SILICON PROCESSING
FOR
THE VLSI ERA

VOLUME 4:

DEEP-SUBMICRON PROCESS TECHNOLOGY

SILICON PROCESSING FOR THE VLSI ERA

VOLUME 4: DEEP-SUBMICRON PROCESS TECHNOLOGY

STANLEY WOLF Ph.D.

LATTICE PRESS

Sunset Beach, California

DISCLAIMER

This publication is based on sources and information believed to be reliable, but the authors and Lattice Press disclaim any warranty or liability based on or relating to the contents of this publication.

Published by:

LATTICE PRESS
Post Office Box 340
Sunset Beach, California 90742, U.S.A.

Cover design by: S. C. Ross Wolf.

Library of Congress Cataloging in Publication Data
Wolf, Stanley

Silicon Processing for the VLSI Era
Volume 4: Deep-Submicron Process Technology

Includes Index
1. Integrated circuits-Very large scale
integration. 2. Silicon. I. Title

ISBN 0-9616721-7-X

9 8 7 6 5 4 3 2 1

PRINTED IN THE UNITED STATES OF AMERICA

MASTER TABLE OF CONTENTS

PREFACE

Chap. 1 - THE EVOLUTION OF THE STRUCTURE OF MOSFETS 1

Chap. 2 - 300-MM SILICON WAFERS 17

Chap. 3 - GATE DIELECTRICS: THIN GATE OXIDES 75

Chap. 4 - HIGH-*k* DIELECTRICS 145

Chap. 5 - THE STRUCTURE OF DEEP-SUBMICRON MOSFETS: Retrograde-Wells; Super-Steep Retrograde Channel Doping; Drain Engineering; Punchthrough-Control Implants 181

Chap. 6 - DEEP-SUBMICRON LITHOGRAPHY I: PHOTORESISTS 227

Chap. 7 - DEEP-SUBMICRON LITHOGRAPHY II: OPTICS AND HARDWARE 259

Chap. 8 - CHEMICAL MECHANICAL POLISHING (CMP) 313

Chap. 9 - SHALLOW TRENCH ISOLATION (STI) 433

Chap. 10 - SILICON-GERMANIUM (Si-Ge) TECHNOLOGY FOR HIGH-PERFORMANCE TELECOMMUNICATIONS ICs 475

Chap. 11 - SILICON-ON-INSULATOR (SOI) TECHNOLOGY 501

Chap. 12 - MULTILEVEL INTERCONNECTS FOR ULSI 573

Chap. 13 - POLYCIDES AND SALICIDES OF $TiSi_2$, $CoSi_2$ AND NiSi 603

Chap. 14 - LOW-*k* DIELECTRICS 639

Chap. 15 - DUAL-DAMASCENE INTERCONNECTS 671

Chap. 16 - COPPER INTERCONNECT PROCESS TECHNOLOGY 711

INDEX 795

DETAILED TABLE OF CONTENTS

PREFACE

Chap. 1 - THE EVOLUTION OF THE STRUCTURE OF MOSFETS 1

1.1 THE STRUCTURE OF DEEP-SUBMICRON MOSFETS: (0.25-μm to 0.13-μm) - COMPARED TO THE STRUCTURE OF "CONVENTIONAL" MOSFETS (2.0-μm to 0.5-μm) 2

1.1.1 Evolution of the MOSFET Gate Stack and Contact Structure
1.1.2 Gate Dielectric Materials in Deep-Submicron MOSFETs
1.1.3 Doping-Concentration Profiles of the MOSFET Channel
1.1.4 Evolution of the Drain Structure of MOSFETs

1.2 DEEP-SUBMICRON CMOS STRUCTURES 10

1.2.1 Substrate Issues for Deep-Submicron CMOS
1.2.2 Well Formation in Deep-Submicron CMOS
1.2.3 Dual-Doped Poly in Deep-Submicron CMOS
1.2.4 Shallow Trench Isolation for Deep-Submicron CMOS
1.2.5 Silicon-On-Insulator (SOI) Technology

1.3 LIMITS TO CONVENTIONAL MOSFET SCALING 15

REFERENCES 16

Chap. 2 - 300-MM SILICON WAFERS 17

2.1 300-mm SILICON CRYSTAL GROWTH 17

2.2 GROWN-IN SILICON DEFECTS: 21 CRYSTAL-ORIGINATED-PARTICLES (COPS) & DISLOCATION LOOPS

2.3 DETAILS OF THE FORMATION OF CRYSTAL-ORIGINATED-PARTICLES (COPS) 23

2.3.1 The Radial Distribution of Grown-In Defects on the Wafer Surface

2.4 THE OXYGEN-STACKING-FAULT RING (OSF-RING) 26

2.5 MITIGATING EFFECTS OF COPS BY USE OF POST-CRYSTAL-GROWTH ANNEALING 27

2.6 ELIMINATION OF COPS WITH "PERFECT-SILICON" 29

2.7 MINIMIZING PERFORMANCE-DEGRADATION CAUSED BY COPS THROUGH THE USE OF HIGH-PULL-SPEED SILICON 30

2.8 GETTERING FOR ULSI PROCESSES 32

2.8.1 Basic Gettering Principles
2.8.2 Extrinsic Gettering
2.8.3 Intrinsic Gettering
2.8.4 Gettering with Oxygen Precipitates
2.8.5 Summary of Gettering

2.9 STATUS OF SILICON WAFER TECHNOLOGY & FUTURE TRENDS 45

2.9.1 Epitaxy-Optimized Substrate (EOS) Wafers
2.9.2 Comparing the Grown-In Defect Characteristics of the "New" Wafers
2.9.3 Gettering Methods & Denuded Zone Formation Techniques Used with the "New" Wafers
2.9.3.1 "Magic DeNuded Zone"

2.10 FROM INGOT TO FINISHED WAFER: SLICING, ETCHING, AND POLISHING 52

2.10.1 Ingot Evaluation
2.10.2 Ingot Surface Grinding
2.10.3 Grinding Flats or Notches on the Ingot for Orientation Purposes
2.10.4 Sawing the Ingot into Slices (Wafers)
2.10.5 Laser Marking the Wafers
2.10.6 Lapping and Grinding the Wafers
2.10.7 Removal of Surface Mechanical Damage by Chemical Etching
2.20.8 Rounding the Wafer Edge (and Notch)
2.10.9 Edge Polishing of the Wafers
2.10.10 Chemical-Mechanical Polishing of the Wafers
2.10.11 Cleaning the Wafers
2.10.12 Depositing Epitaxial Silicon Layers on the Wafers
2.10.13 Shipping 300-mm Wafers: Its Impact on the Evolution of Wafer Shipping Boxes

2.11 SPECIFICATIONS OF SILICON WAFERS FOR VLSI 63

2.12 THE ECONOMICS OF SILICON WAFERS 66

REFERENCES 69

Chap. 3 - GATE DIELECTRICS: THIN GATE OXIDES 75

3.1 REQUIRED CHARACTERISTICS OF GATE DIELECTRICS FOR DEEP-SUBMICRON MOSFETS 76

3.2 THE STRUCTURE OF THERMALLY GROWN SiO_2 AND THE PROPERTIES OF THE Si/SiO_2 INTERFACE 80

3.2.1 The Microscopic Structure of Thermally Grown SiO_2
3.2.2 The Si/SiO_2 Interface
3.2.2.1 Interface Trap Charge
3.2.2.2 Effect of Interface Traps on IC Characteristics
3.2.2.3 Oxide Trapped Charge
3.2.2.3 Effect of Oxide Trapped Charge on Device Characteristics
3.3 DIELECTRIC BREAKDOWN IN SILICON DIOXIDE FILMS 90
3.3.1 Electron Trapping in Silicon Dioxide:
3.3.2 The Electric-Field Driven Model of Oxide Degradation (*E* Model)
3.3.3 The Current-Driven Model of Oxide Degradation (1/E Model)
3.3.4 The Hole-Trapping Model that Describes How Holes are Injected & Trapped in SiO_2
3.3.5 Comparing the Electric-Field Driven and Current Driven Oxide Breakdown Models
3.3.6 Time to Breakdown (T_{BD}) and Charge to Breakdown (Q_{BD})
3.4 LEAKAGE CURRENTS IN SiO_2 FILMS (TUNNELING PHENOMENA) 99
3.4.1 Fowler-Nordheim (FN) Tunneling (Tunneling Into Silicon Dioxide)
3.4.2 Direct Tunneling (Tunneling Through Silicon Dioxide)
3.5 MODELS OF THIN OXIDE GROWTH 104
3.6 SINGLE-WAFER TECHNOLOGY OF THIN OXIDE GROWTH 109
3.6.1 Rapid Thermal Oxidation Tools
3.6.2 Wet RTO Processes
3.7 NITRIDED & FLUORINATED OXIDES AS GATE DIELECTRICS 112
3.7.1 Oxynitridation of Silicon in N_2O
3.7.2 Oxynitridation of Oxides in N_2O or NO
3.7.3 Fluorinated Gate Oxides
3.8 PROJECTIONS OF THICKNESS LIMITS OF GATE OXIDES 121
3.8.1 Minimum Oxide Thickness Due to Defects and Tunneling
3.8.2 Minimum Oxide Thickness: Stress Induced Leakage Current (SILC)
3.8.3 Soft-Breakdown of Oxide Films
3.8.4 Impact of Polysilicon Depletion
3.8.5 Impact of Process Induced Damage
3.8.6 Summary of Oxide Thickness Projections
3.9 MEASURING THIN GATE OXIDES 135
3.10 MANUFACTURING THIN GATE OXIDES 137
3.10.1 Process Control Issues of Growing Ultra-Thin Gate Oxides
3.10.2 Thin Gate Dielectric Stacks Based on Silicon Dioxide

REFERENCES 139

Chap. 4 - HIGH-*k* DIELECTRICS 145

4.1 HIGH-*k* DIELECTRICS AS THE GATE DIELECTRIC IN MOSFETS 145

4.1.1 Integrating High-*k* Dielectrics into MOSFET Structures:

4.2 HIGH-*k* DIELECTRICS AS THE CAPACITOR DIELECTRIC IN DRAMS 150

4.3 HIGH-*k* DIELECTRICS FOR FeRAM APPLICATONS 154

4.4 TANTALUM PENTOXIDE (Ta_2O_5) 154

4.4.1 Tantalum Pentoxide as a DRAM Capacitor Dielectric

4.4.2 Tantalum Pentoxide as a MOSFET Gate Dielectric

4.5 BARIUM STRONTIUM TITANATE (BST) 162
AS A DRAM CAPACITOR DIELECTRIC

4.5.1 Deposition of BST Films by CVD

4.5.2 Microstructure Effects on the Electrical Properties of BST

4.5.3 BST Integration Issues

4.6 USE OF OTHER HIGH-*k* MATERIALS AS GATE DIELECTRICS 169

4.6.1 Atomic Layer Deposition for Depositing Thin High-*k* Dielectric Films

4.6.2 Aluminum Oxide (Al_2O_3) as a Gate Dielectric

4.6.3 Zirconium Oxide (ZrO_2) as a Gate Dielectric

4.6.4 Hafnium Oxide (HfO_2) as a Gate Dielectric

4.6.5 Praseodymium Oxide (Pr_2O_3) as a High-*k* Gate Dielectric

REFERENCES 177

Chap. 5 - THE STRUCTURE OF DEEP-SUBMICRON MOSFETS: Retrograde-Wells; Super-Steep Retrograde Channel Doping; Drain Engineering; Punchthrough-Control Implants 181

5.1 WELL-FORMATION IN DEEP-SUBMICRON CMOS 181

5.1.1 Retrograde-Wells for Deep Submicron CMOS

5.1.2 Process Integration Issues Involving the Use of Retrograde-Wells

5.2 SUPERSTEEP RETROGRADE CHANNEL (SSR) PROFILES 191

5.2.1 Fabricating SSR Profiles for Sub-0.1 *m*m MOSFETs

5.3 SOURCE/DRAIN ENGINEERING IN DEEP-SUBMICRON CMOS 195

5.3.1 Parasitic Resistance of Deep-Submicron Source/Drains

5.3.1.1 Physical Meaning of R_{ac}, R_{sp}, and R_{SDE}

5.3.1.2 Comparing the Resistance of the SDE Region (R_{sp} + R_{SDE}) to R_{sat}

5.3.1.3 Parasitic Resistance of the Deep Contact Region of the Drain R_{co}

5.3.2 MOSFETs with Elevated Source/Drains

5.3.2 MOSFETs with Elevated Source/Drains
5.3.3 Shallow-Junction Formation for Deep-Submicron Source/Drains
5.3.3.1 Ultra-Low-Energy Implants
5.3.3.2 Transient-Enhanced Diffusion
5.3.3.3 Rapid Thermal Annealing of Shallow Junctions
5.3.3.4 Dopant Loss for Sub-keV Implants from Self-Sputtering, Out Diffusion, and Surface Oxide
5.3.4 Effect of the Overlap of the SDE Region and Gate Edge on MOSFET Performance
5.3.5 SDE Junction Lateral Abruptness
5.4 ANTI-PUNCHTHROUGH STRUCTURES IN DEEP-SUBMICRON CMOS 219
REFERENCES 223

Chap. 6 - ADVANCED LITHOGRAPHY I: DEEP-SUBMICRON RESISTS 227

6.1 CHEMICALLY-AMPLIFIED DEEP-UV RESISTS FOR OPTICAL LITHOGRAPHY 227
6.1.1 248-nm Photoresists
6.1.1.1 Amine-Contamination of DUV CA Resists
6.1.1.2 Current State of 248-nm Resist Technology
6.1.2 193-nm DUV Photoresists
6.1.2.1 Problems of 193-nm Resists
6.1.3 157-nm DUV Photoresists
6.2 ANTI-REFLECTIVE COATING (ARCs) 244
6.2.1 Organic BARCs
6.2.2 Inorganic Dielectric BARCs
6.2.3 PVD-Deposited BARCs
6.2.4 BARCs for Dual-Damascene Applications
6.3 PHOTORESIST PROCESSING SYSTEMS 251
REFERENCES 257

Chap. 7 - ADVANCED LITHOGRAPHY II: OPTICS AND HARDWARE 259

7.1 EXCIMER LASER DEEP-UV LIGHT SOURCES 259
7.1.1 KrF Excimer Lasers
7.1.2 ArF Excimer Lasers
7.1.3 F_2 Excimer Lasers
7.2 EXPOSURE TOOLS For DUV LITHOGRAPHY 265
7.2.1 Exposure Tools for 248-nm Lithography

7.2.2 Exposure Tools for 193-nm Lithography
7.2.3 Exposure Tools for 157-nm Lithography
7.2.4 Calcium Fluoride Optical Elements for DUV Lithography
7.2.5 300-mm Lithography Tools
7.2.6 Mix-an-Match Lithography

7.3 RESOLUTION ENHANCEMENT TECHNOLOGIES (RETs) 274
7.3.1 Off-Axis Illumination
7.3.2 Optical Proximity Correction (OPC)
7.3.3 Phase Shift Masks (PSM)

7.4 MASK ERROR FACTOR (MEF, or MEEF) 292

7.5 EXTENDING THE LIMITS OF OPTICAL LITHOGRAPHY 296

7.6 NON-OPTICAL (or NEXT GENERATION) LITHOGRAPHIC TECHNOLOGIES (NGL) 301
7.6.1 Extreme Ultra-Violet Reflective Projection Lithography (EUV)
7.6.2 Electron Beam Projection Lithography (SCALPEL and PREVAIL)

REFERENCES 309

Chap. 8 - CHEMICAL MECHANICAL POLISHING (CMP) 313

8.1 TERMINOLOGY AND EVOLUTION OF PLANARIZATION PROCESSES FOR ICS 313
8.1.1 Defining the Degree of Planarization
8.1.2 The Need for Dielectric Planarization:
8.1.3 Design Rules Related to Intermetal Dielectric-Formation and Planarization Processes
8.1.4 Step-Height Reduction of Underlying Topography as a Technique to Alleviate the Need for Planarization
8.1.5 Planarization through Sacrificial-Layer Etchback
8.1.5.1 Sacrificial-Etchback Process Problems

8.2 INTRODUCTION TO CHEMICAL MECHANICAL POLISHING (CMP) 322

8.3 THE TERMINOLOGY USED TO CHARACTERIZE CMP PROCESSES 325
8.3.1 CMP Removal Rate (RR)
8.3.2 The Degree of Planarization (DOP)
8.3.3 Within-Die Non-Uniformity (WIDNU)
8.3.4 Within-Wafer Non-Uniformity (WIWNU)
8.3.5 Wafer-to-Wafer Non-Uniformity (WTWNU)
8.3.6 Efficiency of Planarization (EOP)

8.5.1 The Mechanical Aspects of Silicon Dioxide Removal by CMP: (Preston's Law)
8.5.1.1 Model of WIWNU Effects in Rotary CMP Tools Based on Preston's Law
8.5.2 Models Which Describe Factors that Impact Oxide RR in CMP Through the Preston Constant
8.5.2.1 Impact on the Oxide RR from the Electrochemical Interactions Between the Slurry Particles and the Oxide Surface
8.5.2.2 Dependence of Oxide RR in CMP on Dielectric Hardness
8.5.3 Models Which Describe Chemical-Mechanical Phenomena that Give Rise to Planarization Phenomena Associated with CMP of Oxides
8.5.3.1 Model of the Mechanism That Produces Planarization of Steps in Oxide CMP
8.5.3.2 Pattern Dependence of the Planarization Rate in CMP Processes
8.5.4 Metal-CMP Mechanisms
8.6 CMP EQUIPMENT 352
8.6.1 CMP Polishing Tools:
8.6.1.1 CMP Polishing Tool Configurations
8.6.1.2 Rotary CMP Tools:
8.6.1.3 Orbital CMP Tools
8.6.1.4 Linear CMP Tools
8.6.1.5 Fixed Table CMP Tools
8.6.1.6 300-mm CMP Tools
8.6.2 Wafer-Carriers in CMP Polishing Tools (and the CMP Edge Effect)
8.6.2.1 CMP Edge-Effect
8.6.3 CMP Consumables (Slurries):
8.6.3.1 Slurries for Oxide-CMP
8.6.3.2 Slurries for Metal-CMP
8.6.3.3 Multi-Step Slurries
8.6.4 Slurry-Distribution Systems
8.6.5 Environmental, Safety, and Handling Issues of CMP Slurries
8.6.6 CMP Consumables (Polishing-Pads)
8.6.7 Pad-Conditioners
8.6.8 Slurry-Free Pads
8.6.9 Endpoint-Detection in CMP
8.7 CLEANING ISSUES IN CMP 392
8.7.1 Mechanisms of Particle Adhesion and Removal from Surfaces
8.7.1.1 Electrical Repulsion of the Particle and the Wafer Surface
8.7.1.2 Chemically Assisted Particle Removal
8.7.1.3 Mechanically Assisted Particle Removal
8.7.1.4 Steric Stabilization for Suspending Particles in a Solution

8.7.2 Practical Particle Removal Processes: Brush Scrubbing
8.7.2.1 Brush Scrubbing to Remove Particles in Post-Oxide-CMP Processes
8.7.2.2 Brush Scrubbing to Remove Particles in Post Metal-CMP Processes
8.7.3 Vibrational Scrubbing to Remove Particles (Megasonic)
8.7.4 Secondary (or Buff) Polish Procedure as a CMP Cleaning Step
8.7.5 Post-CMP Cleaning of Metallic Contaminants
8.7.6 CMP Cleaning Equipment
8.8 CMP METROLOGY 407
8.8.1 Detection of Defects Associated with the CMP Process
8.9 CMP POLISHER TOOL RELIABILITY 411
8.10 CMP SYSTEMS AND PROCESS INTEGRATION 412
8.11 CMP OF VARIOUS MATERIALS 413
8.11.1 CMP of Oxide Interlevel & Intermetal Dielectrics (ILDs & IMDs)
8.11.2 CMP of Tungsten
8.11.3 CMP of Copper
8.11.4 CMP of Aluminum
8.11.5 CMP of Polysilicon
8.11.6 CMP of Photoresist
8.11.7 CMP of Low-*k* Dielectrics
8.12 MISCELLANEOUS ISSUES CMP 421
8.12.1 Dishing and Erosion
8.12.2 Thickness Non-Uniformity Within a Wafer After CMP
REFERENCES 428

Chap. 9 - SHALLOW TRENCH ISOLATION (STI) 433

9.1 EARLY SHALLOW TRENCH ISOLATION STRUCTURES 435
9.2 STI-ENABLING PROCESSES 438
9.3 SHALLOW TRENCH ISOLATION FOR CMOS 439
9.4 DETAILS OF THE PROCESS FLOW TO FORM A BASELINE SHALLOW-TRENCH-ISOLATION (STI) STRUCTURE 446
9.5 ISSUES INVOLVED WITH STI PROCESS INTEGRATION 455
9.5.1 Issues Involved With Etching Trenches for STI
9.5.2 Corner-Rounding Techniques in STI Structures
9.5.3 Trench-Fill Dielectrics for STI
9.5.4 Dishing Problem Associated with the CMP of STI Trench Dielectrics.
9.5.5 Corner Engineering and Its Effects on MOSFET TURN-OFF Characteristics of MOSFETs Fabricated with STI
REFERENCES 472

Chap. 10 - SILICON-GERMANIUM (Si-Ge) TECHNOLOGY FOR HIGH-PERFORMANCE TELECOMMUNICATIONS ICs **475**

10.1 INTRODUCTION TO HETEROJUNCTION BIPOLAR TRANSISTORS 475
10.2 HETEROJUNCTION BIPOLAR TRANSISTORS WITH LARGE-BANDGAP EMITTERS 480
10.3 HETEROJUNCTION BIPOLAR TRANSISTORS WITH SMALLER BANDGAP BASE REGIONS (Si-Ge BASE) 482
10.4 PROCESS TECHNOLOGY FOR FABRICATING SI-GE HBTS 484
10.5 PROCESS INTEGRATION ISSUES FOR BiCMOS TECHNOLOGY WITH Si-Ge HBT 486
10.6 HIGH-QUALITY SI-GE EPITAXIAL FILM GROWTH ON PATTERNED WAFERS 491
10.7 THERMAL STABILITY PROCESS INTEGRATION ISSUES 492
10.8 COMPATIBILITY WITH ION IMPLANTATION PROCESSES 492
10.9 COMPATIBILITY WITH THERMAL OXIDATION PROCESSES 493
10.10 VERTICAL SCALING OF Si:Ge HBTS 493
10.11 BASE-DURING-GATE AND BASE-AFTER-GATE BiCMOS PROCESS FLOWS 495
10.12 EXAMPLES OF THE MODULAR INTEGRATION OF Si-Ge HBTS INTO A BiCMOS PROCESS FLOW 497

REFERENCES 499

Chap. 11 - SILICON-ON-INSULATOR (SOI) TECHNOLOGY **501**

11.1 WHAT IS SILICON-ON-INSULATOR (SOI)? 501
11.1.1 Isolation Issues in CMOS in Bulk Silicon
11.1.2 Isolation in SOI CMOS
11.2 THE "END OF THE ROAD FOR BULK CMOS" SCENARIO PREDICTED BY THE 1999 INTERNATIONAL TECHNOLOGY ROADMAP FOR SEMICONDUCTORS 505
11.2.1 Source/Drain Junction Scaling
11.2.2 Dopant Activation in Shallow Source/Drain Junctions
11.2.3 Doping Profile Abruptness in Shallow Source/Drain Junctions
11.3 WHY SOI? 510
11.3.1 The Path of Using SOI May Allow Extension of the ITRS Roadmap
11.4 HISTORY OF SILICON-ON-INSULATOR (SOI) TECHNOLOGY 516
11.5 SILICON-ON-INSULATOR DEVICES 518
11.5.1 Partially-Depleted Thin SOI MOSFETs (PD-SOIs)

11.5.2 Fully-Depleted Thin SOI MOSFETs (FD-SOIs)
11.5.3 Process Technologies Used to Fabricate Fully-Depleted SOI Devices
11.5.3.1 FD-SOI: Thinning Si-SOI Surface Layer by Thermal Oxidation
11.5.3.2 FD-SOI Fabrication by LOCOS Recess of the MOSFET Channel Region

11.6 FABRICATION TECHNOLOGIES USED TO PRODUCE SOI-STARTING-WAFERS 527

11.6.1 Silicon-on-Sapphire (SOS)
11.6.2 SOI by Separation by Implanted Oxygen (SIMOX)
11.6.2.1 High-Dose SIMOX
11.6.2.2 Low-Dose SIMOX
11.6.3 Wafer Bonding
11.6.3.1 Bond and Etch-back SOI (BESOI)
11.6.3.2 Hydrogen-Implantation-Induced Layer Splitting to Create SOI Substrates (SMART-CUT®)
11.6.3.3 NANOCLEAVE® Wafer Bonding SOI
11.6.3.4 ELTRAN® Wafer Bonded SOI

11.7 LATERAL ISOLATION IN SOI 550

11.8 SOI PRODUCT EVOLUTION 556

11.9 PROBLEMS ASSOCIATED WITH SOI 559

11.9.1 Floating Body Effects in Partially-Depleted SOI MOSFETs
11.9.1.1 Kink Effect in Partially Depleted SOI NMOSFETs
11.9.1.2 Transient I_D Effects
11.9.1.3 Effect of Building Partially Depleted SOI MOSFETs Without Body Contacts
11.9.2 Short-Channel Effects in SOI MOSFETs
11.9.3 Self-Heating Effects in SOI MOSFETs
11.9.4 Reliability Issues of SOI

REFERENCES 568

Chap. 12 - MULTILEVEL INTERCONNECTS FOR ULSI 573

12.1 EVOLUTION OF THE STRUCTURE AND MATERIALS USED IN IC INTERCONNECTS 575

12.1.1 Single-Level Metal Interconnects in Early ICs
12.1.2 Double-Level Metal Interconnects in Early Bipolar ICs
12.1.3 Double-Level Metal Interconnects in CMOS ICs (Partial Planarization of the ILD)
12.1.4 Three-Level Metal Interconnects in CMOS ICs (CMP of the ILD & W-Plugs)
12.1.4.1 Problems of W-plug/Al-line Interconnect Structures
12.1.5 Dual Damascene Cu/Low-*k* Interconnects

12.1.6 Other Geometrical Aspects of Multilevel ULSI Interconnects
12.2 PERFORMANCE REQUIREMENTS OF ULSI INTERCONNECT SYSTEMS 586
12.2.1 Propagation Delays in ULSI Multilevel Interconnects
12.2.1.1 Case When the *k*-Value of the ILD is Constant Throughout the Interconnect System
12.2.1.2 Case When the *k*-Values of the Interlevel and Intralevel ILD are Different
12.2.1.3 Minimizing the *RC* Delay in the Local Interconnect Region
12.2.1.4 Minimizing the *RC* Delay in the Global Interconnect Regions
12.2.2 Power Dissipation Issues in ULSI Interconnects
12.2.3 Crosstalk in ULSI Interconnects
12.2.4 Contact Resistance in ULSI Interconnects
12.2.5 Electrostatic Discharge Protection Issues of Interconnects
12.3 SUMMARY OF USLI INTERCONNECT ISSUES AND FUTURE TRENDS 599
REFERENCES 601

Chap. 13 - POLYCIDES AND SALICIDES OF $TiSi_2$, $CoSi_2$ AND NiSi 603

13.1 POLYCIDES AND SALICIDES 603
13.2 POLYCIDES FABRICATED BY CVD OF TUNGSTEN SILICIDE (WSi_x) 605
13.3 SALICIDE (SELF-ALIGNED SILICIDE) STRUCTURES 608
13.4 THE EFFECTS OF SALICIDES ON MOSFET PERFORMANCE 612
13.5 SALICIDES FORMED WITH TITANIUM SILICIDE ($TiSi_2$) 614
13.5.1 Reaction Kinetics of $TiSi_2$ Formation (and the Problem of *Bridging*)
13.5.2 The $TiSi_2$ Narrow-Line Effect
13.5.3 Proposed Process Modifications to Mitigate the $TiSi_2$ Narrow-Line Effect
13.5.4 Junction Dopant Redistribution and Silicon Consumption During $TiSi_2$ Formation
13.6 SALICIDES FORMED WITH COBALT SILICIDE ($CoSi_2$) 624
13.6.1 Advantages and Drawbacks of Cobalt Salicides
13.6.2 Details of the Cobalt Salicide Fabrication Process
13.6.3 Diode Leakage Issues Involving Cobalt Salicides
13.6.4 Sensitivity of Cobalt Salicide Films to Degradation by Oxygen Ambients: TiN Capping of Cobalt Salicide Films
13.7 EPITAXIAL $CoSi_2$ 630
13.8 $CoSi_2$ AS A SOURCE/DRAIN DIFFUSION SOURCE 631
13.9 NICKEL SILICIDE FILMS FOR SALICIDE APPLICATIONS 634
REFERENCES 635

Chap. 14 - LOW-*k* DIELECTRICS **639**

14.1 INTRODUCTION TO LOW-*k* DIELECTRICS 635

14.1.1 Spin-On versus CVD Methods for Forming Low-*k* Dielectrics

14.1.2 Silicon-Based versus Carbon-Based Low-*k* Dielectrics

14.1.3 Status of Low-*k* Dielectrics in 2000

14.2 DESIRED CHARACTERISTICS OF LOW-*k* DIELECTRIC FILMS 646

14.3 GENERAL PROCESS INTEGRATION ISSUES INVOLVING LOW-*k* FILMS 648

14.4 FIRST GENERATION LOW-*k* DIELECTRICS ($2.8 < k < 3.5$) 650

14.4.1 Hydrogen Silsesquioxanes (HSQ)

14.4.2 Spin-On Methyl Silsesquioxanes (MSQ)

14.4.3 Fluorinated Silicate Glass (FSG) Films

14.5 SECOND-GENERATION LOW-*k* DIELECTRICS ($2.5 < k < 2.8$) 655

14.5.1 2nd-Genereation Spin-On-Dielectrics: ($2.5 < k < 2.8$)

14.5.2 2nd-Generation CVD-Dielectrics: ($2.5 < k < 2.8$)

14.5.2.1 2nd Generation Low-*k* CVD-Films: Organo-Silicate Glasses

14.5.2.2 2nd Generation Low-*k* CVD-Films: *FlowFill* CVD Films

14.5.2.3 2nd Generation Low-*k* CVD-Films: Black Diamond & Coral CVD Films

14.5.2.4 2nd Generation Low-*k* CVD-Films: Parylene CVD Films

14.6 ULTRA-LOW-*k* DIELECTRICS: ($k < 2.3$) 663

REFERENCES 669

Chap. 15 - DUAL-DAMASCENE INTERCONNECTS **671**

15.1 DAMASCENE VERSUS SUBTRACTIVE INTERCONNECT STRUCTURES 671

15.2 DAMASCENE INTERCONNECTS AS ALTERNATIVE DEEP-SUBMICRON INTERCONNECTS 672

15.3 SINGLE-DAMASCENE INTERCONNECT STRUCTURES

15.4 INTRODUCTION TO DUAL-DAMASCENE INTERCONNECT PROCESSES 674

15.5 THE THREE DUAL-DAMASCENE PROCESS SEQUENCES 676

15.6 THE EVOLUTION OF DUAL-DAMASCENE FROM SiO_2 TO LOW-*k* DIELECTRIC STACKS 679

15.7 TRENCH-FIRST DUAL-DAMASCENE PROCESS FLOW 683

15.8 VIA-FIRST DUAL-DAMASCENE PROCESS FLOW WITH AN SiO_2 DIELECTRIC STACK 685

15.9 PROCESS INTEGRATION ISSUES OF THE VIA-FIRST DUAL-DAMASCENE PROCESS FLOW WITH A LOW-*k*-DIELECTRIC STACK 687
15.9.1 Via-First-Damascene Integration Issues: Lithography
15.9.1.1 Via-Patterning Lithography Issues
15.9.1.2 Trench-Patterning Lithography Issues – DUV Resist Poisoning
15.9.1.3 Trench-Patterning Lithography Issues – Trench CD Control Issues
15.9.1.4 Trench-Patterning Lithography Issues – BARC Veils
15.9.1.5 Bi-Layer Resist Process
15.9.2 Via-First-Damascene Integration Issues: Dielectric Etching
15.9.2.1 Via-Etching for Low-*k* Processes
15.9.2.1 Trench-Etching
15.9.2.1 Etching the SiN Barrier at the Bottom of Vias
15.9.3 Via-First-Damascene Integration Issues: Resist Stripping
15.9.4 Via-First-Damascene Integration Issues: Cleaning
15.9.4.1 Post-Dielectric-Etch-and-Strip Residue-Removal
15.9.4.2 Pre-Metal-Deposition Clean
15.9.4.3 Post-CMP Clean
15.9.4.4 Clean of Top-Surface of Cu Prior to Diffusion-Barrier-SiN-Deposition Process
15.9.4.5 Supercritical Fluids as Wafer Cleaners
15.10 VIA-FIRST DUAL-DAMASCENE DIELECTRIC STACK WITH NO EMBEDDED ETCH-STOP LAYER 708
15.11 PROCESS INTEGRATION ISSUES OF DUAL-DAMASCENE DIELECTRIC STACK WITH ULTRALOW-*k* DIELECTRICS 708
REFERENCES 709

Chap. 16 – COPPER INTERCONNECT PROCESS TECHNOLGY 711

16.1 WHY COPPER FOR DEEP-SUBMICRON IC INTERCONNECTS? 712
16.1.1 The Advantages of Copper Interconnects
16.1.2 Technological Challenges of Using Copper Interconnects
16.2 OVERVIEW OF COPPER PROCESS TECHNOLOGY 723
16.2.1 The Transition from Aluminum to Copper Interconnects
16.2.2 The Key Goal of Depositing Copper into Damascene Recesses
16.3 BARRIER LAYERS FOR COPPER INTERCONNECTS 728
16.3.1 The Impact of the Barrier-Layer Deposition Process on the Filling of Damascene Recesses with Copper
16.3.2 Introduction to the Technology of Depositing Barrier-Layers for Cu Interconnects

16.3.3 Ionized Magnetron Sputter Deposition
16.3.4 Hollow-Cathode Magnetron Sputtering Source
16.3.5 Self-Ionized-Plasma Sputtering Source
16.3.6 Material Properties of Tantalum and Refining Tantalum Metal For Semiconductor Applications
16.3.6.1 Material Properties of Tantalum
16.3.6.2 Refining and Forming Tantalum Metal For Sputtering Targets
16.3.6.3 The Properties of Tantalum (and TaN & TaSiN) as Cu-Barrier Layers
16.3.7 Other Candidate Materials for Cu-Barrier Layers
16.4 SEED-LAYER TECHNOLOGY FOR COPPER INTERCONNECTS 740
16.5 COPPER INTERCONNECT FILM DEPOSITION 747
16.5.1 Electrolytic-Plating of Copper
16.5.2 Filling High-Aspect-Ratio Recesses Without Voids Using a Cu Electrolytic Electroplating Process
16.5.2.1 The Electroplating Process Sequence
16.5.2.2 Additives in Plating Baths for Damascene Cu Interconnects
16.5.2.3 Models of How Additives in Plating Baths Impact "Bottom-Up" Filling
16.5.2.4 Electroplating Bath Control
16.5.2.5 Pulsed Plating
16.5.2.6 Copper Thickness Variation Within a Wafer After Electroplating
16.5.3 Post-Electroplating Cu Annealing and Recrystallization
16.5.4 Electrochemical Mechanical Deposition (ECMD™)
16.5.5 Electroless-Plating of Copper
16.5.6 CVD of Copper
16.6 WET-CHEMICAL ETCHING AND ELECTROPOLISHING COPPER 775
16.7 CONTAMINATION CONTROL IN COPPER TECHNOLOGY 778
16.7.1 Sources of Copper Cross-Contamination in a Wafer Fab
16.7.2 Effects of Copper-Cross Contamination at the Back-End-of-Line (BEOL)
16.7.3 Cu Removal from the Wafer Bevel/Edge and Backside
16.8 RELIABILITY OF COPPER INTERCONNECTS 784
16.8.1 Reliability of Copper Interconnects: "Missing-Metal" Defects
16.8.2 Reliability of Copper Interconnects: Plated vs. CVD Cu Films
16.9 ENVIRONMENTAL, HEALTH, & SAFETY ISSUES OF CU TECHNOLOGY 787
16.9.1 Copper Waste Handling
16.9.1.1 Handling Concentrated Waste from Plating and Wafer Reclaim
16.9.1.2 Handling Copper CMP Effluents
REFERENCES 789

INDEX **795**

PREFACE

With the dawn of the new millennium, IC technology-nodes have evolved beyond 0.18-micron. New processes also had to be developed to meet the needs of this *Deep-Submicron Era,* including: copper/low-*k* dual-damascene interconnects; high-*k* MOSFET-gate and DRAM dielectrics; 300-mm wafers; chemical-mechanical polishing (CMP); excimer-laser-based DUV-lithography; chemically-amplified (CA) resists; silicon:germanium (Si:Ge); and silicon-on-insulator (SOI).

The semiconductor industry, however, also experienced an economic decline of unprecedented severity at the start of the 2000 decade. It seams clear that to fuel the recovery, chip manufacturers will have to exploit the above innovations. Such processes are a path to achieving the performance goals of future IC technology-nodes that will be needed for manufacturing new and profitable electronic products.

This is the *first available book dedicated exclusively to describing such deep-submicron process technologies* (including those listed above). It is intended to provide semiconductor engineers and researchers with a comprehensive, state-of-the-art reference about these emerging and leading-edge processes. While many other books have been written about the "traditional" IC fabrication processes, none offer what this volume contains. It's sole focus is those techniques needed to *manufacture chips for 0.18-μm (and smaller) generations*. Thus, it belongs on the shelf of every microelectronic professional whose livelihood depends on having a thorough grasp of the latest advances in IC fabrication techniques. There is no faster way to gain this information. As with our other volumes, owners will find themselves referring to this text again and again!

Readers can still consult Vol.1, 2nd Edition for background material on basic IC fabrication modules, as this book (Vol. 4) does not repeat that information. Instead, it presents over 800 pages of new content about the process technologies of the 2000's. It also has over 900 references (with 200 of these being published since 2000), and more than 500 illustrations. It is written in the same lucid, user-friendly style that has been widely praised in the other volumes of *Silicon Processing for the VLSI Era.*

As with the previous volumes, this text was produced using an accelerated book-production schedule. By employing revolutionary publishing techniques made available in the 1990's, Lattice Press can turn a finished manuscript into a published text in less than 3 months (a task that takes most publishers of technical books 15-18 months). This allows our books to be released faster, and to contain the most-up-to-date information at the time of publication.

SILICON PROCESSING FOR THE VLSI ERA is a multi-volume treatise designed to provide a comprehensive and up-to-date treatment of this important and rapidly changing field. The volume in hand is the fourth in the series. Volume 1 (*Process Technology*, 2nd Edition) was published in 2000 and deals with individual processes employed in the fabrication of silicon ULSI circuits (such as silicon wafering, thermal oxidation of silicon, CVD and PVD of amorphous and polycrystalline films, diffusion and ion implantation of dopants in silicon, microlithography, and patterning technology). Volume 2 (*Process Integration*) was released in 1990, and describes how the individual processes of Volume 1 are combined in various ways to produce MOS and bipolar VLSI and ULSI circuits. The task of integrating these various fabrication processes together is referred to as *process integration.* In Volume 3 (*The Submicron MOSFET,* published in 1995) the topics of submicron MOSFET device physics and the relationship between such device physics and submicron MOSFET fabrication are addressed.

A book of this length and diversity would not have been possible without the indirect and direct assistance of many others. From an indirect perspective, virtually all of the information presented here is based on the research efforts of countless numbers of scientists and engineers. Their contributions are recognized to a small degree in citations of their articles given in the references at the end of each chapter. The direct help was also provided by many people, and the resulting text is thus significantly better. The author expresses heartfelt thanks to those who gave their time, energy, and intellect.

Important direct help came from members of my immediate family. My wife, Carrol, read the manuscript for grammatical and stylistic correctness. My son, Ross, designed the cover and improved many of the graphics. My daughter, Jennifer, helped in marketing the book. Thus, it truly represents a family enterprise.

Stanley Wolf

Chapter 1

THE EVOLUTION OF THE STRUCTURE OF MOSFETS

The silicon MOSFET is the most important active-device in ULSI integrated circuits. Over the past 30 years, MOSFET dimensions have been decreasing at a dramatic rate - from gate lengths of 10-μm in the early 1970s to less than 0.13-μm today. This gate-length scaling is expected to continue, and may reach 0.05-μm (50-nm) at the end of the projected shrink path in 2012. Table 1-1 shows the predicted scaling of MOSFETs according to the 2001 IRTS Roadmap.[1]

MOSFETs are the dominant devices in Si ICs largely because they can be scaled to increasingly small dimensions, and this scaling results in higher performance. However, the structure of the MOSFET has undergone many changes besides the scaling of feature sizes down to the deep-sub-micron regime of today.

Figure 1-1 shows a SEM of a MOSFET with a gate length of 2-μm (Fig. 1-1a, circa 1985), and also of a MOSFET with a gate length of 0.2-μm (Fig. 1-1b). With the exception of being scaled, the basic structure of the MOSFET does not appear to be fundamentally changed. However, several aspects of the MOSFET device structure have been modified. (This can be better illustrated in schematic drawings that depict MOSFET cross-sections with a 1.5-μm gate length and a 0.18-μm gate length, see Fig. 1-2.) In this chapter the structure of deep-sub-micron MOSFETs is compared to those of technology generations in which the gate length exceeded 1.0-μm. In this discussion the term *deep-submicron MOSFET* refers to MOSFETs with a *gate length of 0.25-μm or smaller.* How the structure of CMOS devices must be reconfigured when building ICs using devices whose minimum sizes are in the deep-submicron regime is also covered.*

* By 2001, several novel MOSFET structures were being studied as alternatives for the "conventional" MOSFET when it's scaling limits are reached. Such new structures include the *vertical-*, the *replacement-gate-*, and *double-gate-MOSFETs.* However, by 2002 these structures (and others) were still in their infancy. It is not likely that they will be put into production for many years. The focus of this text, however, is on process technology that is in (or about to be put into) mainstream IC manufacturing lines in the 2001-3 timeframe. Thus, these novel structures will not be described here further. However, readers interested in more data on such futuristic MOSFETs should see such recent reviews as Ref. 8, and the *Technical Digests* of the *IEEE International Electron Devices Meetings (IEDM),* and *Digests of Technical Papers* of the *Symposium VLSI Technology.*

Table 1-1 PRINTED GATE LENGTHS OF MOSFETS ITRS2001[1]

	2001	2002	2003	2004	2005	2006	2007
PRINTED GATE LENGTH (nm)	90	75	65	53	45	40	35

The purpose of this chapter is to show how the MOSFET structure has had to be changed when deep-submicron integrated circuits were fabricated. The device physics details that underlie the need for building such deep-submicron MOSFET structures are not presented here. While some of such information is found in later chapters, a more in-depth discussion of this topic is left to other texts (such as Vol. 3 of this series,[2] and elsewhere.[3])

1.1 THE STRUCTURE OF DEEP-SUBMICRON MOSFETS: (0.25 μm to 0.13 μm) COMPARED TO EARLIER-GENERATION MOSFETS: (2.0 μm to 0.5 μm)

In this section the structures of deep-submicron MOSFETs of the 2001-2002 timeframe (i.e., with gate lengths from 0.25 μm down to about 0.13 μm) are compared to those fabricated in the late 1980s and early 1990s (i.e., with gate lengths that ranged from 2.0 μm down to about 0.5 μm).

The following parts of the MOSFET will be examined, to see how the device structure had to be modified in order to allow high-performance deep-submicron MOSFETs to be implemented:

1. MOSFET gate-stack and contact structure;
2. Gate dielectric materials for the MOSFET;
3. Doping-concentration profiles of the MOSFET channel;
4. Drain structure of the MOSFET.

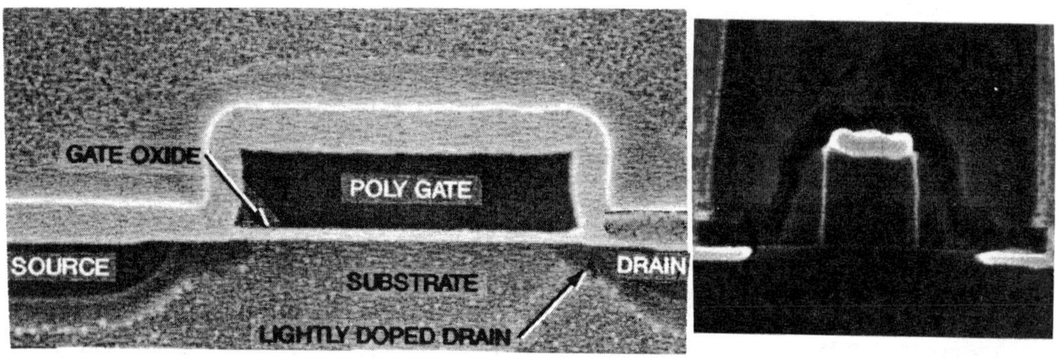

a) b)

Fig. 1-1 SEMs of MOSFET cross-sections with different gate lengths: a) 2.0 μm; b) 0.2 μm.

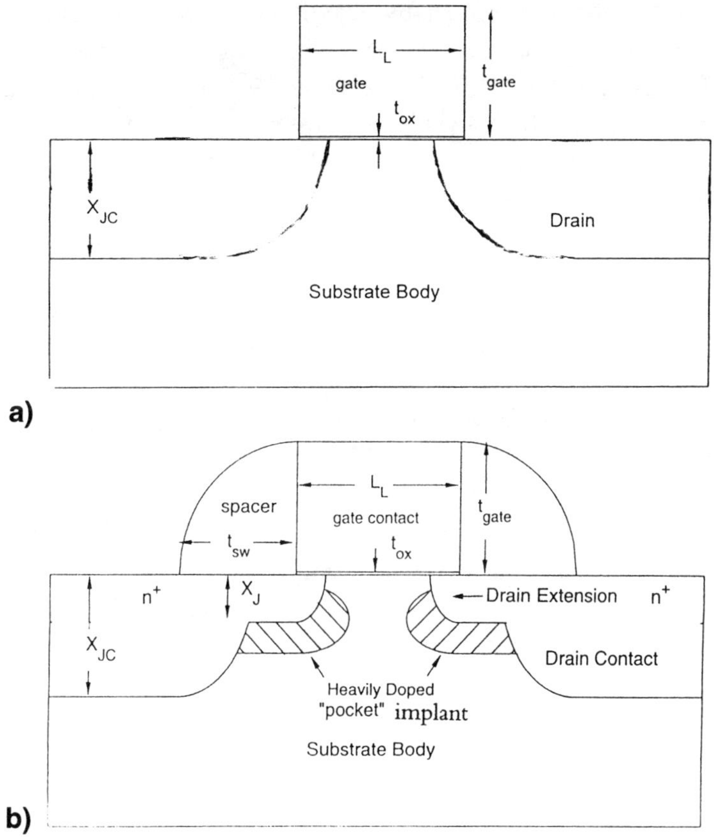

Fig. 1-2 Schematics of MOSFET cross-sections with different gate lengths: a) 1.5 μm; b) 0.2 μm.

Note that in CMOS ICs there are some additional structural aspects of the integrated circuit that must be considered besides those pertaining to the active device (MOSFET) structure. That is, the characteristics of the following two CMOS elements must also be addressed: **1**) the on-chip interconnect system; and **2**) the substrate and isolation structures. The issues relating to the substrate and the isolation structures will be considered in the following section of this chapter, while general issues related to the interconnect system are covered in Chap. 12.

1.1.1 Evolution of the MOSFET Gate Stack and Contact Structure

MOSFETs have used polysilicon as the gate material since the mid-1970s (i.e., when the polysilicon-gate structure replaced metal [Al] gates). The gate material in earlier MOSFETs for both NMOS and CMOS technologies down to about

0.35-μm was n^+-doped poly (usually heavily doped with phosphorus - see also the following section on CMOS technology for more information on poly doping). However, as the MOSFET gate lengths were scaled below about 1.25-μm, the resistance of the polysilicon lines became too large for them to be effective in their secondary role as a local interconnect (even when they were doped as heavily as possible). At that point it became necessary to clad the top surface of the poly film with a metal silicide layer (most commonly WSi_x). This multi-layer gate *stack* was called a *polycide structure* (see Chap. 13 for more details on WSi_x polycides). Use of the polycide gate stack structure allowed the resistance of lines implemented with heavily-doped poly to be reduced by about a factor of four. Implementing WSi_x polycides thus permitted the use of single-level-metal interconnects in MOS-based ICs to be extended for one or two more technology nodes. (Note that the polycide structure is not shown in Fig. 1-2a, but is shown in Fig. 13-1.)

In deep-submicron CMOS, the silicide atop the poly layer is no longer WSi_x. Instead it is replaced with $TiSi_2$ or $CoSi_2$. These two silicides have lower resistivities than WSi_x. In addition, $TiSi_2$ or $CoSi_2$ can be simultaneously formed on the source and drain contact-regions at the same time they are formed on the poly lines. MOSFETs with a silicide on both the poly and the source/drain contact regions are said to contain a *self-aligned silicide* (or *salicide*) structure (as shown in Fig. 1-2b). The salicide structure provides two main benefits to the MOSFET: **1**) the poly lines have a lower resistivity due to the presence of the silicide; and **2**) the contact region exhibits a smaller contact resistance and sheet resistivity as compared to a contact region that does not have such a silicide. The details of the salicide formation processes, and the comparison of titanium silicides and cobalt silicides for this role are described in Chap. 13.

1.1.2 Gate Dielectric Materials in Deep-Submicron MOSFETs

Thermally grown SiO_2 has been used as the gate dielectric material for over 30 years (ever since MOSFETs ICs were first introduced). The thickness of this oxide has been continuously scaled to improve the MOSFET drain current and to control short-channel effects. The gate-oxide thickness has been approximately linearly scaled with channel length, with the ratio of gate length L_E to oxide thickness t_{ox} being maintained at approximately 45 (i.e., L_E/t_{ox} ~45).[4] For example, a 200-Å gate-oxide thickness was used in MOSFETs with a gate length of 1.0-μm, and a 40-Å gate-oxide thickness was used for 0.18-μm MOSFETs.

It is predicted that SiO_2 will continue to be used as the gate dielectric in MOSFETs until about 2004-2005. At that point (for MOSFETs with a gate

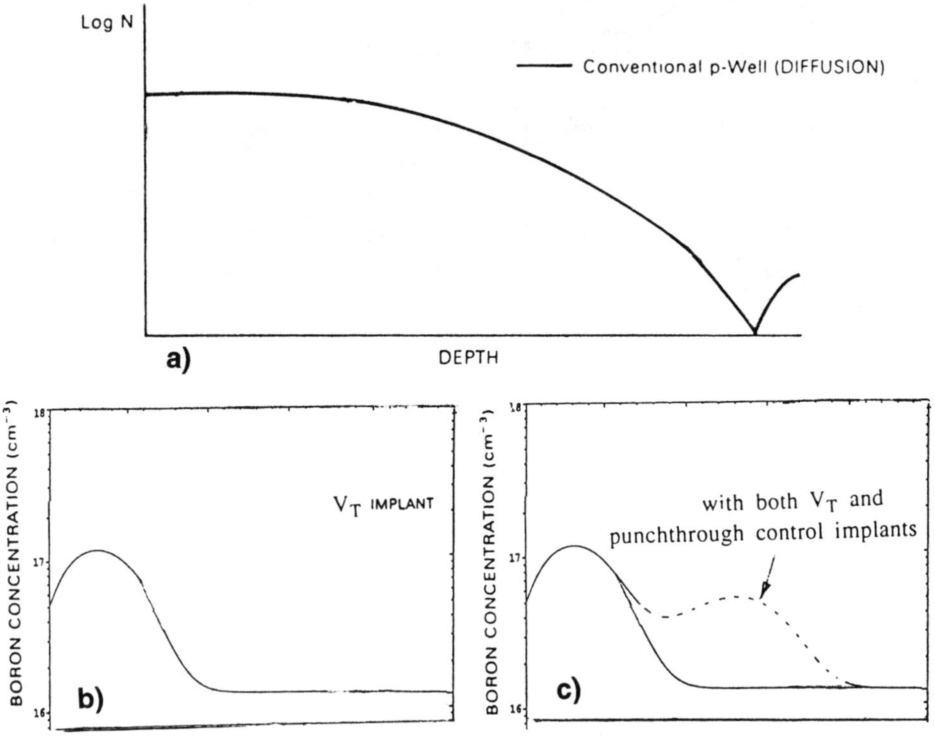

Fig. 1-3 a) Doping profile in a conventional diffused CMOS well; b) Doping profile in the CMOS well with a threshold-adjust implant; c) Doping profile in a CMOS well with a threshold-adjust implant *and* an anti-punchthrough implant.

length of 0.1 μm), the required oxide thickness will be less than 20-Å (assuming the same ratio L_E/t_{ox} is maintained). However, when the oxide reaches a thickness of less than 20Å, the gate-to-channel tunneling leakage-current will become approximately as great as the off-state source-to-drain leakage-current of the MOSFET (~10-nA/μm). As a result, it may not be feasible to build logic MOSFETs (i.e., not even high-performance-logic MOSFETs) with SiO_2 gate dielectric layers thinner than about 20-Å. Other devices (e.g., DRAM capacitors, or oxides used in non-volatile memory structures) will face even more stringent oxide thickness limitations. Such limits to gate oxide thickness are considered in more detail in Chap. 3.

To allow the gate lengths of MOSFETs to continue to be scaled below 0.1 μm, it seems it will be necessary to use alternative high-*k*-dielectric materials instead of SiO_2. With these materials, thicker dielectric layers can be used, and yet the same inversion layer characteristics can be maintained. These thicker layers result in less carrier tunneling, and thus they permit further scaling. The use of high-k dielectrics as alternative gate-dielectric materials is discussed in Chapter. 4.

1.1.3 Doping-Concentration Profiles of the MOSFET Channel

In earlier MOSFETs the channel region had a doping concentration profile (prior to the V_T-adjust implant) that is initially relatively flat with respect to depth below the silicon/oxide interface. In NMOS technology, this flat profile exists because the starting-substrate Si-wafers in which the NMOSFETs are fabricated are uniformly doped. In CMOS technology, the initial flat doping profiles are created from the processes used to create "conventional" CMOS well regions. That is, in such processes the well dopant is introduced into the silicon wafer near its surface using a shallow implant step. Then a long, high-temperature drive-in process diffuses these dopants away from the silicon surface, resulting in a doping profile that is relatively constant with depth near the silicon surface (see Fig. 1-3a). A V_T-adjust implant (using boron for both NMOS and PMOS) is then carried out to provide a doping profile as shown in Fig. 1-3b. As the NMOSFET approached 1-μm in gate length, a blanket anti-punchthrough boron implant was also carried out (typically following the V_T-adjust implant step). This caused a doping profile to exist in the channel as shown in Fig. 1-3c (and the MOSFET structure as in Fig. 1-2a).

In deep-submicron MOSFETs, the doping profile in both the vertical direction (i.e., in depth from the silicon surface) and in the lateral directions (i.e., along the channel) is changed from that which existed in earlier MOSFETs. Consider the profiles in the vertical and lateral directions separately.

In the *vertical direction*, the doping profile in the channel of deep-submicron MOSFETs has a so-called *super-steep retrograde* (SSR) shape (see Fig. 1-4). Such an SSR channel doping is needed to allow the correct value of the threshold voltage of the MOSFET to be achieved, while avoiding a degradation in the mobility of the carriers in the channel (due to excessively high dopant concentrations near the silicon/oxide interface - where the inversion layer exists). That is, it is desired to keep the doping density at the silicon surface to $<5x10^{17}/cm^3$. Such an SSR doping profile can be obtained by implanting a heavier dopant species than B for n-channel devices (typically In is used to replace B), and a heavier dopant species than P for p-channel devices (either As or Sb can be used). These species have a smaller diffusivity than either B or P, respectively, and thus the impurity profile does not get spread out during the

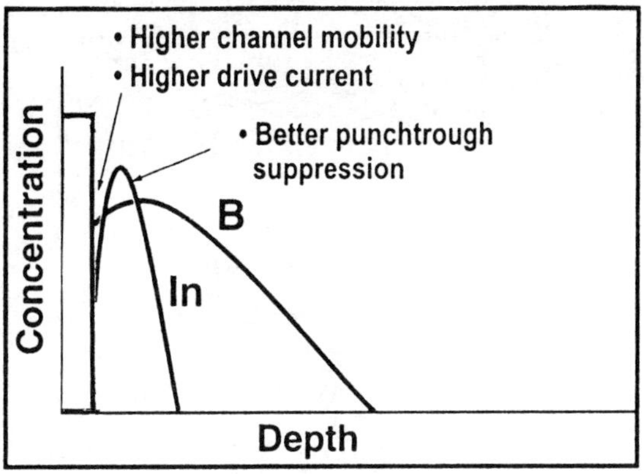

Fig. 1-4 As-implanted doping concentration profiles of threshold-adjust-implants with B and In as the implant species. The In implant produces a super-steep retrograde (SSR) profile.

subsequent heat cycles used to activate the implanted dopants. By choosing the correct implant energy and annealing cycles after implant, an appropriate SSR channel doping profile can be created. Chapter 5 covers this topic in more detail.

In the *lateral direction* (i.e., in the direction along the channel from the source to the drain), there is now also a non-uniform doping profile. That is, in the advanced "conventional" MOSFETs, a blanket implant for improving the punchthrough (PT) performance of the MOSFET is employed. Since this implant is a blanket process, the impurities that it introduces are in uniform concentration in the lateral direction of the channel (as shown in Fig. 1-2a). However, for deep-submicron MOSFETs, such blanket PT implants cause performance degradation.

To avoid these problems (and yet maintain the benefits provided by the blanket anti-punchthrough implant), the same dopant species is introduced in another fashion. As shown in Fig. 1-2b, the dopant is introduced only in selected, local regions; namely at the edge of the drain-extension region. This leaves much of the channel length with a lighter doping that is needed for a low V_T.

As the schematic drawing of Fig. 1-2b depicts, this extra dopant region surrounds the edge of the drain extension much like a halo surrounds the head of an angel in medieval paintings. Hence, this localized anti-punchthrough region has been termed a *halo anti-punchthrough structure*. Such a halo structure was originally produced (ca. 1985) by performing a second implant after the implant that created the lightly doped drain region (LDD - see Sect. 1.4.1). The halo implant is again self-aligned to the gate (and with the beam direction normal to the wafer surface). The same implant species as that already present in the well. is introduced by the halo implant, and it uses an energy that is somewhat higher than that used for the LDD implant. Thus, the range and lateral straggle of this

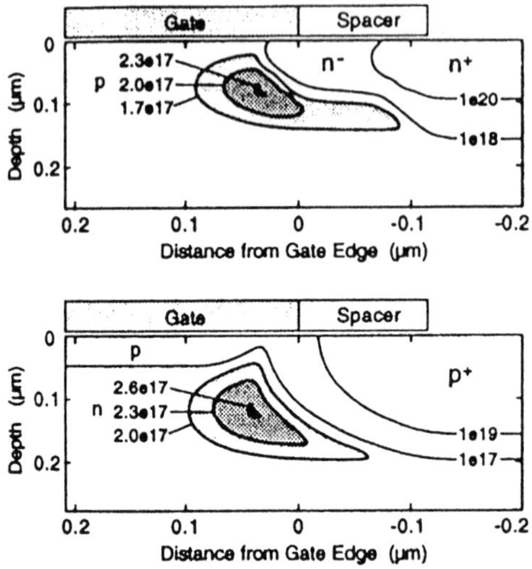

Fig. 1-5 Doping concentration contours resulting from a LATIPS implant.[7]

second implant will be greater than that of the first, resulting in the formation of a halo region as shown in Fig. 1-2b.

In more-advanced-technology generations, this second implant step is carried out with the ion beam directed at a high angle to the wafer surface. This allows the peak concentration of the implanted impurities to be located at a precise depth and lateral distance from the edge of the gate. Such an implant process has been named *large angle tilted implanted punchthrough structure* (*LATIPS*), or alternatively a *pocket implant.* Such lateral tailoring of the channel doping profile provides some additional design space to control V_T and punchthrough. Figure 1-5 shows the doping concentration contours resulting from such a LATIPS process, and more details on LATIPS are given in Chap. 5. These LATIPS (or pocket) implants, however, may be difficult to control at small dimensions.

1.1.4 Evolution of the Drain Structure of MOSFETs

When MOSFETs were first implemented in ICs, gate lengths were on the order of 10-μm. For MOSFETs with such long channel lengths, the drain-side of the *pn* junction at the drain end of the channel consisted of a simple single-region drain structure (Fig. 1-6a). As the gate lengths grew smaller than about 2.0–μm, the electric field in the depletion region of this drain *pn*-junction grew so large that it started to produce severe hot-carrier injection effects (HCI). These HCI effects caused long-term reliability-degradation in MOSFETs, and they had to be curtailed. The technique used to suppress HCI effects was to replace the single-region drain structure with a two-region drain structure (Fig. 1-6b). The two-region drain structure consisted of a region that was in contact with the MOSFET channel (and extended under the edge of the gate) and a second region that was further from the channel (but was connected to the first region). To reduce the electric field near the drain, the first region was more lightly doped (~$10^{17}/cm^3$) compared to the second region (~$10^{20}/cm^3$). The name given to this two-region drain structure is the *lightly doped drain* (or *LDD*).

The process sequence for forming this structure is as follows: **1**) After patterning the gate stack, the lightly doped drain region is formed with an ion implantation step using a medium-current implanter. The gate is used to align the drain dopant with that of the gate edge, forming the LDD region; **2**) A spacer is formed (by CVD of a dielectric film, followed by an anisotropic-etch step that leaves a spacer structure along the sidewalls of the gate); **3**) A high-current implanter is next used to introduce the dopants for the second heavily-doped drain-contact-region (with the spacer preventing these species from being implanted into the LDD region); and **4**) An anneal step activates the dopants implanted in both steps and drives the edge of the lightly doped region the proper distance beneath the edge of the gate.

In deep-submicron MOSFETs, the drain structure has been modified from the 2-region LDD-drain described above. The reason for this change is that the power-supply voltage has been reduced as MOSFET dimensions were scaled (primarily because gate oxide thicknesses have become thinner, and such thin oxides cannot withstand the oxide electric fields that exist at the higher supply voltage values – see Chap. 3). For MOSFETs at 0.18-μm, the power-supply

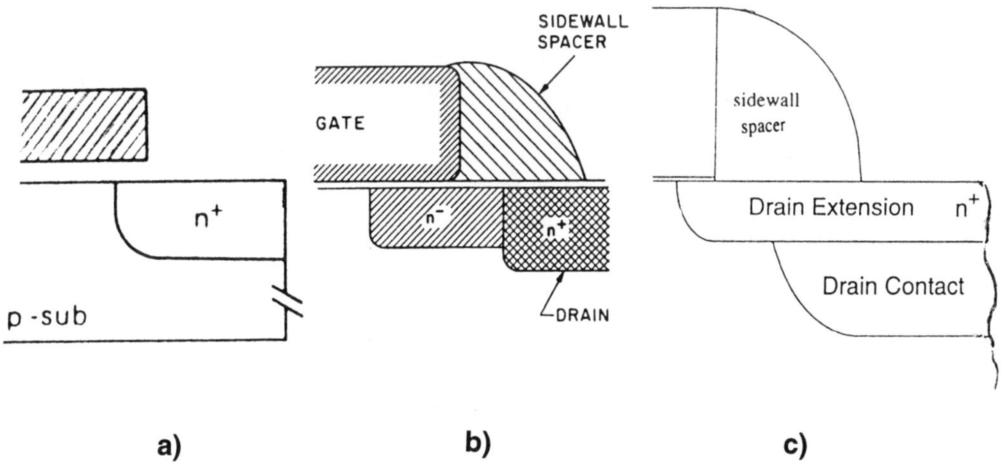

Fig. 1-6 a) Single-region drain in a MOSFET; b) Lightly-doped-drain in a MOSFET; c) Source/drain-extension and contact-drain-regions used in deep-submicron MOSFETs.

voltage has been reduced to about 2.0 V (down from the 5.0 V power supply value that was retained until devices got to the 0.8-μm generation). In any case, for such small drain-voltage values, the electric field near the drain is already reduced to a degree where few carriers can now gain sufficient energy to create interface states or shallow oxide traps. Hence, in deep-submicron MOSFETs HCI-effects no longer need to be controlled by the use of an LDD structure. However, the parasitic resistance contribution to the overall MOSFET ON-resistance due to the LDD region is increased to an unacceptably large degree.

To overcome the above problem, the two-region drain structure is retained. However, the region closest to the channel is no longer an LDD region. Instead, the first region is now termed the *source/drain-extension-region* (*SDE*) because the peak doping is no longer lighter than in the contact region. The increase in peak doping is achieved by raising the implant dose into the SDE region (from $1\text{-}5\times10^{13}/\text{cm}^2$ to $5\text{-}10\times10^{14}/\text{cm}^2$), which produces an extension region of the drain having lower resistivity. The peak doping concentration in the SDE is as high as in the second (contact) region, but its junction depth is shallower. This second drain region is called the *contact region* (Fig. 1-6c). It should also be noted that the penetration of the junction due to the SDE under the gate edge must be more carefully controlled as the gate length gets shorter.

The main challenge is to continue to make the SDE more shallow (to control short-channel effects), while retaining a sufficiently low resistance. Shallow n^+/p S/D junctions can be formed easily by arsenic (As) implantation, because its high mass produces smaller R_p, and the high As dose in the SDE region amorphizes the silicon, suppressing channeling effects. On the other hand, it is more difficult to make shallow p^+/n S/D junctions due to the low mass and susceptibility to TED of boron (the only p-type dopant with solid solubility suitable for this application). This has driven boron implant energies to the ultra-low (0.5-1.0 keV) range, and has forced the development of specialized ion implanters to do these implants economically. These and other issues involved with the formation of the SDE and the contact region are described in detail in Chap. 5.

1.2 DEEP-SUBMICRON CMOS STRUCTURES

Other structures besides just MOSFETs have to be formed when fabricating monolithic ICs. That is, isolation structures and interconnects are also needed.

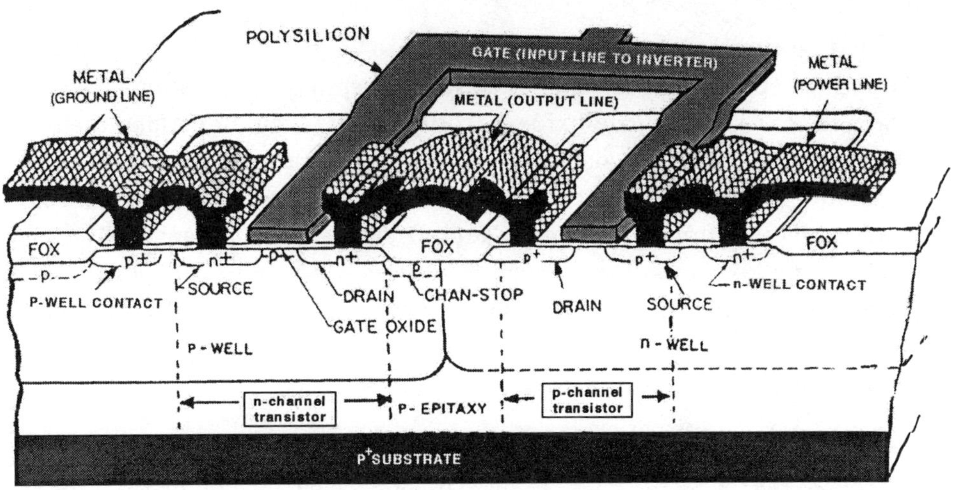

Fig. 1-7 Perspective schematic view of a twin-well CMOS structure.

Furthermore, the IC technology can be MOSFET-based (e.g., NMOS or CMOS), bipolar-junction-transistor (BJT) based (e.g., emitter-coupled-logic – ECL), or one that contains both BJTs and MOSFETs (e.g., BiCMOS). Since CMOS is the dominant silicon IC technology at the start of the 21st century, the focus will be on the CMOS structures within the silicon substrate that must be created when manufacturing deep-submicron CMOS ICs (Fig. 1-7).

1.2.1 Substrate Issues for Deep-Submicron CMOS

Czochralski-grown single-crystal silicon is the starting material on which silicon ICs are fabricated. Such single-crystal silicon is formed into *wafers*. There are two common doping-concentration configurations that ULSI wafers contain: **1**) the wafers are uniformly doped with a relatively light doping concentration of either *p*- or *n*-type dopant (and such wafers are referred to as *bulk* or, *polished wafers*); and **2**) the wafers are heavily *p*-doped during the crystal growing process, and then a lightly *p*-doped epitaxial Si layer is grown on the polished top surface of the wafer (and these wafers are referred to as *epitaxial*, *epi,* or *p-epi-on-*p^+ *wafers*).

Bulk wafers were used for NMOS technologies and some CMOS technologies (especially in chips that have lower selling prices, since the cost of bulk-Si wafers is about one-half that of epi wafers). The epi wafers were used for controlling latchup in CMOS logic circuits that needed high functional density but had a higher average selling price (ASP) than the chips that use bulk-Si wafers.

In deep-submicron CMOS technologies, the *epi wafers* are used for other roles besides just controlling latchup. The use of an epi layer on a heavily *p*-doped Si wafer improves the gate-oxide-integrity (GOI - an important factor that determines the overall yield of the chips), and also the intrinsic-gettering capability of the wafer. That is, some impurities in the Si wafers (e.g., iron) are gettered by the formation of metal-atom-B pairs (Fe-B) in the heavily-B-doped bulk region of the wafer. Chapter 2 discusses both the GOI issue and this gettering capability in detail. It also reviews the new, alternative substrates that are beginning to replace the conventional epi-wafers described above (e.g., such new wafer types as *epi optimized substrates)*.

1.2.2 Well Formation in Deep-Submicron CMOS

As described in Sect. 1.1.3, "conventional" wells in CMOS are formed by introducing the well dopants with a shallow implant process, and then diffusing them deeply into the silicon with a long, high-temperature drive-in step. This creates a well doping-concentration profile that has the highest doping concentration at the silicon/oxide interface, and decreases monotonically with distance into the depth of the silicon. Unfortunately, such a well-doping-profile forces tradeoffs to be made among various MOSFET device-performance characteristics, especially when fabricating deep-submicron MOSFETs.

To avoid this problem, and to enable independent optimization of key MOSFET performance characteristics, a different well-doping-profile must be produced. That is, by creating so-called *retrograde-wells* (instead of "conventional" wells), carrier mobility, S/D capacitance, inter-device leakage currents, V_T-control, punchthrough suppression, parasitic transistor effects, and latchup can be optimized almost independently. Retrograde wells are discussed in Ch. 5.

1.2.3 Dual-Doped Poly in Deep-Submicron CMOS

Conventional CMOS structures (down to about 0.35-μm) used n^+-poly for both NMOS and PMOS devices. Surface-channel MOSFETs are produced when n^+-poly is used as the gate material for NMOS transistors. However, the consequence of using n^+-poly as the gate material for PMOS transistors is that such PMOSFETs are buried-channel MOSFETs (see Vol. 3 for a complete discussion on surface-channel vs. buried channel MOSFETs, and why the use of n^+-poly makes buried-channel PMOSFETs inevitable). However, buried-channel PMOSFETs exhibit OFF-state leakage currents that are unacceptably large once MOSFETs are scaled to 0.25-μm (and smaller) gate lengths.

To overcome this problem, it is necessary to build surface-channel PMOSFETs (as well as surface-channel NMOSFETs) in CMOS technologies at 0.25-μm and smaller. This requires the use of p^+-poly as the poly gate material on the PMOSFETs (Fig. 1-8), and n^+-poly for NMOSFETs (in other words, the poly in the 0.25-μm [and smaller] CMOS circuits must be *dual doped*). Such a change has several implications. First, additional complexity is introduced into the CMOS process flow (e.g., selective dual doping of the poly layer must be carried out instead of single doping, requiring additional masking layers to permit such selective doping). Second, the p-dopant introduced (boron, B), can penetrate the gate oxide into the silicon substrate unless a sufficiently low thermal budget is used after this poly doping step (or unless the gate oxide is somehow made more resistant to B penetration). Such B penetration would cause unacceptable variations in the threshold voltage of the PMOSFETs. Third, B diffusion along the poly lines can counterdope the poly in other parts of the IC. Some of the issues involved with the use of dual-doped poly in deep-submicron ICs are discussed further in Chap. 3, as well as in Vol. 3, Chap. 5 of this series.

1.2.4 Shallow Trench Isolation for Deep-Submicron CMOS

LOCOS isolation was developed in the early 1970s, and it remained the workhorse isolation structure for NMOS and CMOS ICs until these technologies reached the 0.35-μm generation (Fig. 1-9a). However, for IC generations beyond

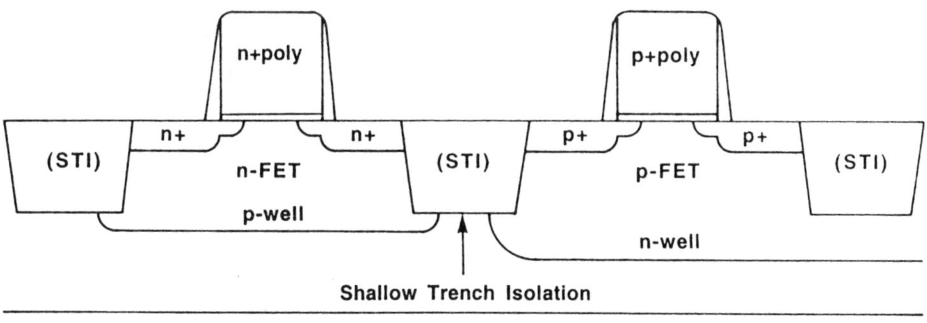

Fig. 1-8 Schematic cross-section of a CMOS structure with STI and dual-doped poly.

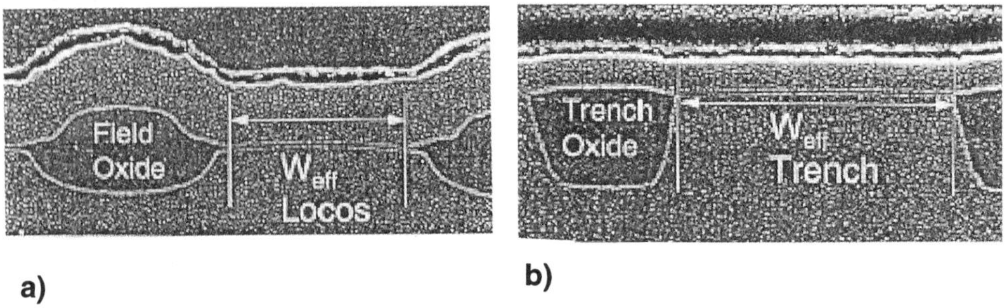

Fig. 1-9 a) Semirecessed LOCOS isolation structure; b) Shallow trench isolation structure.

0.35 μm, shallow-trench isolation (STI) replaced LOCOS. The drawbacks that LOCOS could not overcome were the field oxide bird's beak (that encroaches into the active areas of the chip) and the non-planar topography that results from thermally growing the semi-recessed field oxide.

Both of these problems are avoided by the use of STI (Fig. 1-8 and 1-9b). That is, in STI, trenches are etched and filled with oxide deposited by CVD to form the thick isolation oxide region. This method of forming the isolation oxide does not produce a bird's beak. In addition, by using CMP to planarize the surface in the course of forming this isolation oxide-structure, the wafer surface is flat at the conclusion of the STI-formation sequence. This eliminates the topography that exists at the conclusion of the LOCOS formation process. Chapter 9 covers the details of STI and CMP is discussed in Chap. 8.

1.2.5 Silicon-On-Insulator (SOI) Technology

CMOS technology has largely been built on so-called bulk-Si wafers. In such wafers the devices are isolated from one another by a combination of isolation structures at the silicon surface, and *pn* junctions in the bulk of the Si. However, latent current paths among devices may still be activated in the bulk regions by such phenomena as latchup or transient radiation events. Furthermore, parasitic circuit elements associated with the CMOS structure (such as source-to-substrate and drain-to-substrate capacitances) can become too large in CMOS structures built on bulk-Si wafers. As a result, the use of *silicon-on-insulator wafers* (SOI) instead of bulk-Si wafers for some CMOS applications is being pursued. Device isolation and latchup control are simplified for SOI with fewer processing steps required. SOI technology is described in Chap. 11

1.3 LIMITS TO CONVENTIONAL MOSFET SCALING

Conventional scaling of MOSFETs, with some modifications to the device structures (as described in the previous sections) has allowed MOSFETs with gate dimensions to be reduced from 10-μm in the 1970s down to 0.1-μm. (Figure 1-10 shows a SEM micrograph of 0.18-μm MOSFET transistors in typical logic structures.) Certain MOSFET structures were essentially just scaled (without undergoing any fundamental structural changes) as the minimum device dimension was reduced to about 0.1-μm. These include: the gate-oxide thickness; the source/drain junction depth; and the gate length.

However, below 0.1-μm, MOSFET devices built according to the structure described in this chapter (and depicted in Fig. 1-2), appear to run into some fundamental limits. These limits would prevent further size reductions if we were restricted to using only such traditional scaling approaches. Table 1-2 (as postulated by Thompson *et al.*), catalogs these limits.[4] Consequently, to enable MOSFETs to be scaled even further, new solutions will need to be developed, such as the use of high-k materials for gate dielectrics (see Chap. 4), and novel shallow, ultra-low-resistivity junctions (see Chap. 5). Other issues related to these limits also arise, and will be considered in the chapters of this book. For example, advanced lithography challenges will be covered in Chaps. 6 and 7, and

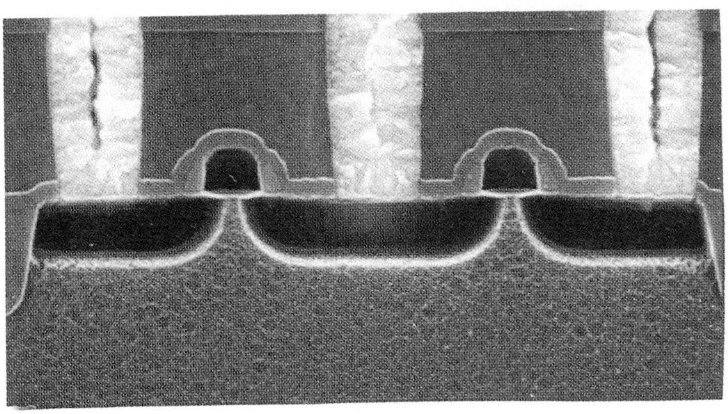

Fig. 1-10 A SEM micrograph of 0.18-μm MOSFET transistors in logic structures.

Table 1-2 SCALING LIMITS OF CONVENTIONAL MOS DEVICES[4]

FEATURE	LIMIT	REASON
Oxide Thickness	2.0 nm	Leakage (I_{gate})
Junction Depth	30 nm	Resistance (R_{SDE})
Channel Doping	$V_T = 0.25$ V	Leakage (I_{off})
SDE Under-Diffusion	15 nm	Resistance (R_{inv})
Channel Length	0.06 μm	Leakage (I_{off})
Gate Length	0.10 μm	Leakage (I_{off})

alternative device architectures, such as Si:Ge technology and SOI CMOS will be discussed in Chaps. 10 and 11, respectively. Issues concerning ULSI interconnect issues are explored in Chaps. 12 to 16. A recent review by Hutchby *et al.*, discusses the novel structures being considered to replace conventional MOSFETs that are still being used in ICs at the start of the decade of 2000.[8] It describes how such nanoelectronic structures for memory, logic, and information processing architectures could possibly extend the time frame of the ITRS beyond CMOS.

REFERENCES

1. Overall Roadmap Technology Characteristics Chapter. The International Roadmap for Semiconductors, Semiconductor Industry Association, 2001 Edition.

2. S. Wolf, *Silicon Processing for the VLSI Era*: Vol. 3 – The Submicron MOSFET, Lattice Press, Sunset Beach CA, 1995.

3. Y. Taur and T.K. Ning, *Fundamentals of Modern VLSI Devices,* John Wiley & Sons, New York, 1997.

4. S. Thompson *et al.*, "MOS Scaling: Transistor Challenges for the 21st Century," *Intel Technology Journal*, 3rd Quarter 1998.

5. S. Parihar *et al.*, "A High Density 0.10 μm CMOS Technology Using Low-*k* Dielectric and Copper Interconnect," *Tech. Dig. IEDM,* 2001, p. 249.

6. S. Thompson *et al.*, "An Enhanced 130-nm Generation Logic Technology Featuring 60-nm Transistors Optimized for High-Performance and Low-Power at 0.7 and 1.4 V," *Tech. Dig. IEDM,* 2001, p. 257.

7. T. Hori, "A 0.1 μm CMOS Technology with Tilt-Implanted Punchthrough Stopper (TIPS)," *Tech. Dig. IEDM* 1994, p. 75.

8. J.A. Hutchby *et al.*, "Extending the Road Beyond CMOS," *IEEE Circuits & Devices Magazine,* March 2002, p. 28.

Chapter 2

300-mm SILICON WAFERS

This chapter deals with 300-mm wafer technology, including the growth of 300-mm Si crystals, 300-mm wafer specifications, and the economics of the 300-mm wafer industry. Readers wishing background information on material properties of silicon, general information on Si crystal growth and wafering can consult *Silicon Processing for the VLSI Era - Vol. 1: Process Technology, 2nd Edition.*

2.1 300-mm SILICON WAFER GROWTH

The fabrication of VLSI and ULSI takes place on Si substrates possessing very high crystalline perfection. The critical importance of utilizing single-crystal material for the transistor regions of microelectronic circuits was first recognized by Teal in 1949.[1] He reasoned that polycrystalline material would exhibit inadequately short minority-carrier-lifetimes, due to the defects at the grain boundaries of the polycrystalline grains. Teal also pointed out the need to define the degree of chemical purity and crystal perfection, the uniformity of structure, and the chemical composition of the single-crystals when specifying starting material on which to fabricate semiconductor products. The method of obtaining such single-crystal Si for IC fabrication, involves several steps (Fig. 2-1):

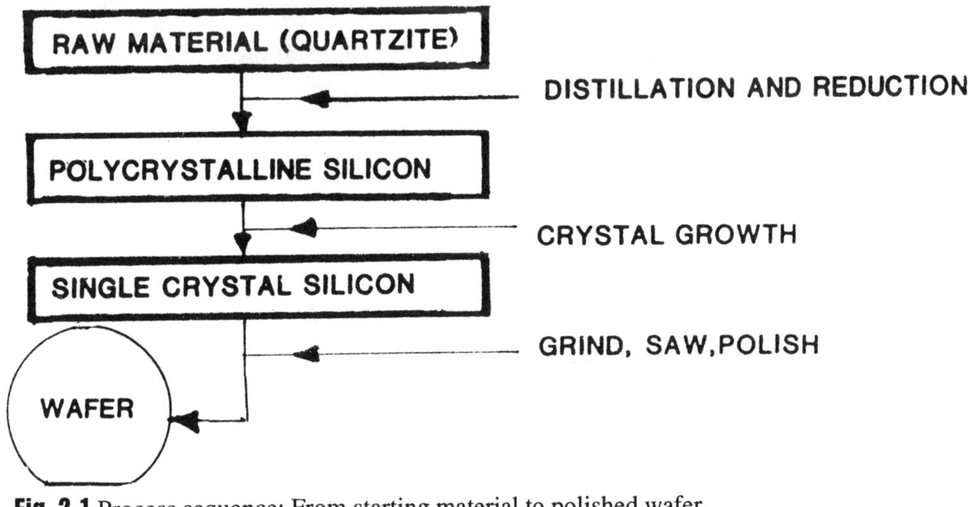

Fig. 2-1 Process sequence: From starting material to polished wafer.

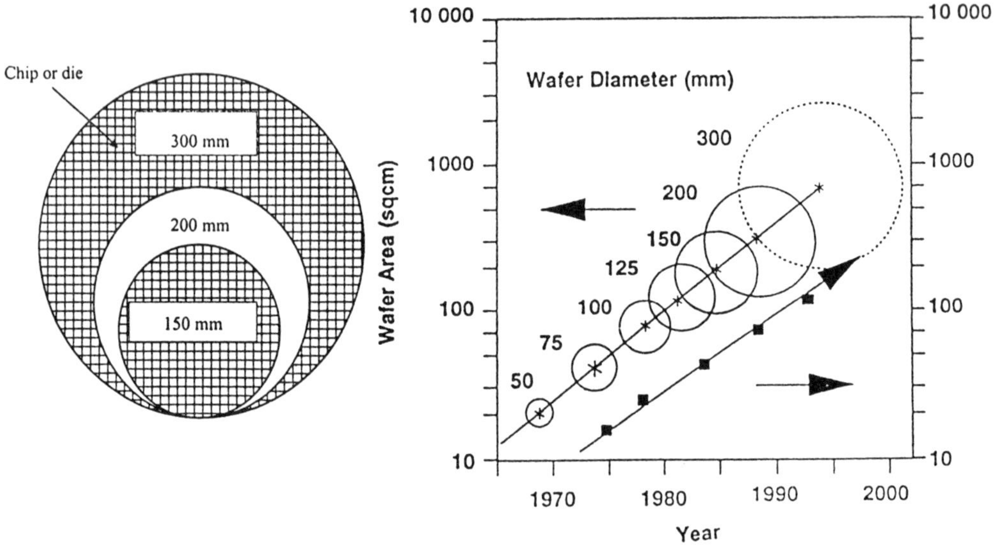

Fig. 2-2 The increase with time of CZ silicon crystal diameter and the percentage increase in wafer area with increase in wafer diameter.

1. Raw material (e.g., quartzite, a type of sand), is refined by a complex, multi-stage process that produces *electronic grade polysilicon*, EGS (Fig. 2-1).

2. This electronic grade polysilicon is used to grow single-crystal silicon by *Czochralski* (CZ) crystal growth or *float-zone* (FZ) crystal growth.

In CZ growth, single-crystal ingots are pulled from molten silicon contained in a crucible. Czochralski silicon is preferred for ULSI applications since it can withstand thermal stresses better than FZ material,[2] and it is able to offer an internal gettering mechanism (not available with FZ crystals) that can remove unwanted impurities from the device structures on wafer surfaces.

The simplicity of the CZ process and the relatively high-degree of crystal purity helped establish it as the dominant silicon-crystal-growing technology in the early years of the semiconductor industry. The development of dislocation-free ingot growth and automatic diameter control in the late 1960s lead to a rapid growth in the sizes of ingot diameters and crystal melts. Further increases in charge size and diameters were driven by the economic advantages of even larger ingots. As shown in Fig. 2-2, by 1998 300-mm diameter crystals had become available, grown with 150-kg charges. Typical commercial ingots are grown 0.5–2.5-m in length, and a 100-kg charge can grow an ingot roughly 1-m long

a) b)

Fig. 2-3 a) 300-mm Si crystal after it has been removed from a 300-mm CZ puller; b) 300-mm ingot cooling on a materials-handling device.

with a diameter of 200-mm. The driving function for going to larger wafer diameters is the increase in Si area available per wafer. Figure 2-2 also shows that the area of a 300-mm wafer is 2.25 times the area of a 200-mm wafer. Figure 2-3b shows a 300-mm Si ingot mounted in a materials handling device. We will only concern ourselves in this section with the issues involving the growth of 300-mm Si-crystals.

The wafer manufacturing process for 300-mm wafers is not a straightforward extension or scale-up of 200-mm processes. Most tools needed to produce 200-mm wafers, including crystal pullers, grinders, polishers, cleaners, and inspection stations are not capable of being cost-effectively utilized for 300-mm wafer production. In addition, the cost of growing 300-mm silicon crystals is the principle challenge facing technologists striving to match the 200-mm unit-area cost performance. That is, to be able to achieve the same yield of 300-mm wafers as is possible with 200-mm wafers, the size of the charges will have be increased from the 150-kg charge being used in 1998 for 200-mm wafers to 300-400 kg for 300-mm wafers. The equipment used to pull crystals from such charges will also have to be able to rotate and lift the crystal while stirring and positioning the

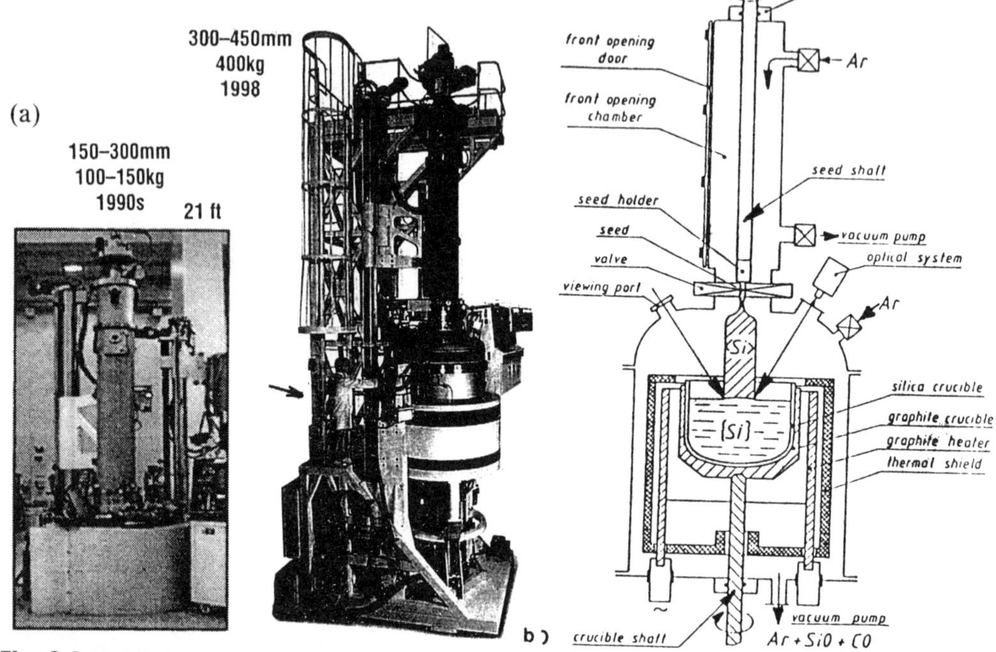

Fig. 2-4 a) Modern computer controlled CZ pullers. Larger silicon wafer sizes and the required manufacturing efficiency have driven crystal puller size. b) Schematic drawing of CZ puller.

melt, doing this all the while without vibration, in a high-vacuum atmosphere, and at temperatures that exceed 1400°C. Figure 2-4 shows how the size of the crystal puller has to be increased with respect to the size of a 200-mm puller to make it possible to handle such larger charges.

The charge must be held in a single-use, high-purity quartz crucible more than 30 inches in diameter. Demands on the crucible increase significantly for a 300-mm process, as the growth cycle time stretches to over 100 hours (~4 days). Any flaw that develops in the crucible during that multi-day crystal-growth time can cause a loss in atomic alignment of the crystal, making it unusable.

Another important factor that impacts the cost of producing the wafer is the crystal pull speed (i.e., the longer it takes to pull the crystal, the higher the cost of producing the wafer). However, in growth chambers of identical design, this speed $v_{pullrate}$ is inversely proportional to the crystal diameter $D_{crystal}$ or, $v_{pullrate} \sim (1/D_{crystal})$. The reason that the pull speed must be decreased as the diameter increases is that the crystal mass increases as the square of the ingot radius, while the surface area of the ingot (which radiates away the heat of fusion), is proportional to the diameter. Thus, a path toward reducing the cost of manufacturing larger diameter wafers is to develop new and innovative puller designs that allow the heat to be removed more quickly than in a conventional puller (a project apparently being pursued by most wafer suppliers).

Increasing the pull speed, however, is only one relatively small aspect of decreasing the cost of manufacturing large diameter wafers because there are many steps that comprise the entire growth cycle. (As shown in Fig. 2-5, by tripling the pull speed from 0.5 mm/min to 1.5 mm/min, the wafer production cost is reduced by only about 5%.) Two other significant factors that increase the cost of producing 300-mm wafers above the cost of producing 200-mm wafers are: **1**) a 300-mm wafer is 50-μm (or 7%) thicker than a 200-mm wafer; and **2**) the back surface of a 300-mm wafer is polished, whereas the back surface of a 200-mm wafer is not. The back-surface polishing process removes an additional 10-20-μm of silicon stock. In total, a 300-mm wafer therefore consumes about 9% more Si-crystal per unit area than a 200-mm wafer, resulting in fewer wafers per unit-length of ingot.

2.2 GROWN-IN CRYSTAL DEFECTS: CRYSTAL-ORIGINATED PARTICLES (COPs) and DISLOCATION LOOPS

As the silicon crystal is being grown in the CZ growth process, several phenomena occur that produce defects in the as-grown crystal. Since these defects arise from the various aspects of the CZ crystal growth sequence (namely, solidification from the melt and cool-down to room temperature), they are termed *grown-in defects*. As the minimum device dimensions in ICs are scaled below 0.25-μm, grown-in defects have been identified as a potential factor that reduces the performance characteristics and decreases the yields of CMOS ICs. The most significant of these grown-in defects are the so-called *crystal-originated defects* (COPs). The COP defects have the form of voids in the Si lattice, and these voids are caused by agglomeration of supersaturated vacancies

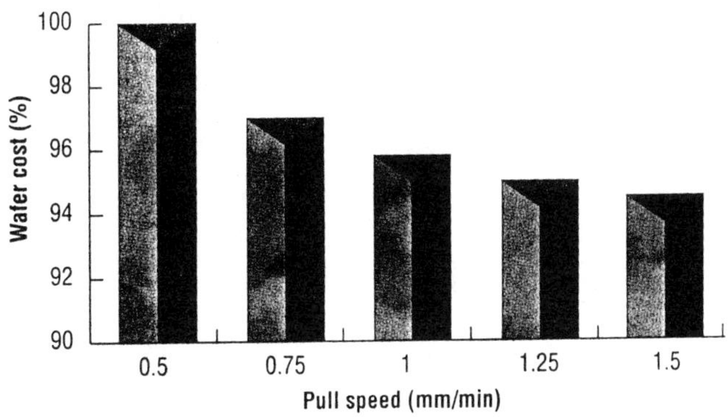

Fig. 2-5 300-mm crystal-growth-cycle-time components.

that are introduced into the crystal as the liquid-phase-silicon solidifies. COPs cause two major problems when such voids intersect the wafer surface: **1**) they degrade the integrity of thin gate-oxide films (GOI); and **2**) they create leakage paths in the isolation structures of DRAMs.[3] In this section we describe the mechanisms that give rise to these COPs and provide details about the problems they cause in the devices fabricated on wafers containing such defects. Finally, we discuss measures that have been employed to mitigate their effects.

Note that another form of defect is also prevalent in CZ Si wafers, namely oxygen precipitates. These defects, however, form during IC processing (instead of during the crystal growth process itself), and thus oxygen precipitates are not considered to be grown-in defects. However, their density in the wafer (number/cm^3) and their size also strongly correlate with prior crystal growth conditions and subsequent heat treatments.

It is believed that the grown-in defects described above (i.e., COPs - as well as their complementary counterparts, dislocation loops) arise from an agglomeration of intrinsic point-defects (namely, vacancies to form COPs and self-interstitials to form dislocation loops), during the cooling of the silicon ingots following the crystallization from the liquid phase. The formation of voids from the homogeneous nucleation (agglomeration) of supersaturated vacancies in silicon was first suggested by Voronkov in 1982.[4]

Intrinsic point defects are incorporated into the silicon when it crystallizes. Estimates from computer simulation indicate that their thermodynamic equilibrium concentrations at the melting point of silicon (1412°C) are ~$1x10^{15}$ cm^{-3} for vacancies, and about 20% less for self-interstitials.[5,6] As the crystal cools continually during CZ growth, the solubility of these intrinsic point-defects decreases, resulting in a condition of supersaturation. Intense recombination of the vacancies and interstitials occurs when the solidified crystal is cooling through the high-temperature range (i.e., from 1412°C to ~1200°C). At the same time, some net out-diffusion of self-interstitials occurs from regions near the surface (outer circumference) of the Si ingot. That is, more self-interstitials out-diffuse from the ingot surface than do vacancies because the self-interstitials diffuse 10 times faster than vacancies near the silicon melting-point temperature.

Computer simulations further indicate that ~10% of the original point-defects in both the vacancy- and interstitial-rich regions survive after the intense recombination events which occurred during the cooling through the high-temperature range, but that these surviving point-defects still remain in supersaturation when the temperature of the ingot drops below ~1200°C. Since the point-defects continue to diffuse rather rapidly at temperatures above

~1150°C, they can come out of solution either by further recombination events, or by agglomerating into clusters of vacancies or self-interstitials. The vacancies have been found to agglomerate into octohedral voids at about 1150-1180°C. As noted earlier, these voids formed by vacancy-agglomeration are called crystal-originated particles (COPs).[7,8] Supersaturated self-interstitials, on the other hand, can form large dislocation loops. CZ-silicon-crystals, however, are pulled under conditions where vacancies are the prevailing type of intrinsic point-defect, and as a result, self-interstitial-agglomerated dislocation loops do not occur in significant numbers in 200-mm Si wafers used in IC production. Thus, our discussion of grown-in defects is primarily focused on COPs.

2.3 DETAILS OF THE FORMATION OF CRYSTAL ORIGINATED PARTICLES (COPS)

In this section we provide additional information about the formation of COPs and the effects they have on IC devices. We begin with a more detailed discussion of how COPs are formed during the cooling of the Si ingot.

The formation of COPs was studied by employing crystal-growth-halting-and-quenching techniques. That is, during CZ-growth the crystal ingot pull-rate was abruptly halted, and the ingot temperature was quickly lowered (quenched) to some specific value. By doing this with a number of ingots and quenching to different temperatures, the defects formed during the cooling of the ingot from solidification to that temperature could be examined. Such studies indicated that the formation of COPs consists of two stages. In the first, voids are rapidly formed as the ingot cools through the narrow temperature range of 1150-1180°C. In the second, an oxide is formed on the inner surface of the voids during the cooling-down process following void formation (down to about 900°C). The resultant voids formed within the bulk have been identified by TEM studies as having a tetrahedral shape, and with sizes between 0.1- and 0.3-μm (100-300-nm), Furthermore, their inner surfaces are covered with a thin oxide layer approximately ~5-nm thick (Figs. 2-6 and 2-7).[9] Such COPs are formed with densities between 10^5 to 10^7 cm^{-3}.[10]

It should be noted that before the nature of grown-in defects was well understood, COPs were identified in the literature under different names, depending on the technique that was used to delineate and detect them. These terms included: *COPs*, *laser-scattering tomography defects* (LSTDs), *flow-pattern defects* (FPDs), and *Secco etch-pit defects* (SEPDs). However, by comparative observation, it was confirmed that they were in fact the same type of defect (namely COPs), that just appear under different guises (depending on the technique used to detect them).

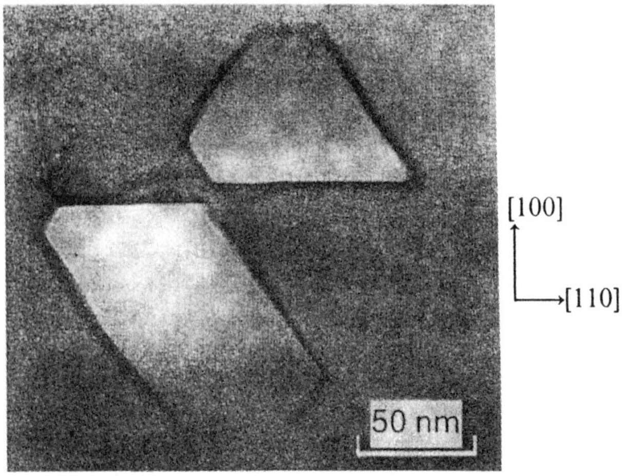

Fig. 2-6 SEM of COPs as they appear on the surface of a Si wafer. Oxide layer at the void:crystal-interface is visible.

In fact, the most common method used to make COPs visible with optical instruments when they intersect a wafer surface is to immerse bare silicon wafers in hot RCA-Clean SC-1 solutions (e.g., 4 h @ 85°C). The SC-1 solution etches Si slowly and isotropically, and this enlarges the lateral and vertical dimensions of any COPs that intersect the surface. They appear with the shapes shown in Figs. 2-6 and 2-7.[11] This shape enlargement also makes it possible for laser-based defect-detecting tools (which do not have as high a resolution capability as do TEMs - which can be used to image COPs directly), to detect the presence of any COPs as *localized light scatterers* (LLDs). By using this delineation technique, the entire wafer surface can be quickly mapped for the presence of COPs with

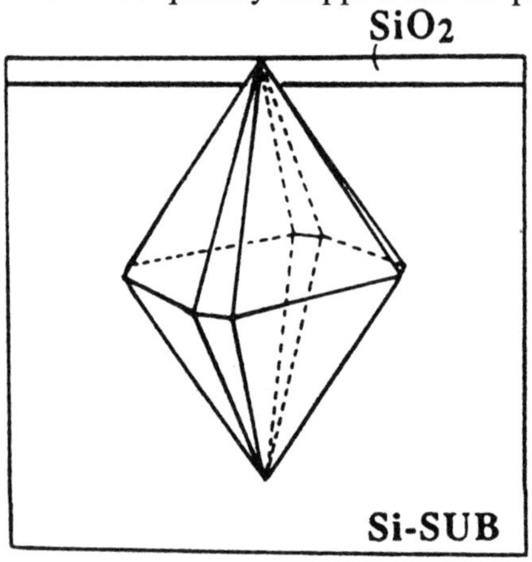

Fig. 2-7 Schematic representation of the octahedral structure of a COP defect.

an automated scanning surface inspection system (such as the KLA-Tencor Surfscan).

2.3.1 The Radial Distribution of Grown-In Defects on the Wafer Surface

As described previously, typical 200-mm wafers being used for ULSI today have a central vacancy-rich region surrounded by a self-interstitial-rich edge. These regions are separated by the vacancy/self-interstitial (V/I) boundary that occurs at a critical value of a parameter, ς_C, defined as:

$$v/G(r) = \varsigma_C = 0.138 \text{ mm}^2/\text{min-K} \quad \textbf{(2.1)}$$

where v is crystal pull rate (in mm/min), G(r) is the axial temperature gradient {i.e., G(r) = dT(r)/dr [in K/mm]} at the crystal/melt interface at a radial position (r) from the center of the ingot, and T is the absolute temperature (in kelvins).[12]

The axial temperature gradient, G(r), at the center of the crystal ingot is usually lower than at the ingot surface. In addition, v is the same across the crystal growth interface at a given pull rate. Thus, in conventional CZ crystal growth, the value of v/G(r) is highest at the radial center of the ingot and decreases continually toward the crystal-ingot edge. The wafer central-region inside the V/I boundary is thus vacancy-rich, characterized by $v/G(r) > \varsigma_C$, while the edge-region outside the V/I boundary is self-interstitial-rich, with $v/G(r) < \varsigma_C$.

The concentration of the vacancies (C_V) and the self-interstitials (C_I) at a given radial position, r, relative to those of the equilibrium levels of (C_V^{eq}) and (C_I^{eq}) at about 1200°C, after passing the intense recombination and out-diffusion regime between the melting-point and ~1200°C, are apparently exponentially proportional to the difference between v/G(r) and ς_C, or:

$$(C_V/C_V^{eq})_r = \exp\{\alpha\,[v/G(r) - \varsigma_C]\} \quad \textbf{(2.2)}$$

$$(C_I/C_I^{eq})_r = \exp\{\beta\,[\varsigma_C - v/G(r)]\} \quad \textbf{(2.3)}$$

where α and β are not known at this time. The supersaturation of vacancies is highest at the radial center, and decreases radially from this value. Thus, more voids are formed at the central-region of the wafer. On the other hand, high supersaturation of self-interstitials is present at the crystal peripheral (edge) region. But, because of out-diffusion of the self-interstitials to the crystal surface (a result of the fact that self-interstitials diffuse 10 times as fast as vacancies near the melting point of Si), high-supersaturation of self-interstitials occurs at some distance from the edge.

In the v/G(r) ratio, the thermal gradient (G[r]) is controlled by the thermal environment in the hot-zone of the crystal puller, and by melt convection. With an appropriately designed crystal-puller, the value of G(r) can thus, in theory, also be controlled. The pull rate v is much more easily controllable. Computer simulations of both the melt convection and the thermal distributions within the puller hot-zone have been extensively employed to modify the design of the crystal puller so that an optimum v/G(r) value can be obtained. The recent rapid progress in the field of simulating crystal-growth processes permit the bulk properties and process stability of the growth processes to be predicted with remarkable accuracy. This should help reduce R&D costs, a requisite for allowing wafer manufacturers to profitably develop the new generations of wafers being called for by the industry.[13]

2.4 THE OXYGEN-STACKING-FAULT RING (OSF-RING)

It is observed that when a typical, polished 200-mm CZ-grown wafer (as described above) is oxidized and then etched in HF and SC-1 to delineate the defects present at the wafer surface, it manifests a multi-ring defect structure. The central region of the wafer contains COPs. The COP-filled region is surrounded by a ring of *oxidation-induced stacking faults*, OSFs (i.e., which emerge after the wafer is subjected to a thermal oxidation step). This feature is also termed an *OSF-ring* (Fig. 2-8). The region of the wafer outside of the OSF-ring is an essentially defect-free region that contains a very low density of dislocation loops. This region extends to the edge of the wafer (Fig. 2-9).[14] The origin of the COPs in the central region has already been described. Here we explain what is believed to be the cause of the OSF-ring.

In the OSF-ring, the delineated defects are oxidation-induced stacking faults (OSFs). These stacking faults, however, arise from the presence of oxygen precipitates.[15] The oxygen precipitates are the actual defects that exist in the as-grown state of the ingot. But since these precipitates act as heterogeneous nucleation sites for the OSFs, the OSFs become the defects that are delineated by the SC-1 etch. Hence, this region is called the OSF-ring. An important aspect of the OSF-ring is that the ring-radius decreases with decreasing pull rate, until at a critical pull-rate-value (that depends on thermal conditions), it disappears into the center of the crystal. The presence and radial location of an OSF-ring in CZ-silicon corresponds qualitatively to the neutral (or boundary) region that separates the vacancy-rich- and self-interstitial-rich-regions of the ingot.[16] (Sometimes anomalous oxygen precipitation is observed between the OSF-ring and the defect-free outer-region, as shown in Fig. 2-9, but we shall not consider this anomalous phenomenon here).

The OSF-ring is thought to be a region in which vacancies are only in slight

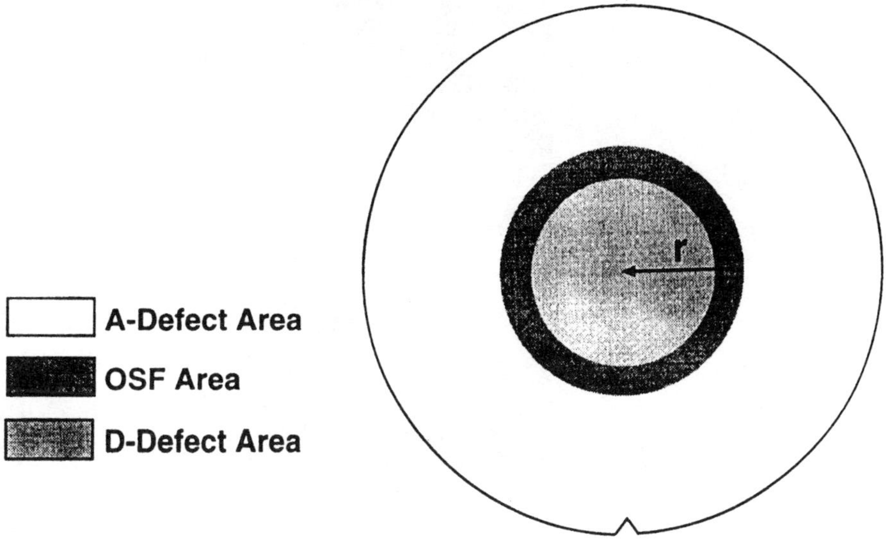

Fig. 2-8 Schematic drawing of the ring regions. An OSF-region after oxidation separates the A-defect-rich region (dislocation loops) from the D-defect-rich region (COPs) .

excess over self-interstitials, so recombination proceeds over a longer period of time, since neither species is strongly dominant. This competition results in a lower vacancy supersaturation and a delay in the onset of nucleation. Smaller voids are formed and the final vacancy concentration in the region of the OSF-ring is actually higher than inside the ring. These vacancies may provide the needed free volume for formation of the oxide precipitates that are observed at the center of the stacking faults in the OSF-ring region.[17]

2.5 MITIGATING THE EFFECTS OF COPS WITH THE USE OF POST-CRYSTAL-GROWTH ANNEALING

As we noted previously, for ULSI circuits it is important that the wafer be free of grown-in defects and oxide precipitates in the surface and sub-surface device regions of the wafer. Typically, a high-temperature annealing step is performed on the wafers to out-diffuse the oxygen near the surface, so that the concentration of oxygen there drops below the solubility limit. Thus, oxygen precipitation can no longer occur in this region, and no oxide precipitates are formed. High-temperature annealing processes have also been investigated as a method of annihilating the COPs near the wafer surface. That is, annealing at high temperatures (typically ≥ 1200°C) in an inert ambient (most commonly hydrogen, but also in argon) has been used to reduce the COP density near the wafer surface. Such hydrogen-annealed CZ-polished-wafers were introduced by

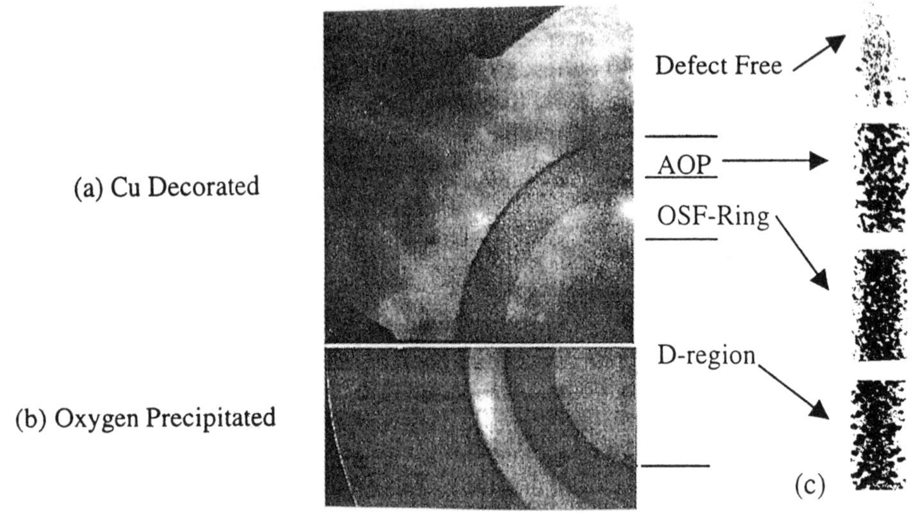

Fig. 2-9 a) Plane x-ray topograph of Cu-decorated CZ-Si wafer; b) Plane x-ray topograph of oxygen-precipitated CZ-Si wafer. The D-region, OSF ring, and defect-free regions are observed. c) Details of the defects delineated in each region.

Toshiba Ceramics in 1986, and they have been commercially available under the name of Hi-WAFER.[TM,18] Other silicon wafer vendors have also offered annealed CZ-polished-wafers (annealed in other gases). It should be noted that it has been reported that the annealing must be carried out in an oxygen-free ambient, otherwise the annealing procedures do not produce optimal benefits.[19]

The mechanism thought to be responsible for reducing the COP density in hydrogen-annealed wafers is as follows: The inner surface of the COP void is covered with a thin oxide. Under high-temperature annealing in a hydrogen ambient, the hydrogen species diffuse into the Si. The oxide on the COP surfaces (in regions near the wafer surface) is decomposed by the reduction effect of hydrogen and oxygen. Eventually the COP surface becomes bare-Si and the surface atoms within the COP-void are mobile enough to reconstruct this surface and cause the void to shrink (by vacancy-diffusion into the Si bulk). As the process of void dissolution proceeds, the void is first transformed into a spherical structure from a regularly octahedron (because this decreases the surface free-energy of the void during annealing).[20] It has also been proposed that a heavily-doped buried layer (created by high-energy implantation of a heavy boron dose) be incorporated into Hi-WAFERs. In that way, the benefits of a heavily-doped boron region in the substrate (as is available with p/p^+ epitaxial wafers) can also be obtained with a Hi-WAFER, at a lower cost than that of such epi wafers.[21]

Hydrogen annealing at 1150°C for several hours to reduce COP levels extends the usefulness of 200-mm polished wafers. However, it adds 15-20% to the wafer cost, and according to some opinions, it will not be a feasible technique for 300-mm wafers.[22]

In another report, wafers were annealed at 1300°C in an inert ambient for one hour. No observable COPs existed (i.e., at the wafer surface) after this annealing.[23] It was reported in another study that very-good GOI was obtained using such wafers. However, these wafers also showed a very high concentration of oxide precipitates after the thermal processes used during the IC fabrication were carried out. The presence of these precipitates may give rise to excess leakage values in such products as high-density DRAMs. As a result, this report concluded that the high-temperature-annealing process needs further optimization.[24]

2.6 ELIMINATION OF COPS WITH "PERFECT SILICON"

A special CZ-silicon-growth technique has been developed that eliminates COPs from silicon wafers by preventing their formation during the ingot growth process. Wafers obtained from ingots grown according to this technique are referred to as *"perfect" silicon wafers*, or *"pure" silicon wafers.*[25] The idea behind this approach is to create growth conditions such that the concentrations of intrinsic point defects remain close to their quasi-equilibrium values all along the radius of the crystal-melt interface. In that way, the silicon is not super-saturated with either vacancies or self-interstitials. Under these circumstances, regions will still exist that are either vacancy-rich, or self-interstitial-rich. But neither COPs nor dislocation loops can form (because even in those regions there is no supersaturation of the point-defect species which dominates in that region). The goal of the "perfect" silicon-growth-process is to the establish the conditions quantified by the following set of conditions:

$$v/G(r) - \varsigma_c \approx 0, \quad C_V \approx C_V^{eq}, \text{ and } C_I \approx C_I^{eq} \qquad \textbf{(2.4)}$$

The name "perfect silicon" was given to such point-defect-equilibrated silicon crystals because wafers from such crystals after immersion in a defect-delineation etch exhibit neither a boundary between vacancy-rich/self-interstitial-rich regions nor an OSF-ring. Instead, the wafer appears "perfect" or "pure." Specifically, a "perfect" silicon wafer will show negligible COP densities, so yield and performance of ULSI devices on such wafers will be enhanced to the levels exhibited by devices fabricated on epitaxial wafers. Growth processes and crystal pullers that can produce "perfect" silicon were being developed in 2000 to be able to produce 200-mm "perfect" silicon wafers in production quantities, and subsequently this effort is expected to be carried out for 300-mmwafers. However, there are some difficulties associated with this approach. The first is that the hot-zone in the crystal puller must be optimized so that it becomes possible to maintain the condition of $v/G(r) = \varsigma_c$ along the whole length of the crystal. The second is that a relatively slow pull-speed is needed to maintain the $v/G(r) = \varsigma_c$ condition. The latter requirement is not favorable in terms of the crystal yield.

The ingot yield from a crystal puller is reduced by this method for two reasons. First, the slower pull-speed reduces the throughput of the puller. Second, because the ingot growth cycle becomes much longer, more time is allowed for a disruptive event to occur during the crystal growth time (which would render the ingot useless). Nevertheless, since the cost to produce "perfect" silicon wafers seems to be less than the cost of producing epitaxial wafers, there is still an incentive to pursue the development of "perfect" silicon growth processes.

A report comparing the behavior of MOS-capacitor structures and *pn*-junction diodes fabricated on conventional polished CZ wafers, "perfect-silicon" CZ wafers, and p/p^+-epitaxial wafers was published by Niwa.[26] Junction leakage currents were smaller on the "perfect-silicon" and p/p^+-epitaxial wafers than on the conventional CZ wafers. GOI characteristics were also superior on the "perfect-silicon" and p/p^+-epitaxial wafers, and the "perfect-silicon" GOI results matched those obtained with the epitaxial wafers.

2.7 MINIMZING COPS PERFOMANCE DEGRADATION WITH HIGH-PULL-SPEED SILICON

Another approach that is being pursued to achieve wafer quality similar to "perfect-silicon" is to use a crystal pull-rate that is much faster than that used to pull a "perfect-silicon" ingot (three or four times as fast). With this faster pull-rate, G(r) is increased to some extent as a result of the higher pull-rate. However, since v is much larger than in the "perfect-silicon" case, the overall v/G(r) value becomes substantially larger than the critical value $\varsigma_c = 0.138\ mm^2/min\text{-}K$. Thus, the crystal becomes completely vacancy-rich and the vacancy concentration at temperatures below 1200°C becomes much larger than the quasi-equilibrium value (i.e., the crystal becomes supersaturated with vacancies), or:

$$v/G(r) >> \varsigma_c, \text{ and } C_V >> C_V^{eq} \tag{2.5}$$

Nevertheless, due to the high pull rate, the crystal passes quickly through the 1150-1180°C COP-forming temperature-range, so only COPs of smaller sizes and at a high density can be formed. Such small-sized voids near the wafer surface can be annihilated by post-growth wafer-annealing. Doping the ingot with nitrogen during crystal growth is another factor that contributes to the mini-

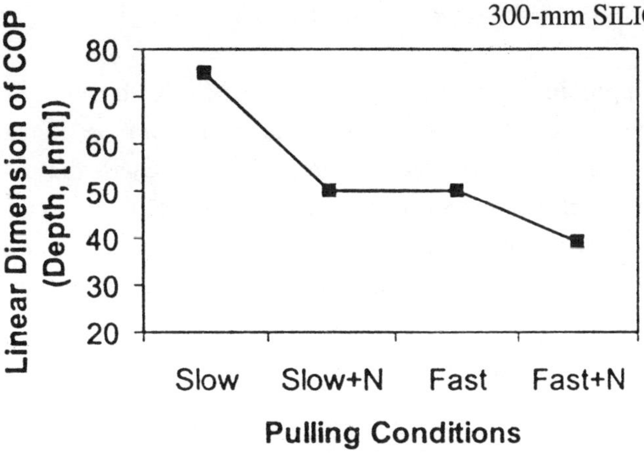

Fig. 2-10 Size of COPs in a silicon wafer as a function of pulling conditions.[28]

mization of void formation.[27] Figure 2-10 shows how the size of COPs is decreased by combining fast-pull and nitrogen-doping of CZ-ingots.[28] The fast-pull-approach offers potential advantages of high throughput and yield, and lower cost.

The process of performing the CZ-ingot-growth process at a high pull-speed and simultaneously doping the ingot with nitrogen (to get small COPs) has been combined with hydrogen annealing of wafers obtained from such ingots. By using infra-red topography, the concentration of COPs near and at the wafer surface (i.e., a technique that can detect the presence of COPs within 20-μm of the surface),[29] it was observed that the concentration of COPs decreases dramatically to depth of over 10-μm when such wafers are annealed in hydrogen at 1200°C.[30] The GOI of these wafers annealed in hydrogen was also significantly improved (Fig. 2-11).

Some concerns exist about the fast-pull CZ-Si growth processes. First, the relatively high vacancy concentrations remain in the wafer (which could cause

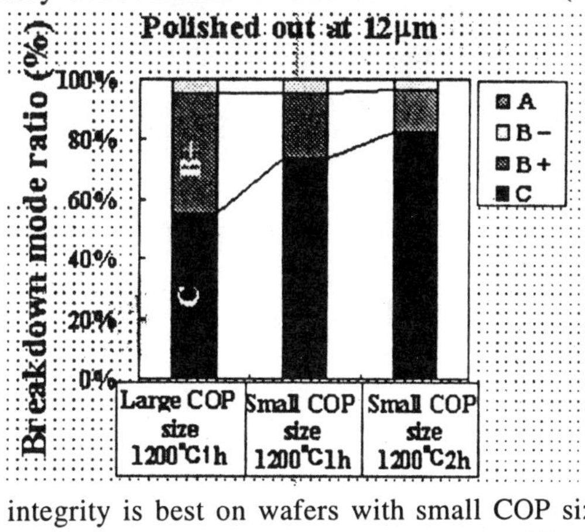

Fig. 2-11 Gate oxide integrity is best on wafers with small COP size ("C" breakdown mode indicates breakdown field of the gate oxide is > 10 MV/cm).[30]

anomalous oxygen precipitation that degrades DRAM refresh time). Second, the formation of proper denuded zones during the high-temperature annealing steps is a question. For applications that require wafers without COPs that intersect the wafer surface, but which cannot afford to use "conventional" epitaxial wafers, a fast-pulled wafer with a thin epitaxial layer on it may be a possibility. Such wafers are termed *epitaxy-optimized substrates* (EOS), and they are described in Sect. 2.9.1.

2.8 GETTERING FOR ULSI PROCESSES

Unwanted crystalline defects and impurities can be introduced during silicon crystal growth or subsequent wafer fabrication processes. The presence of crystalline defects and impurities is undesirable because both can degrade device characteristics and overall yield. A detailed discussion of the degradation of VLSI device properties by *extended crystalline defects* (dislocations, stacking faults, and precipitates) can be found in *Silicon Processing for the VLSI Era* Vol.1, 2nd Ed. Chap. 2. The damaging effects by *impurities* are discussed here.

The impurities that create problems are grouped into two categories: **1**) impurities that participate in the process of nucleating extended crystalline defects (e.g., oxygen and carbon); and **2**) other impurities that decorate pre-existing extended crystalline defects (the latter being primarily metals, and include such elements as Cu, Ni, Au, and Fe). Such metallic impurities are highly mobile and diffuse long distances in the crystal at moderate process temperatures. Thus, there is a high probability they will migrate near extended defect sites, and be captured by them. Extended defects in device regions of the wafer which capture metal impurities give rise to larger leakage currents and lower breakdown voltages.

Efforts undertaken to eliminate the effects of impurities and defects have involved four approaches:[31] **1**) suppression of extended defect nucleation; **2**) annihilation of existing extended defects in the device regions; **3**) preventing device degrading impurities from being introduced into the wafer; and **4**) preventing any remaining metallic impurities from being captured in the device regions. Technique **4**) is referred to as *gettering*, which is a term originally used to describe the process in which an appropriate material (such as cesium), was used to getter (remove) the last traces of gas in vacuum tubes, so that the required high vacuums within the glass tubes could be achieved.

By combining these four approaches, the product yield of ICs can be enhanced. That is, proper application of the first two approaches minimizes the number of extended defects that exist in the active regions, and the third reduces the concentrations of the impurities introduced into the silicon. Gettering is then used to remove any of those impurities that still manage to be incorporated during crystal growth or processing. Thus, gettering reduces the probability of decorating any extended defects formed in the device regions with metallic impurities.

Methods used to suppress extended defect nucleation include: **a**) control of the thermal process conditions during wafer fabrication that lead to excess thermal-stress and resultant dislocations; and **b**) creation of oxygen denuded zones at the wafer surface regions in which active devices are fabricated. The latter procedure eliminates oxygen precipitates in these device regions. *Annihilation of pre-existing* or *process-induced extended defects* such as stacking faults, is achieved by using various techniques, including elevated temperatures (>1200°C)[32] and the addition of HCl to the growth ambient[33] during the growth of SiO_2.

The use of technique **3**), namely *the control of unwanted impurities* (especially metals), involves identification and removal of the sources of contamination. Such approaches entail the thorough characterization of the material and processing equipment used to fabricate devices. Some examples of the contaminant sources that have been identified include: **a**) stainless steel fixtures in ion-implant machines that ended up being sputtered during the ion-implant process;[34] **b**) impurities originating from the sputtering of chamber walls and fixtures of reactive ion etching systems;[35] **c**) the diffusion of heavy metals from the heater coils through the fused silica walls of diffusion furnaces; and **d**) the release of transition metals (e.g., Fe and Ti) by the SiC-coated susceptors of (batch) epitaxial reactors.

A variety of procedures have been implemented to reduce the level of metal contamination arising from these sources. Stainless steel parts (which may release metal atoms as a result of sputtering in ion implanters and reactive-ion etchers) have been replaced with parts made from materials which produce less degradation if they are incorporated into Si than do Fe or Cr. Such materials include aluminum, silicon, and carbon. Suggested procedures to decrease contamination by sources of heavy metals have included the following: **1**) reducing process temperatures (since quartz tubes become increasingly permeable to the metal atoms of the furnace heating coils as the temperature increases); **2**) subjecting the furnace tubes to periodic high-temperature HCl cleaning; and **3**) use of double-wall furnace tubes (with HCl or inert gas flowing in the space between the inner and outer tubes). Contamination from epitaxial susceptors can be reduced by performing the epitaxy process at lower temperatures and reduced pressures, and by *in situ* cleaning of the wafers prior to

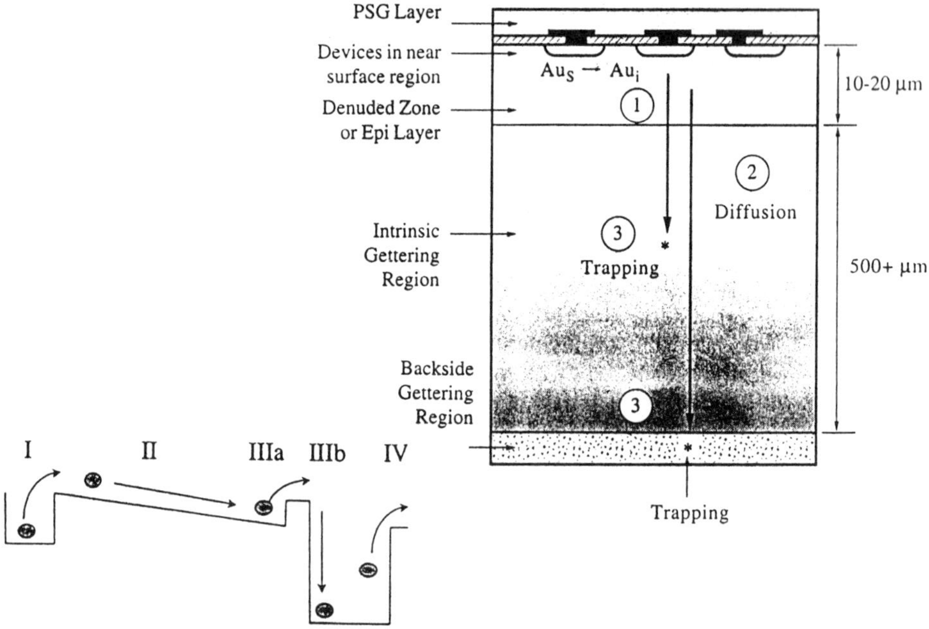

	I	II	IIIa IIIb	IV
Gettering Stage	Release	Diffusion	Capture (over barrier)	Stability (release)
Driving Forces:	•temperature •photons •electric field	•Si_I, V •conc. grad. •charge state	•supersaturation •solubility enhancement.	•temperature •solubility

Fig. 2-12 Four stages of impurity gettering in silicon.

initiating epitaxial growth. The use of cassette-to-cassette operations in transporting of wafers has reduced the possibility of contamination due to manual handling. Use of vacuum wands is now mandated if automatic wafer handling fails or is not yet fully implemented.

2.8.1 Basic Gettering Principles

The mechanism of gettering removes transition metal impurities from device regions by four steps (Fig. 2-12):

1. The metal impurities to be gettered are released into solid solution from any stable-precipitate-form they have assumed;

2. They then diffuse through the silicon lattice;

3. They are "captured" at a position away from the device regions by either: **a**) the formation of chemical pairs (e.g., Fe-B pairs) with other impurities (e.g., boron) present in the crystal; or **b**) by deliberately created extended defects (dislocations or precipitates);

4. The metal impurities are prevented from being re-released into the wafer during subsequent heat treatments.

In general, the transition-metal impurities are initially considered to rapidly disperse themselves uniformly throughout the wafer. This assumption is based on their high diffusivity and long diffusion lengths at typical processing temperatures (see Vol. 1, 2nd Ed., Chap. 2).

Gettering processes are divided into two categories: extrinsic and intrinsic. *Extrinsic gettering* involves the use of external means to create damage or stress on the surface of the silicon wafer that leads to the creation of the extended defects or chemically reactive sites at which the mobile impurities are captured. Normally such capturing sites are generated on the backside of the wafer, although front-side extrinsic gettering has also been explored. Extrinsic gettering can be performed prior to the first oxidation step (*pre-oxidation gettering*), or concurrently with the remainder of the fabrication steps (*gettering during* [or *subsequent to*] *junction formation*). *Intrinsic gettering* involves the localization of impurities at extended defects or regions of high concentration of electrical impurities (such as boron) which exist within the bulk material of the silicon wafer, and whose origin is due to an "intrinsic" property of the starting wafer, such as its oxygen or boron content acquired during CZ crystal growth.

The gettering of gold atoms in silicon by phosphorus diffusion is described, as this process serves to illustrate the basic mechanism of (extrinsic) gettering. A known concentration of gold atoms (e.g., $1x10^{14}/cm^3$) is first diffused uniformly throughout a group of silicon wafers. Phosphorus diffusions at 900°C are then carried out in an attempt to getter the gold. These diffusions produce surface regions having high concentrations of P ($1x10^{19}$–$1x10^{21}/cm^3$). After the P diffusions are completed, the gettering efficiency of the procedure is measured. It is observed that with increasing surface concentration of P, a strongly decreasing concentration of gold in the silicon bulk occurs. At P concentrations of 10^{21} atoms/cm^3, the gold concentration is decreased below the detection limit. Further investigations show that the gold has diffused into the phosphorus-containing layers in the silicon.

A model offered by some investigators describes the kinetics of this gettering process according to the following steps:[36] **a**) P diffusion is accompanied by the generation of phosphorus-vacancy complexes and/or misfit dislocations which become possible capture sites (sinks) for the gold atoms; **b**) during the P-diffusion step the highly mobile gold atoms also diffuse long distances in the wafer; **c**) the gold is captured at the phosphorus-vacancy or dislocation sites, and is thus removed from solution (this maintains a favorable concentration gradient of gold in solution toward the phosphorus-rich zones); and **d**) once a gold atom is captured, the electrostatic interaction energy with the capture site is so high that subsequent heat treatment does not release it back into the wafer. Sufficient phosphorus must be deposited to create enough capture sites to getter the gold present in the wafer.

Diffused phosphorus layers (and other backside gettering methods) are also effective in shrinking OISFs. Oxidation-induced stacking-faults are regions of excess point defects (i.e., Si self-interstitials). Diffused P-layers shrink them for reasons still being argued (i.e., either vacancies are injected, which shrinks the OISFs, or the supersaturated P-concentration causes precipitation to begin and the resulting dislocation loops form sites to soak up the interstitials).

2.8.2 Extrinsic Gettering

Historically, gettering of mobile metallic impurities has been accomplished by *extrinsic gettering*, in which the wafer backside is subjected to a process that creates damage or stress, and subsequently traps mobile impurities. A number of wafer backside treatments have been developed for this purpose. Mechanical damage and polysilicon deposition are the two used most widely.

Mechanical damage by abrasion, grooving, or sandblasting have all been used to create stress fields at the backside of wafers. During subsequent annealing steps, dislocations which relieve these stresses are generated. The dislocations then serve as gettering sites.

A *layer of polysilicon* deposited on the wafer backside (1.2–1.5-μm thick) is another extrinsic gettering technique that has been developed (Fig. 2-13).[37] Start-

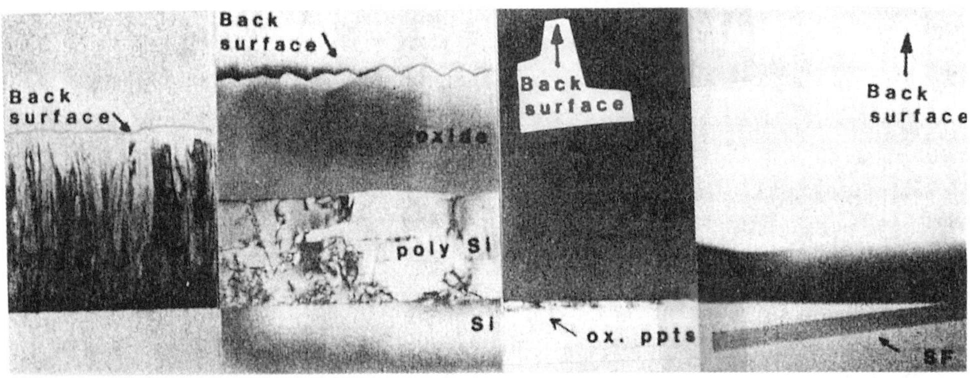

a) b) c) d)

Fig. 2-13 Cross-sectional TEM micrographs of: a) as-deposited poly-Si, then wet oxidized at 1150°C for; b) 100 min; c) 765 min, and; d) 1300 min.[38]

ing wafers up to 200-mm in diameter can be purchased with such a layer already deposited. (However, as we will see, such backside deposition is not compatible with 300-mmwafers.) The grain boundaries, and high degree of lattice disorder in polysilicon act as sinks for mobile impurities.

Unfortunately, extrinsic gettering techniques suffer from a variety of limitations. The primary general limitation of extrinsic gettering procedures is their inherent lack of stability, especially at high temperatures up to 1250°C for bipolar structures. Such instability manifests itself as the dissolution of gettered metal back into the wafer, and the annealing out of dislocations as the wafer undergoes a number of high-temperature steps. Upon being annealed out, dislocations can no longer trap newly introduced impurities, nor recapture the impurities released back into the wafer from getter sites. This limitation is one reason why extensive investigation of intrinsic gettering as a complementary gettering procedure has been undertaken, and why it is being increasingly implemented into ULSI fabrication sequences.

For 300-mm wafers, a *double-sided-polishing* (*DSP*) standard has been adopted. By polishing the wafer backside of the wafer (as well as the frontside), inspection of the entire wafer surface for the presence of particles can be achieved (i.e., small particles cannot be detected on the rough back-surface of a wafer). However, since the backside layer treatments of a wafer for the purposes of external gettering would have to be carried out after the DSP steps, it will no longer be possible to use them for 300-mm wafers. Hence, 300-mm wafers will have to rely on internal gettering processes.

2.8.3 Intrinsic Gettering

Intrinsic gettering is based on the principle that under proper conditions, fast-diffusing impurities are captured at sites existing within the interior of silicon wafer, away from the device regions near the surface. Two mechanisms are employed to implement internal gettering:

Gettering at Oxygen Precipitates: Supersaturated oxygen in silicon wafers will precipitate out of solution and form clusters within the wafer during thermal processing. The stresses that result as these clusters grow into larger precipitates can be relieved by punching out dislocation loops (or by emitting Si-self interstitials that agglomerate to form stacking faults and dislocation loops). These dislocations become sites at which unwanted impurities can be trapped and localized.[36] In an effectively designed intrinsic gettering process, these precipitates are only allowed to form in the bulk regions of the wafer. They are prevented from forming in the active

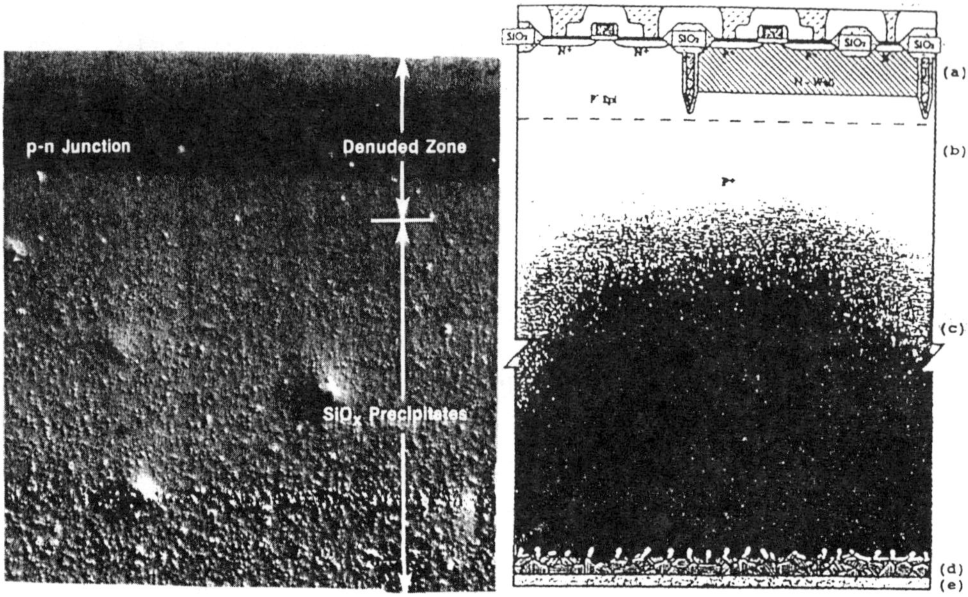

Fig. 2-14 SEM micrograph of a *pn* junction created by a 450°C anneal for 70 hr in denuded-zoned-Si with SiO_x precipitates formed by thermal treatment.[40]

device regions by reducing the oxygen concentrations to levels below the threshold required for precipitation (denuded zone formation). In this way, unwanted impurities are localized (gettered) only in regions not containing active devices (Fig. 2-14).

Gettering by Formation of Impurity Pairs: In this technique some types of metallic impurities are immobilized by chemically pairing them with other impurities deliberately introduced into the Si for this purpose. For example, Fe appears to form immobile Fe–B pairs. Thus, if regions of high B concentration are present in the interior of a wafer, migrating Fe atoms pair with a B atom and become immobilized at that location. Silicon wafers with a high bulk-concentration of B, topped with a lightly doped (e.g., 1×10^{16} cm^{-3}) thin epitaxial layer of Si (e.g., 3-μm-thick) can be used for this purpose. In fact, such silicon wafers have been historically used to control latchup in CMOS. Only recently has it been discovered that they are also very effective at gettering Fe through Fe–B pair formation, as described above. Recent reports indicate that a heavily doped region of B, created by a high-energy ion-implant process may be as effective in gettering Fe in wafers without an epi layer. If this claim is borne out, it may provide a less expensive way of obtaining the benefits of gettering and latchup protection offered by the heavily doped boron region in the Si with an epi layer.

The advantages of intrinsic gettering over extrinsic gettering include the following: **a**) the technique may be employed without any external treatment of the wafer other than heating; **b**) the volume of the wafer provides a sink during intrinsic gettering that is roughly two orders of magnitude larger than the volume of the gettering region on the wafer backside; and **c**) the gettering region in the bulk is much closer to the device regions than the backside of the wafer.[39] The distance required for impurities to diffuse from the circuit regions to the sinks is therefore 25–50 times shorter. Thus, intrinsic gettering allows diffusion cycles which are significantly shorter than those required for extrinsic gettering to remove the same contamination.

2.8.4 Gettering with Oxygen Precipitates

The first requirement for successfully implementing oxygen precipitate gettering techniques is to have appropriate oxygen concentration levels in the starting Si (i.e., 15–19 ppma). The lower limit of this range is somewhat above the threshold concentration of oxygen required for precipitation (~6×10^{17} atoms/cm^3 = 12 ppma). Precipitation does not occur at a measurable rate for concentrations below the 12 ppma level.[40] Wafers from CZ-grown ingots, however, normally exceed this lower limit, with the oxygen being incorporated during crystal growth by erosion of the SiO_2 crucible. On the other hand, if oxygen concentrations exceed 19 ppma, too high a density of precipitates is formed. This may contribute to both wafer warpage and dislocation generation in the vicinity of the active devices (Fig. 2-15).[41,42]

Assume that the wafers contain appropriate concentrations of oxygen, and that the oxygen is uniformly distributed across the wafer diameter. The wafers can then be subjected to a series of temperature cycles to achieve intrinsic gettering (Fig. 2-16). A *high-low-high* sequence of such steps is then:

Step **1**. *Denuded Zone Formation* - a *high-temperature* step produces a region of reduced oxygen concentration near the wafer surface;

Step **2**. *Nucleation of SiO_x Clusters* - a *lower-temperature* step causes homogeneously nucleation of SiO_x precipitation sites;

Step **3**. *Precipitate Growth and Gettering* - a *high-temperature* step grows the SiO_x nuclei formed in Step **2** into larger oxide precipitates. The result of their growth is the formation of dislocation loops, which function as the desired gettering sites.

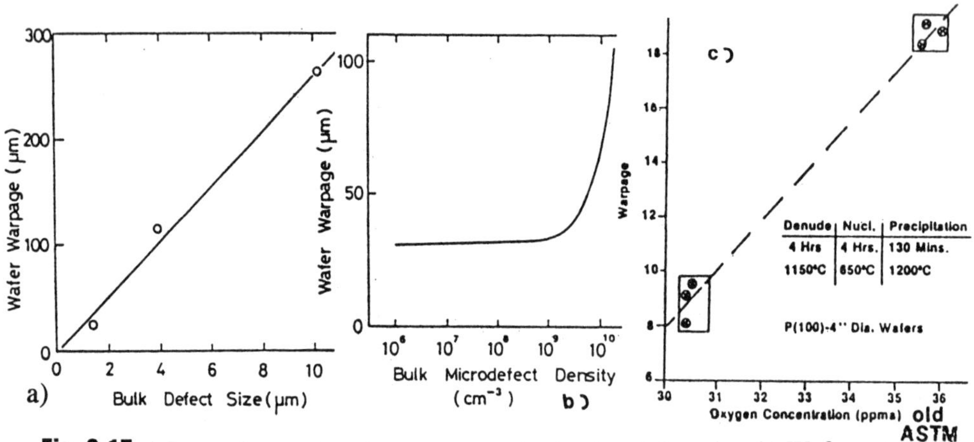

Fig. 2-15 a) Dependence of wafer warpage on bulk microdefect size; b) Wafer warpage as a function of bulk-microdefect-density.[42] c) Effect of Initial Oxygen content of the wafer on warpage.

In *Step* 1, a region near the surface containing less oxygen than is required to form precipitates (*denuded zone*) is created. Thus, such precipitates or their punched out dislocation loops (which could intersect active device junctions or become decorated with impurities near device surfaces), are prevented from forming. Denuded zones are established by subjecting wafers to a high temperature step.[44,45,46] The oxygen near the top and bottom wafer surfaces thus diffuses out, reducing its concentration near the surface to about the solubility level

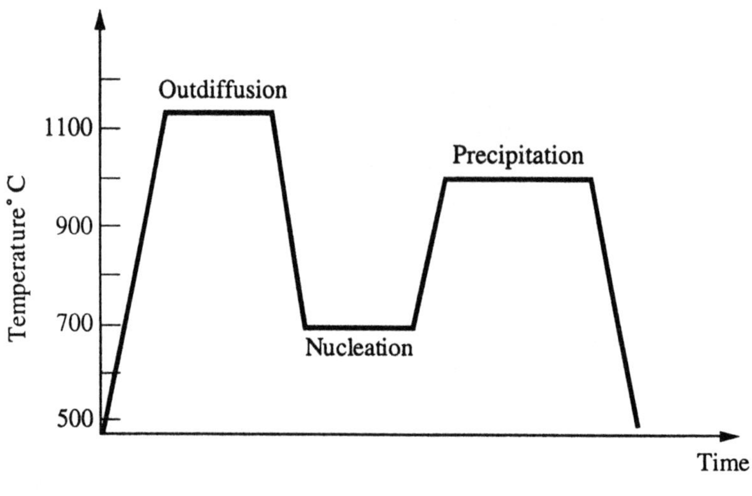

Fig. 2-16 Example of a thermal cycle to form the denuded zone.[48]

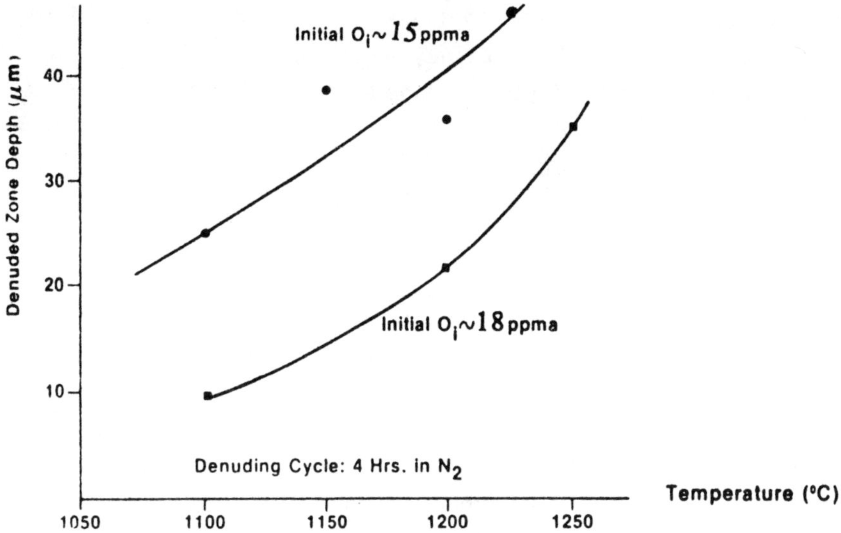

Fig. 2-17 Effect of initial oxygen concentration on the denuded zone depth.[43]

at that temperature. Denuded zones needed for a ULSI process must have a depth exceeding the deepest junction plus its surrounding space-charge region. Since the depletion region thickness depends on the wafer resistivity and the maximum operating voltage of the device, different devices made on the same substrate resistivity may need different denuded zone depths. Therefore, an important question that must be answered is: What is the shallowest denuded zone depth that the design and process can tolerate? There is no real shortcut to determining this information. Functional circuits must be fabricated on wafers containing a variety of initial oxygen levels. These circuits must be tested, and then cleaved and etched to determine the denuded zone depth and precipitate densities. Temperature cycles of 1100–1200°C for 30–240 min reportedly create denuded zones 10–40-μm deep. Figure 2-17 shows the outdiffusion depth for various temperatures and initial oxygen concentrations. Nitrogen and argon are typically suggested as appropriate ambient gases. However, it has been recommended that to avoid pitting, a thin oxide layer be grown before subjecting wafers to the high denuding temperature cycle (≥ 1200°C) in these inert gases.[47]

In *Step* **2**, a lower temperature cycle (600–800°C) is next used to cause interstitial oxygen in the wafer bulk to form the nuclei required for the subsequent precipitation and gettering events of *Step* **3**. The supersaturated (interstitial) oxygen can diffuse and homogeneously nucleate into small clusters if wafers are subjected to temperatures between 650–800°C for prolonged periods (e.g., 4–64 hrs). Note that above 950°C, homogeneous nucleation appears to be negligible. Thus, the first temperature step on starting wafers (the denuding step) should be carried out a temperatures greater than 950°C, to prevent formation of clusters near wafer surfaces during this initial heat treatment.

A large number of relatively small oxygen clusters are formed by this step. This number, for a fixed temperature and time, depends on the oxygen content and increases with initial oxygen concentration. Since the oxygen content of the starting wafers is one important parameter that determines the density of oxygen clusters, it is crucial that this concentration be controlled in CZ silicon. Figure 2-18 shows the nucleation rate and final radius of the clusters that are created at 800°C for an initial oxygen content of 15 ppma.[47] If nucleation cycles are carried out at the lower end of the temperature range (~650°C), significantly longer annealing times are required to achieve a sufficient concentration of adequately sized clusters (16–64 hr). To enable an initial 650°C temperature to be used for this step, and yet allow a reduction in the overall nucleation cycle time, a *ramping up* of the temperature during *Step* **2** has been suggested.[42,49]

The rate of cluster formation can be enhanced by this procedure if the rate of change of the critical size, r_0, with temperature, is smaller than the rate of growth of the clusters during annealing. That is, by the time a growing cluster reaches a higher temperature, it will have grown fast enough that its size exceeds the required r_0 at that temperature.

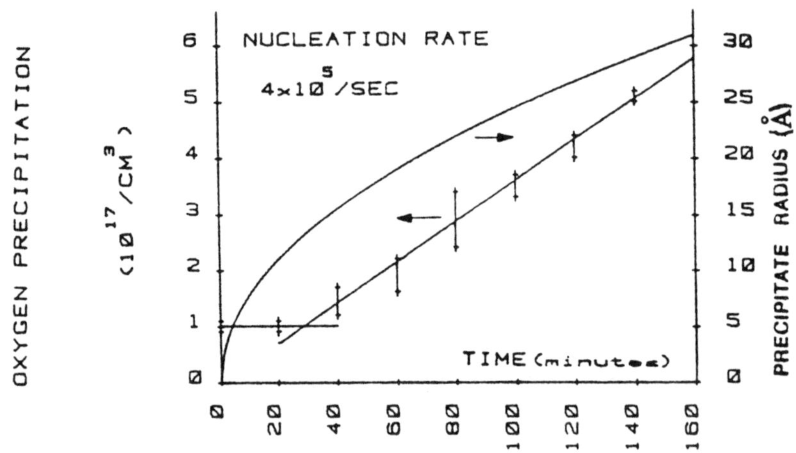

Fig. 2-18 Nucleation rate for oxygen precipitates. The curve is the expected rate of precipitate formation at 880°C.[47]

The time to grow the clusters into precipitates that would be large enough to form the desired dislocations in the bulk is prohibitively long over the range of temperatures of *Step* **2**. Thus, once the clusters have grown to a size where they exceed the r_0 of a higher temperature step, a third thermal cycle at that temperature can be used to grow the clusters into larger precipitates in a shorter time. If a temperature-ramping technique during the nucleation cycle is used, the final temperature can be that called for by *Step* **3**.

In *Step* **3**, wafers are heated to high temperatures (900–1250°C) for extended periods (4–16 hrs). At these temperatures, oxygen clusters of a sufficient size (i.e., $> r_0$) which were formed during *Step* **2**, will grow (e.g., from 3–5-nm to ~50–100-nm in radius). Such growth is caused by diffusion of interstitial oxygen in the bulk to the SiO_x clusters, and reaction with silicon on the precipitate surface. For each SiO_2 molecule thus formed, a Si-self-interstitial is produced to accommodate the volume expansion. The large volume-change associated with the precipitation is not completely relieved by the emission of self-interstitials, and the precipitate builds up a strain field that can punch out prismatic dislocation loops (Fig. 2-19).[50]

The self-interstitials emitted by the precipitate can also form dislocation loops as discussed in the section on *Bulk Defects* in Vol.1, Ch. 2. The dislocation loops contain an extrinsic stacking fault that grows by absorbing the self-interstitials emitted by the growing precipitate. The dislocations formed from both mechanisms act as the gettering sites which can trap fast diffusing impurities.

The ultimate size of the precipitates apparently also plays a role in the warpage resistance of the wafers. Kishino[42] shows that wafer warpage increases

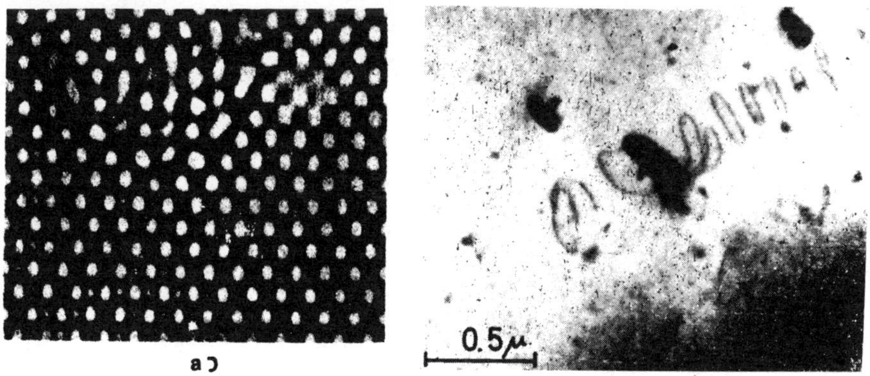

Fig. 2-19 a) Image of the core of a dislocation in Si obtained by high resolution TEM depicts an extra plane of atoms that terminates at the core of the dislocation. A precipitate is at the core.[24] (©1985 IEEE); b) Dislocation loops generated in Si due to prismatic punching by SiO_2 particles.[25]

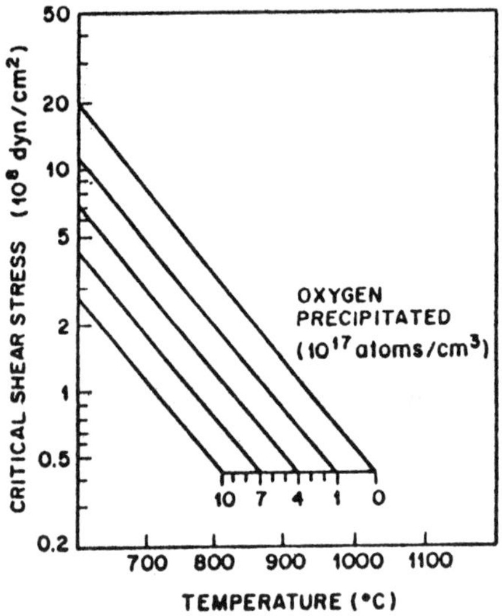

Fig. 2-20 Yield strength of Si showing the influence of oxygen precipitates.[42]

with increasing precipitate size (Fig. 2-20). These data, and that of others,[51,52] suggests that *Step* **3** heat cycles that lead to precipitates larger than 200-nm in size can negatively impact production yield. Several reports have, in fact, indicated that the best wafers are those in which the precipitation occurs slowly and uniformly throughout the wafer (especially with good radial uniformity), and which results in small precipitates with attached precipitation defect complexes.[53,54] The concern for wafer strength has lead most users of internal gettering away from materials with high oxygen levels (e.g., > 19 ppma), to material containing medium levels of oxygen (e.g., 15–19 ppma).

2.8.5 Summary of Gettering

Effective gettering is required to produce high-yielding ULSI. We have described a variety of gettering techniques, both extrinsic and intrinsic. In practice, more than one gettering mechanism has been simultaneously exploited in a wafer during fabrication. That is, the approach to gettering has been to deliberately implement more than one technique. An extrinsic, backside technique and an intrinsic gettering process were employed concurrently. Since typical CZ wafers contain oxygen in the appropriate range of 15–19 ppma, it is likely that many early IC processes had inadvertently benefited from intrinsic gettering. However, as low-temperature processes, and/or RTP, become more extensively utilized, and especially when 300-mm wafers are used, the use of intrinsic gettering can no longer be left to chance. The reduced thermal cycles in some cases, and the inability to use extrinsic gettering with 300-mm wafers will demand use of a gettering technique which allows the placement of gettering sinks in the wafer bulk (and closer to the device regions). The intrinsic gettering method, however, does require the strict control of a number of oxygen-related processes, and this can be a complex task.

2.9 STATUS OF SILICON WAFER TECHNOLOGY IN 2000 AND FUTURE TRENDS

By the year 2000, most 200-mm wafers being used for ULSI IC fabrication were either: **1**) conventional lightly-doped polished CZ wafers (i.e., which exhibit a grown-in defect distribution characterized by central vacancy-rich region surrounded with an OSF-ring); or **2**) conventional heavily *p*-doped CZ-silicon-substrate-wafers covered with a lightly-*p*-doped epitaxial layer. Some of the polished wafers are subjected to a post-growth high-temperature annealing step in hydrogen or argon to decrease the density of COPs at the wafer surface (see Sect. 2.5). The polished-CZ-wafers rely on both external gettering methods (e.g., backside polysilicon) and internal gettering methods which require the presence of oxide precipitates to establish gettering sites in the bulk of the wafer. The *p*-epitaxial/p^+-wafers also employ these two gettering techniques, but are able to rely on the gettering capability of the heavily doped p^+ substrate as an additional gettering mechanism (especially for such impurities as Fe and Cu). The polished-CZ-wafers usually use a deliberate process step (or steps), to create a zone near the surface denuded of oxygen (i.e., a denuded zone), to prevent oxide precipitates from forming in the device regions. Epitaxial wafers do not need to establish a separate denuded zone, as oxygen is not incorporated into the epitaxial layer during the epi deposition step (and thus such wafers automatically have a denuded device region).

As devices are being scaled below 0.18-μm and the wafer sizes are increasing to 300-mm in size, a new set of wafers types with respect to: **1**) the grown-in defect characteristics they exhibit; **2**) the gettering methods they employ; and **3**) the denuded zones they possess, will be used. Let us examine each of these characteristics.

There appear to be four types of wafers that will be offered for processes below 0.18-μm and 300-mm wafers; **1**) "perfect-silicon" wafers; **2**) fast-pulled wafers; **3**) *epitaxy-optimized substrate* wafers (EOS); and **4**) "conventional" epitaxial wafers. The "perfect-silicon" wafers are described in Sect. 2.5. The fast-pulled wafers are described in Sect. 2.7. The EOS wafers are described in Sect. 2.9.1. We compare the advantages and disadvantages of these four types of "new" wafers in Sect. 2.9.2.

2.9.1 Epitaxy-Optimized-Substrate (EOS) Wafers

Epitaxy-optimized-substrate (EOS) wafers typically use a fast-pulled, lightly-*p*-doped wafer as the substrate, and cover it with a medium-*p*-doped, epitaxial layer that is thinner than those used in conventional epitaxial wafers The epitaxial layer in EOS wafers is not deposited thick enough so that it contains the entire device region. Its role instead, is merely to cover the COPs so that they do not degrade the isolation leakage behavior of DRAMs (and also not degrade the GOI of thin gate oxides). As noted, the COPs created in the fast-pulled substrate (i.e., a silicon crystal which may also be doped with nitrogen) are small and are formed with high density.

A thin (~0.5 μm thick), medium-*p*-doped epitaxial Si layer is thus deposited on a lightly-doped substrate. COPs are not formed in epitaxial layers because the epitaxial deposition temperature is much lower than the temperature encountered by the solidifying Si crystal in CZ growth. Thus, the vacancies created during the epitaxial process are in much lower concentrations (too low, in fact, to allow them to aggregate and form COP voids). In addition, it has also been demonstrated that the small COPs that are formed in the fast-pulled silicon which are covered with such a 0.5-μm-thick epitaxial layer are no longer detected with a surface scanning inspection system after an SC-1 delineation etch.[55,56] In fact, more thorough experimental data indicates that substrate COPs are still present after epitaxial deposition, but their morphology is radically changed. Their dimensions significantly decrease after being covered by a thin epitaxial layer (their thickness is around 2-nm, compared to 13-nm before the epi deposition), with no dependence on epi thickness detected for epi layers from 0.5 to 3.0-μm thick. Data also show a 100% GOI yield for epi layers only 0.5-μm thick. Thus, on such EOS wafers, the effects of COPs is dramatically reduced with the use of a fast-pull crystal growth process and the deposition of a very thin epitaxial layer.

Note that since these EOS wafers start with a lightly *p*-doped Si substrate, for some applications a heavily doped *p*-type buried layer (formed with high-energy ion implantation) may also be implemented. This type of buried layer would allow the latchup and gettering benefits of the heavily doped p^+-substrate (present in "conventional" epitaxial wafers) to be obtained on EOS wafers as well. EOS-type wafers are being offered commercially under such names as: fLASH® (Wacker Siltronic AG), and Epi-II® (MEMC).

2.9.2 Comparison of Grown-In-Defect Characteristics of the "New" Wafers

The four "new" types of wafers for near-future generations of ULSI circuits and 300-mm can now be compared, especially with respect to the tradeoffs among them involving low-grown-in defect densities, reliability of devices fabricated on each type, and cost:

1. "Perfect-silicon" wafers are beginning to be offered for mainstream production applications. However, the production of "perfect-silicon" wafers in large quantities presents some serious problems for the wafer manufacturers with respect to developing a cost-effective manufacturing process. That is, the pull-rate required to ensure "perfect-silicon" is slow, and this negatively impacts the throughput of the ingots that can be produced by an (expensive) crystal puller. Furthermore, the ingots must remain within the hot-zone of the puller for much longer periods of time, which increase the likelihood that a growth-disturbing event will occur during the growth cycle. This will further reduce the yield of acceptable ingots produced by a puller. Finally, a "perfect-silicon" growth-process must be developed for 300-mm wafers, which itself is a costly undertaking. In summary, the overall cost of "perfect-silicon" wafers for 300-mm wafer sizes may turn out to exceed the cost of even EOS wafers.

2. The fast-pulled (and perhaps also nitrogen-doped) wafers possess small-sized COPs, present in high densities. However, it has been demonstrated that the presence of COPs no longer degrades the GOI of oxides when the oxide thickness is smaller than 6.5-nm. As an example, it was reported that for MOSFETs fabricated with gate-oxides of 5-nm thickness, almost no COP failures were detected, independent of the size and density of the COPs in different crystals.[57] As a result, the use of fast-pulled wafers may be acceptable for many logic ICs, despite presence of the COPs (provided such circuit applications do not require the extremely low isolation leakage levels required by other products, such as DRAMs). Thus, such wafers might find use for a wide variety of CMOS IC circuit applications that use gate-oxide thicknesses of less than 6.5-nm-thick, built on 300-mm wafers. Of course, for some applications (such as 256-MB [and larger] DRAMS), the presence of a high-density of COPs might be the potential cause of such high leakage currents that this type of wafer would not be a suitable choice for them.

3. As described above, the EOS wafers use a lightly doped, fast-pulled substrate. Manufacturing such substrates can be achieved with higher throughput, and with a much-less-intensive process-development-effort than will be required for developing a "perfect-silicon" process for 300-mm wafers. In addition, the throughput and yield for such wafers will be

significantly greater than for "perfect-silicon" wafers. Finally, the cost of producing such thin epitaxial layers is significantly lower than that to produce the thicker epitaxial layers used in "conventional" epitaxial-Si-wafers (up to 15-μm thick). That is, the production cost of depositing the epitaxial layer on "conventional" epitaxial wafer is a significant part of the overall wafer cost. It is possible that when all of the factors that go into the cost of producing 300-mm "perfect-silicon" wafers are compared to those that go into the cost of producing EOS wafers, that the EOS production costs will be lower.

4. "Conventional" epitaxial wafers will also probably continue to be used in 300-mm wafers, for a number of reasons. First, the p^+-substrate used in such wafers may provide latchup and gettering advantages for some circuit that may not be possible to achieve with either "perfect-silicon" or EOS wafers. The highly doped boron substrate enhances the formation of oxide precipitates as well as providing additional solubility gettering for impurity contamination. Second, a significant reduction in the epitaxial-to-polished wafer price-ratio is expected for 300-mm wafers, compared to that which existed for smaller diameter wafers. Finally, it has been observed that the COP density in highly-p-doped wafers (2×10^{19} cm^{-3}) is very low,[58] and that this contributes to the excellent GOI of gate oxides fabricated on p/p^+ epitaxial wafers. Thus, despite the fact that such "conventional" epitaxial wafers will likely cost more than the other wafer types, they may still be chosen for some circuits built on 300-mm wafers.

2.9.3 Gettering Methods and Denuded Zone Formation Techniques Employed with the "New" Wafers

As noted in Sect. 2.8.3, external gettering processes based on backside wafer treatments are not compatible with the double-sided-polishing standard adopted for 300-mm wafers. Consequently, 300-mm wafers will have to rely on internal gettering processes. However, gettering sinks are even more critical today, when device fabrication increasingly uses copper interconnects (which could potentially introduce a very fast diffusing impurity in both the device region of the IC and into the IC manufacturing process lines). Let us therefore now discuss the forms of intrinsic gettering that will be available with the "new" wafers described earlier.

In the "new" wafers in which the substrate background doping is low, the intrinsic gettering capability will have to rely on gettering produced with the assistance of oxide precipitates in the bulk. In the "conventional" epitaxial wafers, the substrate is heavily-*p*-doped with boron. The formation of oxide precipitates is significantly enhanced by the heavy boron doping (allowing a sufficient number of oxide precipitates to be formed, even if the oxygen content in the silicon is low),[59,60] and some additional solubility gettering will occur due to the heavily-doped substrate that does not occur in the wafers with lightly-doped substrates. Hence, gettering will, in general, be more effective in wafers that have the more heavily-doped substrates (and in particular for gettering Fe). Nevertheless, in one report, it was indicated that equally effective gettering can be obtained in both the lightly-doped and heavily-doped substrates, provided that the oxygen concentration in the lightly doped substrates exceeds $7x10^{17}$ cm^{-3}.[61] By co-doping the Si with nitrogen during crystal growth, the oxide precipitate densities can, however, be significantly increased. Such nitrogen doping also acts to reduce COP sizes, as noted earlier.

Thus, to enable the use of such lower-cost "new" wafers which have a lightly-doped substrate, adequate gettering capability must be exhibited with such a substrate. To ensure that sufficient gettering capability is available, either one of two approaches is being followed: **1**) the wafers are being doped with nitrogen to enhance the number of oxygen precipitate nucleation sites; or **2**) an RTP process that manipulates the excess-vacancy-profile throughout the wafer is employed, as this allows the density of oxygen precipitate nucleation sites to be controlled.

As described in Sect. 2.8.4, part of the gettering process involves forming a denuded zone that extends from the wafer surface to a depth that exceeds the region in which device structures are formed. That is, the denuded zone is a region that is devoid of oxygen precipitates. In the "new" wafers, such denuded zones must also exist. In the case of "conventional" epitaxial wafers, the epitaxial layer is a denuded zone because there is no oxygen incorporated into the epitaxial layer as it is deposited. Hence, no oxygen precipitation will occur within the "thick" epitaxial layer of "conventional" epitaxial wafers. For the other three "new" wafer types, the denuded zone is typically formed as in conventional polished wafers with high-temperature annealing steps (>1000°C) to out-diffuse oxygen from the wafer near the surface (including the EOS wafers, in which the epitaxial layer is so thin that it does not extend deep enough to encompass the entire device region). It has been postulated that due to radical reductions in the total thermal budgets, the "conventional" methods of forming adequate denuded zones may no longer be cost effective.

2.9.3.1 Magic DeNuded Zones^{TM}: A novel way of producing the denuded zone has recently been introduced by MEMC called the *Magic Denuded Zone* (MDZ).[TM,62,63] In this method the denuded zone (and the oxide precipitates within the bulk region of the wafer) are formed with the assistance of vacancy-concentration profiles that are introduced into the wafer with a Rapid Thermal Process (RTP) *prior* to the development of the oxygen precipitates in subsequent heat treatments. (Oxygen diffusion rates are much slower than vacancy diffusion rates in silicon.) The patented MDZ process creates a vacancy-concentration depth-profile that effectively pre-programs the oxide-precipitate-density in the bulk of the wafer, as well as the depth of the oxide-precipitate-free-zone to ideal targets. In late-2000 MEMC licensed this technology to Wacker Siltronic AG, another vendor of Si wafers, in a cross-licensing agreement involving a number of patents.

The physics of the Magic Denuded Zone concept is as follows: An otherwise perfect silicon crystal at any temperature above 0 K (e.g., room temperature) nevertheless contains an equilibrium-concentration of point defects, namely *vacancies* and *interstitials*. (For more details see Vol.1, Chap. 2, 2nd Ed.) The higher the equilibrium temperature, the higher are these concentrations. Consider a region in a bulk-silicon-crystal (i.e., far from any surface). If the crystal is heated above room temperature, thermal generation will create equal numbers of *additional* (or *excess)* vacancies and interstitials (vacancy-interstitial pairs). As long as the silicon is kept at an elevated temperature, such higher equilibrium-concentrations of point-defects will be sustained. However, if the silicon is returned to room temperature, the excess vacancies and interstitials will annihilate each other by a reverse process, i.e., recombination. Such recombination of excess point-defects will continue until the room-temperature equilibrium concentrations are restored.

The events that take place when a Si-crystal *wafer* is heated differ dramatically from when a region deep within the *bulk* of a Si-crystal is heated, due to the presence of *surfaces relatively close to any point within the silicon wafer.* The proximity to such surfaces radically alters the relative concentration of vacancies and interstitials in the heated wafer. This alteration in the relative concentrations of the two point defect types occurs for two reasons: **1**) interstitials are about 10 times as mobile as vacancies; and **2**) when either an interstitial or vacancy arrives at the surface, it is quickly annihilated. That is, when an interstitial arrives at a surface, it will quickly occupy one of the plentiful (vacant) surface lattice sites. Likewise, a vacancy moving onto a lattice site is really just a silicon atom moving from one filled site to an adjacent empty one. Since interstitials can move more quickly to any nearby surfaces, most vacancies trail behind. (Note that since thermal generation of interstitial-vacancy pairs is a continuous process, some vacancies are always able to recombine with interstitials. But, since the new interstitials also migrate more rapidly to the surfaces, there are always fewer interstitials within the bulk of a wafer than vacancies.) In other words, because of the inequality in point-defect mobility, and the presence of nearby surfaces, a net steady-state excess-vacancy-concentration will exist within a heated silicon wafer. The length of time from when a silicon wafer is heated to when the concentration values of interstitials and vacancies reaches their new, steady-state values is very fast – on the order of a few seconds.

If the wafer temperature is reduced to a value below its initial heated temperature (but to a value still higher than room temperature), the concentration of excess vacancies in the silicon will also eventually decrease to a new, lower, steady-state value. However, the process by which this reduction takes place in a wafer is a combination of: **1**) recombination; *and* **2**) the diffusion of vacancies to the surfaces from locations of higher concentration deep within the wafer.

The diffusion of vacancies is also dependent on temperature. At temperatures in the vicinity of 1000°C the vacancies can diffuse rapidly to the surfaces of the wafer (but of course, more slowly than the interstitials), but at room temperature the vacancy diffusion is so slow that the vacancies can be considered essentially immobile. Thus, if a heated wafer is rapidly cooled to room temperature, the concentration profile of excess vacancies within the wafer is "quenched in." In addition, by controlling the intermediate wafer temperature (and the time during which the wafer is kept at that intermediate temperature), the concentration of vacancies that will be "quenched in" at the surface can also be controlled.

MEMC has invented the MDZ process that uses RTP to produce vacancy-concentration-profiles within a wafer that are well-suited for internal gettering techniques. That is, it is known that the rate and oxygen concentration dependence of oxygen clustering can be substantially modified and controlled by the concentration of vacancies present (as was learned from studies of the oxide precipitate formation in the OSF-ring region of the wafer as was described in Sect. 2.3.2). MEMC determined that in regions of the silicon containing high concentrations of "quenched in" vacancies, the rate of formation of oxide nuclei is dramatically increased, whereas in regions where the concentrations are lower, the oxide nuclei formation rate is much slower. Thus, by creating a profile in which vacancies deep in the wafer bulk are high in concentration, the number of oxide precipitates that form there will also be high. Near the surface (and to the desired depth), the vacancy concentration is tailored to be much lower. Thus, in such near-surface regions oxide precipitate formation essentially does not occur, even in the presence of relatively high concentrations of oxygen. Thus, a "true denuded zone" near the silicon surface can be produced with a short RTP process after the wafers are formed - instead of with a long, high-temperature process for out-diffusing the oxygen (and which MEMC argues is not a "true denuded zone" because some oxygen nuclei may still remain from the time the oxygen concentration was high enough to have allowed them to form). On the other hand, in Si wafer regions below the denuded zone at the surface, a high number of oxygen precipitates forms, permitting an effective internal gettering scheme to be implemented. No special processing of the silicon during ingot growth (such as nitrogen doping during ingot growth) is necessary with this method.

By using the MDZ process, the density of oxide-precipitates and their depth distribution are completely decoupled from the oxygen and crystal growth influences and details of the IC thermal processes. The precipitation performance is completely controlled by the engineered vacancy-concentration-profile. In addition, the denuded-zone-depth and bulk-oxygen-precipitate-density may be engineered independently *of one another* by this technique. The denuded zone is also formed without the requirement of oxygen out-diffusion. The entire MDZ process is accomplished in less than one minute, and is therefore much faster if compared to the average oxygen diffusion process (that takes several hours).

MDZ provides internal gettering that is nearly independent of initial oxygen concentration, wafer position along the ingot, and thermal history effects. Since the MDZ process is also independent of the crystal growth-process, accurate MDZ performance may be achieved on any type of wafer. This means that the oxygen precipitation process is effectively decoupled from the crystal-growth-related intrinsic-point-defect issues.

2.10 FROM INGOT TO FINISHED WAFER: SLICING, ETCHING, and POLISHING

After a single crystal ingot has been grown, a complex sequence of shaping and polishing steps are performed to produce starting material suitable for fabricating semiconductor devices (*wafers*).[65,74] The wafering process for 300-mm wafers has been modified to some degree compared to ones used for previous wafer generations. We will emphasize those differences in this section. Figure 2-21 shows a flow chart depicting a generic 300-mm wafer shaping process. Figure 2-22 also illustrates this *wafering* procedure, which includes the steps covered next.

2.10.1 Ingot Evaluation

The grown single crystal ingot undergoes routine evaluation of resistivity, impurity content, crystal perfection, size, and weight. Sections of the ingot that are crystallographically defective, irregularly shaped, or undersized, are cut off and discarded (as are the seed and butt [or *tang*] ends). Loss of the ingot material may approach 50% at this point.

2.10.2 Ingot Surface Grinding

Since ingots do not grow perfectly round, nor with sufficiently uniform diameters, they must shaped to the desired form and dimension. Thus, ingots are deliberately grown larger than the final desired wafer diameter and a grinding operation removes the excess material to reduce as-grown crystals to a cylindrical shape of precise diameter. (Exact diameter dimensions are required to make wafers compatible with automated handling procedures.) Since silicon is a hard, brittle material (registering 72.6 on the Rockwell "A" hardness scale), grinding machines with diamond wheels are used to perform this cylindrical shaping process. The grinding step is followed by an etch step that removes the grinding work-damage. The ingot diameter is reduced 0.25-1.0-cm by the grinding and etch steps.

2.10.3 Grinding Flats or Notches on the Ingot for Orientation Purposes

For wafers up to 150 mm, *flats* are used for aligning and crystal orientation. For larger wafers ≥ 200 mm, *notches* are predominantly used (see Fig. 2-23). For 300-mm wafers, notches are used exclusively. These features are next ground along the length of the ingot. The largest flat, called the *primary flat*, is usually positioned relative to a specific crystal direction. (The primary flat orientation is found by using X-rays to produce a Laue photograph.) The primary flat is used for several purposes. Automated wafer handling equipment utilizes the primary flat to obtain correct alignment, and in addition, devices on the wafer can be oriented to specific crystal directions with this flat as a reference. Smaller flats are called *secondary flats*, and they are utilized to identify the orientation and conductivity type of the wafer. Since automated equipment relies on the flats for correct operation, the flat dimensions must be precisely machined. For example, the flat or notch is placed perpendicular to a <110> direction on a (100) wafer.

2.10.4 Sawing the Ingot into Slices (Wafers)

The sawing operation[66,74] that produces silicon slices from the shaped ingot also defines the surface orientation, thickness, taper, and warp of the slice. The ingot is rigidly mounted to maintain exact crystallographic orientation during the sawing process. Wafers of <100> orientation are normally cut "on orientation" (e.g., within ± 0.5°).

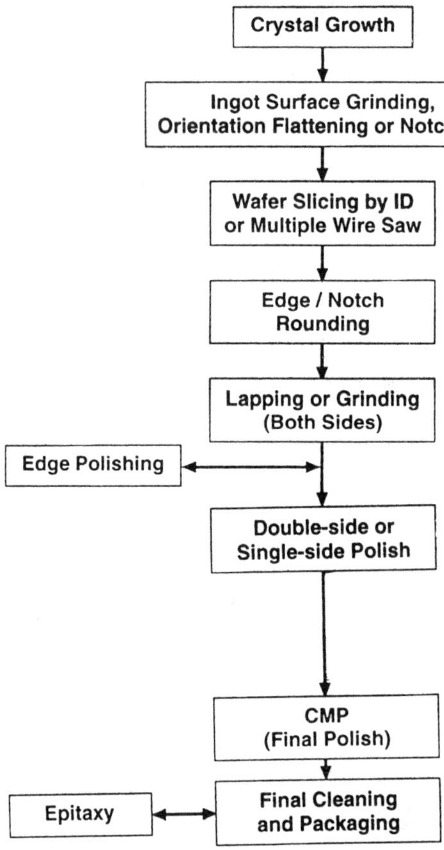

Fig. 2-21 Flow chart depicting the steps involved in wafer preparation starting with a whole Si ingot.

Wafer thickness is primarily set by the sawing operation, although some material is also removed by subsequent operations (e.g., for a 200-mm wafer of final 725-μm thickness, a total of ~1250 -μm of ingot silicon is needed: 725-μm final wafer thickness; 400-μm for kerf loss during sawing; 50-μm lapping loss; 50-μm etching loss; 25-μm polishing loss. Thus, about 8, 200-mm wafers per linear cm of ingot can be produced. As wafers get larger they must be thicker (Table 2-1) to allow them to withstand thermal processes (e.g., epitaxy and oxidation) and handling during fabrication. Thus, fewer wafers per unit length of ingot are possible than for smaller wafers.

The most common mode of slicing has been *inner diameter* (ID) *slicing* (Fig. 2-24) using stainless steel blades with diamond particles bonded to their edges.[74] Continuous monitoring with blade-deflection sensors is required to assure that slices are sawn adequately free of bow, taper, work damage, and saw marks. Cutting speeds are in the neighborhood of 0.05 cm/sec. Since one slice is cut at a time by the saw, this is a relatively slow process.

Wafers with 300-mm diameters are sawn with slurry-coated wires instead of inner-diameter saws. More slices per inch of ingot are possible due to smaller kerf loss (kerf loss is reduced by approximately 30% compared to the ID saw).[67] In addition, wire sawing enables the silicon slicing to be carried out at higher throughput and superior mechanical properties, such as reduced bow and warp. Figure 2-25 shows a schematic of multiple-wire saw. In this arrangement, parallel, equally spaced and properly tensioned stainless steel wires spun across two pulleys are part of single stainless-steel wire winding through a complex set of pulleys. Cutting of multiple slices results when the ingot is pressed against the traveling wires under injection of slurry. Although the cutting rate is much slower than the ID saw (the ordinary ID saw has a 80-100 times higher sawing

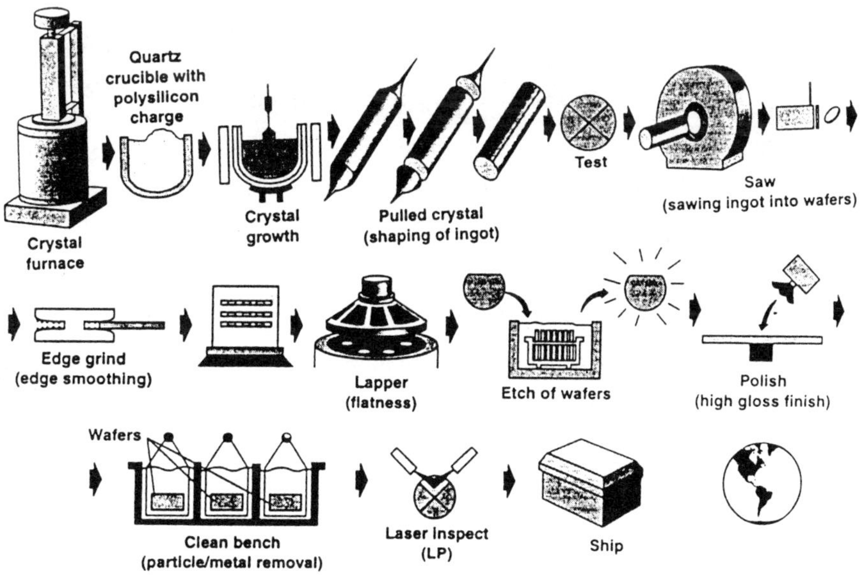

Fig. 2-22 Graphical depiction of the steps in the wafering process.

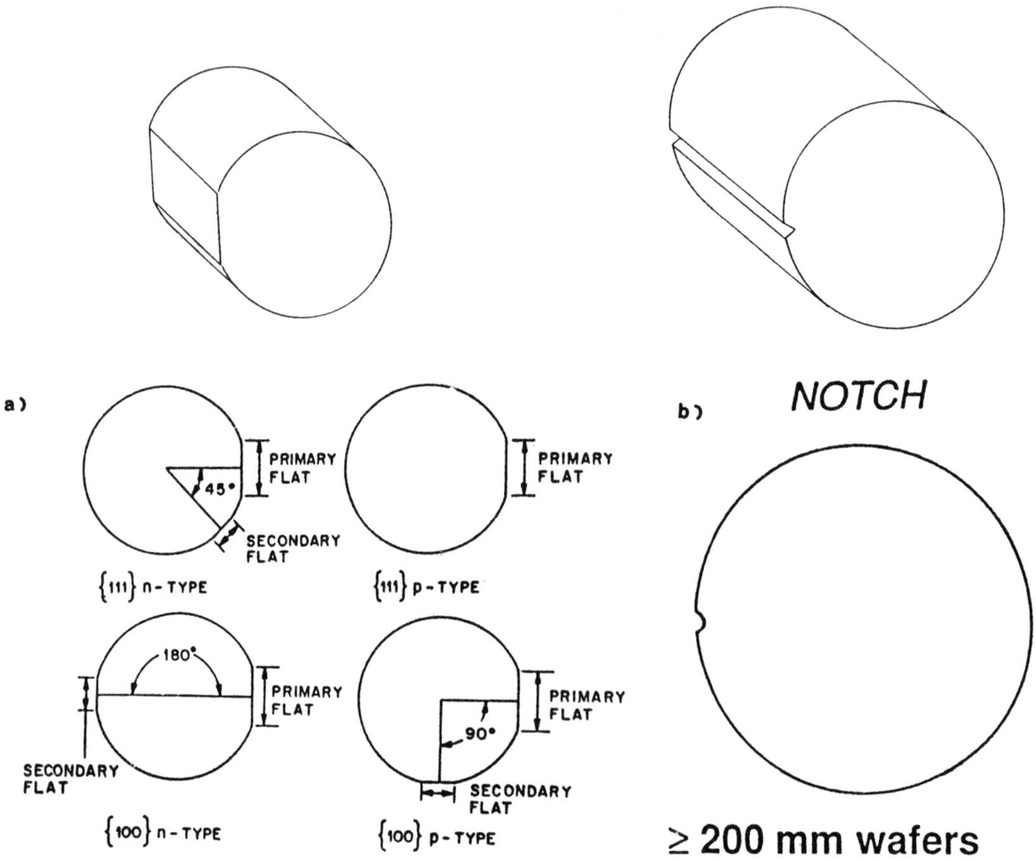

Fig. 2-23 a) Identifying flats on a silicon wafer; b) Notched wafers.

rate), as many as 300 slices can be produced simultaneously. Since slicing by the multi-wire saw is actually the result of low speed grinding action by the slurry, improved bow, warp, TTV, and taper are much more easily obtained than with the ID saw.

2.10.5 Laser Marking the Wafers

A technique using a *laser* to create an alphanumeric (or bar code) identification mark on the front of the wafer near the primary flat or notch (Fig. 2-26) is then employed. *Laser marking* according to SEMI[68] Std. M 1.8, specifies an 18 character field to identify the wafer manufacturer, conductivity type, resistivity, flatness, wafer number, and device type. This allows each wafer to be identified. Wafer traceability, correlation of device characteristics produced on a wafer with the wafer properties, and the ability to prevent in-process rejected wafers from being re-inserted into the fab line are all benefits of utilizing wafer marking.

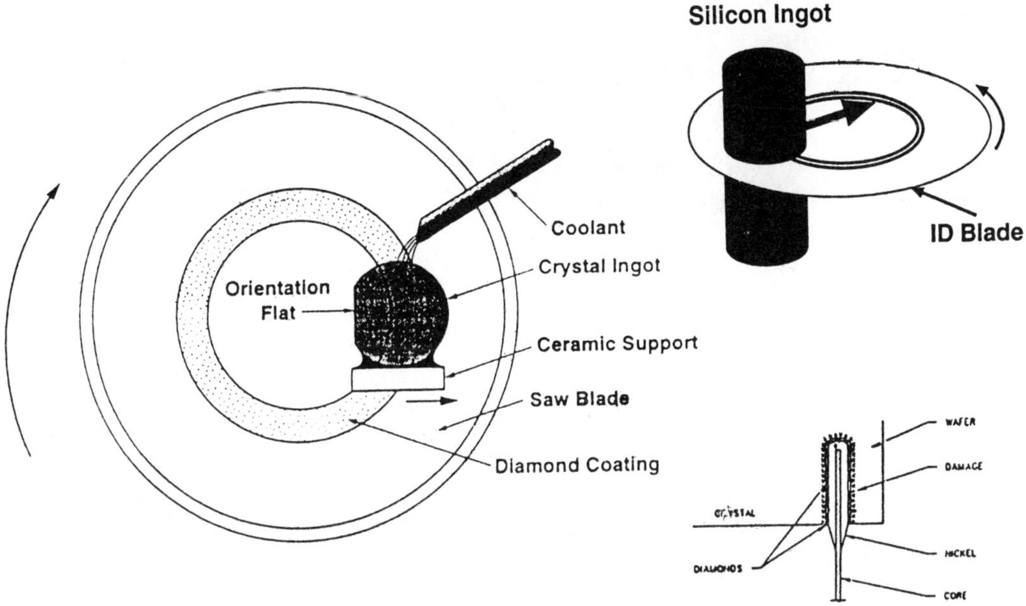

Fig. 2-24 ID saw geometry.[66]

Automatic code readers at various stages of processing can also track wafers and provide data on wafer movement in the fab.[69,70]

2.10.6 Lapping and Grinding the Wafers

The thickness of the as-sawn slices is sufficiently variable that a *lapping* and *grinding* step is next employed to bring all slices to within the specified thickness

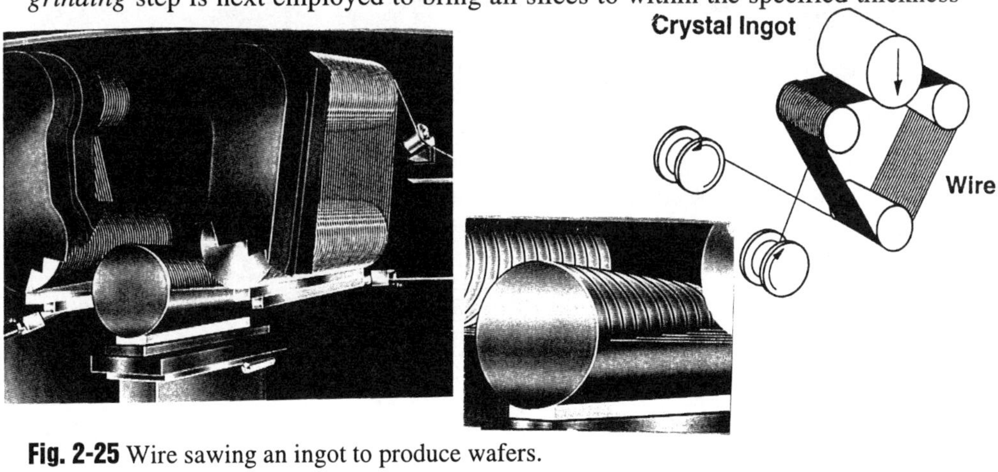

Fig. 2-25 Wire sawing an ingot to produce wafers.

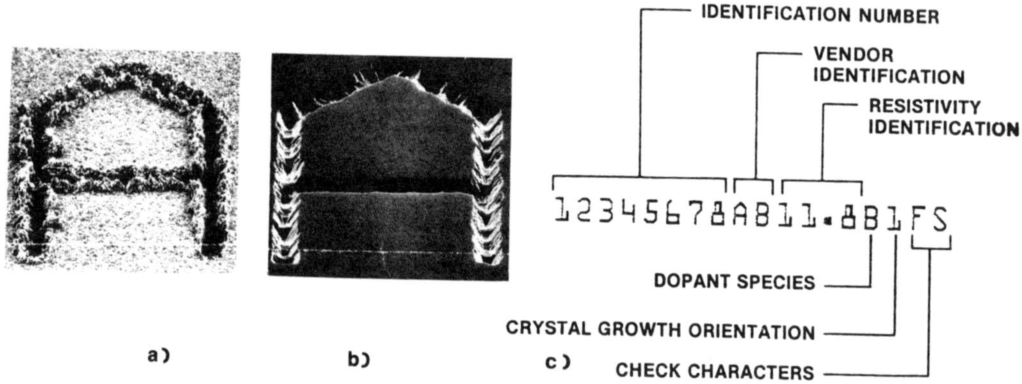

Fig. 2-26 a) SEM of letter "A" marked on wafer after it appears as marked on lapped wafer (left), after wafer has been etched (center), and after polishing. Courtesy of Wacker Siltronic; b) SEMI specification for marking vendor codes on wafers. Dimensions are specified along flat on wafer; c) SEMI specifications for 18-character vendor code. SEMI standard, used with permission.

tolerance. The lapping takes place between two counter-rotating cast-iron plates in the presence of an abrasive slurry, usually a mixture of micron-sized alumina or silicon carbide particles suspended in a solution. This step also serves to reduce bow and warp, and to increase slice flatness. Both sides of the slice are lapped using a mixture of Al_2O_3 and glycerine, and a succession of increasingly finer polishing grits is used in multiple lapping steps. This process can achieve flatness uniformities of 2-μm.

2.10.7 Removal of Surface Mechanical Damage by Chemical Etching

The work damage and contamination caused by the previous mechanical grinding, sawing, and lapping steps must be removed. This is done by chemically etching away the resultant damaged and contaminated layer.[71] The wet etch procedure for this step typically utilizes an etchant solution of hydrofluoric (HF) and nitric (HNO_3) acid (with modifiers such as acetic acid) to attack the silicon. The surface material removal is the result of two-step reactions. The Si surface is first oxidized by the nitric acid to form silicon oxide, and this oxide removed by the HF. A relatively non-porous and clean wafer backside is also produced by this step. Note that etching of the wafer may be repeated after the subsequent operation of edge rounding.

2.10.8 Rounding the Wafer Edge (and Notch)

The edges of the silicon slices are then shaped (rounded) as shown in Fig. 2-27. Edge rounding substantially reduces the incidence of chipping during normal wafer handling, and helps minimize film buildup on the edge of the wafer during photoresist and epitaxial processes. (Edge-chip defect sites are known to nucleate plastic deformation [or *slip*] during subsequent thermal processing steps).[72] The shape of the rounded edge profile usually follows an industry standard (i.e., SEMI standard) in which the edge profile fits within the boundary of a standard template.

2.10.9 Edge Polishing of the Wafers

Particle contamination reduces yield. In advanced, submicron IC processing some of the major sources of particles are the process equipment and the wafer itself. Defects found on the wafers during the device fabrication sequence are often caused by tiny chips of silicon from the wafer, or by particles removed from the process equipment. This abrasive effect is caused by the contact of the silicon wafer with the process equipment, and can be influenced by the smoothness of the contact area, primarily the wafers edge.

Wafer edges of as-sawn wafers are sharp and prone to chipping (Fig 2-27a). As a result, after sawing, wafers are generally subjected to a grinding process that rounds the wafer edge (with much of the edge-grinding damage being later removed during the wafer-surface etching process). This results in a wafer edge that is rounded, but is not perfectly smooth, as shown in Fig. 2-27b.[64] Such relatively rough edges still produce enough particles to reduce the yield of ICs, when compared to the yields obtained when these edges are polished after the edge has been etched to remove the grinding damage. That is, after edge-polishing the wafer edge is perfectly smooth, as shown in Fig. 2-27c, and indications are that the defect density on wafers is reduced if their edges are so polished. For the future generations of ICs that will produced on 300-mm wafers, the benefit of edge-polishing should be even more pronounced. That is, the wafer-edge area increases with increasing wafer radius and thickness. In addition, it appears that other particle sources on the wafer topography (such as the notch and deep laser scribe marks), must also be polished (or in some other way smoothed) to further reduce defect density.

2.10.10 Chemical-Mechanical Polishing of the Wafers

A chemical-mechanical polishing[71] step is used to produce the highly reflective, scratch- and damage-free surface on one or both sides of the wafer. This is

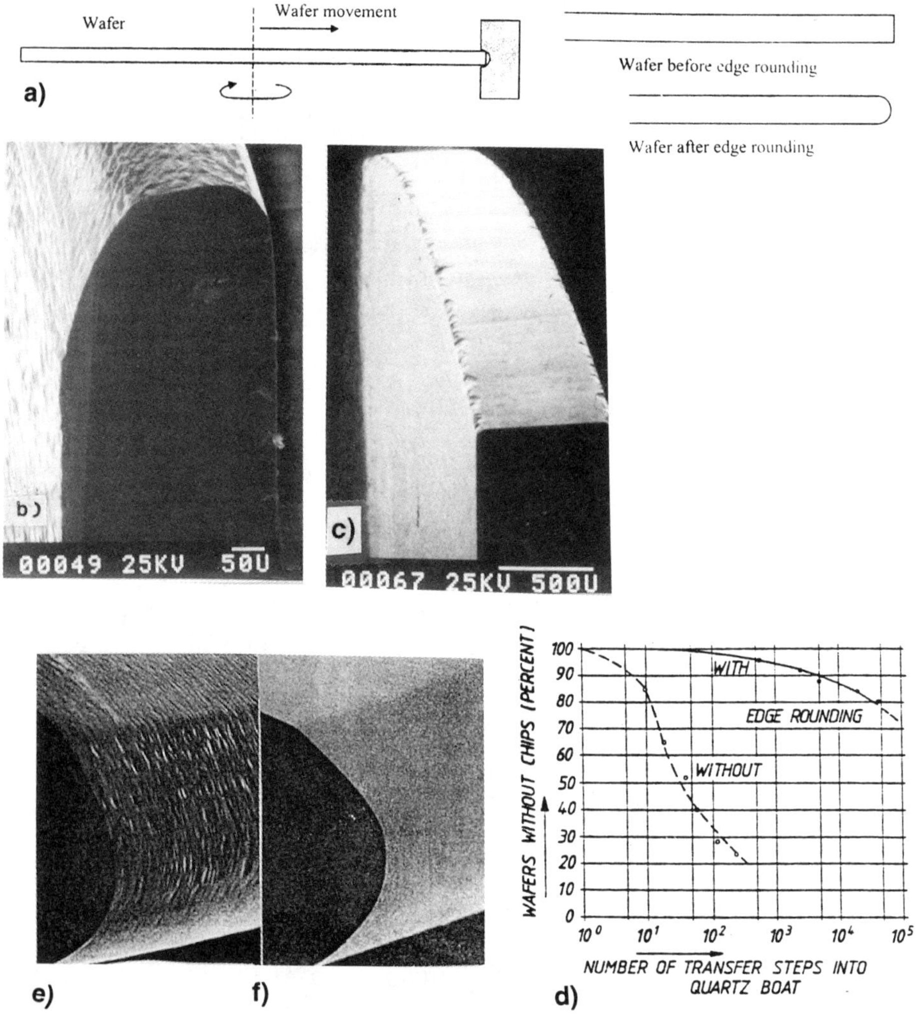

Fig. 2-27 a) Wafer edge-rounding process; b) Edge-rounded wafer; c) Non-edge rounded wafer; d) Mechanical strength of edge-rounded vs. non-edge-rounded wafers; e) Edge-rounded, but non-polished wafer; f) Edge-rounded, polished wafer.

accomplished by mounting unpolished slices onto a carrier, and then putting them on a polishing machine. There, a powered platen drives an appropriate polishing pad material across the wafer surface. A colloidal silica slurry of sodium hydroxide and fine (~10-nm) SiO_2 particles is dripped onto the table. The frictional heat of the sliding mounted wafers causes the sodium hydroxide to oxidize the Si (i.e., the chemical part of the process). The oxide is then abraded away by the silica particles (i.e., the mechanical part). Figure 2-28 shows the wafer thickness and surface roughness changes as it undergoes the various grinding,, lapping and polishing processes.

For 300-mm wafers, double-sided-polishing will be standard (instead of the single-sided-polishing used up to the 200-mm wafer sizes). Double-sided polishing (CMP on both the front and back surfaces simultaneously) has been found to result in superior flatness than single-sided polishing.

The wafer preparation via mechanical methods (i.e., grinding, CMP, etc.) has its limits in the degree of flatness that can be obtained. To supplement and fine tune the local lithography for further improvement in local flatness, tools that can carry out such processes as *plasma-assisted chemical etching* (PACE) have been developed.[73] Such a tool employs a spatially confined plasma with a scanning mechanism to allow material removal to be controlled as desired over the wafer surface. The PACE utilizes low-energy ions (i.e., << 1 eV), instead of the energetic ions that are employed in reactive-ion-etching. Therefore, PACE produces minimum or no subsurface damage.

2.10.11 Cleaning the Wafers

Following the polishing cycle (in which a layer of silicon about 25-μm-thick is removed) the wafers are subjected to series of chemical dips and rinses to remove the polish slurry. A cleaning process concludes this step. This is a very important

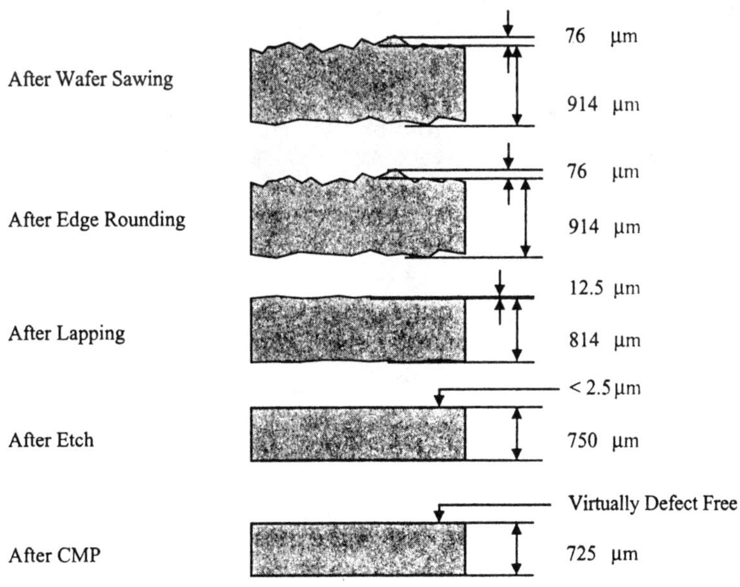

Fig. 2-28 Wafer thickness and surface roughness changes during grinding and polishing.

step as the silicon wafer must be free of contamination before it is shipped to the device fabrication facility. Chemical cleaning is effective in removing contaminants from the wafer surface. The "RCA Clean" (see Vol. 1, Chap. 5) is widely used to clean the wafer before shipping.

2.10.12 Depositing Epitaxial Silicon Layers on the Wafers

Epitaxial layers are required for some ULSI process technologies. Wafer manufacturers are now supplying silicon wafers with a variety of epitaxial layers for CMOS ULSI technologies. Another option being offered is a layer of polysilicon or mechanical damage on the backside of the wafers for extrinsic gettering purposes.

2.10.13 Shipping 300-mm Wafers: Its Impact on the Evolution of Wafer Shipping Boxes

An inspection, followed by packing in a particle-free package finishes the wafering sequence. The box used to ship 300-mm wafers is called a *front-opening shipping box* (FOSB, pronounced *fos-bee*). FOSBs cost approximately 10 times as much as the boxes that are used to ship 200-mm wafers. The main reason for this cost discrepancy is that the 300-mm wafer is more susceptible to breakage because it is *proportionately* thinner than previous diameters. Consequently, the shipping package must provide more protection. In addition, the FOSB must be more rigid in order to interface with automated load ports.

Since the cost is so high, it must be possible to re-use the FOSB. In fact, as shown in Fig. 2-29, if the FOSB is used only once, the cost per wafer to use it is ~\$15, whereas if it is reused 10 times, the cost per wafer drops to ~\$7. Neverthe-

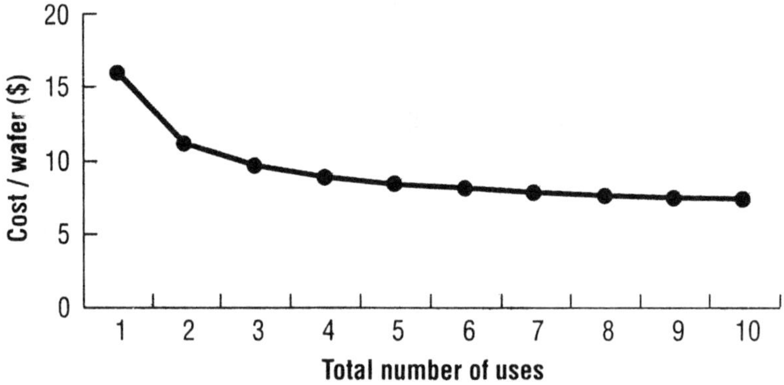

Fig. 2-29 The cost/reuse of a FOSB levels off at \$7/wafer.

less, the overall cost of the box to ship the wafer is at least 55% higher for 300-mm wafers than for 200-mm wafers.

One topic being discussed concerns the issue of FOSB ownership. One proposal before the SEMI 300-mm Wafer Shipping Box Taskforce is that manufacturers become owners of the FOSB. That is, since device manufacturers want a guarantee of wafer cleanliness, the best way to supply this is to maintain completely control the FOSB history. Ownership of the FOSB makes this possible. The protocol for using the FOSB is to remove the wafers upon receiving them from the wafer supplier, to then seal the FOSB in a polypropylene bag and ship it back to the wafer producer. The FOSB is like a wafer. Once it gets contaminated, it becomes hard, or impossible to clean.

2.11 SPECIFICATIONS OF SILICON WAFERS FOR VLSI

There are many properties that a wafer must exhibit (Table 2-2) in order to be an appropriate starting substrate for the fabrication of VLSI and ULSI. Table 2-3 provides some typical specification ranges for 125-, 150-, and 200-mm diameter CZ wafers. Figure 2-30 graphically summarizes the qualitative characteristics desired in 200-mm device quality Si wafers for ULSI circuits.[75] The semiconductor industry has seen quality improvements in Si crystals grown by the CZ technique continually over the years. In today's emerging sub-0.13-μm IC manu-

Table 2-2. COMPARISON OF MATERIAL PROPERTIES AND REQUIREMENTS FOR VLSI AND ULSI

Property	Czochralski	Float Zone	Requirements for ULSI
Resistivity (phosphorus) n-type (ohm-cm)	1–50	1–300 and up	5–50 and up
Resistivity (antimony) n-type (ohm-cm)	0.005–10	–	0.001– 0.02
Resistivity (boron) p-type (ohm-cm)	0.005–50	1–300	5–50 and up
Resistivity gradient (four point probe) (%)	5–10	20	< 1
Minority carrier lifetime (μs)	30–300	50–500	300–1000
Oxygen (ppma)	5–25	Not detected	Uniform and controlled
Carbon (ppma)	1–5	0.1–1	< 0.1
Dislocation (before processing) (per cm^2)	≤ 500	≤ 500	≤ 1
Diameter (mm)	up to 300	up to 100	up to 300
Wafer bow (μm)	≤ 25	≤ 25	< 5
Wafer taper (μm)	≤ 15	≤ 15	< 5
Surface flatness (μm)	≤ 5	≤5	< 1
Heavy-metal impurities (ppba)	≤ 1	≤ 0.01	< 0.001

TABLE 2-3 SPECIFICATIONS FOR SINGLE-CRYSTAL SILICON WAFERS

Parameter	150 mm	200 mm	300 mm
Diameter (mm)	150 ± 0.2	200 ± 0.2	300 ± 0.1
Thickness (μm)	675 ± 25	725 ± 20	775 ± 25
Primary flat length (mm)	55–60	notch	notch
Oxygen content (ASTM '79) (ppma)		≥ 23 ± 2	≥ 22 ± 1.5
Flatness (μm) (GTIR)	1.5	1.5	1.5
Site Flatness (μm) (SFQD)		± 0.17 (1998)	± 0.12 (2001)
Surface orientation	(100)	(100)	(100)

facturing, 200-mm and 300-mm wafers must meet ever tightening wafer quality requirements in terms of the crystalline defects, heavy-metal impurities, oxygen level, oxygen radial uniformity, minority-carrier lifetimes, particles, and flatness.

Note that these specifications have also become tighter with increasing wafer size due to automatic wafer handling and photolithographic requirements. ADE Corporation is the major supplier of tools used to measure many of these specifications. While 300-mm wafers are just being introduced into full-scale IC production, the required quality as updated in the 1999 *Technology Roadmap for Semiconductors* (Table 2-4, p. 69) will be met by further developing and refining crystal growth technology being applied and developed for 200-mm wafers, particularly as it is applied to point-defect and secondary-defect control, and the control of oxygen.

Some questions have arisen, however, about whether some aspects of wafer "goodness" may be specified too tightly, especially with respect to contaminant levels in the silicon. That is, it may be necessary to balance "best wafer possible" against the cost of ownership opportunity of not driving silicon requirements to the detection limit, but to some less stringent value consistent with achieving high yield. This issue is important, as the cost of a 300-mm wafer is projected to account for as much as 30% of the total IC wafer processing cost.[76]

There has been a variety of data in the literature to support such a view.[77] One example, described earlier, is that the specification on the limits of COPS and metals may be relaxed for device technologies in which thinner gate dielectrics are employed. For example, at gate-oxide thicknesses of 5-nm or less, the influence of COPS on gate-oxide-integrity (GOI) has been found to become negligible for CZ polished wafers.[78] (On the other hand, at the 0.1-μm node, the device features will approach the size of the void diameters, and this might cause

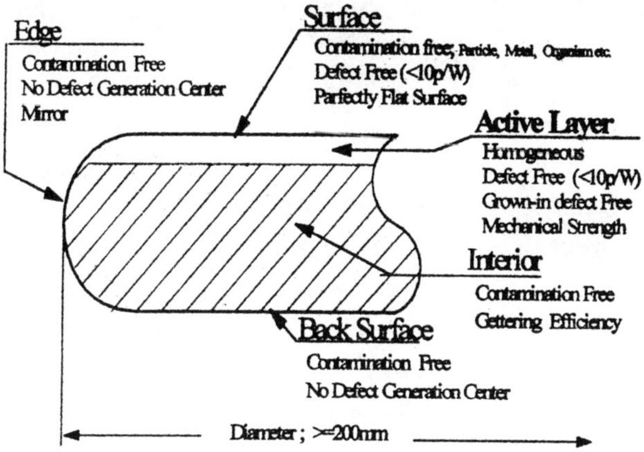

Fig. 2-30 Silicon wafer structure required for ULSI.[65]

shorts, e.g., between adjacent trench capacitors in a DRAM.) Furthermore, gate oxidation is generally performed after a number of prior thermal processes which typically annihilate near-surface COPs, thereby reducing whatever residual influence these defects may have on the GOI. In addition, in epitaxial material, the COPs are covered up to a depth beyond ~100 nm, essentially negating their influence. Accordingly, the necessity of using "perfect-silicon" (while being a remarkable scientific achievement, see Sect. 2.6), apparently needs to be reassessed for future generations of ICs from a cost-of-ownership perspective.

The implications of thin-gate-oxides for the specification of the metal levels in the Si wafer are based on the paradoxical fact that the integrity of thinner gate oxides may be less compromised by metals than is the integrity of thicker oxides.[79] A second reason that the maximum allowed levels of metal in the silicon may be allowed to be increased is that cleaning technologies have been improving, allowing the incoming critical metal level specifications to be somewhat higher than those now in use. This is because the improved cleaning processes drive the initial levels of metal down to an operational level that is an order of magnitude smaller than the initial level.

The utilization of multiple-well structures fabricated by ion-implantation in effect decouples the MOSFETs from the original silicon material characteristics. This reduces the sensitivity of these devices to the initial resistivity- and uniformity-tolerance of both CZ polished and epitaxial wafers. Whereas the nominal substrate doping for epitaxial wafers is typically in the range of ~5-10 mΩ-cm (to ensure latch-up control), shallow-trench-isolation (STI) is rendering classical concerns for latch-up, such as n^+-p^+ spacing moot. Accordingly, resistivity ranges ≥20% may be acceptable, also favorably impacting CoO.

Internal gettering (IG) in CZ-polished-Si wafers will still be required to remove deleterious metallic contaminants inadvertently introduced during IC processing. A ±2 ppma variation in bulk oxygen concentration in the range ~18-31 ppma (old ASTM standard) may be sufficient to ensure stringent bulk defect control with an acceptable homogeneous IG capability. Adequate wafer strength and resistance to warpage, along with denuded-zone width control, is also required. The utilization of heavily-boron-doped substrates for epitaxial wafers also offers an effective IG capability, especially for gettering Fe.

In summary, it appears that calling for further reduction of various contaminants and defects in Si wafers (and the continued drive to reduce certain other wafer specifications) may not necessarily be required to ensure continued improvements in IC performance.

2.12 THE ECONOMICS OF SILICON WAFERS

As discussed earlier, the size of silicon wafers has been continuously increasing. The motivation for using larger wafers is economic, and here we discuss this issue. The economic benefits of larger wafers are gained from the additional silicon area they provide (Fig. 2-31). If the manufacturing cost per wafer remains constant (or if its percentage increase is smaller than the percentage increase in silicon area) a reduction in the manufacturing cost per chip may be possible.

Let us address the issue of the silicon wafer costs. We see from Fig. 2-32 that the cost per square inch of silicon on a given wafer size is highest when the new wafer size is first introduced, but then drops with time.[80] For 125-mm prime

Diameter Increase	Percentage Area Increase
2" → 3"	125%
3" → 100 mm	78%
100 mm → 125 mm	56%
125 mm → 150 mm	44%
150 mm → 200 mm	78%
200 mm → 300 mm	125%

Diameter	Area (Sq. Inches)
2 Inch	3.1
3 Inch	7.1
100mm	12.6
125mm	19.6
150mm	28.3
200mm	50.2
300mm	109.5

Fig. 2-31 The increase in silicon wafer area with increasing wafer diameter.

device wafers the minimum silicon cost reached ~\$1/in^2. For each succeeding wafer size the minimum cost/in^2 is expected to be 1.4x the cost of the previous generation. Thus, for 200-mm wafers the silicon cost is \$2/in^2, resulting in the minimum cost of prime bulk CZ wafers of this size being \$80–\$100. The cost of 300-mm wafers is expected to eventually decline to somewhere above \$3/in^2, with a single bulk-CZ wafer thus eventually costing \$350–\$400. In 1998, the cost of prime 300- mm wafers was ~\$1000. Note that adding an epitaxial layer to the wafer approximately doubles the wafer cost, and the cost of 200-mm epi wafers in 1998 was \$150–\$180.

The cost/in^2 increases with wafer size for a number reasons. First, larger wafers are thicker, meaning fewer wafers per cm of ingot length are produced. Second, a smaller percentage of the electronic-grade polysilicon (EGS) ends up in prime wafers. (For 150-mm wafers this percentage was 30%, for 200-mm wafers it dropped to 17%, and for 300 mm wafers it is expected to be only ~10%.[81]) Third, wafer specs are tighter for each generation, reducing the "sweet spot" in the middle of the ingot that yields prime wafers. Finally, epi may be required for DRAM production on 300-mm wafers, which will create a steep rise in average price of prime wafers.

Note that the consumption of wafers of a given size also varies with time (see Fig. 2-33), and each new wafer size remains in production for about 24 years.[82] For example, production of 150-mm wafers began in the late1970s, and their consumption rate peaked in 1997. Their use in production is expected to be phased out by about 2010. Next, 200-mm wafers were first introduced in 1988, and they are expected to reach their peak consumption rate sometime between 2005 and 2010 (depending on the rate of ramp up of the new 300-mm wafer generation).[72] By 1998, however, full-scale-production IC factories using 300-mm wafers were not yet running. As a result, well-designed 200-mm equipment is predicted to have a long working life before it is made obsolete by future wafer-fabs running larger wafer sizes.

The worldwide consumption of single-crystal silicon has been growing at 15–20% per year. In 1992, about 1.9 billion in^2 of silicon were sold. By 1994 this number was 2.7 billion in^2, and by 1999 it had increased to ~4.5 billion in^2. In 1994 this translated to a total production of about 100 million wafers. These sales generated an income of \$5.9 billion US in 1999.[83,84] There are a relatively large number of silicon wafer suppliers. In the late 1990's the largest were Shin-Etsu Handotai (SEH: Japanese), MEMC (German); Wacker-Siltronic (German); Sumitomo-Sitix (Japanese); Mitsubishi Materials Silicon (Japanese); and Komatsu (Japanese).

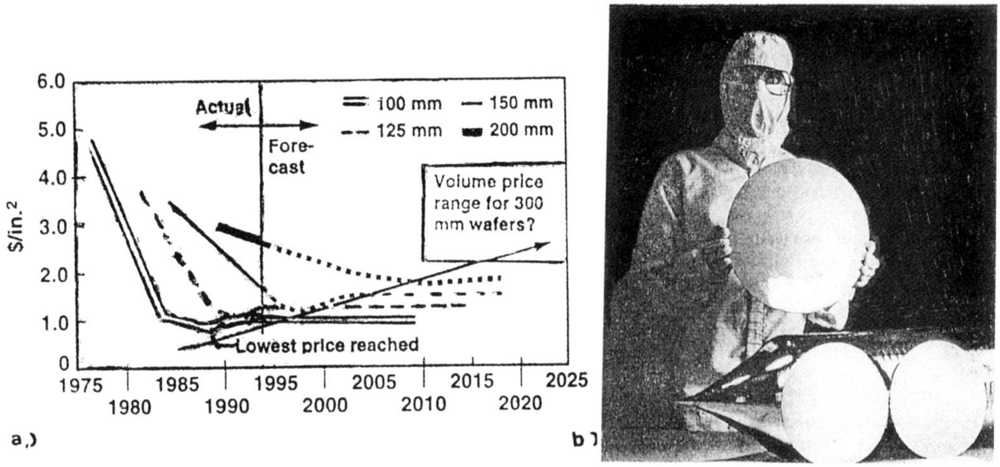

Fig. 2-32 a) Actual wafer pricing cycle by wafer diameter. Reprinted with permission of Semiconductor International.[70] b) Photograph of a 300-mm wafer. Courtesy of MEMC.

The silicon wafer industry historically funded its R&D and capacity expansion for the next generation, large-diameter wafer through the profits and cash flow from the previous-generation wafers. However, in the 2000-year timeframe the industry was being asked to fund the 200-mm and the 300-mm generation at the same time. According to the Rose Associates this is unprecedented, especially in an industry that lost nearly $1 billion in 1998 on sales.[85] These losses resulted from the semiconductor industry downturn, the Asian financial crisis, and excess silicon wafer capacity that combined to cause the average selling price of silicon wafers to fall by 36% over the three year period ending in mid-1999. This caused the industry to focus on cost reduction for survival. While losses in 1999 narrowed, and profitability returned in 2000, the profit outlook was still uncertain. Unless the profitability continues to improve, the funding for 300-mm development will compete with the economics of 200-mm growth and capability requirements.

One possible scenario will be that there will be a shortage of wafers in the 2000 timeframe, especially 300-mm wafers. One vehicle for alleviating this shortage is to use reclaimed wafers, especially for the role of test wafers used in process development applications. That is, during the development and marathon testing of wafer-manufacturing tools, reclaim wafers provide an excellent opportunity to re-use materials. While the maturation of 200-mm processes has greatly reduced the need for test wafers, their use will increase as the 300-mm processes enter the mainstream manufacturing arena. The price of 300-mm

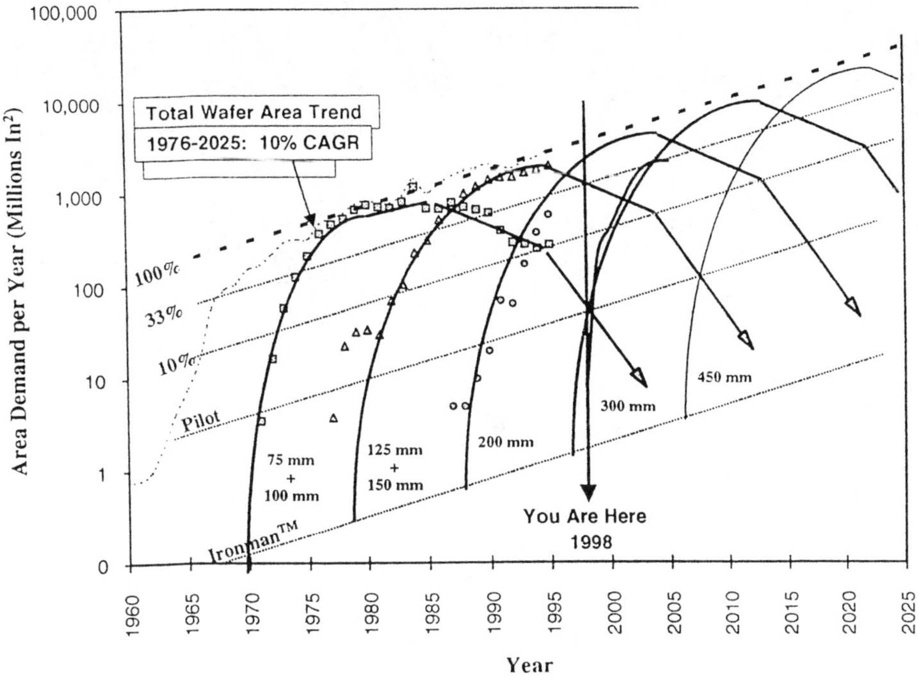

Fig. 2-33 The 24-year single-diameter cycle of silicon wafers.[82]

Selected near-term general characteristics and polished wafer requirements for silicon+

TABLE 2-4

□ = Solutions exist = Solutions being pursued ■ = No known solutions

Year	**1999**	**2000**	**2001**	**2002**	**2003**	**2004**	**2005**
Technology node (nm)	180			130			100
Wafer diameter (mm)*	200	300**	300	300	300	300	300
				Defined limit values			
Edge exclusion (mm)	2	1	1	1	1	1	1
Front surface particle and COP size (nm)	≥90	≥82.5	75	65	60	55	50
Particles and COPs (cm⁻²)***	≤0.13	≤0.12	0.12	0.14	0.13	0.12	0.10
Particles and COPs (#/wf)***	≤38	≤84	80	95	89	84	72
Critical surface metals (at/cm²)	≤$1.8x10^{10}$	≤$1.4x10^{10}$	≤$1.2x10^{10}$	≤$8.8x10^9$	≤$6.8x10^9$	≤$5.8x10^9$	≤$4.9x10^9$
Site flatness (mm)	≤180	≤165	≤150	≤130	≤120	110	100
Oxygen (center point value ±2.0 ppma)	19–31	18–31	18–31	18–31	18–31	18–31	18–31
Total bulk Fe (at/cm³)	≤$1x10^{10}$	<$1x10^{10}$	<$1x10^{10}$	<$1x10^{10}$	<$1x10^{10}$	<$1x10^{10}$	<$1x10^{10}$
OSFs DRAM (cm⁻²)	≤4.4	≤3.9	≤3.4	≤2.8	≤2.5	≤2.2	≤1.9
OSFs MPU (cm⁻²)	≤3.1	≤2.5	≤1.9	≤1.5	≤1.4	≤1.1	≤1.0
Recombination lifetime (μm)	≤305	≤350	≤350	≤350	≤350	≤350	≤350

* Significant gaps in metrology and wafer manufacturing equipment need to be closed on 200mm, especially for the 180nm and 130nm nodes.

** Values are for 300mm, although 200mm will be the dominant wafer diameter.

*** The total allowable, front-surface, light-scattering defect density is the sum of crystal originated pits (COPs) and particles.

+ The complete Table 34A in the 1999 *International Technology Roadmap for Semiconductors* includes other parameters and requirements for epitaxial and SOI wafers.

reclaim wafers is about 60% of the price of virgin wafers. Reclaim wafers in 2000 represented about 10% of the 300-mm market.[86]

REFERENCES

1. G.K. Teal, *IEEE Trans. Electron Dev.*, **ED-23**, 621 (1976).

2. R.N. Thomas *et al.*, "Melt Growth in Large Diameter Wafers: Part II," *Solid State Technology,* April 1990, p. 163.

3. J.G. Park, *et al.,* "Effect of Defects such as COPs, Large Dislocations and OSF-Ring on Device Integrity Degradation. *Mtg. Abs. Electrochem. Soc.,* Spring 1999, Abs. 343.

4. V.V. Voronkov, *J. Crystal Growth,* **59** (1982) p. 625.

5. T. Sinno *et al.*, "Point Defect Dynamics and Oxidation-Induced Stacking Fault Ring in CZ Silicon Crystal," *J. Electrochem. Soc.,* **145**, p. 302 (1998).

6. T. Sinno, R.A. Brown, "Modeling Microdefect Formation in CZ Silicon," *J. Electrochem. Soc.,* **146**, p. 2300 (1999).

7. M. Itsumi *et al., Jpn. J. Appl. Phys.*, **35**, 812 (1996).

8. M. Itsumi *et al., J. Appl Phys*, **78**, 5984 (1996).

9. K.-M. Kim, "Silicon-Pulling Technology for 2000+," *Solid State Technology,* January 2000, p. 72.

10. E. Dornberger *et al.,* "Simulation of Non-Uniform Grown-In Defect Distribution in Czochralski Silicon Crystals," *Mtg. Abs. Electrochem. Soc.,* Spring 1998, Abs. 312.

11. F. Ishikawa *et al.,* "The Formation Mechanism of Coupled Voids in Czochralski Silicon Crystals," *Mtg. Abs. Electrochem. Soc.,* Fall 2000, Abs. 678.

12. W. von Ammon, *et al.*, "The Dependence of Bulk Defects on the Temperature Gradient of Silicon Crystals During CZ Growth" *J. Crystal Growth,* Vol. 151, p. 273, 1995.

13. W. von Ammon, "Silicon Crystal Growth at the Beginning of the 21st Century," *Mtg. Abs. Electrochem. Soc.,* Spring 2000, Abs. 406.

14. K-M. Bae *et al.,* "A Vacancy/Oxygen Based Model Describing Radial Distribution of Voids, OSF-Ring and AOP in Ring Structure of CZ Silicon Crystals," *Mtg. Abs. Electrochem. Soc.,* Fall 1998, p. 644.

15. M. Hasebe, *et al., Defects Control in Semiconductors*, K. Sumino ed., North Holland, New York, p. 157 (1990).

16. T. Sinno *et al.*, "Point Defect Dynamics and Oxidation-Induced Stacking-Fault Ring in CZ-Grown Silicon Crystals," *J. Electrochem. Soc.,* January 1998, p. 302.

17. T. Sinno and R.A. Brown, "Modeling Microvoid Formation in Czochralski Silicon," *Mtg. Abs. Electrochem. Soc.*, Spring 1998, p. 315.

18. Y. Matsushita et al., *Ext. Abs. 18th Intl. Conf Solid State Dev. and Materials,* (Jpn. Soc. Appl. Phys., Tokyo 1986) p. 529S.

19. N. Yamada and H. Yamada-Kaneta, "Comparison of Gate Oxide Reliability Between Ar-Annealed Wafers and H_2-Annealed Wafers," *Mtg .Abs. Electrochem. Soc.*, Spring 1998, Abs. 490.

20. T. Minami *et al.,* "Influence of Void Size on the Formation of Defect-Free Regions in Hydrogen Annealed CZ Silicon Wafers," *Mtg. Abs. Electrochem. Soc,* Spring 2000, Abs. 514.

21. Y. Matsushita *et. al.,* "Hydrogen Anneal of Silicon Wafer Formation of High Quality Device Active Layer," *Mtg. Abs. Electrochem. Soc.,* Spring 1998, Abs. 326.

22. L. Peters, "Striking a Compromise Between Good and Perfect Silicon," *Semiconductor International,* October 1999, p. 19.

23. N. Adachi *et al.*, "Reduction of Grown-in Defects by High Temperature Annealing," *Abs. Mtg. Electrochem. Soc.,* Spring 1998, Abs. 327.

24. R. Winkler *et al.,* "The Gate Oxide Quality of Annealed CZ silicon Wafers and Its Influence on 4-Mb DRAM Device Yield and Reliability," *Mtg. Abs. Electrochem. Soc.,* Fall 2000, Abs. 710.

25. M. Hourai *et al., "Nature and Generation of Grown-in Defects in Czochralski Silicon Crystals,"Mtg. Electrochem. Soc.,* Spring 1998, Abs. 310.

26. H. Niwa, "Reduction of Junction Leakage and Improvement of GOI Using advanced CZ wafers," *Mtg. Abs. Electrochem. Soc.,* Spring 1999, Abs. 478.

27. H. Nishikawa, *et al.*, "Formation of Grown-In Defects During CZ Silicon Crystal Growth," *Jpn. J. Appl. Phys.,* Vol. 36, p. 6595, 1997.

28. R. Schmolke *et al.*, "Advanced Silcon Wafers for 0.18 μm Design Rule and Beyond: Epi and fLASH," *Mtg. Abs. Electrochem. Soc.,* Fall 2000, Abs. 671.

29. K. Moriya *et al., Jpn. J. Appl. Phys.,* 34, 2965 (1995).

30. R. Takeda *et al.*, "Influence of LSTD Size on the Formation of Denuded Zone in Hydrogen Annealed CZ Wafers," *Abs. Mtg. Electrochem. Soc.,* Fall 2000, Abs. 700.

31. G.A. Rozgonyi, "Trends in Defect and Impurity Control for Fine Line VLSI Processing," *Defects in Silicon,* 1983, Electrochem. Soc., Pennington, N.J., 1983, p. 29.

32. S.M. Hu, "Anomalous Temperature Effect of Oxidation Stacking Faults in Silicon," *Appl Phys. Lett.*, **17**, 165 (1975).

33. H. Shiraki, "Stacking Fault Generation Suppression and Grown-In Defect Elimination in Dislocation-Free Silicon Wafers by HCl Oxidation," *Jpn. J. Appl. Phys.*, **15**, 1 (1976).

34. E.W. Haas *et. al.*, *J. of Elec. Matls*, **7**, 525 (1978).

35. L.M. Ephrath and R.S. Bennett, *Extended Abstracts, Electrochem. Soc. Fall Meeting*, Abs. 302, Vol. 80-2, Hollywood, FL., 1980.

36. W. Zuhlehner and D. Huber, "Czochralski Growth Silicon," in *Crystals 8 (Silicon Chemical Etching)*, Springer-Verlag, Berlin (1982).

37. M.C. Chen and V.J. Silvestri, *J. Electrochem. Soc.*, **129**, 1294 (1982).

38. W.T. Stacy *et al*, *Defects in Silicon, Proc. Electrochem. Soc.* 85-9, p. 423, 1983.

39. C.W. Pearce, L.E. Katz, and T.E. Seidel, "Considerations Regarding Gettering in Integrated Circuits," *Semiconductor Silicon 1981*, Electrochem. Soc., Pennington, N.J., 1981, p. 705.

40. D.C. Gupta, R.B. Swaroop, *Solid State Technology*, August 1984, p. 113.

41. J.E. Lawrence and H.R. Huff, "Si Material Properties for VLSI Circuitry," in *VLSI Electronics, Microstructure Science*, Vol. 5 (N.G. Einspruch, Ed.), Academic Press, Orlando FL (1982), p. 51.

42. S. Kishino, "Material Characterization for VLSI Applications," *VLSI Sci-ence & Technology*, 1985, Electrochem. Soc., Pennington, N.J., 1985, p. 399.

43 R.B. Swaroop, "Advances in Silicon Technology for the Semiconductor Industry," *Solid State Technology,* July 1983, p. 97.

44. C.W. Pearce and V.C. Kannan, "Saucer Pits in Si," in *Defects in Silicon*, 1983, Electrochemical Soc., Pennington, N.J. (1983), p. 396.

45. *ibid, Ref. 43.*

46. D. Huber, J. Reffle, "Precipitation Process Design for Denuded Zone Formation in CZ Silicon Wafers," *Solid State Technology,* August 1983, p. 137.

47. *ibid. Ref. 44.*

48. M.B. Shabani *et al.*, "Competitive Gettering of Fe, Ni, and Cu in Si Wafers by Poly Backside Gettering and Internal Gettering," *Ext. Abs. Joint Intl. Mtg. Electrochem. Soc.*, Abs. No. 1959, Fall 1997, p. 2270.

49. R.F. Pinizzotto, *et. al*, "Temperature Ramping for Nucleation of Oxygen Precipitates in Si," *Materials Research Soc. Symp. Proc.* Vol. 36, Materials Res. Soc., 1985, p. 275.

50. T. Ono *et al.*, "Oxide Precipitate-Induced Dislocation Generation in Heavily Boron-Doped Czochralski Silicon," *J. Electrochem. Soc.* September 1999, p. 3461.

51. R.K. Watts, *Point Defects in Crystals*, Wiley and Sons, New York, 1977.

52. *ibid. Ref. 45.*

53. D. Huber, J. Reffle, "Precipitation Process Design for Denuded Zone Formation in CZ Silicon Wafers," *Solid State Technology,* August 1983, p. 137.

54. R. Craven "Internal Gettering in CZ Si," *Matls. Res. Soc .Symp. Proc.* Vol. 36 MRS 1985 p. 159.

55. R. Schmolke *et al.*, "Defect Depth Profile in Si (100) *p/p*$^+$ Epitaxial Wafers," *Abs. Mtg. Electrochem. Soc.*, Spring 1998, Abs. 339.
56. K. Schmolke *et al.*, "Advanced Silicon Wafers for 0.18 μm Design Rule and Beyond: Epi and fLASH!," *Mtg Abstracts Electrochem. Soc.*, Fall 2000, Abs. 671.
57. A. Huber *et al.*, "Reliability of Gate Oxides on Silicon Substrates with 5 – 10 nm Oxide Thickness," *Abstracts Mtg. Electrochem. Soc.*, Spring 2000, Abs. 478.
58. M. Suhren *et al.*, "Crystal Defects in Highly Boron Doped Silicon," *Mtg. Abs. Electrochem. Soc.*, Fall 1996, p. 453, Abs. 358.
59. H. Takeno *et al.*, "Influence of Heavy Boron Doping on Oxygen Precipitation of Czochralski silicon Crystals and Its Computer Simulation," *Abstracts Mtg. Electrochem. Soc.*, Spring 1998, Abs. 351.
60. G. Kissinger *et al.*, "Influence of Boron Doping on Oxide Precipitate Nucleation and Gettering of Iron Impurities in Low Thermal Budget Processing of Czochralski Silicon," *Abstracts Mtg. Electrochem. Soc.* Spring 1999, Abs. 338.
61. D. Graf *et al.*, "300-mm Epi *p/p*$^-$ Wafer: Is There Sufficient Gettering?" *Mtg. Abs. Electrochem. Soc.*, Fall 2000, Abs. 699.
62. R. Falster *et al.*, "The Engineering of Silicon Wafer Material Properties Through Vacancy Concentration Profile Control and the Achievement of Ideal Oxygen Precipitation Behavior," *Mtg. Abs. Electrochem. Soc.*, Fall 1998, Abs. 643.
63. R. Falster *et al.*, *MRS* 510, p. 27 1998.
64. K. Winkler and W. Hanke, "On the Effect of Edge Polishing on the Yield of 4 Mbit DRAMs," *Mtg. Abs. Electrochem. Soc.*, Spring 1998, Abs. 374.
65. W. Helgeland, *et al.*, "CZ Crystal Growth into the 90's," *Microelectronic Mfg. & Test.*, May 1986. p.1.
66. N. Jackson, "Materials and Technology of Wafering," *Solid State Technology*, July 1985, p. 107.
67. G.R. Fisher, "Challenges for 300-mm Polished Wafer Manufacturers," *Semiconductor International*, September 1998, p. 97.
68. Semiconductor Equipment and Materials Institute (SEMI), Mountain View, California.
69. I.A. Morris, "Wafer Identification Increases Yields," *Micro. Manuf. and Test*, Sept. 1985, p. 50.
70. J. Scaroni, "A Close Look at Laser Marking of Silicon Wafers," *Solid State Technology*, July '97 p. 245.
71. D. Biddle, "Characterizing Semiconductor Wafer Topography," *Microelectroncs Manuf. and Testing*, March 1985, p. 15.
72. S. Takasu, "Silicon Wafers for VLSI Technology," *VLSI Technology 1983*, Electrochem. Soc., Pennington, N.J. (1983), p. 490.

74. A. Bonora, "Flex-Mount Polishing of Si Wafers," *Solid State Technology*, **20**, 55 (1977).
75. Y. Matsushita *et al.*, "Hydrogen Anneal of Silicon Wafer Formation of High Quality Device Active Layer," *Mtg. Abs. Electrochem. Soc.*, Spring 1998, p. 326.
76. F. Robertson and A. Allan, *Future Fab*, 1, No. 3, p. 27, 1997.
77. H.R. Huff, *Silicon Wafers for the Mesoscopic Era," Abs. Mtg. Electrochem. Soc.*, Spring 2000, Abs. 405.
78. *ibid. Ref. 13.*
79. K.V. Ravi, *Solid-State Phenomena*, 89-70, p. 103 (1999).
80. W.H. Reed, "Large Diameter Wafer Update," *Semiconductor International,* November 1994, p. 140.
81. D. Rose, "Wafer Makers Challenged by Many Aspects of 300 mm," *Solid State Technology,* February 1996, p. 79.
82. D. Anderson, "Stoking the Productivity Engine with New Materials and Larger Wafers," *Solid State Technology*, March 1997, p. 57.
83. Rose's Annual Forecast for Fab Materials, *Semiconductor International,* April 1995, p. 17.
84. Rose's Annual Forecast for Fab Materials, 1999.
85. J. Highfill *et al.*, "The Cost-Effective Challenges of 300-mm Silicon Wafers," *Solid State Technology,* October 2000, p. S-8.
86. C. Thomson and H. Mynster, "300-mm Wafers: Reclaimers Role and Challenges," *Solid State Technology,* August 1999, p. 40.

Chapter 3

GATE DIELECTRICS: THIN SILICON DIOXIDE FILMS

Thermally grown films of silicon dioxide (SiO_2) play many key roles in the operation and fabrication of silicon integrated circuits. However, the most critical application of such films in deep-submicron CMOS USLI circuits is serving as the MOSFET gate dielectric material. Although the many other roles of SiO_2 are also important (including diffusion masking, surface passivation, and functioning as a field oxide, interlevel dielectric, screen oxide, pad oxide, trench liner in STI, tunneling oxide in non-volatile memory devices, and sidewall spacer), the discussion here will focus on its role as the *gate dielectric of MOSFETs.*

It is commonly believed that crystalline-silicon is the main material responsible for the growth and success of the semiconductor industry. But others maintain that the real magic in silicon technology lies with silicon dioxide and its amazing ability to satisfy the myriad demands placed upon it as the gate dielectric of the MOSFET.

In this chapter we will explore the role of thin gate oxide films in deep-submicron MOSFETs, and address the following issues that relate to this subject:

1. The structure and role of silicon dioxide (and the Si/SiO_2 interface) as the gate dielectric of the MOSFET.

2. Dielectric breakdown and defects in SiO_2;

3. Leakage currents in silicon dioxide films;

4. Models and processes of the growth of ultra-thin oxide films;

5. Strengthening silicon dioxide films through the addition of nitrogen (oxynitrides), and the related problem boron penetration through gate oxides;

6. Technology of growing ultra-thin oxides on large diameter wafers (200-mm and 300-mm);

7. Limits of silicon dioxide as a gate dielectric material.

Despite the fact that we will restrict our discussion to SiO_2-films as gate dielectrics, it is informative to list the other IC applications in which oxide films are employed, and the oxide thickness used in such roles (see Table 3-1).

Table 3-1 RANGES OF THERMAL OXIDE THICKNESS USED IN ULSI PROCESSING

Application	SiO_2 Thickness (nm)
Gate oxide insulators for MOS devices	1.5–10.0
Tunnel oxides in EEPROMs and flash memories	6.0–10.0
Screen oxides for ion implantation	10.0–20.0
Side-wall liners for shallow trench isolation (STI)	15.0–40.0
Field oxide for LOCOS isolation	200–400
Re-oxidation of the etched gate stack side-wall damage	5.0-7.5
Inter-polysilicon dielectric	4.0–20.0
Masking film against ion implantation and diffusion	100–200
Sacrificial oxides for gate applications	6.0–15.0
Capacitor dielectric in DRAM circuits	5.0–10.0

3.1 REQUIRED CHARACTERISTICS OF GATE DIELECTRICS FOR DEEP-SUBMICRON MOSFETS

A dielectric material must satisfy a large number of important demands in order to successfully function as the gate dielectric of a MOSFET. Silicon dioxide has been used as the gate dielectric material in silicon-based MOS ICs ever since they were first introduced in the early 1970's. It continues to serve in this role, even as MOSFETs have been scaled down to the deep-submicron regime. We list the requirements of dielectric material for this application here, and provide more details about them in later sections of the chapter.

1. The most important requirement that the gate dielectric of a deep-submicron MOSFET must meet is the following: *It must be possible to continue to decrease the thickness of the oxide to the degree specified in accordance with the scaling of the gate length L_E of a MOSFET.* We should elaborate upon why being able to scale the gate oxide thickness is so fundamentally important for deep-submicron MOSFETs. Figure 3-1 shows that the ratio of the gate length to the oxide thickness (L_E/t_{ox}) has indeed remained constant at about a value of 45 for process technology at Intel for over 25 years.[1] The gate-oxide requirements for technology nodes for 0.18-μm and beyond (according to the ITRS Roadmap) are given in Table 3-2. At some point, *if SiO_2 cannot be scaled further to meet future demands, another material with higher reliability will be needed.*

   As CMOS IC technology advances, the central focus of scaling the MOSFET device structure involves scaling the gate length L_E. That is, gate lengths of MOSFETs are scaled to obtain the significant benefits offered by such scaling. These benefits include: i) an increase in the drain current of the MOSFET I_D (which usually provides an improvement in the circuit

Table 3-2 Gate Oxide Equivalent Thickness for Deep-Submicron CMOS

	1999 (180 nm)	2001 (150 nm)	2003 (130 nm)	2006 (100 nm)	2009 (70 nm)	2012 (50 nm)
Equivalent Oxide thickness (nm)	3–4	2–3	2–3	1.5–2	<1.5	<1.0

speed); **ii**) a decrease in the gate area of the minimum-size MOSFET (which results in a reduction of the input capacitance that must be charged when logic levels are switched - and thus also improves the circuit speed); and **iii**) an increase in the density of devices that can be fabricated on a chip. (See Vol. 3 for more details on these issues.)

In deep-submicron MOSFETs, however, such reduction of the gate length increases the subthreshold drain current I_{Dst} of a MOSFET (which is one component of the total off-state leakage current of CMOS ICs). That is, in deep-submicron MOSFETs the short-channel effect known as *drain-induced barrier lowering* (DIBL) causes the subthreshold (off-state) drain current to rise significantly as the gate length is reduced (see Vol. 3 for more details). Two measures must be employed to prevent (or at least mitigate) this rise in the subthreshold leakage current as the gate length is

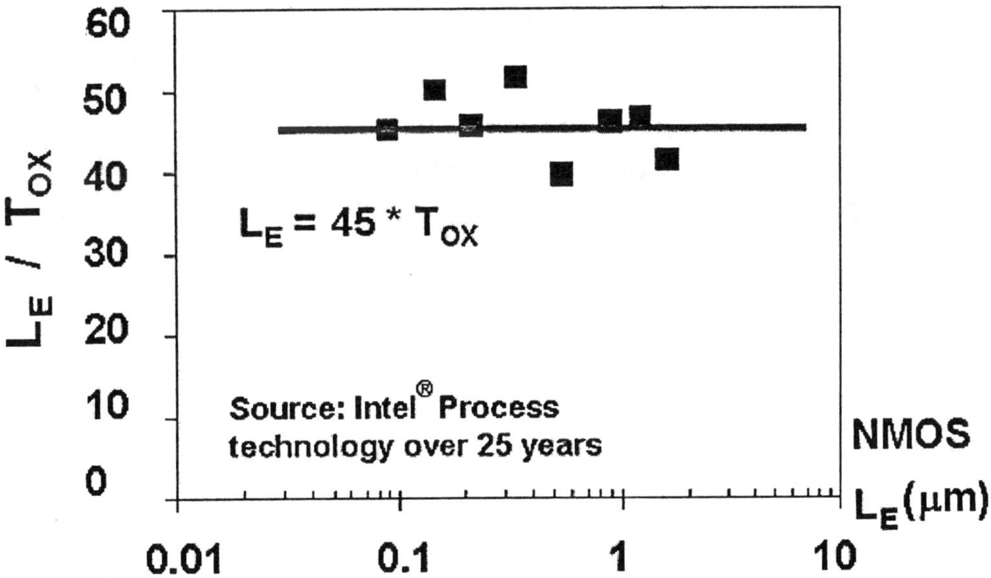

Fig. 3-1 Channel-length divided by gate-oxide thickness (L_E/t_{ox}) versus channel-length L_E for Intel's process technologies for the past 20 years.[1]

reduced. First, the junction depth X_j of the source/drain-extension-regions must be made shallower. Second, the depth of the MOSFET channel-depletion-region must also be made shallower. Both measures serve to keep the electric field of the drain from penetrating along the channel toward the source (which would worsen the DIBL effect).

The second measure (making the channel-depletion-region shallower as the gate length is shortened) is achieved by increasing the doping concentration in the channel (typically with a heavier-dose threshold-adjust-implant). While it is relatively easy to implement such an increase in the channel doping, doing so causes another undesirable effect. That is, if the gate-oxide-thickness of the MOSFET remains unchanged, the threshold-voltage V_T is increased as the doping concentration near the surface of the channel (boron) is increased (e.g., for an NMOSFET with a *p*-doped channel region). Since the threshold voltage is generally kept unchanged (or it may even be slightly *decreased*) as the CMOS technology is scaled, the increase in V_T due to the increased channel doping must be counteracted. This can only be done by reducing the gate-oxide-thickness (see Fig. 3-2). For example, as shown in Fig. 3-2, if the doping concentration of an NMOSFET with a 250-Å gate-oxide is increased from $6x10^{16}/cm^3$ to $1x10^{17}/cm^3$, V_T will increase from 0.6-V to about 1.4-V – which is far higher than the required V_T value of ~0.6 V.

In summary, this shows that in order to continue to scale the gate-length of MOSFETs, it will be necessary to decrease the oxide thickness by about the same scaling factor applied to the gate-length! However, such scaling of the oxide thickness brings with it both benefits and problems. One additional benefit is that the on-state drain current is further increased as the oxide thickness is scaled down (see Vol. 3). However, several other negative factors also ensue: the thinner oxide films cannot tolerate the increased gate electric fields, and this forces smaller power-supply voltages to be used (and such smaller supply voltages reduce the MOSFET on-state drain current); the tunneling leakage-current (gate current) through the oxide increases; the oxide is more prone to penetration by poly dopant (especially boron); and defects and oxide-layer thickness non-uniformities become bigger problems when thinner oxides must be manufactured. Figure 3-3 depicts these concerns for ultrathin gate oxides.

2. It must be possible to form the oxide layer with a thickness that closely matches the design specification of the MOSFET. This oxide thickness

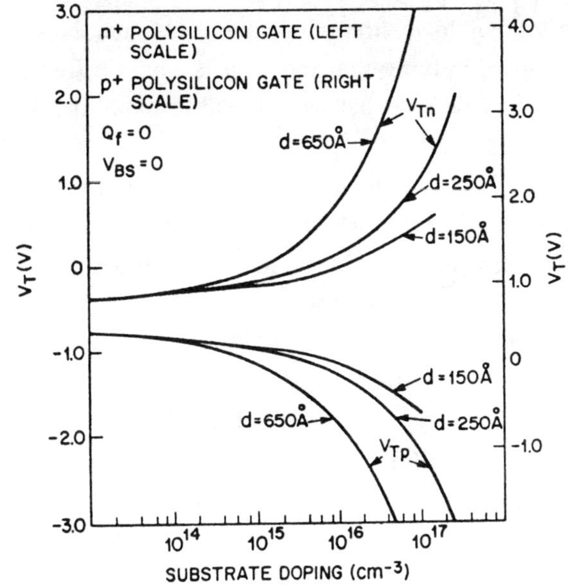

Fig. 3-2 Calculated threshold voltages of *n*-channel (V_{Tn}) and *p*-channel (V_{Tp}) transistors as a function of their substrate's doping, assuming n^+ polysilicon gate (left scale) and p^+ polysilicon gate (right scale). Curves for gate oxide thicknesses *d* of 150Å, 250Å, and 650Å are shown. From S.M. Sze Ed. *VLSI Technology,* 2nd Ed., Chap. 11, "VLSI Process Integration." Copyright, 1988 Bell Telephone Labs. Reprinted with permission.

must also be sufficiently uniform across the entire wafer, and from wafer-to-wafer, and from run-to-run.

3. The oxide film and the Si/SiO_2 interface must maintain stable electrical characteristics. This includes exhibiting adequately small values of charge in the oxide and at the Si-SiO_2 interface (i.e., low Q_f, D_{it}, Q_{ot} and Q_m values).

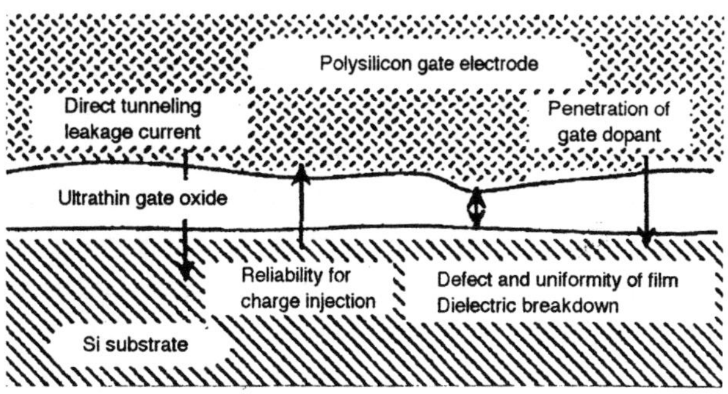

Fig. 3-3 Concerns in ultra-thin gate oxides.

4. The dielectric breakdown strength of the oxide must be sufficiently high (e.g., > 8-MV/cm), implying that the film is pinhole free and contains a negligible number of defects that would lead to oxide breakdown at lower electric fields.

5. The oxide film must exhibit sufficiently long lifetime under normal operating conditions (i.e., t_{BD} is adequate). This characteristic is related to **4**.

6. The dielectric material (oxide) must be compatible with the other materials of the MOSFET structure, and be thermally, electrically, and chemically stable under the processes used to fabricate the ICs.

7. The leakage-current through the gate-oxide must be sufficiently small to meet the off-current leakage requirement of the IC in which it is used.

8. The oxide should exhibit high resistance to hot-carrier damage (i.e., device degradation should be low in the face of hot-carrier stressing).

9. If the oxide is to be used in a symmetric CMOS technology (i.e., in which both p^+ and n^+ poly are used), the oxide film needs to be resistant to the penetration of boron at the process temperatures used after gate doping.

10. The oxide must passivate the silicon surface.

3.2 THE STRUCTURE OF THERMALLY GROWN SiO_2 AND THE PROPERTIES OF THE Si/SiO_2 INTERFACE

3.2.1 The Microscopic Structure of Thermally Grown SiO_2

SiO_2 (silica) can be found in crystalline or amorphous (vitreous) forms. When the atomic structure of silica exhibits long-range order it is in *crystalline* form. Many varieties of crystalline-silica exist, including quartz, crystobalite, coesite, etc. However, the type of silica encountered in silicon ICs (i.e., silicon-dioxide films thermally grown under normal conditions) is not crystalline. Instead, it is silica in its *vitreous* (or *glassy*) form. Vitreous solids do not exhibit long-range order. Instead, their structure is ordered only over short ranges. This glassy state of SiO_2 is also often referred to as *fused-silica.* Since we are interested in thermally grown SiO_2, our discussion will focus on the glassy state of SiO_2 (fused-silica).

The basic structural unit of silica is centered around the *structural formula* (SiO_4). The spatial arrangement of Si and O atoms is due to their respective valence (+4 and -2), their relative sizes, and their electrostatic interactions. These factors give rise to elementary SiO_2 cells of tetrahedric configurations. That is, a silicon atom (with a valence of +4) is located at the center of the tetrahedron, with oxygen atoms (O^{2-}) at each of the corners (Fig. 3-4a).[3] In crystalline-SiO_2 (quartz), each oxygen atom belongs to two tetrahedra and is thus bonded to 2 silicon atoms. Such oxygen atoms are then known as *bridging oxygen atoms*. (Fig. 3-4b).[3] In vitreous- or amorphous-SiO_2 some of the vertices of the tetrahedra have *nonbridging oxygen atoms*, meaning they are not shared between two tetrahedra (Fig. 3-4d). The greater the ratio of bridging-to-non-bridging oxygens, the greater the cohesiveness of the SiO_2 structure.

The interatomic distances (from the center of one atom to the center of its nearest neighbor) have been measured and found to have the following mean values: 1.62-Å for the Si-O bond; 2.27-Å for the O-O distance; and 3.12-Å for the Si-Si distance. Note this implies that a 2.0-nm gate oxide is only 10-12 atomic layers thick!

The structure of fused-silica is a continuous random network of tetrahedra, where the Si-O-Si bond angle varies from 110° to 180°, with the most probable value being 144°. A two-dimensional representation of the vitreous-SiO_2 lattice is shown in Fig. 3-4c.[4]

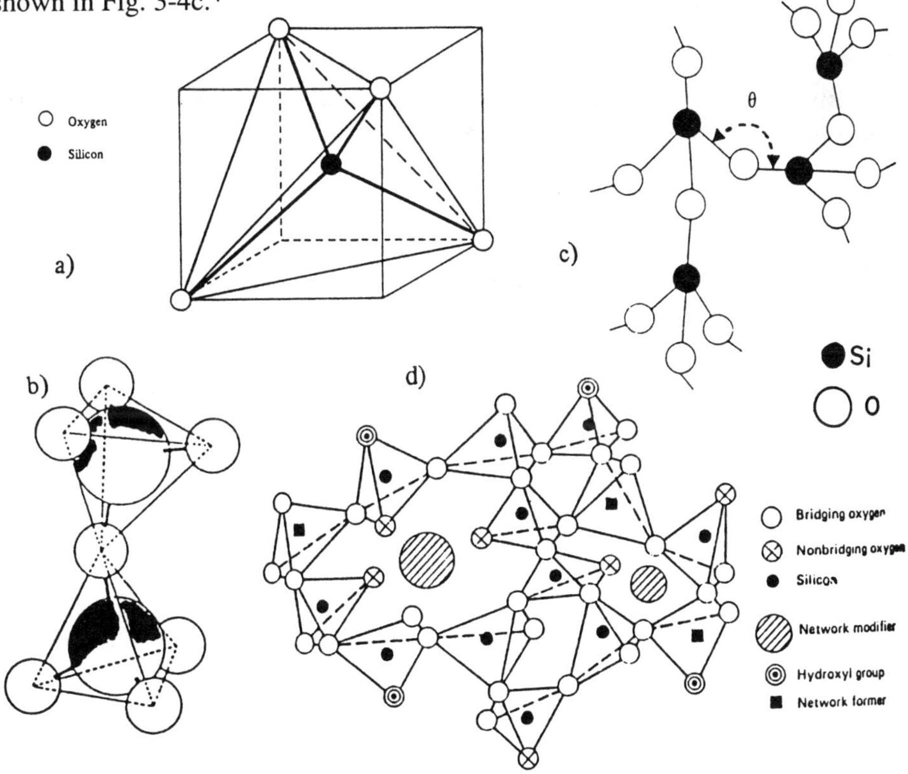

Fig. 3-4 (a) The basic structure of SiO_2.[2] (b) Three-dimensional representation of two neighboring SiO_4 cells, bridged by an oxygen atom. (c) Two-dimensional lattice representing vitreous-SiO_2.[2] (d) The structure of thermally grown SiO_2 showing bridging and non-bridging oxygen atoms and dopant (i.e., network modifier) atoms.[3] (© IEEE 1965)

The density of thermal-SiO_2 (fused-silica) is 2.20-g/cm^3 which is smaller than that of quartz (2.65-g/cm^3). In fused silica only 43% of the lattice space is occupied, making its structure much more open than that of quartz. Consequently, a large variety of impurity atoms can easily enter this oxide network and diffuse through it. Since it is difficult to directly assess the density of thin films, the refractive index of SiO_2 is usually measured instead, and the density is then inferred from it. A refractive index of 1.460 corresponds to a density of 2.20-g/cm^3. Generally, CVD oxides exhibit smaller densities than thermally grown SiO_2, and worse electrical and material properties correspond to less dense films. Hence, measurements of the index of refraction also provide a method for rapidly comparing the characteristics of deposited oxides.

3.2.2 The Si/SiO_2 Interface

In addition to the characteristics of the SiO_2 bulk structure, the characteristics of the Si/SiO_2 interface also play a critical role when SiO_2 functions as the dielectric layer of a MOSFET. Hence, a discussion of these characteristics (and the structure) of the SiO_2 interface is also relevant. First, it is useful to define the types of charges that exist in the oxide and at the Si/SiO_2 interface. (It should be noted that such charges are not expected to exist in an "ideal" oxide or at an "ideal" interface.) Four different types of charges are associated with the bulk oxide and "real" Si/SiO_2 interfaces, as shown schematically in Fig. 3-5. The *effective net* (i.e., *total*) *charge-per-unit-area* due to such charges at the Si/SiO_2 interface and within the oxide bulk (in C/cm^2) is denoted by the symbol Q_{tot}, whereas the *net* (*total*) *number of charges* per unit area (Q_{tot}/q) is given by N_{tot}. The standardized terminology and symbols used to differentiate these various individual charge types are as follows:

1. Mobile ionic charge per unit area Q_m (C/cm^2), and N_m (no. of mobile charges/cm^2);

2. Fixed oxide charge per unit area Q_f (C/cm^2), and N_f (no. of fixed oxide charges/cm^2);

3. Interface trapped charge per unit area Q_{it} (C/cm^2), and N_{it} (no. of interface trapped charges/cm^2); D_{it} - number of interface trapped charges per unit area *and* energy (no. of interface trapped charges/cm^2-eV);

4. Oxide trapped charge Q_{ot} (C/cm^2), and N_{ot} (no. of oxide trapped charges/cm^2).

The issues relating to the reliability of deep-submicron MOSFETs primarily involve Q_{it} and Q_{ot}. That is, the problems caused by Q_m and Q_f are not only well controlled, but their impact does not get any more severe as oxides are scaled. On the other hand, those associated with Q_{it} and Q_{ot} are ongoing, scaling-dependent problems that must be addressed as MOSFETs are pushed further into the deep-submicron regime. Hence, the review provided here will examine only these two categories of interface/oxide bulk charges. This will include the origin of the charges and their effect on device behavior. A more detailed description of the characteristics of Q_m and Q_f can be found in Vols. 1 and 3, and in Refs. 1 and 7.

3.2.2.1 Interface Trapped Charge: The *interface-trapped charges* refer to charges which are localized at sites at the surface of the silicon (i.e., at the Si/SiO_2 interface). Such sites are believed to arise from energy states that exist in the forbidden gap of the silicon, and are referred to as *interface states.* Such surface states are extra allowed energy states (i.e., that electrons can occupy) present at the semi-conductor surface, but not within the bulk.

The term *surface state* was coined by Tamm in 1932.[5] Using quantum mechanical calculations, he showed that new electronic energy states arise in a crystal if the lattice is terminated at a surface. Such new states are confined to the region very close to the surface. He also calculated that these states will arise at interfaces (such as the Si/SiO_2 interface), as well as at free surfaces. Each of the states is associated with a single atom at the surface. Hence, an electron occupying one of these states is localized (i.e., it is forced to remain in a restricted region of space centered on that atom). Since such states thus effectively *trap* free carriers at the surface, they are also referred to as *interface traps.* The charge-per-unit-area stored in these traps is symbolized by Q_{it}.

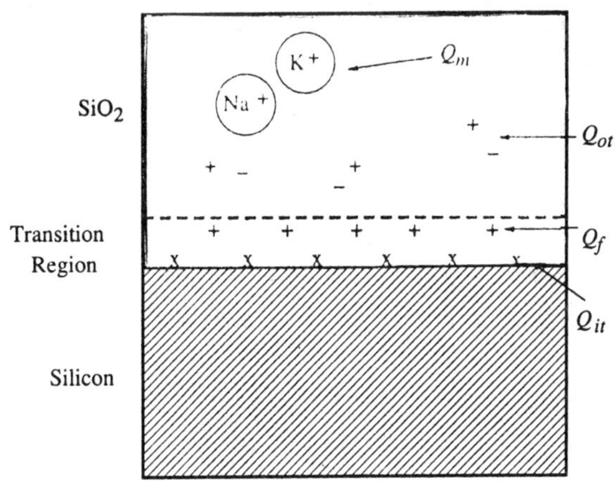

Fig. 3-5 Charges associated with the Si/SiO_2 system.

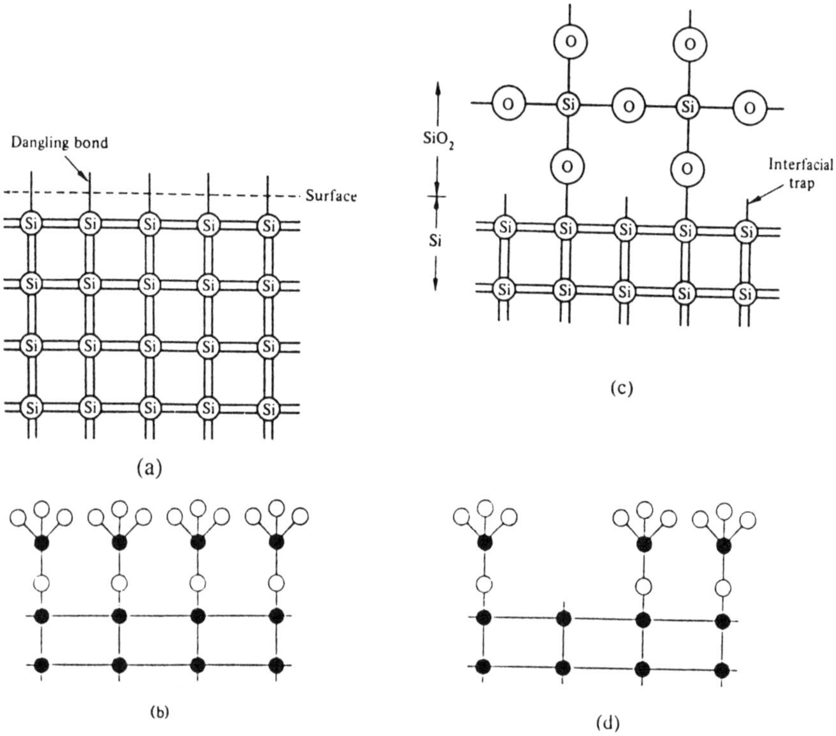

Fig. 3-6 Physical model for the interfacial traps. "Dangling bonds" which occur when the Si lattice is abruptly terminated along a given plane to form a surface. (b) Post-oxidation perfect (ideal) interface. (c & d) Post-oxidation dangling bonds that become interfacial traps.[7] Reprinted with permission of the publisher, the Electrochemical Society.

Although the theoretical basis of interface traps is well accepted and models exist that accurately detail the electrical behavior of the traps, the *physical origin* of these surface states has not been totally determined. The weight of experimental evidence supports the view that surface-states arise primarily from unsatisfied, or *dangling* bonds at the silicon surface (see Fig. 3-6). Note that a Si atom at a surface with a dangling bond can also be viewed as a *trivalent Si atom.*

From a qualitative perspective, the dangling bond model can be visualized with the aid of Fig. 3-6a. Here it is seen that if a Si lattice is abruptly terminated along a given plane to form a surface, one of the four bonds of each Si atom at the surface is left dangling. There will thus be extra states on this surface because the energy field of the crystal is one-sided (i.e., the electrons in the surface region are bonded only from the side directed toward the bulk). It is plausible that thermal formation of SiO_2 can tie up these surface Si bonds, and that along a perfect Si/SiO_2 interface all such bonds could be tied up (Fig. 3-6b). In this case, surface-states at the interface would be suppressed.[6] However, it is more plausible that such oxidation will not tie up *all* the bonds (Fig. 3-6c). If even a very small fraction of the number of dangling bonds is left unsatisfied, a significant number of surface-states could exist. For example, on a (100)-Si-surface there are $6.8x10^{14}$ Si atoms per cm^2. If oxidation left 1/1000 of these bonds dangling, and each of them gave rise to a surface-state, the density of interface trapped charges would be $6.8x10^{11}/cm^2$ (assuming one electronic charge is associated with each energy state). If a gate-oxide-thickness of 20-nm was being used, this would cause a threshold-voltage-shift of 0.63-V. This indicates that if the dangling-bond model correctly describes the origin of interface states, then only a relatively small number of residual-dangling-bonds can significantly perturb MOSFET device characteristics. Crystal-defects near the surface or foreign atoms bonded at the surface are other interface perturbations that have been identified as possible sources of surface-states.

Figure 3-7 represents the bond-stretching and broken-bond model of the Si/SiO_2 bulk of Verwey[2] and of the Si/SiO_2 interface of Sakurai and Sugano.[6] That is, Fig. 3-7a depicts the stretched and broken Si-O bonds in the bulk, and Figs. 3-7b-e at the interface. Figures 3-7f-h illustrate the possible defects in the oxide bulk. In Fig. 3-7b, the dangling-Si-bond at the surface is depicted as an *interfacial-trivalent-Si-atom,* while in Fig. 3-7f, a dangling-Si-bond in the bulk is portrayed as a *trivalent-Si-atom.*

Trivalent-Si-atoms introduce a deep trap-level near the Si midgap when located at the interface, and a deep trap-level in the SiO_2 bandgap when located in the oxide bulk. The weak Si-Si* and Si-O stretched bonds (Figs. 3-7c and 3-7d) at the interface introduce a continuum of deep trap levels, or if such stretched bonds exist in the bulk of the oxide (Fig. 3-7g), they produce shallow trap levels. If a dangling-Si-bond is tied up with an impurity (most commonly H or OH), this is thought to produce an electron-trap, while the oxygen vacancy (or weak Si-Si stretched bond) is considered to be the precursor of a hole-trap.

An anneal in a hydrogen ambient (100% H_2 or 4%-H_2 in N_2 [forming gas]) at approximately 450°C is normally the final step prior to assembly and packaging of an IC. This step is thought to allow hydrogen to penetrate the gate-oxide and then tie up the remaining dangling bonds at the Si/SiO_2 interface not tied up by thermal oxidation (as suggested in Fig. 3-8). However, the Si-H or Si-OH bond can be easily broken by injected hot-electrons, giving rise to interface traps. Thus, hydrogen introduced in the post-metal anneal or other process steps (i.e., during the steam reflow of an interlevel BPSG layer, or during the deposition of silicon nitride), can increase hot-electron degradation in a MOSFET.

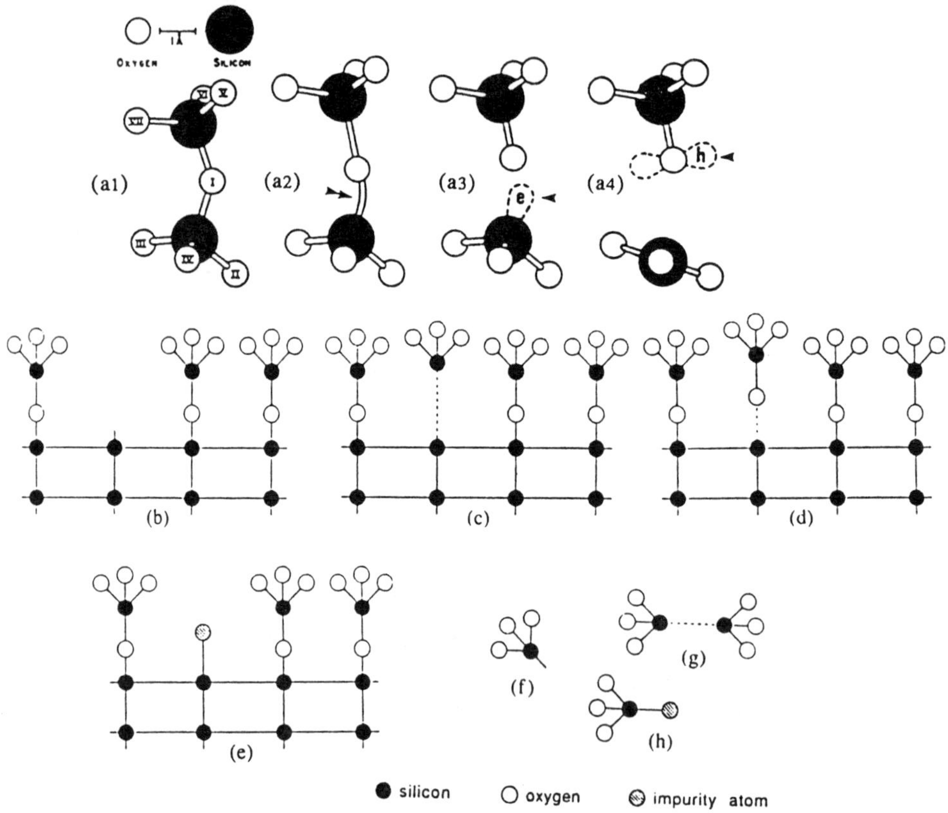

Fig. 3-7 (a) Strained and broken bonds: (a1) Normal Si-O bonds; (a2) Strained bond in vitreous silica; (a3) Broken bond represented with a trapped electron; (a4) Broken bond represented with a trapped hole. Possible interface defects (b-e): (b) Si dangling bonds. (c) Si-Si stretched bond (or oxygen vacancy). (d) Si-O stretched bond. (e) Impurity at interface. Possible oxide bulk defects (f-h): (f) Si dangling bond; (g) Si-Si stretched bond (or oxygen vacancy). (h) impurity in SiO_2.[6] (© IEEE 1988)

Consequently, it has been suggested that the ideal gate dielectric and all subsequent dielectric layers should contain as little hydrogen as possible to make them hot-electron resistant.

3.2.2.2 Effect of Interface Traps on IC Characteristics: The presence of interface traps has four major effects on the characteristics of ICs. First, the charge that is present in the interface traps interacts with the charge in the silicon near the surface and thus changes the silicon charge distribution and the surface potential. Second, interface traps act as generation-recombination centers, giving rise to leakage currents, as described below. Third, since the density of interface trap

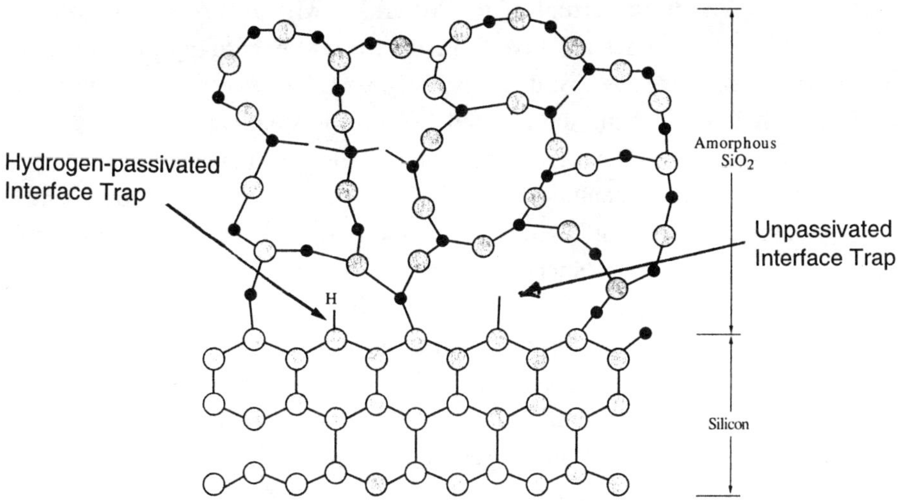

Fig. 3-8 How hydrogen can passivate the dangling bonds at the silicon surface shown in Fig. 3-7c and reduce the interface trap density.

charge changes with the surface potential, it gives rise to an additional capacitance component in parallel with the silicon gate capacitance. Fourth, an increase in interface-state-density has been correlated to weakening of the oxide, which appears to be a prelude to the second-stage breakdown event in gate oxides.

The issue concerning the *effect on the surface-potential of charge being trapped at the interface* is considered first. That is, such charge trapped at the interface has the same effect on the surface-potential as placing a sheet of charge Q_{ox} at the oxide-silicon interface. Such a sheet of charge will impact the threshold voltage, causing it to shift by an amount given by $\Delta V_T = -Q_{ox}/C_{ox}$. To keep this threshold voltage shift as small as possible, the value of Q_{ox} should also be kept as small as possible. The last paragraph of this section explains the measures taken to achieve this goal.

A secondary effect of interface-trap-charge on the surface-potential was introduced earlier. That is, injection of hot electrons into the gate-oxide is thought to produce interface traps (by breaking the bond between a silicon-surface atom and a hydrogen atom, which leaves a dangling bond on the Si atom). Since this process produces additional dangling bonds (and thus interface traps) as time goes by - and since these interface traps can trap electrons - this phenomenon will also cause the value of Q_{ox} to increase with time. In turn, this will cause the value of V_T to change with time. If V_T increases as result of the buildup of Q_{OX}, the drain current of the MOSFET will decrease. If such drain current degradation becomes too severe the device will eventually fail to operate properly. Thus, measures to avoid hot-carrier-injection have been employed in MOSFETs as their gate lengths approached 1.0-μm and smaller.

The second effect of interface states involves the fact that they can serve as generation-recombination centers. Since they are present at the silicon surface, they will exist in the depletion regions of *pn* junctions that intersect the surface of the silicon. There are many such junctions at the Si/SiO_2 interface in every device, and therefore in every IC.

Before the invention of integrated circuits, only discrete diodes, bipolar transistors, and field-effect transistors (FETs) could be fabricated. In the early 1950s, these discrete devices exhibited relatively high reverse-bias junction-leakage and low breakdown-voltage (caused by the large density of interface traps at the surface of single-crystal silicon).

In 1958, a group of workers at Bell Telephone Laboratories, led by Atalla, found that when a thin layer of SiO_2 was grown on the surface of silicon where a *pn*-junction intercepts the surface, the leakage current of the junction was reduced by a factor from 10 to 100. It was later understood that the oxide reduces the number (or stabilizes many) of the interface traps. Not only did such oxide-passivation of the silicon surfaces allow diodes and transistors to be fabricated with significantly improved device characteristics, but the leakage path along the surface of the silicon was also effectively shut off. Thus, one of the fundamental isolation capabilities needed for planar devices and integrated circuits had also been developed. In his report on the evolution of the MOS transistor, C.T. Sah remarks that the successful effort by the Bell Labs group to stabilize Si surfaces was the most important technological advance in microelectronics during the 1950s, and that it blazed the trail that led to the development of the silicon integrated circuit.[8]

The third effect of interface traps involves the interface-trap capacitance. As noted earlier, the *interface-trap capacitance* C_{it} is in parallel with the gate capacitance. However, only those interface traps that can be filled and emptied at a rate faster than the capacitance measurement signal can contribute to C_{it}. Traps too slow to follow the capacitance-measurement signal will not contribute to C_{it}. Therefore, the observed C_{it} is not only a function of the interface-trap density but also depends on the MOS-capacitor gate voltage (which controls the surface potential and hence the probability of occupancy of the traps) and the frequency at which the capacitance measurement is made.

All of these observations point to the conclusion that for a MOSFET to have predictable and reproducible capacitance characteristics, it is important to employ fabrication processes that minimize trap density. Using (100)-oriented silicon wafers, and a post-metallization hydrogen-annealing step (at temperatures around 400°C) at the end of the fabrication process, are the main techniques for keeping the interface-trap-density at a minimum.

It has been observed that the passage of tunneling current through a thin oxide also introduces interface traps at the Si/SiO_2 interface. Furthermore, while breakdown may not occur until about 10 C/cm^2 of charge has passed through the oxide, interface traps are detectable even after 0.001 C/cm^2 of charge passage. Thus, measuring interface traps has been proposed as a more sensitive means of quantitatively measuring oxide damage (due to Fowler-Nordheim stressing of oxides) prior to breakdown.

3.2.2.3 Oxide Trapped Charge: The oxide trapped charge Q_{ot} is due to holes or electrons trapped in the *bulk* of the oxide, and hence can be positive or negative. They become trapped at *trap centers* that are microscopic defects in the oxide structure *in the bulk of the oxide layer.* Such defects may be caused by ionizing radiation, hot-carrier injection, or high currents through the oxide (e.g., due to Fowler-Nordheim tunneling), or other reactions which either bend or break Si-O bonds in the oxide. Figures 3-7f-h show possible defects in the SiO_2 bulk that may give rise to such traps. For example, a trivalent Si atom in the SiO_2 will introduce a deep trap level in the SiO_2 bandgap $(.Si{\equiv}O_3)$, and stretched Si-O or Si-Si bonds [in the latter, a missing bridging-oxygen atom leads to a defect $(O_3{\equiv}Si..S{\equiv}O_3)$]. Figure 3-9 shows schematics illustrating the potential wells of electron traps in silicon dioxide. While, as-grown SiO_2 films contain few such microscopic oxide defects, electron and hole traps can be easily introduced by bombardment with high-energy photons or particles.[9] (The exception to this

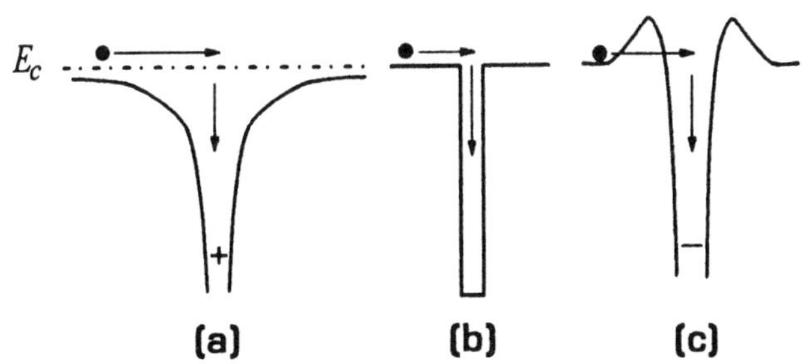

Fig. 3-9 Schematics illustrating the potential wells of electron traps in silicon dioxide: a) Coulomb-attractive trap; b) neutral trap; and c) Coulomb-repulsive trap.

statement is that as-grown wet oxides tend to contain some oxide traps because they also contain OH^- and H_2O species. The density of oxide traps in wet oxides can be significantly reduced with an appropriate post-oxidation-anneal.)

Q_{ot} can be annealed out by low-temperature treatments (above 550°C) although some neutral traps may remain. Therefore, oxide trapped charge was been considered relatively less important than the other oxide charges in just-completed MOSFETs. In early generations of MOSFET ICs, exposure of the devices to ionizing radiation encountered in space flights was the main concern involving Q_{ot}. However, in deep-submicron CMOS ICs, the problems involving Q_{ot} are becoming serious for several reasons. First, processing techniques such as ion implantation, sputter deposition, and plasma etching are potential sources for creating charged and neutral traps in the gate dielectric (all of which may not be completely annealed out at the end of the fabrication process). This may result in long-term reliability problems, especially if hot carriers are injected in these oxides and get trapped. Another reliability concern involves gate-oxide degradation due to high electric-fields in the oxide (and subsequent Fowler-Nordheim tunneling currents in the oxide). This has become more of a concern as oxide thickness has been scaled down.

3.2.2.4 Effect of Oxide Trapped Charge on Device Characteristics: The major effect of oxide trapped charge on MOSFETs is that they are a manifestation of the damage done to the gate oxide during device fabrication or by electrical stressing (either by the application of a high electric-field across the oxide, or by the injection of carriers into or through the oxide). We will discuss this issue in more detail in Sect. 3.3.

3.3 DIELECTRIC BREAKDOWN IN SILICON DIOXIDE FILMS

Dielectric breakdown of the gate-oxide layer is one of the major reliability issues that affects MOSFETs. We provide a review of this phenomenon here, and recommend that interested readers consult Vol. 3, Chap. 7 for more background information on the various breakdown measurements, statistical models, and burn-in tests used to screen out parts with weak oxides. However, significant new information that has been published since 1995 (when Vol. 3 was released) and this new information will be focus of the discussion here.

Figure 3-10 shows the time dependence of the gate current through an oxide under a large and constant voltage stress applied across the oxide layer. One can see that there is a gradual decrease in the current with time until the oxide suffers *dielectric breakdown* (Fig. 3-10). At that point the current rises rapidly (*catastrophic breakdown*).[10] After such breakdown, the oxide is irreparably damaged,

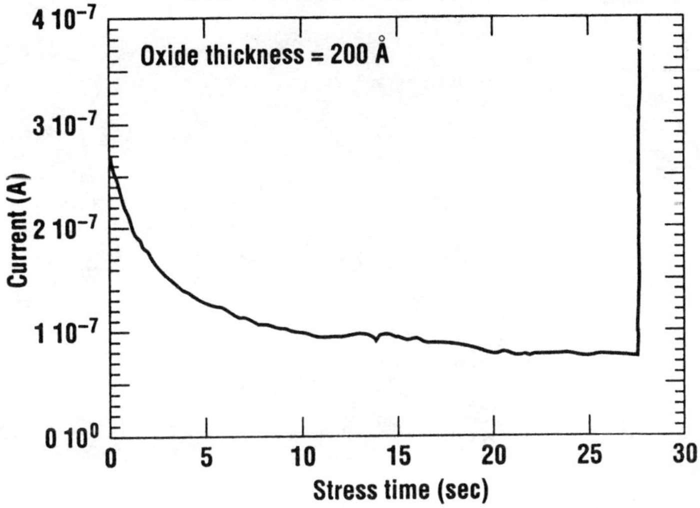

Fig. 3-10 Constant-voltage stress measurements of a 20-nm-thick oxide film. The sharp increase in current near the end of the trace indicates that irreversible breakdown has occurred.

and can no longer function as the dielectric layer in the MOSFET. Since such breakdown occurs after the oxide is subjected to a voltage stress for some extended period of time, this phenomenon is also referred to as *time-dependent dielectric breakdown* (TDDB). Anomalous "soft breakdown" of oxides also occurs for films < 5.0-nm thick, and this will be discussed in Sect. 3.8.

A different metric of the quality of oxide films is expressed in terms of the electric field at which dielectric breakdown occurs. To measure this value, a voltage (that increases linearly with time) is applied between the gate and substrate of a MOS-C test structure until the oxide breaks down. The "best" (apparently defect-free) oxides break down at fields greater than about 10 MV/cm, and failures at such field strengths are referred to as "intrinsic breakdown." Less-than-perfect-oxides (i.e., those that contain some form of defect) may fail at lower voltages (i.e., between 4-6-MV/cm).

In any case, dielectric breakdown in ICs must be avoided. Extensive work has demonstrated that 20-year oxide-lifetimes can be achieved if the oxide field remains below about 8-MV/cm.[11] Thus, for devices used in CMOS logic and memory chips, the maximum electric field during normal device operation is typically restricted to the 5 MV/cm range, and this limit on the electric field is projected to remain at about this value for each technology generation. That is, the 5-MV/cm value is expected to give an acceptable safety margin for 20-30-year oxide-lifetimes. Chips are also normally subjected to burn-in tests to screen out those chips that contain oxides that would fail under the normal operating conditions during their time of service.

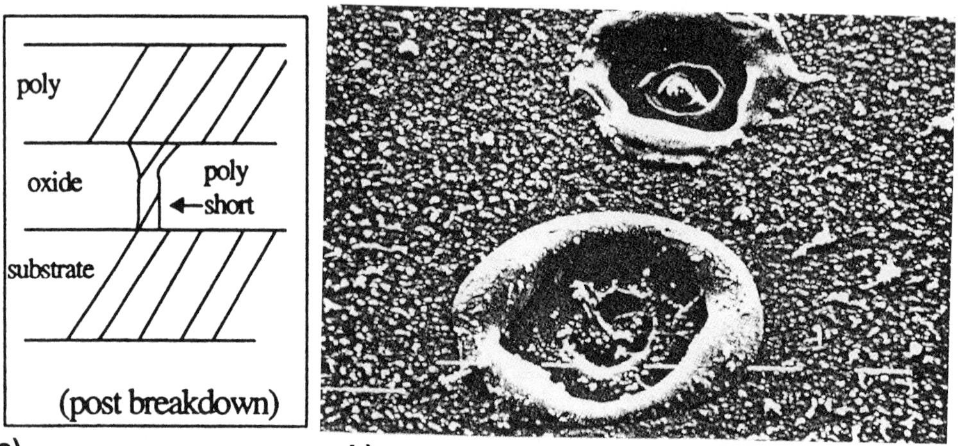

a) b)

Fig. 3-11 a) Localized poly filament shorting the gate and substrate after gate oxide breakdown; b) A SEM photograph of two breakdown spots. The polysilicon electrode has melted away over 20 μm^2. The polysilicon is spread over a large area in the form of dust and little globules. The size of a dash is 1-μm. From D.R. Wolters and J.F. Verwey, "Breakdown and Wear-Out Phenomena in SiO_2 Films," Chap. 6, p. 329, in *Instabilities in Silicon Devices*, G.M. Barbottin and A. Vapaille Eds., © 1986 Elsevier Science.

Because dielectric breakdown is such a critical reliability issue in CMOS ICs, breakdown of thin oxide films has been (and still is) a subject of intense research. The physical mechanisms involved in the dielectric breakdown process are quite complex. We will discuss some of the latest proposed models to shed light on this phenomenon.

In general, the catastrophic breakdown of a dielectric is believed to be a two-stage process. First, the dielectric is slowly degraded over time (perhaps years) due to being subjected to high-electric fields or due to the passing of current through the oxide. This degradation takes the form of creating defects in the oxide (e.g., electron/hole traps and/or bond-breakage). These defects form a localized conductive path through the oxide during this time. The second stage is the very short runaway breakdown process itself (it may occur in microseconds). During the runaway breakdown phase the oxide is shorted (due to severe joule-heating that causes a conductive filament to form in the dielectric) and a very large current flows between the gate and substrate in the MOS device (Fig. 3-11).

There are numerous forms of damage to the oxide that are postulated to occur during the oxide degradation process, including lattice damage, trap creation by electrons, and trap creation by holes. The two physical mechanisms that are blamed for producing these types of damage are: 1) externally applied electric fields; and **2**) current that passes through the oxide. Thus, there are two classes of models that have been developed to predict TDDB: **1**) models that assume the oxide degradation is electric-field-driven (the so-called *E-models*); and **2**) models that assume the oxide degradation is current-driven (the so-called [1/*E*]-models). Here we briefly describe both models. Also note that although the physical basis for oxide degradation is still under debate, definitions and practices have been developed to describe and test the integrity of the oxide.

3.3.1 Electron Trapping in Silicon Dioxide: (and why the Current in the Oxide Decreases with Time During Constant-Voltage Stressing)

As described in Sect. 3.2.2.3, silicon-dioxide layers contain electron and hole traps (either as a result of residual damage inflicted during device processing, or caused by the electric fields applied across the oxide - or gate current that flows - during normal device operation). If these sites trap carriers which are transported in the oxide conduction-band (electrons), they can give rise to oxide-trap charge Q_{ot}. Electrons can enter the conduction band of the oxide in several ways, including: **1**) ionizing radiation can generate electron-hole pairs within the oxide itself; **2**) electrons can enter from the silicon substrate or the gate polysilicon either through tunneling or through hot-electron injection. Some of these electrons may get trapped at oxide trap sites. These trapped electrons will then cause the electric field in the oxide to be modified in the following manner: The field near cathode (i.e., the electrode that acts as an electron source) is decreased, while the electric field near the anode (the electrode that acts as the electron sink) is increased (as depicted in Fig. 3-12). A decrease in the field near the cathode causes Fowler-Nordheim current in the oxide to decrease. Since the gate current is due primarily to Fowler-Nordheim tunneling, this explains why the gate current declines with time under a constant-voltage stress (as shown in Fig 3-10).

3.3.2 The Electric-Field-Driven Model of Oxide Degradation

In the electric-field -driven models of oxide degradation (*E*-models), the cause of low-field (< 10-MV/cm) TDDB is postulated to be field-enhanced thermal bond breakage. That is, at a given temperature, chemical bonds in the SiO_2 structure can be broken by thermal vibrations. Such bond breakage produces local defects that weaken the oxide, and when the concentration of such defects reaches a critical value, breakdown can occur. An externally applied electric field serves to reduce the activation energy required for such thermal-bond breakage, and therefore increases the degradation-reaction rate for failure exponentially. The *time-to-breakdown* t_{BD} which is the inverse of degradation rate, decreases exponentially

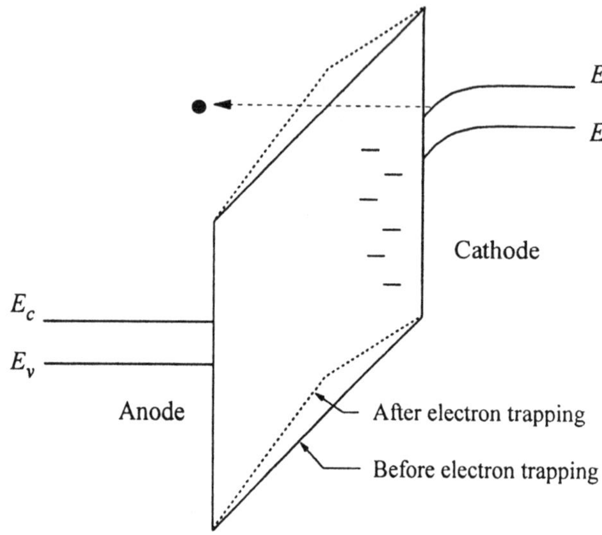

Fig. 3-12 Schematic illustrating the trapping of tunneling electrons. As electrons are trapped, the oxide field near the cathode (electron source) is decreased, while the oxide field near the anode (electron sink) is increased.

with field according to:

$$t_{BD} = A_0 \exp(-\gamma E_{ox}) \exp(Q/k_B T) \quad \textbf{(3.1)}$$

where γ is the field acceleration-parameter, E_{ox} is the electric field in the oxide (usually expressed in MV/cm), and Q is the enthalpy of activation for bond breaking in the absence of an external electric field. Generally, for oxide thicknesses greater than 90-Å, γ ~2.5-3.5-cm/MV.

3.3.3 The Current-Driven Model of Oxide Degradation (1/E Model)

In the so-called 1/E model for TDDB, degradation is postulated to be due to current flow through the dielectric caused by Fowler-Nordheim conduction. Electrons, which are F-N injected from the cathode, may cause damage to the dielectric due to impact ionization as the electrons are accelerated through the oxide. Also, when these accelerated electrons reach the anode, hot holes may be produced (see Sect. 3.3.4 below) which can tunnel back into the dielectric causing damage. Since both the electrons from the cathode and the hot holes from the anode are the result of F-N conduction, the time-to-failure is expected to show an exponential dependence on the reciprocal of the electric field, $1/E$:

$$t_{BD} = \tau_0(T) \exp[G(T)/E_{ox}] \quad \textbf{(3.2)}$$

where $\tau_o(T)$ is a temperature dependent pre-factor, G is the $1/E$-model field acceleration-factor, and k_B is Boltzmann's constant.

The temperature dependence of G has been expressed as a 1/T power series expansion, given by:[12]

$$G = G_o\,[1 + (\gamma/k_B)\{(1/T) - (1/300K)\}] \quad \textbf{(3.3)}$$

where:

$$\gamma = (k_B/G_o)\,[dG/d(1/T)]_{300K} \quad \textbf{(3.4)}$$

and where the derivative is evaluated at 300K. At room temperature, G_o ~350-MV/cm and γ ~0.017-eV.

$\tau_o(T)$ is also usually represented as a 1/T expansion:

$$\tau_o(T) = \tau_o \exp[(-E_a/k_B)\{(1/T) - (1/300K)\}] \quad \textbf{(3.5)}$$

where τ_o ~$1\text{x}10^{-11}$ sec, and E_a ~0.3-eV.

3.3.4 The Hole-Trapping Model that Describes How Holes are Injected and Trapped in SiO_2

The transport of holes through the oxide has been shown to precede breakdown, indicating that holes cause damage to the oxide (namely by creating traps in the bulk oxide).[13] A model (the *anode-hole injection model* [*AHI model*]) has been published that seeks to explain how such holes are generated and injected into the oxide during high electrical field stress.[14] The mechanisms that occur are depicted in Fig. 3-13.

When an electric field is applied across an oxide layer using an MOS-C structure, the energy-band diagram of this situation is depicted in Fig. 3-13. If an electron tunnels from the cathode into the conduction band of an oxide layer (by F-N tunneling), it can then gain kinetic energy from the electric field applied across the oxide, and get accelerated toward the anode. When the electron arrives at the anode it has a large amount of kinetic energy (E_{gain} in Fig. 3-13). Some fraction of such electrons that arrive at the anode are able to transfer this energy to an electron deep within the valence band of the anode silicon near the oxide-anode interface. This valence band electron is excited to the bottom edge of the silicon conduction band in the anode. The excitation event also produces a hole deep within the valence band, and this "hot" hole can tunnel into the valence band of the oxide (again by F-N tunneling, as depicted in Fig. 3-13). These holes can be trapped in the oxide and also create the damage described earlier. This damage accumulates with time, degrading the oxide.[35]

In addition, the trapped holes in the oxide layer cause an increase in the oxide field near the cathode, and a decrease in the field near the anode (Fig. 3-14). A small increase in the oxide field near the cathode can cause a large increase in the

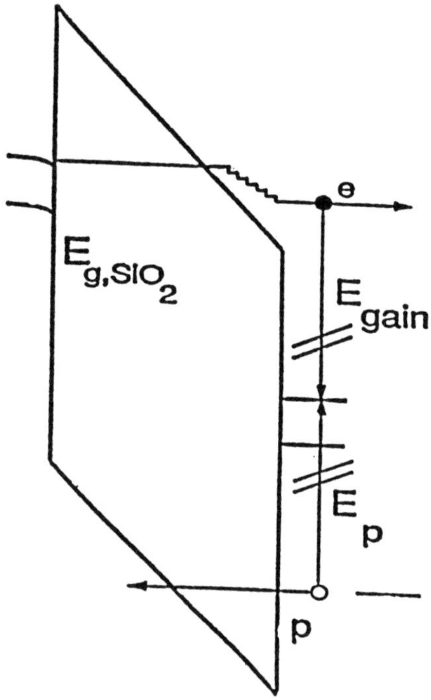

Fig. 3-13 Energy-band diagram of the anode-hole injection (AHI) model, illustrating the generation of an electron-hole pair by a tunneling electron. The hole thus generated can then be injected (by tunneling in this model), into the oxide layer.

tunneling current. Thus, hole trapping in the oxide near the cathode provides positive feedback to the electron tunneling process. It should be pointed out that the increase in tunneling current due to hole trapping is usually not readily observable, except near breakdown, because it is masked by the decrease in tunneling current due to electron trapping.

It should be noted that recent work has called some conclusions of this model into question. That is, until recently it was generally accepted that the substrate current originates from the anode-hole-injection (AHI) mechanism, which was schematically depicted in Fig. 3-13. Part of the holes created by the phenomena postulated by this model tunnel into the oxide, and the oxide degradation is related to the hole fluence (which is found to be constant at breakdown, independent of the stress current or stress voltage). However, the explanation of the substrate current by anode-hole-injection has never been completely verified, and other mechanisms such as minority-carrier generation due to F-N-induced photons have been suggested to be a cause as well. Two recent reports claim that the F-N-induced photons in the gate are the dominant source of the substrate hole

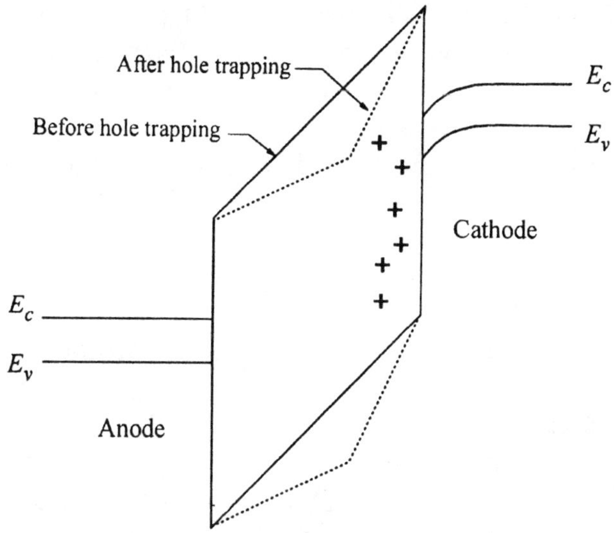

Fig. 3-14 Schematic showing the trapping of holes in the oxide layer. The trapped holes enhance the electric field near the cathode, and decrease the electric field near the anode.

current, and thus the generally accepted explanation of oxide degradation based on anode-hole-injection may have to be revised.[31,32]

3.3.5 Comparing the Electric-Field-Driven and the Current-Driven Oxide Breakdown Models

There has been a great deal of disagreement as to which is the dominant degradation mechanism for low-field TDDB in thin oxide films; i.e., is it field-induced or current-induced? The (1/*E*)-model fits the TDDB data very well at high fields (in which high F-N-currents exist), but falls short at lower fields (< 10-MV/cm). In fact, there have been several low-field/long-term TDDB studies carried out, with each showing that TDDB data were described much better by the *E*-model.[15,16,34] The good fit of the physics-based *E*-model to low-field/long-term TDDB data suggests strongly that it is field (not current), which is the dominant degradation mechanism at low fields because the F-N current is simply so vanishingly small at these lower fields. However, for very-thin oxides (< 40-Å) the direct tunneling-current is very high, which could mean that in these oxides the dominant degradation mechanism may again be current-driven.

3.3.6 Time-to-Breakdown (t_{BD}) and Charge-to-Breakdown (Q_{BD})

The breakdown characteristics of an oxide film are described in terms of the *time-to-breakdown* (as was discussed earlier) and in terms of its *charge-to-breakdown* Q_{BD} (which measures the *integrated* total tunneling current [i.e., total charge] just before breakdown). Recent publications have tended to use the value of Q_{BD}. It appears that it is simpler to develop models relating to the other physical processes, such as hole current, trapping, trap generation, and interface-state generation, than it is to develop models relating time-to-breakdown to these processes. Thus, Q_{BD} needs to discussed further.

As mentioned above, evidence indicates that tunneling-electron-current can lead to hole-injection into the oxide. Thus, Q_{BD} is the sum of the charges passing through the oxide due to electrons and holes. If a MOS-C structure is used to measure Q_{BD}, then due to the two-terminal nature of the MOS-C, only the total charge can be measured. However, if an NMOSFET is used to measure Q_{BD}, then both the total charge and the hole-charge component can be determined. For the case of an NMOSFET, the bias configuration for such measurements is shown in Fig. 3-15. Integration of the gate current gives the total charge, while integration of the substrate current gives the charge due to the holes. For a given oxide film, Q_{BD} is often plotted as a function of oxide voltage. Figure 3-16 is such a plot for oxide thickness in the 2.5-10-nm range.[15] It shows that for a given oxide thickness, Q_{BD} decreases with increasing oxide voltage.

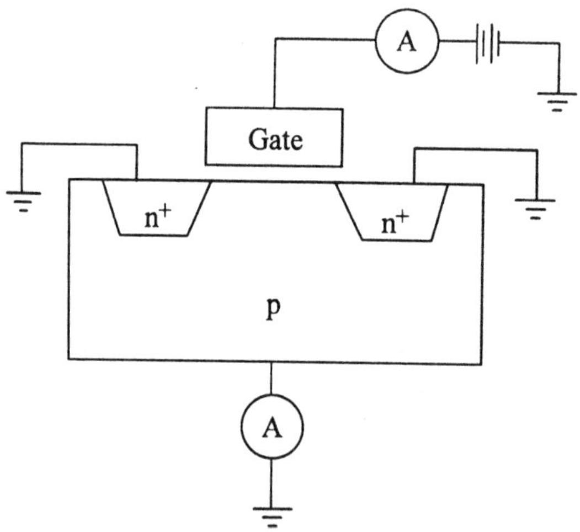

Fig. 3-15 Schematic depicting the bias configuration of an *n*-channel MOSFET for measuring the charge-to-breakdown Q_{BD} and its hole-charge component.

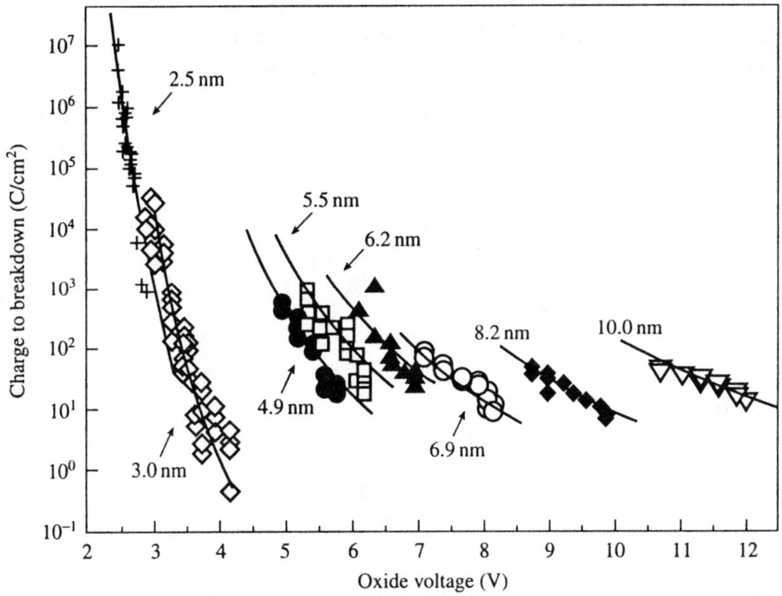

Fig. 3-16 Typical plot of charge to breakdown versus oxide voltage for several oxide thickness values.

3.4 LEAKAGE CURRENTS IN SiO_2 FILMS (TUNNELING PHENOMENA)

When considering the ideal MOS capacitor structure, it is assumed that no current flows through the oxide. However, in real MOS capacitors, a current from the gate to the substrate through the oxide can flow (termed *gate current,* I_g). Although the carriers that constitute such current can enter the conduction band (electrons) and valence band (holes) of silicon dioxide by hot-carrier injection, here we only discuss the other way that they can enter, namely by tunneling. This is because few carriers in deep-submicron MOSFETs enter the oxide as a result of hot-carrier injection during normal device operating conditions; hence they do not play a major role in the observed gate current.

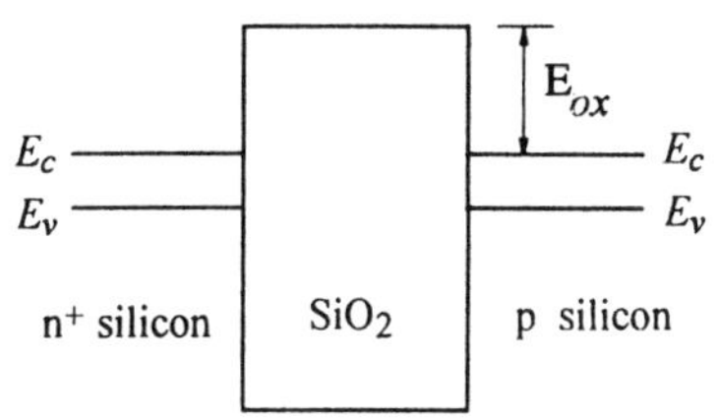

Fig. 3-17 Energy-band diagram of an *n*-type polysilicon-gate MOS structure at flat band.

The phenomenon of carrier injection into the oxide by tunneling is due to the quantum-mechanical nature of the electron. In the classical sense, the oxide represents an impenetrable barrier to injection of electrons into the conduction-band of the silicon if they possess kinetic energies smaller than the interface energy barrier for the electrons (E_{ox} = 3.1-eV). In Fig. 3-17, the energy-band diagram of a MOS capacitor with silicon dioxide as the insulator and heavily *n*-doped-polysilicon as the gate is shown when biased at the flat-band condition. When a large positive voltage is applied to the gate electrode, electrons in the strongly inverted Si substrate surface can tunnel into or through the oxide layer, and give rise to gate current. (Similarly, if a large negative voltage is applied to the gate electrode, electrons from the n^+-polysilicon can tunnel into or through the oxide-layer and again give rise to gate-current.) Such tunneling occurs because the wave nature of an electron (or hole) allows a finite probability of crossing the barrier, even if the electron does not possess sufficient kinetic energy. This probability increases with larger gate electric fields and/or thinner barriers. Electrons injected by tunneling are generally considered to be in equilibrium with the lattice, and are thus characterized as "cold" (i.e., not "hot").

3.4.1 Fowler-Nordheim (F-N) Tunneling (Tunneling *Into* Silicon Dioxide)

The first type of tunneling is called *Fowler-Nordheim tunneling.* Here electrons are injected by tunneling into the conduction band of the oxide through the triangular energy barrier. Once injected into the oxide conduction-band, electrons are accelerated by the oxide field toward the anode (gate), causing a gate current. Figure 3-18a sketches the energy band diagram for this case assuming a positive bias is applied to the gate.

To determine the probability of an electron exhibiting F-N tunneling through a barrier, Schroedinger's equation is solved for a triangular barrier. The WKB approximation (discussed in most quantum mechanical textbooks) is invoked, and the following expression for the tunneling current-density expression is derived:

$$J = A_F E_{ox}^2 \exp(-B/E_{ox}) \quad \textbf{(3.6)}$$

where A_F = 1.25x10^{-6} A/V^2, B ~240-MV/cm, J is the current-density in A/cm^2, and E_{ox} is the oxide field in V/cm.[17] According to this equation, the oxide current is exponentially dependent on the electric field and experimental studies have found that the equation accurately describes the oxide current. This can be seen in Fig. 3-20, which shows the measured gate current and the gate current calculated using Eq. 3-6 in oxides 6-nm- and 8.3-nm-thick. The tunneling probability rapidly rises as E_{ox} increases because the tunneling distance (i.e., the

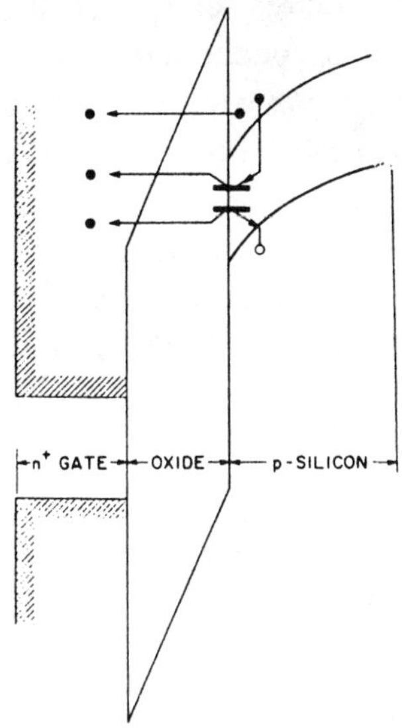

Fig. 3-18 Energy-band diagram for the phenomenon of Fowler-Nordheim tunneling in the MOSFET gate oxide. Also shown are some possible interface-trap assisted leakage paths, with trap levels indicated by short, solid bars.

distance from the Si-surface through the triangular barrier to the oxide conduction-band) becomes smaller as E_{ox} increases.*

It should also be noted that it may be possible for electrons to be "hot" but still not possess sufficient energy to surmount the barrier. In this case, carrier injection may occur by allowing these hot-carriers to tunnel through a smaller triangular-barrier distance. Thus, a synergistic combination of processes which individually would not have caused injection may yet enable carriers to be injected. This effect was modeled by Ning and Yu.[18]

At an oxide electric-field of 8-MV/cm, the measured Fowler-Nordheim tunneling-current-density is about 5×10^{-7} A/cm^2. Thus, for normal device operation, Fowler-Nordheim tunneling does not produce enough gate-current leakage to be significant. Hence, Fowler-Nordheim tunneling does not represent an issue when gate-oxides get thinner as the gate length of MOSFETs are scaled, even to deep-submicron gate-lengths.

3.4.2 Direct Tunneling (Tunneling Through Silicon Dioxide)

If the oxide is very thin (say 4-nm or less), electrons from the inverted silicon

* Because the distance of tunneling between Si surface and the oxide conduction band decreases with increasing electric field.

surface can tunnel *directly* through the forbidden-energy-gap of the oxide layer instead of tunneling into the conduction-band of the SiO_2 layer. In direct-tunneling, electrons tunnel through a trapezoid-shaped barrier as shown in Fig. 3-19. (Because of the differences in the height of barriers for electrons and holes, and because holes have a much lower tunneling-probability in oxide than do electrons, the tunneling limit will be reached earlier for NMOS than PMOS.)

The theory of direct-tunneling is even more complex than that of Fowler-Nordheim-tunneling, and there is no theoretical simple dependence of the tunneling-current-density on voltage or electric field. (While an expression for the tunneling probability can be derived, no analytical expression is available for direct tunneling (as exists for the case of F-N tunneling). Instead, direct-tunneling-current must be calculated from numerical analysis, and the results of such calculations are plotted in Fig. 3-20.[19]) Nevertheless, experimental data indicates that tunneling current exhibits a weaker dependence on oxide fields than is shown by the equation that governs Fowler-Nordheim tunneling (Eq. 3-6). This means that direct-tunneling dominates in thin oxides at low voltages.

For oxide thicknesses smaller than 3-nm, the direct tunneling-current becomes very large. Figure 3-21 is a plot of the measured and simulated thin-oxide tunneling-current versus voltage in polysilicon gate MOSFETs. For the gate-voltage range shown in Fig. 3-21, the current is primarily a direct-tunneling current.[20] This tunneling represents a limit on the oxide thickness that can be used. This limit will be reached approximately when the gate-to-channel tunneling-current becomes equal to the off-state source-to-drain subthreshold

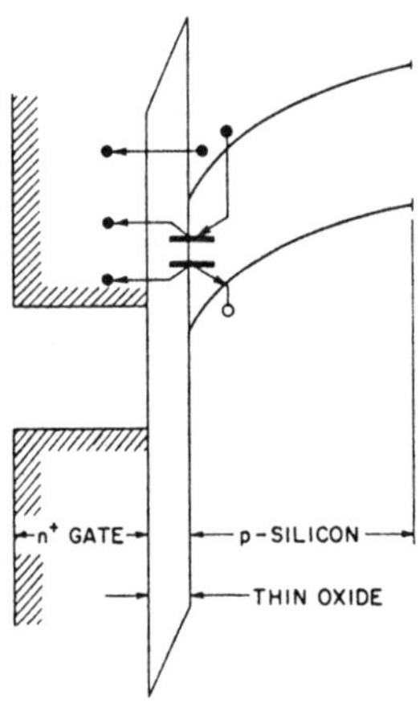

Fig. 3-19 Energy-band diagram for the phenomenon of direct tunneling through the gate oxide for thin oxides. Also shown are both unassisted-tunneling and some possible interface-trap-assisted leakage paths, with trap levels indicated by short, solid bars.

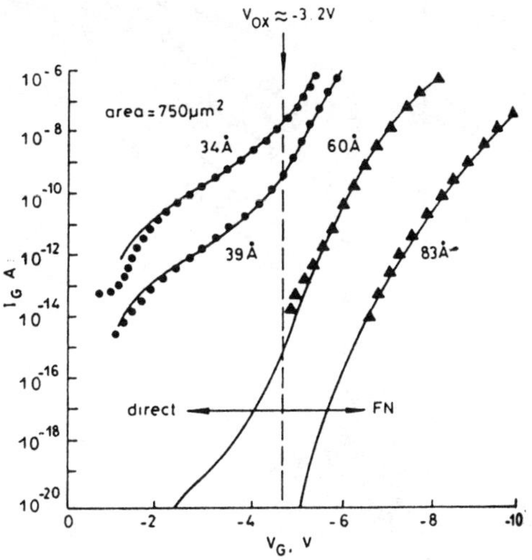

Fig. 3-20 Theoretical (and experimental) tunneling I-V curves of Al-gate *n*-channel MOS structures under negative gate-bias, illustrating the I-V characteristics of Fowler-Nordheim tunneling in 6-nm- and 8.3-nm-thick oxides and the calculated direct tunneling- currents in 3.9-nm- and 3.4-nm-thick oxides.[19] (© IEEE 1983)

leakage current (e.g., $I_{off}/W = 10$ nA/μm). That is, a MOSFET with a 1-μm gate length and width, would have a channel area of 1-μm^2. Thus, to keep its value equal to the subthreshold source-to-drain leakage current, the tunneling leakage current density should not exceed 10^{-8} A/μm^2 (which is equal to 1 A/cm^2).

From Fig. 3-21, the minimum oxide thickness that could be used to achieve a gate current density of 10^{-8} A/μm^2 is 1.4-nm.* Since this is the thickness of the oxide needed for the 70-nm generation (according to the L_E/t_{ox} guideline given at the beginning of Sect. 3.1 of this chapter), it appears that alternative gate-dielectric materials with reduced gate-leakage will be required for 50-nm MOSFETs and beyond. (Another, more detailed analysis of the leakage-currents in CMOS technology in the 100-nm regime arrives at the same oxide-thickness limit due to tunneling-leakage as the one given above [i.e., 1.4-nm].[30]) This topic is considered in Chap. 4. It should also be noted that logic circuits can usually withstand higher leakage currents than can DRAMs. Therefore. the gate oxide

* Note that the data shown in Fig. 3-21 are the results of work by researchers at Agere Systems.[20] They used an *in-situ* cleaning of the silicon surface (with HF gas, followed by ultra-violet [UV] cleaning) before the rapid-thermal-oxidation. The ultra-thin oxides they grew using this cleaning sequence resulted in oxides with very-low gate leakage currents (Fig. 3-21). Ultra-thin oxides grown by wet rapid-thermal-oxidation (RTO) processes have also reportedly shown smaller gate leakage currents (see Sect. 3.6.2).

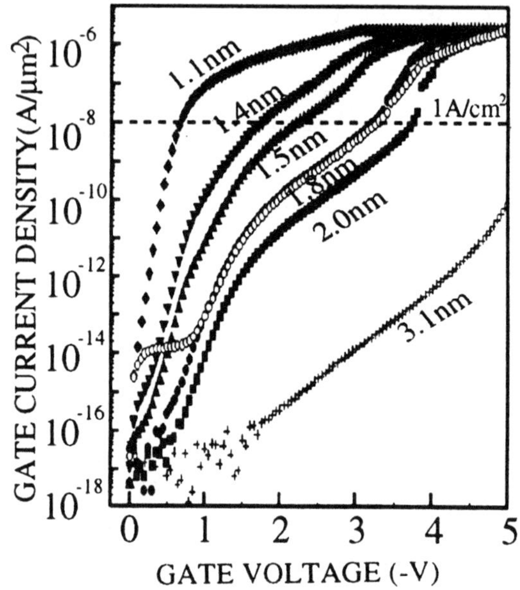

Fig. 3-21 Negative gate-current-density as a function of negative gate-voltage for oxide thicknesses from 1.1-nm to 3.1-nm, showing gate tunneling-leakage-current-densities.[20]

limit for DRAMs is expected to be larger (about 3-nm will be the minimum oxide thickness for DRAMs).

3.5 MODELS OF THIN-OXIDE GROWTH

The Deal-Grove model described in Vol. 1 for the growth of SiO_2 provides excellent agreement with experimental observations for thick oxides.[21] That is, the rate-constants derived in the Deal-Grove model allow the thickness of oxide films >350-Å to be well predicted as a function of temperature, furnace-ambient, silicon-doping-concentration, and silicon-crystal-orientation. However, the model does not give a detailed understanding of the mechanisms that produce these dependencies, nor is it valid when the oxide is thinner than 200-Å. Specifically, the experimental measurements of oxide-growth-rates in dry-O_2 are not accurately predicted by the Grove-Deal for oxides less than about 200-Å (20-nm). This is the so-called *anomalous regime* of the Deal-Grove model. This is particularly troublesome because gate-oxide-thicknesses of $\leq$ 100-Å are used for sub-micron MOSFETs. Figure 3-22 depicts a computer simulation of the formation of the first few layers of oxide being grown on a silicon surface.[22]

Several physical mechanisms have been proposed as models for the enhanced oxidation that is observed when oxides thinner than 200-Å are grown in dry-O_2, including:[23]

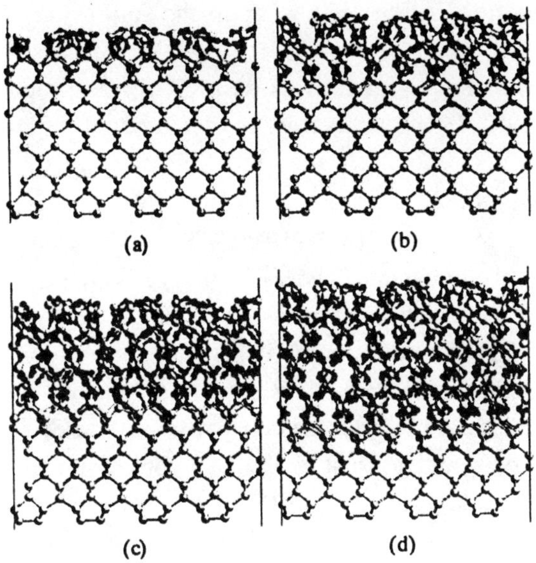

Fig. 3-22 $Si(100)/SiO_2$ interface structures after oxidation of a) 2nd b) 4th c) 6th d) 8th layer.[22]

1. Models based on space-charge effects where the oxidation-enhancement in the thin regime is due to field-assisted oxidant-diffusion (arising from O_2^- coupled diffusion with holes).

2. Models based on the existence of structural defects in the oxide, such as micropores or channels (~10-Å in diameter). These would allow the oxygen molecules to diffuse more rapidly when the oxide is thin.

3. Models based on parallel O_2 and O reactions at the Si/SiO_2 interface.

4. Models which postulate that a blocking layer exists near the Si/SiO_2 interface which forms as the oxide grows and slows down the rate for thicker layers.

5. A model by Massoud *et al.*, based on the presence of a thin surface layer in the silicon where additional reaction sites are available. The concentration of these sites decays exponentially with a characteristic decay length L_2 in Eq. 3-9.[23]

Despite the large number of models proposed to explain oxide growth in this very-thin regime, none has yet been shown to be clearly correct, and thus none have been convincingly accepted as more valid than the others. We are therefore left with several empirical models that yield reasonable agreement with measured values of oxide thickness in the thin regime, but with no completely satisfactory physical explanation. Additional work in the future may help clarify some of these issues.

One *empirical* model for thin oxide growth was proposed by Reisman and Nicollian in 1987.[59] By analyzing a vast amount of data, they derived an empirical model that calculates the oxide thickness versus time using a general power law of the form:

$$t_{ox} = a\,(t_g + \tau)^b \tag{3.7}$$

where a and b are constants, and t_g is the growth time measured in a given experiment, and τ is the time to grow an oxide of thickness t_{ox_i} already present on the surface. Their model could fit all published dry-O_2 data. In fitting this equation to experimental data, they extracted b values between 0.25 and 1.0, depending on temperature and oxidant-partial-pressure. This equation has two fitting parameters (a and b), just as the Deal-Grove model has two (A and B/A). However, there are significant differences between the models. First, the Reisman-Nicollian model is claimed to be able to fit data down to oxide thicknesses of essentially zero thickness. There is no anomalous regime, as exists with the Deal-Grove model. Second, each model has its own physical basis for the oxide growth process. The Deal-Grove model is based on the idea of oxidant-diffusion and an interface-reaction - with each process dominating the growth under different conditions. Reisman suggests that the interface-reaction actually controls the oxidation process at all times, and that the volume expansion necessary at that interface to accommodate the growing oxide was provided by viscous-flow (relaxation) of the oxide layer. The time-dependent viscous-flow of the oxide in the Reisman model, is used to explain the extracted pressure- and temperature-dependence of the parameters in that model (a and b). A third model by Han and Helms (which we will not consider in detail here), has also been proposed to predict oxide growth. We will, however, compare the results of the Reisman and the Han and Helms model to the Grove-Deal model in upcoming paragraphs.

A comparison between the Reisman-Nicollian model of oxide thickness and the Deal-Grove model is shown in Fig. 3-23. Data from Deal and Grove at 700°C are plotted on linear scales and Eq. 3-7 is plotted with measured values of a and b. Shown for comparison is Eq. 3-7 with the appropriate constants obtained by a fractional weighted least squares fit to the data.[24] Kouvatsos extracted the a and b parameters of Eq. 3-7 at 950°C and 1100°C for oxide growth in O_2 diluted in He, and found very good fit to the Reisman-Nicollian power-law model.[25]

Figures 3-24 and 3-25 again compare the predictions of the Reisman and the Han and Helms models with those of the Deal-Grove model. The model para-

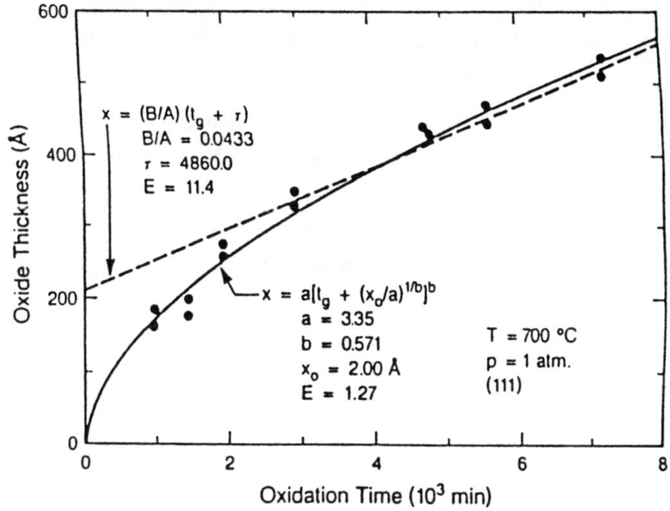

Fig. 3-23 Fits to oxide thickness versus oxidation time using the Deal-Grove model and the Reisman-Nicollian power-law model expression.[24]

meters for these plots for the Deal-Grove model are the standard A and B/A values for 1 atmosphere. For the Reisman model at 800°C, a = 0.302 nm, b = 0.704, and τ = 13.1 min. At 1000°C a = 3.02 nm, b = 0.701, and τ = 1.26 min. At 800°C (Fig. 3-24), the differences in these models for very-thin oxides is clear.

The Reisman (and the Han and Helms) model agree fairly well with each other and are a good match to experimental data in the thin regime. The Deal-Grove model does not do a good job for thin oxides, either with τ set at 8 hours

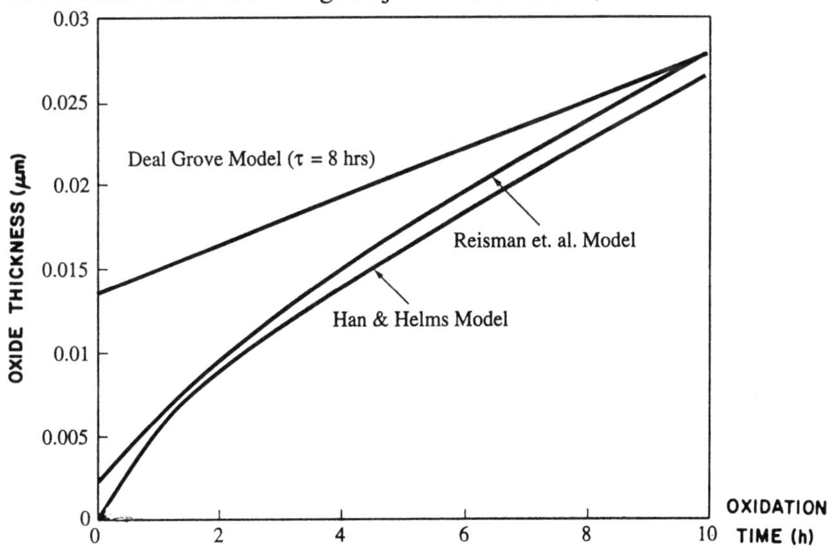

Fig. 3-24 Comparison of three oxidation models for 1 atmos, dry-O_2 oxidation at 800°C.

(the value Deal and Grove extracted in their original work), or with $\tau = 0$ (which predicts even thinner oxide than do the Reisman or Han/Helms models). At 1000°C (Fig. 3-25), the differences in the models in the thin regime are less apparent since much thicker oxides grow at the higher temperature on the same time scale. For the particular parameters chosen for this plot, the Reisman-Nicollian model diverges from the predictions for thicker oxides.

Despite the fact that the Deal-Grove model in its original form does a poor job at predicting oxide growth in dry-O_2 in the thin-oxide regime, their model can be "fixed" to do a much better job. Such a"fix" was developed by Massoud *et al.*, who demonstrated that a much better fit can be accomplished than with the Deal-Grove model alone (by adding an additional term that decays exponentially with thickness to the oxidation rate.). The expression for the oxide growth-rate in the classical Deal-Grove model dt_{ox}/dt (see Vol. 1, Chap. 7) given by:

$$(dt_{ox}/dt) = B/(2t_{ox} + A) \tag{3.8}$$

then becomes

$$(dt_{ox}/dt) = \{B/(2t_{ox} + A)\} + C\exp(-t_{ox}/L_2) \tag{3.9}$$

where

$$C = C^0 \exp(-E_A/k_BT) \tag{3.10}$$

where $C^0 \sim 3.6x10^8$-μm/hr, $E_A \sim 2.35$-eV, and L_2 is the characteristic decay length described in paragraph **5** above. For dry-oxidation of lightly doped substrates in

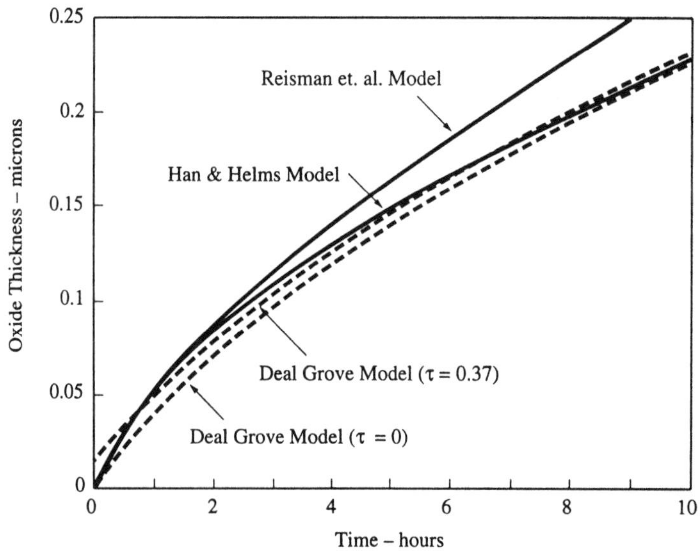

Fig. 3-25 Comparison of three oxidation models for 1 atmosphere, dry-O_2 oxidation at 1000°C. The dashed lines are the Deal-Grove model.

the 800-1000°C range, L_2 was found to be ≅7-nm, independent of surface orientation [i.e., the above numbers apply to either (100)- or (111)-oriented silicon substrates]. Heavy doping was found to only slightly affect the oxidation-rate-enhancement in the thin regime. Equation 3-9 has been inserted into SUPREM III and SUPREM IV as the model for thin-oxide growth in dry-oxygen. With Massoud's addition to the Deal-Grove model, its predictions match the data well over the whole thickness range. Note that it was decided to implement the Massoud model (but not the Reisman model) in SUPREM III and IV. This is because many other oxidation-effects have also been tied to the basic Deal-Grove model, and these effects are also modeled in SUPREM III and IV.

While the debate continues about the physical mechanisms responsible for thin oxide kinetics, the problem has taken on a new importance. This is because deep-submicron MOSFET gate-oxides are routinely grown with thicknesses in the anomalous regime. This is an example of an industrial application outpacing basic scientific understanding. From the industrial view, it is more important to be able to grow thin oxide layers reproducibly and uniformly (and with good electrical properties), than it is to understand the governing physical principles.

3.6 SINGLE-WAFER TECHNOLOGY OF THIN OXIDE GROWTH

Thin oxides can be grown using a batch process or a single-wafer process. Batch processes currently use vertical furnaces (200-mm wafers), and such tools are discussed in more detail in Vol. 1, 2nd Ed. With the advent of 300-mm wafer processing and the need to grow uniformly-thick 20-nm oxides across such 300-mm wafers, the single-wafer process may begin to make inroads against batch-oxide-growth processes. Here we will limit the discussion to such single-wafer oxide processes, which are also based on rapid-thermal-oxidation (RTO) methods. Note that single-wafer RTP processes are also described in Vol. 1, 2nd Edition, Chap. 7, and a more comprehensive review of RTP systems and processes is given in Ref. 26. However, in both of these references, the focus is more on the general aspects of RTP systems. Here our focus is on the details of rapid-thermal-oxidation (RTO) tools and processes. Wet-RTO processes are also examined, a topic not covered in Vol. 1, 2nd Ed.

3.6.1 Rapid Thermal Oxidation Tools

Conventional furnace-based oxidation has been the workhorse of the industry, but it encounters some problems when gate oxides thinner than 2.0-nm must be grown, including: **1**) furnace oxidation is difficult to scale below 2.0-nm thickness; and **2**) the lower temperatures used during such ultra-thin oxide growth processes degrade the oxide quality.

RTO is able to offer better quality ultra-thin oxides when compared to those grown in furnaces for two reasons: **1**) RTO grows the oxides at higher temperatures; and **2**) RTO is more conducive to cluster tooling. RTO can grow thin oxides at high temperatures because the oxidation time is kept down to just a few seconds (whereas the minimum oxidation time in a batch furnace is at least a few minutes).

There are several advantages to growing oxides at higher temperatures. First, oxides exhibit smaller gate-leakage-currents as their growth temperature increases (up to a maximum of 1050°C, beyond which leakage-currents increase).[27] Second, they show higher resistance to boron penetration. In one study, 2-nm-thick oxides grown in a furnace at 800°C were compared to RTP-grown oxides at 1000°C. The results indicated that boron penetration was suppressed better by the RTP-grown oxides.[28] The third advantage is that MOSFETs built with the RTP-grown oxides had higher effective carrier mobilities (for both electrons and holes) as the temperature of the oxide growth process was increased from 800°C to 1000°C.[29] This was attributed to the smoother Si/SiO_2 interface that results from growing the oxides at higher temperatures.

Cluster tools can potentially produce better-quality oxides because they allow process sequences to be carried out in an integrated fashion. For example, in oxide growth processes, a cleaning procedure is done prior to the growth step. If this can be done *in situ,* the quality of the oxides can be improved. Clustering is easily achievable with single-wafer RTP systems, but is impractical with batch furnace tools. By integrating RTO chambers with other processing modules it may be possible to extend the use of oxides to additional technology generations.

As noted above, with the introduction of single-wafer RTO-systems, it becomes possible to carry out *in situ* surface preparation, wafer rotation, ambient control, and process integration. Because such systems can grow oxides at such rapid rates (i.e., more than 1000-Å/min), it also makes RTO capable of competing economically with vertical hot-wall furnaces.

Figure 3-26 is a cross-sectional view of an RTO reaction-chamber depicting the necessary components of a lamp-heated RTO-system. In this reactor, the wafer is inserted into the reaction chamber by a robot wafer handler onto a rotating SiC support ring. The wafer is rapidly heated at rates of 50-100°C to the process temperature (800-1050°C). Process gases are introduced at reduced pressure in an axial flow pattern where they react to form water vapor, hydroxyl ions, oxygen ions, and other species (depending on the H_2:O_2-ratio and whether N_2, NH_3, HCl, or other reactant gases are introduced). At the conclusion of the oxidation and annealing steps, the wafer is cooled rapidly to a temperature where

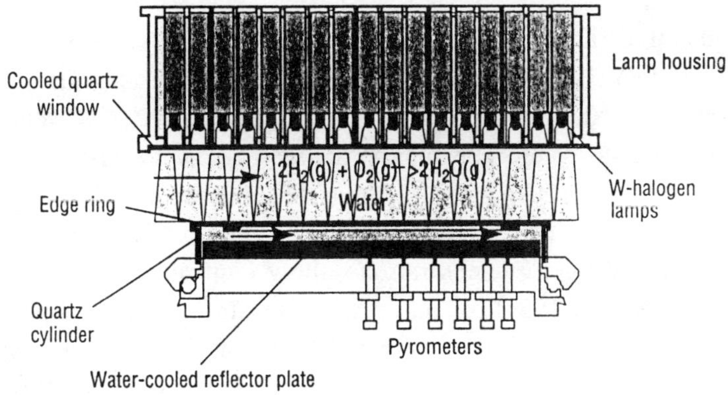

Fig. 3-26 Cross-section of an Applied Materials ISSG RTO reactor.

it can safely be removed by the wafer handler from the reaction chamber to an additional cool-down chamber that can be used to enhance system throughput. Cool-down can be enhanced further by providing a wafer backside cooling gas, such as helium or hydrogen.

3.6.2 Wet RTO Processes

As will be discussed in Sect. 3.8, traditional dry oxides are expected to be limited to minimum thicknesses somewhere between 1.6-nm[36] and 2.5-nm.[37] Various techniques are being explored to reduce this thickness limit. One of these methods employs wet-RTO. A technique pioneered by Applied Materials called *in-situ steam generation* (*ISSG*) has reportedly allowed oxides in the 2.0-nm thickness range to be formed with improved reliability compared to dry oxides of the same thickness.[38] ISSG-oxidation is a low-pressure process (typically below 20 torr, for safety reasons) performed in a cold-wall RTP reactor.

Unlike hot-wall furnaces that employ a pyrogenic torch for growing wet oxides, no hot quartz is used to generate the steam ambient (eliminating the problem of quartz devitrification). Instead, in ISSG-oxidation, H_2O, OH and atomic oxygen are generated directly in the reaction chamber without pre-combustion. The hot wafer acts as an ignition switch, causing reactions among electronic-grade H_2 and O_2 gases (that are pre-mixed in a plenum before being injected into the RTP chamber). This configuration reduces risk of metal contamination, since no metal catalyst is used in the in-situ steam generation method.

The oxidation-rate rapidly *increases* as the pressure is decreased below 20 torr (unlike in conventional oxidation processes where oxidation-rates decrease with decreasing pressure). The growth-rate at 10-torr is nearly twice that of atmospheric dry-rapid-thermal-oxidation. Modeling studies indicate that the oxidation-rate exhibits a strong correlation to the atomic-oxygen-concentration and not to any other atomic or molecular species. The oxidation-rate is also strongly temperature-dependent compared with wet-atmospheric-oxidation.

Oxides grown with the ISSG process show good characteristics. Within-wafer uniformity and wafer-to-wafer repeatability are well below 1% (1σ), both for thin (< 40-Å) and thick (> 100-Å) oxide-films. The faster growth-rates improve productivity while offering a thermal-budget reduction. The wide range of temperatures and hydrogen concentrations allow controlled oxidation for any thickness of oxide. The ISSG-oxides also demonstrate better reliability than comparable furnace-oxides, or those created by dry-RTO. Several independent measurements of charge-to-breakdown of 30-50-Å wet-RTP oxides show a threefold (or more) improvement over furnace-grown-oxides. The wet-RTO-oxides also exhibit about an order-of-magnitude less gate-leakage-current than dry-RTO oxides, which are comparable to furnace-grown-oxides (Fig. 3-27).[76]

3.7 NITRIDED AND FLUORINATED OXIDES AS MOSFET GATE DIELECTRICS

One modification to the basic SiO_2 material that has recently been intensely examined is the addition of nitrogen to the oxide. That is, oxynitrides have been studied as replacements for silicon dioxide as the gate dielectric of MOSFETs for thicknesses below 4.0 nm. The advantages of oxynitrides over silicon dioxide for this application are the following: **1**) they exhibit improved boron penetration resistance; **2**) they have better breakdown characteristics; **3**) they show greater immunity to hot-carrier-injection effects; and **4**) MOSFETs made with such layers exhibit improved high-field channel-electron-mobility. Improved performance of nitrided silicon-dioxide gate-dielectrics has been associated with the accumulation of a small amount of N at the Si/SiO_2 interface. Interfacial N-accumulation is thought to reduce interfacial defects by relieving strained Si-O bonds and "dangling" bonds. An interfacial peak-concentration of about 1-2 atom % is typically associated with higher breakdown electric fields and lower fixed-oxide charge densities. But, a higher N-concentration (> 5 atom %) is needed to effectively suppress boron penetration through the dielectric layer.

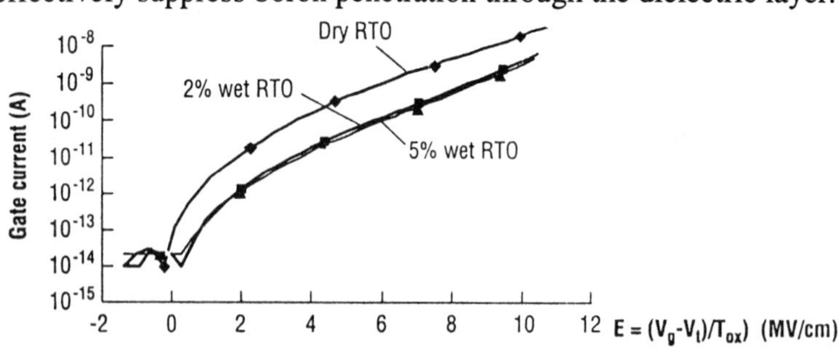

Fig. 3-27 Gate leakage current comparing dry-RTO to 2% & 5% ISSG RTO for 30Å oxides.[76]

Nitridation of oxides is considered to be a variation of silicon oxidation, and such films are formed mainly though the nitridation of silicon oxide or the oxidation of silicon in a nitrogen containing ambient such as NH_3, N_2O, NO (or mixtures of O_2 and NO). Such films have been formed in high-temperature furnace processing or rapid-thermal-processing tools. Fluorine can also be incorporated into gate oxides as a method for improving its dielectric characteristics. This subject is also discussed in a following section.

3.7.1 Oxynitridation of Silicon in N_2O

Early oxynitride films were formed by nitriding a thermally grown oxide film by exposing it to an NH_3 ambient at high temperatures, and then re-oxidizing the nitrided oxide to reduce the amount of hydrogen that remains in the film after NH_3 nitridation. Such oxynitrides were thus called *reoxidized nitrided oxide (ROXNOX)* films.

Although ROXNOX films exhibit better hot-carrier resistance than do pure oxides, some practical issues make them troublesome to manufacture. First, the ROXNOX formation sequence is complex and difficult to optimize, especially for CMOS applications. For example, the combination of nitridation and reoxidation steps used in the ROXNOX process may leave excess hydrogen in the film. In that case, the density of electron traps will be higher than in a pure oxide film. Since electron trapping is the dominant degradation mechanism in PMOS devices, PMOSFETs with such films will exhibit worse hot-carrier degradation.[39] On the other hand, if the reoxidation step is excessive, the film can lose its strength against interface-state generation by hot-carriers. Thus, NMOSFETs containing such films would be prone to more hot-carrier degradation. The gate dielectric layer might also grow too thick. Another drawback is that the best NH_3-nitrided films always exhibit higher densities of positive fixed charge (Q_f) than do the best SiO_2 films. The fixed charge arises as a result of the nitridation process. Finally, the ROXNOX process generally involves multiple high-temperature steps, which are undesirable for ULSI CMOS applications. As a result, ultra-thin oxynitrides are not formed in production by using NH_3 as the nitrogen source.*

A solution to some of the problems of NH_3-nitrided films is to avoid the incorporation of hydrogen in the film in the first place. One way to accomplish this is to replace NH_3 with N_2O as the nitriding ambient. Such N_2O-based techniques have indeed proved to be more fruitful because of the hydrogen-free nature of the processing.

* Interested readers can find more details on ROXNOX films in Vol. 3.

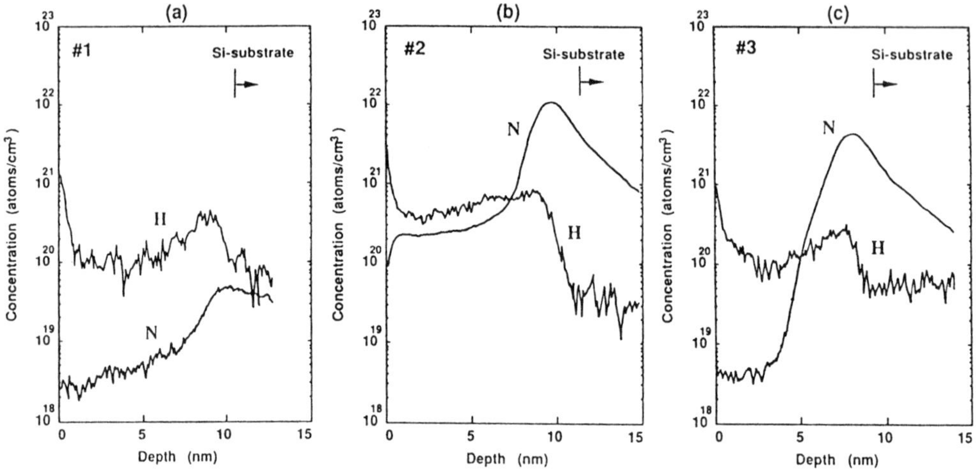

Fig. 3-28 SIMS depth profiles of N and H in three kinds of oxide films: (a) pure SiO_2 (#1); (b) NH_3-nitrided SiO_2 (#2); and (c) N_2O-nitrided SiO_2 (#3).

This approach was first used to grow a gate dielectric by directly reacting N_2O at the silicon surface in an RTP system.[40,41,42,43] Since there is no hydrogen in N_2O, this process results in films with significantly less incorporated hydrogen than NH_3-nitride films. Figure 3-28 shows SIMS profiles of dielectric films prepared by: (a) thermally oxidizing silicon; (b) nitriding an oxide film in NH_3; and (c) growing a film on silicon by RTP in an N_2O ambient.[41] It can be seen that the hydrogen concentration in the N_2O film is significantly smaller than in the NH_3 film. It is also evident that the nitrogen concentration in the N_2O film is highest at the Si-SiO_2 interface. Early studies of such films concluded that they are self-limiting in thickness, as is shown in Fig. 3-29a.[40,44] The build up of the nitrogen concentration was thought to block the oxidation reaction that occurs simultaneously at the interface. Later work showed that the thickness was not self-limiting (see Fig. 3-29b),[45,46] and that the thickness in fact, increases approximately as $(\text{time})^{1/2}$. This seems to indicate that the growth is instead limited by the diffusion of N_2O through the oxide bulk (rather than through the nitrogen-rich layer).[47] The discrepancy in the apparent growth mechanisms may have been explained by Okada, who observed significant retardation of the growth only at relatively high (~0.9 at%) interfacial nitrogen concentrations, which is higher than observed in some N_2O furnace processes.[59] However, the interfacial reactions may still dominate growth rates for short times and thin oxides.

Nevertheless, the nitrogen profile in the N_2O film is similar to that in the ROXNOX film, yet this is achieved with only a one-step process (i.e., no

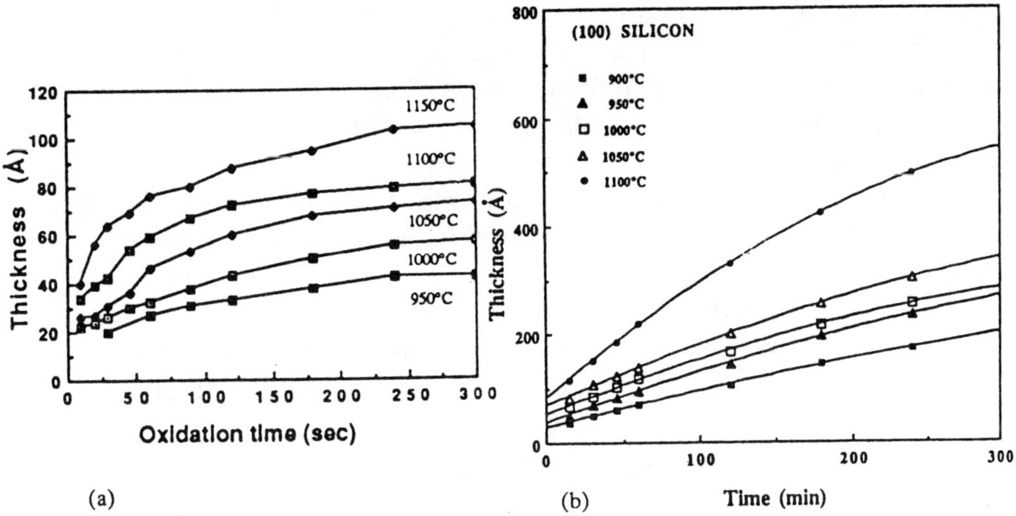

Fig. 3-29 (a) N_2O oxynitride thickness versus oxidation time using RTP for various process temperatures. Ramp-up rate was 200°C/sec. Oxide thickness was measured at a fixed refractive index (1.46)[40] (© IEEE 1990); (b) N_2O oxynitride thickness versus time in a furnace-anneal process. (© Electrochemical Society 1993)[45]

reoxidation is necessary). The peak nitrogen concentration (at the Si/SiO_2 interface) is ~0.5%, which is less than that found (2-10 atom%) in SiO_2 furnace-nitrided in NH_3 under atmospheric pressure. Furthermore, residual hydrogen incorporation into the film is no longer a concern. This is evidenced by the low electron trapping density exhibited by such films. NMOSFETs fabricated with such films also exhibited less tendency toward hot-carrier degradation than those made with control oxides (i.e., they showed smaller values of ΔD_{it}), and were also less susceptible to the creation of electron traps throughout the oxide under hot-carrier stressing.[48]

However, several limitations were still observed in these N_2O-grown oxynitrides. First, larger initial values of D_{it} were observed than in pure oxide films (Fig. 3-30), even though the rate of their generation under hot-carrier stressing was much smaller (see also Fig. 3-30).[48] Second, the initial Q_f values were also somewhat higher. Third, the slow growth-rate of such films make them only practical for ULSI applications which use dielectric films thinner than 100Å (see Fig. 3-29). Fourth, the maximum mobility of electrons in MOSFETs made with such films is ~5% less than in MOSFETs with oxide gate dielectrics (but the mobility is actually higher at large lateral electric fields - see Fig. 3-31).[49] The most severe limitation, however, is that such RTP N_2O-grown films exhibit unacceptably large nonuniformities in composition and thickness across a wafer.[50] Such non-uniformities would limit the use of such films in ULSI MOS devices. Temperature nonuniformities in the RTP chamber, gas depletion, and heat-transfer effects were some of the phenomena suspected of leading to these nonuniformities.

Nevertheless, it was found that growth of such N_2O dielectrics in a furnace (rather than by RTP) resulted in films with thickness uniformities comparable to those of pure oxide films, and with adequate compositional uniformity as well.[51,52] The process described in these reports was a 950°C furnace step in N_2O, which grew 60-Å-thick dielectric films having a nitrogen concentration at the Si/SiO_2 interface of ~3%. Hot-carrier degradation of PMOSFETs with such films was excellent since electron-trap-densities in these films was also low.

3.7.2 Oxynitridation of Oxides in N_2O or NO

Another method used to form thin oxynitride layer involves a two-step operation. Oxides are first grown in O_2, and are then subjected to an oxynitridation step in N_2O or NO. When N_2O is used as a post-oxidation-annealing ambient, a wide range of final oxide thicknesses is achievable with much smaller thermal budgets. Processes using this two-step N_2O sequence have been described using an atmospheric-pressure-furnace,[46,53,54,55] a low-pressure-furnace,[56] and RTP conditions,[57] Figure 3-32 shows an example of such a two-step process performed in a conventional furnace at atmospheric pressure.[54]

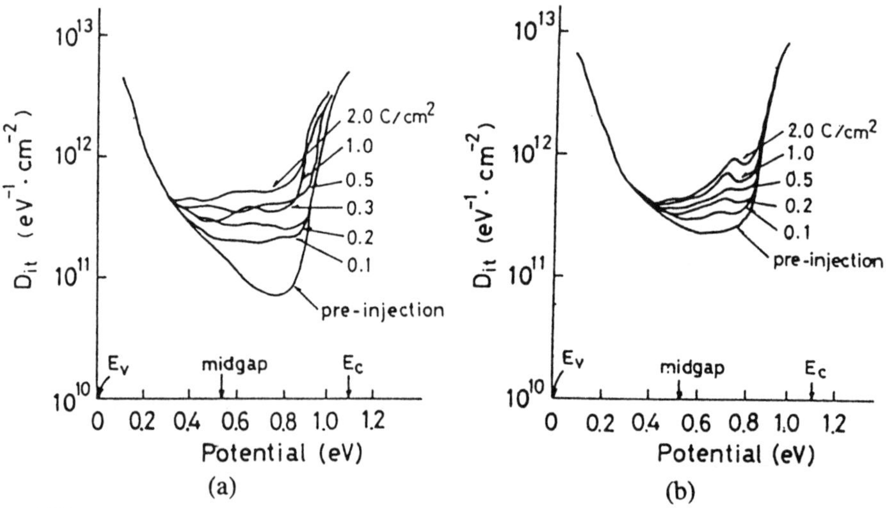

Fig. 3-30 Changes in the interface trap densities of MOS capacitors with: a) pure SiO_2; and b) SiO_xN_y by Fowler-Nordheim electron injection. Electrons were injected from the gate electrodes into the oxide films at a current density of ~10 mA/cm^2.[42] (© IEEE 1991)

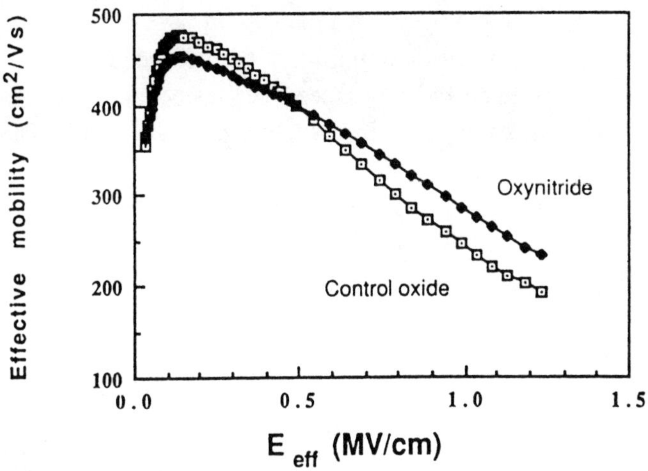

Fig. 3-31 Effective electron mobility versus effective normal field of NMOSFETs with a 65-Å-thick gate dielectric. The mobility values were calculated using the linear region I_D versus V_{GS} curves (V_{DS} = 100 mV), with a *Z*/*L* ratio of 75-μm/60-μm. The maximum effective mobility of the oxynitride is 5% lower than that of the control oxide. However, the high-field mobility of the oxynitride shows a 10% improvement.[49] (© IEEE 1991)

Typically, the oxide is grown in dry-O_2 in a furnace at 850-900°C. The N_2O step is usually carried out at 800-850°C for 5-40 minutes. In the process described in Ref. 55, 50-Å SiO_2 films are grown in dry-O_2 at 850°C for 30 minutes. During the second step in N_2O at 950°C, an additional 35-Å of oxide is grown to form films with a final thickness of 85-Å.

PMOSFETs made by oxynitridation of SiO_2 in N_2O also exhibit enhanced

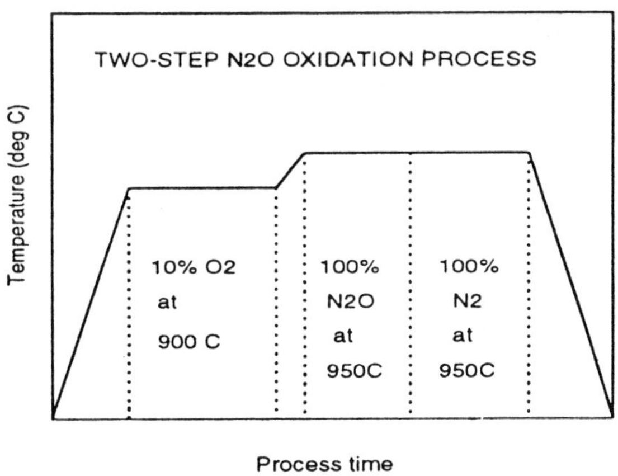

Fig. 3-32 Two-step oxidation & N_2O nitridation process in a conventional furnace.[54] (© IEEE 1993)

hot-carrier immunity compared to devices made with pure oxide. This is thought to be due to the fact that electron trapping is suppressed in devices with such N_2O dielectrics.[58] Consequently, it has been suggested that the optimum N_2O oxynitride process is oxidation in O_2 followed by nitridation in N_2O, rather than oxidation in N_2O alone.

In the low-pressure approach[56] it was reported that a minimum pressure of between 1 and 50 torr of N_2O was needed before a measurable amount of nitrogen was incorporated into the oxide. However, at low pressures of N_2O (i.e., just beyond the minimum needed for the onset of nitrogen incorporation), the concentration of nitrogen at the interface could be well controlled, but at the same time, minimal additional oxide was grown during the oxynitridation step. Thus, tight control of the final film thickness could be maintained.

As noted in a previous paragraph, the oxide film continues to grow during the second (i.e., N_2O) step. However, if re-oxidation of N_2O oxides is carried out after they are formed, unusual behavior occurs. That is, new SiO_2 grows at the Si/SiO_2 interface, and the existing N-rich layer remains intact and is displaced away from the interface by growth of the new oxide (see Fig. 3-33).[59]

Using N_2O as the nitrogen source during the second step of this procedure was found to have some disadvantages. The concentration of N-incorporation may be

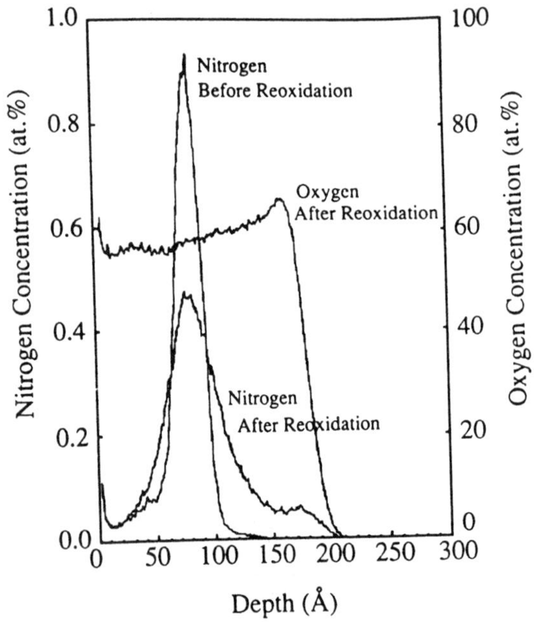

Fig. 3-33 SIMS profile of the oxynitride film grown from 100% N_2O at 1100°C before reoxidation, and after 85 min reoxidation.[59] By permission of the Electrochemical Society.

too low for some applications, especially suppression of boron penetration, unless a much higher thermal budget is permitted.

More recently oxynitrides formed with the two-step process using NO as the nitrogen source have been reported. Generally, the results for NO oxynitrides overcame the problems of using N_2O as the nitrogen source, in that they offer a lower thermal budget process for incorporating higher concentrations of N, as well as better thickness control for scaled dielectrics.[60,61] It is thought that nitrogen incorporation in the N_2O process is driven by NO species, which is a product of N_2O molecule dissociation. The N_2O molecule likely dissociates into NO (4.7%), N_2 (64.3%), and O_2 (31%), according to calculations by Tobin *et al.*[62] Such by-products act in competition. That is, while the NO incorporates nitrogen, O_2 continues the oxidation by reacting with the silicon substrate, and N_2 reduces the partial-pressure of the nitridation species (which increases the thermal budget). The parallel oxidation explains the observed increase of oxide thickness during annealing in N_2O. The high temperature needed for N_2O dissociation (> 850°C) mandates the high thermal budget needed to get acceptable results. Good quality gate-dielectric-films have reportedly been produced with RTP processes using NO + O_2 mixtures.

The percent nitrogen incorporation also depends on the relative amount of NO in the gas mixture. Figure 3-34 shows the dependence of N incorporation on NO concentration for both direct (1-step), and 2-step oxynitride growth carried out in a NO + O_2 mixtures using RTP processes. This indicates that NO can produce

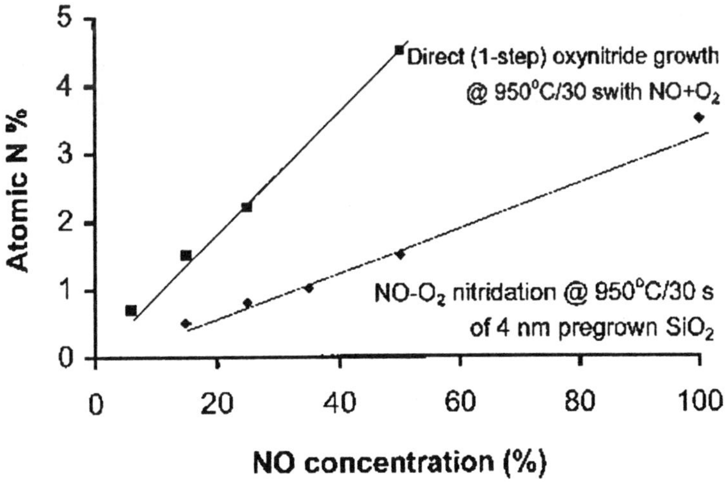

Fig. 3-34 Dependence of nitrogen incorporation in oxide on NO concentration in NO + O_2 mixtures.

oxynitride films with a wide range of incorporated N, allowing the films to be tailored for specific applications.

In another reported process, 4.0-nm-thick oxynitrides were formed in a two-step furnace operation. First, 4.0 nm oxides were grown in O_2 at 800°C with a 6% HCl, and a post-oxidation anneal in N_2. The second step involved annealing this oxide in a diluted NO ambient at 800°C for 45 min.[60] Finally, a process that uses a mixture of N_2O and NO in the reaction chamber has been described. The postulated advantage of this method is that the vertical positions of the nitrogen concentration peaks within the oxynitride layer can be adjusted by changing the N_2O + NO mixture.[64]

We should also mention that adding nitrogen to gate oxides has also been reported by several other novel techniques, including implantation of nitrogen through the polysilicon gate and drive-in annealing,[64] reoxidation of films formed by growing the oxynitride in N_2O or NO,[65] and nitridation of the Si surface by using a nitrogen remote plasma.[66,67,68]

3.7.3 Fluorinated Gate Oxides

The presence of fluorine in the gate oxide has also been reported to increase the hot-electron resistance of devices fabricated with such oxides.[69,70] Such fluorine may inadvertently be introduced from a BF_2^+ source/drain implant, or from an LPCVD-W or LPCVD WSi_x process. (In the W-CVD and WSi_x-CVD processes fluorine is produced during the reaction of WF_6 and SiH_4.[71]) Other reports confirm that deliberate incorporation of fluorine into the gate oxide (from implanting fluorine into a polysilicon film, and then diffusing it into the gate oxide) produces a more hot-electron resistant interface.[72,73] The increased hot-carrier-resistance of the F-containing-oxides is thought to be due to fluorine atoms which form Si-F bonds. These are stronger than Si-H bonds. The presence of the fluorine implies that some of the Si-H bonds at the Si/SiO_2 interface are replaced by Si-F bonds, thus yielding an interface that has better hot-carrier-resistance.

Unfortunately, while the above benefits make fluorinated oxides attractive for NMOSFETs, the presence of fluorine in the gate oxide may increase the penetration of boron (from p^+-poly gates) though the gate oxides, and this represents a potential problem for deep-submicron PMOSFETs. That is, BF_2^+ is used when the shallow source/drain-extension-region is formed, and this implant enters the poly gate. Thus, the F from this implant will diffuse through the poly and into the gate oxide during the implant activation steps. However, a recent study indicated that if the fluorine dose implanted into the poly is restricted to a medium dose ($1x10^{14}/cm^2$), no noticeable enhanced boron penetration effect is observed. In addition, the charge-to-breakdown (Q_{BD}) characteristics of the PMOSFET capacitors fabricated with 4-nm-thick oxides are also significantly improved.[74]

Recently, a new form of "anomalous" leakage was discovered in oxides used in Flash-memory devices. Such leakage occurs at temperatures below 150°C, and thus cannot be screened by a traditional 250°C high-temperature retention bake. However, by incorporating a high concentration of fluorine in the tunnel oxide (introduced in a WSi_2 capped-control-gate process), a much higher resistance to this anomalous leakage is obtained.[75]

3.7.4 "Dual-Gate-Oxide-Thickness" Structures

An increasing number of mixed-signal and system-on-a-chip applications require dual on-chip power-supply-voltages. The higher-voltage may be used for the analog and input/output (I/O) circuitry, and the lower-voltage for the core digital-logic-devices. However, a serious reliability problem would be created if the same thin oxide needed for the high-performance logic MOSFETs was used in the devices subjected to the higher voltage (i.e., the analog and I/O circuitry MOSFETs). That is, the analog and I/O devices require thicker gate oxides than the core logic devices. (The thicker oxide also makes the I/O devices more robust against transient, off-chip electrostatic-discharge-events.) Consequently, it must be possible to fabricate gate oxides with different thicknesses on such ICs.

Early "dual-gate-oxide-thickness" processes had two oxidation steps, with a masked etch-step in between to remove the first oxide from regions intended to have thinner oxide. Newer processes use a different, simpler approach. That is, it has been discovered that the rate of thermal-oxidation on silicon is reduced by nitrogen implants.[88] Hence, nitrogen implantation into the silicon substrate prior to the oxidation process can be used to facilitate the implementation of such "dual-gate-oxide" ICs. Namely, selective implantation of nitrogen will cause oxides grown on such areas to be thinner than those simultaneously grown on non-implanted areas (Fig. 3-35a). Typical implant conditions for the nitrogen implant are energies of 10-30 keV, at doses of $2x10^{14}/cm^2$-$1x10^{15}/cm^2$. Figure 3-35b shows how the oxide thickness from an 800°C, dry-O_2 oxidation is reduced with increasing nitrogen dose.[89] It is suggested that nitrogen suppresses oxidation by forming Si-N bonds at the interface of the growing oxide, thereby constraining further growth.

3.8 PROJECTIONS OF THICKNESS LIMITS OF GATE OXIDES

In the late 1970's it was projected that the scaling limit of MOSFETs would be 0.5-μm. In the mid-1980's, this limit was reduced to 0.25-μm. In 2001, there

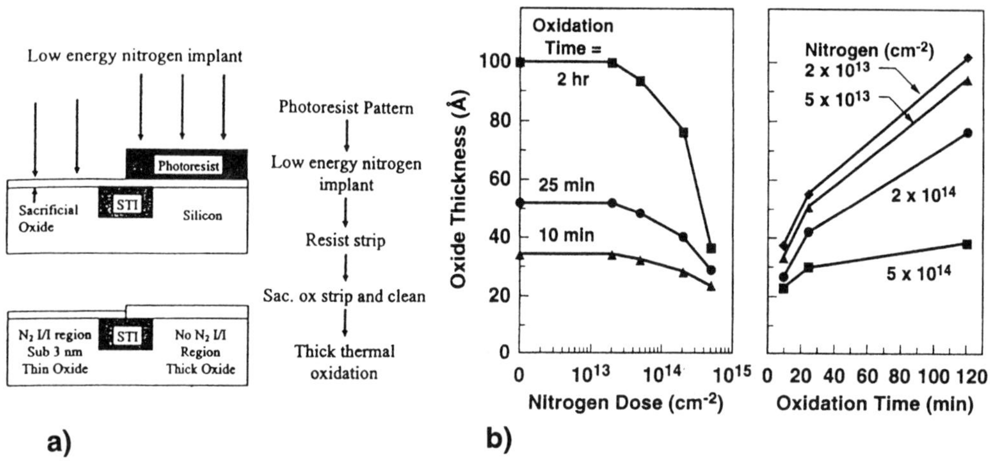

Fig. 3-35 a) Schematic of the process flow for sub-3-nm gate oxide formation on N_2-implanted substrates. b) Oxide thicknesses obtained in dry-O_2 at 800°C for various nitrogen implant doses (energy = 25 keV) and oxidation times.[89] (© IEEE 1997)

does not seem to be any device physics barrier for scaling at least to 0.05-μm (50-nm). The factor that kept the predicted limit on MOSFET gate lengths in previous decades higher than it is today was the uncertainty in the minimum thickness of the gate-oxide-layer that could be successfully implemented in MOSFETs as they were scaled. Thus, the question of the minimum-oxide-thickness that can be implemented in MOSFETs remains an important one. Here we address the predictions of the minimum-oxide-thickness as viewed in 2001.

There are a number of factors that play a role in the minimum-oxide-thickness that can be used, including: oxide-reliability, tunneling-leakage-current, stress-induced leakage-currents (SILC), polysilicon-depletion, process-induced oxide-damage, and hot-carrier-induced damage. The type of application in which the oxide is to be used also plays a role. That is, different minimum-oxide-thickness limits exist for logic devices, DRAMs, and nonvolatile memory devices. We will discuss each of these topics here, with the exception of hot-carrier-induced damage. The latter phenomenon is not treated because hot-carrier-reliability is expected to greatly improve as the gate voltage is decreased below 3-V (since at such low voltages, few carriers can gain sufficient energy to create interface-traps or shallow oxide-traps). This is one of the positive aspects of scaling devices below 0.25-μm.

3.8.1 Minimum-Oxide-Thickness Due to Defects and Tunneling

The first two (and probably the most important) phenomena that impact the minimum-oxide-thickness that will be usable in full-scale IC production are: **1**) defect densities in the thin-oxide films and; **2**) the direct-tunneling-current through these films. We discuss these two effects together here because approximately the same minimum-oxide-thickness limit is set by either of them.

In 1996, Hu predicted what the minimum-oxide-thicknesses would be in production ICs, depending on the power-supply voltage.[11] He based these values on the following assumptions: First, the oxides will be used in ICs designed to operate for at least 10 years. Second, the time-to-breakdown (t_{BD}) that oxides of various thicknesses will exhibit can be extrapolated to the 10-year limit by using the anode-hole-injection (AHI) model discussed in Sect. 3.3.4. Figure 3-36 shows a plot of such t_{BD} data for oxides of various thicknesses and intrinsic failure behavior (i.e., the oxides depicted in this plot are assumed to be defect-free, and thus fail according to intrinsic-breakdown failure mechanisms). For such defect-free oxide films, the following thicknesses could be expected to exhibit a 10-year time-to-breakdown behavior for respective power supply voltages: **a**) 80-Å for 5.5 V (5 V power supply + 10%); **b**) 45-Å for 3.6 v (3.3 V + 10%); **c**) 33-Å (3.3-nm) for 2.75 V (2.5 V + 10%).

To allow for defects in the oxide that would weaken them (i.e., the B-mode defects, which locally thin, but do not create catastrophic pinholes in the oxide), Hu suggested that a safety margin be employed. That is, he used a 30% increase

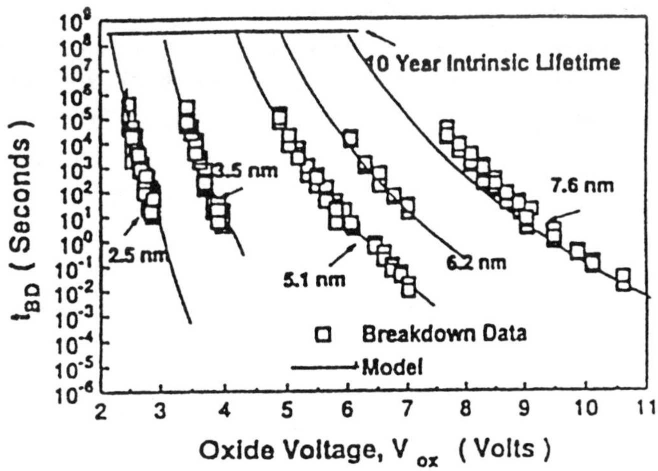

Fig. 3-36 Oxide lifetime as described the anode-hole-injection model.[11]

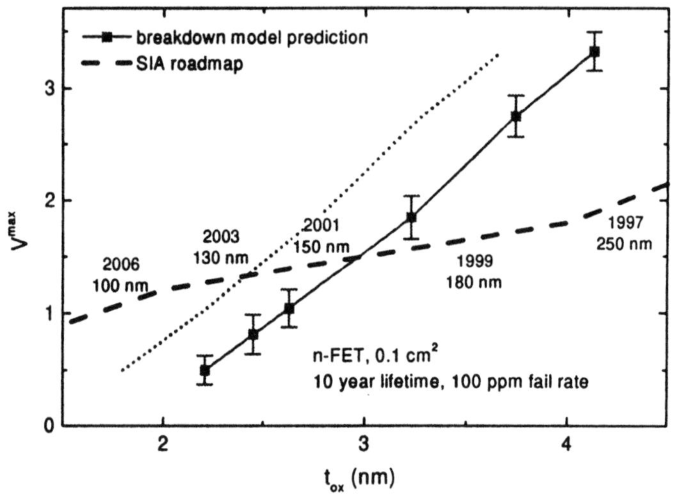

Fig. 3-37 The largest allowable supply (gate) voltage, V^{max}, as a function of oxide thickness, consistent with a failure rate of 100-ppm for a total gate area of 10 cm^2 on a chip operating at room temperature for 10 years. Solid and dotted curves correspond to t_{ox} measured by CV-extrapolation or quantum mechanical methods, respectively. The heavy dashed line is the SIA Roadmap (1997 version) for t_{ox} and V_{DD}, corresponding to the indicated year of first manufacture and technology generation.[37]

in the oxide thickness to allow for such defects. Thus, the minimum oxide thicknesses that he predicted that could be used in production were: a) 110Å for 5 V; b) 65Å for 3.3 V; and c) 45Å for 2.5 V.

Since that time, however, MOSFET scaling has been projected to even smaller gate lengths, mandating the use of even thinner gate-oxides (and smaller power-supply voltages). That is, projections for MOSFETs scaled to 50-nm (and using oxides below 2.0-nm) have been subsequently published. In 1998, Stathis and Di Maria of IBM published a study in which they projected the minimum oxide thickness that could be used in production was 2.6-nm for a 1.0-V power supply in order to meet the 10-year operating lifetime.[37] This was based on the conclusion that intrinsic oxide-degradation and breakdown would become the most significant reliability concerns, and therefore the dominant limiting factors that would set the allowable oxide thickness. Figure 3-37 shows the graph reported in that study, indicating that at 1.0 V, the minimum oxide thickness to meet a 10-year lifetime would be about 2.6-nm. This plot also implies that a 2.2-nm-thick gate oxide would no longer meet the ITRS Roadmap requirements.

In 1999, however, another study by Weir *et al.*, of Bell Labs (Lucent Technologies) indicated that if the oxide uniformity could be significantly imp-

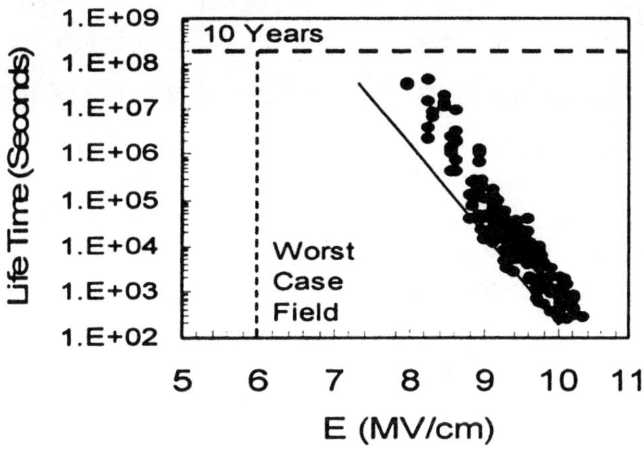

Fig. 3-38 TDDB data for 2.0-n- thick oxides at 125°C from MOS-capacitor data.[79]

roved, the reliability projections of an oxide at a given thickness can also be increased. That implies that thinner oxide films than were projected by Stathis and Di Maria may be possible. In fact, Weir showed that oxides with a thickness of 1.6-nm could meet the 10-year operating lifetime goal with a power-supply of 1.2 V - if they were produced with a process that allowed the film-thickness-uniformity to be increased.

In 1999, a 100-nm CMOS-technology that uses an oxide with a thickness of 2.0-nm and a power-supply voltage of 1.5 V was reported by Intel.[79] Figure 3-38 indicates that this oxide is able to meet the 10-year operating lifetime. In 2000, another report by the same group indicated it could fabricate 130-nm CMOS ICs with a gate oxide that was 1.5-nm-thick and with a 1.3-V power supply. These recent results appear to exceed the predictions of the minimum gate oxide thicknesses made just 2 years ago. At this point (circa 2002), the limit of oxide scaling does not yet seem to have reached.

The tunneling gate-leakage-current is another factor that has been predicted to put a limit on the minimum-oxide-thickness that can be used. In Sect. 3.4.2 we mentioned that at oxide thicknesses below 1.4-nm, the tunneling-leakage-current would become excessive.* This about the same thickness of oxide that has been reported to be used in the most advanced CMOS processes by 2000. Thus, both gate oxide reliability and excessive gating tunneling currents both limit the gate

* At some point this leakage will grow to be too large. That is, for SiO_2 at a gate-bias of ~1 V, the leakage-current changes from $1x10^{-12}$ A/cm^2 at 35-Å, to 1x10 A/cm^2 at ~15-Å (see also Fig. 3-21). This represents a change of *twelve orders of magnitude* in leakage-current for an oxide-thickness-change of little more than a factor of 2!

oxide thickness at about the same value. (Although, it has been reported that even high levels of tunneling current do not seem to affect device reliability.[87])

3.8.2 Minimum-Oxide-Thickness Due to Stress-Induced Leakage-Current (SILC)

When thin oxides are exposed to high electric-field-stress (and consequently Fowler-Nordheim gate current), it causes them to subsequently exhibit low-field leakage. This effect, termed *stress-induced leakage current* (SILC), was first reported in 1987 (Fig. 3-39).[80] Apparently, such high-field stress generates neutral oxide-traps that facilitate electron tunneling. In addition, SILC is a transient phenomenon in thicker oxides, but is a continuous current in thinner oxides.[81,82] That is, in thicker oxides the SILC is observed to decay as traps are filled without significant tunneling out of traps. In thinner oxides, a steady-state current flows when there is an equilibrium between trap-filling and -emptying processes.[83]

The SILC is a particular problem for non-volatile memory devices because such oxide leakage reduces data retentivity. As shown in Fig. 3-40, it also tends to increase under the high-field program/erase cycles that the tunnel-oxides of such devices experience. The leakage makes nonvolatile-memory tunneling-oxide scaling much below 8-nm difficult (and perhaps impossible), unless the 10-year retention requirement is relaxed.

3.8.3 Soft-Breakdown in Thin Gate Oxides

In Sect. 3.3 we described how oxide films undergo sudden, catastrophic breakdown in which the I-V characteristic becomes essentially ohmic after being

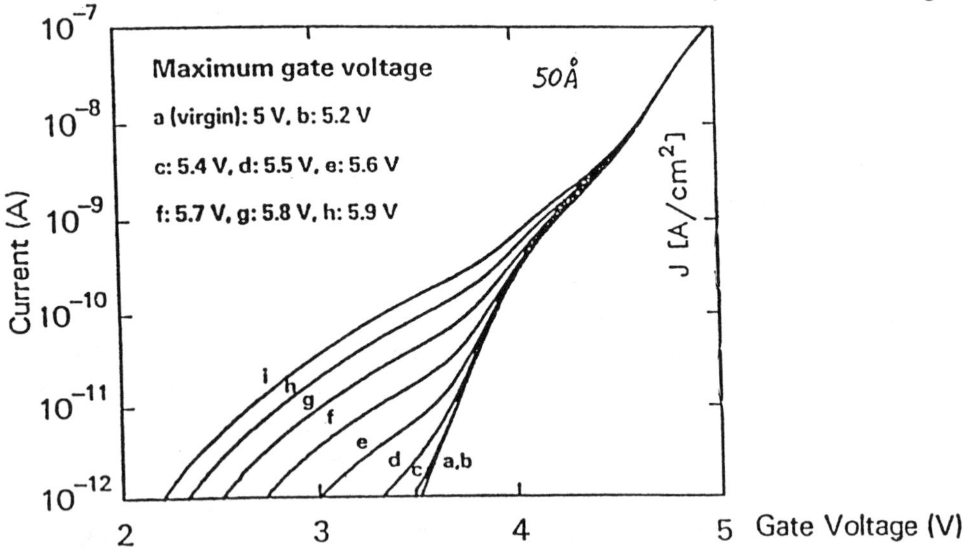

Fig. 3-39 Stress-induced (gate) leakage current (SILC) in 50Å gate oxides.[80]

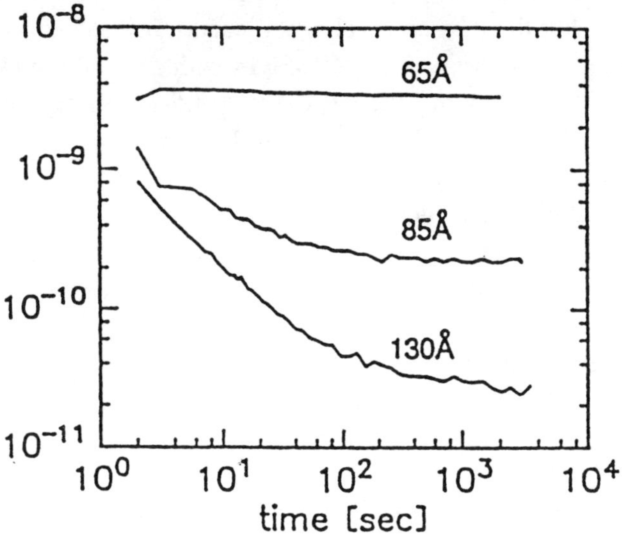

Fig. 3-40 Stress-induced leakage current is a transient current in thicker oxides but is a continuous current in thinner oxides, preventing nonvolatile memory scaling.[81]

subjected to sufficient electrical stress. However, for oxide films thinner than 5.0-nm, the breakdown mechanism may be fundamentally different. That is, another "anomalous" breakdown mode, termed *soft breakdown* or *quasi-breakdown,* occurs. Soft breakdown is detected by a much smaller change of voltage or current during stress, and a post-breakdown I-V characteristic that is given by a power law. In fact, while hard breakdown dominates for thicker oxides, it is often not observed in thin gate oxides. Instead, thin oxides appear to undergo soft breakdown. Also, soft breakdown is exhibited when oxides are stressed using lower and more realistic voltages or current densities. Soft break-down becomes "softer" and even less abrupt as the stress and oxide thickness is decreased.[90]

Figure 3-41a shows how the gate current in MOSFETs with thin (2.5-nm) oxides changes with time during voltage stressing (V_G = 4.1V and T = 140°C). Two breakdown regimes are observed: **1**) soft breakdown - where the current rises suddenly to an intermediate level of ~10^{-4} A in some devices, and **2**) hard breakdown – where the current rises suddenly to about 10^{-2} A. The I-V curves of a pre-breakdown oxide (before), and a post-breakdown oxide (after) are shown in Fig. 3-41b. In fact, when the thin oxide is first stressed with excess voltage, it undergoes SILC. However, further stress beyond a critical oxide-field-value causes the SILC to transition to soft breakdown. That is, soft breakdown is observed to be triggered once the oxide electric-field exceeds 10.7-11.0-MV/cm.[91]

After the oxide has undergone soft breakdown, the gate voltage becomes noisy (Fig. 3-42), but the threshold voltage V_T and transconductance g_m do not change significantly (as observed in MOSFETs with an oxide thickness of 2.5-nm). While MOSFETs with thicker oxides (5.5-nm-thick) undergo hard breakdown, their V_T and g_m also undergo little change. However, the current through the gate becomes so large that the transistor characteristics are distorted (making the MOSFETs that undergo hard breakdown unusable). Nevertheless, the only measurable effect on MOSFETs after soft breakdown is an increase in noise level in the gate voltage signal. While such noise fluctuations can couple into the drain current, for many applications this effect will not prevent MOSFETs from functioning properly. This suggests the possibility of a low-voltage technology which does not have gate-oxide breakdown as a failure mode.[90]

A later report, however, indicated that the soft breakdown mode in MOSFETs with channel lengths smaller than 1.0-μm is not observed. That is, such small MOSFETs undergo hard-breakdown.and immediate device failure.[92] Thus, being able to operate deep-submicron MOSFETs in a mode in which gate oxides do not undergo hard breakdown is not a likely scenario. This same study also demonstrated that hard- and soft-breakdown events are triggered by the same type of defect paths. The impact of soft breakdown on the reliability of ICs fabricated with deep-submicron MOSFETs is still being actively examined.

3.8.4 Impact of Polysilicon Depletion

Polysilicon depletion becomes a significant effect in MOSFETs when gate

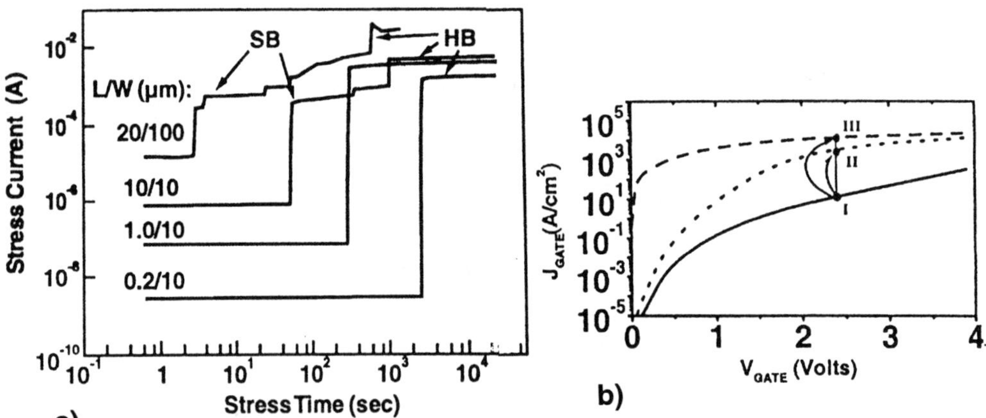

Fig. 3-41 a) Observed soft and hard breakdown in stress current curves vs. time for t_{ox} = 2.5-nm with V_G = 4.1 V at 140°C.[93] b) Oxide I-V characteristics before and after soft-breakdown.[94] (Pre-breakdown – solid line; Post-soft (short-dash); Post-hard (long-dash).

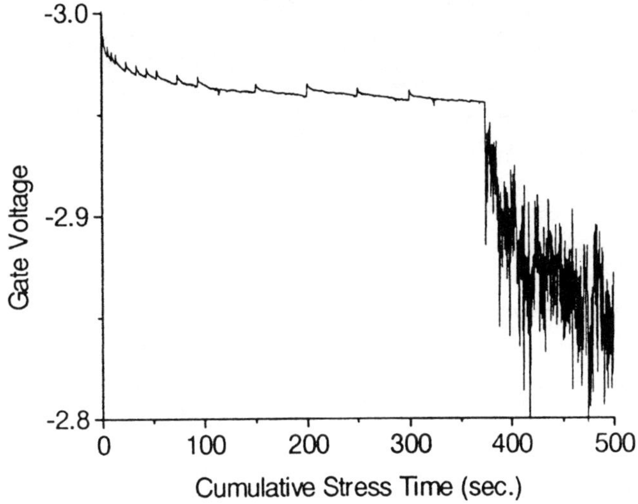

Fig. 3-42 Gate voltage measured during sequential –10 mA/cm^2 constant-current stresses of a 2.0-nm gate oxide. Soft breakdown occurs after about 360 seconds. Spikes are visible in the pre-breakdown trace because the stress was removed periodically.[90]

lengths are scaled below 0.25-μm. However, while the effect is widely discussed as being a problem, its physical basis on a fundamental level has not been carefully addressed in most texts. Thus, it is useful to explain the origin of this effect before describing its ramifications on the characteristics of deep-submicron MOSFETs.

While the polysilicon-depletion-effect has recently become a topic of attention in the technical literature, in truth it is an effect that *always* exists when a MOS-C (or MOSFET) is biased into depletion or inversion. Although the polysilicon-depletion-effect always occurs under such biasing conditions, it can be ignored by treating the polysilicon gate layer as being an ideal, equipotential region. In fact, this idealized perspective is *merely an approximation, that can only be invoked as long as doing so has insignificant effects on the device behavior.* Once the effect becomes significant, this idealization must be removed. (Note that most book authors are probably oblivious to the fact that this approximation is even invoked, as the correct energy-band diagram is only mentioned in one or two of the dozens of device physics texts that purport to treat MOSFETs.)

Let us consider an MOS-C which has an n^+-doped polysilicon gate and a p-doped silicon substrate. When a positive bias is applied between the gate and substrate, this causes depletion or depletion/inversion in the silicon substrate (depending on the bias voltage), and the charges induced in the silicon substrate by this biasing are *negative.* In any case, an equivalent *positive* charge must exist

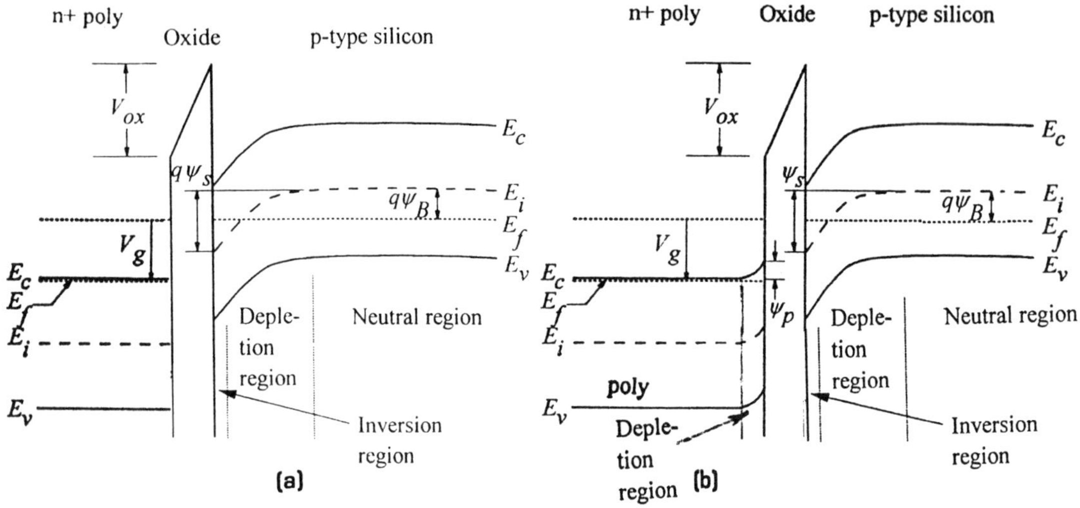

Fig. 3-43 a) Energy-band diagram of an NMOS-C biased to inversion which ignores the depletion-region formed in the polysilicon gate; b) Energy-band diagram of an NMOS-C biased to inversion showing the polysilicon depletion-region that actually always forms.

on the gate (polysilicon electrode). Donor ions in the n^+-doped poly constitute this positive charge (i.e., some of the mobile, negatively charged electrons leave the poly gate when a positive bias is applied to it). In other words, the biasing causes a *depletion region* of some finite thickness to be formed in the poly gate (at the poly/oxide interface). This depletion region in the poly is the basis of the *polysilicon-depletion-effect* (*PDE*). While most energy-band diagrams depict the situation of a MOS-C under depletion (or inversion) as in Fig. 3-43a, this is not strictly correct. That is, this diagram ignores the polysilicon depletion-region. The more correct energy-band-diagram is shown in Fig. 3-43b. Here, the presence of the poly depletion-region is indicated by band-bending (which implies the presence of an electric field) in the poly gate near the oxide.

The existence of an electric field means that a voltage-drop exists across the polysilicon depletion-region (which represents some fraction of the applied gate voltage). This means that the remaining voltage (i.e., the sum of the voltage across the gate oxide and the silicon substrate) is smaller than if no voltage was dropped across the poly depletion-region. The reduced remaining voltage will induce a smaller inversion-layer charge-density than if there was no voltage dropped across the polysilicon depletion-layer. (The effect is the same as if the gate oxide thickness was increased; i.e., PDE increases the "equivalent oxide thickness" $t_{ox_{eq}}$ – see Chap. 4, Sect 4.1). The decreased inversion-layer charge-density results in a degradation of the drain-current I_D of the MOSFET.

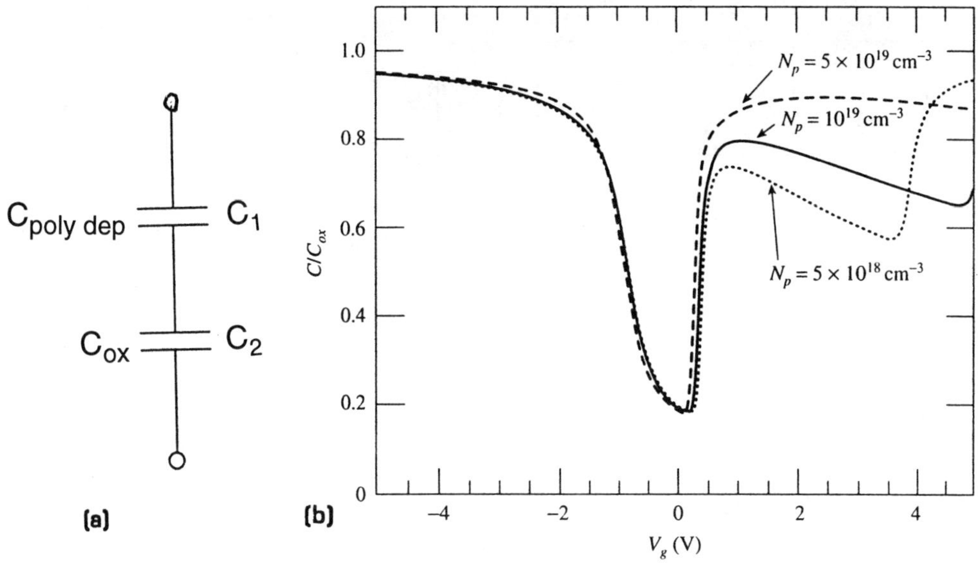

Fig. 3-44 a) Equivalent-circuit diagram of the polysilicon depletion-capacitance in series with the gate-oxide-capacitance. b) Low-frequency C-V curves of *p*-type MOS capacitor with an n^+-polysilicon gate doped at several different concentrations.[96]

The polysilicon-depletion-effect (PDE) becomes more significant as device scaling continues because the oxide capacitance increases as the oxide is made thinner, but the capacitance due to the polysilicon-depletion-layer remains constant (assuming there is no increase in the poly doping). At some point (i.e., at about when the gate-oxide thickness is scaled below 10.0-nm and the gate voltages are 2.5-V [or lower]), the depletion effect starts to adversely impact the device performance of MOSFETs made with polysilicon gates to a significant degree. We will now explain quantitatively why this is so.

The depletion-capacitance due to the PD-region is in series with the gate-oxide-capacitance (and the silicon-substrate depletion-region-capacitance, as well). An equivalent circuit diagram of the poly-depletion-capacitance in series with the gate-oxide-capacitance is shown in Fig. 3-44a.* The total capacitance of two capacitors in series is given by:

$$1/C_{tot} = (1/C_1) + (1/C_2) \qquad \textbf{(3.11)}$$

Thus, the smaller capacitor largely determines the total capacitance value. For the

* Note that although no capacitor electrodes physically exist at the poly/SiO_2 interface, such "virtual" electrodes are shown in Fig. 3-44a. This is nevertheless a valid equivalent-circuit representation because the two "virtual" electrodes each have the same amount of charge (although of opposite polarity). Thus, the effect of one cancels that of the other.

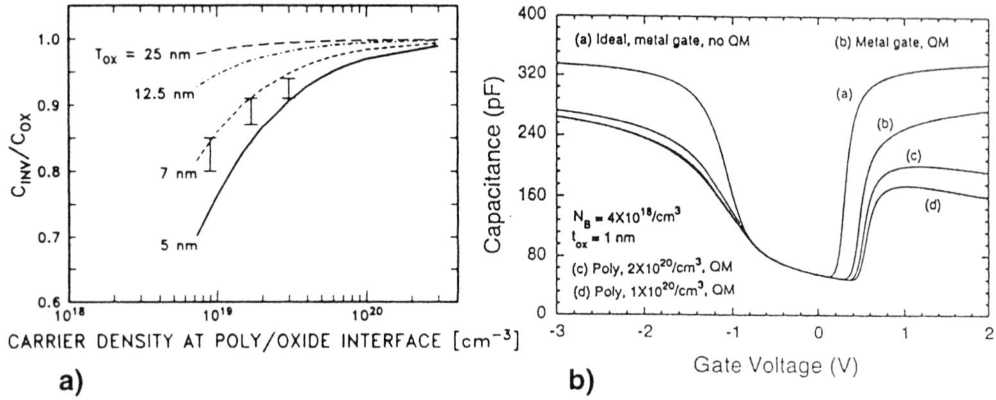

Fig. 3-45 a) Theoretical plot of C_{inv}/C_{ox} as a function of carrier concentration at the polysilicon/oxide interface for different gate oxide thickness.[101] b) Theoretical C-V curves for a thin gate oxide (1-nm-thick) including quantum mechanical effects and polysilicon depletion.[96]

case of the polysilicon-depletion-capacitance and the gate-oxide-capacitance, the gate-oxide-capacitance is much smaller than the polysilicon-depletion-capacitance until the oxide thickness gets below about 10.0-nm. For such thin oxides (assuming the poly layers doping remains constant), the oxide-capacitance starts to get large enough that the poly-depletion-capacitance can begin impacting the total capacitance. This is also shown in Fig. 3-44b in which the low-frequency C-V curves are shown for an MOS capacitor (with a 7-nm gate oxide, a *p*-substrate, and an n^+-poly gate with several doping concentrations).[86] It can be seen that the total capacitance at inversion (C_{inv}) does not return to the full oxide capacitance (i.e., the total capacitance in inversion is reduced by the PDE). The degree of reduction in C_{inv} depends on the poly doping concentration (Fig. 3-45a). This is because the higher the doping concentration, the narrower is the PD region, and thus the smaller is effect of PD. For the 7-nm gate-oxide thickness used in Fig. 3-44b, the poly-depletion effect becomes negligible only when the poly doping reaches about $1x10^{20}/cm^3$ (C_{inv} ~.98C_{ox}, see Fig. 3.45a).

As the oxide thickness gets smaller, the problem gets worse.[95] Figure 3-45b depicts computer simulated C-V curves for an MOS-C with a 1-nm-thick oxide, and an n^+-poly on a *p*-type substrate having a doping density of $4x10^{18}/cm^3$ (to allow for a V_T of several tenths of a volt). Curve a) is the C-V curve for an ideal (infinite conductivity) gate material, ignoring quantum effects. Curve b) is for a metal gate but including quantum effects. Curves c) and d) are for n^+-poly gates doped to $2x10^{20}/cm^3$ and $1x10^{20}/cm^3$, respectively, and including quantum effects. Curves c) and d) illustrate the effects of finite polysilicon doping.[96]

The above discussions indicate that if the polysilicon doping density could be increased without limit, the polysilicon depletion effect could be minimized. However, it appears that it will be very difficult to get electrically active doping densities above $10^{20}/cm^3$ for *n*-type polysilicon and above the mid- to upper-$10^{19}/cm^3$ for *p*-type polysilicon. This will result in reductions in drive-current capability that might not be acceptable.

To summarize, the impact of polysilicon-depletion on the minimum-oxide-thickness is the following: As the oxide is made thinner, the capacitance of the MOS gate structure is more impacted by the polysilicon-depletion-layer capacitance (assuming the doping in the poly remains constant). The effect is to reduce the total MOS-gate-capacitance to a lower value than if there was no poly depletion-layer-capacitance. For the same gate voltage, the potential at the Si surface is thus reduced, which in turn decreases the inversion-layer charge-density. Thus, the drain current actually decreases as the oxide is made thinner, causing the speed of the circuits made with such MOSFETs to decrease. Figure 3-46 shows how the propagation-delay in a ring oscillator increases with decreasing oxide thickness due to this effect.[84] This implies that it may not be appropriate to use the thinnest oxide thickness that is allowed by the 5-MV/cm breakdown limit.

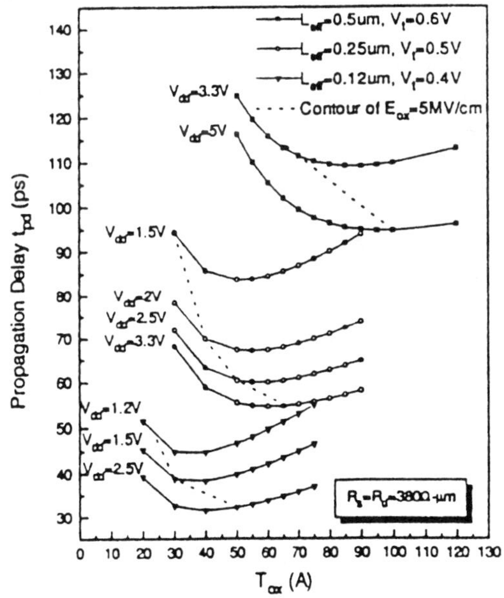

Fig. 3-46 At low V_{DD} (e.g., 2 V) and 0.2- μm, due to the polysilicon-depletion-effect, optimum speed may require a thicker t_{ox} than that allowed by the 5-MV/cm limit.[84]

3.8.5 Impact of Process Induced Damage

In 1996, Hu suggested that process -induced damage would have less impact on gate-oxide-reliability as oxides less than 10-nm in thickness were used.[78] That is, oxides that were 6.4-nm-thick showed an order of magnitude smaller interface-trap-density than did oxides that were 11.6-nm-thick (Fig. 3-47). A model that sought to explain this effect treats plasma charging-current as a constant-current-source, rather than a fixed voltage-source. Thus, for the same plasma process conditions, the same amount of current will pass through oxides as they are scaled below 10-nm, regardless of their thickness. In addition, the thinner oxide (e.g., 6.4-nm) is more tolerant of the same current stress than the thicker oxide (11.6-nm-thick). This data led Hu to conclude that plasma-process damage would not become a more serious issue with oxide scaling below 5-nm.

However, more recent information was not as optimistic. In 2000, at the 5th International Symposium on Plasma Process-Induced Damage, P.W. Mason of Lucent Technologies reported that plasma damage did, in fact, increase as the gate oxide thickness was decreased from 11.5-nm to 5.5-nm. The main damage source was the rf-backsputter via-clean process used to prepare vias for metal deposition. The plasma damage was found to correlate with long-term oxide reliability degradation, although not with short-term reliability problems. Another report from K.P. Cheung of Lucent at the same conference indicated that currently available measurement techniques cannot determine whether or not a low-level of plasma-induced damage is occurring.

The fact that plasma-induced damage in gate oxides may still remain a relia-

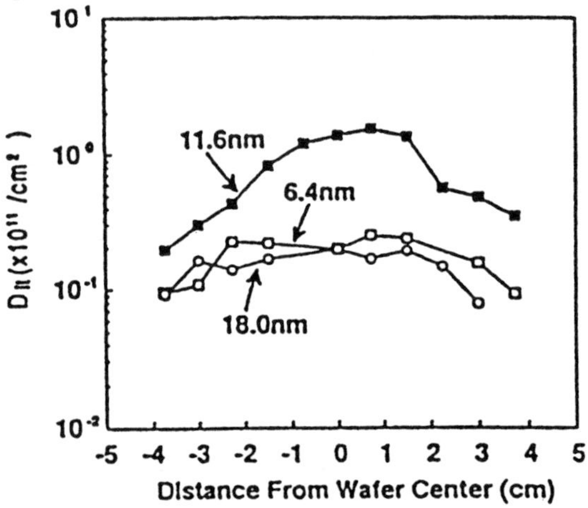

Fig. 3-47 Plasma-process-antenna-effect actually causes less damage to 6.4-nm oxides than to 11.6-nm oxides.[11]

bility problem, has led work to continue on ways to limit such damage. In July 2000 it was reported that using 3.0-nm-thick oxides nitrided in N_2O produced CMOS transistors which showed significantly improved resistance to charging damage, especially for PMOS devices.[85]

3.9 MEASURING ULTRA-THIN GATE OXIDES

Ultra-thin oxide films can be measured in a number of ways, as described in detail in Ref. 98. Two of the more common methods are *ellipsometry* and *high-resolution transmission electron microscopy* (HRTEM). Ellipsometry is a powerful and accurate optical method. The technique uses plane-polarized monochromatic light to illuminate the oxidized silicon surface at an angle. (The name *ellipsometer* comes from the use of elliptically polarized light to measure film properties.) An elliptically polarized laser beam is directed at the Si surface at an angle, and is then reflected from both the silicon substrate and oxide surface (Fig. 3-48). The technique makes use of the change of state of polarization when light is reflected from a surface. If the index of refraction (n_1) and the extinction coefficient (κ) of the substrate are known, and if the film is non-absorptive at the wavelength being used (so it's $\kappa = 0$), the state of polarization of the reflected beam then depends only on the thickness of the transparent film (and n_1).

The reflected beam passes through an analyzer drum and onto a detector. The analyzer drum is rotated to produce a minimum value in the light intensity reaching the photodetector. By reading the polarizer and analyzer settings, the

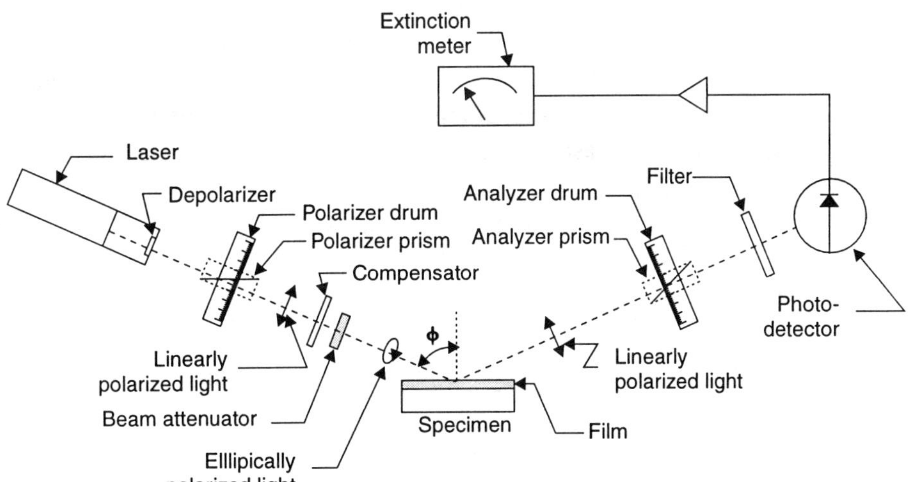

Fig. 3-48 Schematic of an ellipsometer.

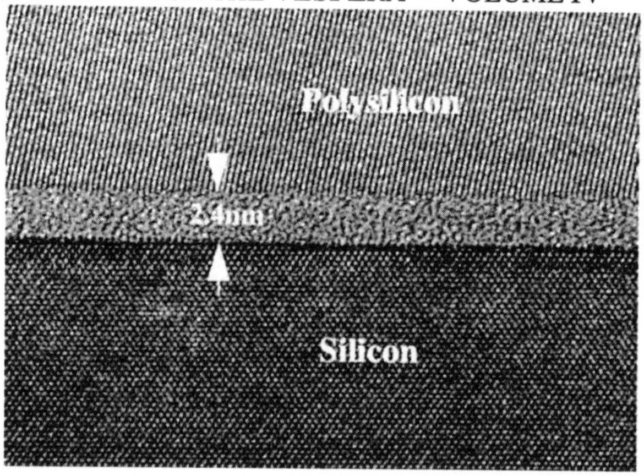

Fig. 3-49 HRTEM of 2.4 nm-thick oxide.[90] (© 1997 IEEE)

film thickness and index of refraction can be determined. The direct outputs of an ellipsometric measurement are the angles ϕ and ψ, which are related to the refractive index of the silicon substrate, the refractive index of the silicon oxide, and the thickness of the film.

If a single reflectance angle is used, multiple film thicknesses can produce a null, and the user must thus start with some idea of the expected film thickness to select the correct value. However, if multiple angles are used, an unambiguous single thickness value can be obtained. In any case, ellipsometry can be used to measure oxide thicknesses down the very thin regime (< 50-Å). However, inaccuracies in this regime may be present. Therefore, more complicated methods, such as multiple-wavelength ellipsometry or multiple-angle ellipsometry may be used to improve the accuracy of these measurements.[97] One recently described technique for measuring ultra-thin oxides involves a modification to conventional ellipsometry. In this method, an oxide *under* a polysilicon gate film is measured with a multiple-wavelength, multiple-angle, laser ellipsometer (the Rudolph Technologies S200-*ultra*® model). This allows the thickness of oxides fabricated *in-situ* in the single-wafer cluster tools described in Sect. 3.8.6, to be determined.[99]

Thin-oxide film-thickness can also be measured from micrographs obtained by HRTEM (Fig. 3-49). Preparing a sample for HRTEM requires the tedious tasks of slicing the wafer, mounting the slice in epoxy, and mechanically polishing and ion milling it to the proper thickness. However, unlike other techniques, no physical values such as index-of-refraction or dielectric constant are needed.

As gate-oxides scale to the ultra-thin regime, issues associated with measuring these films becomes increasingly critical. A transition region is thought to exist between the crystalline-silicon and bulk-silicon-dioxide that extends from several angstroms up to about 10-Å from the silicon. For ultra-thin oxide films this could constitute 20-50% of the film. The structural differences in the transition region can change the properties of the film, including its refractive index, static dielectric constant, band gap, and effective mass. These property changes can effect many measurement techniques used to determine the oxide thickness. Thus, it is necessary to examine the extent to which this interlayer must be included in optical models to obtain the required measurement precision for ultra-thin oxide films. Surface roughness will also play a role since many of the measurements assume an atomically smooth transition between silicon ad silicon dioxide. Leakage and tunnel currents through the oxide will also affect many metrology aspects, including capacitance and conductance measurements.

The presence of contamination on the surface of the dielectrics to be measured is another complicating factor. Such airborne contaminants can form an ad-layer on the surface (see Sect. 3.10.1). Under these circumstances, the ellipsometric response will include the effect of the ad-layer. To eliminate these effects, it is necessary to develop a reproducible and easily implemented cleaning procedure. The analysis of ultra-thin stacked dielectrics consisting of individual layers of oxide and nitride may also be made more complex by parameter correlation effects.

3.10 MANUFACTURING ULTRA-THIN GATE OXIDES IN EARLY-2001

3.10.1 Process Control Issues of Growing Ultra-Thin Gate Oxides

In the 2001 timeframe, gate oxides were still largely being grown with batch processes that employ vertical furnaces. Details of such batch oxide-growth processes (and the tools used to carry them out) are found in Vol.1, 2nd Ed., Chap. 7. It has been demonstrated that these types of processes can grow oxide films as thin as 2.5-nm, and with a uniformity of ±0.5-Å across a 200-mm wafer.[100]

However, when oxide thicknesses have to be made even thinner (≤ 2.0-nm), it becomes very difficult to maintain this kind of control using such conventional processing methods (i.e., in which the gate oxide is grown in an oxidation furnace and the polysilicon gate electrode is subsequently deposited in a separate processing tool). In the time between gate-oxide-formation and polysilicon-deposition, the gate oxide is exposed to the fab environment, allowing adsorption of *molecular airborne contamination* (MACs – see also Chap. 6).[102] Such MACs have been shown to have a negative effect on gate-oxide integrity, and they are also known to cause the measured thickness of the gate oxide to increase over time. In traditional batch processing, wafers may be stored for hours (or even days) after the gate oxide is grown, before the poly gate electrode is deposited.

Newer production methods thus form both the gate oxide and the polysilicon in the single vacuum environment of a cluster tool (with separate process chambers linked by an isolated vacuum robot that transfers wafers between chambers). This not only speeds the formation of the gate stack but also does not MACs the opportunity to become incorporated MACs between the two layers.[99]

3.10.2 Extending the Use of Ultra-Thin Silicon Dioxide Layers as Gate Dielectrics: Stacked Oxide Films Which Incorporate Nitrogen

As argued earlier, growing gate dielectrics with a thickness of < 2.0-nm in a manufacturing environment may not be possible, due to several limits: reliability; tunneling leakage-current; and boron penetration. However, work has continued to try and extend the use of oxides into this thickness regime. The main thrust in the 2001 timeframe for doing this involves the formation a dielectric *stack*, still mainly based on SiO_2. However, nitrogen is incorporated into the oxide such that it is present in several places: **1**) in a very-thin layer near the Si/SiO_2 interface at a concentration of < 1atom%; and **2**) in another very thin layer at the poly/SiO_2 interface, at a percentage of greater than 5 atom% (or, more commonly, in the form of a very thin SiON layer on top of the SiO_2 layer). To make such a gate-dielectric stack-structure effective, it is critical to be able to control the amount, position, and concentration-profile of the nitrogen throughout the stack.

Excess nitrogen in the bulk of the film may harm the oxide properties, and thus it is desirable that little (or none) exists in the bulk of the oxide. However, nitrogen at a concentration of < 1% close to the Si/SiO_2 interface has been found to be useful in strengthening the bottom Si/SiO_2 interface against hot-carrier degradation and interface-state generation, without degrading the electrical properties of the stack. On the other hand, a nitrogen concentration of at least 5 atom% is needed at the top oxide/poly interface to prevent boron penetration into the oxide bulk during subsequent high-temperature processing steps. Such boron penetration into the oxide (i.e., from the p^+-doped poly of deep-submicron PMOSFETs) has two detrimental effects: **1**) If not prevented from doing so, the boron can enter the Si substrate, causing unacceptable shifts in the V_T of the PMOSFETs; and **2**) If the boron does enter the oxide, but is prevented from entering the silicon substrate, it will nevertheless pile up in the oxide. Such boron-pileup in the oxide has been shown to degrade the oxide reliability, likely by causing defects that weaken the oxide structure.

Thermal oxidation with O_2, N_2O, and NO (as described in Sects. 3.7.1 and 3.7.2) can tailor the nitrogen profile such that reasonable control of the nitrogen density and profile at the Si/SiO_2 interface and within the bulk oxide can be maintained. However, it may not be possible to form a layer with enough nitrogen at the oxide/poly interface to prevent B penetration (i.e., > 5 atom% is needed) using thermal techniques in oxide films ≤ 2.0-nm thick. Instead, a low-temperature plasma-technique (e.g., using a remote nitrogen plasma) can be used to form a very-thin nitride film on the top oxide surface (actually a heavily-nitrided oxide-film is formed with this method).[103] Such thin SiON-layers may be enough not only to reduce or eliminate boron penetration, but also to increase the stacks permittivity. The result is that by adding an oxynitride film to the stack may cause an increase in its effective thickness of approximately 30%. This, in theory, may also make it possible to reduce the gate leakage-current. The nitride film formed by such a remote plasma nitridation process has also recently been shown to improve the reliability of the dielectric stack.[104] These improvements reportedly arise from reduced leakage and higher voltage acceleration factor.

3.10.3 Use of Ultra-Thin Oxide Layers Under High-*k* Gate-Dielectric Films

Even when it comes time to use high-*k* dielectrics in MOSFETs, an ultra-thin silicon oxide layer may still need to be formed on the silicon surface under the high-*k* dielectric film (to produce a reliable interface between the high-*k* material and silicon substrate). Such an SiO_2 layer would preserve the interface state characteristics and channel mobility. However, since the interfacial layer contributes to the equivalent oxide thickness, the oxide thickness will still have to be minimized the without compromising reliable interface properties. Thus, it seems that processing ultra-thin SiO_2 films may remain a part of silicon IC fabrication for yet a long time!

REFERENCES

1. S. Thompson *et al.,* "MOS Scaling: Transistor Challenges for the 21st Century," *Intel Technology Journal*, 3rd Quarter 1998.

2. D. Wolters and J.F. Verwey, "Breakdown and Wearout Phenomena in SiO_2 Films," Chap. 6, in *Instabilities in Silicon Devices - Vol. 1,* Ch. 1 Eds. G. Barbottin and A. Vapaille, (Elsevier) North-Holland, Amsterdam, 1986.

3. A.G. Revesz, "The Defect Structure of Grown Silicon Dioxide Films," *IEEE Trans. Electron Dev.,* **ED-12**, p. 97 (1965).

4. S. Rigo, "Silica Films on Silicon," in *Instabilities in Silicon Devices - Vol. 1,* Ch. 1 Eds. G. Barbottin and A. Vapaille, (Elsevier) North-Holland, Amsterdam, 1986.

5. I.E. Tamm, *Phys. Z. Sowjetunion* **1**, p. 733 (1932).

6. P. Roblin, *et al.*, "Simulation of Hot-Electron Trapping and Aging in *n*MOSFETs," *IEEE Trans. Electron Dev.,* **ED-35**, December 1988, p. 2229.

7. B.E. Deal, *J. Electrochemical Society.,* **121:** 198C (June 1974).

8. C. T. Sah, "The Evolution of the MOS Transistor," *Proceedings of the IEEE,* October 1988, p. 1280.
9. J.C. Burgoin, "Radiation-Induced Defects in Si/SiO_2 Structure," Ch. 17, *Instabilities in Silicon Devices,* Vol. 2, Ed. Barbottin, North Holland, 1989.
10. E. Harari, "Dielectric Breakdown in Electrically Stressed Thin Films of SiO_2," *J. Appl. Physics,* Vol. **49**, No. 4, p. 2478, 1978.
11. C. Hu, "Gate Oxide Scaling Limits & Projections," *Tech. Dig. IEDM* 1996, p. 319
12. R. Moazzami, *et al.*, "Temperature Acceleration of Time-Dependent Dielectric Breakdown," *IEEE Trans. Electron Dev.,* **ED-36**, p. 761, 1989.
13. I.C. Chen *et al.*, "The Effect of Channel Hot-Carrier Stressing on Gate Oxide Reliability of N-MOSFETs," *IEEE Trans. Electron Dev.,* **ED-35**, p. 2253, 1988.
14. K.F. Schuegraf and C. Hu, "Hole Injection SiO_2 Breakdown Model for Very Low Voltage Lifetime Extrapolation," *IEEE Trans. Electron Dev.,* **ED-41**, May 1994, p. 761.
15. J. Suehle and P. Chaparal, "Low Electric Field Breakdown of Thin SiO_2 Films Under Static and Dynamic Stress," *IEEE Trans. Electron Dev.,* 1997, p. 801.
16. J. McPherson and H. Mogul, "Underlying Physics of the Thermochemical *E*-Model in Describing Low-Field Time-Dependent Dielectric Breakdown in SiO_2 Thin Films," *J. Appl. Phys.* **84**, p. 1513, 1998.
17. Y. Nissan-Cohen, *et al.*, "Measurements of Fowler-Nordheim Tunneling Currents in MOS Structures under Charge Trapping Conditions," *Solid-State Electronics,* **28** (7) p. 717, (1985).
18. T.H. Ning and H.N. Yu, "Optimally Induced Injection of Hot Electrons into SiO_2," *J. Appl. Phys.*, vol. **45**, p. 5373, (1974).
19. C. Chang et al., "Carrier Tunneling Related Phenomena in Thin Oxide MOSFETs," *Tech. Dig. IEDM*, 1983, p. 194.
20. T. Sorch *et al.,* "Ultra-Thin, 1.0-3.0 nm Gate Oxides for High Performance Sub-100 nm Technology," *Dig. Tech. Papers IEEE Symp. VLSI Tech.* 1998, p. 222.
21. B.E. Deal and A.S. Grove, "General Relationship for the Thermal Oxidation of Silicon," *J. of Appl. Phys.* **36**, p. 3770, 1965.
22. T. Watanabe and I. Ohdoman, "Large-Scale Modeling of SiO_2/Si(001) Interface Structures by Using a Novel Inter-Atomic Interaction Model," *Abs. Mtg. Electrochem. Soc.,* Spring 2000, Abs. 473.
23. H.Z. Massoud, *et al.*, "Thermal Oxidation of Silicon in Dry Oxygen: Growth-Rate Enhancement in the Thin Regime I Experimental Results, II Physical Mechanisms," *J. Electrochem. Soc.,* **132**, 1985, p. 2685 and 2693.
24. A. Reisman *et al.*, "The Modeling of Silicon Oxidation from $1x10^{-5}$ to 20 Atmospheres," *J. Electron. Mat.* **16**, p. 45, (1987).
25. D.N. Kouvatsos and M.K. Hatalis, "Thin Silicon Dioxide Films Grown at Very Low Oxygen Partial Pressure in Oxygen Diluted by Helium," *Abs. Mtg. Electrochem. Soc.,* Fall 1992, p. 399.

26. P.J. Timans *et al* “Rapid Thermal Processing,” in *Handbook of Semiconductor Manufacturing Technology,* Y. Nishi and R. Doering, Eds., Chap. 9, pp. 201-86, Marcel Dekker, New York, 2000.
27. G. Eftekhari, *J. Electrochem. Soc.,* **140**(3), p. 787 (1993).
28. M.L. Green *et al., Proc. 6th Intl. Conf. on Advanced Processing of Semiconductors, RTP*, 1998, p. 72.
29. A. Joshi and D.L. Kwong, *IEEE Electron Dev. Letts.,* **12**(1), p. 28 (1991).
30. W. Kirklen Henson *et al.*, “Analysis of Leakage Currents and Impact on Off-State Power Consumption for CMOS Technology in the 100-nm Regime,” *IEEE Trans. Electron Dev,* **47**, July 2000, p. 1393.
31. M. Rasras *et al.*, “Photo-Carrier Generation as the Origin of F-N-Induced Substrate hole Current in Thin Oxides,” *Tech. Dig. IEDM,* 1999, p. 465.
32. P. Palestri *et al.*, “Impact Ionization and Photon Emission in MOS Capacitors and FETs,” *Tech. Dig. IEDM*, 2000, p. 97.
33. C. Lam *et al.*, “Anomalous Low Temperature Charge Leakage Mechanism in ULSI Flash Memories,” *Tech. Dig. IEDM*, 2000, p. 335.
34. J. McPherson *et al.*, “Comparison of E and 1/E TDDB Models for SiO_2 Under Long-Term/Low-Field Test Conditions,” *Tech. Dig. IEDM*, 1998, p. 171.
35. J.D. Bude *et al.*, “Explanation of Stress-Induced Damage in Thin Oxides,” *Tech. Dig. IEDM,* 1998, p. 179.
36. B.E. Weir, “Gate Oxides in 50 nm Devices: Thickness Uniformity Improves Projected Reliability,” *Tech. Dig. IEDM* 1998, p. 167.
37. J. H. Stathis and D.J. Maria, “Reliability Projection for Ultra-Thin Oxides at Low-Voltage,” *Tech. Dig. IEDM* 1998, p. 167.
38. S. Kuppurao *et al.*, “In Situ Steam Generation: A New Rapid Thermal Oxidation Technique,” *Solid State Technology,* July 2000, p. 233.
39. B. Doyle and A. Philipossian, "*p*-Channel Hot-Carrier Optimization of RNO Gate Dielectrics Through the Reoxidation Step," *IEEE Electron Dev. Letts.*, **EDL-14**, April 1993 p. 161.
40. H. Hwang *et al.*, "Electrical and Reliability Characteristics of Ultrathin Oxynitride Gate Dielectric Prepared by Rapid Thermal Processing in N_2O, *Tech. Dig. IEDM*, 1990, p. 421.
41. A. Uchiyama *et al.*, "High Performance Dual-gate Sub-Half-Micron CMOSFETs with 6 nm-thick Nitrided SiO_2 Films in an N_2O Ambient," *Tech. Dig. IEDM*, 1990, p. 425.
42. H. Fukuda, T. Arakawa, and S. Ohno, "Highly reliable thin Nitrided SiO_2 films formed by rapid thermal processing in an N_2O Ambient," *Jap. Jnl. of Appl. Phys.*, **vol. 29**, p. 2333, 1990.
43. *bid.* Ref. 40.
44. Z. Liu *et al.*, "The Effects of Furnace Annealing on MOSFETs," *Tech. Dig. IEDM,* 1992, p. 625.

45. S.C. Sun, T.S. Chen, and H.Y. Chang, "Thermal Oxidation of Silicon in Pure N_2O," *Ext. Abs. Mtg. Electrochem. Soc.*, Spring 1993, Abs. #749, p. 1080.
46. N.S. Saks *et al.*, "Characteristics of Oxynitrides Grown in N_2O," *Ext. Abs. Mtg. Electrochem. Soc.*, Spring 1994, Abs. #162, p. 251.
47. Y. Okada *et al.*, "Evaluation of Interfacial Nitrogen Concentration of RTP Oxynitrides by Reoxidation," *J. Electrochem. Soc.,* **140**, p. L87, June 1993.
48. H. Fukuda *et al.*, "Novel N_2O - Oxynitridation Technology for Forming Highly Reliable EEPROM Tunnel Oxide Films," *IEEE Electron Dev. Letts.*, **EDL-12** (1991) p. 587.
49. H. Hwang *et al.*, "Improved Reliability Characteristics of Submicrometer MOSFETs with Oxynitride Gate Dielectric Prepared by Rapid Thermal Oxidation," *IEEE Electron Device Letters*, **EDL-12** (1991), p. 495.
50. T.Y. Chu *et al.,* "Thickness and Compositional Nonuniformities of Ultrathin Oxides Grown by Rapid Thermal Oxidation of Silicon in N_2O, *J. Electrochemical Soc.*, **138**, June 1991, p. 113.
51. W. Ting *et al.*, "MOS Characteristics of Ultrathin SiO_2 Prepared by Oxidizing Si in N_2O," *IEEE Electron Device Letters*, **EDL-12** (1991), p. 416.
52. G.Q. Lo *et al.,* "Improved Performance and Reliability of MOSFETs with Ultrathin Gate Oxides Prepared by Conventional Furnace Oxidation of Si in Pure N_2O Ambient," *Symp. VLSI Tech.* p. 43, 1991.
53. J. Ahn, W. Ting and D.-K. Kwong, "Furnace Nitridation of Thermal SiO_2 in Pure N_2O Ambient for ULSI," *IEEE Electron Dev. Letts.*, **EDL-13**, 1992, p. 117.
54. H.G. Pomp *et al.*, "Lightly N_2O nitrided dielectrics grown in a conventional furnace for E^2PROM and 0.25 μ m CMOS, *Tech. Dig. IEDM*, 1993, p. 463.
55. G.Q. Lo, J. Ahn, and D.L. Kwong, "Improved Hot-Carrier Immunity in CMOS Analog Device with N_2O-Nitrided Gate Oxides," *IEEE Trans. Electron Dev.*, **EDL-13**, September 1992, p. 457.
56. M.S. Soo *et al.*, "Very Light Nitridation of Thin Gate Oxide in Low Pressure N_2O Ambients," *Ext. Abs. Mtg. Electrochem. Soc.*, Spring 1994, Abs. #168, p. 262.
57. Y. Okada *et al.*, "The Performance and Reliability of 0.4 Micron MOSFETs with Gate Oxynitrides Grown by Rapid Thermal Processing Using Mixtures of N_2O and O_2," *IEEE Trans. Electron Dev.*, **ED-41**, p. 191, February 1994.
58. G.Q. Lo *et al.*, "*p*-Channel MOSFETs With Ultrathin N_2O Gate Oxides," *IEEE Electron Device Letters*, **EDL-13** (1992), p. 111.
59. Y. Okada *et al.,* "Evaluation of Interfacial Nitrogen Concentration of RTP Oxynitrides by Reoxidation," *J. Electrochem. Soc.,* **140**, June 1993, p. L87.
60. R.I. Hegde *et al.,* "Surface and Interface Roughness of Ultrathin Nitric Oxide Oxynitride Gate Dielectric," *J. Electrochem. Soc.,* January 1998, p. L13.
61. C. Gerardi *et al.,* "Nitridation of Thin Gate or Tunnel Oxides by Nitric Acid," *J. Electrochem. Soc.* , August 1999, p. 3958.
62. P. Tobin *et al., J. Appl Phys.,* **75**, 1811 (1994).

63. S. Haddad and M.S. Liang, "Improvement of Thin GOI Using Through-Silicon-Gate Ion Implantation," *IEEE Electron Dev. Letts.*, **EDL-8**, p. 58, 1987.
64. G.D. Leonarduzzi and D.-L. Kwong, "Improving Performance with Oxynitride Gate Dielectrics," *Semiconductor International*, July 1998, p. 225.
65. R.I. Hegde *et al., J. Electrochem. Soc.,* **144**, 1081 (1997).
66. S.V. Hattangady *et al., Tech. Dig. IEDM* 1996, p. 495.
67. D. Kapila *et al.,* "Modeling and optimization of Oxynitride Gate Dielectric Formation by Remote Plasma Nitridation of Silicon Dioxide," *J. Electrochem. Soc.,* **146**, 1111 (1999).
68. P.E. Nicollian *et al.,* "Extending the Reliability Scaling Limit of SiO_2 through Plasma Nitridation," *Tech. Dig. IEDM,* 2000, p. 545.
69. E. F. Da Silva, *Tech. Dig. IEDM*, 1987, p. 848.
70. P. Wright *et al.*, "Hot-Electron Immunity of SiO_2 Dielectrics with Fluorine Incorporation," *IEEE Electron Device Letters*, **EDL-10**, August 1989, p. 347.
71. V. Jain, *et al.*, "Improved Sub-micron CMOS Device Performance in CVD Tungsten Silicide," *1991 Symp. VLSI Tech.*, p. 91.
72. Y. Nishioka *et al.*, "Hot-Electron Hardened Si-Gate MOSFET Utilizing F Implantation," *IEEE Electron Device Letters*, **EDL-10**, April 1989, p. 141.
73. N. Kasai, P. Wright, and K.C. Saraswat, "Hot-Carrier Degradation Characteristics for Fluorine Incorporated NMOSFETs," *IEEE Trans. Electron Dev.*, **ED-37**, June 1990 p. 1426.
74. C.-C. Chen *et al.*, "Improved Ultrathin Gate Oxide Integrity in p^+-Polysilicon-Gate p-Channel MOS with Medium-Dose Fluorine Implantation," *Electrochem. & Solid-State Letts.*, **3**(6), June 2000, p. 290.
75. C. Lam *et al.,* "Anomalous Low Temperature Charge Leakage Mechanism in ULSI Flash Memories," *Tech. Dig. IEDM*, 2000, p. 335
76. K.G. Reid *et al.,* "Dilute Steam Rapid Thermal Oxidation for 30Å Gate Oxides," *Ext. Abs. Mtg. Electrochem. Soc.*, Spring 1999, Abs. #489, Seattle WA.
77. S.C. Everist, "Characterization of Oxynitride Dielectric Films Grown in Nitric Oxide:Oxygen Mixtures by Rapid Thermal Oxinitridation," *Abs. Mtg. Electrochem. Soc.*, Spring 200, Abs. #455, Toronto, Canada.
78. C. Hu, "Gate Oxide Scaling Limits & Projection," *Tech. Dig. IEDM* 1996 p. 319.
79. T. Ghani *et al.,* "100-nm Gate Length High Performance/Low-Power CMOS Transistor Structure," *Tech. Dig. IEDM* 1999, p. 415.
80. T.N. Nguyen, *Internat. Reliability Physics Symp.* 1987, p. 66.
81. M. Moazzami *et al., IEEE Trans. Electron Dev.* p. 1643, 1990.
82. H. Massoud *et al., 5th Intl. Conf. Adv. Thermal Process of Semiconductors*, p. 3.
83. M. Moazzami *et al.,* "Stress-Induced Current in Thin Silicon Dioxide Films," *Tech. Dig. IEDM,* 1992. p. 139.
84. C. Chen, *et al., IEEE Electron Dev. Lett.,* p. 202, 1996.
85. C.- C. Chen *et al., IEEE Trans. Electron Dev.,* July 2000, p. 1355.

86. R. Rios and N.D. Arora, "Determination of Ultra-Thin Gate Oxide Thickness for CMOS Structures Using Quantum Effects," *Tech. Dig. IEDM* 1994, p. 613.

87. Caillet *et al., Dig. Tech. Ppers, Symp. VLSI Technology*, 1999, p. 89-90.

88. H.R. Soleimani *et al.*, "Formation of Ultrathin Nitrided SiO_2 Oxides by Direct Nitrogen Implantation into Silicon," *J. Electrochem. Soc.* **142**, p. L132, 1995.

89. C.T. Liu *et al.*, "Light Nitrogen Implant for Preparing Thin-Gate Oxides," *IEEE Electron Device Letters,* **18**, March 1997, p. 105.

90. B.E. Weir *et al.*, "Ultra-Thin Gate Dielectrics: They Breakdown, But Do They Fail?" *Tech. Dig. IEDM* 1997, p. 73.

91. W. Mizubayashi *et al.*, "Soft Breakdown Mechanism in Ultrathin Gate Oxides," *Abs. Mtg. Electrochem. Soc.* Spring 2000, Abs. 481.

92. J. Sune *et al.*, "Understanding Soft and Hard Breakdown Statistics, Prevalence Ratios, and Energy Dissipation During Breakdown Runaway, *Tech. Dig. IEDM* 2001, p. 117.

93. E. Wu *et al.*, "Structural Dependence of Dielectric Breakdown in Ultra-Thin Gate Oxides and Its Relationship to Soft Breakdown Modes and Device Failure, *Tech. Dig. IEDM* 1998, p. 187.

94. M.A. Alam *et al,* "A Study of Soft and Hard Breakdown Part I (p. 232), and Part II (p. 239), *IEEE Trans. Electron Dev.* February 2002.

95. C.-H. Choi *et al.*, "Gate Length dependent Polysilicon Depletion Effects," *IEEE Electron Device Letts.*, **23**, April 2002, p. 224.

96. W.C. Holton *et al.*, "Overview of Semiconductor Devices," Ch. 1 in *Handbook of Semiconductor Manufacturing Technology,* Y. Nishi and R. Doering, Eds., Marcel Dekker, New York, 2000, p. 1.

97. H.G. Tomkins, *Spectroscopic Ellipsometry and Reflectometry: A User's Guide,* John Wiley & Sons Inc., New York, 1999.

98. C. Hayzelden, "Gate Dielectric Metrology," Chap. 2, pp. 17-47 in *Handbook of Silicon Semiconductor Metrology*, Ed. A. Diebold, Marcel Dekker, 2001.

99. D. Leet *et al.*, "Process Metrology for Ultrathin Gate Dielectrics," *Semiconductor International,* August 2001, p. 79.

100. D.A. Buchanan, "Scaling the Gate dielectric: Materials, Integration, and Reliability," *IBM J. Res. & Dev.*, Vol. **43**, No. **3**, July 21, 1999, p. 265.

101. C.Y. Wong *et al.*, "Doping of n^+ and p^+ Polysilicon in a Dual-Gate CMOS Process, *Tech. Dig. IEDM* 1988, p. 238.

102. MM. Heyns *et al.*, "Cost-Effective Cleaning and High-Quality Thin Gate Oxides, *IBM J. Res. & Dev.*, Vol. **43**, No. **3**, July 21, 1999, p. 293.

103. Y. Wu *et al.,* "Performance & Reliability of PMOSFETs with Stacked Gate Dielectrics with Nitrided Si/SiO_2 Interfaces Prepared by Remote PECVD," *IEEE Trans. Electron Dev.* July 2000, p. 1361.

104. P.E. Nicollian *et al.*, "Extending the Reliability Scaling Limit of SiO_2 Through Plasma Nitridation," *Tech. Dig. IEDM* 2000, p. 545.

Chapter 4

HIGH-*k* DIELECTRICS

For IC applications, high-*k* dielectrics are defined as those with a dielectric constant greater than that of silicon nitride (i.e., > k = 7). High-*k* dielectric materials are being investigated for use in three main IC applications: **1**) as a gate dielectric material to replace SiO_2 in advanced MOSFET structures; **2**) as a dielectric material in the capacitor structures of dynamic RAMS (DRAMs); and **3**) as a dielectric in the capacitor structure of non-volatile ferroelectric random access memories (FeRAMs). In this chapter we will primarily focus on the first two of these applications. That is, since FeRAMS are not yet a mainstream IC product, we will only give a brief description of high-*k* FeRAM structures.

There are currently a variety of materials with dielectric constants greater than that of SiO_2 being studied as such high-*k* replacement dielectrics. These include: silicon nitride (Si_3N_4); aluminum oxide (Al_2O_3), tantalum pentoxide (Ta_2O_5); zirconium oxide (ZrO_2); hafnium oxide (HfO_2), barium strontium titanate (BST), and lead-lanthanum-zirconium-titanate (PLZT - $PbLaTiO_3$). Table 4-1 lists the dielectric constant and bandgap of these materials.

4.1 HIGH-*k* DIELECTRICS AS THE GATE DIELECTRIC IN MOSFETS

Let us begin by describing the need for high-*k* dielectrics in the first of the above three applications, namely, as a gate dielectric in advanced MOSFET structures. Metal-oxide-semiconductor field-effect transistors (MOSFETs) have been continually scaled since they were first introduced in the early 1960s. The gate dielectric employed in all of these scaled generations has been silicon dioxide (SiO_2). As each generation was scaled, this gate oxide also had to be made progressively thinner. For the most advanced CMOS technologies being fabricated in 2001 (0.13-μm), the gate oxides had thicknesses of ~3.0-nm. According to the *International Technology Roadmap for Semiconductors - 1999* (ITRS-99),[1] when MOSFET devices used in CMOS are scaled to gate lengths below 0.1-μm (100-nm), they will require gate-oxide-equivalent thicknesses, t_{oxeq} < 1.5 nm. However, because direct tunneling increases exponentially with decreasing thickness, it may not be feasible to use SiO_2 films with thicknesses

TABLE 4-1 IMPORTANT HIGH-*k* DIELECTRIC MATERIALS

	Dielectric Constant	Bandgap (eV)	Crystal Structure
PbLa $ZrTiO_3$	1000		
$BaSrTiO_3$	300		
Ta_2O_5	26	4.5	Orthorhombic
ZrO_2	25	7.8	Monoclinic
HfO_2	24	5.7	Monoclinic
Al_2O_3	9	8.7	Amorphous
Si_3N_4	7	5.1	Amorphous
SiO_2 (Reference)	3.9	8.9	Amorphous

smaller than 1.5 nm. Thus, it appears that SiO_2 may become unsuitable as the gate insulator in sub-100-nm MOSFETs if future device scaling requirements are to be met. Of course, process engineers will do everything possible to keep using SiO_2 for this role. This is because SiO_2 is compatible with silicon (and polysilicon) and is well behaved and well understood. Furthermore, it has also been used as the gate dielectric since the inception of MOSFET technology some 40 years ago. Finally, it has been widely predicted that the substitution of high-*k* dielectrics for SiO_2 as a gate dielectric will be markedly difficult due to intrinsic chemical bonding effects and difficult process integration challenges.[2] *Some have predicted that the greatest challenge to the continued scaling of the MOS transistor may be the replacement of SiO_2 with a high-k dielectric!*

In any case, when (and if) SiO_2 is replaced, the replacement material will have to be thicker than oxide to reduce the direct-tunneling current and yet maintain the same MOSFET drive-current capability. This can be achieved by using an insulator that has a higher dielectric constant ($\varepsilon_{high\text{-}k}$) than does oxide ($\varepsilon_{ox} = 3.9$). That is, it must be possible to achieve the same equivalent thickness as required for SiO_2, where the oxide equivalent thickness t_{oxeq}, is given by:

$$t_{oxeq} = t_{high\text{-}k}\,(\varepsilon_{ox}/\,\varepsilon_{high\text{-}k}) = t_{high\text{-}k}\,(3.9/\varepsilon_{high\text{-}k}) \qquad (4.1)$$

For example, a high-*k* gate insulator with a thickness $t_{high\text{-}k}$ = 10-nm and a dielectric constant ($\varepsilon_{high\text{-}k}$) of 20, will behave in the same way in a MOSFET as an SiO_2 layer with a thickness of ~2.0 nm (see also Fig. 4-1).

Of course, an interfacial layer of SiO_2 (or some other lower-*k* material) will limit the highest possible gate stack capacitance or, equivalently, the lowest achievable t_{oxeq} value. That is, when a gate-dielectric structure consists of a stack of several dielectric materials, the layer with the lowest-*k* value dominates the overall capacitance and also sets a limit on the minimum achievable $t_{ox_{eq}}$ value.

The capacitance of two dielectrics in series is given by:

$$1/C_{tot} = (1/C_1) + (1/C_2) \quad \textbf{(4.2)}$$

For example, a dielectric stack structure with an SiO_2 bottom layer and a top layer having a higher-*k* value, will thus have a $t_{ox_{eq}}$ that is calculated from:

$$t_{ox_{eq}} = t_{SiO_2} + t_{high\text{-}k} (\varepsilon_{ox}/\varepsilon_{high\text{-}k}) \quad \textbf{(4.3)}$$

Since the minimum achievable $t_{ox_{eq}}$ value will never be less than that of the lower-*k* layer, a bilayer approach necessarily compromises the performance of the high-*k* dielectric film.

4.1.1 Integrating High-*k* Dielectrics into MOSFET Structures

The use of high-*k* dielectrics as the gate dielectric in submicron MOSFET structures also improves other device characteristics. First, as a MOSFET is scaled below about 0.3 microns, the threshold voltage V_T begins to decrease as the gate length is decreased (for a constant channel doping value and gate dielectric thickness). This is seen in Fig. 4-2a, where the effect of channel length reduction on the threshold voltage value is plotted for deep-submicron MOSFETs (i.e., gate lengths smaller than 1.0-μm) fabricated with gate dielectric films of

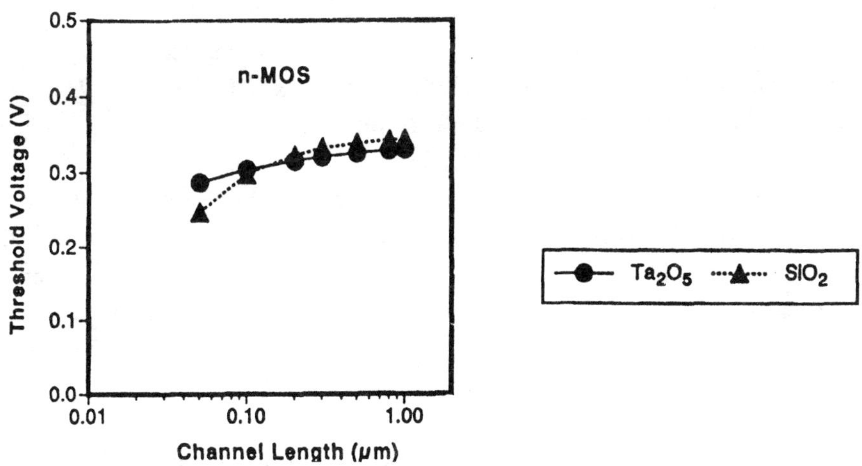

Fig. 4-1 As shown in this hypothetical transistor gate stack, a high-*k* material can be physically thicker yet provide an oxide-equivalent thickness that is electrically thinner ($t_{ox_{eq}}$ = 10-Å for both) than would be possible using oxynitride dielectrics.

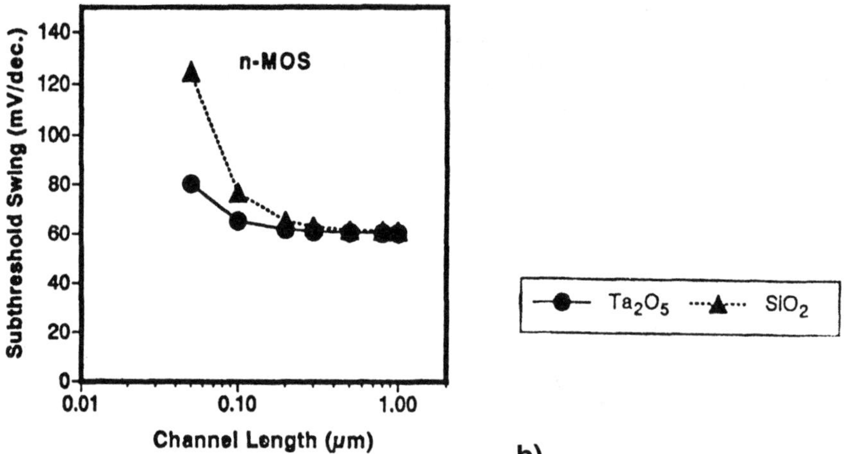

a) b)

Fig. 4-2 a) The effect of channel-length-reduction on the threshold voltage V_T for deep-submicron MOSFETs fabricated with SiO_2 and Ta_2O_5 as the gate dielectric. The use of Ta_2O_5 reduces the V_T degradation with channel length scaling.[3] b) The effect of channel length reduction on the subthreshold swing S_t for deep submicron MOSFETs fabricated with SiO_2 and Ta_2O_5 as the gate dielectric. The use of Ta_2O_5 allows an improved S_t value with channel length scaling.[3]

SiO_2 and Ta_2O_5. This data shows that for a specific gate length (e.g., 0.1-μm), the change in V_T from that of a long channel MOSFET (e.g., with a length > 0.5 μm, in this example) is greater for MOSFETs built with a gate oxide than with a Ta_2O_5 gate dielectric.

The second improvement in device performance observed when a high-*k* gate dielectric is used involves the subthreshold-swing of a MOSFET, S_t. This is a measure of how effectively a MOSFET is turned off as the gate voltage is decreased below V_T (see Vol. 3 for more information on S_t). As shown in Fig. 4-2b, a value of 80 mV/decade can still be obtained for MOSFETs built with a high-*k* gate dielectric (e.g., Ta_2O_5). Note that S_t values smaller than 100 mV/decade are considered very good, and the smaller the S_t value the better. It has been shown that for a dielectric thickness of 5-nm, the optimum dielectric constant is 25, which corresponds to the *k*-value of Ta_2O_5.[3] For MOSFETs, there is a specific dielectric constant for a given dielectric thickness for which the circuit performance is optimum.

One of the drawbacks of using high-*k* dielectrics is that if they are deposited directly on Si they tend to have a high-level of interface states, D_{it}. The interface-state-density can be reduced by growing a thin layer of oxide (1-2 atomic layers) on the silicon to passivate the surface. We will describe this issue and other

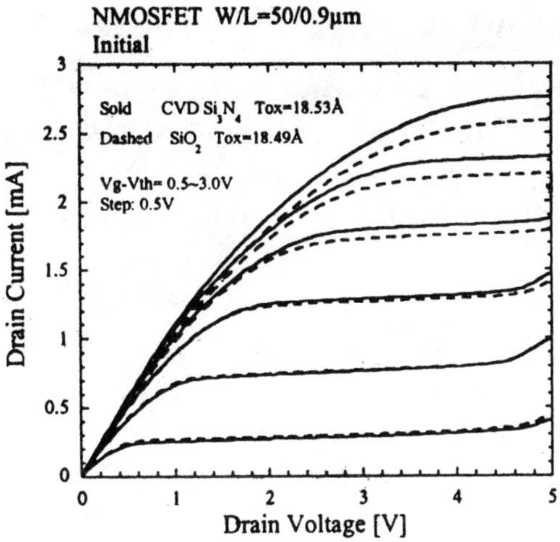

Fig. 4-3 The current-voltage (I_D-V_{DS}) characteristics of a 0.9 μm NMOSFET using Si_3N_4 and SiO_2 as the gate dielectrics.[4]

processing challenges related to the deposition and etching of various high-k films with respect to their suitability as alternatives to SiO_2 as a gate dielectric.

Note that initial results of using an alternative dielectric instead of SiO_2 in MOSFETs has produced some promising results, although none are clearly yet ready for full-scale IC production. Several examples of MOSFETs fabricated with alternative gate dielectrics have been reported. Figure 4-3 compares the current-voltage (I_D-V_{DS}) characteristics of a 0.9-μm NMOSFET using a 20-nm-thick Si_3N_4 gate dielectric to one using a 20-nm-thick SiO_2 gate dielectric. The device structure using the RTCVD-Si_3N_4 gate dielectric exhibits greater drain current.[4] A second MOSFET used a thermal-CVD process to deposit a Ta_2O_5 gate dielectric. The current-voltage characteristics of 150-nm MOS transistors using such a 4.0 nm-thick Ta_2O_5 film showed an improved drain current value compared to those fabricated with a gate oxide.[5] A third MOSFET used a Ta_2O_5/SiO_2 gate dielectric with a 2.8-nm equivalent-oxide thickness.[6] A fourth MOSFET used a gate dielectric stack consisting of $SiON/Ta_2O_5/TiN$.[7]

According to some researchers,[54] the ideal gate dielectric stack may end up having an interface consisting of a few atomic layers of Si-O (and possibly N), containing a pseudo-binary oxide. This oxide layer could preserve the critical high-quality characteristics of the SiO_2 interface while providing a higher-k value. Then a different, higher-k material could be used on top of this interfacial layer.

4.2 HIGH-*k* DIELECTRICS AS THE CAPACITOR DIELECTRIC IN DRAMS

A DRAM capacitor is a stack of materials consisting of a dielectric sandwiched between two electrodes. Since the invention of the DRAM nearly 30 years ago, either SiO_2 with a dielectric constant (k) of 4, or SiN with $k = 7$, or a combination of these materials has been used as the capacitor dielectric. Polysilicon is used as both the bottom and top electrode. Structures containing these materials are therefore called *silicon-insulator-silicon capacitors* (SIS).

As progress is made to put more memory bits on a chip, the surface area of silicon available for each individual memory cell in a DRAM shrinks rapidly (Fig. 4-4). At the same time operating voltages have been falling from 5V a few years ago, to below 3V today (and probably to about a volt in a few years). However, the charge-storage-capacity for each cell remains at about 30 fF/cell (corresponding to about $1\text{-}2 \times 10^6$ electrons/cell). There are three reasons that the amount of stored charge per cell has not been decreased. First, this charge-storage-capacity is needed to prevent the occurrence of soft errors. Soft errors are caused by alpha particles emitted from naturally occurring trace radioactive impurities in packaging materials or from cosmic rays. Such energetic particles generate minority carriers in the silicon, whose charge can alter memory states. By maintaining the capacitance-per-cell at about 30 fF/cell, the soft-error-rate (SER) can be kept acceptably low. Second, as DRAMs have evolved, the detection capability of the DRAM sense amplifiers has remained about the same. Thus, each cell still needs to have about the same charge-storage-capacity to enable the state of the cell to be read. Third, the data-retention time has remained constant, again requiring about the same charge-storage-capacity per cell.

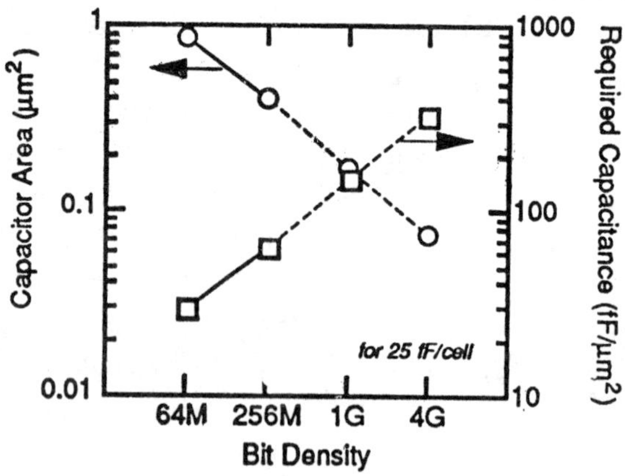

Fig. 4-4 Capacitor area and the required capacitance density as a function of DRAM bit density.[27] (© IEEE)

Until recently, three approaches have been used to maintain this constant value of charge storage per cell. The first involves the use of three-dimensional structures to increase the cell surface area (such as trench capacitors and stacked capacitors). The second involves using ever-thinner layers of dielectric in the capacitor. The latter approach entails increasing the surface area of the capacitor by about a factor of 2x by roughening the surface of the polysilicon bottom electrode. This is achieved by growing the bottom-electrode polysilicon with larger grains, and the resulting films are called *hemispherical silicon grain* (HSG) or *rough-surface polysilicon* (RSP).

However, as the silicon area per cell has continued to decrease, the three-dimensional structures can no longer be scaled further to increase surface area. Thus, in the face of shrinking capacitor area and reduced capacitor voltage, even thinner layers of SiO_2 (or multilayers of *oxide/nitride/oxide* - ONO) have been implemented. Nevertheless, in ONO dielectrics, the minimum thickness is about 50-Å (5-nm). In thinner films, tunneling currents produce excessive leakage. However, the combination of such ONO films, together with HSG is currently used in 0.18-μm and 0.15-μm DRAM production.

But, as memory chips surpass about 256 MB this approach will no longer be adequate because unacceptably high values of leakage current occur. Furthermore, the storage capacitors suffer premature catastrophic breakdown, even with reduced operating voltages. Thus, by examining the capacitor equation, the only remaining option for increasing the capacitance per unit area is to incorporate a dielectric with a higher *k*-value. This enables use of a capacitor dielectric with a large enough thickness that low leakage and high operating reliability to be maintained - while still providing sufficient charge storage with a continuously shrinking plate area and storage voltage. By using such high-*k* dielectrics, the DRAM cell structures could also be significantly simplified.

When the list of high-*k* dielectric candidates for DRAM capacitors is examined one notes that Al_2O_3 (with a *k*-value of 9), does not provide a sufficient increase in *k* to maintain the required charge storage/cell. However, Ta_2O_5, with a *k*-value of 25 can meet the requirement for the next generation or two of DRAMs. The IC industry has had some experience with Ta_2O_5, and it can be deposited in a variety of ways, including CVD, metal-organic CVD, sputtering, and anodization. However, Ta_2O_5 does not provide a sufficiently large improvement in charge storage for the memory generations above 4 Gb that are expected to begin production within the next few years (~2006). The long-term solution appears to lie in the various ferroelectric or near-ferroelectric dielectric materials, such as the various combinations of barium and strontium with titanate (e.g., barium/strontium-titanate - BST), lead/zirconium-titanate (PZT), and other perovskite metal oxides. These types of materials exhibit *k*-values from a few hundred to several thousand, enabling them, in principle, to provide sufficient capacitance for DRAMs into the foreseeable future. Such ultra-high-*k* materials can be deposited using almost any method by which a thin film can be formed (and they come in a wide variety of compositions and morphologies). The dielectric constants tend to be a strong function the specific elemental ratios and crystal structures which, in turn, are a function of the deposition method. The selection of a particular dielectric material and the design of the DRAM cell depend on a number of complex factors, including the process simplification leading to cost reduction.

There are, however, several challenging process integration issues involving the use of high-*k* materials in DRAM capacitor structures. First, note that in DRAM capacitors the bottom capacitor electrode is the *storage node* and the top capacitor electrode is termed the *plate electrode* (Fig. 4-5). Doped-polysilicon has traditionally been the material used to form the bottom electrode of the DRAM capacitor, and either polysilicon or metal has been used to form the plate electrode. (As noted earlier, if polysilicon is used for both electrodes, the capacitor is a called a *silicon-insulator-silicon* capacitor [SIS]. If the top electrode is made of metal, the structure is called a *metal-insulator-silicon* capacitor [MIS].) In any case, polysilicon has been used to form at least one of the capacitor electrodes. However, polysilicon has the problem that it tends to form a

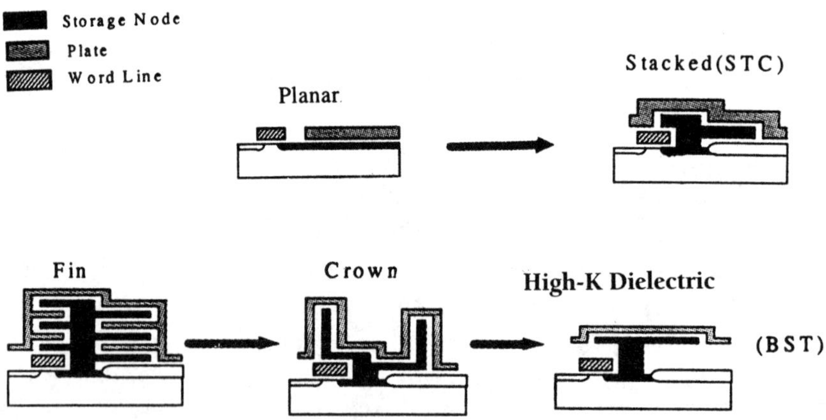

Fig. 4-5 The evolution of the capacitor-electrode structure in stacked-capacitor DRAMs.

silicon oxide layer at the polysilicon/high-*k* interface, either as a native oxide, or during the processing steps used to form or anneal the high-*k* dielectric layer (or even during the subsequent chemical-vapor-deposition of interlevel dielectric layers). These interfacial SiO_2 layers cause an increase in the SiO_2-equivalent thickness of the overall capacitor dielectric layer (which, in turn, reduces the cell capacitance). As a result, a number of other bottom-electrode materials are being considered for the manufacture of cell capacitors, including aluminum (Al), iridium (Ir), platinum (Pt), ruthenium (Ru), tantalum nitride (TaN), titanium nitride (TiN) and tungsten (W). Note that if *both* the bottom and top electrodes of the DRAM capacitor are made of metal, the structure is called a *metal-insulator-metal capacitor* (MIM). Table 4-2 shows the requirements of DRAM capacitors as the DRAM size increases above 1Gb/chip.

The Platinum Group Metals (PGM) - Pt, Ir, and Ru - have been actively studied due to their low resistivity, chemical corrosion resistance, thermal stability, and diffusion barrier properties. Each, however, has unique problems with respect to serving as an electrode. For example, Pt has adhesion and etch residue problems, while Ru forms RuO_4 as a toxic by-product of the etch process (and Ru is also susceptible to pinholes and notching because of its ease of oxidation). We discuss these issues in more detail in later sections. Note that PVD and CVD are the most common deposition methods for these materials. Each has its advantages and disadvantages. CVD is capable of much better conformality at lower processing temperatures. However, obtaining high-purity precursors of the PGMs is a rather difficult process. PGM deposition by CVD also presents a significant challenge in the development of volatile precursors.

Cost is another significant factor. For example, to sputter deposit a 50-nm (500-Å) film of Pt onto a 200-mm wafer with a 40% deposition efficiency, the cost is roughly $1000 per wafer. At 20,000 wafer-starts per month, this cost is

TABLE 4-2 DRAM Stack Capacitor Films Technology Requirements

Year	1999	2002	2005	2008	2011	2014
Technology Node	180 nm	130 nm	100 nm	70 nm	50 nm	35 nm
DRAM Capacity	1 Gb	4 Gb	8 Gb	24 Gb	64 Gb	192 Gb
Capacitor Structure	Cylinder MIM	Pedestal MIM	Pedestal MIM	Pedestal MIM	Pedestal MIM	Pedestal MIM
Capacitor Dielectric Material	Ta_2O_5	Ta_2O_5	BST	Epi-BST	???	???
Capacitor Dielectric Constant	22	50	250	700	1500	1500

is significant. In CVD, only 20% of the precursor is deposited in the chamber, and roughly 80% of the precursor is exhausted as waste. After subsequent processing, on the order of only 1% of the original material remains on the finished wafer. The remaining deposited-99% is removed by plasma etch or chamber cleaning. For expensive metals, this remaining 99% must be reclaimed.[8] One report describes an MOCVD deposition process for Pt films.[9]

4.3 HIGH-*k* DIELECTRICS FOR FeRAM APPLICATONS

The use of high-*k* dielectric materials in non-volatile memory FeRAMs is an emerging area. Ferroelectric dielectrics such as *lead zirconium titanate* (PZT - whose *k* value is 1000-1500), or *strontium bismuth titanate* (SBT), have a perovskite structure and are unique in that this crystal structure retains its "memory" without power. That is, since ferroelectric polarization retention is nearly perpetual, refresh is not needed. The capacitor structure in FeRAMs controls read/write speeds, data-storage lifetime, device layout, and packing density. Figure 4-6 shows a cross-section of an FeRAM cell and an example of a process flow used to incorporate such a cell into a conventional IC structure.[25] The early FeRAM device (manufactured by Ramtron Corp.) is being used in game cartridges, where memory is maintained without the use of a battery. The challenges of implementing ferroelectric dielectrics in FeRAMs are similar to those encountered in integrating high-*k* dielectrics into DRAMs. These include: controlling the interfacial reactions between the dielectric and electrode materials; maintaining thermal stability of the electrodes of the capacitor; and developing plasma-etching processes that provide critical dimensions of the devices. A review of the current status of FeRAMs has recently been published.[25]

4.4 TANTALUM PENTOXIDE (Ta_2O_5)

Tantalum pentoxide, Ta_2O_5, has a *k*-value of 25 and the IC industry has been developing techniques for forming thin films of this material since the early 1980s. It can be deposited in various ways including: **1**) LPCVD; **2**) metalorganic CVD; **3**) sputtering of Ta films followed by oxidation; and **4**) anodic oxidation of deposited Ta films. For use as a high-*k* dielectric layer in DRAMs, the most common deposition method of Ta_2O_5 films is LPCVD. Vendors that supply CVD tools to deposit Ta_2O_5 include Lam Research and Applied Materials.

4.4.1 Tantalum Pentoxide as a DRAM Capacitor Dielectric

We will discuss the use of Ta_2O_5 as a DRAM capacitor dielectric first because it

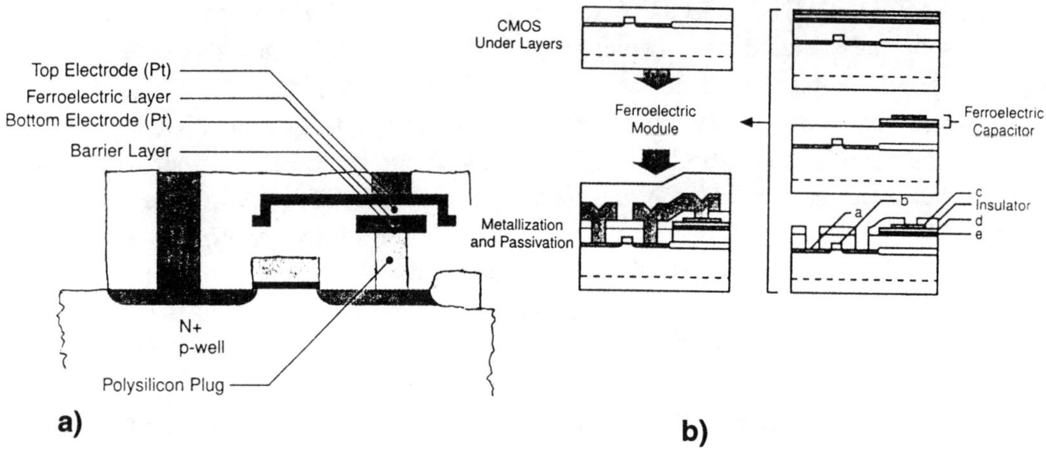

Fig. 4-6 a) The cross-section of a ferroelectric RAM cell (FRAM); and b) A typical FRAM process flow.[25]

is more likely to be implemented in this role than as a gate dielectric in MOSFETs. However, we will also discuss the application of Ta_2O_5 as a gate dielectric in the next section. The integration of Ta_2O_5 as a DRAM-capacitor-dielectric has been pursued because its k-value of 25 is high enough to result in adequately high charge-storage values per cell for DRAMs after the 256-Mb generation (i.e., perhaps for the 1-Gb and 4-Gb DRAM generations). In addition, Ta_2O_5 capacitors can simply replace ONO-based capacitors without major process integration problems. (The other change in the DRAM capacitor structure that accompanies the shift to a Ta_2O_5 dielectric is the replacement of the top plate polysilicon electrode with TiN – forming a TiN/Ta_2O_5/polysilicon structure [which is an MIS capacitor]). Here we describe the various processing issues associated with implementing Ta_2O_5 as a DRAM capacitor dielectric material. While Ta_2O_5 films exhibit many outstanding characteristics, their biggest drawback for DRAM applications is their relatively high leakage-current densities.

As noted above, the most common technique for forming Ta_2O_5 films for DRAM dielectrics is LPCVD.[10] Such films provide good step coverage. The film is formed by the oxygen-assisted thermal decomposition of an organometallic precursor in the presence of oxygen, typically tantalum ethylate (i.e., $Ta[OC_2H_5]_5 + O_2$), or tantalum tetraethoxy dimethylaminoethoxide (i.e. $Ta[OC_2H_5]_4[OCH_2CH_2N(CH_3)_2]$ TATDMAE and O_2) at 450°C, and at a pressure of 1-4 torr. Tantalum ethylate is also referred to as tantalum ethoxide – or TAETO. TAETO was the first precursor used for CVD of Ta_2O_5 and has become the standard one. Typically, a 10-nm-thick Ta_2O_5 layer is deposited, whose SiO_2-equivalent thickness (t_{oxeq}) is 2.5-nm.

Note that both of these precursors are liquid at room temperature, and so must first be vaporized at 170°C before being flowed into the CVD chamber with Ar as the carrier gas. In addition, the freezing point of TAETO is 18°C and that of TATDMAE is 9°C. Thus, it is a challenge to deliver a fully vaporized precursor to the heated wafer surface without any condensation or decomposition anywhere in the liquid delivery system. To circumvent this problem, TAETO is typically mixed with ethanol. TAETO with 2.5% ethanol has a freezing point of <10°C and thus does not require a heated ampule or heated liquid lines.

The deposition rate of TAETO in the 450-485°C range is reaction-rate limited and is thus sensitive to temperature. (The activation energy of the deposition process is 1.3 eV, which implies that the deposition rate doubles every 25°C in this region.) Above 485°C the process transitions into a mass-flow limited regime. Since excellent step coverage and film conformality are needed in DRAM capacitors, the deposition process must be carried out in the reaction-rate limited regime. Figure 4-7 is a SEM showing that excellent step coverage and conformality is achieved in such high-aspect-ratio structures as found in DRAM devices. The smaller SEMs in this figure show the film details at the top and bottom of the opening.

Since these films are deposited on the doped-polysilicon bottom electrode of the DRAM capacitor, it is undesirable to have an interfacial oxide form on the polysilicon either as a native oxide or during thermal process steps that are performed after the Ta_2O_5 layer has been deposited. (As noted earlier, such an interfacial oxide layer would cause the effective thickness of the capacitor dielectric layer to increase.) To eliminate the possibility of such an oxide layer from existing on the poly, a rapid-thermal nitridation (RTN) step is thus carried out on the polysilicon prior to depositing the Ta_2O_5 film (e.g., at 900°C for 60 sec in NH_3). This RTN step both nitrides any native oxide layer already existing on the poly, and the nitride layer that is thus formed acts as a diffusion barrier to oxygen, preventing any new oxide from forming on the poly during later processes.

The as-deposited Ta_2O_5 films, however, exhibit poor leakage characteristics. It is believed that this occurs because the as-deposited LPCVD-Ta_2O_5 films are oxygen deficient and amorphous. Thus, such Ta_2O_5 layers are densified using an anneal in O_2 to improve their leakage characteristics. Annealing in oxygen is

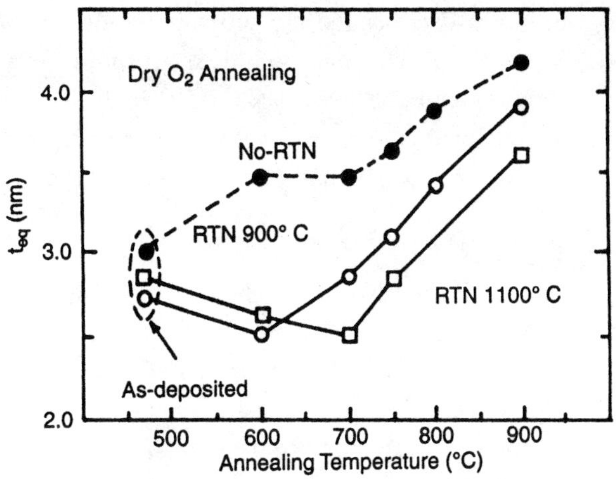

Fig. 4-8 An oxide grows on Ta_2O_5 during post-deposition O_2 annealing. This graph depicts the dependence of $t_{ox_{eq}}$ of Ta_2O_5 on the annealing temperature in a dry-O_2 atmosphere.[11]

thought to balance the stoichiometry of the films by filling vacancies in the undensified film, thus causing the leakage current to decrease.

The conventional oxygen-annealing method is performed in a furnace at 500°C in the presence of UV-generated O_3.[27] However, more recent methods use a rapid-thermal-process in oxygen (RTO). The RTO process minimizes the impact on the thermal budget and reduces the time during which oxygen can react with the substrate beneath the electrode. The oxygenation process in one approach is carried out at 600°C in an oxygen plasma and the temperature is then ramped quickly above 700°C to achieve crystallization of the Ta_2O_5 film.[26] The maximum temperature of the RTO step must still be kept below 800°C because the interfacial oxide under the Ta_2O_5 will grow too thick at higher temperatures (Fig. 4-8)[11] - making the SiO_2-equivalent thickness of the Ta_2O_5 film too great. (That is, at higher temperatures than 800°C, Si can diffuse through the SiN barrier on the polysilicon surface and react with the Ta_2O_5 to form SiO_2.) Up to a six-order-of-magnitude improvement in leakage current has been reported at 1V using such an RTO-annealing procedure.[26]

The densification step also crystallizes the amorphous as-deposited film, and this is also thought to reduce the leakage current through the Ta_2O_5 layers. Finally, since the precursor $Ta[OC_2H_5]_5$ is an organometallic material, the as-deposited Ta_2O_5 layer is not carbon-free. The densification step desorbs some of the hydrocarbon and water molecules, which also shrinks the film thickness. The decrease in film thickness also decreases the $t_{ox_{eq}}$ of the film.

Another issue involves keeping the top polysilicon electrode from forming a silicide with the Ta_2O_5 layer. This can be prevented by inserting a barrier layer between the Ta_2O_5 layer and the poly. Materials such as TiN and WN have reportedly be employed for this purpose. Another way around this problem is to use a plate electrode material other than polysilicon. For example, as mentioned earlier TiN (deposited by PVD), is typically used as a top (or plate) electrode material with Ta_2O_5.

In addition, even the nitride layer formed on the bottom polysilicon electrode will degrade the overall capacitance to some degree, reducing the benefit of using a high-*k* dielectric. As a result, an alternative material for the bottom electrode has also been studied. In one report, TiN was used for the bottom electrode, and Pt for the top electrode.[12] In another report, reactively sputter-deposited RuO_2 was used for the bottom electrode. (RuO_2 is a conductive oxide with a conductivity of 35 $\mu\Omega$-cm. Since it is already an oxide, it will not form a dielectric oxide layer at the RuO_2/Ta_2O_5 interface during post-Ta_2O_5 thermal processes.) In one other publication, a $Ru/Ta_2O_5/Ru$ capacitor structure was described. The Ru electrodes were deposited by sputtering. In this report, the Ta_2O_5 film after annealing was found to have a *k*-value of 65. It was speculated that the Ta_2O_5 layer crystallized on the Ru material during the annealing step in such a way that its *k*-value was significantly increased compared to layers used with other electrode materials. Significantly lower leakage-currents through the Ta_2O_5 layers were also observed after annealing them in forming-gas at 400°C.[13]

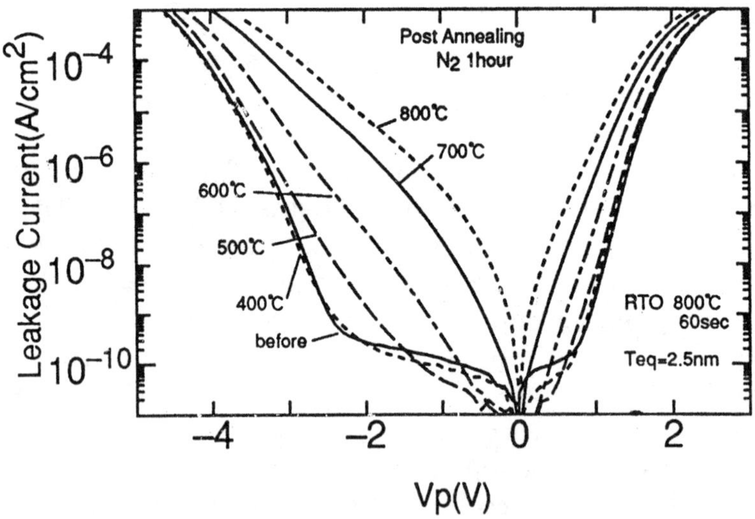

Fig. 4-9 Annealing temperature dependence of leakage current density for 2.5 nm Ta_2O_5 capacitors. A rapid thermal oxidation (RTO) is performed after Ta_2O_5 film deposition.[19]

Subsequent high-temperature process steps, however, cause the leakage currents through the Ta_2O_5 layers to significantly increase (see Fig. 4-9).[14] In this figure, it is observed that if the Ta_2O_5 layers are subjected to furnace-annealing temperatures of 800°C for 1 hour (or more), the leakage current rises precipitously. Two such high temperature steps that are commonly used in post-Ta_2O_5 layer deposition are those used to reflow the BPSG ILD layer, and to activate the dopants after the source-drain implant step. In one paper, this leakage current was reduced by using a novel low-thermal-budget process that exposed the Ta_2O_5 layers to temperatures no higher than 500°C after the Ta_2O_5 layers are deposited. That is, instead of using a conventional BPSG film, the interlevel dielectric layer (ILD) was formed as follows: An SOG film was first spun on, and this was followed by the deposition of an SiO_2 capping-layer using a low-temperature-CVD process. The source/drain regions were formed with an optimized implant step that employed only a 500°C activation anneal.[14] The application of a Ta_2O_5 layer to a DRAM capacitor formed with rugged polysilicon as the bottom electrode and TiN as the top electrode was described in Ref. 15.

Since leakage through the Ta_2O_5 layers is one of the biggest issues with such films for DRAM applications, work has been done to try and solve this problem. One factor that seems to influence leakage is the surface roughness of the underlying electrode.[16]

Another technique for depositing Ta_2O_5 layers is by *metalorganic solution deposition* (MOSD). This also allows the Ta_2O_5 layer to be doped with Al (a factor that improves some of the Ta_2O_5 layer dielectric properties). For example, thin films with $0.9Ta_2O_5$-$0.1Al_2O_3$ composition exhibited an increased *k* value of 42 and a lower leakage-current density (6×10^{-8} A/cm^2 up to an applied electric field of 1 MeV/cm).[17]

Etching of Ta_2O_5 layers can be carried out in variety of fluorocarbon gases (e.g., C_2F_6, CH_4, CF_3Cl, CHF_3, and CCl_3F_2),[18] as well as in discharges of Cl_2, SF_6, N_2, or CH_4/H_2. Selectivities of > 6 for Si over Ta_2O_5 were obtained for SF_6/O_2 discharges.[19]

In summary, Ta_2O_5 seems to be the DRAM capacitor dielectric material that will replace ONO. Combined with such polysilicon-electrode techniques as hemispherical-grain poly (HSG), Ta_2O_5 capacitors will be suitable for DRAMs up the 1-Gb and 4-Gb generations (0.15-μm node). The minimum thickness of Ta_2O_5 is estimated to be about an equivalent oxide thickness of 1.5 nm. This implies that even by using HSG and the minimum possible Ta_2O_5 thickness, the limits of Ta_2O_5 technology may be reached at the 1-Gb integration level.

However, for the 0.13 μm technology node it may be possible that Ta_2O_5 will be used with a MIS capacitor structure (i.e., RuO_2/Ta_2O_5/poly). At the 0.1 μm node it may be necessary to use metal-insulator-metal (MIM) structure (i.e., $RuO_2/Ta_2O_5/RuO_2$). At the 0.05 μm node (50 nm) it is likely that the insulator in an MIM structure will be BST.

4.4.2 Tantalum Pentoxide as a MOSFET Gate Dielectric

Tantalum pentoxide has also been considered as an alternative gate-dielectric material to SiO_2 for MOSFETs with gate lengths of less than 0.1-μm. Since oxide films for such devices would have to be thinner than 1.5-nm, excessive leakage-current through the oxide film would make such MOSFETs impractical for use in most circuit applications. Due to the fact that direct-tunneling leakage currents decrease exponentially with the thickness increase of the dielectric layer, a dielectric with a higher k-value could provide the same electrical behavior as SiO_2 but it could be used with a larger thickness. For the case of Ta_2O_5, which has a k-value of 25, a Ta_2O_5 layer would have a thickness 25/4.0 times larger than that of an equivalent SiO_2 film. Many of the process integration issues of incorporating into a MOS structure were discussed in the previous section. Here, we describe the challenges specific to using Ta_2O_5 as a gate-dielectric layer.

The first problem that must be addressed involves the fact that when Ta_2O_5 is deposited directly on silicon (or when it is subsequently annealed in oxygen), SiO_2 grows under the Ta_2O_5. As much as 2-nm of oxide can grow, thereby increasing t_{oxeq} (see Eq. 4-3). This was demonstrated in a report in which Ta_2O_5 was deposited directly on Si and NMOSFET transistors were subsequently fabricated.[20] An interfacial SiO_2 layer was found to be under the Ta_2O_5 at the conclusion of all processing. A way to prevent uncontrollable growth of such an oxide during subsequent thermal processing steps is to form an interfacial layer between the Ta_2O_5 and the Si substrate. In the first report of fabricating MOSFETs with a Ta_2O_5 gate dielectric, a 2-nm-thick oxide layer was in fact deliberately grown on the Si before the Ta_2O_5 was deposited. This layer was then annealed in N_2 at 800°C for 30 min (nitriding the oxide). A diffusion-barrier is thus created which prevents oxygen from reaching the underlying Si substrate and growing additional oxide. A TiN gate electrode was employed instead of poly (Fig. 4-10). The leakage current was reported to be equivalent to that of a MOSFET with an SiO_2 gate dielectric of identical thickness.[21]

In a second report describing MOSFETs Ta_2O_5 with a gate dielectric layer, the gate dielectric consisted of a stack (i.e., 1.5-nm SiO_2/3.0-7.5-nm Ta_2O_5/0.5-1.5-nm SiO_2). This dielectric stack was electrically equivalent to 20-35-nm

SiO_2.[22,31] A later report described the use of RTP to grow an interfacial layer on the Si substrate with NO at 800°C for 20 sec - which when combined with a CVD Ta_2O_5 layer - produced a gate dielectric with a t_{oxeq} = 2.3-nm thick. The main benefit of using this approach was that the leakage current through the dielectric film was 1000x smaller than when an oxide was used as the interfacial layer.[23] A similar process flow reported use of a RTN step (nitridation of the silicon substrate surface by RTP in an NH_3 ambient at 800°C for 30 sec), to form an interfacial layer on the poly before depositing the Ta_2O_5 film.[24]

Related to the oxide growth under the Ta_2O_5 is the issue of unacceptably large values of interface traps when using a dielectric other than SiO_2 is in contact with Si. In the Kizilyalli study above, a Ta_2O_5 gate-dielectric-stack (with SiO_2 in contact with both the underlying Si and the overlying poly gate) exhibited interface trap densities comparable to those in MOSFETs fabricated with SiO_2. Charge trapping (a major problem of layered dielectrics) was reportedly not an issue.

A final limitation of using Ta_2O_5 as a gate dielectric is that a barrier over the Ta_2O_5 is needed to prevent silicide from forming due to a chemical reaction of the Ta_2O_5 and the polysilicon gate during subsequent thermal processes. This barrier must be conductive, and TiN, WN, or PGMs used in DRAM top electrodes can be used. However, the difficulty with using such a metallic gate barrier is that the threshold voltage of MOSFETs is increased by about 0.6V. Obviously, this high threshold voltage value is not compatible with low-voltage, low-power technologies. Hence, proper threshold voltage adjustment with counterdoping in the active areas must be employed. However, this produces buried-channel MOSFETs, which degrades their short-channel behavior.

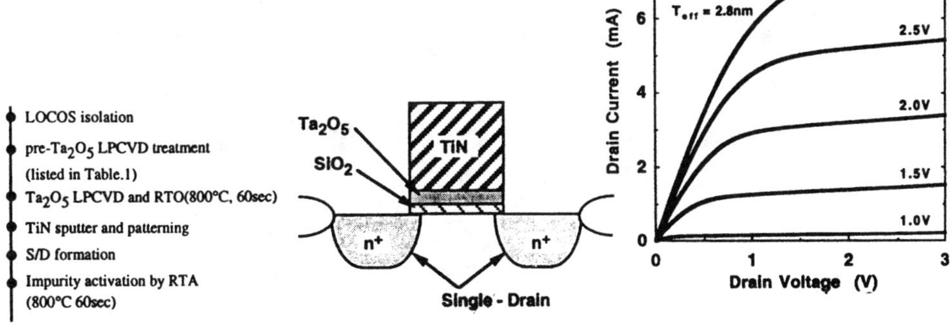

Fig. 4-10 Process flow, cross section, and I-V characteristics of an NMOSFET fabricated with a with a Ta_2O_5/SiO_2 gate dielectric. A TiN gate electrode was also employed.[21]

4.5 BARIUM STRONTIUM TITANATE (BST) FILMS FOR DRAM CAPACITOR DIELECTRICS

Barium:strontium-titanate (BST) is the candidate material of most promise for use as a capacitor dielectric in future generations of high-density DRAMs. BST films are polycrystalline films, with the crystalline grains having a perovskite oxide structure. The chemical formula of BST is $(Ba_xSr_{1-x})TiO_3$. Common compositions for DRAM use are $0.5 < x < 0.7$. BST films are the leading candidates for advanced DRAM capacitor dielectrics because of their high dielectric constant (k = 88-90 in the film thickness range used in DRAM capacitors), and because the value of k in BST films exhibits a relatively small frequency dependence. That is, the permittivity of perovskite oxide films tends to decrease at high frequencies. For BST films deposited by MOCVD k decreases by < 2% from 100 Hz to 10 MHz and by < 7% from 200 MHz to 2GHz.[30] This is sufficient to meet the requirements of future DRAM generations. Its large dielectric constant (like those of many other perovskites) is the result of the strong ionic polarization in the material.

Many believe that dielectric films with such high-k values will be needed for DRAM capacitors following the 130-nm generation. For example, the required capacitance density in a DRAM technology with a minimum feature size of 0.15-μm with a DRAM capacitor having a stack structure of 0.3-μm in height (i.e., a 2:1 aspect ratio) and a rectangular area of 0.15-μm by 0.45-μm is ~80 fF/μm^2. Figure 4-11 shows a schematic representation of a simple stack-capacitor DRAM cell incorporating a high-k dielectric in the stack capacitor. The high-k dielectric

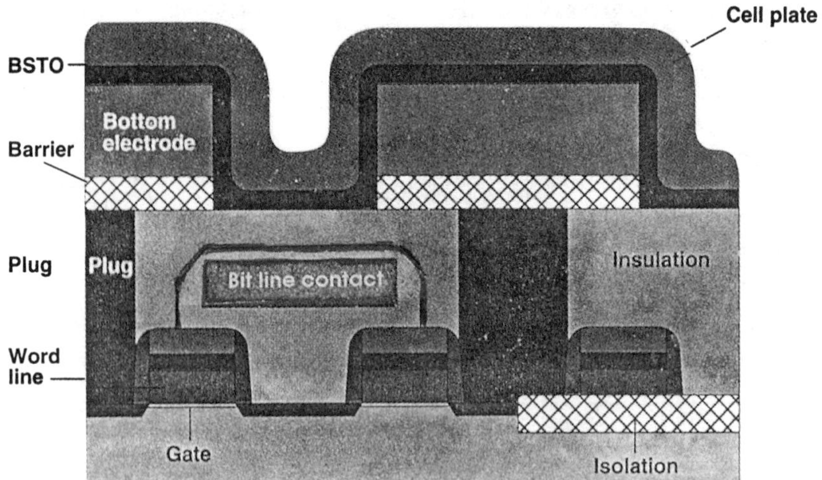

Fig. 4-11 A schematic representation of a simple stack capacitor DRAM cell incorporating a high-k dielectric layer.[47] Courtesy of Semiconductor International.

material (e.g., BST) covers the top and sides of the patterned bottom electrode. The top electrode forms the common plate electrode, and the word lines and bit lines are located below the capacitor. The structural simplicity in the capacitor of this cell is apparent. In addition, the vertical height of the bottom electrode can be reduced to provide improved control over the critical dimension by reducing depth-of-focus issues.

Note that perovskite materials exhibit *ferroelectric behavior* when held below their Curie point, and *paraelectric behavior* when above that temperature.[28] In conventional DRAMs, the perovskite materials used as the capacitor dielectric are chosen so that they exhibit paraelectric behavior. On the other hand, perovskites that exhibit ferroelectric behavior are used as the dielectrics of FeRAM capacitor structures. BST is a perovskite that has a Curie point that varies linearly with composition, with a value of about 1207°C when x = 1, and about –2207°C when x = 0.[29] The strontium is added to reduce the Curie point below room temperature, which ensures paraelectric operation. The dielectric constant also depends on the film composition, and barium/strontium composition is an important parameter in the use of BST in memory capacitors.

To obtain the high-*k* values necessary for high-density DRAMs, the films must be free of impurities, have correct stoichiometry, and be crystalline.

4.5.1 Deposition of BST Films by CVD

The deposition of BST films can be carried out by a number of methods, including PVD by rf sputtering, thermal MOCVD, sol-gel, and pulsed laser ablation. However, BST film deposition is most commonly done with MOCVD. One reason for this is that rf sputtering requires an appropriately designed composite target. Since sputtering rates for the different elements vary, the useful lifetime of a single sputtering target is too short. BST films deposited by sputtering and sol-gel methods also have a conformality which is typically smaller than 60%, while the conformality of BST films deposited by thermal MOCVD processes can exceed 80%. Because of the high conformality which can be realized, the MOCVD method is likely to be used to deposit the high-dielectric layers for DRAM capacitors. Consequently, we will restrict our discussion of deposition of BST films to the MOCVD processes.

Most precursors of BST are solids at room temperature and are dissolved in an organic solvent for delivery as a liquid to a vaporizer. The liquid is flash-vaporized (to minimize the time precursors spend at elevated temperatures, which minimizes precursor decomposition), and is then transported in a gaseous state to the CVD chamber. This liquid delivery method enables the use of solid precursors. However, it chemically complicates the deposition process by introducing a solvent. Most of the published work on CVD BST uses precursors that are adducts of barium and strontium β-diketones along with organometallic titanium compounds, primarily: $Ba(THD)_2$·tetraglyme, $Ba(THD)_2$·PMDTA, $Sr(THD)_2$·tetraglyme, $Sr(THD)_2$·PMDTA, and $(THD)_2Ti(i\text{-}Pr\text{-}O)_2$. (Note that: THD = 2,2,6,6-tetramethylheptane-dionate, and (i-Pr-O) = isopropoxide). These diketones and organometallic-Ti compounds are usually dissolved in butyl acetate. BST films of good quality have been produced using this precursor scheme at deposition temperatures around 550°C. However, at these temperatures, conformality of the deposited films is typically poor and may not be acceptable for three-dimensional capacitor fabrication. A different set of precursors has therefore been developed, which appear promising for BST film growth at lower temperatures. In this precursor scheme, non-adducted versions of the same barium and strontium β-diketones: $(Ba\{THD\}_2)_4$ and $(Sr\{THD\}_2)_4$, along with $(THD)_2$ $Ti(i\text{-}Pr\text{-}O)_2$, are dissolved in tetrahydrofuran (THF). Reports in the literature state that BST films with capacitance densities (planar capacitors) of over 100 fF/μm^2 and leakage currents in the 10^{-8} A/cm^2 range can be grown at wafer temperatures of 480°C using these precursors. Other work has demonstrated conformality of 100% at 480°C using non-adducted precursors.

The low-temperature CVD processes for depositing BST films give better conformality than do the higher temperature deposition processes. This is due to the fact that different mechanisms are limiting the deposition process. An Arrhenius plot of the deposition rate of BST growth is shown in Fig. 4-12. It can

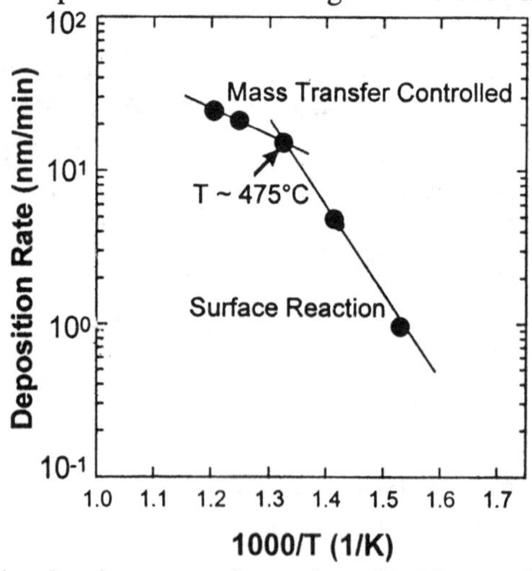

Fig. 4-12 Arrhenius plot showing two regimes of rate-limiting mechanisms for CVD BST deposition using non-adducted precursors. Courtesy of Marcel Dekker Publishing.

be seen that two distinct regimes are apparent with a crossover at ~480°C. In the higher temperature regime (where the deposition rate is approximately constant with increasing temperature), film growth is rate-limited by mass transport from the vapor phase to the film surface. In the lower temperature regime, film growth is limited by the chemical reaction rate on the surface. Films grown at the higher temperatures exhibit poor conformality because the precursor material reacts immediately upon arriving at the surface and does not have time to diffuse into narrow patterned features before being incorporated into the growing BST film. In the surface-reaction-limited regime, the precursor molecules can diffuse a long way on the wafer surface (i.e., into the narrow patterned regions) before reacting and being incorporated into the growing film.

In BST deposition, a tradeoff occurs between needing a high enough deposition temperature to ensure good crystallinity of the perovskite phase, and needing a low enough temperature to achieve adequate conformality. The adducted precursors form highly crystalline perovskites at high temperatures, but such films deposited at low temperatures are mostly amorphous (i.e., they do not exhibit the perovskite crystalline structure). On the other hand, non-adducted precursors can produce perovskite BST films in the low-temperature regime. The low-temperature deposition is also favored when Ru bottom electrodes are used (see Sect. 4.5.3), to prevent the Ru surfaces from being oxidized during the BST deposition in an O_2 atmosphere.

Note, however, that since the BST films are deposited by CVD at low temperatures (400-500°C) to meet the stringent step-coverage requirements, these films must be subsequently annealed to achieve the desired capacitance and leakage-current behavior. Annealing is usually done after the top electrode has been deposited. Such a post-top-electrode deposition anneal is especially beneficial if the top electrode is deposited using a technique that bombards the BST surface with energetic species (such as PVD). That is, the anneal then serves the additional purpose of healing the damage done to the BST film by the PVD process that deposits the top electrode. Reported annealing parameters vary widely. Temperatures between ~500°C and 700°C have been used, and both oxidizing (N_2O, O_2) and inert (N_2) ambients have been observed to improve the BST electrical properties. Typically, films grown at higher temperatures (550°C) using the adducted precursors require lower annealing temperatures – around 550°C in oxygen is commonly used. Low-temperature depositions require higher annealing temperatures, closer to 700°C.

The main benefit of the annealing processes is a reduction in the leakage characteristics of BST capacitors. Before annealing, films are leaky (Fig. 4-13),

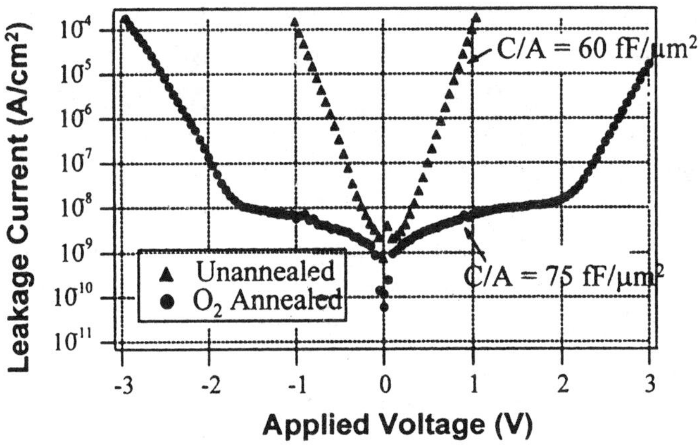

Fig. 4-13 Leakage curves from unannealed and 550°C oxygen-annealed BST films deposited at 600°C. Labels on the leakage curves give the corresponding capacitance densities of the films.

although the capacitance is quite high. After annealing leakage improves to an acceptable level and capacitance increases by roughly 25%, although the improvement is not as dramatic as that seen in the leakage-current properties.

4.5.2 Microstructure Effects on the Electrical Properties of BST Films

A key factor that impacts the electrical properties of BST films is the amount of titanium in the film. Titanium content is referred to as a percentage of all cations in the film, i.e., Ti% = {[Ti]/([Ba] + [Sr] + [Ti])} x 100%. The best properties (e.g., highest capacitance-density) are found when there is a slight excess of titanium, ~51% (compared to the stoichiometric perovskite composition of 50%). Above this amount, the dielectric constant decreases. The sensitivity of the value of k to the titanium content is quite high (e.g., the dielectric constant drops by ~40% when the titanium concentration increases from 51% to 53%). Hence, it is critical that the amount of titanium incorporated during deposition be tightly controlled. A low-temperature, the non-adducted precursor process has been shown to provide such control of incorporated titanium content.

Another important parameter that impacts the BST electrical properties is the surface roughness. Increased surface roughness correlates with lower dielectric constants and higher leakage currents. Roughness is often observed as a hazy film. That is, films that are rough appear hazy to the naked eye under bright light. Roughness can be caused by depositing the film on a rough underlayer, as well as film growth occurring in an island mode. Roughness resulting from film growth mode is very sensitive to deposition process parameters, and such parameters must be optimized to find a haze-free process window. Typically less haze is observed at higher deposition temperatures, above 600°C.

4.5.3 BST Integration Issues

Integrating BST into a DRAM process poses many difficult challenges. First, BST is incompatible with the materials that have been used for electrodes up to this point, namely: polysilicon, TiN, and aluminum. This is because BST films contain significant amounts of oxygen and also require high-temperature post-deposition annealing in oxygen ambients. Thus, aluminum films cannot be used as the bottom electrode of a BST DRAM capacitor because they cannot tolerate the high-temperatures used to anneal BST films after deposition (> 500°C). Likewise, polysilicon will oxidize either by reaction with the oxygen that can bleed out from the BST film or with the oxygen present in the annealing ambients. Finally, TiN is no longer stable against oxidizing species at temperatures above 500°C. Hence, alternative electrode materials must be employed.

The most suitable electrode materials are based on the metals found near the middle of the periodic table - such as platinum (Pt), ruthenium (Ru), and iridium (Ir). Oxides of these metals are also conductive, making them candidates for inclusion in the electrode stack. As we will discuss in more detail, one of the major problems of using these materials is the problem of electrode patterning, as these materials produce minimal volatile species during dry etching of the noble metal electrodes (with Ru and RuO_2 being the exceptions).

Platinum is the most commonly used electrode for BST in R&D work, as it is non-reactive with BST, and the growth characteristics of BST films on Pt films are good. However, using platinum is also difficult. Since it is a chemically-inert noble metal, it is difficult to dry-etch. But, it does form silicides, and hence it must be deposited on a diffusion barrier to prevent Si-Pt reactions with the Si wafer. (An example of such a film is stacked layer of Ti/TiN, which is deposited on the silicon substrate. Then the Pt film is deposited on the Ti/TiN film.) Since Pt has poor adhesion to most surfaces, the barrier layer must also serve as a "glue layer." This entire multi-layer bottom electrode structure must then be oxidation resistant and thermally stable enough to withstand the high-temperature, oxidizing environments employed in the processes used to deposit and anneal BST films. Finally, platinum can catalyze reduction in the BST during a forming-gas anneal (instead of protecting the BST against such reduction). Thus, Pt is not a good candidate for the top electrode.

Other electrode materials can also be used, especially for the top electrode. These include ruthenium metal (Ru), ruthenium oxide (RuO_2), iridium oxide (IrO_2), or conducting perovskite oxides such as $SrRuO_3$ or (La,Sr)CoO_3. Ruthenium and RuO_2 are relatively easy to etch, although one of the products of the etching process, RuO_4, has a melting point of 25.4°C, and is both a flammable and highly toxic gas. Figure 4-14 shows the process flow for the fabrication of Ru/BST/Ru capacitors. However, none of these materials is standard in semiconductor processing, and hence all will require significant process-development for incorporation in DRAMs.

Dry-etching of metal oxides, perovskites (including BST), and refractory metals will also provide some serious challenges because they do not have volatile by-products. The etching mechanism for these materials must therefore be more physical, like ion milling. Ion milling has been used for patterning these kinds of materials, but the small feature sizes hinder this technique. Dual-frequency plasma etching is also being examined as a way to etch materials such as platinum, metal oxides, and perovskites. The energy of the ions striking the wafer is controlled to maximize the physical effect while taking advantage of whatever chemistry effect there is. Work has also been carried out on developing high-temperature etch processes for such materials (i.e., > 350°C), as a way to improve the etch-profile and etch-rate of such materials. Masking for the etch process is another challenge. The material to be etched must be physically attacked, while the materials that are used for the mask are etched more easily.

The exceptions, as noted earlier, are Ru and RuO_2, both of which can be patterned by dry-etching (in O_2/Cl_2 plasmas) to give totally anisotropic profiles through chemical reactions which produce RuO_4 in oxygen-containing plasmas. Thus, the profile and feature size of the storage electrodes can be precisely controlled by short gas-residence-time techniques to suppress the deposition of reaction products from the plasma onto the sidewalls. This will permit a 3-D stacked capacitor structure with vertically sided Ru bottom electrodes and BST dielectrics to be used down to 4-16-Gbit DRAMs.

It should also be noted that a diffusion barrier under the bottom electrode is also needed to keep the electrodes from reacting with the polysilicon plug during BST deposition and post-deposition annealing (see Fig. 4-14). Several kinds of materials have been investigated for this role, including: TiN, TiAlN, and TiSiN.

The introduction of high-dielectric materials for DRAM fabrication will also require additional tooling, including:

1. A sputtering system for electrode deposition;

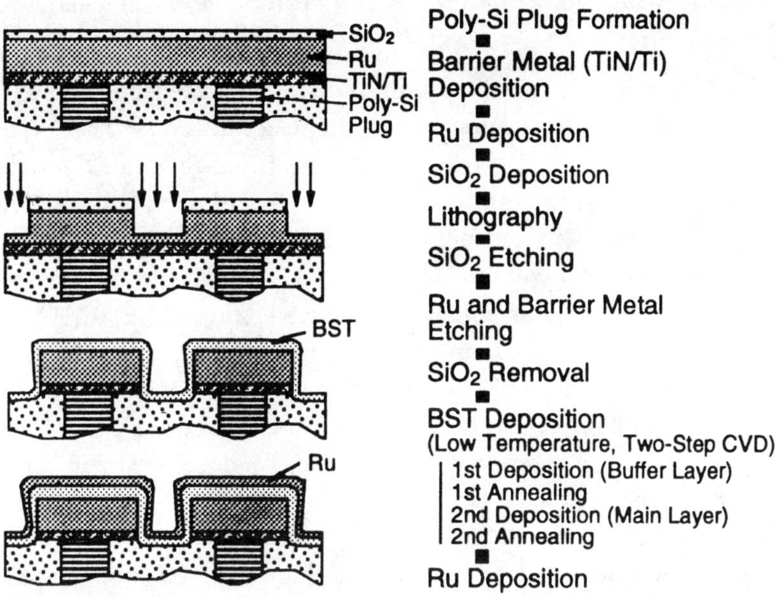

Fig. 4-14 Process flow for the fabrication of Ru/BST/Ru capacitors.[48] (© 2000 IEEE).

2. An MOCVD reactor for the deposition of the high-*k* dielectric layer, and;
3. A reactive-ion-etch system to pattern the electrode.

Numerous companies manufacture sputtering systems capable of depositing the electrode materials. Applied Materials and Novellus have developed MOCVD reactors capable of depositing BST on 200-mm wafers. Tegal Corp. and Lam Research (along with other suppliers of etch equipment) have demonstrated etch processes for Pt electrodes.

4.6 USE OF OTHER HIGH-*k* MATERIALS AS GATE DIELECTRICS

An extensive worldwide research effort is being undertaken to develop alternative high-*k* dielectrics for MOSFETs to replace silicon dioxide. By the end of 2000 workers in the field were not optimistic that despite intensive efforts to find a high-*k* material that works well, such a material would readily be available for IC manufacturing applications. Their view is that it is going to take at least five years (i.e., until 2006) before a suitable high-*k* material can be introduced into manufacturing. A review of research on the use of binary metal oxides and silicates up to 2000 was published by Wilk *et al*,[50] and by Gusev *et al*,[51] Here we describe some of the candidates for that application, as of 2001.

4.6.1 Atomic Layer Deposition for Depositing Thin High-*k* Dielectric Films

Before describing the individual high-*k* materials that are candidates for gate dielectrics in MOSFETs, it is useful to discuss a new deposition technique that may be employed for forming such thin dielectric layers. That is, a promising method for depositing very-thin films has been developed for a variety of applications, including gate dielectrics, DRAM capacitor dielectrics, and diffusion barriers. This technique is called *atomic layer deposition (ALD)*, or *atomic layer chemical vapor deposition* (ALCVD). This technique was pioneered by Tuomo Suntola who invented *atomic layer epitaxy* (ALE) in 1980. The appeal of ALE was that it could deposit very-thin layers of material (<10 nm-thick) with tight composition control. However, at that time the thinnest layers in IC devices were about 100-nm thick, and so ALE was not immediately seen as a technique for the semiconductor market. (It was initially developed for depositing ZnS films for electroluminescent devices.)

In the ensuing 20 years, however, things have changed. Aggressive scaling of critical feature sizes led to lateral dimensions approaching 100-nm, and the thinnest layers approached atomic dimensions. For example, in the gate-dielectric layer, only about 1-nm, or about 4 atomic layers in silicon-oxide equivalents are needed. Clearly, the aggressive scaling has now made applications of the Suntola invention appropriate for integrated circuits. The ALE technique has therefore been modified and commercialized as the ALD technique. The ALD process appears to have several advantages over traditional CVD: it can be carried out at lower temperatures (typically < 400°C); it uses a wider range of precursors; it produces very-thin films; and it inherently obtains 100% step coverage and excellent conformality even over the most challenging topography, and with lower impurity levels.

However, deposition rates are limited by the ALD mechanism. By definition, the thickness of each cycle cannot exceed a few angstroms as determined by the lattice constant of the materials. ALD is therefore targeted at very-thin films having the highest quality demands. The thickness of these films can be controlled simply by counting the cycles. The composition can also be modified and controlled at the atomic level, thus providing an opportunity to create new dielectrics that cannot be easily obtained by other means. Further scaling of critical IC feature sizes will inevitably lead to more applications into the realm of ALD. We will show how ALD can be used to deposit various films in the upcoming sections.

Several equipment vendors have introduced ALD tools. The first is ASM International (Bilthoven, Netherlands) which, through its acquisition of Microchemistry Ltd. (Finland) now has several systems capable of ALD for semiconductor and flat-panel display applications. (Note that Atomic Layer CVD and ALCVD are trademarks of ASM International). Genus (Sunnyvale, CA) also announced a new ALD capability on its Lynx2 platform. Other equipment suppliers are also apparently developing tools with ALD capability. The nature of the ALD process is such that it can be done in a simplified reactor, where gas flow and temperature uniformity will be relatively unimportant. Instead, the hardware design will focus on preventing deposition on the reactor interior and the wafer backside. Development efforts will be focused on creating ALD tools with high throughput, long mean times between cleans, and low defect density. Reliable and precise control of reactants, quick removal of extra reactants and reaction by-products, and short cycle-time will also be critical to the success of this emerging technology.

ALD works quite differently from conventional CVD techniques. Instead of mixing two or more reactant gases inside the deposition chamber - where they either react in the gas phase and fall on the wafer surface, or they react on the wafer surface - in ALD one reactant gas is introduced at a time. Through a process called "chemisorption," a monolayer of the first gas is adsorbed on the wafer surface. That is, individual precursors are pulsed onto the wafer surface in a sequential manner, without mixing in the gas phase. Each individual precursor reacts with the surface to form an atomic layer in such a way that only one layer can form at a time. This is because the surface reaction occurs such that the reaction is complete, and permits no more than one layer at a time to be deposited, no matter how many molecules are applied to the surface in an overdosing mode. Films are built up by introducing short bursts of gases in cycles.

As an example, ALD can be used to deposit a thin silicon-oxide layer by using silane and oxygen. At room temperature, silane gas is introduced into a chamber containing the wafer. Silane molecules will be adsorbed onto the wafer surface until a monolayer of silane molecules exists over the entire exposed surface of the wafer. Thereafter, no more molecules will be adsorbed by the wafer surface. If the temperature is kept low enough, a stable chemisorbed monolayer will form over the exposed wafer surface. Then, the excess silane gas is pumped from the chamber, and it is replaced with oxygen or water vapor. These molecules react with the silane to form an oxide, and the reaction is completed when one monolayer of oxide has been formed. Then, the excess oxygen is removed, and a second burst of silane gas is fed into the chamber to form the next monolayer of silane, and so on. This discussion demonstrates that ALD is a technique best suited for depositing binary materials such as SiO_2 and metal oxides (e.g., Ta_2O_5, Al_2O_3, ZrO_2, HfO_2, etc).

By quickly switching the gases on and off (in less than 1 sec), it is possible to obtain reasonable growth rates, up to 200-Å/min. For those films that are deposited in the thickness range of 10-50-Å (or less), such growth rates should produce acceptable throughputs. In addition, since it is only necessary to saturate the wafer with the precursor, the chamber does not have to be designed for uniform reactant coverage. This allows the chamber to be kept very small (e.g., only 4 mm high), which in turn, allows as many as 4 cycles per second to be run. The claim made earlier that ALD deposition temperatures are generally low can now be understood in light of the fact that this process depends on chemisorption.

4.6.2 Aluminum Oxide (Al_2O_3) as a Gate Dielectric

Aluminum forms a stable oxide (aluminum oxide, Al_2O_3), and it is one of the materials that is being studied as a high-*k* gate dielectric. The *k*-value of Al_2O_3 is about 9, which is about three times as large as that of SiO_2. The effective *k*-value of the dielectric layer which employs Al_2O_3 is, in fact, even somewhat smaller (i.e., it is ~5.5-6.0). This reduction in the effective *k*-value occurs because a thin interfacial SiO_2 layer inevitably forms between the Al_2O_3 layer and the silicon substrate during Al_2O_3 deposition. This extra, unwanted, interfacial layer reduces the effective *k*-value, and increases the equivalent oxide thickness (EOT) of the Al_2O_3 layer. (But this interfacial layer also plays a beneficial role in reducing the interface state density leakage current.) Thus, while Al_2O_3 promises to provide a dielectric layer that has an improved *k*-value, this improvement may only be adequate for a single technology node.

Aluminum oxide is attractive as a gate dielectric for future MOSFETs not only because it has a higher *k*-value than SiO_2, but because it also exhibits excellent thermal stability when it is in contact with silicon, and because it has a wide energy gap (~8.3 eV). Furthermore, it is compatible with polysilicon without requiring any barrier materials.

Recent work has described the fabrication of CMOS devices with an Al_2O_3 gate dielectric.[37,38] The Al_2O_3 layer is deposited by atomic layer deposition (ALD - as described in Sect. 4.6.1), using trimethyl-aluminum (TMA) and ozone as the precursors, at 450°C. (Figure 4-15 shows how an ALD process can be used to deposit an Al_2O_3 layer using the above precursors.) Polycrystalline silicon is then deposited and patterned, junction implants are performed, and a rapid ther-

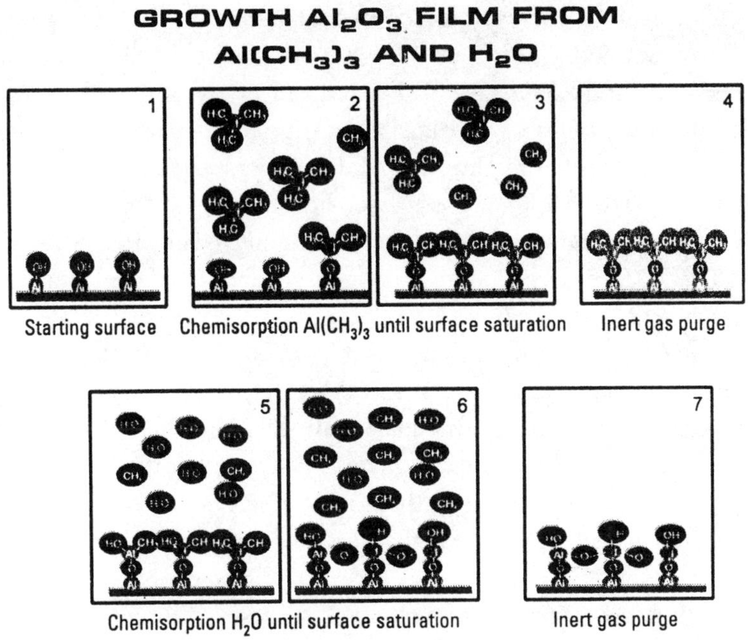

Fig. 4-15 Atomic layer CVD works by introducing one reactant gas at a time. The first gas is "chemisorbed" onto the wafer surface. Excess gas is then purged and the second gas (in this case, water vapor) is introduced. This gas reacts with the chemisorbed layer, creating a monolayer of deposited film. Courtesy of Semiconductor International.

mal anneal at T = 1000°C is performed. The leakage current density exhibited by MOS capacitors with an Al_2O_3 layer is 2-3 orders of magnitude lower than that observed in MOS capacitor built with an SiO_2 layer with the same EOT.

One major drawback of MOSFETs built with an Al_2O_3 dielectric layer is that the carrier mobility is smaller than in MOSFETs built with SiO_2 dielectric layers, probably due to electron trapping by the Al_2O_3 material. By the end of 2000, the reason why such electron trapping occurs in Al_2O_3 layers (and other high-*k* dielectric materials) is still unknown.

4.6.3 Zirconium Oxide (ZrO_2) as a Gate Dielectric

Zirconium oxide (ZrO_2) is considered to be one of the most promising high-*k* dielectric candidates to replace the SiO_2 gate dielectric. ZrO_2 has a dielectric constant of ~20-25 and is thermodynamically stable in contact with silicon.[44] In addition, ZrO_2 has a band gap of ~5.1 eV and a lattice mismatch with silicon of only 2.1%. It has been shown that reactively sputtered ZrO_2 (i.e., a Zr sputter target is used in an Ar + O_2 ambient) has a low equivalent-oxide-thickness

(EOT), low leakage, and good reliability characteristics.[40,41] (Another way that ZrO_2 can be formed with a PVD process is by oxidizing a sputtered Zr film in an RTP process.[45]) ZrO_2 films can also be etched in HF solutions. It has also been reported that doping the ZrO_2 films with Al produces amorphous films with better uniformity, but with a slightly lower dielectric constant. This doping can be achieved in a PVD deposition process by co-sputtering with an Al target.[41]

ZrO_2 films can also be deposited using an atomic layer deposition process (ALD).[42,43] The ALD process has the inherent advantage of conformality over PVD deposition processes. Such an ALD process for ZrO_2 is illustrated in Fig. 4-16. Generally, ALD requires that the metal compound precursor and nonmetal compound precursor ($ZrCl_4$ and H_2O, respectively, for the case of ZrO_2) are introduced alternately instead of simultaneously as in a standard CVD process (see Sect. 4.6.1). This results in the layer-by-layer growth that creates the characteristics of ALD films, making them appropriate for several applications. In the case of Zr-based dielectrics, the material may consist of zirconium, aluminum, and oxygen.

In any case the ZrO_2 films are amorphous, and a very thin interfacial silicate layer (9-Å thick) exists at the silicon/ZrO_2 interface. Post-deposition annealing in oxygen or inert gases (at 400-500°C) is found to be necessary to reduce the leakage current of deposited films. After high-temperature annealing, this layer grows and converts to a more stoichiometric SiO_2 layer. Significant SiO_2 growth is observed at the ZrO_2/Si interface annealed at 400-600°C even under low partial pressures of oxygen. Reaction with the Si substrate and the polysilicon gate also occurs at higher temperature ($\leq$ 1000°C) processing steps.

Films of zirconium silicate doped with aluminum (Zr-Al-Si-O) have also been studied as candidate high-*k* gate materials.[42] It was noted that doping ZrO_2 films with Al improves the resultant ZrO_2 dielectric layer. That is, the undoped ZrO_2 films have high diffusivity of oxygen, and so oxygen can diffuse through the ZrO_2 during the post-deposition anneal and react with the underlying Si to form an SiO_2 interfacial layer. The addition of Al reduces the rate of oxygen diffusion through the ZrO_2 film. Another report described the use of yttria-stabilized ZrO_2 (YSZ) films as a high-dielectric constant layer in MOSFETs. The advantage of the YSZ film is that it can be epitaxially grown on a silicon substrate by pulsed laser ablation without the formation of an interfacial silicon oxide layer.[46]

At the end of 2001, Intel announced that a new transistor structure that included a ZrO_2 gate dielectric layer, deposited by ALD. They stated that this new MOSFET structure would be used for devices built by Intel from 2005 and beyond.[55]

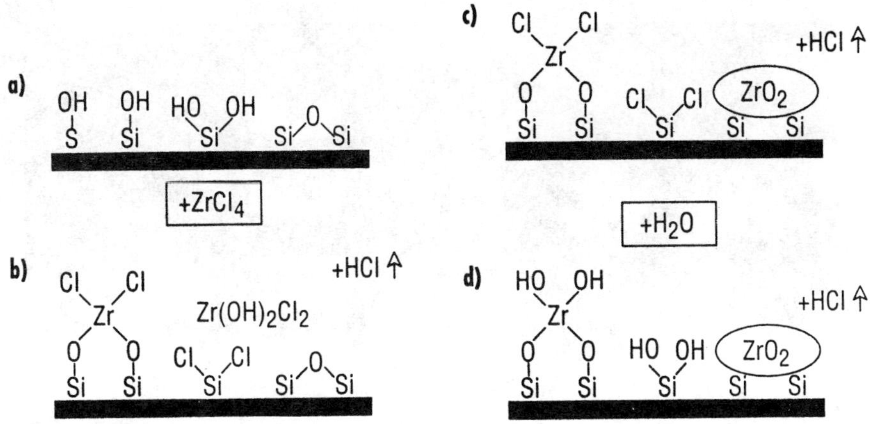

Fig. 4-16 Atomic layer CVD growth mechanism of the high-*k* dielectric ZrO_2 from $ZrCl_4$ and H_2O. Courtesy of ASM International.

4.6.4 Hafnium Oxide (HfO_2) as a Gate Dielectric

Hafnium oxide (HfO_2) has been investigated as a high-*k* dielectric for MOSFET devices. Early reports described fabrication of MOSFETs with n^+ polysilicon gates,[32] and more recent ones have discussed the characteristics of MOSFETs with p^+ polysilicon gates as well.[33] Hafnium oxide exhibits excellent materials properties for these applications, namely: a high dielectric constant (k = 24 - 40); good thermal stability when in contact with silicon; and an energy gap of 5.65 eV. (Hf forms the most stable oxide with the highest heat of formation [ΔH_f = 271 Kcal/mol] among the elements in the IVA group of the periodic table [Ti, Zr, Hf].) In addition, HfO_2 is very resistive to impurity diffusion and intermixing at the silicon/dielectric interface because of its high density (9.68 g/cm^3).[39]

The HfO_2 films described in the reports were deposited either by PVD (reactive sputtering),[32,33] or by *in situ* RTCVD.[34] MOS capacitors were fabricated with an equivalent oxide thickness (EOT) of 12-Å (and a physical thickness of 38-Å) by optimizing the polysilicon deposition process and the subsequent annealing steps that would be used to activate the MOSFET source/drain implants. That is, no interfacial layer was found to be formed between the HfO_2 and the polysilicon gates if the polysilicon deposition temperature was kept below 600°C, and the RTA anneals were carried out at 1000°C. However, the same report indicated that an unwanted interfacial SiO_2 layer IL does form between the HfO_2 and the silicon substrate during the HfO_2 deposition (Fig. 4-17). In addition, some crystallization of the HfO_2 is observed to occur during the RTA anneal steps. A more recent study claims that HfO_2

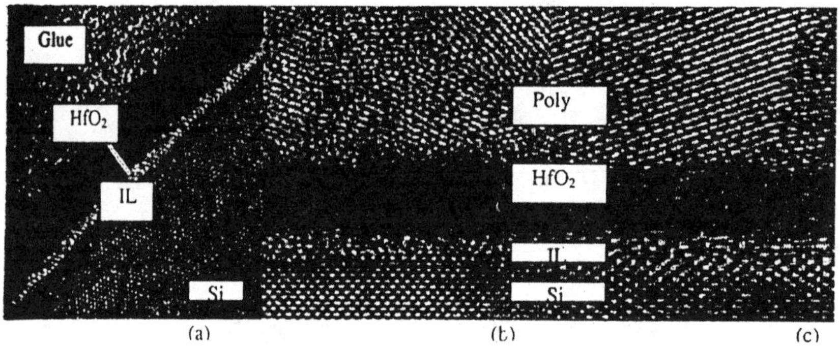

Fig. 4-17 HR-TEM showing HfO_2 (~38Å) and IL (~6Å): (a) after PDA, and (b) and (c) RTA 900°C and 1000°C for 1 min, respectively.[33] (© 2000 IEEE).

exhibits more thermal stability than does ZrO_2.[51] Two other papers published at the same time also states that better MOSFETs were fabricated with HfO_2 than with ZrO_2 as the gate dielectric.[52,53]

The leakage current characteristics of such MOS capacitors was about four orders of magnitude smaller than a MOS capacitor fabricated with SiO_2 having the same 12-Å EOT (as a result of the HfO_2 layer having a physical thickness more than double that of the SiO_2 layer). However, like other high-*k* dielectrics, the HfO_2 films also tend to trap electrons and thus degrade carrier mobility in the MOSFET channel. In another report, scalability of the EOT of the HfO_2 layer to less than 10-Å was demonstrated by using a TaN gate electrode to replace the polysilicon gate electrode.[35] That is, the polysilicon-depletion effect is avoided if a TaN gate electrode is used instead of polysilicon, and therefore an EOT as small as 8-Å can be obtained.

A recent study was published on the use of HfSiON as a high-*k* gate dielectric. This is an amorphous material and exhibits smaller leakage currents than does silicon dioxide at equivalent oxide thicknesses. HfSiON films scaled below 13-Å showed good reliability performance.

4.6.5 Praseodymium Oxide (Pr_2O_3) as a High-*k* Gate Dielectric

Another exotic material that has been studied as a high-*k* gate dielectric is *praseodymium oxide* (Pr_2O_3).[36] The key advantage that is exhibited by MOS capacitors made with Pr_2O_3 layers is that they exhibit ultra-low leakage current densities. That is, leakage currents of 5 nA/cm^2 at 1V have been demonstrated for Pr_2O_3 layers with EOT of 14-Å. This is four orders of magnitude lower than the best reported results for other high-*k* materials. Such Pr_2O_3 films also exhibit an effective dielectric constant of 31. In addition, no significant hysteresis is observed in the C-V measurements, provided the material is furnace annealed only briefly at 600°C. These layers can also survive an RTA anneal of up to 1000°C for up to 15 seconds with no degradation in electrical properties.

The Pr_2O_3 layers are polycrystalline. Ordinarily, the dielectric layers of MOS capacitors have an amorphous structure, and if the structure begins to become polycrystalline, the films become leaky. High-temperature post-deposition anneals tend to crystallize the dielectric films, and this can cause an increase in the leakage currents through other dielectric layers. Apparently this is not the case with Pr_2O_3. It should be noted, however, that although it is possible to deposit polysilicon films on these layers, the poly/Pr_2O_3 interface is rough. (In this work, instead of using polysilicon as the top electrode, a top electrode to the Pr_2O_3 was formed by evaporating gold onto the Pr_2O_3 through a shadow mask.)

The Pr_2O_3 layers in this study were deposited by molecular beam epitaxy (MBE) at temperatures between 625 and 725°C at a growth rate of 0.1-Å/sec. They were then subsequently furnace-annealed at 600°C for 5 minutes in dry nitrogen. A 1000°C/15 sec RTA anneal in N_2 was shown to cause no structural change in the films, and no change in their electrical properties.

REFERENCES

1. *International Technology Roadmap for Semiconductors*, 1999 Edition, Semiconductor Industry Association, San Jose, CA.
2. G. Lucovsky, "Integration of Alternative High-*k* Gate Dielectric Aggressively-Scaled CMOS Si Devices: Chemical Bonding Constraints at Si-Dielectric Interfaces," *Mtg. Abs. Electrochem. Soc.*, Spring 1999, Abs. 495.
3. H. Shimada and T. Ohmi, "Current Drive Enhancement by Using High-Permittivity Gate Insulator in SOI MOSFETs and Its Limitation," *IEEE Trans. Electron Dev.*, **43**, p. 431 (1996).
4. S.C. Song *et al.*, "Ultra-Thin (<2 nm) CVD Si_3N_4-Gate Dielectric for Deep-Sub-Micron CMOS Devices," *Tech. Dig. IEDM,* 1998, p. 373.
5. T. Devoivre *et al., Proc. 1998 IEEE/SEMI Advanced Semiconductor Manufacturing Conference*, p. 434 (1998).
6. Y. Momiyama *et al.*, "Ultra-Thin Ta_2O_5/SiO_2 Gate Insulator with TiN Gate Technology for 0.1 μm MOSFETs," *Digest of Symp. VLSI Technology,* p. 135, 1997.
7. D. Park *et al.,* "SiON/Ta_2O_5/TiN Gate-Stack Transistor with 1.8-nm-Equivalent SiO_2 Thickness," *Tech. Dig. IEDM,* 1998 p. 381.

8. C. Hoover, "Metal Electrode Challenges with the Emergence of High Permittivity Dielectric Materials," *Mtg. Abs. Electrochem. Soc.*, Spring 2000, Abs. 251.
9. C.V. Srividya *et al.*, "MOCVD Processing of Pt Films for High-*k* Bottom-Electrode Applications," *Mtg. Abs. Electrochem. Soc.*, Spring 2000, Abs. 253.
10. S. Kamiyama *et al.*, "Highly Reliable 2.5-nm Ta_2O_5 Capacitor Process Technology for 256 MBit DRAMs," *Tech. Dig. IEDM*, 1991, p. 827.
11. K.S. Tang *et al.*, "Trends in DRAM Dielectrics," *IEEE Circuits and Devices Mag.*, May 1997, p. 29.
12. M.G.-L Tai *et al.*, "Thermal stability of Ta_2O_5 Films in Capacitor Systems Using a TiN Metal Electrode," *Mtg. Abs. Electrochem. Soc.*, Spring 2000, Abs. 254.
13. J.-W. Kim *et al.*, "Development of Ru/Ta_2O_5/Ru Capacitor Technology for Giga-Scale DRAMs," *Tech. Dig. IEDM*, 1999, p. 793.
14. Y. Takaishi *et al.*, "Low-Temperature Integrated Process Below 500°C for Thin Ta_2O_5 Capacitor for Giga-Bit DRAMs," *Tech. Dig. IEDM*, 1994, p. 839.
15. I. Asano *et al.*, "1.5-nm-Equivalent-Thickness Ta_2O_5 High-*k* Dielectric with Rugged Si Suited for Mass Production of High Density DRAMs," *Tech. Dig. IEDM*, 1998, p. 755.
16. Y.S. Kim *et al.*, "The Influence of Surface Roughness on the Electric Conduction Process in Amorphous Ta_2O_5 Thin Films," *J. Electrochem. Soc.*, September 1999, p. 3398.
17. P.C. Joshi and M.W. Cole, "Enhanced Electrical Properties of Al-Doped Ta_2O_5 Thin Films for Microelectronic Applications," *Mtg. Abs. Electrochem. Soc.*, Fall 2000, Abs. 812.
18. Y. Kuo, "Reactive Ion Etching of Sputter Deposited Tantalum Oxide and Its Etch Selectivity to Tantalum," *J. Electrochem. Soc.*, February 1992, p. 579.
19. K.-P. Lee *et al.*, "Inductively Coupled Plasma Etching of Ta_2O_5," *J. Electrochem. Soc.*, October 1999, p. 3794.
20. G. Bersuker *et al.*, "Experimental Observation of Ta_2O_5 Gate Dielectric Transistors," *Mtg. Abs. Electrochem. Soc.*, Spring 2000, Abs. 243.
21. Y. Momiyama *et al.*, "Ultra-Thin Ta_2O_5/SiO_2 Gate Insulator with TiN Gate Technology for 0.1 μm MOSFETs," *Dig. Symp. VLSI Tech.*, 1997, p. 135.
22. I.C. Kizilyalli *et al.*, "Stacked Gate Dielectrics with Ta_2O_5 for Future CMOS Technologies," *Dig. Symp. VLSI Tech.*, 1998, p. 216.
23. H.F. Luan *et al.*, "Ultra-Thin High-Quality Gate Ta_2O_5 Dielectric Prepared by *In-Situ* Rapid Thermal Processing," *Tech. Dig. IEDM*, 1998, p. 609.
24. *ibid., Ref. 7.*
25. G.F. Derbenwick and A.F. Isaacson, "Ferroelectric Memory: On the Verge of Breakthrough," *IEEE Circuits & Device Magazine*, January 2001, p. 20.

26. V. Sivaramakrishnan *et al.*, "High-*k* Dielectrics for Advanced DRAM Applications," *Semiconductor Fabtech*, 12th edition, 2000, p. 217.
27. K.W. Kwon *et al.*, "Ta_2O_5 Capacitors for 1 Gbit DRAM and Beyond," *Tech. Dig. IEDM*, 1994, p. 835.
28. C. Kittel, *Introduction to Solid State Physics, Sixth Edition*, John Wiley & Sons, New York, 1986.
29. F. Jona and G. Shirane, *Ferroelectric Crystals*, Dover Publications, 1993 and references cited therein.
30. R.B. Laibowitz *et al.*, *American Physical Soc. Meeting*, Abs. R28.05 (1996).
31. I.C. Kizilyalli, R.C. Huang, and P.K. Roy, "MOS Transistors with Stacked SiO_2-Ta_2O_5-SiO_2 Gate Dielectrics for Giga-Scale Integration of CMOS Technologies," *IEEE Electron Device Letts.*, November 1998, p. 423.
32. L. Kang *et al.*, "Single-Layer Thin HfO_2 Gate Dielectric with n^+ Polysilicon Gate," *Dig. Symp. VLSI Tech.*, 2000, p. 44.
33. L. Kang *et al.*, "MOSFET Devices with Polysilicon on Single-Layer HfO_2 High-*k* Dielectrics," *Tech. Dig. IEDM*, 2000, p. 35.
34. S.J. Lee *et al.*, "High Quality Ultra Thin CVD HfO_2 Gate Stack with Poly-Si Gate Electrode," *Tech. Dig. IEDM*, 2000, p. 31.
35. B.H. Lee *et al.*, "Characteristics of TaN Gate MOSFET with Ultrathin HfO_2 (8-12Å)," *Tech. Dig. IEDM*, 2000, p. 39.
36. H.J. Osten *et al.*, "High-*k* Dielectrics with Ultra-Low Leakage Current Based on Praseodymium Oxide," *Tech. Dig. IEDM*, 2000, p. 653.
37. D.A. Buchanan *et al.*, "80 nm Poly-silicon Gated *n*-FETs with Ultra-Thin Al_2O_3 Gate Dielectric for ULSI Applications," *Tech. Dig. IEDM*, 2000, p. 223.
38. J.H. Lee *et al.*, "Effect of Polysilicon Gate on Flatband Voltage Shift & Mobility Degradation for ALD- Al_2O_3 Gate Dielectric," *Tech. Dig. IEDM*, 2000, p. 645.
39. B.H. Lee *et al.*, "Ultrathin Hafnium Oxide with Low Leakage & Excellent Reliability for Alternative Gate Dielectric Application," *Tech. Dig. IEDM*, 1999, p. 133.
40. W.-J. Qi *et al.*, "MOSCAP and MOSFET Characteristics Using ZrO_2 Gate Dielectric Deposited Directly on Si," *Tech. Dig. IEDM* 1999, p. 145.
41. Y. Ma *et al.*, "Zirconium Oxide Based Gate Dielectrics with Equivalent Oxide Thickness of Less Then 1.0 nm and Performance of Submicron MOSFETs using a Nitride Gate Replacement Process," *Tech. Dig. IEDM* 1999, p. 149.
42. L. Manchanda *et al.*, "Si-Doped Aluminates for High Temperature Metal-Gate CMOS: Zr-Al-Si-O, A Novel Gate Dielectric for Low Power Applications," *Tech. Dig. IEDM*, 2000, p. 23.
43. C.H. Lee *et al.*, "MOS Characteristics of Ultra Thin Rapid Thermal CVD ZrO_2 and Zr Silicate Gate Dielectrics," *Tech. Dig. IEDM*, 2000, p. 27.
44. K.J. Hubbard and D.G. Schlom, *J. Mater. Res.* 2757 (1996).

45. R. Nieh *et al.,* "Processing Effects and Electrical Evaluation of ZrO_2 Formed by RTP Oxidation of Zr," *Mtg. Abs. Electrochem. Soc.,* Spring 2000, Abs. 247.
46. B. Dance, "Alternative Gate Dielectric Material Shows Promise," *Semiconductor International*, May 2001, p. 58.
47. D. Kotecki, "High-*k* Dielectric Materials for DRAM Capacitors," *Semiconductor International* November 1996, p. 109.
48. K. On *et al.,* "(Ba,Sr)TiO_3 Capacitor Technology for Gbit-Scale DRAMs," *Tech. Dig. IEDM* 2000, p. 803.
49. A. Shanware *et al.,*"Reliability Evaluation of HfSiON Gate Dielectric Film with 12.8Å SiO_2 Equivalent Thickness," *Tech. Dig. IEDM* 2001, p. 137.
50. G.D. Wilk *et al.,* "High-k Dielectrics: current Status and Materials Properties Considerations," *J. Appl Phys.* (Review) **89** (2001) 5243.
51. E.P. Gusev *et al.,* "Ultrathin High-*k* Gate Stacks for Advanced CMOS Devices," *Tech. Dig. IEDM* 2001, p. 451.
52.. Y. Kim *et al.,* "Conventional *n*-MOSFET Devices Using Single Layer HfO_2 and ZrO_2 as High-*k* Gate Dielectrics with Polysilicon Gate Electrode," *Tech. Dig. IEDM* 2001, p. 451.
53. C. Hobbs *et al.,* "80 nm Poly-Si Gate CMOS with HfO_2 Gate Dielectric," *Tech. Dig. IEDM* 2001, p. 651.
54. R. Wallace and G.D. Wilk, "Identifying the Most Promising High-*k* Gate Dielectrics," *Semiconductor International*, July 2001, p. 227.
55. R. Chau and G. Marcyk, "New Transistors for 2005 and Beyond – The TeraHertz Transistor," Intel Press Release, Nov. 26, 2001.

Chapter 5

THE STRUCTURE OF DEEP-SUBMICRON MOSFETS: Retrograde-Wells; Super-Steep Retrograde Channel Doping; Drain Engineering; Punchthrough-Control Structures

In Chapter 1 we discussed the evolution of MOSFET structures down to the sub-micron technology nodes in production at the time publication of this text. In this chapter we delve further into the details of the MOSFET regions that are fabricated in such deep-submicron MOSFETs, namely: **1**) retrograde wells for CMOS twin-well technology; **2**) super-steep retrograde (SSR) channel doping profiles; **3**) drain engineering issues in deep-submicron MOSFETs; and **4**) punchthrough-control structures.

5.1 WELL-FORMATION IN DEEP-SUBMICRON CMOS

Twin-well CMOS is now the dominant CMOS technology in IC production. In twin-well CMOS each well is formed separately by implanting the well dopants into a lightly doped substrate region. Having two well implants allows each well-type to be tailored separately, permitting independent device performance adjustments for both PMOS and NMOS transistors. When first introduced, twin-well CMOS structures were fabricated by using moderate-energy implants to place both types of well dopant near the wafer surface. These dopants then had to be driven into the substrate by a long (12-24 hour), high-temperature (>1100°C) diffusion process (Fig. 5-1). Such a deep drive-in step was needed to prevent punchthrough from the MOSFET drain regions at the surface to the substrate under the well. This type of CMOS-well-structure is referred to as a *diffused-well.* However, use of diffused-well structures gives rise to some problems which limits their use when CMOS is scaled below 0.5 μm:

1. Lateral (as well as vertical) diffusion occurs during the diffusion step, which has the effect of reducing packing density. For example, since the lateral diffusion is ~0.7 times the junction depth, a typical diffused well having a vertical junction depth of 3-4-μm, results in a lateral spread of well dopant of 2-3-μm. Consequently, the maximum aspect ratio of a diffused well is limited to about 1.6:1 (i.e., wells narrower than about 1.6 times their depth would intermingle with the dopant of adjacent wells).

2. If LOCOS isolation is used, boron from the *p*-well-dopant implant-step segregates into the oxide during the long, high-temperature field-oxide growth step. Compensating for this effect requires a separate field-threshold-adjust implant (also referred to as a *channel-stop-implant*), prior to the field-oxidation step. This additional step adds process complexity.

3. Diffusion of the channel-stop-implant dopant into the active regions during field oxidation adversely affects both the junction-capacitance and the turn-off characteristics of the NMOSFETs (see Vol. 3).

4. The lengthy, high-temperature drive-in-step increases process cycle time, and increases the risk of silicon wafer warpage, bow, and flatness degradation (and the likelihood of metal contamination from furnace heating elements).

5. The diffused-well profile for a particular technology node is the net result of compromises between the need to control threshold-voltage, to suppress hot-carrier effects, to control punchthrough and soft-error effects, and to provide latch-up immunity. For example, a high dopant concentration is desirable deep in the well to strengthen latch-up immunity. This cannot be achieved in a diffused-well because doping at the bottom of such a well must be lower than at the surface, and the maximum surface doping is set by limits on junction breakdown voltage and junction capacitance.

For all of the above reasons, the diffused-well approach to forming a twin-well CMOS process becomes less effective for each smaller technology node. As a result, an alternative method of forming the wells was sought. The technique adopted to replace the older diffused-well structure when CMOS technologies are scaled below about 0.5-μm involves *retrograde-well formation* with *high-energy ion-implant steps*.

5.1.1 Retrograde-Wells for Deep-Submicron CMOS

The medium-energy implant steps used to introduce well-dopants into the substrate in the diffused-well approach (50-100-keV), can be replaced with high-energy-implant steps (300-keV to 1-Mev). Such high-energy implants cause the peak of the implant to be located considerably deeper within the silicon substrate (at roughly 1-μm for the energies most frequently used in high-energy implant processes). Since the implant peak is now far beneath the silicon surface, the impurity concentration thus decreases as the doping profile approaches the wafer surface. Since the well-profile in this case is different from that of a conventional

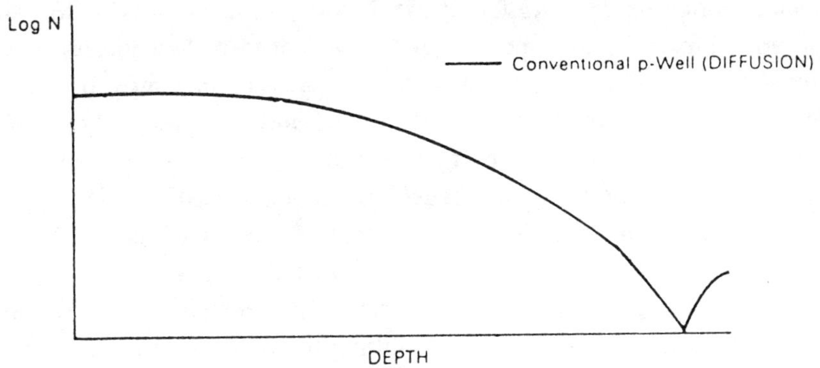

Fig. 5-1 Doping profile in a conventional well formed by a shallow implant and drive-in step.

diffused well (in which the doping concentration after drive-in is highest at the surface, and decreases monotonically with depth, as seen in Fig. 5-1), such deeply implanted wells are known as *retrograde-wells.** Comparison of the doping profiles in conventional and implanted *p*-well profiles (obtained by simulation) is shown in Fig. 5-2.[2,3] The retrograde-doping-profile is retained by minimizing the temperature-cycles of later process steps (e.g., by using RTP instead of furnace anneals).

The retrograde-well approach to forming the twin-well CMOS-structure has several major advantages over the diffused-well process:

1. Much less lateral-dopant-spread occurs for a retrograde-well having the same junction depth as a diffused-well. This is because ion implantation is a more anisotropic dopant-distribution process than diffusion (i.e., the lateral-straggle for implant energies of 500-keV is only ~0.7-μm). In addition, little additional diffusion occurs after the implant is performed, since only a short anneal at a relatively low-temperature is needed to activate the implant. Thus, the total lateral-dopant-spread barely exceeds 1-μm after all thermal processing (compared to at least 2-μm for the same well-depth using a diffused-well process).[2] Figure 5-3 depicts the doping concentration profile of a high-energy implanted *p*-well. In this example, the implanted *p*-well is formed by boron implantation at 400-keV.

In addition, high-energy implanters can place dopant beneath a shallow-

* In order to avoid excessive spread of the active-device electrical parameters, the surface concentration after the retrograde implant should be at least five times smaller than its final value - which will be set by subsequently performed, shallower low-dose implants specifically designed for threshold and punchthrough control.

trench-isolation oxide or a LOCOS field oxide, allowing the CMOS twin-well structure to be implanted *after* the dielectric isolation features are formed. (In diffused-wells, the well-dopants must be implanted before the field-oxide, and driven-in either before - or during - the field-oxidation step). Thus, since the high-energy implanted wells are formed after the long, high-temperature cycle needed to grow the LOCOS field-oxide, lateral-spread of the well-dopants is further reduced. (A higher threshold voltage can also be achieved in the LOCOS field-regions of the *p*-wells if the boron implant is done following field-oxidation, because the boron does not segregate out into the oxide during the field-oxide growth-step.[1]) Hence, the 7-9-μm n^+-to-p^+ spacing required for a twin-well 1.2-μm CMOS process which uses conventional wells, can be reduced to ~5-μm if retrograde wells are used instead. Such overall reduced lateral-dopant-diffusion permits higher circuit-packing-densities.

2. By combining the high-energy-implant step with the other lower-energy MOSFET channel implants (V_T-adjust and punchthrough stopper) and since minimal thermal diffusion occurs, control of dopant levels throughout the depth of the well becomes possible with fewer tradeoffs (a task that has been dubbed "Well Engineering"). This permits optimization of such key

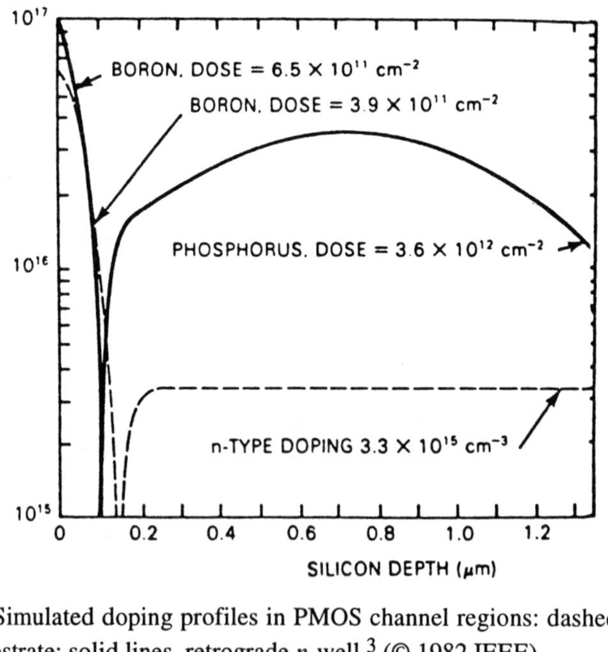

Fig. 5-2 Simulated doping profiles in PMOS channel regions: dashed lines, lightly doped substrate; solid lines, retrograde *n*-well.[3] (© 1982 IEEE).

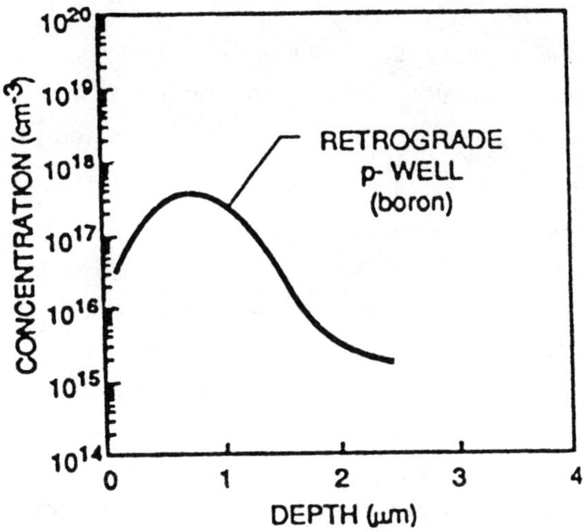

Fig. 5-3 Example of the doping profile of a retrograde *p*-well (boron) implant (400 keV).

MOSFET performance characteristics as: device speed (carrier mobility and, S/D capacitance), inter-device leakage-currents, V_T-control, punch-through suppression, parasitic-transistor effects, and soft-error and latch-up resistance. That is, low surface-doping levels increase mobility and decrease capacitance. Heavier doping below the surface (retrograde doping) suppresses minority carrier collection by decreasing depletion widths; thus reducing soft errors and latchup susceptibility. This degree of flexibility in optimizing device parameters is not possible with diffused-wells.

Besides the primary benefits of increased packing-density and greater process flexibility, retrograde-wells offer the following secondary advantages:[5]

1. A retarded electric field is created in the parasitic-vertical-bipolar-transistor, thereby reducing its current gain. This affords some extra protection against latchup.

2. Susceptibility to vertical-punchthrough is reduced without having to significantly raise the surface-doping-concentration.

3. The conductivity in the bottom of the well is increased, which also provides additional resistance against latchup.

4. The shallower well depth should also make the drain of the parasitic n^+-to-n-well (or p^+-to-p-well) parasitic-FETs shallower. This contributes to improvements in both the short-channel threshold-roll-off-effect and the in the punchthrough characteristics of these parasitic field-region devices (see Vol. 3, Chap. 6).

5. The formation of retrograde-wells with high-energy implants results in a reduced-doping-concentration under the bottom of the source and drain regions, which reduces the source/drain-to-body junction-capacitance.

However, some concerns associated with the retrograde process should also be mentioned. First, the process requires a thick photoresist as a masking material for the high-energy implant. Typical resist thicknesses are 2.0-2.5-μm, and special exposure and development techniques are required for this process. Implant shadowing arising from the use of such thick resists must also be addressed. (We will address the challenges associated with the use of thick resists as high-energy-implant masks in more detail at the end of this section.) Questions have also been raised about the possibility that the high-energy implants could cause damage that may degrade device and circuit performance.

5.1.2 Process Integration Issues Involving the Use of Retrograde Wells

Advanced twin-well CMOS processes with both retrograde *n*- and *p*-wells (using short [~5 min] RTP well-drive-in steps), were reported for submicron applications as early as 1991.[6] Further advances continued throughout the decade. Scaling CMOS to 0.5-μm and below reduces the well depths required to prevent vertical punchthrough and latchup. In turn, as device and isolation dimensions shrink with each device node, retrograde-well implant-energies for both DRAM and logic circuits are also gradually being reduced. However, the retrograde-well implants used for DRAM devices typically require higher doses than those used in logic devices (i.e., to achieve better minority-carrier suppression [soft-error rate]). This, in turn, requires a higher well-implant-energy (to avoid excessive dopant in the near-surface [channel] region). In addition, the well-depth and well-dopant-concentration can be reduced if additional, lower-energy implants are precisely located with respect to the transistor structure (to suppress punch-through of the MOSFET channels and enhance the field-isolation characteristics).

The high-energy implant step used to form retrograde-wells in deep-submicron-CMOS is typically performed together with the V_T-adjust, punchthrough, and channel stop (isolation) implants. These implants are all done in the same implanter, without removing the wafers from the implant process chamber in between (unless channel engineering implants [that use high wafer-tilt-angle for precision dopant placement] are required - see Sect. 5.4). Such a "chained" implant sequence is performed by doing the highest-energy implant first, with the lowest-energy implant being done last. This ensures that no ions traverse the end-of-range (EOR) damage of a previous implant. Note that because of economic requirements for high equipment utilization, this sequence must be carried out in an implanter that has the capability of doing not only high-energy implants (e.g., up to about 1-MeV for the retrograde *n*-well implant), but also lower-energy boron implants (down to about 10-keV, for the NMOS V_T-adjust implant).

An example of a "chained implant" retrograde twin-well CMOS process is given in Ref. 7. Here, a retrograde *n*-well is formed by a 600-keV, 3×10^{13} cm^{-2} phosphorus implant. Then, a second phosphorus implant (150-keV, 4×10^{12} cm^{-2}) is used to obtain both punchthrough prevention and intra-field isolation (channel stop), and a third phosphorus implant (70-keV, 6×10^{12} cm^{-2}) sets the threshold voltage. In an identical manner, three selective boron implants are placed in the *p*-well region. The retrograde *p*-well implant is boron at 300-keV, 3×10^{13} cm^{-2}. Then a punchthrough/channel-stop-implant at 150-keV, 4×10^{12} cm^{-2} and a threshold-adjust-implant at 30-keV, 5.5×10^{12} cm^{-2} are added. The process displays very good inter-well isolation characteristics, which allows a minimum p^+-*to*-n^+ spacing 40% lower than the same process without retrograde wells. It also achieves very good intra-well isolation, allowing n^+-to-n^+ (and p^+-to-p^+) device separation of 0.5-μm, without an additional channel-stop implant.

From a manufacturing perspective, the well-formation sequence thus appears to be a single implant step, reducing manufacturing complexity and cost. Figure 5-4 compares an advanced diffused-well process with a high-energy implanted process up to gate oxidation, showing the process simplification possible with "well-engineering." Eliminating process steps has other benefits besides reducing cycle time. These include a reduction in work-in-progress (WIP) inventory, less wafer handling, and fewer defects.

The cycle-time and cost savings obtained from such a "chained-implant" sequence was assessed by SEMATECH. They sponsored a comparison of a baseline diffused-well process with a high-energy retrograde-well process at two different chip manufacturers. The high-energy processes were engineered to achieve minimum cycle-time and maximum cost savings, while maintaining the same basic device and circuit parametric data (and thus were identical to the baseline processes after gate-oxidation). Significant improvements in cycle-time, cost, and device-performance were achieved by both companies.

By chaining these implants, the doping-profile in the well-regions of the MOSFETs has a *modulated* characteristic as depicted in Fig. 5-5 (where the term "modulated" means that the doping-concentration-profile has several maxima and minima).[7] The incorporation of such multiple implants to form a modulated well has become a firmly established processing approach in chip production.

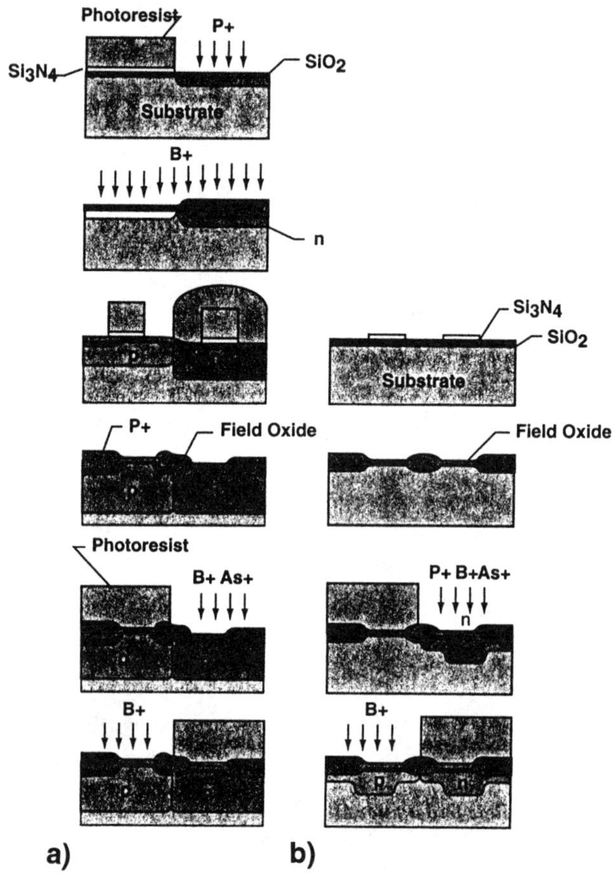

Fig. 5-4 Fabrication sequences for: a) conventional-diffused-well and; b) high-energy-implanted-well processes.

Another issue of using high-energy implants to introduce the well and isolation dopants involves implant shadowing. That is, as shown in Fig. 5-6a, thick photoresist films must be used as masks against the high-energy implant (e.g., 2.0-2.5 μm thick). The doping asymmetry resulting from high energy implants with a small (7°) tilt angle can have significant effects on retrograde well structures and inter-well isolation. Figure 5-7 depicts the simulated doping profiles from a 1 MeV phosphorus implant with a 7° tilt.[8] The simulation accounted for the fact that opposite corners of the photoresist will wear down and carbonize differently because of the asymmetry in the incoming ion beam. As the sidewall shape of the photoresist changes with time, this will affect the implanted profile in the silicon. The doping profiles in Fig. 5-7 clearly show a bulge of lightly doped material on the shadow side of the profile. More seriously, on the

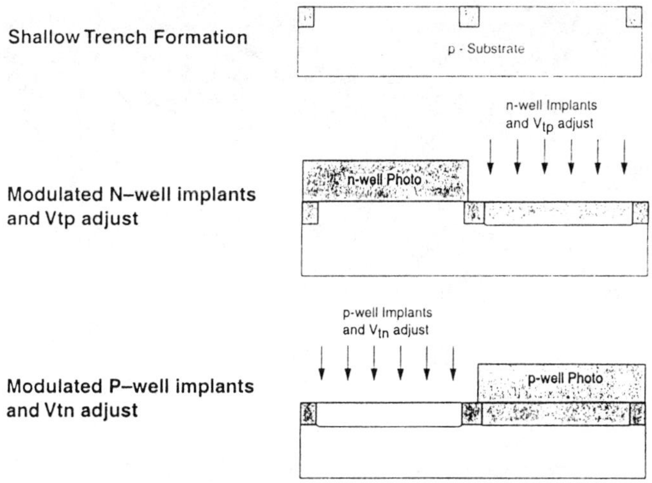

Fig. 5-5 Modulated *n*-well structure for advanced devices, showing the functions of each implant.[7]

incoming side, a channel doped to 20% of the peak-concentration extends from the projected range up to the surface. Doping variations such as these can adversely affect the mobilities and junction-capacitances of the resulting devices. This will also make it harder to maintain adequate isolation structures as isolation widths scale below 400-nm. These problems can be mitigated by reducing the tilt-angle to the 2-4° range, using 0°-tilt. However, using 0°-tilt-implants is not always possible, due to channeling effects (which are severe at high energies).

An alternative solution to the implant shadowing problem during high-energy

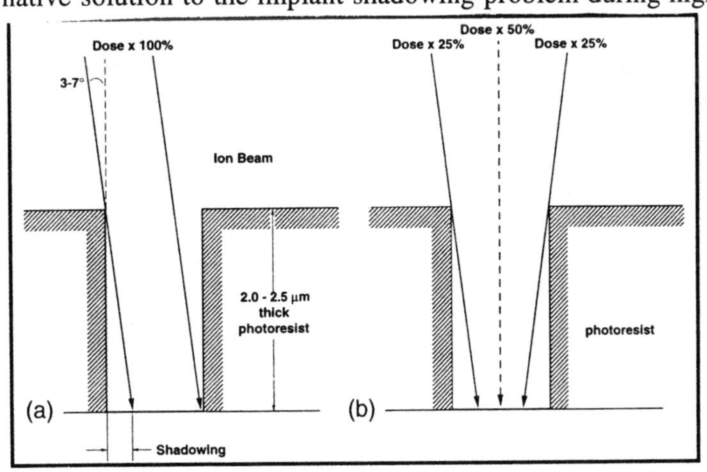

Fig. 5-6 The geometric effects of ion implantation into a narrow active region: a) low-tilt-angle shadowing and well misalignment; b) low-tilt quad implant provides both lateral and vertical (no channeling) uniformity.

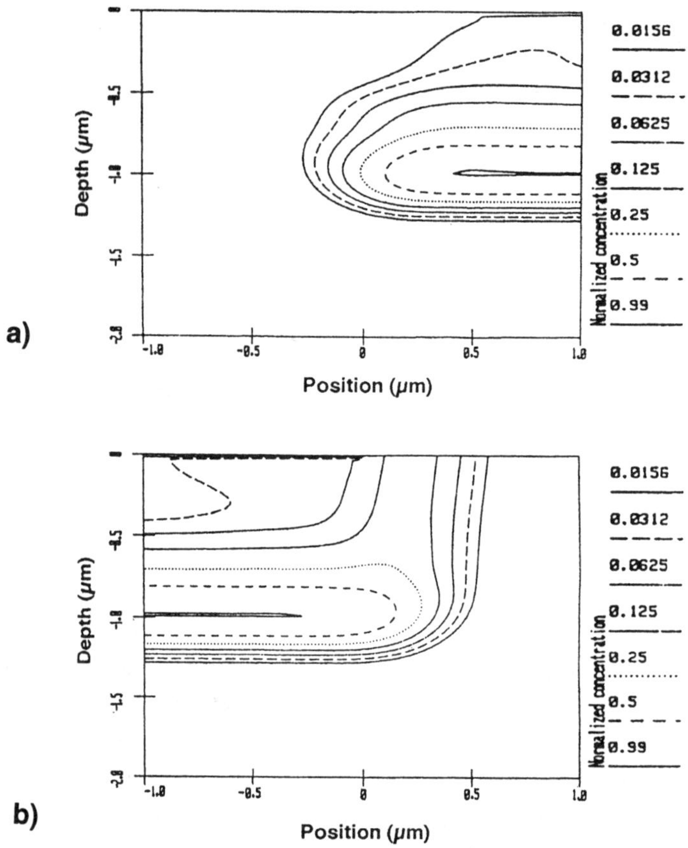

Fig. 5-7 Normalized doping profile of 1 MeV phosphorus implantation. a) On the shadow side (a) a lowly doped bulge is found (according to Monte Carlo simulation). b) On the incoming side, Monte Carlo simulation shows a trunk.[8]

7°-tilt implants is to reposition the wafers part-way through the implant process. If the wafers are repositioned four times, and 25% of the total dose is implanted with the wafer in each position (*quad implant*), all the areas of the exposed silicon surface will receive the same implant dose, eliminating the implant-shadowing problem (Fig. 5-6b). That is, well profiles are symmetric, both laterally and vertically, and shadowing effects near the surface are compensated.

A high-energy batch-implanter capable of performing low-tilt quad-implants without unloading the wafers from the disk addresses the isolation scaling problem with high productivity. Since it is not practical to scan a high-energy ion beam, implanters with fixed beams and batch end-stations using simultaneous disk spinning and scanning is the method-of-choice for process uniformity. The

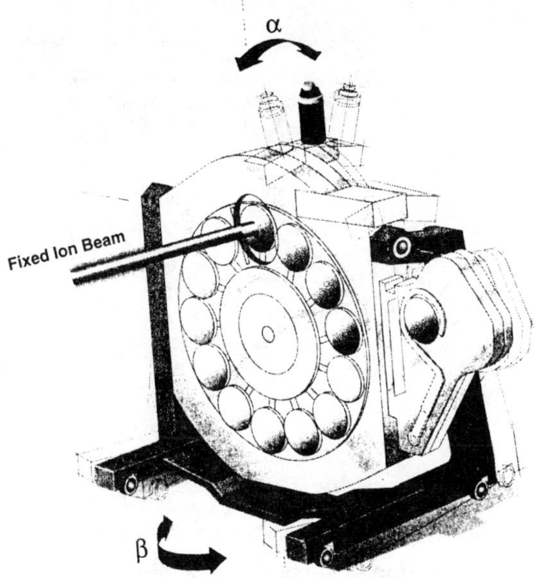

Fig. 5-8 *In situ* quad implants can be run on a batch end-station with orthogonal tilt axes (designated α and β).

quad-implant capability is achieved with a disk chamber as shown in Fig. 5-8.[9] Here the disk can be independently tilted in two orthogonal directions. Such batch high-energy implanters are sold by Axcelis Corp. Single-wafer high-energy implanters are also commercially available. A model offered by Applied Materials also permits tilting the wafer and repositioning it several times during an implant step. Higher tilt-angle control is claimed for such single-wafer high-energy implanters compared to the batch types.[10]

5.2 SUPERSTEEP RETROGRADE (SSR) CHANNEL DOPING PROFILES

The appropriate choice of substrate doping and engineering of doped layers under the gate are critical to optimum performance of deep-submicron MOSFETs (with respect to drain-to-source punchthrough, threshold voltage, and channel mobility). Specifically, the doping-density in the region between the source and drain must increase as the gate-length is reduced to prevent punchthrough. As shown in Fig. 5-9, the mean channel-doping-density needed to control punch-through when gate lengths scale below about 0.13-μm begins to exceed $5x10^{17}/cm^3$. However, to avoid reduction of carrier mobility in the channel, the doping at the surface must be kept below $5x10^{17}/cm^3$ to obtain the universal mobility curve.[11] To allow the surface concentration to be kept below this value, and still have a mean doping concentration in the channel above this number, a vertical *retrograde doping concentration profile* is needed in the channel region.

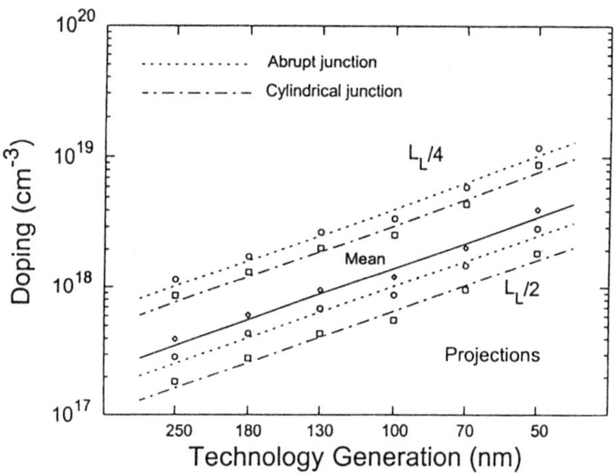

Fig. 5-9 Doping density required for punchthrough control.[11]

Assume that a lighter doped surface layer with a concentration of $1 \times 10^{17}/cm^3$ exists down to a depth X_I (taken to be approximately 1/2 of the source/drain junction depth X_j), and that at that depth the doping abruptly changes to the value needed for the heavily doped punchthrough-control layer. Figure 5-10 shows the scaling required for X_I at each technology generation. Because the transition in doping concentration from the lighter doped region near the surface to the heavily doped punchthrough-control region below must occur over such a short distance, the gradient in the doping concentration must be very steep. Hence, such a doping concentration profile is termed a *super-steep retrograde* (SSR) profile.

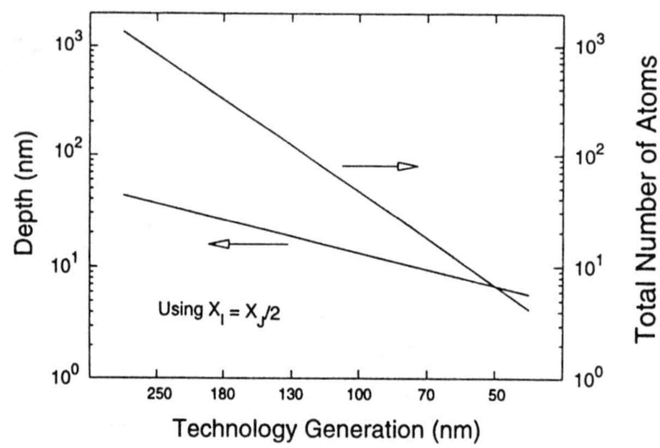

Fig. 5-10 Typical thickness of lighter-doped layer and total number of impurity atoms for $10^{17}/cm^3$ surface doping density.[11]

The formation of SSR channel-doping-profiles for technology nodes from 0.18-μm down to about 0.1-μm is done using ion implantation (in place of the conventional V_T-adjust implant step). In these implant processes, however, higher-mass dopants are substituted for the dopants typically used to carry out the V_T–adjust implants. That is, indium (In) is used instead of boron (B) for the NMOS devices, and either arsenic (As) or antimony (Sb) replace phosphorus (P) in PMOS devices. Heavier implant species provide steeper as-implanted profiles (due to their smaller straggle), and the heavier ions have reduced diffusivity (keeping the as-implanted profile from spreading out during subsequent thermal processes). Thus, well-profiles implanted with In and As (using implant energies in the 100-200-keV range) can be fabricated with sufficiently low doping (~$5x10^{16}/cm^3$) at the gate-oxide interface, and a higher peak doping level (~$1x10^{18}/cm^3$) 80-nm below the gate oxide. (Note that peak doping concentrations of indium cannot be made higher than about $10^{18}/cm^3$ because of the solubility limit of indium atoms in silicon.)

Figure 5-11a shows indium doping-profiles for various annealing temperatures, up to 1000°C.[12] It is evident that little change occurs in the as-implanted indium doping profile for annealing temperatures up to 1000°C, making it a promising candidate for forming SSR channel-doping-profiles. Indium implants for retrograde-channels in NMOSFETs have been reported in some cases to *improve* subthreshold-slopes relative to conventional boron implanted-channels,[12,13] while in another case it was found to have *degraded* the subthreshold-slope.[14] It is generally agreed that SSR doping-profiles are effective at reducing short-channel-effects and drain-induced-barrier-lowering, but it is not equally certain whether this translates into overall improvement in device performance (as this also depends on many other variables, including gate-oxide-thickness, gate-length, and junction-doping). Figures 5-11b and 11c, respectively, show the vertical concentration-doping-profile for an SSR and a conventional process for PMOSFETs (Sb for SSR, and P for the conventional process), and NMOSFETs (indium for SSR, and boron for the conventional process).[15]

The use of SSR channel-doping-profiles is entering some mainstream process flows, despite disagreements over their advantages compared to conventional channel-profiles. In late 1999, the successful fabrication of 100-nm-gate-length devices was reported. They were built using SSR channel-profiles, source/drain-extensions, and deep source/drain-regions for contact. A 100-nm-process yielded devices with 1.04-mA/μm (NMOS) and 0.46 -mA/μm (PMOS) on-currents at 1.5 V with 3-nA/μm off-current. This process uses indium and arsenic as the implant

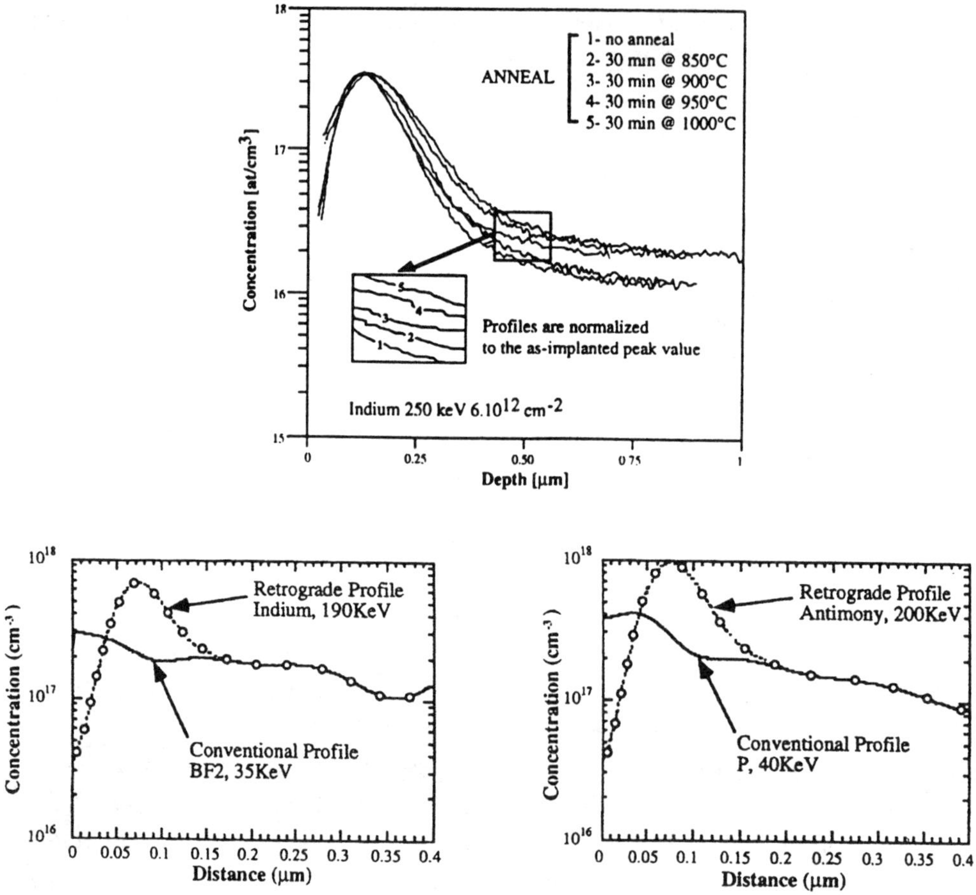

Fig. 5-11 a) SIMS data of indium profiles over a large range of annealing temperatures. Conventional and retrograde doping profiles (simulated) for b) NMOS and c) PMOS .[12]

species for the SSR-channel-doping-profiles, low-energy boron and arsenic source/drain-extensions, and high-angle pocket-implants.[16]

The source for indium ions in implant processes is $InCl_3$, a solid material at room temperature. The production of ion beams from solid $InCl_3$ presents some special challenges for ion implanters. First, $InCl_3$ is hygroscopic (i.e., it readily absorbs moisture), and thus may require excessive outgassing if not handled properly outside of the ion-source vaporiser. In addition, although $InCl_3$ is a solid, it has a low melting-point. Thus, indirect heating from a second vaporiser in the ion-source may cause unintentional vaporization. Both of these effects can lead to premature loss of the source charge, as well as the problem of indium coating in and around the source area. Finally, even during normal operation, indium tends to drift down the beamline, covering the beamline components. This creates a risk of cross-contamination, since the indium atoms can be dislodged from these surfaces if they are struck by ion beams of other species.

Two ion-implanter features can adequately suppress such effects. First, active cooling in the source eliminates indirect heating, decreasing indium-loss and excess beamline-coating. Second, an electrostatic (or other) filtering element near the end-station greatly reduces the magnitude of such cross-contamination, allowing all species to be run on a single implanter. The latest generations of single-wafer, medium-current-implanters incorporate such features.[17]

5.2.1 Fabricating SSR Profiles for Sub-0.1 μm MOSFETs

When gate-lengths scale below 0.1-μm, it probably will no longer be possible to create SSR channel-doping-profiles using ion implantation (even if In and As are used as the implanted species). For example, if a lightly-doped surface-layer is established for a 50-nm-generation MOSFET (with a doping concentration of $1x10^{17}/cm^3$), the heavily-doped punchthrough-control layer ($10^{19}/cm^3$) will have begin at a depth of just 7-nm beneath the silicon surface. Since the width of the surface-inversion-layer is typically thinner than 7-nm, the channel electrons will still be located primarily in the lightly-doped surface-region. However, thin surface-layers with such an abrupt doping-concentration-increase cannot be fabricated with ion implantation (and its ancillary annealing processes). Other doping techniques, such as low-temperature selective-epitaxy (of perhaps Si-Ge layers[19]) will likely be needed for 50-nm and beyond MOSFETs.*

5.3 SOURCE/DRAIN ENGINEERING IN DEEP-SUBMICRON CMOS

In deep-submicron MOSFETs, the typical source/drain contact-structures consist of two regions: **1**) A shallow source/drain-extension-region (SDE), with a junction-depth X_j; and **2**) A deeper contact-source/drain-region, with a junction-depth X_{jC}. As also described in Chap. 1, when MOSFETs were first designed with such two-region source/drain-structures, the first-region had lighter doping than the second-region (and it was in fact called a *lightly-doped drain* (*LDD*) re-

* Figure 5-10 also shows a graph of N_T (the total number of impurity atoms in the lightly doped layer) versus technology generation. For 50-nm and below, this value approaches a small number (< 10), so that for cases where the doping density is important to device operation, any device parameter depending strongly on N_T will show large statistical fluctuations from device-to-device. If all parameters depend only very weakly on N_T then the layer may as well be undoped,[18] so that V_T-control relates to the positioning of the step (i.e., the increase in doping) and the doping in the heavily doped layer.

gion. This region was added to the single-region MOSFET drain-structures at about the time when MOSFETs were scaled below 1.5-μm. Its function was to reduce the peak electric-field (and thus hot-electron-injection) near the drain-end of the channel. However, as MOSFETs gate-lengths were scaled to the deep-submicron regime, the power-supply voltages also had to be reduced, and this resulted in a significant decrease in the peak electric-field near the drain. Hence, the LDD region was no longer needed to suppress hot-carrier-injection effects.

However, in deep-submicron MOSFETs the SDE-region remains a part of the drain structure. But, its role now is to reduce short-channel-effects. The drain contact-region is required to accommodate the thickness of the salicide contact-layer and minimize junction-leakage. Nevertheless, the junction depth X_{jC} of the contact-region should not be larger than that needed to meet these conditions. That is, the greater the junction-depth, the larger must be the width of the sidewall-spacer. This, in turn, will result in a longer SDE-region (which increases the value of the parasitic SDE-resistance, R_{SDE}). In Sect 5.3.1.3, we will see that to enable a MOSFET to deliver improved performance as the gate length is scaled, it is critical that the source/drain contact design should minimize both the parasitic series-resistance-effects and the salicide-substrate contact-resistance. In addition, to ensure that short-channel-effects are effectively controlled as the MOSFET is scaled, both X_j and X_{jC} will also have to appropriately scaled,.

Figure 5-12 shows the depths of the source/drain-region junctions and undoped channel-region as MOSFETs are scaled below 200-nm (where L_L is the lithographically-drawn gate-length), as taken from the ITRS 1999 study.[20] The design rules used in ITRS 2001 to calculate these same values have been relaxed slightly from those in the 1999 study: **1**) the source/drain-contact junction-depth is 1.1 L_L (±33%), compared to 0.8 L_L; **2**) the source/drain-extension junction-depth is 0.55 L_L (±25%), compared to 0.4 L_L; and **3**) the undoped region near the Si/gate-oxide interface in a super-steep retrograde doped channel (SSR) is 0.3 L_L (±25%), compared to 0.2 L_L. Creating and maintaining these ultra-shallow junctions remains a fundamental challenge for doping and metrology techniques.

In this section we will discuss two subjects related to the formation of the source/drain-structure in deep-submicron MOSFETs, namely: **1**) the issues involved in keeping the value of the resistance of the drain region R_S (which is the sum of the parasitic resistance components and contact resistance), to a value that is sufficiently low to meet the requirements of scaled-MOSFET device-design, and **2**) the process technology needed to form shallow SDE-junctions.

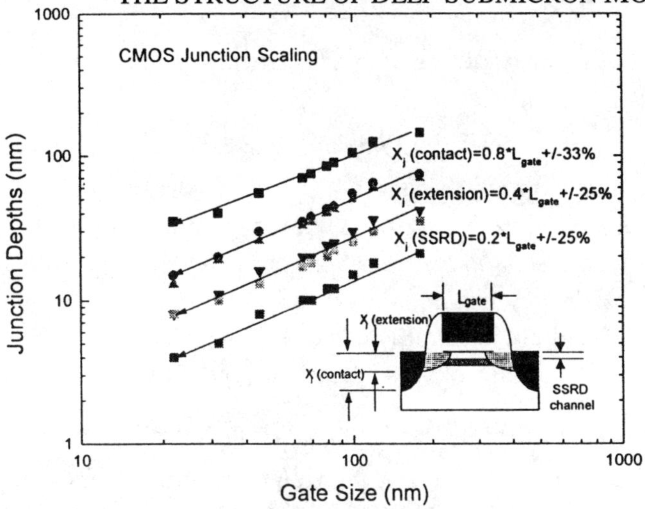

Fig. 5-12 Source/drain junction and undoped-channel depths for planar CMOS on bulk-Si.

5.3.1 Resistance of Deep-Submicron Source/Drains

In the planar MOSFET, the device-region in which transistor-action occurs is the *channel* under the gate. The current coming from the metal-interconnect lines must flow from the metal-Si-contact to the edge of the channel through the source or drain regions (Fig. 5-13). In the MOSFET, current enters the contact perpendicular to the wafer-surface, and then travels parallel to the surface in order to reach the channel. The total parasitic series-resistance, R_S, of the current path from the contact to the edge of the channel of a single-region drain (i.e., one

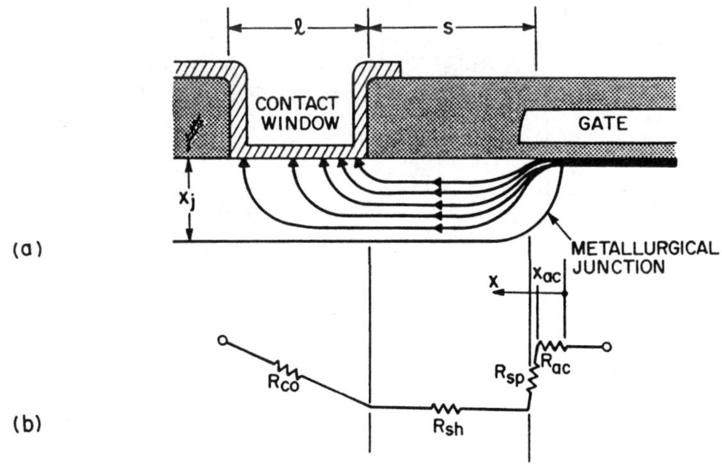

Fig. 5-13 Schematic drawing showing: a) current paths within the source and drain regions, of a planar MOSFET and; b) their corresponding representative series resistance components.[21] (© 1987 IEEE).

that does not have an LDD or a SDE) can be modeled by the sum of four components;[21]

$$R_S = R_{co} + R_{sh} + R_{sp} + R_{ac} \tag{5.1}$$

where R_{co} is the *contact resistance* between the metal and the source/drain region, R_{sh} is the *sheet resistance* of the bulk region of the source and drain, R_{sp} describes the *resistance of the current lines crowding near the channel end of the source*, and R_{ac} is the *accumulation-layer resistance.*

If the MOSFET drain has an SDE, the R_{sh} component of R_S can be replaced by R_{SDE}, as indicated in Fig. 5-14, so that R_S is now given by:

$$R_S = R_{co} + R_{SDE} + R_{sp} + R_{ac} \tag{5.2}$$

In any case, the value of R_S must be viewed from the perspective of how it affects the performance of a MOSFET. Specifically, its value must be compared to that of the channel-resistance R_{sat} of the MOSFET in its ON state:

$$R_{sat} = V_{DD}/I_{Dsat} \tag{5.3}$$

When the external resistance of the source + drain regions is equal to approximately 10% of the value of R_{sat}, the saturation-drain-current I_{Dsat} is reduced by about 8%. In most MOSFET design approaches, the goal is to keep I_{Dsat} from being reduced by no more than 10% due to such series-resistance of the source and drain regions. Thus, it is useful to analyze the components of R_S, so that it can be understood what factors can be adjusted to keep the value of R_S to the value which will meet the I_{Dsat}-degradation design goal.

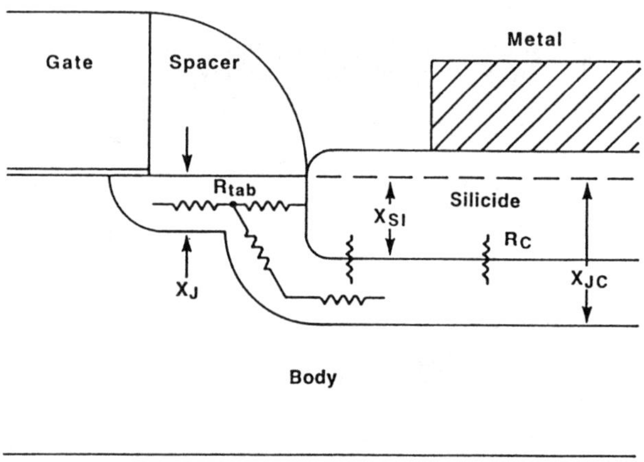

Fig. 5-14 Schematic drawing of the components of source or drain resistance in a deep-submicron MOSFET.

For future scaling, the value of R_{sat} (actually $R_{sat}W$, normalized per unit device width) is given by:

$$R_{sat}W \cong V_{DD}t_{ox}/(0.47v_{sat}\varepsilon_{ox}\,[V_{DD} - V_T]) \cong t_{ox}/(0.376v_{sat}\varepsilon_{ox}) \quad \textbf{(5.4)}$$

Equation 5-4 indicates that since t_{ox} will continue to be made smaller as MOSFETs are scaled, the value of $R_{sat}W$ will also decrease as the devices are scaled. This implies that the value of R_S will also have to be made lower with each technology node. Before we examine in detail the implications of a decreasing R_{sat} (and therefore also a decreasing R_S), we first need to give a more thorough discussion of the physical meaning of the parasitic resistance components of the drain contact region, namely R_{ac}, R_{sp}, and R_{SDE}.

5.3.1.1 Physical Meaning of R_{ac}, R_{sp}, and R_{SDE}: Let us discuss the physical origins of the three parasitic resistance components of the drain/source-regions, with respect to deep-submicron NMOSFET structures. Figure 5-15 depicts the doping gradient of an n^+-p junction at the source (or drain) end of an NMOSFET. The dotted lines indicate contours of constant doping concentration. We define the end of the MOSFET channel as the location where the metallurgical junction occurs (i.e., where $N_D = N_A$ at the surface). When the MOSFET is turned ON, an inversion layer is induced along the length of the channel (i.e., in the substrate-region - where for this NMOSFET the doping is *p*-type). However, because the gate is an equipotential region, the positive bias needed to turn the MOSFET ON, will have a different effect on the silicon region under the gate if it is doped *n*-type. That is, in the region where the *drain doping* overlaps under the gate edge, an accumulation-layer (instead of an inversion-layer) is formed (dark region in Fig. 5-15).

The length of the accumulation-region is determined by the point where the induced (accumulated) carrier-concentration at the surface equals the background doping level. For example, a typical bias induces ~10^{19}/cm^3 carriers in the accumulation layer while the maximum doping level in the source/drain region is in the 10^{20}-10^{21}/cm^3 range. Therefore, the accumulation-layer has a more negligible effect in the more heavily doped region of the drain (i.e., further away from the channel edge). Nevertheless, there is still a component of resistance due to the resistance of the accumulation-layer in series with the channel-ON-resistance, and this component is referred to as the accumulation-resistance, R_{ac}. Both the accumulation-layer and the inversion-layer have about the same thickness (which is given by X_C,and to which we will refer again in the discussion on R_{sp}). *In any case, for deep-submicron MOSFETs, the length of the accumulation-layer is so short, that R_{ac} does not play a significant role in causing the value of R_S to*

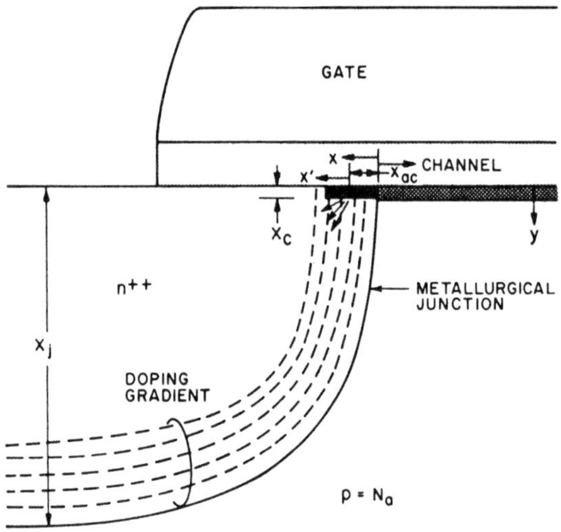

Fig. 5-15 Details of structures near the vicinity of the channel end. The dotted lines are contours of constant concentration. Dark region represents the accumulation layer.[21] (© 1987 IEEE).

rise as the devices are scaled. Hence, we will not consider this component any further in discussing the problem of the increase in R_S as the MOSFET is scaled.

Let us consider R_{sp} next, as this parasitic resistance component does play a role in the increase of R_S as the devices are scaled in the deep-submicron regime. Qualitatively, for a non-abrupt junction (i.e., where the doping contours are spaced over some finite distance, as depicted in Fig. 5-15), R_{sp} arises because after leaving the channel, the current does not immediately spread into the bulk of the source-drain regions (where the resistance effect is given by R_{SDE}). Instead, the carriers leave the accumulation layer and spread out as they move from the thin accumulation layer into the (laterally) uniformly-doped SDE-region of the source or drain SDE (Fig. 5-16).[22] This resistance component is called the spreading resistance, R_{sp}.

An analytical approximation for calculating the value of R_{sp} is given by:

$$R_{sp} \sim (2\rho/\pi W) \ln (\beta X_j/X_C) \tag{5.5}$$

where ρ is the resistivity of the SDE-region, and X_C is the inversion-/accumulation-layer thickness, and β is a constant that varies from 0.37 to 0.9 (depending on the method of derivation). For the purpose of this discussion, an intermediate value of 0.58 will be used.[21]

The value of the series-resistance component R_{SDE} of the SDE-region as shown in Fig. 5-14 can be estimated from (i.e., $R = \rho l/A$):

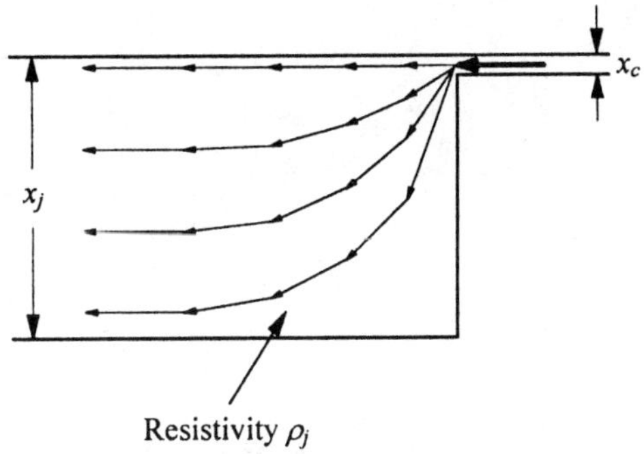

Fig. 5-16 Schematic diagram showing the resistance component associated with the injection region where the current spreads from a thin surface layer into a uniformly doped source or drain region.[22] (© 1983 IEEE)

$$R_{SDE} \cong \rho\, l_{SDE}/(WX_j) \tag{5.6}$$

Without resorting to a two-dimensional analysis to calculate the length of the SDE region l_{SDE}, a reasonable estimate of its value is $L_L/2$ (where L_L is the lithographically drawn gate-length). Thus, since $X_j \sim 0.55\, L_L$ (as described in the introduction of this section), with scaling the ratio of l_{SDE} to the SDE junction-depth X_j will remain approximately 1.2.

The sum of R_{sp} and R_{SDE} will give the value of the parasitic series resistance component of the SDE region of the drain contact structure. That is, when Eq. 5.5 and 5.6 are added together we get:

$$(R_{sp} + R_{SDE})W \cong \rho\,(l_{SDE}/X_j)\left[1 + (2X_j/\pi\, l_{SDE}) \ln (\beta X_j/X_C)\right] \tag{5.7}$$

From Eq. 5-7 it can be seen that, in effect, R_{sp} multiplies the resistance of the SDE region by the quantity given in the brackets of Eq. 5-7. This quantity is expected to range from a factor of 2.09 at the 250-nm generation to about 1.48 at the 50-nm generation, which indicates the R_{sp} plays a significant role in the parasitic resistance of deep-submicron MOSFETs, and it would be beneficial if methods could be developed that could reduce its value.

5.3.1.2 Comparing the Resistance of the SDE Region (R_{sp} + R_{SDE}) to R_{sat}: With the equations derived in Sect. 5.3.1.1 it becomes possible to estimate the values of (R_{sp} + R_{SDE}) that will be required for MOSFETs scaled down to 50-nm. (Note that this analysis was adapted from Ref. 23.) The center curve of Fig. 5-17 shows the calculated values of the channel resistance $R_{sat}W$ (in Ω-cm) of MOSFETs from the 0.25-μm (250-nm) node down to the 50-nm node. The upper and lower curves show what the value of $20(R_{sp} + R_{SDE})W$ would be if the SDE region had sheet resistance values of 500-Ω/sq and 1000-Ω/sq, respectively. Note that

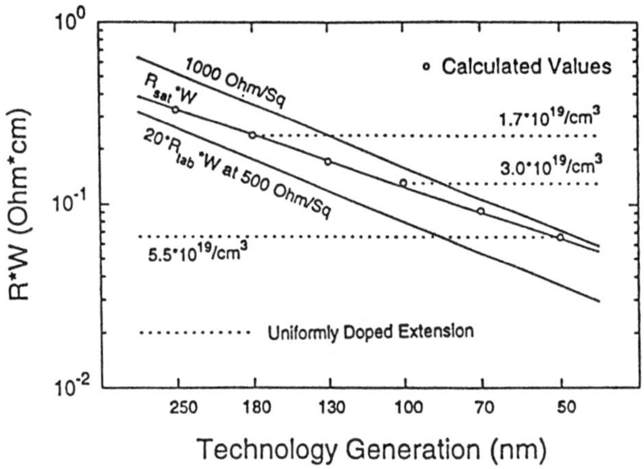

Fig. 5-17 Allowed values of drain (or source) extension resistance considering both bulk extension resistance R_{SDE} and extension spreading resistance R_{sp}.[23]

values are shown for *n*-channel devices, *p*-channel devices would be about 20% higher. In addition, $20(R_{sp} + R_{SDE})W$ is used because the source and drain contact regions each contribute to the resistance in series with $R_{sat}W$.

The data in Fig. 5-17 indicates that the sheet resistance values of $(R_{sp} + R_{SDE})$ will need to be about 650-Ω/sq for 250-nm MOSFETs and about 900-Ω/sq for 50-nm MOSFETs if the value of $20(R_{sp} + R_{SDE})W$ needed to limit the degradation in I_{Dsat} to 8-10% is to be achieved. However, since the junction-depths of the SDE-regions decrease as the gate-lengths are reduced, in order to keep the sheet-resistance-values in the 650-900-Ω/sq range, the average doping-densities in the SDE-regions will have to be increased. As shown in Fig. 5-17, the peak values of doping will have to exceed $10^{19}/cm^3$ (e.g., for the 50-nm node, the peak doping will have be $5.5x10^{19}/cm^3$).

By 2001 it was determined that junctions whose depth was 35-nm could be fabricated with 0.5-keV implants and spike-anneals, but that this was about as low as could be practically achieved (see Sect. 5.3.3). Thus, for 0.1-μm technologies, it will be possible to fabricate SDE-regions with adequately shallow junctions and sufficiently low-values of sheet-resistance. However, it is not clear if it will be possible to do this for 50-nm devices, since the junction-depth will have to decreased to about 20-25-nm, and this will mean that the sheet resistance will have to be significantly increased above 350-Ω/sq.

Also note that the model for R_{sp} given by Eq. 5-5 does not include the effect of the steepness of the lateral-doping-concentration-gradient of the SDE-implant.

As MOSFETs are scaled below 100 nm, this effect becomes the dominant factor that controls the value of R_{sp}. We discuss this issue further in Sect. 5.3.5.

Since increasing the concentration of dopants in the shallow SDE-regions is a difficult task, other ways to reduce the value of the resistance of the SDE-regions have also been sought. From Eq. 5-6 it can be seen that if the length of the SDE region could be reduced, the value of its resistance would also be decreased. While this can be accomplished by decreasing the width of the sidewall spacer, this is not practical in the conventional deep-micron MOSFET-structure. That is, since a deep contact-region is needed (to accommodate the reacted salicide-layer, see Sect 5.3.1.3), the lateral-straggle of the deeper contact-implant would penetrate under the gate edge too far if such a narrower sidewall-spacer is used. However, as we will discuss in Sect. 5.3.2, incorporating elevated-source/drains into the MOSFET structure would permit the use of such narrow sidewall-spacers, thus allowing the resistance of the SDE-region to be made smaller.

5.3.1.3 Parasitic-Resistance R_{co} of the Deeper Contact-Region of the Drain: Although the contact-region of the drain-structure has a deeper junction-depth than the SDE, it can also contribute significantly to the parasitic series resistance of the drain structure, namely through the *contact resistance*, R_{co}. That is, a salicide ($TiSi_2$ or $CoSi_2$) is formed on the surface of the contact-region by sputter-depositing a layer of Ti (or Co) onto the wafer. Then the Ti (or Co) is reacted with the silicon at elevated temperatures, forming the salicide layer. This salicide-layer (and the kinetics of its formation) play a role in the value of R_{co}.

A simple model for R_{co} (or $R_{co}W$) is given by:[21]

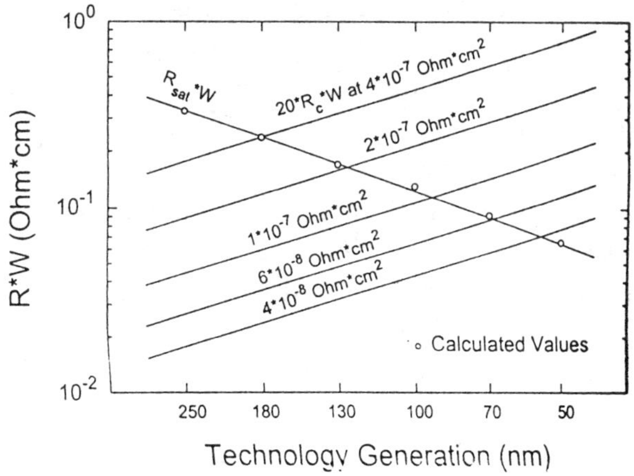

Fig. 5-18 Allowed values of drain (or source) contact-resistance.[23]

$$R_{co}W \geq \rho_c/2\,L_L \qquad \textbf{(5.8)}$$

where ρ_c is the specific contact resistivity of the silicon-to-silicide interface (in Ω-cm^2). This can again be compared with the channel-resistance R_{sat}. Figure 5-18 gives values of $20R_{co}W$ for a 10% contribution of (source + drain) contact resistance. To have a 10% or less contribution, the data in Fig. 5-18 indicates that ρ_c must be in the range of 10^{-7}-Ω-cm^2 for 0.25-μm MOSFETs, but must get down to around $4\text{x}10^{-8}$-Ω-cm^2 for 50-nm MOSFETs. It has been demonstrated that for a barrier height of about 0.5-eV, the required doping-density needed to achieve such a low value of ρ_c is above 10^{20}/cm^3. This means that the silicon regions in contact with the silicide will have be very-heavily doped, a condition that may not be possible to attain in 50-nm MOSFETs with conventional structures, for reasons we describe next.

When the silicide is formed on the source/drain contact-regions by the reaction of Ti (or Co) and silicon, some of the silicon in the substrate is consumed (i.e., when $TiSi_2$ is formed, the silicon consumed is about 75% of that of the thickness of the final silicide film, while for $CoSi_2$, even more silicon is consumed – about equal to the thickness of the $CoSi_2$ that is formed). This has two repercussions. First, the contact-region of the drain must deep enough so that the junction is not entirely lost as a result of the silicide-formation. To keep the implanted dopants from negating the benefits of the shallow SDE-junction (as a result of lateral-straggle that causes the implanted species to penetrate laterally under the implant-mask edge) a sufficiently wide gate-sidewall spacer must be used to mask the contact-region-implant. This impacts the resistance of the SDE-region, through an increase in l_{SDE}. Second, the silicon consumed is that nearest the surface, just where its doping concentration is the highest. The interface between the resultant silicide and the silicon is far below the initial silicon surface, where the implanted doping-concentration is considerably lower (which will impact the value of R_{co}). Furthermore, the dopant in the silicon (especially boron) is known to segregate into the silicide layer, further depleting the doping-concentration at the silicide-silicon interface. Since the contact-region of the drain cannot be made arbitrarily deep (due to the fact that lateral-straggle also increases as the projected-range of the implant is increased), it is not clear that it will be possible to make contact-regions with doping concentrations that are adequately high to achieve the required low-values of ρ_c. The use of elevated-source/drain regions (Sect. 5.3.2) would also allow a way around this obstacle.

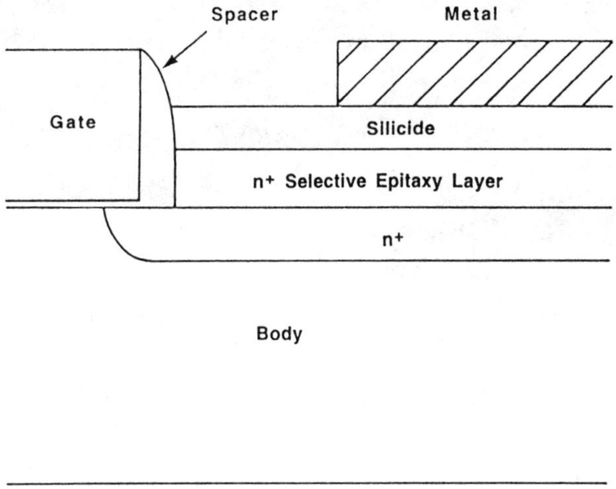

Fig. 5-19 Schematic of elevated-source/drain contacting structure.

5.3.2 MOSFETs with Elevated-Source/Drains

One proposed method for reducing the series-resistance of the source/drain structures in deep-submicron MOSFETs is to modify the basic device structure by adding so-called *elevated* (or *raised*) *source/drains* (Fig. 5-19). In this method, the heavily-doped contact is elevated above the original substrate surface by forming a selectively-grown epitaxial-Si (or perhaps Si-Ge) layer.

The MOSFET with elevated-source/drains has several advantages for devices when they are scaled to 50-nm and beyond. First, the width of the sidewall-spacer can be made much smaller. This can greatly reduce the resistance of the SDE region by decreasing its length l_{SDE}. Second, the n^+ contact layer can be sufficiently thick to contain the salicide layer without junction-leakage problems. Third, the selective-epitaxy-layer can potentially be heavily-doped during growth to obtain a lower contact-resistance. However, two selective-epitaxy processes will be needed since both n^+ and p^+ contacts must be made. The Si-epitaxial-growth steps occur after the SDE-implant. To keep the dopant of the SDE-region from diffusing too-far under the gate-edge, the selective-epitaxial-processes will have to be carried out as sufficiently low temperatures (i.e., probably below 850°C). A technique involving such a low-temperature epitaxial-process for forming elevated-source/drains was described by Miyano *et al.*[24]

In late-2001, Intel announced that the deep-submicron-MOSFET structure it intends to use for the next decade is designed with elevated-source/drains (and an SOI-structure that also employs a high-*k* dielectric [ZrO_2], SSR-channel-doping, and pocket-implants for punchthrough-control).[58]

5.3.3 Shallow-Junction-Formation for Deep-Submicron Source/Drains

The main thrust for achieving shallow SDE region junction depths is to decrease the energy of the SDE-implant. For the source/drain-structures of the 0.13-μm and 0.1-μm technology nodes, single-digit keV energies are expected to be needed for the implants that will form the drain contact regions, and sub-keV energies will be needed for the SDE-region-implants.

5.3.3.1 Ultra-Low-Energy Implants: There are three reasons that reducing the energy of the implant leads to a decrease in the junction-depth. First, the projected range R_p of the implant becomes smaller as the energy is reduced. Second, the straggle and ion-channeling effects both scale with R_p (which is especially important for B implants). Third, TED that occurs during the implant-annealing step is reduced (see Sect. 5.3.3.2). It turns out that for sub-keV B-implants the channeling-effects become relatively unimportant (compared to the broadening of the implant doping profile caused by TED during the annealing step). For this reason, elaborate measures to reduce channeling (such as performing a pre-amorphization-implant prior to the low-energy SDE-implant) are not likely to be used for MOSFETs with gate lengths of 0.1-μm and smaller.

The difficulty with implementing such low-energy-implant processes in manufacturing applications is achieving the productivity that the industry has been accustomed to getting at higher-implant-energies. For example, at energies greater than 10-keV, a high-current-implanter with a beam-current between 10- and 25-mA can implant 230 wafers per hour with the high dose of $4x10^{15}/cm^2$. (Note that beyond 0.18-μm, the doses of the SDE will be in the $1\text{-}2x10^{15}/cm^2$ range, and the doses of the contact-regions of source/drains will be even higher.) At sub-keV energies, the maximum beam-current available on commercial implanters in the mid-1990s timeframe was in the 1-2-mA range. Even after a reduction in dose (made possible by the decreasing energy), the productivity is only fraction of what it previously was. The decreased beam-current at lower energies follows from Child's law, which states that the maximum ion-current that can be extracted from any cathode, drops off as $V^{3/2}$ where V is the extraction voltage. While this basic constraint can be somewhat circumvented (by extracting a larger current at a higher energy and then decelerating the beam to a final lower energy; i.e.., what implant vendors refer to as "decel"), the effective beam-current still decreases with the final energy. With decel, care must also be taken to minimize energy-contamination caused by ion-neutralization before (or in) the decelerating field. That is, ions neutralized while being transported in the beamline will not be counted as part of the dose. Furthermore, they will be implanted at much higher energy, possibly causing deeper junctions (thus defeating the central purpose of using an ultra-low-energy implant).

There are two commercially available high-current, low-energy-implanter configurations: **1**) *batch* (or *multiwafer*) implanters; and **2**) *single-wafer* implanters. Batch type implanters are sold by Axcelis (GSD/200E® model) and Applied Materials (Quantum LEAP®). A single-wafer, high-current, low-energy-implanter (with a scanned ribbon beam) is offered by Varian Semiconductor (VIISta-80®). The vendors of the "batch" type low-energy-implanters argue that the throughput is significantly higher than for single-wafer machines. The batch systems use a fixed beam, while the single-wafer systems use a scanned beam. The tuning time for a fixed beam is always shorter than that needed for a scanned beam, Hence, the set-up time for a batch system is smaller. Capacity modeling studies have shown that fewer batch low-energy-implant-systems are required (e.g., 3 to 4 for a fab that starts 5000-wafers per week with a 0.18-μm process, versus 5 to 9 single-wafer-implanters), primarily because of the fast tuning in the batch machines. This reduces the overall capital-cost and cost-per-wafer.

The advantages claimed for the low-energy, single-wafer-implanters are the following: **1**) the angle of the implant with respect to the wafer surface can be controlled more accurately in a single-wafer-implanter; **2**) sputtering of metals from the spinning disk (the largest source of contamination in batch-implanters) is eliminated; and **3**) a 70°-angle analyzing-magnet allows the VIISta-80® to essentially eliminate the problem of energy contamination.

While a significant effort was being expended to improve the productivity by increasing the beam current,[25,26] the basic capability of being able to implant at an ultra-low-energy (i.e., sub-keV) already existed by 2000. (That is, by 2000 implanters were available that could provide 4-mA at 0.2-keV.) This allowed researchers to investigate the limits of shallow SDE-junction-depths with sub-keV implants and conventional RTA methods. Drain extensions with junction-depths that meet 0.1-μm technology requirements were demonstrated.[27] It appears, however, that it may not be possible to improve on these results unless either the extension-resistivity-specification is significantly relaxed or new annealing technologies are developed which provide a smaller thermal budget

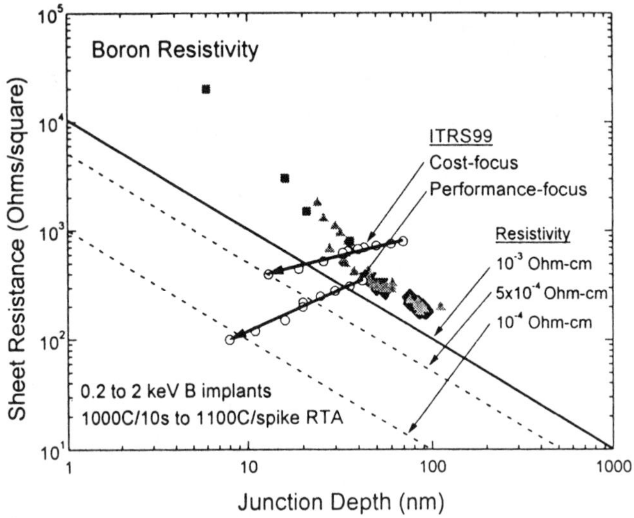

Fig. 5-20 Sheet resistance and source/drain junction depth expectations for CMOS transistors according to the ITRS99 study for cost and performance-focused product and gate sizes from 180 nm towards deeply-scaled devices with gates ~20 nm. Also shown are experimental results for sub-keV B^+ implants with "spike" RTA in CZ-Si materials.[59]

and are compatible with CMOS processing. That is, in order to meet the decreased resistivity requirements of the scaled MOSFETs, it is necessary to implant a larger dose or increase the annealing temperature to activate more of the smaller dose implant. The issue is illustrated in Fig. 5-20, where the ITRS99 expectations for the source/drain-extension junction-depth and sheet-resistance are compared to experimental results for boron-doping with sub-keV B-implants and a "spike"-RTA process (with a high-temperature thermal-cycle of 1050°C for 1 sec). All of the experimental results are above the ~10^{-3}-Ω-cm resistivity "wall" (the dark downward-sloping line in this figure). The shaded squares and triangles in Fig. 5-20 are actual experimental data, whereas the open circles are expected results that have not yet been attained. The resistivity requirements for "high performance" MOSFETs are ~10^{-4}-Ω-cm for 20-nm MOSFETs.

As compared to the SDE-regions of the source/drain, higher doping-concentrations are needed for the contact source/drain-regions. This is because these regions are used to contact the silicide-layer, and high doping-concentrations help to reduce the value of R_{co}. Since these regions are not adjacent to the channel, they do not have as great an impact on short-channel behavior as the SDE-junctions. Consequently, doses and implant-energies can be higher than those permissible for SDE-regions, even when B-implants are used.

By 2001 the trend of using lower implant energies for the SDE-implant for each generation appears to have reached a practical limit in terms of device performance (especially for *p*-channel transistors). That is, 0.5-keV boron implants appear to be the lower useful limit for shallow junctions, not just for 0.13-μm but for all technologies.[28] Below 0.5-keV, the implant becomes less productive, more costly, and the final junction performance (i.e., the simultaneous achievement of junction-depth and resistivity) is the same. The performance limit with a 0.5-keV boron-implant and spike-anneal is a junction-depth of about 35-nm and a sheet resistance of about 350-Ω/sq. It appears that to further reduce either value, the other must be increased.

5.3.3.2 Transient-Enhanced Diffusion: The second key reason for developing dopant implant processes with energies in the sub-keV range (besides reducing the as-implanted depth), is to minimize the transient-enhanced diffusion (TED) of the dopant. When high-energy ions enter the silicon lattice, they displace silicon atoms from their lattice sites, creating pairs of vacancies and interstitials. When the post-implant anneal is initiated, most of these interstitials and vacancies recombine, but some interstitials remain behind (in an amount about equal to the quantity of implanted atoms – most of which now occupy the sites that were formerly filled by the still-displaced silicon interstitials). The fact that each implanted atom produces about one residual interstitial after a very short anneal is called the "*+1 approximation.*" TED can be explained by relating its enhanced-diffusion effects to the presence of these excess interstitials for a wide range of low- and medium-dose implants, and low- and medium-energies.

The residual excess interstitials rapidly coalesce into such unstable extended-defects as {311}'s and also into more-stable extended-defects, such as dislocation loops. While this coalescence occurs because there is a lower system-energy when the interstitials populate extended-defects (compared to when they exist as individual interstitials), the defects are still metastable. Thus, they dissolve with continued annealing (especially the {311}-defects). During their dissolution they emit interstitials into the silicon. The rate at which these defects dissolve depends on the activation-energy and annealing-temperature. Nevertheless, as long as these defects exist, an interstitial supersaturation is maintained (i.e., a higher concentration of interstitials is present than would exist in the absence of the extended defects). Since boron diffuses by an interstitial mechanism, its diffusivity is enhanced by the excess interstitials - with the time-averaged diffusivity-enhancement equal to the time-averaged interstitial-supersaturation. In any case, both the interstitial-supersaturation and the diffusivity-enhancement soon end after the defects have dissolved. The enhancement is thus transient.

The *increase* in depth of a shallow SDE-junction ΔX_j due to TED can be expressed in terms of the initial-supersaturation of interstitials and the time of the interstitial defect dissolution as:

$$\Delta X_j^2 \propto N R_p \exp\left[-(-1.4\ \text{eV})/kT\right] \qquad \textbf{(5.9)}$$

where N is number of interstitials initially trapped in the defects (approximately equal to the implanted dose, as noted above) and R_p is the projected ion range (which is where the peak concentration of the excess interstitials is initially located). The linear dependence of N simply represents the fact that the larger the dose, the larger the supersaturation, and hence the greater the diffusivity enhancement. The linear dependence on R_p has been demonstrated experimentally. The nearer the excess interstitials are to the surface, the faster they reach the surface (where they are annihilated). The linear dependence on R_p explains why use of ultra-low-energy implants reduces TED effects. That is, if an ultra-low-energy implant is used, R_p will be small. Thus, the location of the interstitials will be closer to the surface, allowing them to escape there faster. Thus, the duration of the transient is shortened, making the value of ΔX_j smaller.[29] (Later work indicated that for implants below 1-keV in energy, TED is even further reduced because there is less lattice damage per ion.[30])

Another important dependence contained in Eq. 5-9 is the *negative* activation-energy of ΔX_j^2, which arises from the fact that the interstitial-supersaturation due to the presence of the extended-defects is larger at a lower temperature. The negative-activation-energy implies that if the defects are annealed out at a low temperature, the final junction will be deeper than if the defects are annealed out at a higher temperature. This is one of the main reasons why the S/D anneals are currently carried out in an RTA tool rather than in a conventional batch furnace (that has a ramp-up rate of a few degrees per minute). In an RTA chamber, significantly less time is spent during the temperature ramp-up at a lower temperature (at which the diffusivity-enhancement is larger).

5.3.3.3 Rapid-Thermal-Annealing of Shallow Junctions: To form the shallowest junction possible using ultra-low-energy implantation, it is necessary to minimize the thermal-budget after implant. (Thus, as noted above, annealing such implants is carried out in RTA tools to realize good-quality shallow junctions.) In the ideal case, the as-implanted-dose would spend "zero" time at the temperature that would electrically-activate the dopants and heal the lattice damage from the implant. (Exposure to a high-temperature for a very-short time is effective because the activation-energy for dopant-activation is typically higher than that for diffusion, thus electrical activation is favored while limiting diffusion. Reported values for boron are 5.4-eV and 3.4-eV for activation and diffusion, respectively.) Attempts to reach this ideal annealing condition have led to the development of so-called "spike" anneals, which consist of very fast ramp-up and ramp-down of temperature, with virtually zero-time at the peak temperature. Experiments have shown that the difference in the percentage-of-activation and sheet-resistance between spike-anneals and those that keep the wafer at temperature for 10-sec is negligible. On the other hand, for the 10-sec-at-temperature anneals, the junction-depth increases with temperature.[88] This indicates that spike anneals are more effective at higher peak-temperatures.

In practice, spike anneals have an effectively finite-time at temperature. Also, in reality, while ramp-up rates as fast as about 200°C/sec are possible, the ramp-down is limited to approximately 50-80°C/sec. As a result of these deviations from ideality, it is useful to determine what is a sufficiently-fast ramp-up rate. The negative-activation-energy of TED seems to indicate that as high a ramp-rate that can be attained, should be used during a spike anneal. However, experiments show that for such ultra-fast ramp-rates, the advantage gained by going from moderately fast (e.g., 100°C/sec) to ultra-fast ramp-up rates (200-400°C/sec) is relatively small for ultra-low-energy implants. Such a small advantage must be balanced against the loss of process control (e.g., across-wafer thermal-uniformity and temperature-repeatability) which inevitably occurs when such ultra-fast ramp-up-rates are used. The reason the advantage is not very great is that TED is already sufficiently minimized at 100°C/sec.[32] This reduction in thermal-budget during ramp-up is negligible compared to the portion of the thermal budget which cannot be reduced (the finite-dwell-time at temperature and the relatively low ramp-down-rate).

The data shows that the largest reduction in junction-depth with increased ramp-up-rates occur for the lowest-energy-implants (see Fig. 5-21).[31] Since TED decreases with decreasing implant-energy, it can be inferred that the advantage is primarily due to the overall thermal-budget and not just of the TED. It is therefore evident that without the ability to increase the ramp-down-rate (when it is relatively smaller than the ramp-up-rate), the advantage from increasing the ramp-up-rate will quickly saturate at ramp-up-rates greater than the ramp-down-rate. In addition, it has been determined that the ramp-rate is not as good a measure of the quality of a spike-anneal as is the time spent diffusing the dopant. That is, since about 90% of the dopant-activation occurs in the short-duration from approximately 50°C below the peak-temperature to the peak-temperature

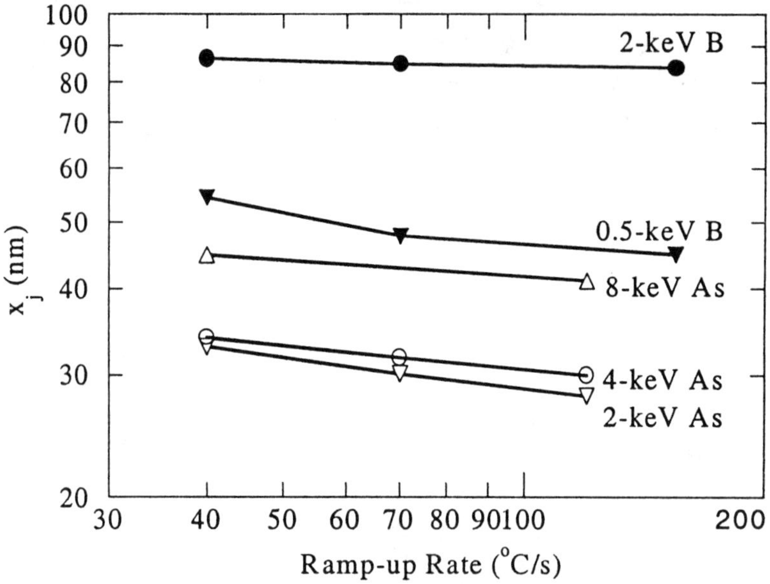

Fig. 5-21 Junction-depth, measured at a concentration of $1x10^{18}/cm^3$ after spike-annealing at various ramp-up-rates. The ramp-down-rate was the same in all cases. For both As and B the advantage from a faster ramp-up is largest for the shallowest implant. This implies that the reduction in TED (known to decrease with implant-depth) is not the dominant mechanism responsible for the larger reduction in junction-depths observed for the shallower-implants with increasing ramp-up-rates.[31]

and back to 50°C below, this interval of time is really a better specification of the quality of a spike-anneal.[28]

Figure 5-22 is a survey of published spike-anneal data that was presented in Ref. 32. It can be seen that many of the data points are coincident, despite having been taken from different implant-and-annealing conditions (energy, dose, and ramp-rate). It is evident that more than one set of process conditions can be used to arrive at the same junction properties.

A variety of RTA-tool configurations are commercially available, including the following: **a**) linear lamps, double-sided heating; **b**) axisymmetric lamps, single-sided heating; **c**) arc lamp; **d**) hot-plate; and **e**) hot-wall RTP. Several general references are available that give detailed descriptions of these types of RTP tools.[33,34,35] Data on shallow junctions formed with the various RTP-tool types can be found in the following references: lamp-based RTP systems (36); hot-plate (susceptor – 37); and hot-wall RTP (38).

New ways to increase the value of the electrically-active-dopant in shallow-junction regions are also being actively being researched. Some of these novel techniques include: **1**) laser anneals;[39] **2**) strain-compensation implants

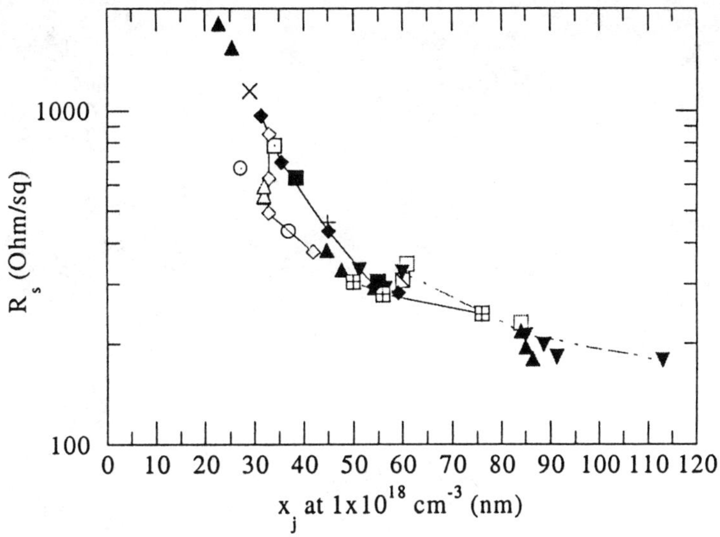

Fig. 5-22 Comparison of *p*-type sheet-resistance and junction-depth data for spike-anneals using various implant-and-annealing conditions. The data set includes a 400°C/sec ramp-up-rate and 0.2-keV implant-energy. Note that all the data points fall in a rather narrow band. No one has been able to make a 30-nm junction with a 300-Ω/sq sheet-resistance using ion-implantation and conventional rapid-thermal-annealing.[32]

(e.g., by adding Ge-atoms to the silicon lattice, their larger size compensates for strain caused by the smaller B atoms. This increases the solubility of the boron atoms in the silicon.);[40] and **3**) using RTP-tools which have a short lamp-response-time, which increases the ramp-down-rate.[41]

5.3.3.4 Dopant Loss for sub-keV Implants from Self-Sputtering, Out Diffusion, and Surface Oxide: Several factors associated with ultra-low-energy implants give rise to loss of some of the species that are implanted, and they are discussed in this section. The effects that lead to such dopant loss are: self-sputtering, out-diffusion, and the presence of a surface oxide during the implant process. When the dose-loss from all three of these effects are combined, the loss can be large enough to pose a significant problem for ion-implant processes carried out at energies of 0.5-keV or lower.

The first phenomenon associated with dopant loss in ultra-low-energy implant processes is *self-sputtering*. That is, when any solid-surface (the target) is struck by an ion-beam, sputtering of atoms from the target will occur. The number of atoms sputtered for each bombarding ion rises as the energy of the ion decreases. In particular, when the ions being implanted have energies in the 0.2-1-keV range, significant sputtering of the target will occur. For example, when 0.5-

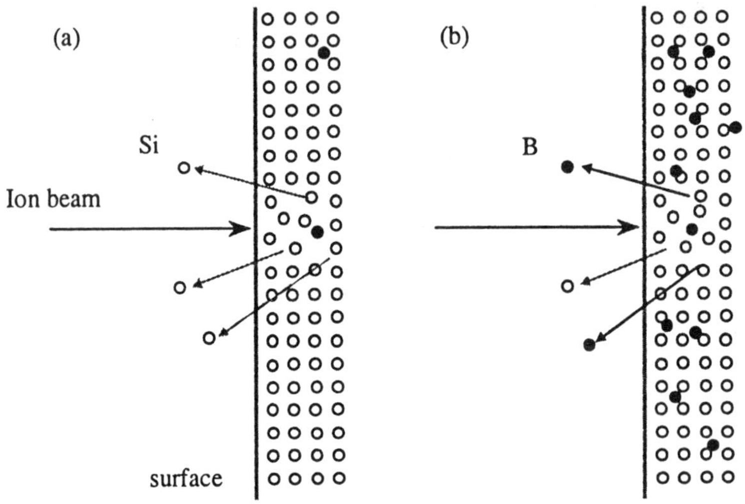

Fig. 5-23 The dose-loss due to self-sputtering depends on how much dopant has already been implanted. (a) At the start of the implant most of the sputtered atoms are silicon. (b) As the composition of the near surface layer becomes altered, with an increasing fraction of them being boron, more boron atoms are sputtered.[59]

keV B-ions strike a silicon target, about one Si-atom is sputtered for every four implanted B-atoms. Since at such low-energies the implanted species also come to rest near the Si surface, some of these implanted B-atoms themselves will be subject to sputtering. In fact, it is expected that with increasing B-dose, more and more of the implanted B-atoms will be sputtered from the surface. (This phenomenon is depicted in Fig. 5-23.) Since additional B-implanted-atoms will be sputtered with increasing dose, this implies that the fraction of implanted dose that is retained in the silicon wafer will decrease with increasing dose. Given the difficulty associated with achieving a large beam-current at such low-energies, a further reduction in the retained dose is very undesirable. That is, while the dopant-loss could be compensated by increasing the implant dose, this will further decrease throughput (which is already too low).

Figure 5-24 plots the retained-dose compared with the nominal-dose for various energies.[32] It indicates that at 2-keV essentially the entire dose is retained. But at implant-energies of 0.5-keV and 0.2-keV, a significant fraction of the implanted-dose has been sputtered out, with the missing fraction rising with increasing dose. For a nominal-dose of $1 x 10^{15}/cm^2$, sputtering causes a 10% reduction in the retained dose at 0.5-keV, and a reduction of 20% of the retained dose at 0.2-keV.

If BF_2^+ is used to implant B at low-energies, the sputtering problem is expected to be even worse than if B is implanted. That is, in BF_2^+-implantation, for each implanted B-atom that contributes to the retained dose, there are now 3 atoms (one B and 2 F) which can cause sputtering (compared to just one B-atom for an implant process that uses B^+-ions instead). For example, for a nominal dose of $1x10^{15}/cm^2$, we stated that a 0.5-keV B^+-implant will cause a 10% reduction in retained dose. However, if a BF_2^+ implant is carried out so that the B-atom in the BF_2^+-ion has the same energy as in the 0.5-keV B^+-process (i.e., the BF_2^+-ions have energies of 2.2-keV), the reduction in retained-dose increases to ~20%. The above circumstances imply that caution must be exercised when comparing junction data from the literature for different sub-keV implants or for different species. A correction needs to be applied for the different retained-doses to allow meaningful comparisons to be made.

The second effect that gives rise to dose-loss is *out-diffusion.* Out-diffusion of implanted species occurs during the annealing steps that are performed after the implant. As the projected-range R_p of the dopant is decreased (as a result of implanting at lower energies), the out-diffusion grows worse. The reduction in retained-dose after a 1000°C/15-sec-anneal is approximately 10% for an implant energy of 0.5-keV, but increases to 15% for 0.2-keV energies.

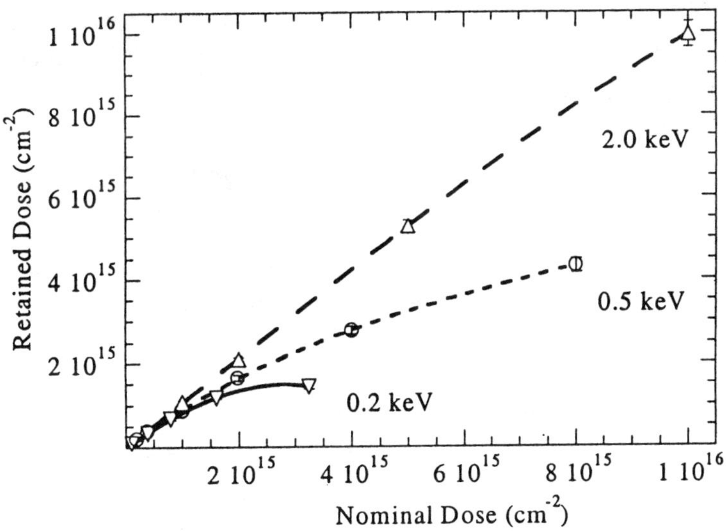

Fig. 5-24 Retained B-dose as a function of nominal B-dose implanted into (100)-silicon wafers with an as-received native oxide. The trend lines are 3rd order polynomial spline-fits. For a nominal B-dose of $1x10^{15}/cm^2$, sputtering leads to a reduction of ~10% in the retained dose at 0.5-keV, and a reduction of ~20% at 0.2-keV.[52]

The final cause of dopant-loss in sub-keV implants is due to the presence of an *oxide on the silicon-surface during implant.* Such an oxide is often deliberately grown to protect the silicon surface from metal contamination during the implant process (see Chap. 10, Vol. 1, 2nd Ed.). While this protective oxide layer can be made quite thin, it nevertheless has a finite thickness. It is clear that as the implant energy is reduced, a larger fraction of the as-implanted dose will be located in the protective oxide. Since the oxide is stripped off at the end of the implant step, the dopant that remains in the oxide cannot contribute to the formation of the junction. Approximately 25% of the B implanted at 0.5-keV is lost by removal of a 15-nm-thick surface-oxide, while more than 40% is lost for 0.2-keV implants. With these results, it is possible to estimate the equivalent-dose needed at lower energies (such as 0.2-keV) to make up for the larger loss as compared to 0.5-keV implants. For example, it has been calculated that to successfully implant a dose of $4x10^{14}/cm^2$ B-atoms at 0.2-keV into the silicon beyond the protective oxide, the beam-current would have to be as much as 60% greater than the beam-current at 0.5-keV to obtain an equivalent throughput process.[32] To circumvent the oxide dose-loss problem, some ultra-low-energy implant processes are being carried out without the use of a protective oxide.

In summary, the dose-loss effects are significant for sub-keV implant processes, and the loss increases as the implant-energy is reduced. For the extreme case of 0.2-keV implants, as much as 40% of the B-dose can be lost due to the combined effects of self-sputtering, out-diffusion, and native-oxide removal for the high doses needed for an SDE-implant.

5.3.4 Effect of the Overlap of the SDE-Region and Gate-Edge on MOSFET Performance

In addition to being able to carefully set the junction-depth X_j and length of the SDE region l_{SDE}, it has also been reported that the amount of overlap of the SDE-region with the gate-edge is also critical for optimum device performance.[42,57] That is, if the overlap of the SDE-region and the gate-edge is too great, the deep-submicron MOSFET will suffer excessive leakage (due to larger short-channel effects that arise if the channel-length is too small). Excessive leakage means that the off-state current exceeds its maximum design value (e.g., 1 nA/μm). In addition, the gate-to-(drain/source) overlap-capacitance increases, which negatively impacts device switching speed.

On the other hand, if the SDE-to-gate-edge-overlap is too small, the ON-state saturated-drain-current I_{Dsat} will decrease because the SDE-region and the end of the channel are not electrically well-coupled. Thus, an excessively-high-resist-

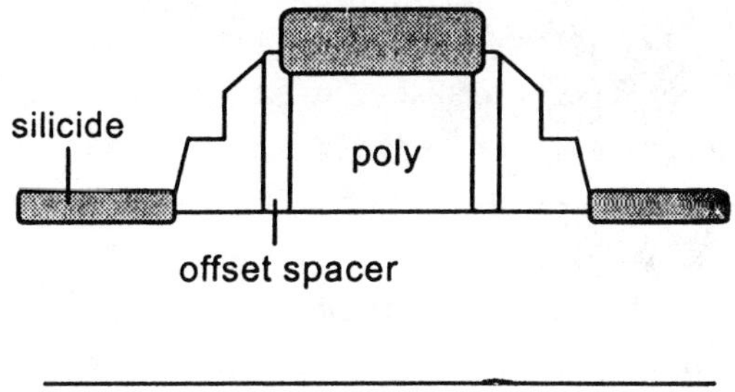

Fig. 5-25 Schematic of a deep-submicron MOSFET with an offset-sidewall spacer.[57]

ance-region separates them (and this increases the parasitic resistance of the source/drain regions, which leads to a decrease in the value of I_{Dsat}).

Controlling the position of the edge of the SDE region with respect to the gate edge can be done in the following way. First, the implant-and-annealing processes for forming the SDE (and contact) region of the drain structure are determined, and the drain structure is formed using these process conditions. However, an offset-spacer is fabricated on the side of the gate after the gate-poly is etched (but before the SDE implant step is carried out, Fig. 5-25).[57] The width of this off-set spacer is varied from zero (no-spacer) to some upper-bound (e.g., 50-nm). Measurements are then made for the MOSFET having different offset-spacer-widths. The offset-spacer-width that yields optimum device characteristics (i.e., the best combination of off-current and I_{Dsat}) is taken to be the value that produces the correct overlap of the SDE-region and the gate edge.

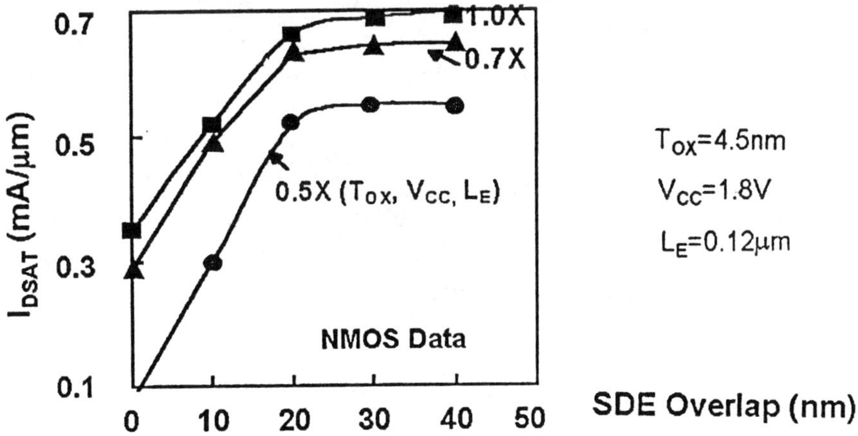

Fig. 5-26 I_{Dsat} versus SDE overlap (NMOS).[42]

The results of such a study are shown in Fig. 5-26. In this case, the effects of SDE-overlap on 0.25-μm NMOSFETs were analyzed. The arsenic-implanted SDE junction-depth X_j was measured to be 50-nm. Since the lateral diffusion of the SDE junction under the gate-edge is estimated to be 0.6-0.7 times the vertical depth (about 30-nm), minus the width of the offset spacer, the SDE-overlap for each offset spacer width can be determined. Fig. 5-26 shows that if the SDE-overlap is less than 20-nm, the value of I_{Dsat} falls off rapidly. Thus, a minimum overlap of ~20-nm is recommended by this study. Information as to how to form the offset spacer was not given, but a method (such as re-oxidizing the poly-silicon gate-structure after it has been etched) is one possible way of doing so.

5.3.5 SDE-Junction Abruptness

The SDE-region must not only have the correct junction-depth, average-doping-concentration, and position under the edge of the gate, but recent studies have shown that there must also be a highly-abrupt lateral-doping-gradient at the interface of the SDE and the MOSFET channel-regions.[43,44] Not only does such an abrupt-doping-gradient help limit the lateral-encroachment of the extension-depletion-regions into the channel, but more importantly it minimizes R_{sp}. The design rules for lateral-junction-abruptness are that the doping-falloff must be at least one decade in a distance equal to ~3% of the size of the gate-length (for a spreading-resistance less than 5% of R_{sat}). For 100-nm MOSFETs this means that the junction-abruptness must be ~3 nm/decade, which is about the profile slope of as-implanted B at 0.2-keV.[45] Since the abruptness will have to increase even further as smaller MOSFETs are designed, this implies that new techniques will have to be developed that result in final junction-doping-profiles that are even steeper than the as-implanted profiles.

The design of the implanter can play a role in the lateral-abruptness of the SDE as-implanted doping profile. That is, unlike mid- to high-energy ion-implant applications, channeling-effects are minimal for low-energy, high-current implants (because of increased channeling acceptance angle). Nevertheless, there is still a significant difference between single-wafer and batch-implanters due to implant shadowing effects. Ideally, a 0°-implant will give the most-abrupt lateral-junction because any tilt-angle larger than 0° induces shadowed areas at the vicinity of the gate pattern (leading to laterally-graded doping-profiles). If the incident beam has an angular-spread deviating from the ideal parallel beam, each of the ions having different incidence angles will make different shadows. The result is a loss of dose- and junction-abruptness at the outskirts of the gate. A single-wafer, parallel-beam-implanter minimizes the incident-angle-variations.

Spinning-disk batch-implanters use a stationary-spot ion-beam. The wafers pass through the beam due to the rotation of the disk. Such batch-implanters use a concave-shaped disk to secure wafers in their respective sites during the high-speed motion of the disk. Centrifugal force exerted on the disk surface holds them in place. Since the conical disk has a curved surface and the wafer-surface is flat, this leads to an incident-beam angle-variation across the wafer during the implant process. Depending on the dimensions of the disk, cone-angle, and wafers, this can amount to 1.5°-variation across a wafer.[46]

Recently, the abruptness of the SDE-lateral-doping was measured using both scanning capacitance microscopy (SCM)[47] and scanning tunneling microscopy (STM).[48] The latter measurement detected a tail-region of dopant at the end of the doping profile, spreading from the extension-region in the horizontal direction near the gate oxide. The cause of this anomalous diffusion tail was not understood.

5.4 ANTI-PUNCHTHROUGH STRUCTURES IN DEEP-SUBMICRON CMOS

Lateral tailoring of the channel doping profile can provide some additional flexibility to control V_T and punchthrough. Recently, such lateral tailoring of the channel profile has been carried out with the use of *large-angle-tilt implanted punchthrough stopper* (LATIPS), which are also referred to as *pocket-implants* (and are depicted in Fig. 5-27). Such a pocket-implant puts additional dopants in the channel close to the source/drain-junctions for punchthrough control, and leaves much of the channel region with a lighter doping, as needed for a low V_T.* The peak doping values of the pocket-implants must be somewhat larger than the doping-concentration values needed for a punchthrough-control layer that is uniform in the lateral channel direction between the source and drain.

These punchthrough-suppression pocket-implants are done after the source/drain-extension-implant (and before the formation of the spacer used to define the edge of the deeper contact-source/drain-implant). This ensures that the pocket-implant-dopant is self-aligned to the gate. The dopant is placed just below the active channel, adjacent to the source and drain, and is of the same dopant-type as the well. Precise control of the placement and dose of this implant within the transistor-structure is needed to obtain the intended enhancements. As noted

* As pointed out in Chap. 1, if an implant with a beam directed normally to the wafer surface is used to introduce dopants to create a punchthrough-control structure, instead of one whose direction is tilted at a large angle to the wafer surface, the resulting punchthrough-control structure is termed a *halo* structure. Such a halo structure is described in Chap. 1, and is depicted in Fig. 1-2b.

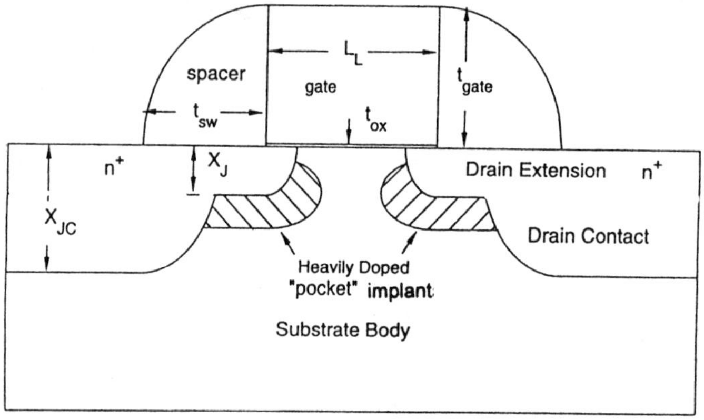

Fig. 5-27 MOSFET with heavily-doped pocket-implants for punchthrough-control.

above, the use of high wafer-tilt-angles when performing such pocket-implants offers an additional, controllable processing parameter for producing complex channel structures, without diffusion or additional masks. (The presence of the pocket-implanted-species also helps keep the source/drain-extension [SDE] junction-depths shallow, because they compensate some of the dopant in the SDE region [which, in effect, squeezes the SDE-junction closer to the surface].)

As introduced by Hori, high tilt-angles are needed for achieving greater

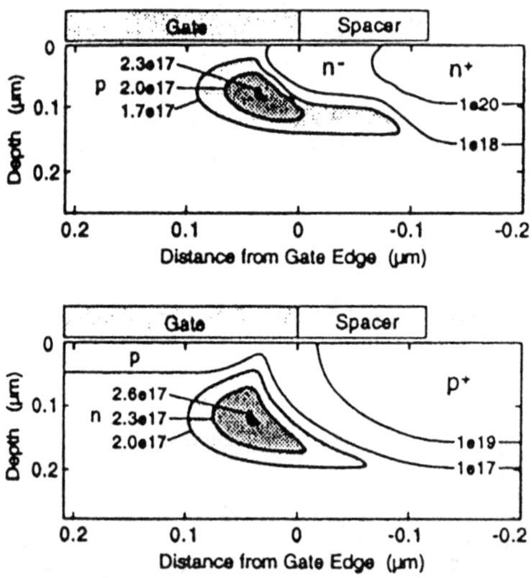

Fig. 5-28 Simulated 2-D impurity profiles of typical LATIPS structures for (a) NMOSFET with *n*-LDD region and; (b) Buried-channel PMOSFET.[49]

lateral-penetration of the pocket-implant than the adjacent source/drain-extension implant without thermal diffusion.[49] That is, the species implanted by the pocket-implant should reside outside of the final source/drain-extension-doping-profile after implant and annealing. Typically, tilt-angles are 20°-45° from the wafer normal. Figure 5-28 shows the resulting (simulated) doping profile in the channel region and in the adjacent well of a LATIPS device.[49]

To maintain device symmetry with high tilt-angle, the implants must be split into segments. Since most circuits contain transistor gates placed 90° from each other, high tilt-angle implants are typically done in four segments of equal dose ("quad-implants") with the wafer rotated 90° between segments. (As gate-widths shrink, less undercutting of the gate is required to produce the punchthrough stop effect. Consequently, tilt-angle requirements for this application are steadily decreasing below the 0.25-μm node, and are expected to reach the "low-tilt-angle" threshold of 10° by the 0.1-μmtechnology node.)

High-tilt-angle punchthrough-suppression-implants improve device performance. Junction-capacitance can be reduced by up to 65%; while the subthreshold slope of the device becomes steeper (which suppresses short-channel effects).[50] The dramatic improvement in punchthrough-resistance does not depend on extreme scaling of the S/D junction-depth or n-well background-doping. The reduced subthreshold-slope and junction-capacitance directly result from carefully controlled doping underneath the channel at the gate edge and the near elimination of all other undesired doping that results from less controllable doping methods.

Because of their lower-diffusivity and as-implanted-straggle, higher mass dopants can be placed more precisely in the channel. Consequently, indium (In) and antimony (Sb) are being increasingly used for this application.[51,52] Punchthrough-stop implants were the primary production application of In and Sb in ion-implantation in the 2000 timeframe. The lower susceptibility of these species to TED also helps reduce the reverse-short-channel-effect (the slight increase in threshold-voltage with decreasing effective-channel-length that occurs before punchthrough-effects dominate the smallest channel-lengths).

One proposed structure for MOSFETs with an effective gate length (L_E) of 0.25-μm is shown in Fig. 5-29.[53] This device has a very heavily doped source/drain next to a very heavily doped pocket-implant. Since the gate-length is so small, the pocket-implanted-region actually doubles for the traditional retrograde-well-implant. This aggressive device design shows the challenges of doping a 25-nm MOSFET. Very-high doping levels and very-steep doping gradients are required – with a minimum of thermal processing.

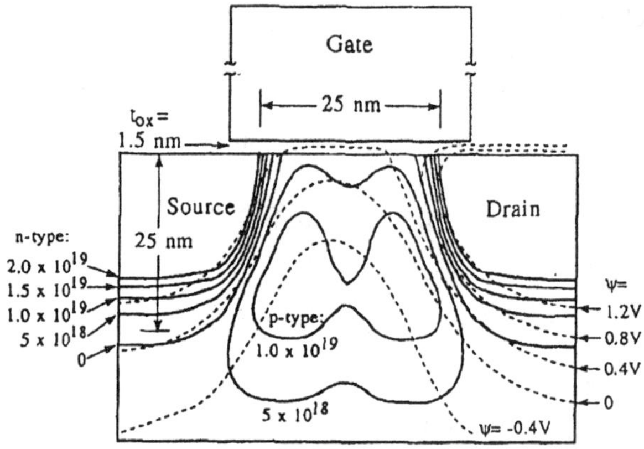

Fig. 5-29 Doping contours in 25-nm PMOSFET.[53]

An issue associated with the use of In as a pocket-implant species has recently been reported. That is, an In implant at the doses needed for the pocket-implant process causes a buried amorphous layer to form. This layer has an interface with the single-crystal silicon. The silicon at these interfaces is still crystalline, but is heavily damaged. The indium segregates to these damaged regions and becomes

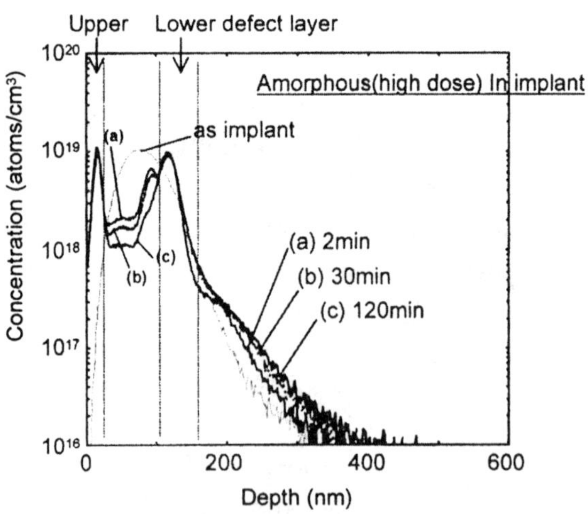

Fig. 5-30 Indium SIMS profiles for $1x10^{14}/cm^2$ implants at 200 keV annealed at 750°C for 2, 30, and 120 min in nitrogen. This figure shows large dose loss for indium. The depth of the defect layer (measured by XTEM) is shown (upper & lower defect layers).[54]

electrically de-activated (negating its ability to provide punchthrough control).[54,55] Figure 5-230 shows a SIMS profile of the In as-implanted, and after annealing. The loss of indium from its peak is seen. A double-implanted extension region (As + Sb) has been reported to be able to control the damaged region near the In pocket, mitigating the indium segregation and de-activation.[56]

REFERENCES

1. R.D. Rung, C.J. Dell'Oca, and L.G. Walker, "A Retrograde *p*-Well for High Density CMOS," *IEEE Trans. Electron Dev.,* **ED-28**, p. 1115, 1981.
2. D.M. Brown *et al.,* "Trends in Advanced CMOS Process Technology," *Proceedings of the IEEE,* December 1986, p. 1646.
3. G.J. Hu *et al.*, "Design and Fabrication of *p*-channel FET for 1-μm CMOS Technology," *IEDM Tech. Dig.* 1982, p. 710.
4. R. Martin and J.Y. Chen, "Optimized Retrograde *n*-Well for 1 μm CMOS," *Proc. Custom Integ. Circuits Conf.,* 1985, p. 199.
5. J.Y. Chen, "CMOS - The Emerging Technology," *IEEE Circuits and Devices Magazine,* March 1986, p. 16.
6. Y. Mochizuki, "Special Report: High Energy Ion Implantation Technology," Nikkei Microelectronics (Dec. 1991), p. 92.
7. K. Tsukamoto *et al.,* "High Ion Energy Implantation for ULSI: Well Engineering and Gettering," *Solid State Technology,* June 1992, p. 35.
8. R. Wijberg *et al.,* "Strong Asymmetric Doping Profiles at Mask Edges in High-Energy Ion Implantation, "*IEEE Trans. Electron Dev.,* **37-1**, January 1990, p. 79.
9. T.M. Parrill and M.J. Namaroff,: "Ion Implantation Productivity for Advanced Applications," *European Semiconductor*, July 1999, p. 44.
10. B. Adibi and V. Meniailenkov, "Transistor Performance: The Impact of Implant Doping Accuracy," *Solid State Technology,* January 2001, p. 88.
11. S. Takagi *et al.,* "On the Universality of the Inversion Layer Mobility in Silicon MOSFETs: Part I-Effects of Substrate Impurity Concentration," *IEEE Trans. Electron Dev.*, **41**, 2357 (1994).
12. P. Boullion *et al.,* "Re-Examination of Indium Implantation for a Low Power 0.1 μm Technology," *Tech. Dig. IEDM*, 1995, p. 897.
13. S.C. Williams *et al.,* "Scaling Trends for Device Performance and Reliability in Channel-Engineered NMOSFETs," *IEEE Trans. Electron Dev.*, **45**, January 1998, p. 254.
14. I. De and C.M. Osburn, "Impact of Super-Steep-Retrograde Channel Doping Profiles on the Performance of Scaled Devices," *IEEE Trans. Electron Dev.*, **46**, August 1999, p. 1711.

15. S. Venkatesan *et al.,* "Device Drive Current Degradation Observed with Retrograde Channel Profiles," *Tech. Dig. IEDM* 1995, p. 419.
16. M. Harlan and K. Petry, *Ion Implantation Technology,* 1998, June 22-26, Kyoto, Japan.
17. T. Ghani *et al.,* "100 nm Gate Length High Performance/Low Power CMOS Transistor Structure," *Tech. Dig. IEDM,* 1999, p. 415.
18. T. Mizuno, *IEEE Trans. Electron Dev.,* **47**, 756 (2000).
19. Ohguro *et al.,*"Undoped Epitaxial Si Channel in NMOSFET Grown by UHV-CVD with Preheating," *IEEE Trans. Electron Dev.,* **45**, March, 1998, p. 710.
20. International Technology Roadmap for Semiconductors, 2001 Edition, Semiconductor Industry Association, San Jose, CA.
21. K.K. Ng and W.T. Lynch,"Analysis of the Gate-Voltage-Dependent Series Resistance of MOSFETs," *IEEE Trans. Electron Dev.*, July 1986, p. 965.
22. G. Baccarani and G.A. Sai-Halasz, "Spreading Resistance in Submicron MOSFETs," *IEEE Electron Dev. Lett.,* **EDL-4,** 1983, p. 27.
23. W.C. Holton *et al.,* "Overview of Semiconductor Devices," Ch. 1 in *Handbook of Semiconductor Manufacturing Technology,* Y. Nishi and R. Doering, Eds., Marcel Dekker, New York, 2000, p. 1.
24. K. Miyano *et al.,* "Low Thermal Budget Elevated Source/Drain Technology Utilizing Novel Solid Phase Epitaxy and Selective Vapor Phase Etching," *Tech. Dig. IEDM* 2000, p. 433.
25. D.W. Duff and L.M. Rubin, "Ion Implant Equipment Challenges for 0.18 μm and Beyond," *Solid State Technology,* June 1998, p. 83.
26. T. Parrill *et al.,* "A New Era for High-Current, Low-Energy Ion Implantation," *Solid State Technoloy,* November 2000, p. 103.
27. H.J. Gossmann *et al.,* "Junctions for Deep-Sub-100-nm NMOS: How Far Will ion Implant Take Us?" *Symp. on Front End Processing* at the *Spring 2000 Meeting of the Materials Research Soc.,* April, 2000.
28. L. Peters, "Ramping the 0.13 μm Generation," *Semiconductor International,* January 2001, p. 56.
29. A. Agarwal *et al.,* "Reduction of Transient Diffusion from 1-5 keV Si^+ Ion Implantation Due to Surface Annihilation of Interstitials," *Appl. Phys. Letts.*, **71**, p. 3141 (1997).
30. A. Agarwal *et al., Mat. Res. Soc. Symp. Proc.*, **568**, p. 19 (1999).
31. A. Agarwal *et al.,* "Ultra-Shallow Junction Formation Using Ion Implantation and Rapid Thermal Annealing: Physical and Practical Limits," *Abs. Mtg. Electrochem. Soc.,* Spring 2000, Abs. 526.
32. A. Agarwal *et al.,* "Ultra-Shallow Junctions by Ion Implantation and Rapid Thermal Annealing: Spike Anneals, Ramp-Rate Effects," *Mater. Res. Soc. Symp.*, **19** p. 568 (1999).

33. P.J Timans *et al.*, "Rapid Thermal Processing," Ch. 9 **in** *Handbook of Semiconductor Manufacturing Technology,* Y. Nishi and R. Doering, Eds., Marcel Dekker, New York, 2000, pp. 201-86.

34. *Advances in Rapid Thermal and Integrated Processing,* F. Roozeboom Ed., Klewer Academic Publishers, Dordrecht, NL, 1996.

35. R.B. Fair, *Rapid Thermal Processing,* Academic Press, San Diego, CA, 1993.

36. M.A. Foad and D. Jennings, "Formation of Ultra-Shallow Junctions by Ion Implantation and RTA," *Solid State Technology,* December 1998, p. 43.

37. A. Atanos *et al.,*"Susceptor-Based Rapid Thermal Processing for Ultra-Shallow Junctions," *Solid State Technology,* April 2000, p. 50.

38. M. Ameen and J. Hebb, "Implanter, RTP Systems Issues for Ultrashallow Junction Formation," *Solid State Technology,* September 2001, p. 77.

39. B. Yu *et al.,*"70-nm MOSFET with Ultra-Shallow, Abrupt, and Super-Doped S/D Extension Implanted by Laser Thermal Process (LTP)," *Tech. Dig. IEDM* 1999, p. 509.

40. S. Banerjee,"High Ramp Rate Thermal Annealing for Ultra-Shallow Junctions," *Abs. Mtg. Electrochem. Soc.,* Spring 2000, Abs. 521.

41. D.M. Camm and M. LeFrancois,"Spike Thermal Processing Using Arc Lamps," *Abs. Mtg. Electrochem. Soc.,* Spring 2000, Abs. 567.

42. S. Thompson *et al.,* "MOS Scaling: Transistor Challenges for the 21st Century," *Intel Technology Journal,* 3rd Quarter 1998.

43. C. Osburn *et al.,* "Design and Integration Considerations of End-of-the-Roadmap Ultrashallow Junctions," *J. Vac. Sci. Technol.***B-18**, p. 338, 2000.

44. T. Ghani *et al., Dig. Tech. Papers, Symp. VLSI Technol.* 2000,

45. M.I. Current *et al.,* "Process Integration Issues for Doping of Ultra-Shallow Doping Junctions," *J. Vac. Sci. Technol.* **B-18**, p. 468, 2000.

46. U. Jeong *et al.,* "Devices Dictate Control of Implant-Beam Incident Angle," *Solid State Technology*, October 2001, p. 79.

47. R.N. Kleiman *et al., J. Vac. Sci. Technol.*, **B-18**, p. 2034 (2000).

48. H. Fukutome *et al.,* "Anomalous Diffusion in the Extension Region of Nanoscale MOSFETs," *Tech. Dig. IEDM,* 2001, p. 67.

49. T. Hori, "A 0.1 μm CMOS Technology with Tilt Implanted Punchthrough Stopper (TIPS)," *Tech. Dig. IEDM* 1994, p. 75.

50. M. Rodder *et al.,* "A Scaled 1.8 V, 0.18 μm Gate Length CMOS Technology: Device Design and Reliability Concerns," *Tech. Dig. IEDM* 1995, p. 415.

51. K. Miyashita *et al.,* "Optimized Halo Structure for 80-nm Physical Gate CMOS Technology with Indium and Antimony Highly Angled Ion Implantation," *Tech. Dig. IEDM* 1999, p. 645.

52. Y. Momiyama *et al.,* "Indium Tilted Channel Implantation Technology for 60 nm NMOSFET," *Dig. Tech. Papers, Symp. VLSI Technol.* 1999, p. 67.

53. Y. Taur, "25 nm CMOS Design Considerations," *Tech. Dig. IEDM,* 1998, p. 789.
54. T. Noda, "Modeling of End-of-Range (EOR) Defects for Channel Engineering," *Tech. Dig. IEDM*, 2001, p. 839.
55. T. Noda *et al.*,"Effects of End-of-Range Dislocation Loops on Transient-Enhanced Diffusion of Indium Implanted in Silicon," *J. Appl Phys.*, **88**, 4980 2000.
56. H.C.-H. Wang *et al.,*"Antimony Assisted Arsenic S/D Extension Engineering for Sub-0.1 μm NMOSFETs: A Novel Approach to Steep and Retrograde Indium Pocket Implants," *Tech. Dig. IEDM* 2001, p. 63.
57. S. Narasimba *et al.,* "High Performance Sub-40 nm CMOS Devices on SOI for the 70 nm Technology Node," *Tech. Dig. IEDM* 2001, p. 625.
58. R. Chau *et al.,* "A 50-nm Depleted Substrate CMOS Transistor (DST)," *Tech. Dig. IEDM,* 2001, p. 621.
59. M.I. Current *et al.*, "Ultrashallow Junctions or Ultrathin SOI?" *Solid State Technology*, September 2000, p. 66.

Chapter 6

ADVANCED LITHOGRAPHY I - PHOTORESISTS

Lithography is the most heavily used process technology for fabricating ICs. In 2002 some ULSI process flows use more than 20 mask levels. Thus, the entire lithography procedure is applied more than 20 times when fabricating many modern ICs. The semiconductor industry is also continuing to drive toward smaller feature sizes, and it relies on progress in lithography to help maintain this effort. Here many of the changes taking place in lithography processes in the 2001-2002 timeframe are discussed. While the basics of the lithography process are not presented in this text, such information can be found Vol. 1 (2nd Ed.) of *Silicon Processing for the VLSI Era,*[1] and elsewhere.[2,3]

Figure 6-1 shows the lithography-scaling roadmap as it was envisioned in 1998.[4] Not all of the predictions shown in this figure have come to pass (e.g., proximity-X-ray-lithography, predicted to be inserted into manufacturing by 2000, never got commercialized). Nevertheless, the data in the figure clearly indicates that many changes have occurred in lithography, and many more are occurring today. The purpose of this chapter (and Chap. 7) is to cover the changes depicted on the right-hand side of Fig. 6-1, including some that are not shown in this chart (e.g., 157-nm lithography, based on F_2-laser light-sources).

6.1 CHEMICALLY-AMPLIFIED DEEP-UV RESISTS FOR OPTICAL LITHOGRAPHY

As device dimensions were scaled to less than 1.0-μm, advances were occurring in both the DNQ/novolac resists and the lithography hardware. To achieve sub-wavelength resolution, advances in lithography were made (i.e., improvement in stepper optics and inspection techniques). Such progress allowed DNQ/novolac resists to continue to be used to fabricate device patterns in the sub-micron realm. Specifically, by using *i*-line illumination to replace *g*-line, by using *i*-line lenses with larger numerical aperture (NA) in the steppers, and by improving the novolac resin and DNQ materials, the resolution of *i*-line lithography kept being improved. Eventually, it became possible to print the features used in 0.35-μm CMOS technology with DNQ/novolac resists. Nevertheless, it was understood that new photoresists would be needed for the generations beyond 0.35-μm. Thus, a major industrial effort was launched in the early 1980's to develop photoresist systems for shorter wavelengths. This effort spanned about 10 years.

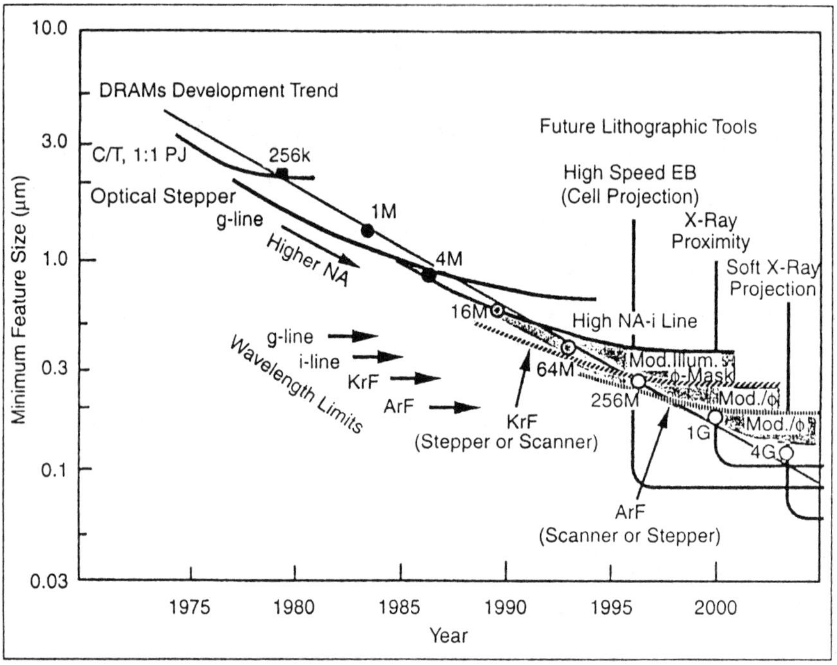

Fig. 6-1 Lithography Scaling Roadmap

6.1.1 248-nm DUV Photoresists

Eventually, by the early 1990s, when the 0.25-μm generation was reached (and used to make such ICs as 256-Mbit-DRAMs), it became clear that *i-line* lithography could no longer print 0.25 μm features (even when using such resolution-enhancement-techniques as phase-shift-masks and off-axis illumination). A shift from *i*–line to deep-UV illumination had to be made. Since both novolac resin and DNQ strongly absorb UV light at around 250 nm, DNQ/novolac resists cannot operate properly in the DUV region. Thus, the new resist chemistry developed during the 1980s (based on *chemical amplification* - CA) was adopted. This CA-chemistry also offered increased photospeed and contrast. Since CA-resists were compatible with DUV illumination, they became the resist materials of choice for advancing optical lithography toward higher resolution.[5]

In chemical amplification the important chemistry is not driven directly by the photons absorbed during the exposure step. Instead, the absorption of photon energy causes the decomposition of a *photoacid generator* (*PAG*). This causes a small amount of acid to be formed throughout the resist. (In early CA-resists such PAGs were *onium salts*.) This acid induces a cascade of chemical transformations in the resist film, typically during the *post-exposure bake* (*PEB*) step. These transformations increase the solubility of the resist in the developer. Since the latter reactions are catalytic, the acid is regenerated after each chemical reaction. Thus, the same acid molecule can participate in further reactions. From 500-1000 chemical reactions can result from a single photon-absorption event. In summary, imaging in CA resists occurs in two steps: **1**) acid generation during exposure; and **2**) acid-catalyzed reactions during the PEB step.

This technology was pioneered by IBM in the early 1980s in an attempt to accommodate the low intensity of illumination produced by Hg-arc lamps at 248-nm wavelengths. Very fast resists were needed, and this led to the development of the CA reactions. Both positive- and negative-tone DUV-resists were developed. Details concerning the *chemical compositions* of these resists are described in Ref. 5. Information about the *chemical reactions* occurring during the various resist processing steps of 248-nm DUV-resists (i.e., exposure, PEB, and develop) can be found in Ref. 6.

One of the first commercially available CA-DUV-resists was a negative-tone resist (Shipley MEGAPOSIT SNR200). Its introduction allowed lithographic practitioners to gain experience with CA-resists and to witness first hand their high resolution capability. However, the SNR200-resist did not find widespread use because it is difficult to print contact holes with negative resists. Production-worthy positive-tone CA-DUV-resists soon followed, with the Shipley APEX-E resist being developed by the IBM/Shipley DUV-resist alliance. The IBM-developed CA-DUV resists have phenolic sites on poly hydroxystyrene (pHOST) which have been protected with t-butoxy-carbonate (t-BOC) groups. The IBM t-BOC resist works in positive-tone with aqueous developer, and alternately in negative-tone using an organic developer. Other major resist vendors now also offer CA-DUV-resists, including Clariant, Olin, JSR, and TOK. Positive-tone materials have become the preferred DUV resists. Some of the newer resists are compatible with TMAH developer and have been made more resistant to airborne contamination. Line/space patterns down to 165-nm, with a depth-of-focus (DOF) of 1.2-μm can be printed with such resists. Negative-tone CA-DUV-resists are also available, although not as many types are offered. Because of the improved aerial image of dark-field masks versus bright-field masks, the ability to print isolated-lines with negative resists is superior to that of positive resists. Thus, for some pattern types, negative resists are preferable. CA-DUV resists, however, are expensive. Their cost in 1998 was \$1500–\$2000 per gallon, which is about 2–3 times higher than DNQ/novolac resists. Future CA-DUV resist formulations are predicted to be even more costly.

6.1.1.1 Amine-Contamination of DUV CA Resists: While the biggest advantage of CA-resists is their high sensitivity (i.e., fast photospeed, in the range of 10–50 mJ/cm^2), the catalytic nature of the CA process makes reproducibility a concern. If the catalytic chain of reactions is somehow interrupted, many of the lithographically important reactions will not occur, leading to catastrophic resist failure. That is, prior to post-exposure-bake (during which the photoacids perform the task of deprotection), these acids are susceptible to neutralization by bases.

The most common mechanism that quenches the catalytic reactions occurs as a result of *ambient contamination.*[7] Specifically, the culprits are trace quantities (on the order of 10 ppb) of vapors of volatile base compounds (such as amines, ammonia, ammonium hydroxide, HMDS, or N-methylpyrrolidone [NMP - a common ingredient in strippers]). Such trace levels of amines in the ambient air can significantly affect the photochemistry because (as noted above) only a small quantity of photo-acid is produced by the exposure dose. If the ambient contains any of these contaminant vapors, the top surface of the resist will absorb them during the time that elapses between the exposure and post-exposure-bake (PEB) steps. These base molecules will then react with and neutralize the photo-generated acid. The surface region of the resist is now devoid of acid, and thus becomes less soluble than the acid-containing regions below. This less-soluble skin develops more slowly, resulting in the formation of a T-top structure (Fig. 6-2).

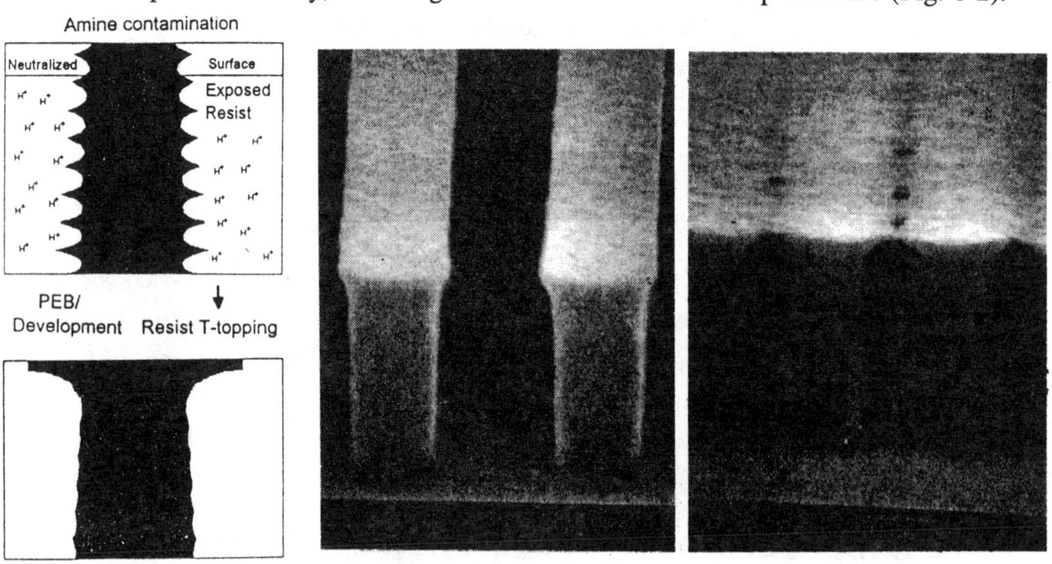

Fig. 6-2 (a) Environmental amine contamination of a chemically amplified resist, resulting in "T-top" formation after development.

This causes linewidth variations from wafer-to-wafer, as the extent of the skin formation is not controlled. In addition to airborne contamination, the acids also decay with time. Thus, any delay between exposure and PEB exacerbates the linewidth control problem. After PEB, resist susceptibility to amine contamination drops to negligible levels.

Let us elaborate on the problem of T-top formation. As noted above, the resist is most sensitive after the coated wafer is exposed in the exposure tool and transported back to the PEB/development tool, until the time it is baked (PEB). Typically the time delay between flash and post-exposure bake may be less than 3 minutes, depending on tool loading, but on occasion it may be 10 minutes or longer. A 10-minute pause can easily cause a linewidth variation on a wafer that may have just finished exposure but is not yet baked. If this event happens when a wafer is in an area where the amine contamination level has reached 5 ppb (or more), it is possible to cause a CD shift of as much as 10-20-nm. If the CD shift is great enough, changes in the resulting linewidth can cause the electrical characteristics and performance of the devices to change (possibly to the point where the device will not perform to specification and its speed or function is diminished). The length of time the wafers are exposed relative to the contamination concentration determines the potential risk to linewidth control. The effect of amine contamination on an example DUV resist is shown in Fig. 6-3 for three conditions: **1**) no time delay between exposure and PEB; **2**) 45-minute delay between exposure and PEB (with the wafer being held within the scanner enclosure, in which the amine concentration was about 1 ppb); and **3**) 45-minute

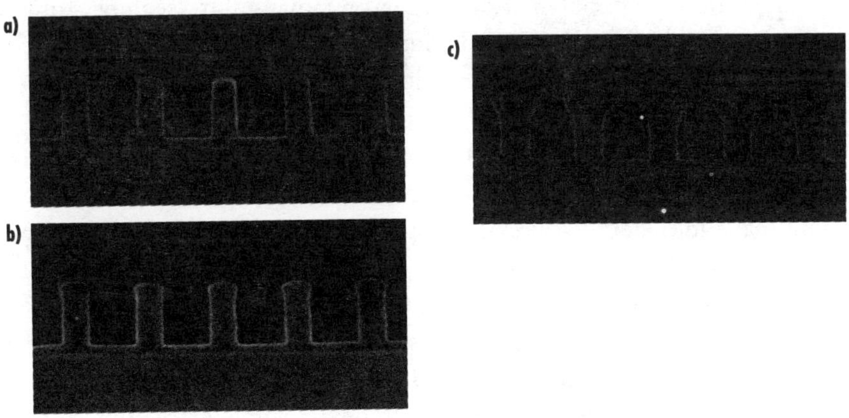

Fig. 6-3 The effect of molecular bases on a DUV-resist is shown with: a) No delay between exposure and PEB; b) after a 45-min delay in the exposure tool; and c) after a 45-min delay in a cleanroom ambient.

delay between exposure and PEB (with the wafer being held in the cleanroom-ambient which has an ambient-amine-concentration of about 5-ppb). Significant T-topping occurs as a result of allowing the wafer to remain in the cleanroom-ambient for 45 minutes.

As shown above, the environmental contamination problem can be quite serious, as trace quantities of basic (i.e., as opposed to acidic) vapors are ubiquitous in the semiconductor fabrication ambient (typically in the 5-10-ppb range). Several methods have thus been established to combat this problem. The first involves filtration of the cleanroom air with activated charcoal. This was the practice first adopted by the industry when DUV-resists were initially being used, but it only reduces the nominal amine concentration in the cleanroom ambient to the 5-10-ppb range. (Environmental base contamination can also be neutralized and further reduced by adding weak acids to the filters.) More recently (as noted above), scanners and track systems (where the resist-coating, exposure, and development steps are performed) are housed in enclosures, and the air that enters these enclosures is purified with charcoal filtering and monitoring systems. Such airborne-amine-filtration and real-time monitoring systems are built by Extraction Systems (Franklin, MA) and other firms (see Figs. 6-4 and 6-5). According to the ITRS Roadmap, the total airborne contamination levels inside such DUV-lithography-enclosures must be kept below 1-ppb (molar) by such filtration systems as long as CA-resists are used. Another approach is to have the wafer track place the wafers in "safe holding" locations after exposure. Other ways to controlling this problem include: **a**) coating resist layers with a protective water-soluble transparent polymeric film (and removing this "sealing" film prior to development with a water rinse); **b**) incorporating stabilizing additives to the resist; and **c**) reducing the rate of absorption of contaminants by "densifying" the resists through annealing.

Nevertheless, even use of all of the above techniques is not completely effective in curing the problem of amine-poisoning of CA resists. Therefore, it is important that additional measures be employed to minimize the levels of amine in the cleanroom ambient. This is a challenging task since there are many potential sources of amines in a fab. Amine levels in wafer fab areas can easily exceed the nominal 10-ppb concentrations when small quantities of developer or adhesion promoters (HMDS) are spilled. (Note that HMDS application is an ammonia- producing process that therefore requires ample exhaustion to prevent

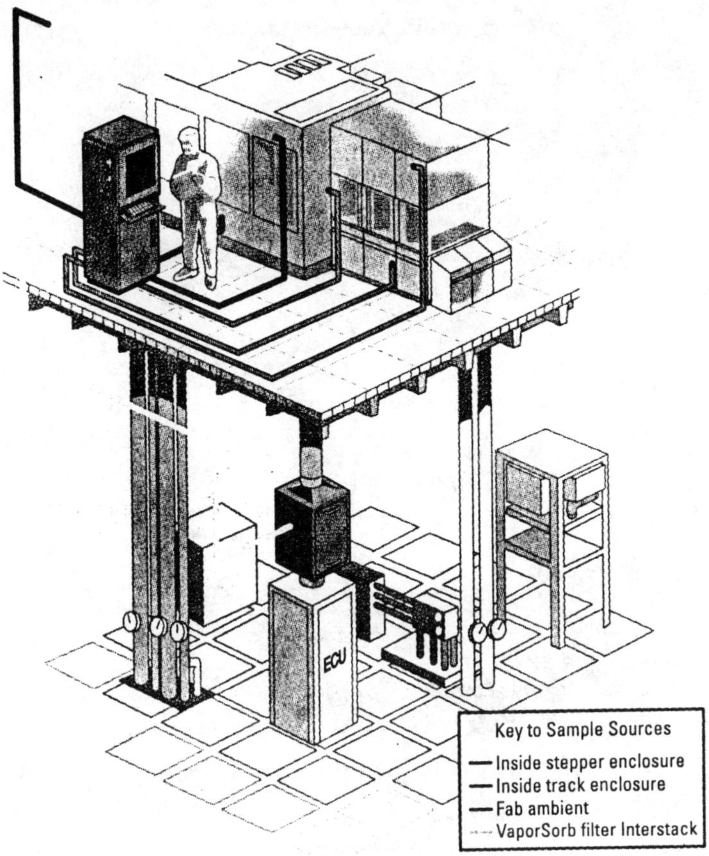

Fig. 6-4 Schematic of filtration units connected to photo-island clusters to avoid amine contamination of chemically amplified resists. Courtesy of Extraction Systems, Inc.

leakage into the track environment, causing amine levels within the track system enclosure to rise above 1-ppb.) Many paints, plastics, and epoxies also outgas amines. Materials that initially evolve large quantities of volatile bases, but quickly reach low-outgassing rates can be used in the fabrication of lithography tools. However, their use can be problematic in operating a wafer fab. Some materials continue to evolve amines over long periods of time and should not be used in wafer fabs, nor in the construction of lithography equipment.

6.1.1.2 Current State of 248-nm Resist Technology: Devices fabricated at 0.35-μm use a mixture of optical lithographies, including traditional *i*–line and 248-nm-DUV (the latter employing single-layer CA-resists). The conversion from *i*–line to 248-DUV was completed in the transition to 0.25-μm devices. The

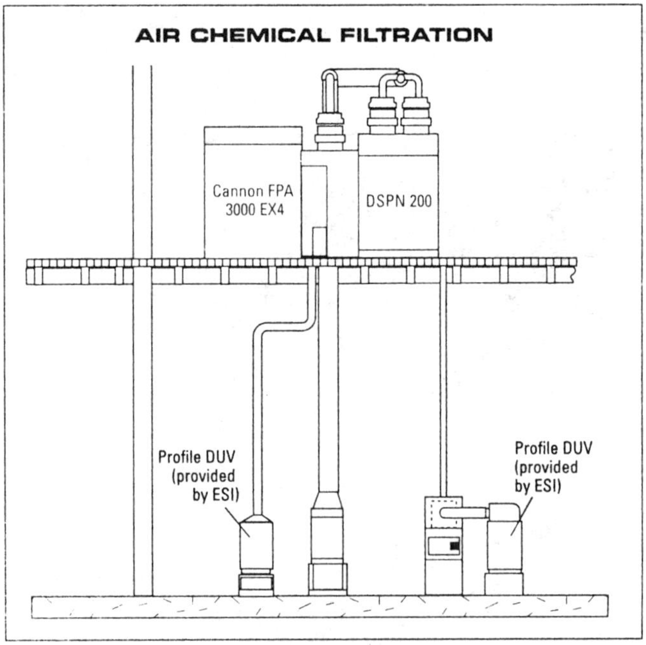

Fig. 6-5 Cabinet-style chemical air filtration was added to the existing integrated filters to provide protection from possible spills and elevated room concentrations.

resolution limits for 248-nm resists continues to decrease. By adding such enhancements as phase-shifting-mask technology and/or more complex multi-layer resist schemes, the use of 248-nm-lithography is predicted to be extended to features as small as 0.13-μm. In fact, some claim that positive resists for 248-nm can image down to 0.1-μm (100-nm) for isolated lines in logic applications, and 130-nm in dense arrays using the latest high-NA (0.8) scanners. (Smaller isolated lines can be imaged than can features in dense arrays because scattering-bar and OPC technologies are more readily applied to isolated structures.) However, at these critical dimension limits, it is very hard to maintain adequate process windows in a production environment and process optimization is engineering-intensive. The applications in which 248-nm will be pushed to 100-nm will therefore be very limited.

With manufacturers supplying fifth and sixth generations of 248-nm resist, these resist types are maturing rapidly. For example, their etch resistance is now comparable to *i*-line resists. The 248-nm resists are also optimized for specific challenges, resulting in resists for isolation layers, contact holes, metal lines, etc. Although such layer-specific resist specialization causes complexity in the fab line, it is mandatory in order to achieve sub-wavelength lithography.

TABLE 6-1 Linewidth Sensitivities of Commercially Available DUV KrF Resists

Resist	Supplier	Linewidth sensitivity (nm/°C)
Apex E	Shipley	16.0
UVIIHS	Shipley	7.5
UV6	Shipley	2.6
TM-461	JSR	2.6
DP-024	TOK	1.8
ARCH 2	Arch	~0

Another issue involving 248-nm DUV-resists is their sensitivity to post-exposure-bake temperature. The earliest commercially available 248-nm resists (e.g., APEX, from Shipley) exhibited large linewidth variations with fluctuations in the temperature of the PEB (16.00 nm/°C). However, as shown in Table 6-1, resist suppliers responded to industry requests to provide better linewidth control. (For example, Shipley later produced the UVII/HS resist - and thereafter the UV6 resist - which progressively improved this parameter.) However, these linewidth sensitivities have driven hot-plate specifications, since the resists with essentially no post-exposure linewidth-variation may not be universally applicable to all feature types, nor are they necessarily able to provide the best resolution.

In order to satisfy the extraordinarily tight linewidth-control requirements of today's and future processes, hot-plates must therefore exhibit excellent across-plate uniformity – significantly less than 0.5°C (range) – and must be matched to tighter levels. This is a challenge not only in terms of achieving the required capability, but it also poses a formidable metrology challenge. Resistance temperature detectors are the best available thermometers today. These are usually bonded to a silicon wafer, which is then placed on the hot plate. This placement is also difficult, since the wires must go between the wafer and the measurement electronics.

As noted above, printing isolated lines is easier than printing nested features. Consequently, contact-hole layers have rapidly become the ones that are most difficult to image (and may be the first to transfer to 193-nm). One example of how layer-specific resist specialization is being used to tackle the contact layer imaging problem is the so-called *RELACS* process.[8] That is, this is a process developed by Mitsubishi Electronic Corp. for shrinking KrF-imaged via holes

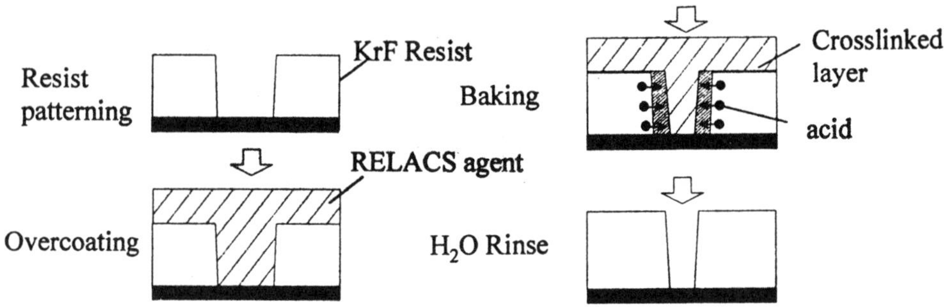

Fig. 6-6 Process flow of RELACS.[8] (© IEEE 1998)

from 0.2-μm to 0.1-μm using a coat, diffusion-bake-and-rinse sequence, following resist-exposure and develop. That is, after the coating step, acid from the unexposed resist diffuses into the wall of the via hole covered by the coating – and then into the coating material – during the bake step. The coating material is crosslinked by this acid (see Figs. 6-6 and 6-7). After a rinse step, some of the coating material remains behind, thereby causing the opening size of the via to shrink. Clariant AZ has licensed the materials from Mitsubishi (R200 coating and R2 developer) and commercializes the process as RELACS (Resist Enhancement Lithography Assisted by Chemical Shrink).

Once the performance of single-layer 248-nm DUV-photoresists based on pHOST chemistries becomes exhausted, bilayer-resists may be adopted to extend the 248-nm-based lithography process. Bilayer processes decouple requirements for imaging, performed by the thin (200-nm thick, or less) top resist, from needs

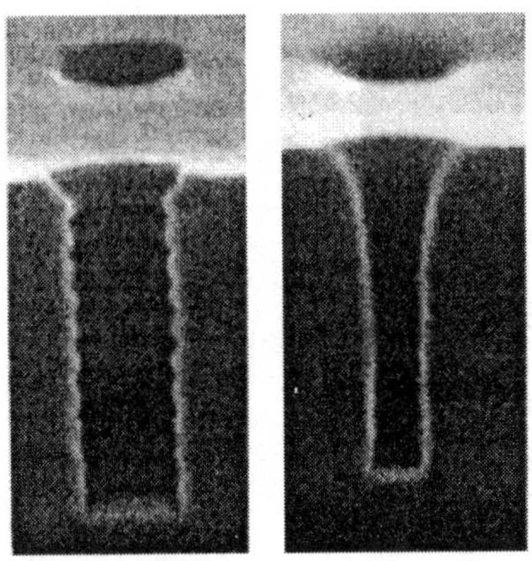

Fig. 6-7 The RELACS process shrinks these KrF-imaged 0.22 μm contact holes (left) to 0.11 μm (right) using a coat, diffusion bake, and rinse step after wafer patterning. (Courtesy of Clariant AZ)

for etch resistance, managed by the thick (400-1000-nm thick) bottom resist. The top layer uses standard single-layer resist imaging, and the image is transferred down anisotropically using O_2/SO_2 plasma etching. The bottom layer acts as an anti-reflecting coating (ARC), reducing the pattern-width degradation from underlying topography. Importantly, bilayer-image performance tends to be more limited by the capability of the scanner. At least one bilayer approach has been implemented in manufacturing. The CARL (chemically amplified resist lines) developed by Infineon Technologies, is being evaluated at 248-nm and 193-nm. (See Vol. 1, 2nd Ed. for more details on the CARL process.)

Another bilayer approach is being proposed for dual-damascene patterning with 248-nm lithography. Arch Chemical offers a thin imaging system (TIS-2000®) optimized for dual-damascene patterning. The top resist layer provides patterning and the bottom organic-layer acts as the BARC, while also planarizing via layers for trench imaging in a via-first approach (Fig. 6-8, see also Sect. 15.9.1.5).[9] One advantage to this process is its dual use at 248-nm and 193-nm.

6.1.2 193-nm DUV Photoresists

The insertion point for 193-nm lithography is predicted to be the second-generation 0.13-μm or even 0.10-μm devices (late in 2002 or early 2003). Successful extension of 248-nm resist is stalling the 193-nm transition. In addition, the compressed timeline for arriving at the 0.13-μm node on the SIA Roadmaps is also part of the reason. That is, if there were four years allowed for developing the 0.13-μm technology, then 193-nm lithography would have a good chance of being accepted into manufacturing from the beginning of that technol-

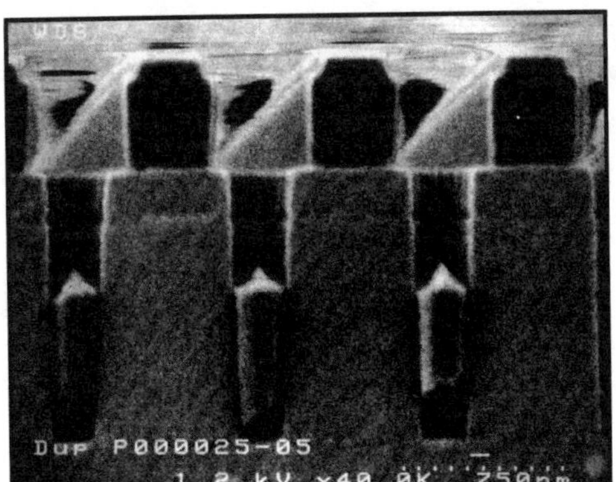

Fig. 6-8 In the via-first damascene approach, the underlayer resist serves the purposes of planarization, BARC and uniform via fill to protect the via bottom during the trench etch. (Courtesy Arch Chemicals and IMEC)

ogy node. But, since there is not that amount of time, there is more of an emphasis being placed on extending 248-nm lithography.

At 0.13-μm there is substantial benefit to transferring to 193-nm using standard binary reticles, rather than using 248-nm and enhancing the reticles with RET. However, the decision to make this transition will depend on the maturity of 193-nm resist materials and the performance of 193-nm exposure tools with very high NA (i.e. > 0.68).[10] In addition, at the 0.13-μm node, economics (rather than technology) may influence the insertion point of 193-nm lithography. That is, some customers (who manufacture a large number of parts from a single mask set) may try to extend the use of 248-nm lithography by any means possible, including the route of using more expensive reticles. Others (such as foundries, which on the average, run fewer wafers per mask) may prefer the ‘simpler" approach of driving to a shorter (193-nm) wavelength. It is estimated that possibly 10 mask levels will require the new systems (enhanced-248-nm or 193-nm). Critical masks will include gate, active area, contact, first metal and first via..

One of the most significant aspects of 193-nm resist development is that after nearly a decade of R&D (primarily performed empirically, due to the complexity of reactions within the resist), a single superior polymer backbone material has not yet been identified (even though the insertion point of 193-nm lithography is drawing closer). That is, no single polymer platform material seems to be able to provide the best process for every application. Instead, there are three seemingly-viable resist polymer types, each exhibiting some benefits – but none of which have all the benefits needed to be used across all production lines. What is predicted to occur is that users will select different chemistry platforms for different applications.

That is, despite the fact that process engineers prefer one resist/one supplier, at 193-nm there are three useful polymer platforms: **1**) *acrylates* (acrylate platforms with attached ali-cyclic structures); **2**) *pure cyclic olefins*: and **3**) *cyclic olefin/maleic anhydrides (COMAs)*.[10] (These are all also different than the hydroxy-styrene-based resists used for 248-nm lithography.) The acrylates for example, reportedly provide the best dense-line performance (particularly with the adamantane methylacrylate chemistry developed by Fujitsu), but they generally exhibit poor etch-resistance. On the other hand, the cyclic-olefin polymers offer better etch—resistance. IBM (in collaboration with Arch Chemicals) is pursuing this group of materials (which reportedly show intrinsically higher robustness toward fairly harsh etch steps). Finally, Clariant AZ licensed a free-radical copolymer of func-tionalized cyclo-olefins and maleic anhydride from Hyundai. This polymer appears to be very cost-effective and may migrate back into 248-nm to reduce resist cost.

The partial benefits that each polymer offers has caused resist suppliers to pursue multiple platforms for 193-nm resists. That is, by 2000 many resist suppliers were working on two or three of these different polymer families (whereas in 1997 each 193-nm resist-supplier had its own, single platform). In addition, a lithography consortium headed by IMEC has been established to evaluate various 193-nm resists from Arch Chemical, Clariant, JSR, and Shipley on a common lithography tool (the ASML 5500/900 scanner with integrated TEL track). This will simplify comparisons of resists whose performance is often measured under different conditions. By 1999 several promising 193-nm resists were emerging (although at that time they still remained 1-2 years from being production worthy). Nevertheless, using a single-layer 193-nm resist, a 0.6-NA scanner and alternating phase-shift mask, 70-nm gates could be patterned (Fig. 6-9).[21]

Beyond selecting the most appropriate polymer platform, resist chemists are charged with finding the photoacid generator (PAG) best able to deliver strong dissolution inhibition, high photospeeds, thermal stability, and tailorable photoacid size (the last factor correlates strongly to imaging performance).

6.1.2.1 Problems of 193-nm Resists: As noted above, the 193-nm photoresist will need to be selected by chip manufacturers in 2001-2002 to begin developing 193-nm lithographic processes. Together with the issues of selecting resists based on their polymer platform, a set of other resist performance characteristics also needs to be considered. We discuss those issues here.

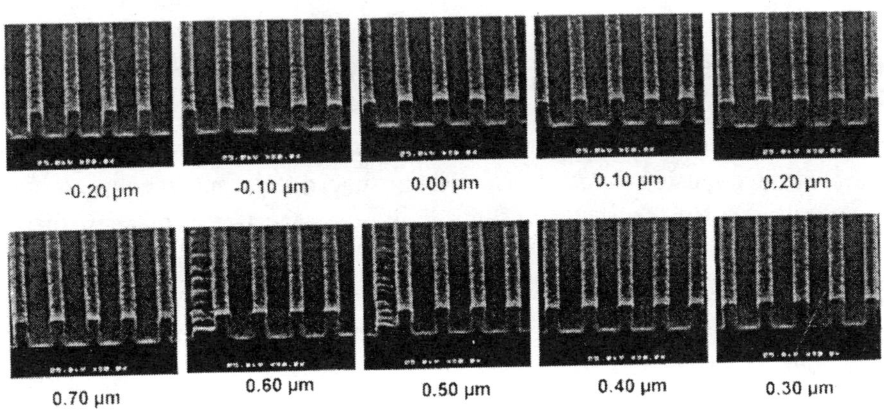

Fig. 6-9 70-nm gates patterned with 193-nm resist and alternating phase-shift mask. Courtesy of Semiconductor International.[21]

The *primary performance factors* involving the still-immature 193-nm resists are those involving process latitude, namely: exposure-latitude and DOF (i.e., an exposure latitude of ~20% and a DOF of 0.5-μm are needed at 193-nm). The are also numerous *secondary performance characteristics* including: etch-resistance; degree of linewidth variation with fluctuation in post-exposure-bake temperature; line-edge-roughness; resist-collapse; resist adhesion (lifting); resist shrinkage during irradiation with e-beams during CD inspection; and resist outgassing.

Etch resistance becomes more critical for 193-nm resists because thinner resist layers will have be used than in previous lithography generations. Resist thickness is expected to be reduced to 400-600 nm in 193-nm (down from about 1000-nm in prior generations of lithography). The advances in etcher design and the use of high-density-plasma systems are helping to improve 193-nm resist thinning during etch steps. However, in cases of insufficient dry-etch-resistance, use of a hard-mask instead of resist-masking may need to be considered.

The *degree of linewidth-variation* in the face of post-exposure-bake temperature-fluctuation is another inherent problem of the immature 193-nm resists. By 2001 193-nm resists exhibited linewidth variations of 10 nm/°C, which is significantly higher than that exhibited by the latest generation of 248-nm resist at that time (see Sect. 6.1.1). But, the same situation existed when the early 248-nm resists were first released. The target value for 193-nm is 3 nm/°C.

Line-edge-roughness is much more pronounced at 193-nm than at 248-nm. Complex interactions occur between lithographic performance, line-edge roughness, and etch-resistance. Resist chemists are seeking the right compromise. Some researchers claim that line-edge-roughness (LER) can be reduced to the 5-7-nm range. The effect of LER on device performance is still being quantified.

Resist-collapse becomes a major issue as the resist features are brought closer together. That is, as developer and rinse water are being removed from developed wafers, surface-tension forces cause closely-spaced adjacent lines to be pulled together (Fig. 6-10). This phenomenon limits the aspect-ratio of resist features (another reason why resist films have smaller thicknesses as features are scaled). Resist-collapse becomes common when resist-feature aspect-ratios exceed about 4:1. There is some concern that resist-collapse may be one factor that could delay the deployment of 193-nm lithography.

The immature 193-nm resists are also prone to *lifting* since their formulations are not yet optimized for adhesion (as are those of *i*-line and 248-nm resists). Adhesion can be optimized with respect to wafer-temperature and HMDS-delivery for each resist type. Lifting can also be addressed by reducing the resist thickness. Using a multilayer resist is another potential remedy for resist-lifting.

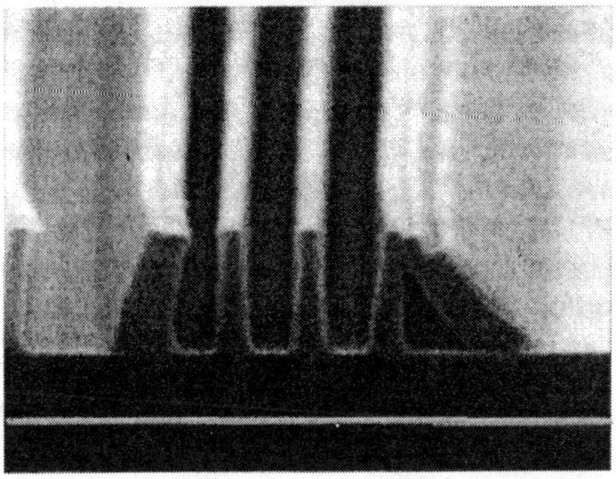

Fig. 6-10 Closely spaced, high-aspect-ratio lines of resist that have collapsed after development.

The 193-nm resists also exhibit significant *shrinkage* (both in the vertical and lateral directions) when they are irradiated by electron breams (as occurs when CD inspections are performed using an SEM). Such resist shrinkage was also observed in early 248-nm resist, but modified resist formulations helped to overcome that problem.[11] Some 193-nm resists exhibit more shrinkage than others, but the tradeoff is that those that have better shrinkage behavior tend to have worse photospeeds and are more prone to resist collapse. Additional work is needed to make 193-nm resists able to meet all of these needs.

Another significant problem with 193-nm resists is that they *outgas organic compounds* that can then build up on scanner optics. This issue may be addressed by PAG selection or by using various additives. Aromatic fragments such as the iodobenzene liberated by iodonium salt sensitizers are among the most critical outgassing species. While the outgassing problem is evident with 248-nm systems as well, the working distance between the lenses and resist at 193-nm is shorter, and the adsorption of vapors is greater at shorter wavelengths. Thus, it is predicted that the outgassing problem will further increase for 157-nm systems.

6.1.3 157-nm DUV Photoresists

The existing deep-UV lithographic technologies and associated optical extensions combined with early 193-nm-based systems are expected to provide sufficient resolution to allow them to be used for the 130-nm device generation outlined in the 1999 SIA Roadmap. However, extension of optical lithography beyond that point may require that 157-nm DUV-illumination systems be adopted by 2005. If this time-schedule is to be met, the first pilot-line quantities of 157-nm resist will need to be delivered to SEMATECH by the first-half of 2002 to begin process development efforts toward the 100-nm device node. Volume production capability using 157-nm illumination systems should be then become available by the 1st quarter of 2005.[12]

The question of "Will 157-nm lithography be ready by 2005?" was debated at the SPIE Microlithography Conference in March 2001. However, there was no consensus as to whether the schedule will (or even, should) be met. Nevertheless, considering that the development cycle for 248-nm DUV resist materials spanned at least 10 years (1980-1990), and the 193-nm resist development program was approximately 8 years long (1991-1999), the 157-nm resist development efforts are faced with a much more abbreviated and challenging development schedule. In addition, the cost of developing a 157-nm resist is also expected to be very high (estimated to be $50-60M). Since such a financial burden is not feasible for any single company, the International SEMATECH formed a University-based consortium to perform fundamental research on 157-nm resists. Just as significantly, IMEC and ASML have begun a cooperative 157-nm lithography development effort, with funding support from many top device manufacturers, including AMD, Infineon, Intel, Micron, Philips, Samsung, ST Microelectronics, and UMC. The First International Symposium on 157-nm Lithography (sponsored by 43 companies), was held in Dana Point, CA in May 2000, and the Second Symposium was held in the same location in May 2001. In any case, researchers are confident that the industry will design workable resists for 157-nm lithography, but they are less sure that they will be able to do it by the 2003 deadline (i.e., set by the possibility that this will be time at which it will be necessary to pattern critical levels in the 70-nm device generation).

The biggest difficulty facing 157-nm lithography is that the illuminating radiation consists of such high-energy photons (7.9 eV). Most polymers used to make up the backbone of previous generation resists strongly absorb such high-energy photons. Thus, it is difficult to find polymers for 157-nm resists that have sufficient transparency (as well as adequate etch resistance, and good adhesion to substrates). Fortunately, however, two promising candidates for the *backbone* (or *platform*) polymers in 157-nm resists have been identified: **1**) highly fluorinated hydrocarbons (fluoropolymers based on norbornane); and **2**) silicon-based siloxanes. A key challenge to developing 157-nm resists based on the latter group of materials (siloxanes) is the ability to achieve a sufficiently high glass-transition temperature.

Fundamentally, the design of materials for developing 157-nm resists is identical to that used to develop resists for the previous DUV lithographic generations. That is, the materials must satisfy the following set of requirements common to all CA-photoresist materials: **1**) high transparency at the exposure wavelengths; **2**) sufficient resistance to plasma-etching processes; and **3**) functional units that are capable of undergoing efficient photochemical transformations that change their solubility in developer solutions. For positive-tone resists, this is typically accomplished by designing the matrix-resin (backbone material) with an acidic functionality that is masked by an acid labile protecting group. Strong acids generated within the exposed regions of the photoresist-film catalyze the thermolysis of the protecting groups, rendering these regions soluble in the aqueous-base-developer solutions. The resists should also have glass-transition-temperatures higher than the processing temperatures, acceptable shelf-storage lifetime, and minimum toxicological risk. A single-layer-photoresist process for imaging at 157-nm is also clearly preferred over a multilayer-resist approach.

When attempting to develop the next-generation of resists, it is often convenient to adopt a modular approach for the design of new photoresist materials. This allows the polymer design to be broken down into several functional sub-units that may be addressed in a somewhat independent fashion (Fig. 6-11).

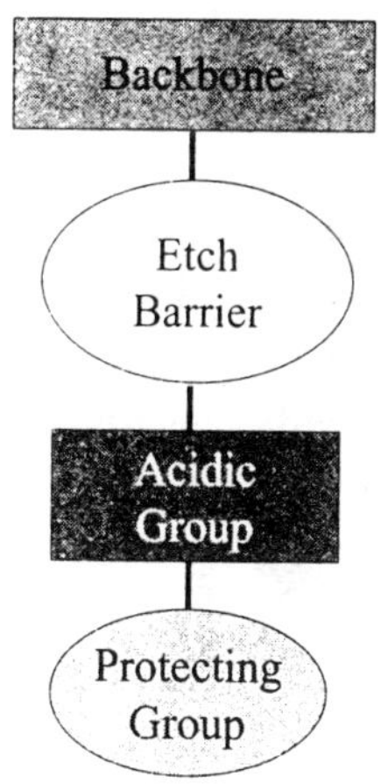

Fig. 6-11 A modular approach in the design of new photoresist materials provides flexibility in design, but transparency among all subunits must be maintained.

Since the 248-nm and 193-nm resist materials are too-absorbing at 157-nm, and since the 157-nm resists were still in their earliest stages of development in 2001, a two-pronged approach was chosen to develop the technology of 157-nm lithography. The first prong involved using existing resists, but at a greatly reduced thickness (60-70-nm-thick). This provides a pathway for tool-testing and for developing early baseline-processes. The second-prong pursues development of the 157-nm resist itself (i.e., exploring the use of novel polymers - such as the fluorinated variations of the resins used in 248-nm resists). In these research efforts, attempts are also made to optimize the formulations and imaging conditions required to print sub-100-nm features. This is a complex task, since it is necessary to explore a variable space that includes the photoacid-generator, base, dissolution-inhibitor, base-polymer, exposure, bake, and development. Figure 6-12 shows some preliminary results of 70-100-nm-features printed using early formulations of 157-nm resists that have evolved from that development effort. Imaging is performed at 0.6-NA, 50-mJ/cm^2, with an alternating phase-shift-mask, using 148-nm-thick resist on a 70-nm antireflective-layer (DUV30), followed by a 60-sec 140°C PAB, a 90-sec 130°C PEB, and development for 20-sec in a 0.26-N TMAH developer. A good image quality is obtained, with sidewall angles between 85-89°.

Beyond having to meet the performance requirements for reduced line-edge-roughness, fast-photospeed, dry-etch-resistance, and solubility in aqueous-base - developer, 157-nm resists have a strong tendency to outgas materials, which settle on the lens. At 157-nm, even a monolayer of oxygen or water can significantly change the transmission of light through the lens and even attenuate light. Fluorocarbons are less of a concern than silicon dioxide, because the user can bleed oxygen into the system with the laser on, making the fluorocarbon residues volatile. With siloxanes, however, SiO_2 deposits are more difficult to remove. Attention to designing effective gas (nitrogen) flows between the wafers and lenses in 157-nm lithography tools will help combat this problem.

6.2 ANTI-REFLECTIVE COATINGS (ARCs)

Interference effects caused by the reflection of light that is transmitted through the photoresist and reflected off the substrate and back, can interfere with incoming light and cause the resist to be unevenly exposed. (See Vol. 1, 2nd Ed., p. 517-24 for a discussion of such effects.) Instead of reducing such interference effects through the modification of the resist itself (as also discussed in Vol. 1), methods that reduce reflectivity at resist interfaces can provide linewidth control with minimal loss of resist performance. That is, to dampen these reflections, *anti-reflective coatings* (*ARCs*) can be used.[13] By minimizing standing waves and reflective notching from substrate reflections, ARCs provide improved resolution and better CD control.

ARCs suppress the reflections either by: **1**) *attenuating light* that passes through the ARC or; **2**) by *matching the index of refraction* of the ARC to the resist-system at the exposure-wavelength employed (such that destructive interference of the light passing through the ARC occurs). These layers can be deposited either on top of resist layers (forming a *Top AR* coating or *TAR*), or on the surface of the wafer beneath the resist (to form a *Bottom AR Coating* or *BARC*). For 0.25-μm lithography (and below), BARCs must be able to suppress more than 99% of the substrate-reflected-light to provide sufficient CD control. Although it significantly increases the process complexity, use of *both* a BARC and a TAR have been reported for layers that require the tightest CD control (i.e., poly gate patterning).

Bottom anti-reflective coatings are materials applied between the substrate and the resist. They are used extensively in applications such as gate definition and for opening contacts, where it is important to keep tight control over critical dimensions. They are not used much at the interconnect level, since the layer of TiN typically used to control electromigration and prevent hillocking on aluminum lines has the extra benefit of being a good anti-reflective coating. As the industry goes to shorter wavelengths, the challenges of minimizing reflections increase. That is, the reflectivity of the substrate becomes higher with

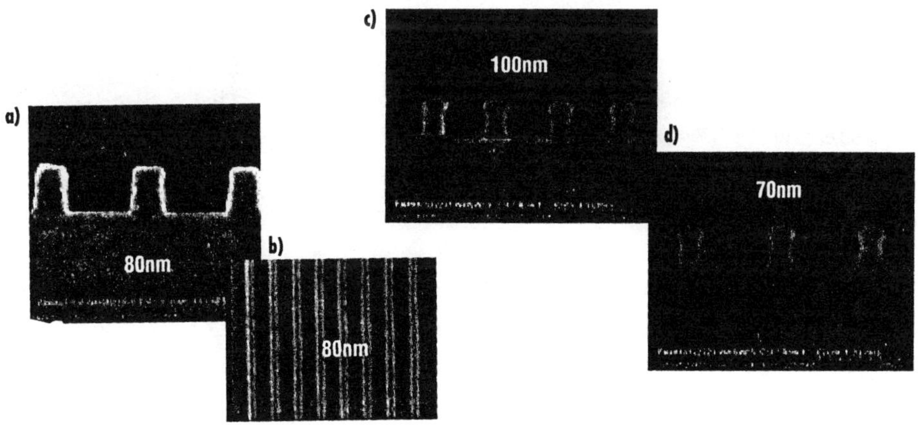

Fig. 6-12 Using 157-nm exposures, phase-shift masks, and 148-nm thick resist on a 70-nm AR layer, various resist recipes resolve at linewidths of: a) and b) 80 nm; c) 100 nm; and d) 70 nm.

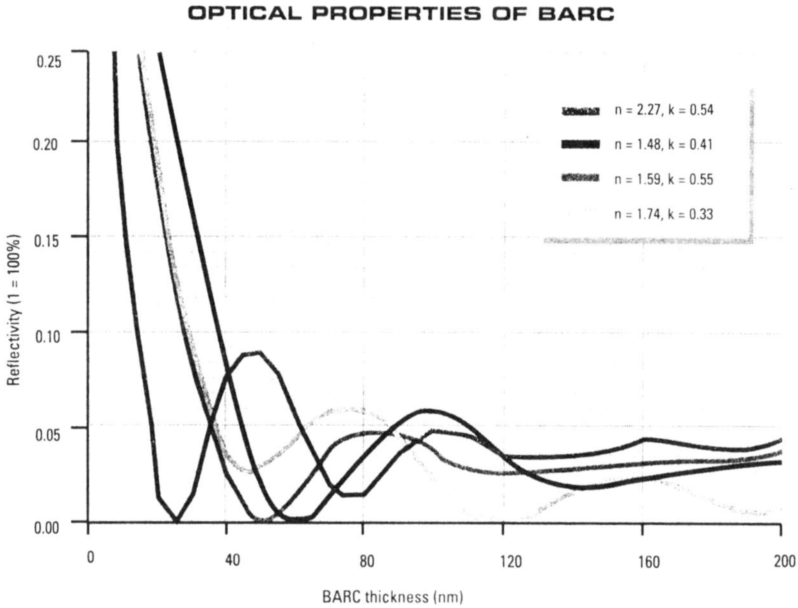

Fig. 6-13 Reflectivity is a function of film thickness, refractive index (n) and extinction coefficient (κ).

decreasing wavelength, and thus interference problems also become more severe. At the same time, advances such as phase-shift-masks, off-axis-illumination, and optical-proximity-correction have allowed features to be printed that are smaller than the wavelength of light. This has made the elimination of CD variations caused by reflected light to be even more critical.

BARCs reduce the problems involving reflectivity by decreasing the value of the reflection coefficient of the resist-substrate interface. They can accomplish this in two ways. First, the BARC may absorb the incident light that manages to penetrate to the bottom of the resist (*absorption*). This requires that a material with a high extinction-coefficient κ be chosen for the BARC, and/or that a thick BARC layer be used. Unfortunately, if the value of κ is too high, the reflectivity of the resist/BARC interface increases – which degrades the effectiveness of the BARC. Thus, useful values of κ are typically in the range of $0.22 \leq \kappa \leq 0.5$ in order to maintain reflectance back into the resist below 10%. For processes that used *i*-line resist, a fairly thick BARC-layer was used (60-90-nm thick, see Fig. 6-13).[13] But DUV-lithography mandates the use of thinner resist and BARC layers because of decreased depth of focus. One of the major challenges for BARCs is maintaining a small enough reflectivity as the BARCs get thinner with decreasing feature size (see also Fig. 6-13).

Thus, a second approach which decreases the value of the reflection coefficient of the resist-substrate interface is favored for 0.25-μm lithography (and smaller). That is, a BARC material whose index of refraction and thickness match the incident light wavelength is sought (*index-matching*). To implement index-matching, a structure is formed that has an optical-path-length of $\lambda/4$ for the wavelength of light being employed. Such a $\lambda/4$-optical-path-length will exist in a thin film (the BARC) made of a material having the correct index-of-refraction and thickness (in the range of 50-80-nm). Light strikes the BARC/resist interface, and some is reflected back toward the air/resist interface, but the remainder enters the BARC.

The light entering the index-matched BARC layer will pass through the BARC layer twice (i.e., once after reflecting from the BARC/substrate interface). These two passes through a quarter-wavelength optical-path-length will produce a light wave with a half-wavelength phase-shift directed toward the resist/BARC interface. This phase-shifted wave interferes destructively with the part of the light originally reflected from the BARC/resist interface. Thus, reflectivity at this interface is decreased. Complete destructive interference only occurs if the amplitude values of the interfering waves are identical. Tight control of BARC thickness (to ensure the two-pass optical-path-length is exactly $\lambda/2$) is an obvious requirement for index-matching by the BARC layer. If thicker BARCs can be tolerated, further reduction in reflectivity will occur by making the index-matching BARC thicker (i.e., to a two-pass optical-path-length of $3\lambda/2$).

In fact, all BARCs employ some of both the absorber and index-matching mechanisms. In any case, if the BARC is effective, little of the light incident on the resist/BARC interface is reflected back toward the resist/air interface, and as a result, the *clearing-dose swing ratio* (see Vol. 1, 2nd Ed., p. 522) is smaller. Standing-wave-effects are reduced because the smaller amplitude of the reflected light decreases the standing wave intensity. Since less light scatters off the bottom surface, the severity of notching is also mitigated. BARCs thus compensate for variations in wafer topography, substrate reflectivity, and surface characteristics. One price paid for these benefits is increased process complexity (and, of course, increased process costs, in terms of equipment and materials). For example, a separate process may be needed to remove the BARC edge-bead. Another issue which arises when BARCs are employed is that their use may also significantly extinguish the light signals that come from alignment marks (these signals arise from the scattering of light from the mark). A variety of deposition techniques have been used to deposit BARCs including spin-on, PVD, and CVD processes.

6.2.1 Organic BARCs

Spin-on organic BARCs are primarily absorber-type polymer films that contain highly absorbing dyes (introduced into the polymer at concentration levels that deliver appropriate extinction-coefficient values). They must also be non-bleaching at the exposure wavelength. Such *organic BARCs* are applied directly to the substrate to a thickness in the range of 60-120-nm. The resist is then spun on top of the BARC, using conventional track equipment. Suppliers of organic spin-on BARCs include Brewer Science, Clariant, Shipley, Hitachi, and Tokyo Ohka. Proponents of the organic approach point out that organic BARCs are advantageous in terms of cost-of-ownership, refractive-index-reproducibility, planarization-capability, film-thickness-tolerance, rework-capability, and surface-control.

Early organic BARCs were about 200-nm-thick, and were wet-etched (in a step often called "ARC open"). Since wet etching is an isotropic-removal process, the resulting undercutting of the BARC layers made them unsuitable for optical lithography smaller than 0.5-μm. Newer organic BARCs are designed to be developed by an anisotropic dry-development process using an oxygen plasma. This extends their use to DUV lithography. However, the self-planarizing nature of spin-on films does not make the dry-developing process a complete solution to the above undercutting problem. That is, when spun-on over topography, some regions of the BARC will be thicker than others, and this causes a problem when organic BARCs are dry-developed. An anisotropic dry-etch-step can be used to etch the BARC in the flat areas of the wafer, and while this may erode some of the resist thickness, it should not attack the resist laterally. However, the thicker residual-BARC-regions must still be removed, and this may be done with a second, isotropic over-etch-step. The isotropic BARC removal process will also laterally attack the resist feature, and this can give rise to linewidth-control problems. ARC-open may be carried out in a separate etch chamber prior to the oxide, metal, or poly etch, or the ARC may be etched *in-situ* prior to the film being etched. Using a separate chamber approach leads to a larger COO penalty. Hence, dry-etch-tool vendors (such as Lam Research), offer *in-situ* ARC-open processes integrated with the etch process of all film types (Al, W, poly, and oxide). Organic BARCs are typically etched with O_2 chemistries (together with appropriate additives such as Ar, He, and CF_4).

Another issue involving organic BARCs is their (in)compatibility with chemically-amplified (CA) resists. That is, the acidity levels of organic BARCs can negatively interact with CA resists (resulting in acidic or basic by-products being formed during thermal hardening of the resist). These by-products can create an intermixing layer at the BARC/resist interface. Such intermixing must not be allowed to occur. Researchers have developed organic BARCs for KrF and ArF resists that have chromophore and cross-linkable functional groups in the same copolymer form as the photoresist to avoid this problem.[14] The key difficulty of implementing the solution of closely matching the chemistry of the organic BARC to that of 197-nm resist is that by 2001 there was not yet a mature 197-nm resist. Until a baseline product is chosen, it will be hard to closely match the organic BARC and resist. Only when the industry standardizes to one or two 197-nm resists, will it be possible to apply this solution successfully.

A manufacturing-related challenge of using organic BARCs is caused by the fact that the liquid, spin-on BARC materials coat spinning bowls and coagulate in drain lines. A new material introduced by Nissan Chemicals (DUV44) reportedly uses 70% less catalyst, which results in less spin-bowl coagulation. It retains the optical properties of its predecessors so that there is minimal process variation with respect to depth of focus or exposure latitude.

6.2.2 Inorganic Dielectric BARCs

Inorganic dielectric BARCs are deposited by PECVD. These are typically films of silicon oxynitride (SiON) or silicon nitride (SiN) that function both as index-matching and absorbing BARCs. The thickness, index-of-refraction, and extinction-coefficient of such silicon-oxynitride layers can be tuned to produce an effective BARC layer. The inorganic-CVD-BARCs are thinner (30-50-nm-thick) than the organic BARC layers. These layers also conformally coat the underlying topography as a result of the CVD process. This allows the entire film to have the correct thickness (i.e., the film does not self-planarize). Inorganic BARCs have chemistries similar to the underlying layers, which can make it tricky to remove them after resist-stripping. Thus, some processes leave the inorganic BARC material in place. On the other hand, inorganic BARCs can have a much higher selectivity to the material being etched (e.g., poly-Si or Al) than can organic BARCs (or the photoresist itself). Consequently, the "ARC open" step is not as much of an issue as when an organic BARC is used. In fact, in some processes that use inorganic BARCs, no separate ARC-open step is used. A smaller overetch time is also needed to clear the BARC all over the wafer and large feature sizes than for organic BARC clearing. Dedicated production tools for depositing inorganic BARCs have been developed by such suppliers as Applied Materials and Novellus Systems.

For 193-nm lithography, the composition of the inorganic SiON films must be altered, since most silicon-based inorganic materials are opaque to 193-nm light.

The BARC films for 193-nm lithography will have to change from SiON films (which are silicon-rich films, where 40+% of the film composition is silicon) to films that are much more like an oxide. However, the PECVD process allows such tuning of the composition of silicon, oxygen, and nitrogen of the film over a relatively wide range. By adjusting the CVD process conditions, the BARC film can be tuned to be suitable for *i*-line lithography, 248-nm lithography, or 193-nm lithography.

Inorganic BARCs can cause adverse reactions between the ARC film and the photoresist (akin to the problems that can occur when organic BARCs are used). The culprits are again amine groups (NH_3) that are left on the surface of the SiON films after processing. Deep-UV resists are more sensitive to such interactions than older *i*-line and *g*-line resists. One technique to avoid this interaction is to *in-situ*-deposit a thin oxide on top of the BARC before spinning-on the resist. This oxide becomes an inert barrier between the resist and BARC and has no impact on the properties of the BARC itself. Another way of preventing such interaction is to use an oxygen treatment after the SiON film has been deposited, but before putting down the resist. This effectively removes any amine groups from the BARC surface. The Novellus PEARL® (Plasma-Enhanced Anti-Reflective Layer) includes an *in-situ* nitrous-oxide treatment before the wafer is taken out of the deposition chamber.

6.2.3 PVD-Deposited ARCs

The most common PVD-deposited ARCs are TiN films. TiN films act primarily as absorber-type BARCs, with TiN having a κ-value of 1.1. Their main application is for patterning Al films with *i*-line lithography. Because TiN ARCs are difficult to remove, they are normally left behind and become part of the device (where they may serve in the role of a diffusion barrier). TiN films are reactively sputter-deposited to a thickness of ~80-nm for this purpose. Their use in DUV-lithography is limited by their high-DUV reflectivity (20–30%).

6.2.4 BARCs for Dual-Damascene Applications

When BARCs need to be used with dual-damascene, low-*k*/copper interconnect systems, another set of complex challenges are encountered. The biggest obstacle is patterning transparent dielectric layers (in which the trenches and via holes of the damascene structures must be etched). These dielectric layers may lie over the rough topography of a reflective film such as copper or polysilicon. Unlike the case of applying a BARC directly onto a reflective layer (such as a poly or Al film), in this situation the intermetal dielectric is transparent. Below it are all kinds of reflective surfaces, at multiple levels. This makes the conventional method of controlling reflection by destructive interference inadequate. That is, a reflection minimized for one local film stack may result in a high reflection for another local film stack. To solve the problem, multilayer ARCs, termed *universal ARLs (UARLs)* are deposited on the dielectric to be patterned (Fig. 6-14). Either two-layers (or sometimes even three layers) are used to create such structures. The bottom layer is opaque to UV light (and reflective), to mask the underlying oxide, while the upper layer cancels out the reflected light by destructive interference. The underlying layer has a higher index of refraction and extinction coefficient so that it is more reflective than absorbing (in the same way that poly and aluminum films behave). Such UARLs are deposited by CVD, in tools from vendors including Novellus Systems and Applied Materials. The UARL formed in the Applied Materials process is termed *Double–DARC.*® Note that the third layer shown in Fig. 6-14 is an oxide film to prevent chemical interaction between the first two BARC layers and the DUV resist.

6.3 PHOTORESIST PROCESSING SYSTEMS

In the early days of IC fabrication the pieces of equipment that performed the individual resist-processing steps were stand-alone tools. As each step was

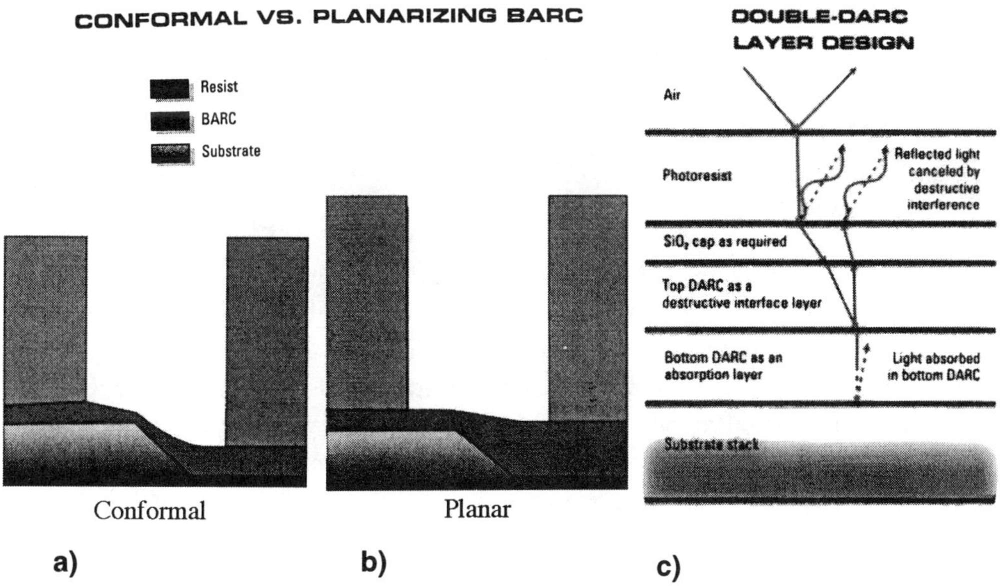

Fig. 6-14 a) Conformal BARC; b) Planar BARC; c) three-layer universal AR layer (UARL).

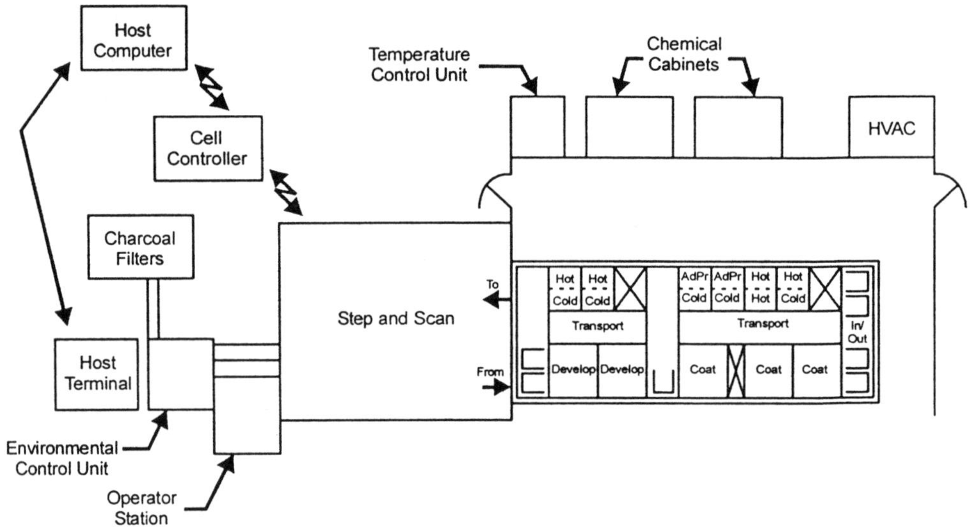

Fig. 6-15 DUV Photo-island.

completed, the wafers would be manually carried from one tool to the next. By about 1990, these tools were integrated into resist-processing systems that automatically moved the wafers from one tool to another without human intervention. Such systems are now configured as either wafer tracks or wafer clusters. In either case, such resist-processing systems are then linked to a stepper to form an integrated stepper-track/photo-island (Fig. 6-15).[15] A cassette of wafers is loaded onto the input station of the photo-island, and 30–60 minutes later they emerge as patterned wafers, ready to be sent to the next process (e.g., ion-implantation or etching).

Wafer tracks originally consisted of two side-by-side lines of resist processing tools (i.e., the prime, spin, and soft-bake modules on the line that fed the stepper, and the PEB, develop, and hard-bake modules on the other). The wafers were transferred sequentially from one module to the next using a belt transport mechanism. In today's *track systems*, the individual tools are still arranged in two lines, but a central robotic transfer shuttle moves the wafers from one tool to another, offering the capability to randomly access the tools (Fig. 6-16). In *wafer clusters* the individual resist-processing tools are arranged in a circular fashion around a central robot that moves the wafers among the tools, again allowing flexible access to each of the process stations.[16,17] The wafer-cluster-system approach was pioneered by Texas Instruments in the mid-1980's and was introduced commercially by FSI in 1991. It has an anthropomorphic, 6-axis

Fig. 6-16 Clean Track ACT platform.® Courtesy of Tokyo Electron America.

wafer-handling system that provides more reach and fewer, larger robots per cluster. SVG's ProCell wafer-cluster system offers a hexagonal platform (Fig. 6-17). Wafers follow a circular path from process module to coating, stepper interface, develop cell, and back to the cassette interface. By scheduling processes with more than 40 wafers simultaneously with varying positions, time requirements and sequencing, the lithography process can be completed in 12 balanced moves. This is done with a patented synchronous scheduling algorithm called *advanced guidance system* (AGS), which controls wafer movement by calculating the optimal wafer path before movement is initiated. System-induced delays are minimized through time management that automatically builds-in scheduled intervals at user-defined non-critical steps, allowing consistent process timing.

The stepper-track (or stepper-cluster) photo-islands are among the most complex systems in a wafer fab. They may consist of up to 20 individual process tools (besides the stepper) that must all work together. At any one moment, the track may be transferring 15-20 wafers through multi-process levels, putting resist on some, developing some, and baking or chilling others. In addition to handling more wafers, the newer lithography processes require the track to perform more steps per wafer. These additional steps include: edge-bead removal,

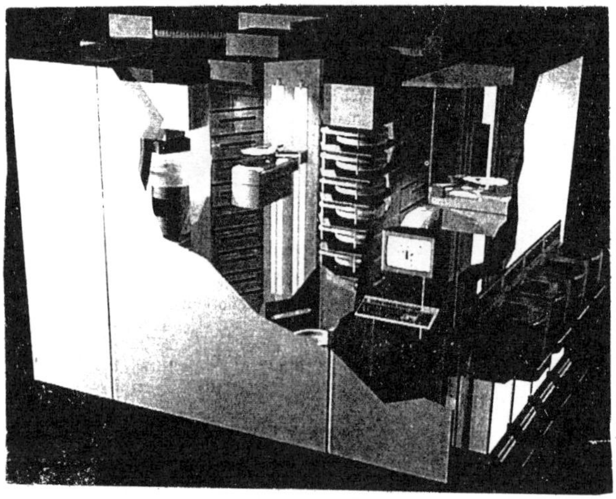

Fig. 6-17 Pro-Cell cluster.® Courtesy of Silicon Valley Group.

applying anti-reflective layers, and processing multi-layer resists. Track systems must also ensure that each wafer (and each lot) has a consistent photoresist coating. The largest manufacturers of such track and cluster systems are Tokyo Electron Ltd., Dainippon Screen, FSI International, and Silicon Valley Group (roughly in that order). We will now describe some of the issues associated with the functioning of photo-islands.[20] They include: capacity, flexibility, reliability, cost of ownership, and environmental/safety features.

The *capacity* of the wafer track is one of its most important characteristics. Advanced wafer stepper/scanners are both fast and expensive ($5M to $10M). To get maximum use from the stepper/scanner, the track must be able to supply wafers as quickly as the stepper can accept them. By 2001, some stepper/scanner throughputs exceeded 100 wafers/hour (200-mm). The wafer track thus requires multiple tools for each process to keep up. For example, as many as six spinners may be needed to support one stepper.

Another critical feature is *flexibility*. The track system must be capable of being modified to meet not only today's, but also tomorrow's needs. It should be possible to add, upgrade, or replace process stations without replacing the entire system. Flexibility is so important because deep-UV resist processing is still evolving, and both hardware and software must be adaptable to "yet to be determined" queue times, chemistry management, and process steps. However, since 200-mm tracks do not accommodate upgrades to 300-mm wafer sizes, the flexibility show-stopper will be the change to the 300-mm wafer fabrication.

The flexibility capability must also extend to the system software. The software should also be modular, so that it can be reconfigured if needed. The software must also be able to handle multiple process recipes so that an array of processes can be run on the same track (e.g., both *i*-line and DUV resists). The software must meet the SEMI SECS or GEM standards for process-tool-to-host-computer communication. In addition, the operator interface should be well designed so that chances for operator errors are minimized. Ultimately there will be an expert system in the cluster controller that will collect and analyze data from the system using statistical techniques that will determine the overall lithography process status.

The software will also have an impact on throughput. For example, if a new lot cannot be started until the previous lot has finished, the *actual* throughput will be significantly lower than the *steady-state* throughput. To eliminate this problem the software controlling the track must be able to start a new lot while the previous lot is still being processed (a procedure termed *lot cascading*). Hence, it must be possible to start a new lot as soon as the input station is empty. The software must recognize the boundary between the two lots and change the recipe when the first wafer of the new lot arrives at each processing station. Capabilities such as on-line recipe editing and graphical user interfaces also help maximize actual wafer production time.

Next, the system must be highly *reliable*. All wafer track manufacturers specify at least 95% uptime, but even this value means that the system will not be available for 24–36 hours per month. Use of "intelligent redundancy" in subsystems can improve track reliability (for example, by adding an extra hot-plate, or by having qualified standby modules).

The cost of the track system itself, its footprint, and the amount of resist and developer consumption are other factors that impact the *cost-of-ownership*. In the move from 200- to 300-mm wafer tracks, users are demanding that 300-mm tracks be no more than one-third larger than 200-mm systems. Use of vertically stacked modules should allow such system footprint goals to be met. Resist consumption is one of the major costs of operating a semiconductor fab. To control this cost, the dispense volume of resist per wafer has been decreased with time . The design of the resist pump and spinning equipment plays a major role in achieving reduced resist consumption.

The tracks must also be *clean* and *safe to operate.* Both external particles and contaminants must be suppressed, and the track tools and processes cannot generate particles. Point-of-use filtration of resists and developer liquids and advanced wafer transport techniques (that employ gentle handling of wafers) reduces particle generation. Photo-islands are also enclosed in Class-1 environmental chambers. Safety issues involve the control of toxic fumes from resist and developer solutions, and measures to ensure that wafers do not fly off the high-speed spinning chucks.

The environment within the photo-island enclosure must be strictly controlled, including the temperature, relative humidity (RH), and vibration (and for DUV resist processing, the trace quantities of amines).[18,19] The temperature control issue is challenging because the users want an enclosed system in a small footprint. Yet they also require that the temperature throughout this volume (which contains bake modules and motors) also be uniform and tightly controlled. The RH is chosen to be some value in the range from 35–50%, but it must then be maintained at that selected value to ±0.2%. The temperature of the resist and developer chemicals used in the system must also be well controlled.

The goal of the lithographic process is to produce resist features with consistent CDs. In fact, the acceptable CD variation is getting smaller faster than the actual CD size.[20] At 0.18-μm and beyond, good CD control will become the most important challenge of resist processing. Many factors (already mentioned) impact the value of the CD including the resist thickness, the use of an anti-reflection coating, and the exposure and develop processes. For example, in *i*-line lithography, across-wafer and wafer-to-wafer tolerances of 50-Å were acceptable. But sub-0.18-μm processes require < 20-Å ranges. To achieve such tighter film thickness tolerances, the track system must rigorously control cup humidity and temperature, resist temperature, resist dispense rate, and wafer temperature. (Some sub-0.13-μm processes will require ±0.1°C temperature control and humidity within ±0.2%.) The thermal control in baking, as well as the time elapsed between process steps, are becoming additional variables that play critical roles in determining the CD value, especially when DUV-resists are used. In DUV processes, the coating step and the PEB hot-plate-temperature are key, but so is the lag between the time PEB is finished and when the wafer goes on the chill plate. Sequences will have to be timed within 1 or 2 seconds at every step. If the system cannot ensure this transfer timing, CDs will vary from wafer to wafer. (In the past the purpose of the PEB step was to stabilize the image. Now it is the step that basically creates the image. So any variation in the PEB process can significantly impact the CDs for 0.25-μm features and below.)

REFERENCES

1. S. Wolf and R.N. Tauber, *Silicon Processing for the VLSI Era* – Volume 1: Process Technology, Lattice Press, Sunset Beach, CA, 2000, Chaps. 12 & 13.

2. H.J. Levinson, *Principles of Lithography*, SPIE Press, Bellingham WA, 2001.

3. *Microlithography: Science and Technology,* J.R. Sheats & B.W. Smith, Eds., Marcel Dekker, New York, 1998.

4. P.K. Vasudev and P.M. Zeitzoff, "Si-ULSI with a Scaled-Down Future," *IEEE Circuits & Devices*, March 1998, p. 19.

5. H. Ito, "Deep-UV Resists: Evolution and Status," *Solid State Technology*, July 1996, p. 164.

6. C. Garza *et al.,* "Photoresist Materials and Processing," in *Handbook of Semiconductor Manufacturing Technology,* Y. Nishi and R. Doering, Eds., Marcel Dekker, Inc., New York, 2000. Chapter 17, p. 515-528.

7. D. Ruede *et al.,* "The Impact of Airborne Molecular Bases on DUV Photoresists," *Solid State Technology,* August 2001, p. 63.

8. T. Toyoshima *et al.,* "0.1-μm-Level Contact-Hole Pattern Formation with KrF Lithography by Resolution Enhancement Lithography Assisted by Chemical Shrink (RELACS)," *Tech. Dig. IEDM,* 1998, p. 333.

9. L.J. Peters, "Resist Developers Try to Beat the Clock," *Semiconductor International,* February 2001, p. 77.

10. L.J. Peters, "Resists Join the Sub-Lambda Revolution," *Semiconductor International,* September 1999, p. 68.

11. B. Su *et al.,* "Analyzing and Characterizing Resist Shrinkage," *Solid State Technology,* May 2001, p. 52.

12. "Developments in Materials for 157-nm Photoresists," *Solid State Technology*, October 2001, p. 63.

13. P. Singer, "Anti-Reflective Coatings: A Story of Interfaces," *Semiconductor International*, March 1999, p. 55.

14. L. Peters, "Progress in KrF Antireflective Coatings," *Semiconductor International,* May 2000, p. 25.

15. P. Burggraaf, "Stepper-Track Integration: Solving the Problems," *Semiconductor International,* August 1992, p. 30.

16. R. DeJule, "DUV Cluster Tools Go Mainstream," *Semiconductor International,* August 1999, p. 44.

17. M. Krishna *et al.,* "Improve Yields and Enhance CDs. With Integrated DUV Resist Track," *Solid State Technology,* October 2000, p. 135.

18. B. Demandante *et al.,* "Protect DUV Processes with Real-Time Molecular Monitoring," *Semiconductor International,* September 2000, p. 115.

19. L. Peters, "Maintaining a Clean Environment in DUV Lithography Bays," *Semiconductor International,* May 1999, p. 52.
20. A.E. Braun, "Track Systems Meet Throughput and Productivity Challenges," *Semiconductor International*, February 1998, p. 63.
21. L. Peters, "Readying Resists for the 90-nm Node," *Semiconductor International,* February 2002, p. 63.

Chapter 7

ADVANCED LITHOGRAPHY II: OPTICAL HARDWARE

In this chapter the topics related to microlithography optics and hardware pertinent to IC technology at the start of 2001 decade are discussed. These subjects include: excimer-laser light-sources; optical-alignment-tools (stepper-scanners) for the 248-, 193-, and 157-nm lithography generations; resolution-enhancement-techniques; and next-generation lithography (NGL).

7.1 EXCIMER-LASER DEEP-UV LIGHT SOURCES

Excimer lasers are used to produce the light needed for deep-UV optical lithographic exposure systems. Deep-UV laser sources will dominate what many industry experts believe will be optical lithography's final arena before an industry jump to the post-optical era. Such laser-light sources are essential enabling elements for lithography for the next three critical-dimension nodes of the industry's roadmap. The term *laser* is an acronym for *light amplification by stimulated emission of radiation*. The krypton-fluoride (KrF) excimer laser is used to produce DUV light with a wavelength of 0.248-μm (or 248-nm). Argon-fluoride (ArF) excimer lasers are employed to produce DUV light with a wavelength of 193-nm, and F_2 excimer lasers for 157-nm DUV light. No fundamental changes are necessary to allow a KrF-laser to operate as an ArF-laser, except to change the gas composition in the laser's plasma tube and the mirrors that make up the cavity.

By 2001 the KrF-excimer-laser had become the light source of choice in mainstream, high-volume, advanced IC production (i.e., for 0.35-, 0.25-, and 0.18-μm-CMOS technologies). At the same time, ArF-excimer-lasers were being installed as the light source for 193-nm stepper/scanners that were being used for advanced process development applications, as well as for some pilot-line production (i.e., for sub-0.2-μm-CMOS process-development lines). It is conjectured that ArF lasers will displace KrF between 130-nm and 100-nm, albeit not until ArF maturity is reached in production. ArF has the potential to be the dominant source down to the 100-nm to 70-nm technology-nodes. Below 100-nm, the F_2 laser appears as a possibility.[1]

An *excimer* is an exotic molecule formed by one atom each of a noble gas and halogen atom. This *dimeric* molecule is bound only in a quasi-stable excited state.

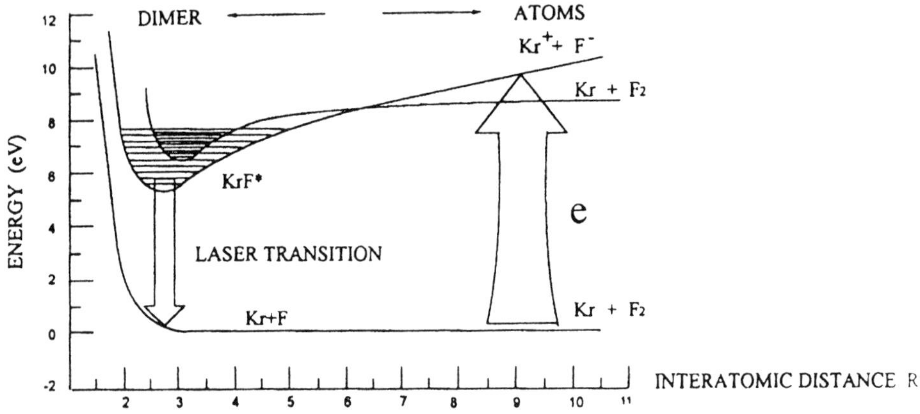

Fig. 7-1 Energy diagram for a KrF excimer laser. KrF* is formed via two reaction channels. It decays to the ground state via dissociation into Kr and F while emitting a photon at 248-nm.

That is, if the noble gas and halogen gas atoms are in the ground state they will not react, but if one (or both) are in an excited state, a chemical reaction can occur to form a *dimer.* The term "excimer" is thus a contraction of the phrase "*excited dimer.*" (Note that true dimer*s* are molecules containing two atoms of the *same* element, but this term is also used here to describe a molecule with two different atoms.) When the excimer decays back to the ground state, the dimer falls apart into its two constituent atoms (with the emission of a DUV-photon accompanying the decay process). When such decay is stimulated by absorption of energy from another light photon, DUV-laser light is emitted by an ensemble of KrF* dimers. The spontaneous (i.e., not stimulated) dissociation time of an excimer (e.g., KrF*) is relatively long (i.e., in the nanosecond to microsecond range). Therefore, population inversion (higher population of the excited state than the ground state) can readily be achieved by exciting a high-pressure mixture of Kr and F gases in a high-voltage pulse discharge (see Fig. 7-1).

A capacitor (that will serve as the transient power supply to produce the plasma discharge) is first charged to 12-kV. Then it is discharged in a very short time (75 ns) across electrodes present in the laser cavity. The plasma discharge created by this injection of energy produces Kr^+ and F^- ions, and these ions then combine to form the KrF* excimers (with the process that creates the excited population of dimers called *pumping*). The laser light is produced by stimulating this created population of excimers to decay within a very short time. Some of the excimers begin to decay spontaneously. The light they emit passes through the gas containing the excimer population, and this stimulates additional excimer decay (which creates more radiation). This is the *amplification* process alluded to in the acronym term *laser*. The laser cavity also uses mirrors to reflect radiation back into the gas. Thus, the stimulated decay rate (and laser-light emission) continues to increase, until almost all of the excimers have undergone stimulated decay. In fact, in a properly designed laser cavity, the stimulated decay process is completed very quickly after it is initiated. Thus, the light of an excimer laser is emitted in the form of short pulses of energy (i.e., these excimer-laser sources do not emit laser-light continuously, as do continuous-emission lasers). The laser-light exits the laser cavity though one of its ends, at which a partially-reflecting surface is located. As soon as the capacitor can be charged again to a high-enough voltage (making it able to act as the excitation source for creating the discharge) the excitation, pumping, and laser-emission cycle can be repeated.

One of the biggest problems in using excimer lasers as lithographic light-sources, however, is the fact they emit their energy in such a pulsed mode. Excimer lasers with a power output that is useful for lithographic applications (2-20 W) typically produce pulses of laser energy at a rate of 200-2500 Hz, with each pulse being only 5-20-nsec in length and containing about 10-mJ of energy. Since each pulse is so short, and there is a relatively long time between pulses, the peak power within each pulse is extremely high (even when the average power of the laser appears to be quite modest). Such peak powers can cause damage to the optical coatings and materials of the optical lithography system if it is concentrated in too small an area (e.g., the stepper lens). Thus, design features must be employed to keep the power-density low enough at all points within optical elements. An increase in the repetition rate of the laser is also beneficial, since this produces the same average power with a lower power per pulse (and exposure-dose-control is also improved).

Excimer lasers are attractive as DUV light-sources for several reasons. First, KrF-lasers can deliver much more usable power at 248-nm than can a mercury-arc lamp. The radiance of the laser is several orders-of-magnitude greater than the arc lamp because the laser emits a highly-collimated beam of light, whereas the arc lamp emits light in all directions (isotropically). Excimer-lasers also have a low degree of coherence compared to most other kinds of lasers, as well as a relatively broad linewidth (which results in low temporal coherence). Low coherence is generally a desirable characteristic of a lithographic light source.

On the other hand, the 0.300-nm (300-pm) *spectral width* of the 248-nm laser is not narrow enough to satisfy the bandwidth requirements of the refractive optical systems of step-and-repeat aligners. (That is, the dispersion [the change of optical index with wavelength] of optical components in the DUV region is high. Thus, spectral bandwidths even in the picometer range, can produce significant chromatic aberration in systems with refractive optical components. This makes it difficult or even impossible to focus and image lithographic patterns properly.) Consequently, since the lenses of these steppers have no chromatic correction, this forces the illumination source to have a spectral bandwidth of less than 0.001-nm (or < 1-pm). It is therefore necessary to put dispersive optical elements (such as prisms and diffraction gratings) within the laser cavity to reduce the bandwidth of the KrF laser-light to this value. These elements somewhat reduce the total power of the laser, but more importantly they also decrease the stability of the power level. A feedback system must be employed to keep the center wavelength of the line-narrowed laser-light-beam constant to about the same level of accuracy as the bandwidth. This control is needed because if the laser wavelength drifts by a few thousandths of a nm, the chromatic aberrations of the lens will cause an excessive shift in the focus of the lens. For each shift of the wavelength by 1-pm (0.001-nm), the lens-focus will shift by about 0.15-μm.

7.1.1 KrF Excimer Lasers

By 2001, excimer-laser wavelength-stability of much less than 1-pm was being achieved (i.e., ±0.25-pm). For example, the Cymer ELS-6010 KrF laser (introduced in 2000) has a spectral bandwidth: ≤ 0.5-pm full-width-at-half maximum (FWHM – see Fig. 7-2a) and a 95% energy integral of ≤ 1.4-pm (this latter specification defines the spectral range that includes 95% of the energy of the pulse, see Fig. 7-2b). Also, the repetition rate of this laser has been increased to 2500 Hz. However, beyond 180-nm lithography, even a 0.5-pm spectral bandwidth and a 95% energy integral of 1.4-pm will not be sufficiently small. These specification will need to be narrowed to <0.4 pm and <1 pm, respectively. Using double-pass Etalon (reflecting surface, Fig. 7-4b) optics with a 10W, 2000

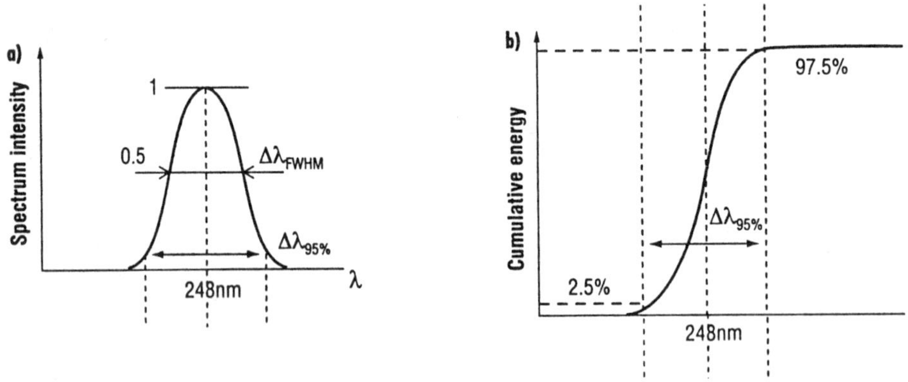

Fig. 7-2 a) Definition of "full width at half maximum," FWHM; b) Definition of 95% integral pulse bandwidth.

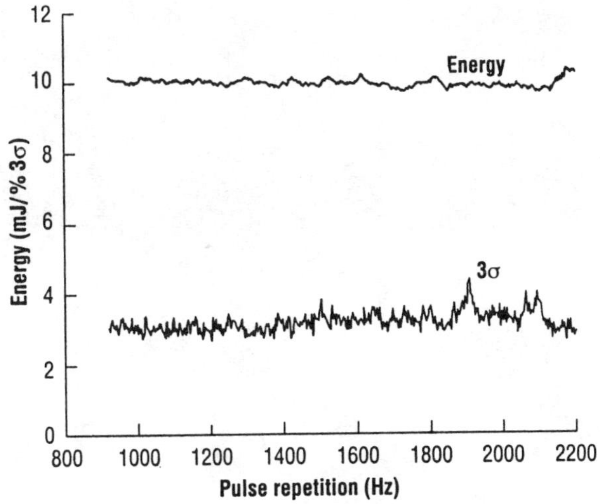

Fig. 7-3 Pulse energy and stability vs. repetition rate for a KrF laser running in the constant charge voltage mode.

Hz laser, laser pulses with an actual spectral bandwidth centered at 248.3-nm, a spectral bandwidth of 0.27-pm and 95% energy integral of 0.93 were produced.[2]

Another issue related to pulse power is exposure control. The method for controlling exposure in a stepper illuminated with a pulsed laser is to integrate the pulsed energy until the required exposure dose is reached (at which point a signal is sent to stop the laser from pulsing). To achieve a 1% control of accuracy, at least 100 pulses must be used. Early excimer lasers had lower pulse-repetition-rates (1000 Hz), and thus had slower exposure times. By doubling the pulse-repetition-rate to 2000 Hz, the delivered exposure energy within the same unit of time was also doubled, and this also cut the exposure time approximately in half. By 2002, the newest laser models have overcome this limitation (repetition-rates are up to 4000 Hz), and the exposure speeds of excimer laser steppers now match (or exceed) those of arc-lamp-illuminated systems. Photonic energy output is a related critical parameter for lithography. The recently improved output characteristics of KrF systems are such that the relationship between output power and laser repetition rate is essentially linear (Fig. 7-3).

In addition to the issue of pulsed power, excimer lasers have some other drawbacks. They are much larger than an arc-lamp illumination source, typically occupying a clean-room footprint of 2x5 feet (Fig. 7-4). The excimer laser is also expensive ($500,000–$1,000,000), which in 2001 added 10-20 percent to the cost of the stepper. Gas used by the stepper is also costly and toxic and the plasma cavity in which it is contained, is not sealed. Thus, it must be periodically refilled

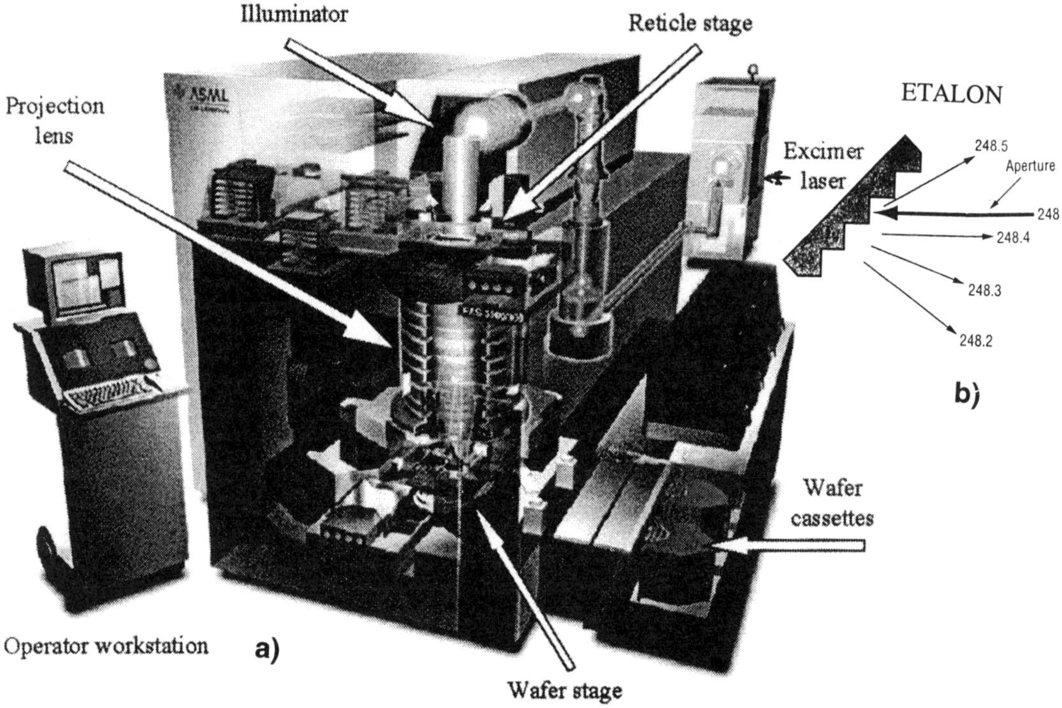

Fig. 7-4 a) Configuration of excimer laser light source and stepper (Courtesy ASML); b) Etalon

with this gas (about every 5 days). This requires a gas-handling system that can fulfill the rigorous industrial safety requirements for toxic gases. (The interval between gas refilling has been increased by more than a factor of 10 since the earliest lasers were used for lithography, and improvement is continuing.)

The laser pulses also sputter materials onto the mirrors and windows of the cavity. Thus, they must be cleaned after a number of pulses. The laser must be shut down during such servicing. Eventually, the entire laser-cavity must be replaced (which represents the largest cost of operating an excimer laser). The number of pulses between cleanings (and replacement) has been continuously increased. In 1998, servicing was required after 1 billion pulses and laser-cavity replacement after more than 2.5-billion pulses. Since laser usage during production is about $2x10^9$ pulses in a year, in 1998 window cleaning had be done about once every six months, and laser-cavity replacement about once a year.

The laser system is also relatively inefficient in terms of electrical-power usage, and thus it consumes about 10-kW while the laser is operating. Finally, the DUV-light of the laser is very hazardous to human skin and eyes. Hence, stringent safety precautions must be enforced when maintenance work is performed on the laser. The laser enclosure and beamline must also be interlocked to prevent anyone operating the system from being exposed to the laser-light. Nevertheless, as excimer-laser technology has matured the reliability of lasers has been greatly increased, and the frequency-of-repair rates have dramatically declined. KrF lasers are commercially available from Cymer, Inc., Lambda Physik, and Komatsu, with Cymer, Inc. at the turn of the century having the largest market share.

7.1.2 ArF Excimer Lasers

The demands of ArF-laser technology are even more stringent than those of KrF-lasers. The power output and repetition-rate must be increased, line-narrowing must be smaller, and pulse-uniformity must be improved to allow accurate dose-control. Finally, the entire set of issues dealing with accelerated degradation of materials and system components caused by 193-nm radiation must be addressed. It appears that a basic space-line resolution of less than 105-nm will be possible using ArF illumination. By 2002, Cymer introduced the Nanolith-7000™ ArF-laser with a 4-kHz rep-rate, an average power of 20-W, and high spectral purity. The optical-component degradation under 193-nm radiation was also found to be lessened as the peak-power of the laser pulse is decreased. Thus, efforts to reduce the peak-power (by increasing the pulse-width) were being pursued for ArF-lasers. Some success toward achieving this goal have been described. Progress reported by 2000 indicates that the manufacturing performance of ArF-lasers will be comparable to that exhibited by KrF-lasers currently being used in production.

7.1.3 F_2 Excimer Lasers

There are many challenges that must be met to build production-worthy F_2 excimer lasers, but these also appear to be tractable. As with the ArF-laser, reducing peak power, but increasing average power (10-40-W) will be important. F_2-lasers operating at multi-kHz pulse rates will be needed. Spectral linewidth demands at 157-nm will depend on lens design, but it seems likely that a spectral bandwidth of 0.2-pm will be required. F_2-lasers emit two bands of light (at 157.63-nm and 157.52-nm), but 87% of the energy is located at the longer wavelength. A prototype 2000-Hz F_2-laser, based on a current ArF design was described in 1999, in which single-line operation at 157.63-nm was achieved. The output of this F_2-laser scaled linearly with repetition rate. Its energy stability was better than 6% (3σ) over 100 bursts and better than 4% over more than 80% of the burst range. Gas lifetimes appear comparable to current ArF lifetimes.

7.2 EXPOSURE TOOLS FOR DUV LITHOGRAPHY

The wafer stepper was initially commercialized by the GCA Corporation of Andover, MA. GCA's first stepper, the DSW4800 (Direct Step on the Wafer) was equipped with a 10X reduction capability, a 0.28-NA and *g*-line lens that could pattern a maximum square field of 10 mm x 10 mm. The lens was supplied by Carl Zeiss of Oberkochen, Germany. The stepper had a resolution of 1.25-μm lines and spaces over a 1.5-μm depth-of-focus and could achieve an overlay of ±0.7-μm. Its list price was about $300,000.

Several other companies soon began selling wafer steppers, and the major suppliers today are Nikon, ASM Lithography (ASML), Canon, and Ultratech. Modern steppers have considerably greater capability than the original GCA DSW4800, but the operating principles are essentially the same. Leading-edge systems in 2001 come in the step-and-scan configuration. (The stepper and step-and-scan configurations are described in more detail in Vol. 1, 2nd Ed., Chap. 13.) The characteristics of 248-nm step-and-exposure tools are listed in Table 7-1. To show how their imaging and overlay capabilities have been improved, these are listed both for the modern step-and-scan systems and the GSW4800 in that table.

7.2.1 Exposure Tools for 248-nm Lithography

Optical lithography systems continue to reach further and further beyond their previously perceived limits. By 2001 248-nm scanners – which use a KrF-laser light-source – were routinely printing 0.18-μm features. Industry experts believe that 248-nm lithography will get them to at least the 0.13-μm node. The pressure to push 248-nm lithography past what had been earlier considered its limits is largely a matter of economics. Chipmakers must weigh the trade-offs between the cost of the advanced reticles necessary to extend the current wavelength, and the systems costs and risks that come with the next wavelength. Manufacturers that run large volumes of devices per reticle are more willing to put effort into the current wavelength, spending the money needed on the reticles. On the other hand, those making logic or application-specific ICs (ASICs), where only a few hundred wafers may be manufactured with a reticle set, face a much higher cost of ownership if forced to pay the price of advanced reticles. Advances in resist technology and the ability to make lenses with higher numerical aperture are two of the reasons higher resolutions are possible with 248-nm lithography.[3] (Note that 248-nm resists are discussed in Chap. 6, Sect. 6.1.1.)

Lenses can now be built with higher numerical aperture (NA) and lower aberrations than was previously thought possible (Fig. 7-5). Even in 1999 it was not considered likely that lenses with an NA greater than 0.6 could be manufactured. Yet by 2001 various steppers were being introduced with even higher NA. For example, Nikon offers a stepper with a lens having an NA of 0.75, and the ASML PAS 5500/800 uses a lens with an NA of 0.8.

TABLE 7-2 Commercially Available KrF Step-and-Scan Systems - Compared to the First Commercially Available Stepper

	GCA	Canon	SVGL	Nikon	ASML
Model	DSW4800	ES3	Micrascan III+	NSR-S205C	5500/800
M:1	10×	4×	4×	4×	4×
Wavelength	436 nm	248 nm	248 nm	248 nm	248 nm
Maximum lens NA	0.28	0.73	0.60	0.75	0.80
Field size	10 mm × 10 mm	26 mm × 33 mm	26 mm × 34 mm	25 mm × 33 mm	26 mm × 33 mm
Overlay, Machine-to-itself	500 nm	25 nm	45 nm	30 nm	30 nm
Lens distortion	250 nm	20 nm	25 nm	20 nm	20 nm
Overlay, machine-to-machine	700 nm	45 nm	55 nm	45 nm	45 nm
Wafer size	3," 4," 5"	200, 300 mm	150, 200 mm	200, 300 mm	150, 200 mm
Throughput (200 mm wafers, except for GCA)	20 wafers per hour (4" wafers)	100 wafers per hour	90 wafers per hour	140 wafers per hour	120 wafers per hour

7.2.2 Exposure Tools for 193-nm Lithography

The ASML PAS 5500/900 ArF step-and-scan tool was introduced in 2001. It is based on the ASML 5000 architecture and uses a Carl Zeiss refractive Starlith 900® lens with CaF_2 elements for partial achromatization. The lens has an NA that is adjustable from 0.45 to 0.6. There is a large working distance from

k_1 - factor vs. NA and Resolution

NA	210	200	190	180	170	160	150	140	130	120	110	100	90	80
0.65	0.55	0.52	0.50	0.47	0.45	0.42	0.39	0.37	0.34	0.31				
0.70			0.54	0.51	0.48	0.45	0.42	0.40	0.37	0.34	0.31			
0.75				0.54	0.51	0.48	0.45	0.42	0.39	0.33	0.30			
0.80					0.55	0.52	0.48	0.45	0.42	0.39	0.35	0.32	0.29	

RESOLUTION NODE

k_1 > 0.5 Binary mask, off-axis illumination

k_1 > 0.4 Add attentuated PSM, OPC and Quadrupole

k_1 > 0.3 Add alternating PSM, TSi

Fig. 7-5 The 248-nm lithography tools have seen vast improvements in lens aberrations in recent years. (Courtesy of ASML)

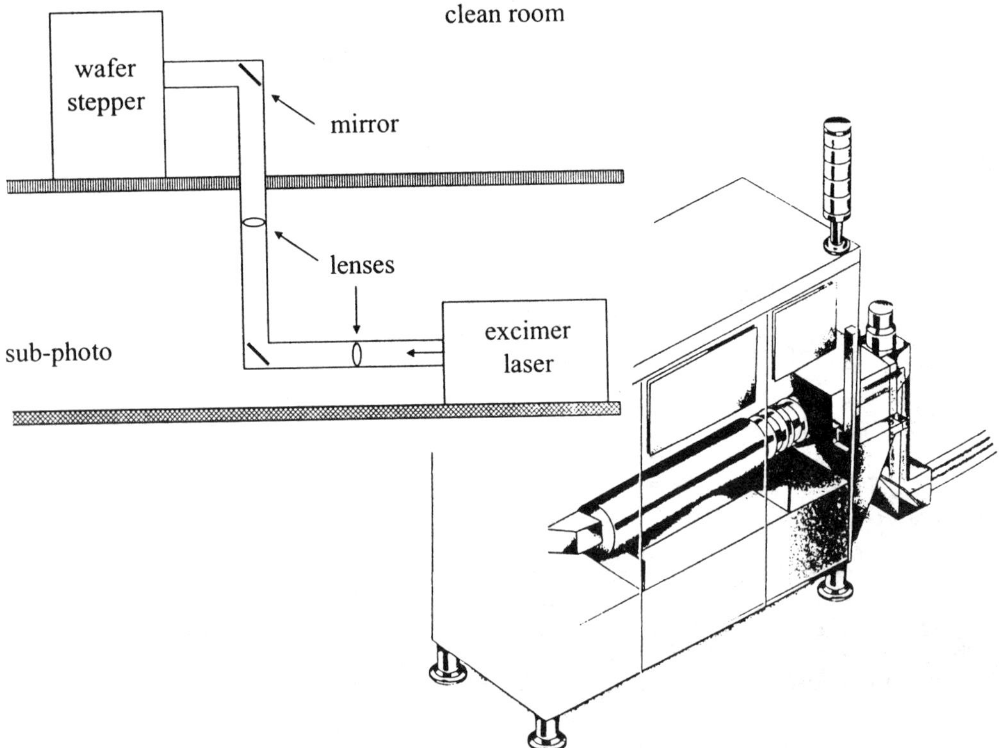

Fig. 7-6 The ASML PAS 5500/900 step-and-scanner. (Courtesy of ASML)

the final lens element to the photoresist surface, so that the lens contamination caused by outgassing from 193-nm photoresist during exposure is minimized. Figure 7-6 shows a picture of the PAS 5500/900 ArF step-and-scan system.

Canon's 193-nm scanner the FPA-5000-AS1 (introduced in 2000) has a 0.6-NA lens, and has demonstrated 600-nm DOF and can print 130-nm lines/spaces using annular illumination. The wafer throughput is projected to be 80-90 wph if the resist sensitivity is 10-mJ/cm^2. Challenges include availability of CaF_2 and improvements in resist performance.[4]

SVG offered two 193-nm exposure tools in 2000, the Micrascan V model (with an NA of 0.75) and Micrascan 193 (with an NA of 0.60). Both of these scanners use a catadioptric design, which requires only 1.7 kg of CaF_2 (as opposed to the 48 kg required for the chromatic correction of a refractive system).

7.2.3 Exposure Tools for 157-nm Lithography

Amid concerns about the technical risks and cost for all next-generation lithography (NGL) technology at and below the 70-nm node, work continues to develop 157-nm lithography technology.[5,77] This technology, however, is beset by a number of obstacles new to the evolution of lithography processes. First, the material that has been used for the lenses and reticles of previous generations of lithography, namely fused-silica, will have to replaced by calcium fluoride (CaF_2). Second, the optical path will need to be a purged system requiring a nitrogen environment. Third, the optical system is likely to be of a catadioptric design, since such optical systems are smaller in overall design and require smaller optical elements than do comparable refractive optical systems. For example, due to the way the design is folded, the catadioptric design is approximately one third the size and weight of an equivalent all-refractive system, and the lens elements may be half that of all refractive optics in both size and numbers. While catadioptric lens elements could range from 50 to 200 mm, all-refractive lens elements as large as 400-mm might be required, depending on design.

The beamsplitter, the heart of the catadioptric optical systems, is a cube the size of which depends on the NA. The larger the NA the larger the beamsplitter that is needed. Fabricating beamsplitters from CaF_2 is a difficult challenge. As we will describe in the next section, the handling and polishing of this thermally sensitive and easily chipped material is a delicate operation. In 1999, however, SVG (in cooperation with Schott) announced that they have been able to successfully produce a 100-mm CaF_2 cube (Fig. 7-7).[6] Though not large enough for a 0.7-NA system, it meets the specifications for a 0.45-NA scanner.

At the First Symposium on 157-nm Lithography in 2000, ASML said it expects to ship its first 157-nm scanner in the 3rd or 4th quarter of 2003, and Canon expects to be ready about a year later. Estimates for the cost of a single 157-nm scanner range from $12m to $20M (for a system with a high NA).[7]

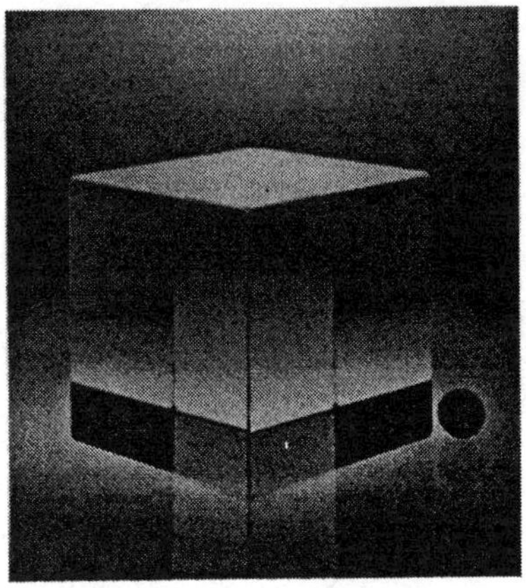

Fig. 7-7 SVGL has demonstrated the ability to grow and make a beam-splitter cube 100-mm on a side for use in its 157-nm catadioptric lithography tool. (Courtesy SVGL)

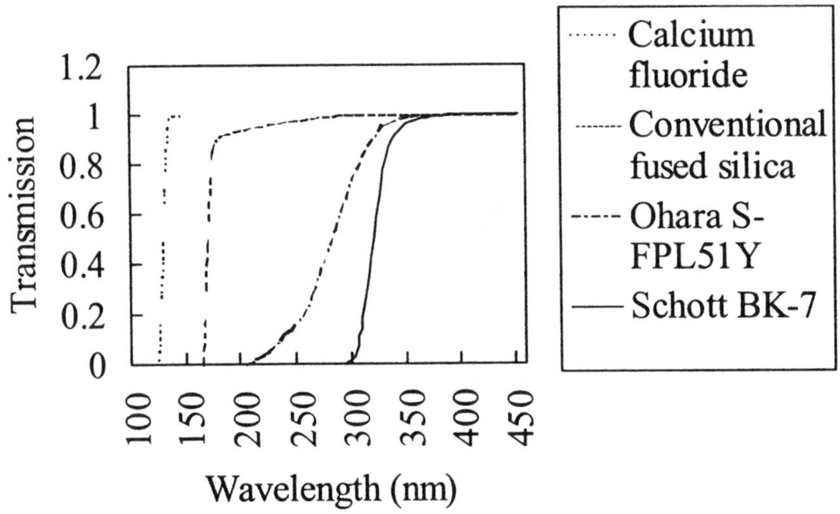

Fig. 7-8 Transmission of 10-mm-thick samples of materials used in lenses of wafer steppers.

7.2.4 Calcium Fluoride Optical Elements for DUV Lithography

Although the optical elements of lithography systems down to 248 nm can be made with fused silica, refractive optical systems at 193-nm often need to use at least some CaF_2 elements. At 157-nm, CaF_2 is virtually the only available useful transmissive material (with BaF_2 as the other possible candidate). The transmission properties of various types of optical glass are shown in Fig. 7-8. Lenses for 193-nm usually include CaF_2 as an optical material for some of their elements, especially at locations where the light intensity is especially strong. CaF_2 elements are used because damage is a concern under the high intensity levels of light delivered by excimer lasers. That is, one type of such damage is compaction of the lens material. Compaction occurring in the projection optics will degrade the image quality, and this compaction will also increase light absorption by the lens (and such increased absorption will therefore also reduce light intensity at the wafer, thereby reducing stepper throughput). Fused-silica undergoes compaction (becomes more dense) and thus also more absorbing under the intense light from excimer lasers. Consequently, CaF_2, which seemingly does not compact because it is a stable, single-crystal, is used in the illuminator and projection optics wherever high-levels of light intensity is expected. In 193-nm steppers, the compaction of fused-silica that occurs after extensive use is sufficiently severe so that it can render 193-nm lithography unsuitable for high-resolution lithography. In the illuminators of 193-nm systems, where such high intensities occur, appreciable amounts of CaF_2 are thus needed.

However, short wavelength applications also put very stringent requirements on materials and fabrication. To minimize susceptibility to optical damage (as well as scattering and absorption losses that degrade throughput), bulk CaF_2 must be extremely pure and homogeneous. To reduce image degradation stemming from optical fabrication errors, the component surfaces must be flat and smooth to the level of tens or hundreds of angstroms rms. CaF_2 is hygroscopic, soft, and very prone to chipping, particularly at the edges. As a result, fabrication time CaF_2 for is typically 50% to 100% longer than for glass.

Besides transparency, there are other properties that a material must have in order to be used to make optical elements for DUV exposure tools. One such property is that the material does not exhibit excessive birefringence. (*Birefringence* is the optical property of a material in which a different index of refraction exists along various crystal axes.) However, many materials do exhibit significant birefringence, and such materials are generally not suitable for use in high-resolution optical systems. For example, SiO_2 in its purely crystalline form (quartz), is not easily incorporated into lenses because it has dissimilar behavior among different optical axes. On the other hand, amorphous fused-silica is a standard optical material because its non-crystalline structure causes it to have only a small degree of birefringence. Some materials, like CaF_2 have (until recently) been thought to exhibit little *intrinsic* birefringence. However, CaF_2 can exhibit low levels of birefringence induced by mechanical stresses. This type of *extrinsic* birefringence is termed *residual-stress birefringence.*[8]

Residual-stress birefringence occurs from the mechanical stress and strain formed during the production of an optical element (for example, if there are large thermal inhomogeneities in the furnaces in which the crystals [e.g., of CaF_2] are grown). A high level of residual-stress birefringence in optical components can lead to the aberration of light rays in an imaging system, and thus affect the imaging quality of a step-and-scan system. Residual-stress birefringence currently limits the yield of CaF_2 crystals suitable for use in stepper lenses.

Recently (May 2001), it was discovered that crystalline CaF_2 does, in fact, possess *intrinsic birefringence.* Lens designers were familiar with the residual-stress birefringence in CaF_2 described above, and were prepared to compensate for this effect with relatively complex lens designs. But, intrinsic birefringence – which may inflict about 6x more birefringence - presents a more sticky problem. However, a group of technologists gathered in mid-2001 to assess the issue, and they concluded that it will not be a showstopper for 157-nm lithography. Furthermore, intrinsic birefringence is not a concern for 193-nm lithography because corrections can compensate for any effects it causes at this wavelength. While it certainly will make 157-nm lens design more complex, the fact that this phenomenon was discovered before such lenses had been fabricated is a good thing. It also implies that lithography systems will have to manage more carefully the polarization of the light as it enters a lens.[78] The lens design software (e.g., CODE V from Optical Research Associates) will also have to be modified to compensate for intrinsic birefringence.[9]

There are also questions about whether it will be possible to supply the quantity and quality of CaF_2 needed for 157-nm lithography. Nikon plans to use its own production for some of its optics supply. While, its major supply will not come from in-house, the company plans to develop the intellectual property for CaF_2 and optical coatings to help it get the quality needed. Corning Inc. recently acquired Optovac and Tropel Corp., to beef-up its capabilities to tackle CaF_2 production. It also has a facility in North Brookfield MA for producing CaF_2.

Note that while conventional fused-silica is not transmissive enough at 157-nm to be used as a 157-nm reticle-blank, its transmission can be improved by preparing the fused-silica with low-OH content (and by adding fluorine to the glass network). This increases the transmission to 82%, allowing fused silica to meet the International SEMATECH targets for 157-nm mask materials.[77]

7.2.5 300-mm Lithography Tools

Development of 300-mm tools began in about 1995. Before the end of 1997 the first 300-mm lithography stepper was installed (the Canon FPA-3000EX3L KrF-stepper) at I300I (now International SEMATECH). Initially, vendors were advised to develop tools that could handle only 300-mm wafers, but a few tool vendors took the bridge-tool approach (which allows a tool to be used on either 200-mm or 300-mm wafers). By late 1998, however, the industry was still not ready to embrace 300-mm production. Instead it stayed with technology-driven cost-reduction at 200-mm. By 2001, 300-mm had still not been widely adopted, as the industry went into a deep downturn. Nevertheless, mature 300-mm tools were ready, but the insertion node was predicted to be 130-nm, not at the 250-nm node, as originally envisioned.

It should be noted that stepper-scanners expose each field and then step to the next one. Since there are 2 1/4 times as many fields on a 300-mm wafer as on a 200-mm wafer, it could take twice as long to expose one. Achieving the desired throughput requires the wafer and reticle stages to move further and faster, while the overlay and resolution demands of 130-nm (and smaller) imaging reduce the tolerable errors and vibrations. All of the exposure tool manufacturers announced new products or platforms to meet these needs.[10,11]

ASML chose a "reaction-free-mass" concept for its new "Twinscan" platform, intended for operation both at 200-mm and 300-mm. To allow rapid acceleration of the relatively heavy wafer stages, the linear stage motors push against a countermass, which moves in the opposite direction, rather than against the fixed stepper frame. Since the countermass is free to move, it absorbs the impulse from the acceleration without inducing vibrations in the exposure tool structure. The center of gravity remains fixed, eliminating gravitational torques. The stages accelerate and decelerate according to Newton's third law, just like a rocket in vacuum (which spews reaction mass behind it). According to ASML, the new systems reduce moving-average alignment error to 3-nm and moving-standard-deviation to < 2 nm, a 10-fold improvement. The initial "Twinscan" tools were specified at 62-wph (in 2000), with throughput expected to double by 2002.

Canon chose a more conservative approach for 300-mm production, introducing a tool set that uses a high-NA KrF-scanner for critical levels, and a matching *i*-line-stepper for non-critical levels. All are designed for high throughput, specifically 88, 300-mm wph for the scanner, and 100 wph for the stepper. The matched fields allow for a mix-and-match strategy with the lowest CoO tool applied to each level. An ArF-tool is expected to be available in 2001. While Canon did not reveal its stage design, countermasses and improved aerodynamics allow 350-mm/sec wafer speeds with overlay accuracy below 23-nm.

Nikon has had a 300-mm-capable KrF-scanner in the field since 1998 (the NSR-S203B), and the newly introduced 0.75-NA NSR-S205C promises 83 wph, along with 130-nm resolution (using RET) and < 30-nm overlay. Nikon expected to ship 100 of these systems in 2001. The NSR-SF100 *i*-line-scanner also matches the 25x33-mm field of the excimer scanners for mix-and-match applications. A new 0.68-NA ArF-scanner was promised for 2001.

Silicon Valley Group (acquired by ASML in 2001) introduced the Micrascan IV and V systems in 2000. These incorporate innovative low-mass precision wafer stages, without interferometer retroreflector mirrors. The interferometer systems of the Micrascan IV and V are inside out, compared to previous designs. That is, the mirrors are on the frame, and the laser-beam launching and collection optics are on the vertically oriented stages. The lightweight stages allow scan speeds above 300-mm/sec. The reticle stage, however, must move 4x-faster. The quick, four wavelength HPAS alignment system has 1σ overlay accuracy of 3-nm. The 193-nm Micrascan V tools are designed to expose 130-nm CDs with standard mask illumination and 100-nm half-pitch using RETs. These tools have a large NA (0.75) and catadioptric projection lenses. The 248-nm Micrascan IV with 0.60-NA optics is intended for non-critical applications.

7.2.6 Mix-and-Match Lithography

One way to lower the cost of lithography is to use the most advanced (and thus expensive) machines only for the critical levels, and employ less-expensive tools for the less-critical levels. This has been widely practiced for many years, and is referred to as *mix-and-match lithography.*

Vendors of exposure tools generally offer both high-resolution (expensive) and lower resolution (lower cost) models. The exposure tools intended for use on those levels with relatively lower-resolution requirements are normally optimized for throughput and price (instead of resolution and overlay). Lower-NA lenses are cheaper to design and build than higher-NA lenses. Such less-expensive systems deliver high throughput by exposing larger field sizes. In addition, in the 2001 timeframe all lower-cost models are steppers (i.e., not stepper-scanners).

Another strategy for lowering the capital cost of lithography is to utilize exposure tools from several generations of technology. As exposure tools get older they can no longer image the smallest feature sizes of the next technology nodes. Yet these steppers can still be used for imaging the less critical levels. This is an effective cost-saving method since it is not necessary to purchase new tools, and those in use may have already been fully depreciated. Smaller-NA steppers are also useful for patterning implant masks (that generally have larger feature sizes, but also often thicker resist films). In such cases, high-resolution steppers may actually do an inferior job, due to their limited depth-of-focus.

While mix-and-match lithography appears economically attractive, it is not without its drawbacks. It may involve using steppers from more than one supplier, or steppers that employ dissimilar operator interfaces. This necessitates extra training for the lithography workers, and a separate spare-parts inventory for each stepper model.

7.3 RESOLUTION ENHANCEMENT TECHNOLOGIES (RETs)

By 2001 it appears that the conventional lithographic processes that have enabled the descent of the IC feature sizes over the past 30 years, are approaching the limits of their usefulness. That is, even by using the shortest possible illumination wavelength, λ, and the largest numerical aperture (NA) lenses, the last factor, k_1, in the basic resolution equation:

$$\text{Minimum resolution} = k_1\,(\lambda/\text{NA}) \quad \textbf{(7.1)}$$

presents a formidable obstacle. (Note that k_1 represents the measure of lithographic aggressiveness.)

That is, to obtain acceptable yield with a conventional lithography process (i.e., one which does not use any resolution enhancing techniques), the value of

k_1 is restricted to values larger than ~0.75 (because unacceptable image degradation occurs below this value). However, in the past two decades, various *resolution enhancement techniques* (*RETs*) have been developed, and their use permits lower values of k_1 to be achieved.[12,13] These RETs include the use of anti-reflective coatings and top-surface imaging (as is discussed in Chap. 6), and off-axis-illumination (OAI), optical-proximity-correction (OPC), and phase-shift-masking (PSM), which are discussed in this section. Use of these techniques can possibly reduce k_1 from 0.75 to 0.35, which represents more than a 50% decrease in critical dimension compared to conventional lithography. Thus, we have gained the capability to print features that are one-half the size of those that can be printed with conventional lithography. Hence, by employing RETs, optical lithography has been able to maintain its vitality - allowing it to be used much longer than was ever thought possible 10 years ago.

7.3.1 Off-Axis-Illumination

Off-axis-illumination is a method for improving the resolution of a stepper without resorting to a higher NA lens or a shorter wavelength of light. The method was long employed in optical microscopes to enhance image contrast, and was introduced to microlithographers by Mack[14] and Fehrs[15] in 1989. Researchers at Canon, Inc.,[16] and Nikon Corporation[17] incorporated it into the optical systems of steppers in 1992. The off-axis-illumination technique improves the resolution capability of the stepper by modifying the direction of incidence of the illumination on the reticle. That is, it directs the beam of light from the illumination system of the stepper through the reticle in a direction that would make it strike the projection lens at the edge of the entrance pupil (rather than at the center of the projection lens entrance pupil, as is done in the conventional illumination technique). The term *off-axis-illumination* is somewhat of a misnomer. This is because conventional partially-coherent illumination also includes a range of angles of incidence on the mask, both on-axis and off-axis. Thus, when the term "off-axis-illumination" is used to describe the *post-1992 techniques*, it is intended to mean an illumination scheme *without any on-axis components*.

The principle of the new "off-axis illumination" techniques can be understood by considering the imaging process of optical lithography from a Fourier optics perspective (see, for example, Vol. 1, 2nd Ed., Chap. 13), as applied to imaging the pattern of a grating of lines and spaces. As shown in Fig. 7-9, in conventional stepper optical systems the light beam is directed from the illumination system through the reticle in a direction which would coincide with the center of the entrance pupil of the projection lens. Recall that for a grating pattern, the light is

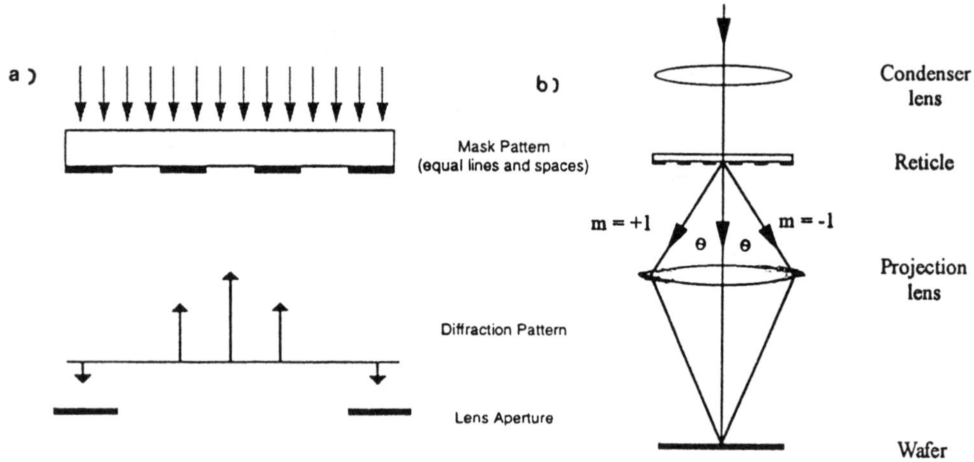

Fig. 7-9 Principle of on-axis illumination: a) Light illuminating the mask from an on-axis direction, and the diffraction pattern from an on-axis coherently illuminated grating mask. The pattern is centered on the entrance pupil of the objective lens. b) Aperture with on-axis illumination showing the 0-order (undiffracted beam) and the two first order diffracted beams, m = +1 and m = -1.

diffracted into a set of beams which propagate away from the grating only at specific angles (corresponding to the orders of the diffracted beam). The undiffracted beam (0-order) propagates at a direction perpendicular to the reticle plane. The first-order diffracted beams (m = +1 and m = -1) propagate at angles $\theta_1 = \sin^{-1}(\lambda/p)$ and $\theta_1 = -\sin^{-1}(\lambda/p)$, respectively (see Fig. 7-9) where p is the pitch of the grating (i.e., the distance equal to the sum of the line and the space). The third-order diffracted beams propagate at larger angles than the first-order beams (i.e., $\theta_3 = \sin^{-1}[3\lambda/p]$ and $\theta_3 = -\sin^{-1}[3\lambda/p]$). Since no imaging is possible unless at least one additional beam is accepted by the lens besides the 0-order beam, the NA of the lens must be large enough to accept both the +1 and the −1 first order beams. For a given NA and λ, there is a minimum pitch below which the first order beams will fall outside the collection area of the lens. For this pitch and smaller, no imaging of the grating pattern is possible. An optical system that collects the three beams (0-order and +1 and -1 order beams) is sometimes called a system that employs *three-beam imaging*. Note that if at least the first order beams are captured within the pupil of the lens (and for the case of coherent illumination), this leads to the simple resolution limit of 0.5 λ/NA, or in equivalent terms, a k_1-limit of 0.5. For gratings with pitches below this limit, no pattern is formed by an optical system with this λ and NA, using coherent light.

If an *off-axis* (or *oblique*) illumination scheme is employed (Fig. 7-10a), such that the illuminating beam is directed so the 0-order beam strikes the entrance pupil of the projection lens just at its edge, and only one of the other first-order beams strikes the other edge of the entrance pupil (i.e., either the m = +1 or the m = −1 beam) only two-beams are collected (hence the name *two-beam imaging,* see Fig. 7-10b). In this case the angle, θ, can have a value greater than in three-beam imaging, implying that the pitch of the grating can be smaller (and in turn a higher resolution can be achieved for an optical system employing the same NA and λ). In the extreme case of this type of off-axis illumination, a doubling of the resolution-limit capability of the imaging system is obtained, compared to a coherent fully on-axis-illumination scheme. Such resolution-enhancement is most pronounced for the case of dense-patterns of equal lines and spaces (such as occur in the regular array patterns of DRAM and SRAM interconnect lines).

Although resolution-enhancement may seem to be the primary benefit of off-axis illumination, in fact it is the impact of illumination on DOF that provides the more significant advantages. Defocus causes a phase error for each diffraction order that is proportional to the square of the radial position within the pupil. Consider the case of off-axis-illumination where the zero-order and one of the first-orders are the same distance from the center of the objective lens pupil. In this case, both the zero-order and the collected first-order will have phase errors due to defocus, but these errors will both be the same, and the relative phase difference between the zero- and first-order beams will thus be zero. Such a circumstance will result in an image that is much less prone to defocus, resulting

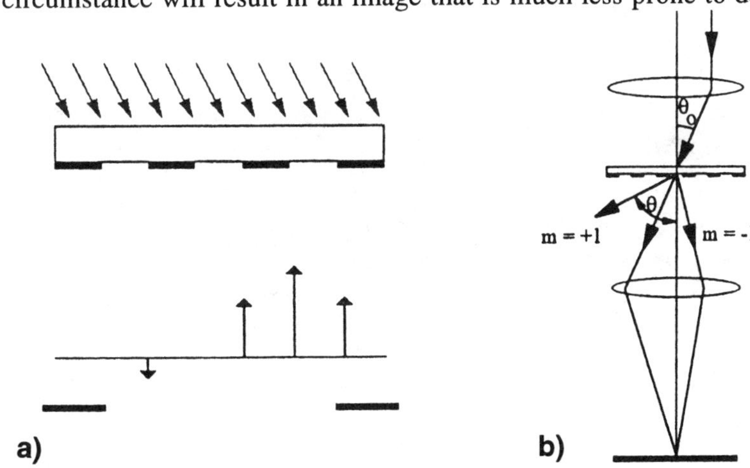

Fig. 7-10 Principle of off-axis illumination: a) Light illuminating the mask from an off-axis direction, and the diffraction pattern from an off-axis coherently illuminated grating mask is no longer centered on the entrance pupil of the objective lens, but is instead shifted. b) Aperture with one off-center hole that produces the illumination shown in (a).

in an important improvement in DOF. Thus, the proper choice of an off-axis illumination angle for a given equal line/space pitch will result in a dramatic increase in the DOF of that feature.

One limitation of off-axis illumination is that it does not provide the same resolution enhancement for isolated lines as it does for dense lines and spaces. This is because discrete diffraction-orders do not exist for isolated features. Instead an isolated feature generates a continuous diffraction pattern. Hence, isolated line performance is only minimally improved by off-axis illumination. When both dense and isolated features are considered together in a field, a dense-to-isolated feature-bias will arise if off-axis-illumination is employed.[18] To circumvent this problem, off-axis-illumination tricks have been proposed. One involves adding small (sub-resolution) auxiliary patterns close to an isolated feature to establish the required interference effects (see Sect. 7.3.1).[19,20,21] Another issue that must be addressed is that some illumination energy is lost because one of the 1st-order beams is not collected. As a result, it is necessary to increase the exposure time to compensate for this loss of illumination.

If the illumination of the mask only comes from one beam (that is, through an aperture with only one off-center hole, as shown in Fig. 7-10b), the +1st-order will fall completely out of the pupil, but the 0th and -1st-orders will be collected to form an image (Fig. 7-10b). However, the asymmetry between the 0th and -1st orders leads to severe telecentricity errors. Symmetry can be restored by illuminating with two sources, placed on opposite sides of the pupil, using an off-axis illumination aperture as shown in Fig. 7-11a (*dipole illumination*). This is an improvement over the single-beam aperture, but it will only improve imaging for gratings oriented in one direction. It will degrade them for gratings oriented in the perpendicular direction (see Fig. 7-11a1). Nevertheless, if geometries are oriented in only the vertical direction, such dipole illumination is optimal. *Quadrupole illumination* (Fig. 7-11b) works well when all features are oriented left-right or up-down, but distorts features of all other orientations (e.g., a line at a 45° angle). The latter limitation can often be tolerated because critical features are rarely oriented at odd angles to the sides of the chip. The *annular aperture* shown in Fig. 7-11c improves features of all orientations, but at the price of milder improvements in the resolution and DOF.

Note that a specific off-axis illumination setting will improve imaging performance for only a narrow range of pitches, but this is still an advantage for many applications, such as memories, where many structures are highly periodic.

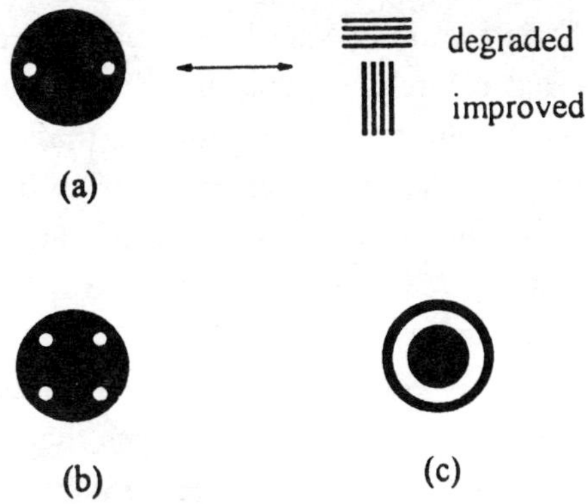

Fig. 7-11 Different types of off-axis illumination apertures. The aperture with holes at the edges or corners - (a) and (b) – will improve lines oriented in one direction (see a1), but will degrade imaging of lines in the orthogonal direction. (c) Annular off-axis illumination aperture.[23]

However, an off-axis illumination setting optimized for one pitch may work poorly for others. For patterns with more than one pitch, this becomes a problem. In such cases, an off-axis-illumination source that includes some on-axis illumination (with lower intensity than off-axis light), reportedly gives better results.[22]

Off-axis-illumination is relatively easily implemented. Stepper manufacturers can provide any of the illumination-aperture types described above. A series of these apertures can even be mounted on a turret, and the particular aperture desired for any mask can be automatically supplied by the stepper control program. More sophisticated illuminators can also be used to transmit a larger fraction of the available light into the annulus of illumination.

7.3.2 Optical Proximity Correction (OPC)

The techniques developed to extend optical lithography beyond the range of conventional imaging are based on the idea that the 2-D aerial image formed by the projection lens is a structure to be optimized. One way to optimize the aerial image is to pre-compensate the reticle pattern to account for expected pattern distortion due to diffraction effects. One important consequence of diffraction is the so-called *proximity effect*, in which the exact size and shape of the features patterned on the wafer depend on their proximity to other features. These proximity effects manifest themselves in the form of distorted geometries (and in some extreme cases, even unprinted patterns). For designers, this phenomenon translates into line-width variations and other distortions that can significantly decrease performance – or even worse - cause missing, incomplete, or shorted structures that can result in circuit failure.

One technique for determining such expected pattern-distortion involves computer simulation. That is, the 2-D aerial image of a particular reticle feature (or a group of features) is simulated by calculating the effects of diffraction on the image.[24] Based on the results, the reticle pattern can be altered and simulated again to determine if the change has improved the 2-D aerial image.[25] The reticle patterns are altered by adding (or subtracting) sub-features to the original pattern, and these features are called *optical-proximity-correction-features.* Additional aspects of the optical-proximity-correction procedure is discussed next.

One effect long observed in microlithography is that when patterns smaller than the exposure wavelength are being imaged, individual pattern features do not image independently. Instead, they interact with neighboring patterns. A simple qualitative argument for why this occurs is based on the light intensity pattern of the Airy function. That is, overlap of the secondary diffraction peaks of the Airy function with adjacent pattern features leads to increased or decreased exposure intensity at any point in the image compared to a purely geometrical image model. For example, the linewidth of isolated- and densely-grouped-lines eventually varies by a significant amount from that of the predicted linewidth (based on the pattern size on the reticle) when the feature sizes approach the

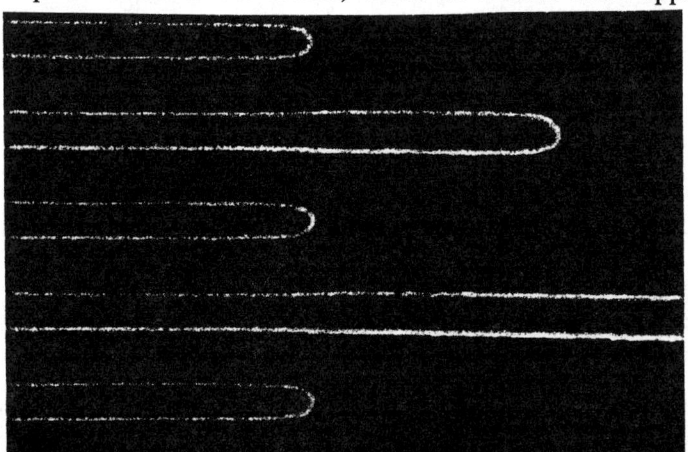

Fig. 7-12 Optical proximity effect. All pattern features are the same size on the reticle, but the final image is larger for the isolated line extension than for the densely packed lines.

wavelength of the illuminating light. However, very little difference between two such aerial images is observed if the feature sizes are large, i.e., > 1 μm. For example, assuming the use of positive photoresist, 0.35-μm-wide *isolated* lines are 0.09 μm wider than *closely spaced* 0.35-μm-wide lines (i.e., which might be part of a diffraction-grating pattern). Figure 7-12 shows an example of this effect. Since the linewidth depends on the proximity of other features, this effect is called an *optical-proximity-effect.* Such proximity-dependent linewidth variation (specifically termed *iso-dense print-bias*) may become significant in the interconnect-lines of a memory chip. That is, lines within the memory-array region of the chip would be regularly and closely spaced, but in the peripheral sense-amplifier and logic-circuitry they could be spaced farther apart. If this dimensional-difference can be predicted, the width of the chrome lines on a reticle can be adjusted to make isolated and densely-spaced printed lines on the wafer all the same width.[26] Such adjustments to the sizes of patterns on a reticle are termed *optical-proximity-corrections* (*OPC*), and in this case such an OPC is performed as a *one-dimensional, pattern-dependent width correction* to lines on a reticle. (Note that biasing the size of features has been used historically for many years to assist in lithography-process centering and optimization, but until recently the same bias was applied to the entire reticle. On the other hand, the OPC-bias is custom-tailored for each line in its own environment.)

Originally, OPC was considered to be just a method to correct the line-size variations as a function of the proximity of other features (as described above).[27] However, other pattern-modification techniques have been developed to improve the images of such two-dimensional IC features as contact holes (or vias) and lines (e.g., of poly or metal). That is, corner-rounding and general loss of shape fidelity in small features is caused by the inability of the projection lens to

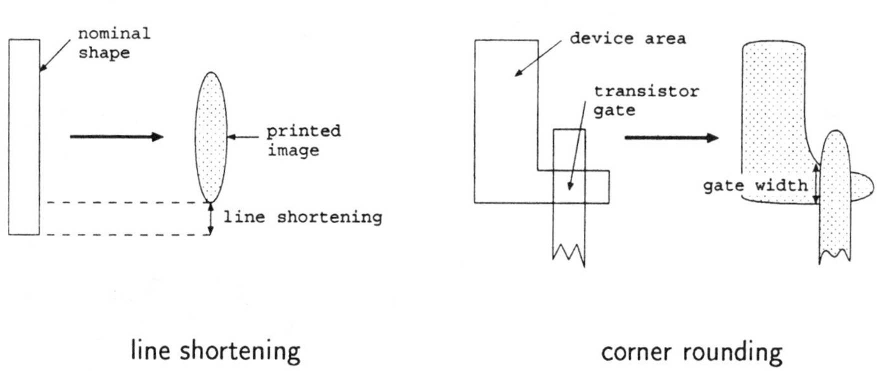

Fig. 7-13 Example of line shortening due to image distortion.

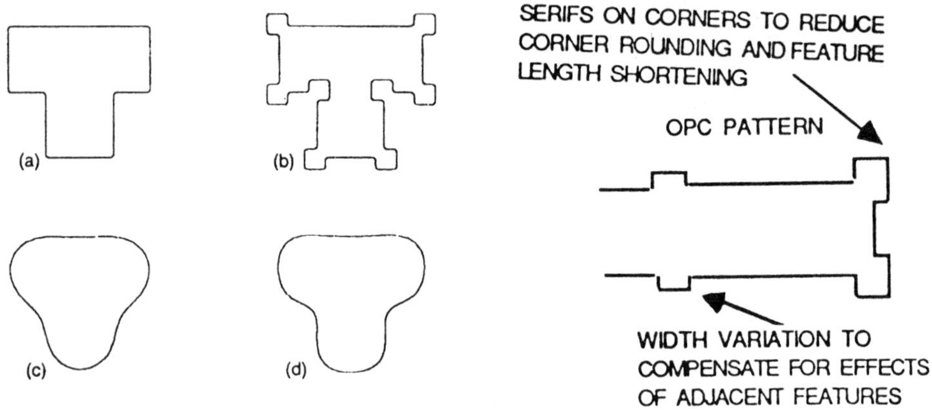

Fig. 7-14 A T-shaped feature (a) with dimensions near the resolution limit of the projection lens is printed as a rather featureless blob (c) on the wafer. Addition of serifs (b) brings the printed image (d) closer to the shape originally designed. Reprinted by permission of Marcel Dekker. (e) Adjustments to the linewidth can correct for the proximity of nearby features, and serifs on the end of a line can correct line shortening effects.

resolve details smaller than the diffraction limit of the lens. For example, a square contact-hole pattern with a size close to the minimum resolution limit of the optics will print more nearly like a circle. Likewise, a line (which should print with a rectangular shape - see Fig. 7-13a), instead prints as a feature with an elliptical shape (and an even shorter length). These effects have therefore also been classified as another form of optical proximity effect.

The OPC techniques used for modifying the 2-D reticle patterns to alleviate such deviations in the aerial images involve appending small, secondary patterns to the primary contact hole or short poly- and metal-line patterns. As noted above, such added-on auxiliary features are intended to be below the resolution capability of the projection optics. Several types of OPC features are employed. The most common are secondary patterns (called *serifs*) which are added to the corners of square hole patterns (as shown in Fig. 7-14b & e). They make the aerial image (and hence also the printed image) of such small contacts become more square-like. This occurs because the serifs enhance the amount of light transmitted though the corners of transparent mask features (e.g., the pattern of a contact hole being imaged in a positive resist). For short lines, serifs (together with linewidth changes) may be also added to improve the fidelity of the printed feature. The second type of OPC feature is the *line-jog*. These are used for lines that are isolated along part of their run across the wafer surface, but are near

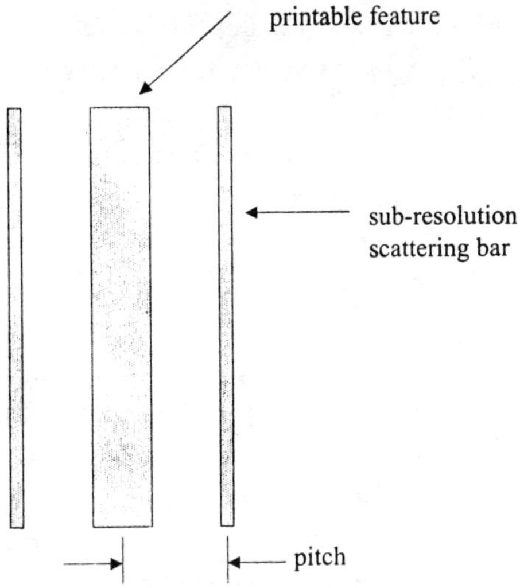

Fig. 7-15 Sub-resolution scattering bars applied to isolated features.

other lines at other places. For such lines, incremental linewidth changes can be made to the linewidth by employing such line-jogs (see Fig. 7-14e).

A third type of OPC feature is the *scattering bar,* whose role is to improve the imaging of isolated lines. Scattering bars are lines having a sub-resolution width that are placed adjacent to isolated lines on the same pitch as dense features on the same reticle (Fig. 7-15). The purpose of scattering bars is to create a dense environment around an isolated feature. However, when properly designed, these scattering bars do not print. The isolated line thus appears dense to the projection system, but prints as a fine, isolated line.*

While OPC techniques provide significant benefits, they are not gained without some costs.[18] The adoption of OPC includes the following drawbacks. First, serifs add a large number of additional features to the database of the reticle pattern. This increases the time needed to write the reticle (and therefore also its cost). Second, the sizes of the serifs are typically much smaller than the minimum feature size of the reticle, which creates problems during the inspection phase of the reticle manufacturing process. Likewise, when sub-resolution features first appeared, they represented a decrease of minimum feature sizes by factors of three and four. Such large changes in minimum feature size necessitated rapid advances in mask patterning. Third, this technique may not be feasible in all situations. That is, if the spacing between two contacts is already at

* As noted earlier, off-axis-illumination is an effective method used to enhance the patterning of periodic structures, but it provides no benefit for isolated lines.

the minimum allowed dimension, adding serifs to the square contact-hole pattern would bring parts of the two contact hole patterns to a distance below the minimally allowed value, and in such cases they cannot be used. Finally, it must be remembered that OPC is intended to correct for proximity effects. It does not actually improve the resolution or other quality metrics of the lithography process.

Since OPC is not performed until the design data are prepared for mask fabrication, OPC is typically transparent to circuit designers (although in some cases the design team may elect to get involved in deploying OPC in their layouts). However, in the first case, designers must still select a chip-manufacturer and mask-maker capable of accommodating OPC adjustments to the design. In both cases, some method must nevertheless be employed to determine which types (and where) OPC features have to be added to circuit layouts. Two approaches have been developed for carrying out this latter task.

The first is the so-called *rules-based* method. This uses a pre-determined set of rules together with a pattern-layout database program. The database program systematically determines the proximity environment of every pattern feature-edge, and then uses a lookup-table algorithm to apply the appropriate rules. These rules can generate biasing, serifs, scattering bars, etc. The rules may be developed directly from experience in the wafer fab, or they may be determined from image and resist simulations. Such rules-based approaches to determining the addition of OPC features to mask patterns are generally quite fast (requiring from several minutes to several hours on a typical workstation, per pattern level). However, the process of generating and validating the rules themselves can be a lengthy and tedious process. Since the accuracy and precision of the OPC features are completely determined by the extent and quality of the rules, this task must be effectively performed.

The second method is the so-called *model-based OPC approach.* In this case, each feature and its proximity environment are modeled by an image analysis program. A set of mathematical model parameters is used within the program to determine the appropriate corrections to the pattern layout. The model-based method is computationally intensive and may require many hours (or even days) of processing per pattern level. The model parameters are very important, but there are a limited number required. Manufacturers fabricate special test chips to generate OPC models of their lithography process. The chief advantages of the model-based approach is that the correction accuracy and precision can be readily controlled, and that corrections can be successfully applied to any sort of pattern layout. Software companies that offer OPC programs include Mentor

Graphics, Numerical Technologies (NumeriTech), and MaskTools (an ASML Company).

7.3.3 Phase-Shift Masking (PSM)

The most dramatic improvement in resolution limits imposed by diffraction involves the use of the phenomenon of *phase shifting*. For example, if a lithography process can attain a resolution of 0.3-μm using a 248-nm KrF light source with a stepper having an NA = 0.6, then by incorporating phase-shifting techniques on the reticle, the resolution can be improved to as much as 0.18-μm. The concept of increasing the resolution of a lithographic image by modifying the optical-phase of the mask transmission was first suggested in 1982 by Levenson, Viswanathan, and Simpson.[29] While a number of different types of phase-shifting masks have subsequently been developed, they all employ the same basic concept, as illustrated by the original version. This concept will be used here to demonstrate the principle of *phase-shift masking* (PSM). Several references review the topic of PSMs in more detail than is provided here.[30,31]

The phase-shifting of light as it passes through the clear part of a mask is accomplished by adding an extra layer of transmissive material to the optical path (Fig. 7-16). For PSM applications, a phase shift of 180° is usually desired, and this corresponds to an optical path-length difference of $\lambda/2$ at the stepper wavelength. Assume now that a grating (representing a *binary imaging mask*, or *BIM*, Fig. 7-17a1) is being imaged by optics with a particular numerical aperture

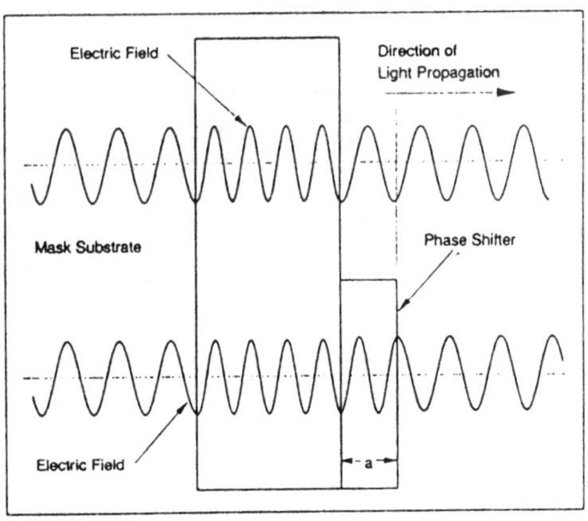

Fig. 7-16 The phase-shifting layer introduces a difference in optical-path-length. Here the difference is $\lambda/2$.[30] © IEEE (1993)

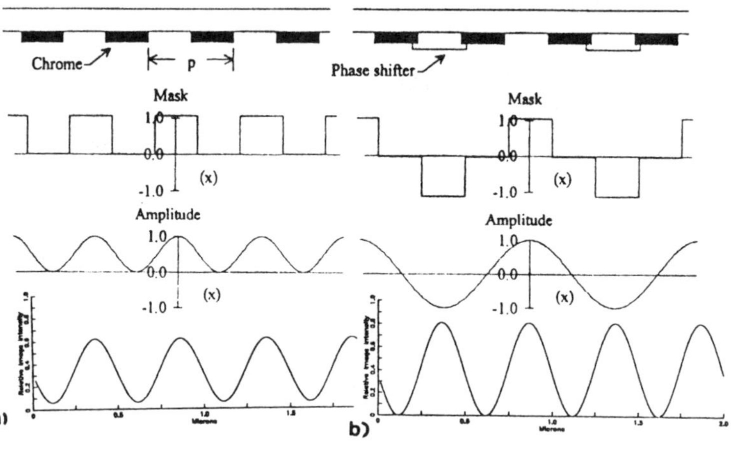

Fig. 7-17 Schematic of (a) a conventional binary mask (b) an alternating phase shift mask, (ALT PSM). The mask electric field, amplitude, and image intensity is shown for each type of mask.[28] Reprinted with permission of Marcel Dekker.

and partially coherent light. The (normalized) amplitude of the electric field (*E*-field) at the mask is either +1 or 0, as shown in Fig. 7-17a2. However, the imaging is degraded at the wafer because light from the clear areas on the mask is diffracted so that it strikes regions on the wafer that would ideally be completely dark. The nominally-dark-regions have light diffracted onto them from the clear spaces on both the right and left of them. The *E*-field amplitude on the wafer as a result of such diffraction is given in Fig. 7-17a3, and the light-intensity (which is the square of the *E*-field-amplitude) is plotted on 7-17a4.

If a phase-shifting material is placed on *alternating* clear spaces of a mask (as shown in Fig. 7-17b1), this forms a type of PSM called an *alternating-PSM,* or *ALT-PSM* (or sometimes even a *Levenson-PSM*). It should be noted that producing a phase shifter on a clear region of a mask may be done in one of two ways. In the first, an appropriate film of material can be deposited over the entire mask, and then etched away in the desired areas. In the second, the fused-silica glass itself can be etched in the appropriate clear-areas to produce recesses in the glass. The latter technique requires well-controlled etches so that recesses of the appropriate depth can be produced with only a timed etch. However, since deposition of a film different than fused-silica is not necessary, the use of this second approach to produce ALT-PSMs avoids the potential problem of the phase-shifter affecting either the light amplitude or phase.

In an ALT-PSM the light is phase-shifted by 180° when it passes through the spaces containing the PSM material, and the *E*-field amplitude on the mask is shown in Fig. 7-17b2. Note that the *E*-field amplitude at the mask spaces containing the phase shifters is -1, but at those without a phase shifter has a value of +1. The *E*-field intensity imaged on the wafer using such a PSM is shown in Fig. 7-17b3, and the corresponding light intensity on Fig. 7-17b4. It can be seen that the contrast in the light-intensity distribution of the image on the wafer is greater when the PSM is used. Contrast is increased because the light diffracted into the nominally dark area from the left-clear-area will interfere destructively with the light diffracted from the right-clear-area. All types of PSMs use this same principle, in which the destructive interference of light of opposite phases is used to improve image contrast.

While ALT-PSMs give the greatest improvement in resolution for the case when closely spaced clear lines are separated on an opaque background, this type of mask is not effective for some other types of patterns. One example is the case when tightly packed features are layed-out in a pattern that resembles the stacking of bricks in a wall (Fig. 7-18). Since such a pattern has alternating rows offset from one another, it is not possible to assign a phase value to each pattern that will make it 180° out of phase with all its neighboring features. The resolution of isolated features (such as contact holes) is also not improved by ALT-PSMs. In general, non-repetitive patterns cannot be given phase assignments that meet the ALT-PSM approach.

Another problem with ALT-PSMs occurs when opaque lines are created on a clear background. At the edge where the clear area covered with a phase shifter abuts an area of clear-glass without a phase-shifter, interference between the two phases will make such boundaries print as a narrow dark line (Fig. 7-19a). A possible cure for this problem is to put another layer of material at such boundaries which shifts the phase by only 90° (see Fig. 7-19d), but in practice such a cure is not completely satisfactory. That is, it requires additional steps in an already complex reticle-fabrication process. Another approach to solve this problem involves the use of negative photoresist to produce masks for IC levels that produce line patterns. By using negative resist, if patterns on a mask which define *lines* are clear-features (and those defining spaces and field-regions are opaque, Fig. 7-20), phase-shifter material never abuts a clear area. This is one of the reasons that negative resists are again being considered for deep-submicron lithography applications that call for PSMs. A lower-cost version of ALT-PSMs (the *sidewall-chrome alternating aperture* [*SCAA*] PSM) specifically tailored for small production runs (< 500 wafers), was recently described by Levenson.[32] In this type of PSM, the phase-shifting regions are patterned into the mask substrate before the chrome is deposited onto the mask blank. Then the chrome is pattern-

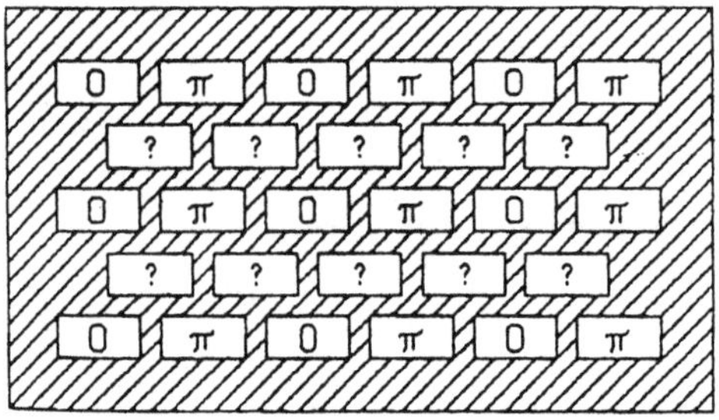

Fig. 7-18 Limitation of ALT-PSMs. There is no way to assign phases so that a 180°-phase difference occurs between adjacent transparent features In this example, the phases have been assigned to the odd-numbered rows, but there is no way to assign phases consistently to the even rows.[36] Reprinted with permission of Marcel Dekker.

ed with a second exposure. To make the masks less costly, about 1000 mask blanks would have to be layed out with pre-determined phase-shift locations. Special EDA tools would then be used to lay out chips so all that all fine-features lie at these predetermined phase-shift locations. Levenson opines that this situation would be compatible with >90% of ASIC designs.

Other types of PSMs have also been developed in an attempt to overcome the drawbacks of ALT-PSMs for some applications. Before describing these other types of PSMs it should be mentioned that the various types of PSMs are grouped into so-called *strong* and *weak* PSMs. Therefore, these terms must be defined. The distinction is based on the maximum degree of improved resolution and depth of focus that each type can provide. Strong-PSMs (such as ALT-PSMs) can provide up to a 40% improvement in resolution, but a lesser improvement in depth of focus. Weak-PSMs (such as the *rim-shift-PSMs* and the *attenuated* [or *halftone*] *PSMs* which are described next) offer less improvement in resolution, but more improvement in depth of focus. The benefits each group offers are different, and this implies that the "best" type of PSM depends on the particular application (as described in more detail later).

One of such alternative type of PSMs is the *rim-shift-PSM*, which falls into the group of weak-PSMs. The rim-shift-PSM was developed to address the problem of overlay in PSM structures by creating a self-aligned phase-shifting structure. In the fabrication of such structures, a phase-shifter film is deposited onto the chrome layer and then resist is spun onto the reticle. The resist is then patterned by e-beam exposure and developing.

This is followed by an anisotropic etching of the phase-shifter film. Finally, the chrome film is etched isotropically so that it undercuts the phase-shifting film

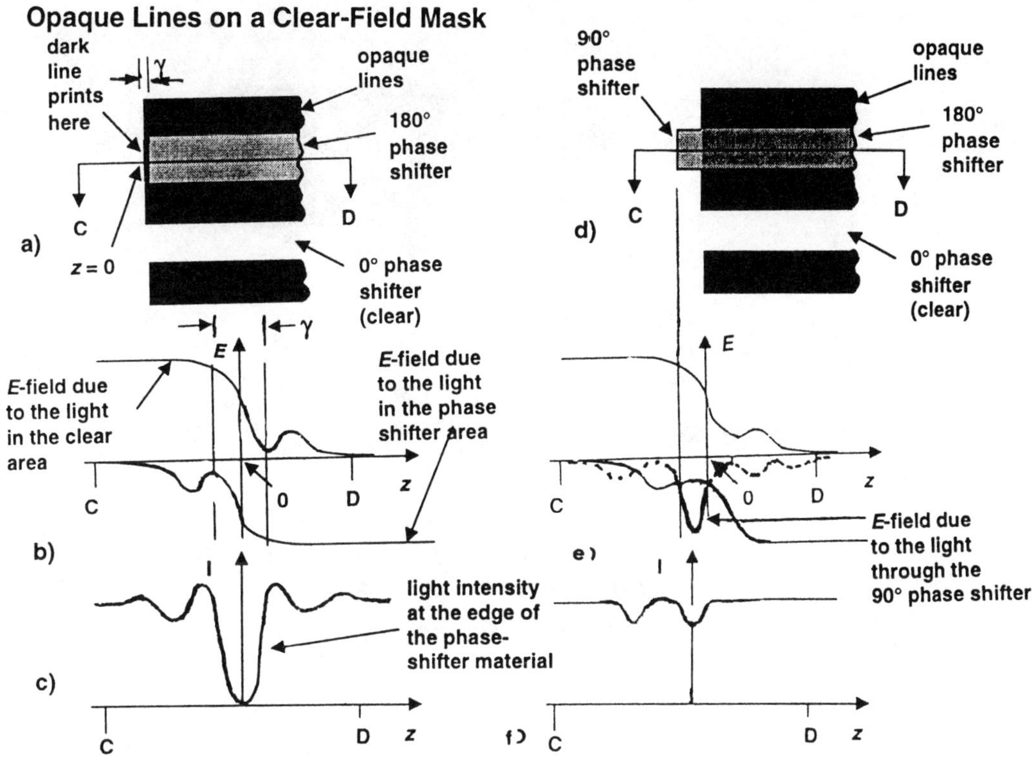

Fig. 7-19 Problem at the edge of a phase-shifter at a clear boundary: (a) ALT-PSM used on lines and spaces pattern with a clear field. Dark lines will print at the boundaries between the edge of the phase shifter and non-shifted clear areas; (b) *E*-fields along the *z*-axis between C and D for light incident on the phase shifter and clear area; (c) These two *E*-fields interfere in the vicinity of $z = 0$, causing the intensity of the *E*-field there to drop, resulting in a dark line printed as shown in (a); (d) Adding an extension of phase shifting material (with $\phi = 90°$) at the end of the phase-shifting region (as shown in [d]), helps mitigate the problem; (e) The amplitude of the *E*-fields of the light for the three phase regions versus *z*; (f) Intensity of light versus *z* when the $\phi = 90°$ extension is present. Note that the intensity drop-off is much smaller when interference occurs, and the dark line may no longer print.

(see Fig. 7-21). In rim-shift -PSMs the phase -shifting takes place only at the rim of each mask feature. As a result, such PSMs do not provide as much resolution improvement as do the ALT-PSMs. However, they do not suffer from the pattern restrictions that hinder ALT-PSMs. Thus, they can be used for arbit-

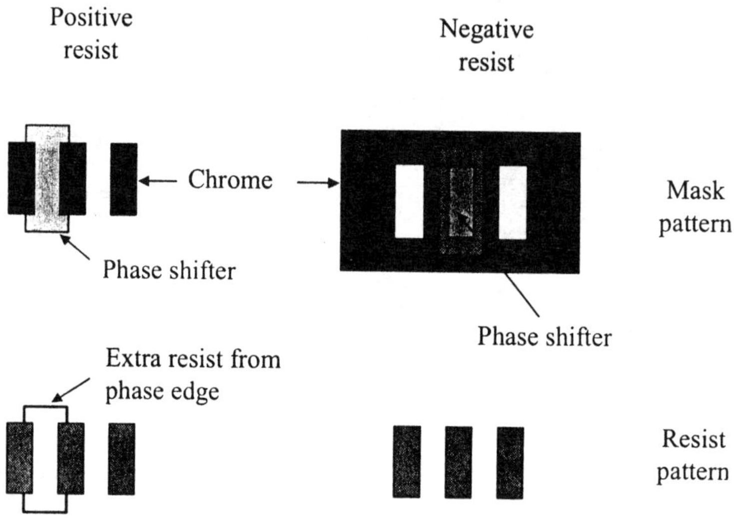

Fig. 7-20 On the left, extra resist printed by the phase edge, for positive photoresist. The right side of the figure shows what occurs with negative resist.

rary reticle patterns, including isolated features. Among the issues of concern are rim-size optimization and the lack of inspection and repair of the phase-shifting materials of the mask.

Another type of weak-PSM that can also be applied to arbitrary layout patterns is the *attenuated* (or *halftone*) *PSM*.[33,34] Such *attPSMs* are fabricated by replacing the opaque part of a conventional reticle with a halftone film (i.e., one that is partially transmissive, about 10%). For example, for *i*-line-light such a film can be made from a very thin layer of chromium oxynitride or Mo-Si-ON (molybdenum silicon-oxynitride) having the proper thickness to shift the phase of the light it transmits by 180° (Fig. 7-22). While some light will pass through the halftone film, it will be too weak to expose the photoresist to the degree necessary for it to be washed away during development. But the negative amplitude of this weak phase-shifted light will destructively interfere with the non-phase-shifted light so that a useful (if weak) improvement in the resolution is achieved. The attPSMs are generally considered better than rim-shift PSMs. This

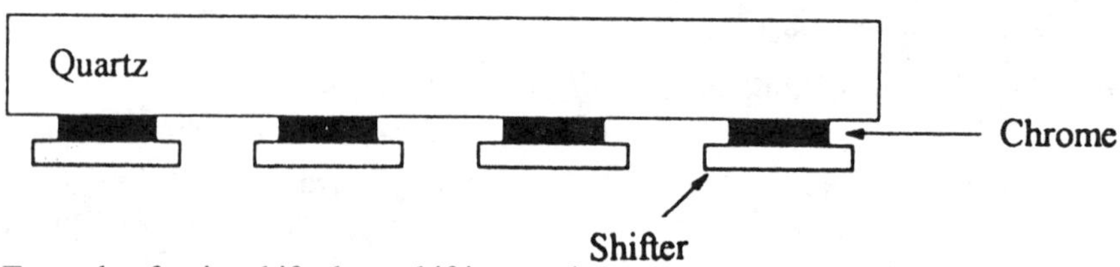

Fig. 7-21 Example of a rim-shift phase-shifting mask.

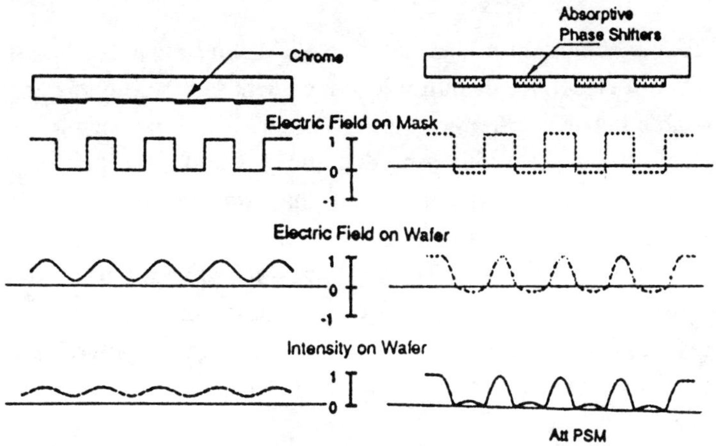

Fig. 7-22 Principle of the attenuated-PSM (attPSM).[31] © IEEE (1993)

is because attPSMs are much easier to fabricate (since the halftone absorber layer can be patterned just as if it was the opaque layer of a conventional mask). Isolated defects (either clear or opaque) can also be repaired with conventional repair techniques. Because of their relative simplicity, attPSMs have been the first type of PSM to be adopted in commercial IC fabrication. One limitation of attPSMs is that they are themselves subject to optical-proximity-effects. One approach to reducing this problem is to combine OPC with attPSMs. The reticle in this case must be made with three tones (clear, attenuated chrome, and opaque chrome). This technique is called *ternary attenuated PSM*, and MicroUnity provides software that allows such OPC features to be designed into the attPSM reticle pattern.[35]

Although PSMs were introduced in 1982, there was scant interest in the technique until about 1988. At that point, a large surge in research about PSMs began. While this resulted in dozens of papers dealing with this topic being published over the next ten years, PSMs were not widely introduced into production until the late 1990s. Even by 1997 only about 1% of the world's mask output contained PSMs. However, as 0.18-μm-CMOS was being moved into production in 1999 it became apparent that resolution-enhancement-techniques including PSMs would have to be used to enable optical lithography to print such small features. By 2000 attPSMs were being used extensively in manufacturing. Such weak attenuating (or halftone) PSMs were being used on 0.18-μm contact-hole levels. Strong-ALT-PSMs were being used on levels containing gates and interconnect lines. ALT-PSMs were also being used in memory circuits, in which interconnect lines are closely and regularly spaced. Reticle inspection techniques for ALT-PSMs were introduced by 1999, but repair techniques still lagged at that time. The combination of attenuated PSMs and off-axis illumina-tion provides an especially effective resolution-enhancement technology, as each method alone is only able to partially reduce the zero-order light component.

From the descriptions provided above – and from that of previous sections - we can see that a particular type of phase-shifting-mask or modified-illumination is applicable as a RET in specific situations, but not in others. It was noted that ALT-PSMs are effective for dense lines and spaces but not for isolated contacts. On the other hand, attPSMs are applicable for contacts and isolated features. Likewise, off-axis-illumination is found to enhance imaging of dense lines and spaces, but does not particularly improve the patterning of isolated-features. These general trends appear to be true for many of the newly developed methods of enhancing optical lithography. That is, as the improvement is restricted to specific types of features; there is no single solution for them all.

7.4 MASK ERROR FACTOR (MEF, or MEEF)

Masks generally include geometries of various dimensions. The fact that a mask contains a fairly wide range of feature sizes implies that lithographic imaging tools and resist processes should exhibit *linearity*. By this we mean that imaged patterns on the wafer should have the same size as the original patterns on the reticle (after applying the lens reduction factor). In other words, the printed features should all be approximately the size of the corresponding feature on the mask (divided by the lens reduction factor), over a wide range of feature sizes (i.e., a 4X reduction stepper should print a 10-μm-wide line on the wafer from a line on a reticle which is 40-μm wide, and it should print a 1-μm-wide line on the wafer from a 4-μm-wide line on a reticle).

However, as the printed feature sizes approach the resolution limit of the lens, *non-linearity* is found to occur.[37] This effect is shown in Fig. 7-23, which shows the results for a 248-nm stepper. The data in the figure indicates that linearity between the reticle-feature-size and printed-feature-size is obtained for printed-feature-sizes down to about 150% of the exposure-wavelength. That is, one can see that for the case of a 248-nm stepper, as long as the reticle dimensions are larger than 1300-nm (which is 4 x 325 nm), the reticle-dimensions are transferred to the printed-features in a 4:1 ratio. However, for smaller features, the dimensions on the wafer are transferred at a ratio greater than 4:1. This means that when critical geometries are being printed, corrections on the mask must be made to account for this non-linearity. This type of print-bias is a function of feature-size, and therefore can be corrected by appropriately resizing the features on the mask.

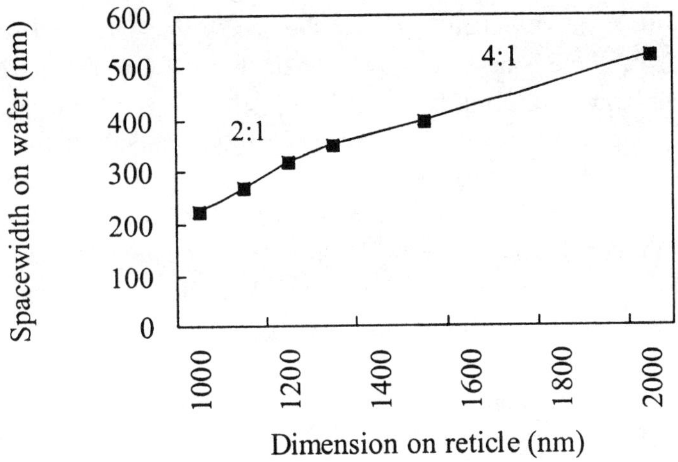

Fig. 7-23 Transfer of features from a reticle to the wafer. The features are reduced in size by 4x for large feature size, but the reduction changes for features near the resolution limit of the lens.

Since the quality of exposure tool lenses has improved and resolution enhancement technology has provided the means to push imaging performance below the wavelength of the exposure light (i.e., to less than 100% of the exposure wavelength), the smallest features that we now expect to be able to print are in the regime where such non-linearity becomes significant. Hence, this non-linearity effect has become very important from a practical point of view.

Another consequence of the non-linear printing of features near the resolution limit of a lens is that there is a *change in the magnitude by which mask linewidth errors* are transferred by the lens. That is, for larger printed feature sizes (i.e., those which exceed about 150% of the exposure wavelength), mask errors are transferred to the printed feature in the same ratio as the lens reduction factor. For example, in a 4X lens, a 20-nm error in the reticle CD for a 400-nm (4-μm) mask feature, would result in only a 5-nm printed error in the 1-μm printed line (because the transfer is still linear for this feature size). However, what is implied by the data of Fig. 7-23 is that *variations in linewidth about their nominal size are not reduced by the lens reduction factor when small features are imaged.* Instead, the effective mask-dimension-error is magnified. The severity of this amplification is described by the so-called *mask error factor (MEF).* Note that this factor is also called the *mask error enhancement factor (MEEF)*, but we will use the abbreviation MEF here. The MEF is defined as:

$$MEF = M\,(\Delta CD_{printed}/\Delta CD_{mask}) \quad (7.2)$$

Where M is the nominal magnification of the mask compared to the wafer image, and $\Delta CD_{printed}$ is the change in the CD of the printed feature, and ΔCD_{mask} is the change in CD on the mask. If MEF > 1, the errors in the mask CDs are transferred to the printed features on the wafer by an amount that is greater than would occur if MEF = 1, even if the mask CD gets geometrically reduced by the proper ratio (e.g., 4:1). This makes it more difficult to control wafer CDs in the presence of inevitable mask errors. For example, for the case described in the previous paragraph, the MEF = 1. However, if the MEF in that case was instead = 2, the error in the CD of the printed line would increase from 5-nm to 10-nm for an error of 20-nm on the reticle (see Fig. 7-24).

The primary factor that causes the values of MEF to rise above 1.0 is the degradation of the image contrast as the factor ($CD_{printed}$ λ/NA) is decreased below 0.6. In an imaging system with no aberrations and no defocus, it has been found that MEF rises to over 4 when ($CD_{printed}$ λ/NA) = 0.35. (The 0.35 value corresponds to a resolution of 125-nm for a KrF-stepper with an NA of 0.7.) In such circumstances (i.e., if MEF = 4), any reticle linewidth variations will transfer 1:1 on 4X reduction exposure tools. In such cases, the entire benefit of lens reduction (whose role is to reduce the impact of reticle variations on wafer linewidth control) is completely absent.

It has been found that exposure-dose has little significant effect on the MEF. The MEF is related to the mask size and the imaging system, not the resist size.

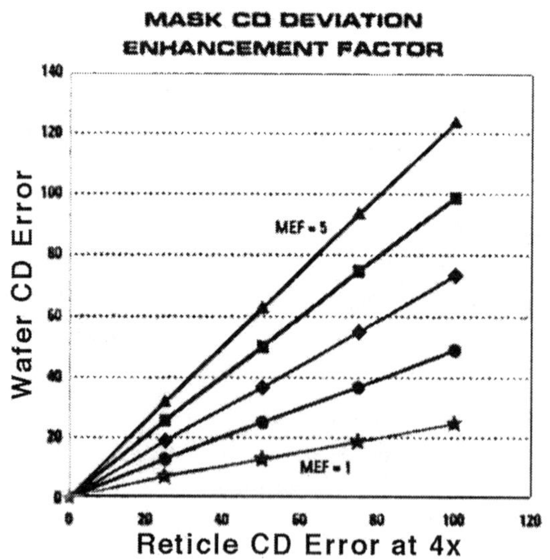

Fig. 7-24 For a MEF >1, a small CD-error on a 4X mask can transfer a larger magnitude of error onto the wafer. (Courtesy KLA-Tencor)

Therefore, reticle biasing does not really help to reduce MEF, although it may be desired for other reasons. The MEF degrades somewhat with defocus, but the degradation is small for defocus smaller than the Rayleigh unit of defocus, $\lambda/2NA$.

The most significant factor that can control the MEF is improvement in the image contrast. This improvement can be achieved in various ways, including the use of more advanced exposure tools (shorter wavelength, higher NA, better aberration control), and through the use of some of the resolution enhancement techniques. Of particular benefit, at small cost, is the use of annular or quadrupole illumination. The disadvantage is that the proximity effects tend to be emphasized and the isolated-to-dense feature bias control suffers.

The MEF also has a significant impact on reticle costs. That is, as long as MEF = 1, improvements in reticle linewidth control only had to match the decrease in minimum feature size (in order to keep constant the reticle contributions to wafer linewidth variations). However, as noted above, with MEF values as large as 4 now being reported, the challenges facing reticle makers have been increased. That is, multiple improvements in the quality of photomasks will be necessary to limit the effects of MEF. Even if these improvements should prove technically feasible, they will only be achieved with increased cost.

These reticle improvements include tighter CD-control and metrology, and tighter enforcement of defect inspection parameters. Characterization must be performed in each area of the wafer because MEF is not uniform (isolated lines generally have a lower MEF than do equally sized dense lines, and near the diffraction limit, contacts have ~2x the MEF of lines and spaces of similar width), and defects near the feature edge are magnified by MEF (just like CD errors). This creates a paradigm shift in sample size and will also require new developments in defect inspection tools. Typically, reticles have been specified with 5 to 10 measurement-sites using white light or *i*-line-metrology. However, with tighter specifications in both CD and defects, the 1-μm optical metrology limit and such small sample sizes are no longer adequate. Generally used at the wafer level, CD SEMs with resolutions ~4 nm, are now being applied to manufacturing the mask sets for 0.15 μm memory cells (whose CDs are 0.6 μm on the reticle). Inspection-tool improvements are allowing such higher resolution limits to be achieved on 5 to 15 times more sites – with overall measurement times remaining the same (~5 min/mask). Another path toward reducing MEF, as described by Kim *et al.*,[38] is to widen the exposure latitude using all means available (including improving aerial image contrast with strong phase-shift masks).

7.5 EXTENDING THE LIMITS OF OPTICAL LITHOGRAPHY

In the late 1970's, the demise of optical lithography was predicted to occur when IC feature sizes reached about 1-μm. Since optical lithography is being used for 0.13-μm IC technologies, this prediction was obviously incorrect. Nevertheless, some limits to optical microlithography do exist, and the consensus today is still that some other lithographic technology will replace optical lithography at a future date.

The resolution limit of optical lithography was also once considered to be about the same dimension as the wavelength of the exposure light. However, as noted in Sect. 7-3, by employing *resolution enhancement technologies* (RETs) the minimum feature size that can now be practically imaged is about 50% smaller than the exposure wavelength (e.g., 0.13-μm features can be produced with 248-nm KrF light).[39] As a result, the rate of shrinkage of the minimum feature size has been faster than the reduction of the exposure wavelength. (For 0.18-μm technology, KrF excimer-laser-based lithography is employed, with the KrF wavelength being 248-nm [0.248μm]).

To image 0.13-μm features without having to resort to such resolution enhancement techniques as PSM, off-axis-illumination, OPC, and multi-layer resists, however, a shorter wavelength of light is needed (e.g., ArF-excimer-laser light that has a wavelength of 193-nm). Since ArF technology (including appropriate light sources, lens materials, and resists), did not become available in time for full-scale production of 0.18-μm ICs, adoption of such RETs onto KrF has been the way that 0.18-μm technology is being manufactured. In 1999 the first generation 193-nm ArF 4:1 stepper/scanners from ASML, Canon, and SVG Lithography were installed on 0.13-μm development lines. The average price of ArF steppers is $10-$15 million. ArF resists from Olin, and others (based on acrylates and cyclic olefin polymer materials – see Chap. 6), were available for pilot-line use. But it appears that ArF excimer-light (or some other shorter wavelength energy source) will be needed to obtain adequate resolution with sufficient depth of focus for the next generation (0.1-μm-CMOS). Note that even at 0.1-μm, the dimension of the feature being printed is smaller than even the wavelength of ArF excimer-light. Thus, RETs will also be needed. (Another financial drawback is that ArF light appears to damage the fused-silica used for ArF lenses. Thus, the $1.5-million lens in ArF stepper/scanners may need to be replaced after three years of use.) Since (for reasons we will describe next), it appears that the limit of optical lithography is now predicted to be 0.05-0.07-μm (50-70-nm), non-optical techniques will eventually be needed.[45] Such non-optical alternatives are considered in the next section.

We noted above that in the 1970's it was predicted that it would not be possible to use optical lithography for IC technologies below about 1 μm. There were several reasons why this estimate proved to be incorrect, as follows:

1. In the decade starting in 1985, lens-making technology made significant and unexpected advances. For example, as noted in Sect. 7.2, 0.3-NA, *g*-line lenses being manufactured in 1980 had a resolution limit of 1.25-μm and a depth-of-focus of 1.5-μm. By 1996 Nikon was offering lenses with the same NA, but with much improved performance (i.e., these Nikon *i*-line-lenses with 0.3-NA had a resolution limit of 0.7-μm and a depth of focus of 5.0-μm). The reason for this improvement was that the lenses being manufactured in the mid-1985's were not yet diffraction limited (but were still being limited by lens aberrations). By 1996, lens-making technology had improved to the degree that the lenses actually were diffraction limited. In addition, it was learned how to make lenses with much larger NA than in 1980. For example, by 2001, lenses with an NA as large as 0.8 were being offered commercially. (However, since the NA of lenses cannot exceed 1, and since diffraction-limited lenses are as perfect as they can be, it will not be possible to obtain the gains provided to optical lithography from such ingenuity and innovation in lens-making again.)

2. Optical lithography technologies were developed that could use shorter wavelengths than the mercury *g*-line (436-nm). The ability to use such shorter wavelengths not only reduced the resolution limit, but also brought with it an increased depth of focus.

3. Photoresists also were improved, allowing the k_1 values to be reduced. For example, in 1980 $k_1 = 0.8$, but by 2001, k_1 was decreased to < 0.5.

4. The depth of focus requirements (DOF) were decreased (down to about ±0.2-μm), primarily due to the development of chemical-mechanical polishing (CMP).

5. Resolution enhancement technologies (RETs) were developed.

Despite these advances, there are several physical limits that are being reached that imply that the optical lithography will need to be replaced when the minimum feature sizes on ICs reach 50-70-nm. Let us list those limits, which have been stated by such industry experts as H. Levinson to justify the 50-70-nm resolution limit of optical lithography.[40]

1. To get to 70-nm resolution with optical lithography, shorter wavelengths of light will be needed. UV lasers operating at 157-nm (F_2) and 126-nm (Ar_2^*), are possible illumination sources for such shorter wavelengths.

Since virtually all materials absorb strongly in the range of 20–193-nm (except for some, like calcium fluoride [CaF_2] that still transmit 157-nm and 126-nm light), it may be necessary to use all-reflective optical systems at these wavelengths. (Extreme UV [EUV] lithography, discussed in Sect.7.6.1, uses such all-reflective optical systems.) However, as discussed previously, by 1999 advances in F_2-laser technology, in the growing of CaF_2-crystals, and in 157-nm resist chemistry made 157-nm lithography a possible candidate to fill the 0.1-μm "lithography gap" that exists.[41]

However, the thermal stability of the mask-substrate material may make it impractical to go to 126-nm illumination. That is, fused-silica can be still used for the mask substrate material because it still sufficiently transparent at 157-nm. However, it becomes too absorbing just a few nanometers below 157-nm, and thus it is not a material that can be used for masks at 126-nm. For such masks, a material such as CaF_2 must be used (since CaF_2 is still sufficiently transparent at 126-nm). However, the coefficient of thermal expansion of fused-silica is much smaller than that of CaF_2 (0.5-ppm/K vs 19 ppm/K). When a reticle is being exposed with an e-beam (to create the patterns on the reticle), substantial energy is absorbed by the mask-substrate material, and this causes the temperature of the mask blank to rise. Assuming that the temperature-rise can be limited to only 0.1K, there will be a 5-nm registration error across 100-mm on a substrate made

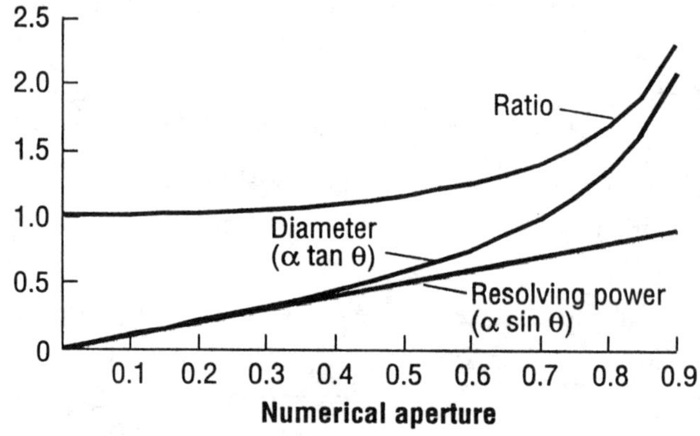

Fig. 7-25 Diminishing returns of higher NA for higher resolution. For lens with an NA greater than 0.80 NA, each incremental increase in lens-resolving power requires disproportionately larger increases in lens diameter. The ratio represents the difference of lens diameter to resolving power at a given NA.

of fused-silica (a magnitude of mis-registration that can easily be accommodated within overlay budgets). However, if the mask blank is made of CaF_2, the same temperature-rise will cause a mis-registration error of 100-nm across 100-mm, making it impractical to use CaF_2 as the mask substrate material. Thus, it appears that 157-nm will be the shortest wavelength that can be used for an optical lithography process.

2. As noted earlier, lenses cannot be made any more perfect than when they are diffraction limited. Thus, the remaining limitation imposed by diffraction remains and becomes dominant. In addition, it is unlikely that lithography will be carried out in a medium other than a vacuum or gas (such as air or nitrogen). This limits the NA of a lens to < 1. Since the NA of lenses available in 2001 is already at 0.8 in the most advanced scanners, increases in NA in the future can no longer be a key driver to higher resolution that it has been in the past. Thus, the great strides of lithography due to improvements in lens fabrication are no longer possible (Fig. 7-25).[39]

3. The decrease in depth-of-focus requirements down to ±0.2-μm was brought about by the use of CMP. However, it will not be possible to reduce this requirement further, as the wafer topographies are now about as flat as they can get.

4. The degree to which optical lithography can be extended toward the 50-nm limit depends on how effectively all of the RET methods described earlier can be implemented. The ability to implement a particular RET technique also depends on the application. As an example that was described earlier, the use of off-axis-illumination is highly effective for imaging highly repetitive structures (such as those found in memories), but the same technique is of not much benefit for imaging isolated features.

In summary, there are many challenges to be met to enable optical lithography to be extended to 70-nm (0.07-μm), although there is optimism that this goal will be achieved (Fig. 7-26).[40] However, for the reasons given above, 50-nm (0.05-μm) appears to be a clear lower limit for optical lithography.[41]

It is also probable that a mix-and-match strategy will be employed when ICs are fabricated with features smaller than 0.1-μm. That is, a non-optical lithography process may be used for the layers with the smallest minimum dimension, and optical lithography techniques for the non-critical layers. Figure 7-27 shows the relationship between resolution and exposure wavelength for various types of microlithographic technologies.[42]

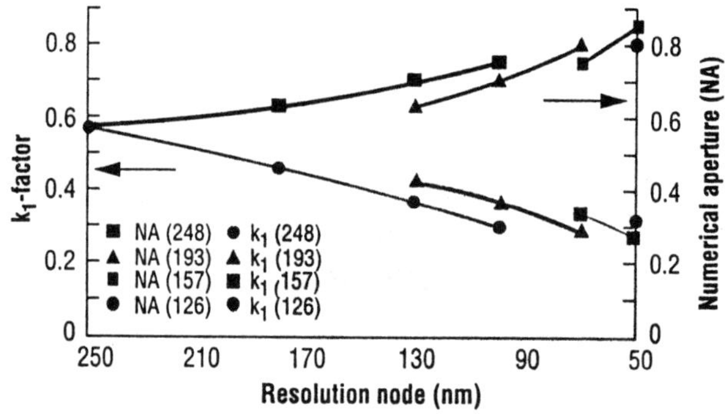

Fig. 7-26 Wavelength, NA, and k_1 for 250 nm to 50 nm resolution nodes. (Courtesy ASML)

Overlay limits are another issue to be considered with optical systems. Stage precision is the biggest practical concern, but work is reportedly being carried out to solve this problem.[43] Another overlay issue involves alignment target acquisition. Metal and grainy polysilicon have long presented difficulties for stepper

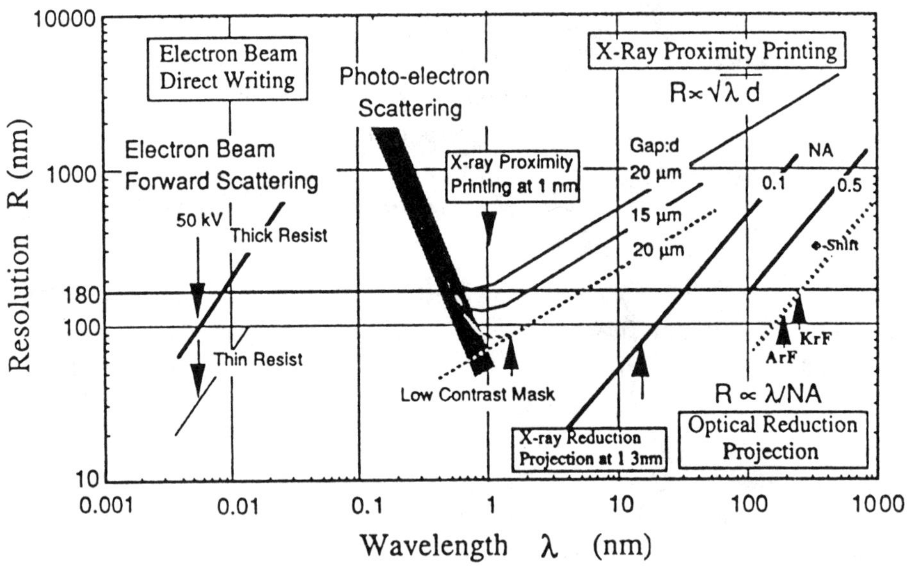

Fig. 7-27 Relationship between resolution and exposure wave length for various microlithographic technologies.[42] © 1996 (IEEE).

alignment systems. More recently, additional problems have been encountered on wafers that have undergone chemical-mechanical polishing. One solution has been to align all layers to a "zero layer" having consistent alignment targets.

The best level-to-level overlay-capability of optical steppers is estimated to be about 30-nm, which will be adequatc for 0.1-μm technology, but not for 0.07-μm. For 0.07-μm technology, it may be necessary to dedicate individual steppers to a particular wafer lot. This will reduce overall stepper productivity and introduce another source of additional costs.[44]

7.6 NON-OPTICAL (or NEXT GENERATION) LITHOGRAPHIC TECHNOLOGIES (NGL)

In order to image features smaller than 70-nm it will probably be necessary to employ some form of non-optical technology (or *next generation lithography, NGL*).[52] Two such NGL candidates will be discussed in this section: **1**) extreme ultra-violet (EUV) lithography; and **2**) electron-beam projection lithography (EPL). It should be noted, however, that by 2001, extreme ultra-violet (EUV) lithography had become the leading NGL technology. The others had become extinct (proximity x-ray and direct-write e-beam), or on put on the endangered species list (EPL). The EPL approach suffered a punishing blow in late-2000 when Agere Bell Laboratories cancelled it's research effort. However, since it remains alive under the auspices of an IBM-Nikon led research effort, we will provide a brief description of this candidate NGL technology as well.

It should also be noted that in 2001 the consensus was that non-optical techniques were still behind the overall production capability of optical micro-lithography (especially with respect to productivity).[45] Two additional difficulties are shared by non-optical lithographic techniques: **1**) the cost of a single mask is expected to be in the \$50,00–\$100,000 range; and **2**) no pellicles can be mounted on the masks to protect them from particles. The latter problem makes masks for non-optical technologies much more vulnerable to particles and other defects.

7.6.1 Extreme Ultra-Violet Reflective Projection Lithography (EUV)

An attempt is being made to extend the principles of projection lithography into the soft x-ray spectrum. (The region of the spectrum with photon energies of about 100-eV and with wavelengths between 10-15-nm [or 0.01–0.015-μm] can be considered either at the low end of the x-ray spectrum (*soft x-rays*), or the short wavelength limit of the *extreme ultra-violet* spectrum [EUV].) The chief reason EUV seems to be emerging as the winner in the NGL race is that EUV demands the fewest changes in manufacturing from the current methods, and

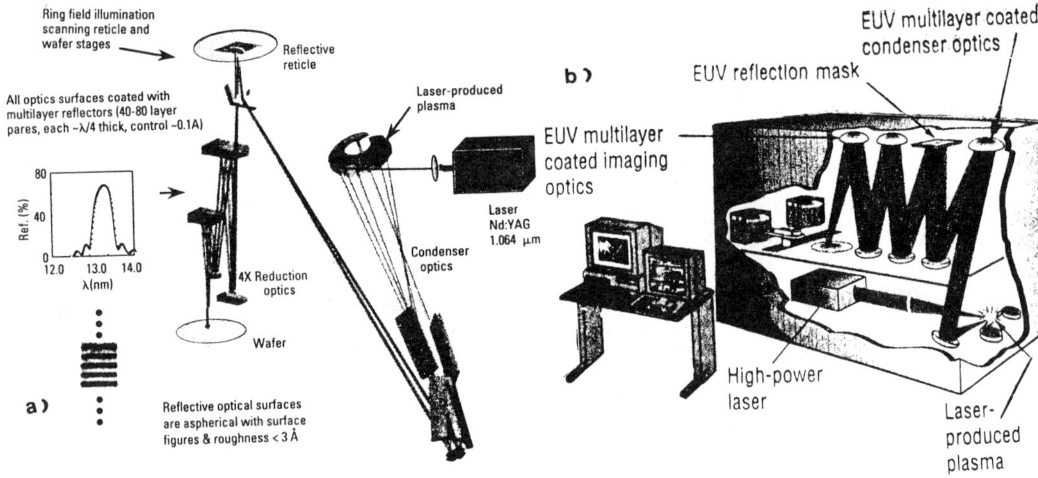

Fig. 7-28 (a) Schematic drawing of the optics of an EUV system. (b) An artists drawing of an EUV lithography system. A laser generates EUV radiation by exciting a laser produced plasma. This radiation is collected by a multilayer condenser system. The radiation illuminates a multilayer coated reflection mask, which is imaged and demagnified onto a resist-coated wafer.[46] Reprinted with permission of Solid State Technology, published by PennWell.

therefore appears to involve the least risk. (In an industry where downtime at an average fab costs $4 million a day, risk is not taken lightly.) For example, EUV radiation is more like light than any of the other alternatives. That is, it still involves photons, rather than electrons or ions, and the industry has grown comfortable with photons.

Furthermore, EUV photons can be focused with reflective lenses (which are actually curved mirrors). Use of these mirrors enables reduction optics to be used, which in turn allows photomasks to be fabricated with much larger dimensions than the features being imaged. Such masks are much easier to fabricate than masks containing patterns of the same size as the imaged features.

A number of interference coatings that exhibit relatively good reflectivity of such EUV light have been developed (i.e., silicon/molybdenum composites - deposited by sputtering), and they can be used to produce EUV mirrors. These coatings make it feasible to design projection systems that use all-reflective lithographic components to image patterns with EUV. The microlithographic technology based on this approach is called *extreme ultraviolet lithography*

(*EUVL*), or *soft x-ray projection lithography.*[47] Figure 7-28 shows a schematic of an EUV lithography system. A short survey of EUV technology is provided here. For more details, readers are referred to the excellent book by Attwood.[53]

EUV radiation can be generated in a number of ways. *Laser plasma sources* have shown promise as EUV light sources. They utilize pulses of high-intensity laser-light (normally from a Nd-YAG laser) to strike a target material (typically a supersonic jet of xenon gas). This produces a plasma that emits light with a broadband radiation that contains a significant component of EUV radiation. Xenon gas is the most common target material because it produces relatively strong EUV radiation, and being inert, does not directly generate debris that would contaminate and degrade the EUV optics. Unfortunately, these lasers may not be robust enough to be able to provide reliable operation in a production environment.

Synchrotrons (such as those used in proximity x-ray lithography) can also serve as the source of EUV light. Their main drawback is that they are not standalone light sources. Chip makers would prefer a standalone light source. With many standalone light sources in a fab instead of one synchrotron, the risk of having to shut down an entire fab if the light source fails is avoided. Some other alternative light source technologies were reportedly under development, but by the end of 2001 public information about them was not yet available.

In any case, the EUV radiation from the light source is collected and shaped into a narrow arc by a system of mirrors coated with EUV interference films. This radiation is then used to illuminate an EUV reflection reticle. The pattern on the reticle is then de-magnified (by 4X or 5X) and imaged onto a resist-coated wafer. The entire reticle pattern is exposed onto the wafer by synchronously scanning the reticle and the wafer (i.e., a step-and-scan exposure is performed). By using Rayleigh's resolution equation, it is predicted that a resolution of 32-nm can be achieved using a wavelength of 13.4-nm in a system having an NA = 0.25, with a k_1-value of 0.6. EUV lithography also offers the benefit of a large depth-of-focus for a system of such resolution (i.e., a DOF of greater than 1-micron). As noted earlier, since reduction optics are possible, many of the mask problems associated with proximity x-ray lithography are overcome.

However, there are a number of formidable challenges that must be met to make this technology practical. First, a modestly-priced, highly-reliable light source is still needed. Second, the stepper and the EUV beam-delivery system must be operated in a vacuum to prevent absorption and scattering of EUV rays. Third, the absence of transparent materials at EUV wavelengths implies that pellicles cannot be used (especially since the EUV light must pass through a pellicle twice in the reflection optics of EUV systems). Finally, since there are no high-transmission materials in the EUV, not only must all optical elements be reflecting, so must the mask. There are numerous issues involved in the fabrication of such reflecting optical elements and masks, and these will be discussed in the following paragraphs. While all of the above challenges appear daunting, significant progress was reportedly being made by 2001 to meet them.[55,56]

Fabrication of the reflective optical elements and masks for EUV requires a technology that can not only create surfaces with a relatively high reflectance, but that can also shape the surfaces of such reflective optical elements to extremely high optical tolerances. High mirror reflectance is important for EUV exposure tool productivity because such tools contain many reflecting surfaces (e.g., six mirrors in the projecting optics, two mirrors in the illuminator, and a reflection mask). Since the EUV light reflects nine times in such a system, the intensity of light at the wafer plane will be nearly 30% greater if each reflector has a 70% reflectance - compared to the case where each reflector has a 68% reflectance. Hence, even small increases in reflectance are significant.

Reflection occurs at interfaces between materials of different indices of refraction. However, at wavelengths below 50-nm all materials exhibit comparable indices of refraction (~1). Thus, it is difficult to make *single* interfaces that are good reflectors of EUV light. However, a technique that uses *multiple* interfaces can produce moderately effective EUV reflectors (i.e., with 60-70% reflectance). That is, such reflectors consist of a multilayer structure having a stack of many (e.g., 40 pairs) thin films (i.e., each with a thickness of tens of angstroms). This stack consists of alternating thin layers of high-Z and low-Z materials (where Z is the atomic number of the element being used in the film). This arrangement is used because there is a small but effective difference between the refractive indices at a high-Z/low-Z interface (see Fig. 7-29). Furthermore, by making the periodicity d of the multilayer satisfy the Bragg condition ($\lambda/2\sin\theta$), the reflected light from each interface adds to the reflected light from the other interfaces. The net effect is that by combining the small reflectivity at each interface it is possible to achieve a moderately high reflectivity overall (assuming there are a sufficient number of layers).

The materials for each layer must also be weak absorbers of EUV light, since the light must be able to penetrate to the lower levels of the film stack (which also explains the need to use as thin a film as possible for each layer). Most efforts have focused on using Mo and Si as the high-Z and low-Z materials for each layer of the EUV reflector, respectively. A 68% reflectance has been

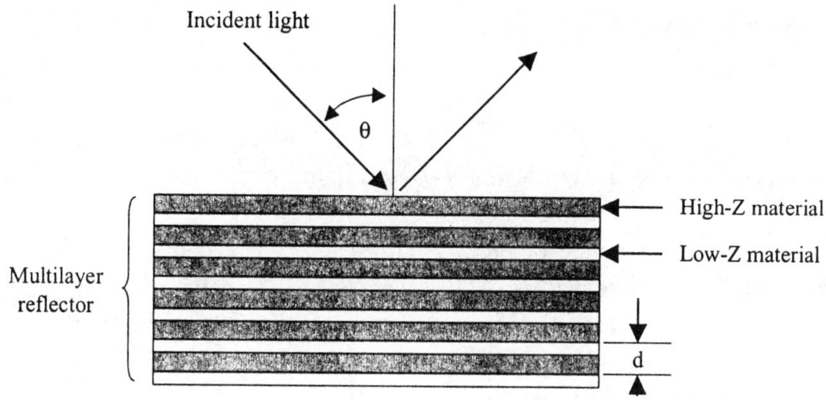

Fig. 7-29 Principle of a multilayer reflector for EUV radiation.

achieved for reflectors made with such Mo/Si multilayers for an EUV wavelength of 13.4-nm. However, the reflectivity will need to be improved to >70% to enable high-throughput cost-effective lithography. These layers are deposited by physical vapor deposition (sputtering), and each layer must be deposited to a thickness accuracy of less than an angstrom. Note that since most elements are strong absorbers at EUV wavelengths, even sub-monolayers of contaminants can reduce mirror reflectance. As a result, mirror (and mask) contamination is an important issue for EUV lithography. Finally, these interference films must be able to withstand exposure to strong EUV radiation without damage over long periods of time.

The fabrication of the aspheric (non-spherical) mirror substrates (the curved surfaces onto which the multilayer stacks must be deposited) is difficult because of the extremely tight tolerance (i.e., better than 2-nm) on the *surface figure* of these elements. Optics producers will need new metrology techniques to measure figure and surface roughness of these mirrors to the required accuracies. It was reported by workers at the Lawrence Livermore Laboratories in March 2001 that it was taking a full year to make a mirror set for EUV at that time. Industrial insertion would require manufacturing 300 such sets/year, a daunting challenge.

The *EUV masks* are fabricated by depositing the reflecting multilayers described above, on a flat substrate. Absorbing materials are then deposited and patterned on the multilayer, completing the mask fabrication. Since nearly all materials absorb EUV radiation, various films can be used for the absorber, including chromium, TiN, and TaN. The choice of absorber material is based on their ability to be deposited, etched, and cleaned. One of the problems of using Mo/Si multilayers on EUV masks is that they begin to significantly intermix at temperatures above 150°C. Such intermixing reduces the reflectivity of the stack. This limits the maximum process temperatures used to deposit and etch the absorber layers on EUV masks. By 2001, it was reported that EUV developers had devised a way to repair masks when defects are inadvertently incorporated during their fabrication.

Another consequence of the high absorptivity of all materials at EUV is the need for ultrathin resists (< 200-nm-thick). The resist-collapse problem that occurs when resist features have high aspect-ratios also dictates the use of thin resists (Chap. 6). However, it has been demonstrated that lithography processes suitable for manufacturing are feasible with such ultrathin resists. Moreover, resists quite similar to those used for DUV lithography appear to be suitable for EUV lithography.

EUV development is being carried out by a unique U.S. government-industry consortium known as the EUV Limited Liability Corporation (LLC). Under the terms of the agreement, three national laboratories, Sandia, Lawrence-Livermore, and Lawrence-Berkeley – are doing most of the development work, while industry members supervise and foot the bills. The core of the industry group consists of five chip makers: Intel, AMD, Infineon, Micron Technology, and Motorola. In April, 2001, the first full-scale prototype EUV lithography system was unveiled (Fig. 7-30).[54] This system, called the Engineering Test Stand (ETS) measures about 10x10x10 ft. The light source is a 1500-watt laser plasma source that generates an EUV beam with 10-W of power. (More than 50-W will be needed by the time the EUV systems reach production, as this is the power expected to be needed to allow throughputs of 70 300-mm wafers per hour.) The EUV LLC announced in March 2001 that the ETS was used to successfully print lines at 100-nm spacings. It was predicted that by the end of 2001, the ETS would be printing 70-nm lines and spaces at full field sizes. Debates are occurring at the end of 2001 as to when EUV will be ready for high-volume production. The optimists say that this will take place by 2005. Others argue that while early production units may be ready by then, high-volume manufacturing will take 2 to 3 years longer.

7.6.2 Electron Beam Projection Lithography (SCALPEL and PREVAIL)

Another NGL technique based on *e-beam projection* has been pursued (EPL). The system developed by Lucent Technologies (now Agere), is called *Scattering*

Fig. 7-30 A worker inspects the Engineering Test Stand, a full-scale prototype EUV lithography system – as assembled at Sandia National Laboratories. (Courtesy Sandia Labs).

with Angular Limitation in Projection Electron beam Lithography (or *SCALPEL*).[47,48] In this approach, a special scattering mask is used to produce a high-contrast image with a technique commonly used in transmission electron microscopy. That is, an electron beam (coming down from the top of Fig. 7-31) is scanned across a special mask that consists of a thin silicon nitride (SiN) blank (100–150-nm thick) having mask patterns formed of very thin (i.e., 25–50-nm-thick) tungsten or gold features. The electrons from the scanned beam can either: **1**) strike regions of the mask covered with one of the metal patterns; or **2**) strike a region of the mask *not* covered with metal. Electrons striking the SiN film pass straight through it (i.e., SiN only weakly scatters the beam electrons). On the other hand, electrons striking the metal patterns are strongly scattered as they pass through the heavy metal of the mask features. Those that have passed unscattered through the silicon nitride are then focussed by a lens so that they pass through the aperture of the *back focal-plane filter*. But most of the electrons that are scattered as they travel through the metal patterns get blocked by the back focal-plane filter (and thus do not travel though its aperture). The electrons that make it through the filter aperture are demagnified and strike the wafer surface (whereupon they form a high-contrast, 4:1-reduced image of the scattering mask).

The resolution limits of SCALPEL can be extended below 0.1-μm. In addition, because the scattering patterns are so thin, the demagnification of the mask can be made small enough that chips can be fabricated with as many as 2×10^{10} pixels (defined here as a small square whose side is equal to the minimum printable linewidth). Since 1-Gigabit DRAM patterns, for example, need roughly 10^{10} lithographic pixels per chip, the SCALPEL technique is suitable for such chip designs. The thin silicon nitride mask-blank substrate is supported on a silicon wafer with periodic silicon support struts. These struts are not imaged onto the wafer since the patterns are shifted into place as they are illuminated. While the mask structure is similar to that used in x-ray masks, the silicon support struts provide greater dimensional stability of the patterns on the

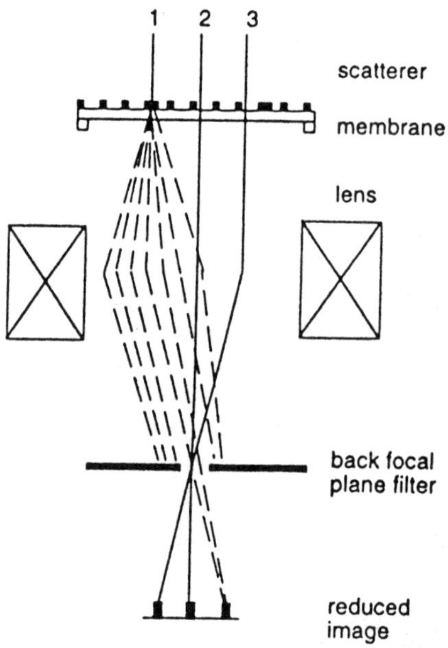

Fig. 7-31 Schematic of the SCALPEL technique. Electrons (1) that hit the scatterer (the metal patterns on the mask) are strongly scattered, and most are filtered out by the aperture. Electrons traveling through the silicon nitride membrane (2, 3) are only weakly scattered and pass through the aperture. These are demagnified to form a high contrast image on the substrate. Courtesy of Lucent Technologies, Inc.

SCALPEL mask, and the use of 4:1 reduction optics makes mask fabrication of the 4:1 reduction masks simpler. The initial price of a SCALPEL system (1998) was $8 million.

One advantage of SCALPEL over *x*-ray lithography is that it does not require a synchrotron orbital radiation (SOR) ring. Such an SOR ring is expensive, and is designed to be used with multiple steppers. Instead, a good e-beam source for SCALPEL is available. For SCALPEL, if another stepper is needed, only the stepper need be purchased. Another advantage is that the scanning e-beam that passes through the mask produces little heat in the thin nitride mask membrane and this reduces distortion of the imaged pattern on the wafer. Finally, SCALPEL's depth-of-focus has been measured to be ~50 times greater than for 0.25 μm optical lithography techniques. On the other hand, the traditional problems of e-beam lithography, such as low throughput, proximity effects, radiation damage, charging, and lack of a good resist are limitations of SCALPEL. Some other concerns involve keeping the membrane clean, and finding ways to repair it. A goal of producing working SCALPEL steppers that can deliver up to 50 300-mm wafer per hour at 0.13-μm resolution by 2003 has been established.

In 2001, however, EPL appeared to be losing some of it's momentum, with EUV gaining support instead. Motorola dropped out of the group of companies backing EPL, and Applied Materials and ASML disbanded their eLith LLC venture that was to have commercialized the SCALPEL technology. However, several of the EPL backers remained committed to developing the approach. Agere and TI switched their allegiance to the EPL approach being developed by Nikon and IBM (as did the semiconductor companies belonging to the Selete equipment consortium in Japan). The Nikon-IBM approach is called PREVAIL. These companies argue that EPL is better-suited to small-volume ASICs than the EUV system, which requires more expensive masks. Nikon is preparing to offer the PREVAIL e-beam column for e-beam projection lithography by 2004 (and perhaps earlier).

REFERENCES

1. O. Semprez, "Excimer Lasers for Future Lithography," *Solid State Technology,* July 2000, p. 255.

2. A. Ershov *et al.,* "Feasibility Studies of Operating KrF Lasers at Ultra-Narrow Spectral Bandwidths for 0.18 μm Linewidths," *Proc. SPIE.,* Vol. 3334, Feb. 1998.

3. A. Hand, "KrF Lithography Not Finished Yet," *Semiconductor International,* February 2001 p. 58.

4. M.D. Levenson, "Is There Light at the End of NGL Tunnel?" *Solid State Technology,* May 2001, p. 51.

5. R. DeJule, "157-nm Lithography Moves Forward," *Semiconductor International,* July 2000, p. 82.

6. R. DeJule, "NGL's Critical Bridge: 157 nm," *Semiconductor International,* March 1999, p. 36.

7. A. Hand, "Ready or Not, When Will 157-nm Lithography Really be Needed?" *Semiconductor International,* May 2001, p. 42.

8. B. Wang, "Birefringence in Fused Silica and CaF_2 for Lithography," *Solid State Technology,* February 2000, p. 77.

9. A. Hand, "Intrinsic Birefringence of CaF_2 Won't Halt 157-nm Lithography," *Semiconductor International,* September 2001, p. 42.

10. "Lithography Tools for 300-mm," *Solid State Technology,* September 2000, p. 40.

11. P.M. Ware, "300-mm Lithography Enters Reality," *Semiconductor International,* July 2000, p. 128.

12. A. K.-K. Wong, *Resolution Enhancement Techniques in Optical Lithography,* SPIE Press, Bellingham, WA, 2001.

13. R..DeJule, "Update on Resolution Enhancement Techniques," *Semiconductor International,* August 2000, p. 54.

14. C.A. Mack, "Understanding Focus Effects in Submicron Optical Lithography: A Review," *Optical Engineering*, Vol. 32, No. 10 (October 1993, pp. 2350-2362).

15. C. Mack, "Optimum Stepper Performance By Image Manipulation" *KTI Microelect. Sem.* p. 209 (1989).

16. D.L. Fehrs *et al.,* "Illuminator Modification of an Optical Aligner" *KTI Microelect. Sem.* p. 217 (1989).

17. M. Noguchi *et al.,* "Subhalf Micron Lithography System with Phase-Shifting Effect," *Proc. SPIE*, Vol. 1674, p. 92, (1992).

18. W. Partlow *et al.,* "Depth-of-Focus and Resolution Enhancement for *i*-line and Deep-UV Lithography Using Annular illumination," *Proc. SPIE*, Vol. 1927, p. 137 (1993).

19. S. Asai *et al., J. Vac. Sci. Technol.* **B 10(6)**, 3023 (1992).

20. K. Toh *et al,* "Three-Dimensional Simulation of Optical Lithography," *SPIE,* Vol. 1463, p. 402 (1991).

21. K. Tounai *et al..,* "Optimization of Modified Illumination for 0.25 μm Resist Patterning," *Proc. SPIE,*Vol. 2197, p. 31 (1994).

22. T. Ogawa *et al.,* "Effective Light Source Optimization with the Modified Beam for Depth-of-Focus Enhancements," *Proc. SPIE,* Vol. 2197, p. 19 (1994).

23. H.J. Levinson and W.H. Arnold, "Optical Lithography," Chap. 1, pp. 11-138, in *Handbook of Microlithography, Micromachining, and Microfabrication,* Vol. 1, Microlithography, Ed. P.Rai-Choudhury, SPIE Optical Engineering Press, Bellingham, WA., 1997.

24. R. DeJule, "OPC Simulation," *Semiconductor International,* September 1997, p. 56.

25. C. Mack, "Evaluating Proximity effects Using 3-D Optical Lithography Simulation," *SemiconductorInternational*, July 1996, p. 237.

26. C.A. Mack and P.M. Kaufman, "Mask Bias in Submicron Optical Lithography," *J. Vac. Sci. Technol.,* Vol. **B6** (Nov./Dec. 1998) p. 2213.

27. N. Shamma *et al.,* "A Method of Correction of Proximity Effect in Optical Lithography," *KTI Microlithography Seminar Interface 1991*, p. 145.

28. *Microlithography: Science and Technology,* J.R. Sheats & B.W. Smith, Eds., Marcel Dekker, New York, 1998.

29. M. Levenson *et al.,* "Improving Resolution in Photolithography with a Phase-Shifting Mask," *IEEE Trans. Electron Dev.*, **ED-29**, 1828 (1982).

30. B.J. Lin, "Phase-Shifting Masks Gain an Edge," *IEEE Circuits & Devices Mag.,* March 1993, p. 28.

31. *ibid.*, Ref. 24, p. 74-82.

32. M.D. Levenson, "Strong Phase-Shift Mask Designs with Lower Cost," *Solid State Technology,* January 2001, p. 104.

33. T. Terasawa *et al.,* "Imaging Characteristics of Multi-Phase-Shifting Mask," *SPIE,* Vol. 108, 25 (1989).

34. B.J. Lin, "The Attenuated Phase-Shift Mask," *Solid State Technology,* January 1992, p. 43.

35. R. DeJule, "OPC for Attenuated Phase-Shift Masks," *Semiconductor International,* May 1998, p. 44.

36. M.S. Hibbs, "System Overview of Optical Steppers and Scanners," in *Microlithography: Science and Technology*, Eds. J.R. Sheats and B.W. Smith, Marcel Dekker, New York, 1998, Ch. 1, pp. 1-107.

37. W. Maurer *et al.,* "Pattern Transfer at k_1 = 0.5: Get 0.25 μm Lithography Ready for Manufacturing," *Proceeding of SPIE*, Vol. 2726, p. 113, 1996.

38. B.-G. Kim *et al.,* "Beneath the MEEF," *Solid State Technology,* August 2000, p. 107.

39. S. Okazaki, *Proc. of SPIE*, Vol. 2440, p. 18 (1995).

40. H.J. Levinson, *Principles of Lithography*, SPIE Press, Bellingham WA, 2001.

41. J.A. McClay and A.S.L. McIntyre, "157 nm Optical Lithography: The Accomplishments and the Challenges," *Solid State Technology,* June 1999, p. 57.

42. S. Okasaki, "Lithography Prospects for 0.18 μm Technology & Beyond," *Tech. Dig. IEDM,* 1996, p. 57.

43. S.A. Lis, "An Air-Turbulence Compensated Interferometer," *Proc. SPIE*, Vol. 2440, 9891 (1995).
44. *ibid.*, Ref. 23, p. 124.
65. T. Brunner, "Pushing the Limits of Lithography for IC Production," *Tech. Dig. IEDM,* 1997, p. 9.
66. A.M. Hawryluk *et al.*, "EUV Lithography," *Solid State Technology*, July 1997, p. 151.
67. S. Berger *et al.*, SPIE Proceedings, Vol. 2322, p. 434 (1994).
68. L.R. Harriott *et al., J. Vac. Sci. Technol.,* **B14(6)**, 3825 (1996).
69. P.M. Ware, "Why Optics is Forever in Lithography," *Solid State Technology,* May 2000, p. 192.
70. C. Wagner *et al.*, "The Technical Considerations of Extending Optical Lithography," *Solid State Technology,* September 2000, p. 97.
71. J.S. Petersen, "Integrated Imaging a Key Approach to Optical Lithography at 100-nm+," *Solid State Technology,* January 2001, p. 80.
72. R. DeJule, "Next Generation Lithography Tools," *Semiconductor International* March 1999, p. 48.
73. D.T. Attwood, *Soft X-rays and Extreme Ultraviolet Radiation, Principles and Applications,* Oxford University Press, 2000.
74. Hand, "EUV Lithography Makes Serious Progress," *Semiconductor International,* June 2001, p. 54.
75. J. Canning, "Challenges of Next-Generation Lithography," *Semiconductor International,* September 2000, p. 107.
76. M.D. Levenson, "Is There Light at the End of the NGL Tunnel?" *Solid State Technology,* May 2000, p. 51
77. A.K. Bates *et al.*, "Review of Technology for 157-nm Lithography," *IBM J. Res. & Dev.,* Vol. **45**, No. 5, 2001.
78. A. Hand, "157-nm Optics Demand a Bag of Tricks," *Semiconductor International,* February 2002, p. 50.

Chapter 8

CHEMICAL MECHANICAL POLISHING (CMP)

As circuit features are scaled to submicron dimensions and additional levels are added to multilevel-interconnection schemes, the required degree of planarization is increased. In fact, the complexity of current microelectronic devices demands global planarity of the substrate after shallow-trench isolation and at all interconnect levels. Thus, such planarization needs to be implemented in both conductor and dielectric materials. In this chapter the technique of *chemical mechanical polishing* (the only known technique that provides global planarization for ULSI applications) is described. The discussion will begin by defining terms through which planarization is characterized, and then a brief recount of the evolution of planarization techniques that preceded CMP will be given. After concluding this introductory matter, the remainder of the chapter is devoted to the CMP process.

8.1 TERMINOLOGY & EVOLUTION OF PLANARIZATION PROCESSES FOR ICS

The term planarization is employed quite frequently (and often loosely) in other technical literature. Thus it is useful to define the term more completely. The example case used for this discussion is that of a dielectric layer deposited after Metal-1 is patterned (i.e., an *intermetal dielectric* [or IMD] film). In the case where no planarization exists (Fig. 8-1a), the step heights on the IMD surface closely approximate the step heights of the Metal-1-layer and the underlying topography. Furthermore, in this case the steps on the IMD surface generally also have steep slopes.

8.1.1 Defining the Degree of Planarization

The steps on the surface of the IMD can be made less severe through various planarization methods. The degree to which this can be successfully accomplished differs according to the technique used. The degree of planarization is classified according to the following qualitative planarization criteria:

1. The first degree of planarization (smoothing) involves a lessening of the step slopes at the IMD surface (Fig. 8-1b). The step heights in this case, however, are not significantly reduced in magnitude.

2. In the second degree of planarization (partial- or semi-planarization) the step heights are reduced (but not eliminated), and the slopes of the steps are also smoothed (Fig. 8-1c).

3. In the third degree of planarization (complete local-planarization), the steps at the surface of the IMD are completely eliminated wherever the spaces in the underlying topography are relatively close together (e.g., < 10-μm

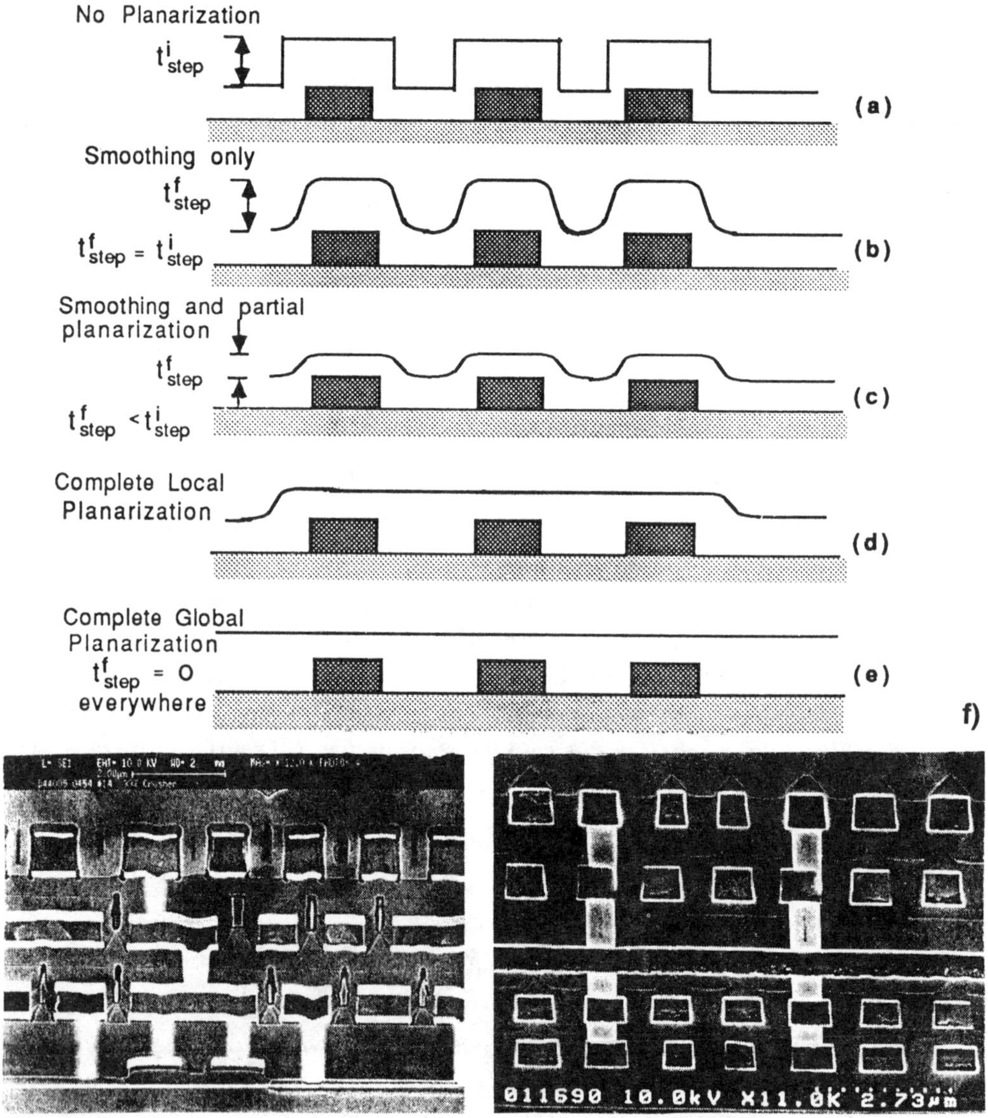

Fig. 8-1 (a) through (e) - Degrees of dielectric planarization; (f) SEM of 6-metal-level interconnect structure, with complete global-planarization realized at each level.

apart), but the steps at isolated, wide features still exhibit a step (Fig. 8-1d).

4. In the fourth degree of planarization (complete global-planarization), the surface is completely planarized over arbitrary topography (Fig. 8-1e).

Figurc 8-1f shows a SEM of a 6-level-metal interconnect-system in which complete global-planarization has been realized in each of the metal levels (i.e., using CMP).

A quantitative measure of the step-height-reduction, referred to as the *planarization factor*, β, is given by[1]

$$\beta = 1 - (t^{f}_{step}/t^{i}_{step}) \tag{8.1}$$

where t^{f}_{step} and t^{i}_{step} are the final- and initial-step-heights, respectively. In complete-planarization cases, $\beta = 1$. If no planarization (or only smoothing) exists, then $\beta = 0$.

8.1.2 The Need for Dielectric Planarization

As the number of levels in an interconnect technology is increased, the stacking of additional layers on top of one another produces a more and more rugged topography. Consider, for example, a two-level-metal, single-poly CMOS technology. Assume the step-height of the semi-recessed field-oxide is 0.3-μm, and the thicknesses of the poly and the first and second metals are 0.4-, 0.5- and 1.0-μm, respectively. The maximum height of the steps on the wafer surface after each of these processes will be 0.3-, 0.7-, 1.2-, and 2.2-μm, respectively (Fig. 8-2). It is apparent that the surface of the wafer must be planarized in some fashion to prevent the topography roughness from growing with each level. Without such planarization, the microscopic canyons that result on the wafer surface from stacking of device features can lead to topography conditions that would eventually reduce the yield of circuits to unacceptably small values.

A nonplanar wafer topography results in three conditions that prevent the fabrication of reliable electrical connections in multilevel interconnect systems. These conditions are as follows:

1. Poor step-coverage of metal lines as they cross over the high and steep steps. A measure of how well a film maintains its nominal thickness is expressed by the ratio of minimum thickness of a film as it crosses a step, t_s, to the nominal thickness of the film on horizontal regions, t_n (Fig. 8-3a). This property is referred to as the *step-coverage* of the film, and is expressed as the percentage of the nominal thickness that occurs at the step:

$$\text{Step coverage (\%)} = (t_s/t_n) \times 100\% \tag{8.2}$$

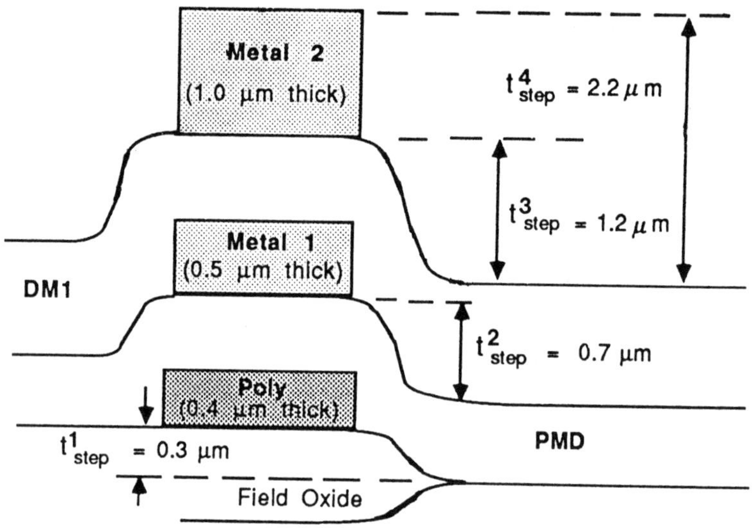

Fig. 8-2 As the number of levels in an interconnect technology is increased, the stacking of additional layers on top of one another produces steps of progressively greater heights.

Step coverage of 100% is ideal for a conductor, but in general the height of the step and the aspect-ratio of the features being covered impact the expected step-coverage. (The *aspect-ratio* is defined as the height-to-spacing ratio of two adjacent steps.) The greater the step-height or the aspect-ratio, the more difficult it is to attain coverage of the step without a corresponding thinning of the film that overlies it. Figure 8-3b shows an example of poor step-coverage. Hence, worse step-coverage is expected under these conditions. In addition to the step-height and aspect-ratio, step-coverage depends on two other topological-factors: the contour and slope of the step. In general, the smoother the contour of a step and the less vertical its slope, the better will be the step-coverage of an overlying film.

2. Metal stringers remain behind at the foot or sides of steep steps when anisotropic-etching is used. This condition is shown in Fig. 8-3c. In addition, the resist is typically eroded by the same gases that etch the metal. The resist must therefore be thick enough in all locations that it will not be completely removed before the etch-process is completed. If the dielectric surface has steps, in certain locations (i.e., over the upper corners of the steps in the dielectric, see Fig. 8-3d) there may not be sufficient resist thickness to survive during the time of overetch that must be used for removal of the stringers. The resist-erosion-rate during the overetch time

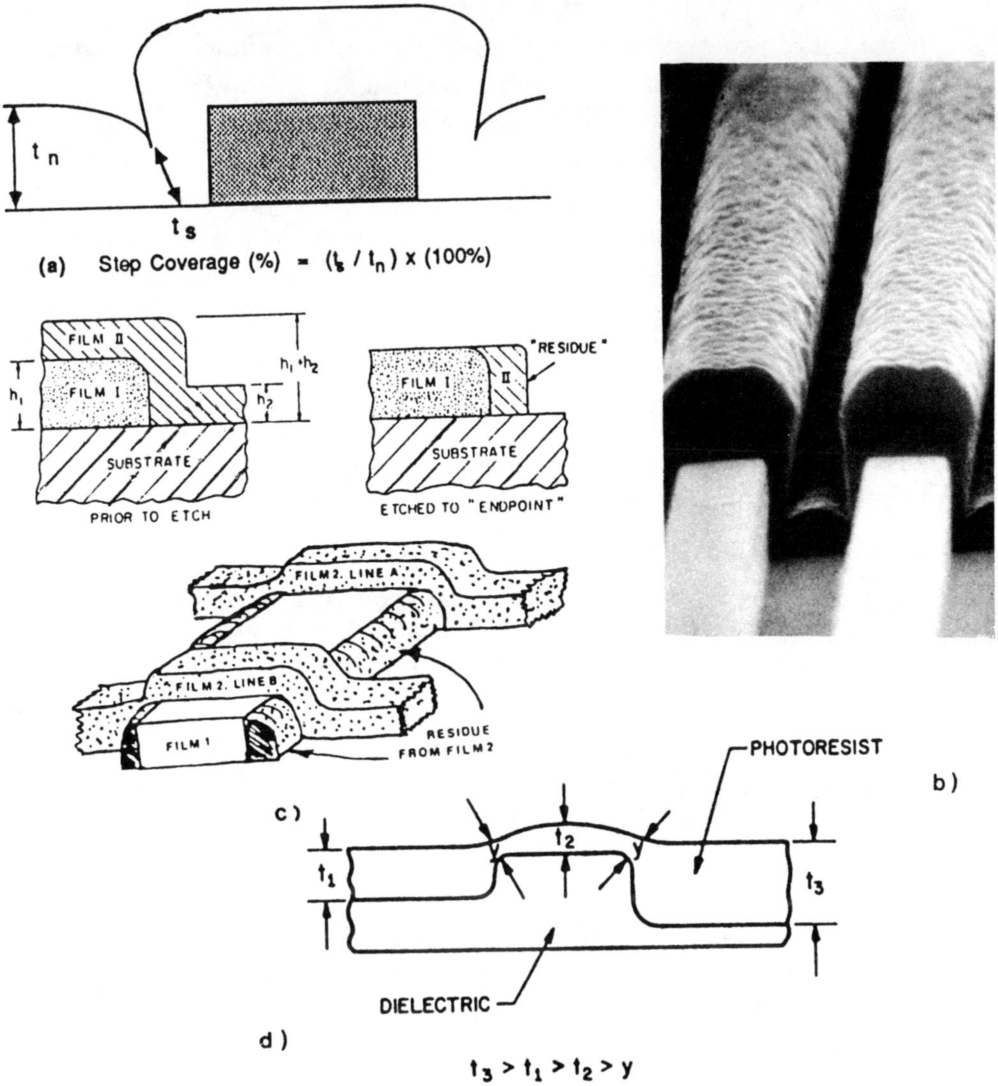

Fig. 8-3 a) Definition of step-coverage. b) SEM micrograph showing poor step-coverage. c) Metal stringers remain behind at the foot or sides of a steep step when anisotropic-etching is used. d) Problem of resist-thinning as film crosses underlying steps.

is also generally enhanced. Because most of the exposed-metal has been etched away, the reactant gases that were being consumed during etching of the metal are now present in greater concentrations. In cases where the resist is completely eroded at thin spots, the metal beneath the resist will be exposed to the reactive species and may be etched away to an unacceptable degree.

Even if the above two limitations could be overcome, the following obstacle would eventually force planarization to be implemented:

3. The depth-of-field limitations of submicron optical-lithography tools will require surfaces to be planar within ±0.5 μm. As a result, if steps larger than 0.5 μm exist on the surface of the IMD (or any other layer on the wafer), it will not be possible to pattern features in Metal-1 to the maximum resolution of the stepper. Thus, planarization will be mandatory if optical lithography is to be usable for fabricating ICs with such submicron feature sizes. (This topic is discussed in more detail in Vol.1, 2nd Ed., Chap. 13.)

Planarization of the intermetal dielectric layers is one of the chief approaches that has been taken to alleviate the problem of rough surface topography.

8.1.3 Design Rules Related to Intermetal Dielectric-Formation and Planarization Processes

The type of intermetal dielectric material, its thickness, and the method of deposition and planarization can all have an impact on the design rules that must be applied when laying out an integrated circuit with a particular multilevel-metal system. Here a two-level-metal system with an underlying poly-silicon level is considered. Assume that the topography beneath the polysilicon is completely planar. In this system, design rules dealing with the following conditions must be specified: **1**) minimum polysilicon-to-Metal-1 spacing, assuming no metal overlap of the poly; **2**) minimum distance between coincident edges when Metal-1 overlaps polysilicon; and **3**) minimum spacing of Metal-1-to-Metal-1.

If polysilicon and Metal-1 come too close together, a gap will exist between them that will lead either to void formation in the CVD oxide PMD layer following deposition (Fig. 8-4a) or to the formation of a deep, narrow crevice that cannot be covered by the next level of metal, even after some planarization procedures have been utilized (Fig. 8-4b).

In the second case (Fig. 8-4c), if polysilicon and metal edges are coincident, the total step height may be too high for adequate coverage by Metal-2 to be achieved. Figure 8-4d illustrates the case in which the problems of coincident edges and a narrow crevice are exacerbated by their simultaneous occurrence.

In the third case, the spacing between adjacent Metal-1 lines must be large enough to prevent void or crevice formation in the deposited dielectric film. The step that exists in the LOCOS field oxide must also be considered in real situations when such design rules are being formulated. One possible way to prevent the formation of voids, crevices, and excessively high steps, is to allow

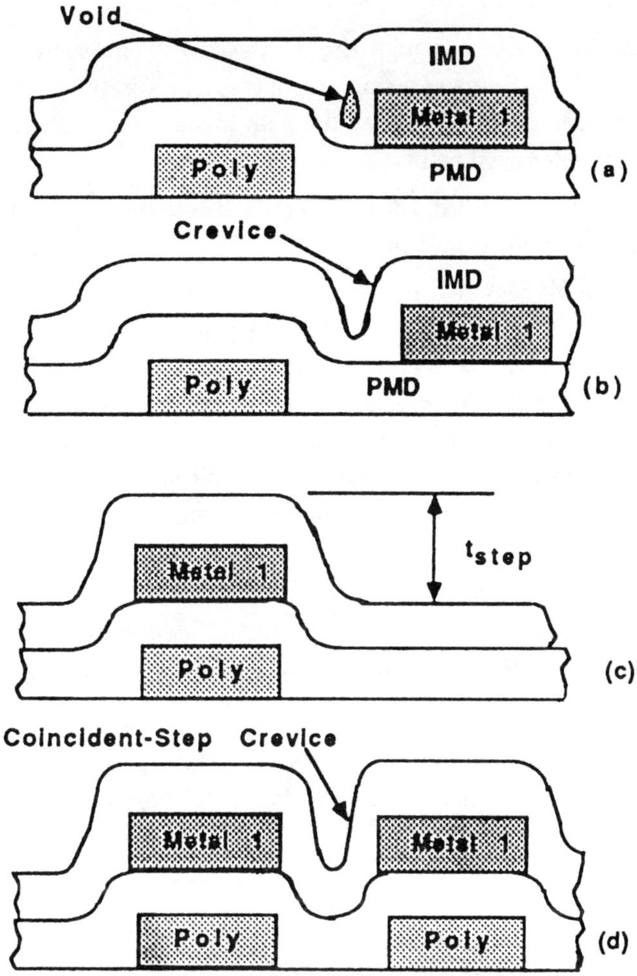

Fig. 8-4 Conditions associated with the relative spacing of polysilicon and Metal-1 that lead to layout design rules of these layers.

the spacings between Metal-1 and underlying features to be sufficiently large so any possible combination of misalignment cannot lead to the problems listed. However, this alternative is impractical, because it increases chip size and causes excessive complexity in layout design.

The trend in advanced IC designs today is toward the elimination of all such restrictive dielectric-layer-related design rules. The development focus is on dielectric planarization processes that allow all spaces between any adjacent or overlapping features (and all possible step heights), to be covered with a dielectric layer whose surface does not exhibit any voids, crevices, or excessively high steps. Even though this is an actively pursued goal, it has not yet been attained in all processes and technologies. Various planarization approaches that have been developed along the path toward the condition in which no restrictive design rules need be invoked will be described.

8.1.4 Step-Height Reduction of Underlying Topography Alleviates the Need for Planarization

In two-level-metal processes it may be possible to avoid performing any planarization procedures with respect to the IMD if the topography underlying IMD is planarized by the process steps that precede IMD deposition. Although this approach may appear to be obsolete for ULSI circuits, several of the step-height-reduction techniques turn out to be useful for providing a planar surface under Metal-1. Such a condition will ultimately be required for three- and four-level-metal processes. The survey of planarization techniques of dielectric layers begins by considering this approach. In this section one older step-height reduction technique is described, namely *planarization through sacrificial-layer etchback.*

8.1.5 Planarization through Sacrificial-Layer Etchback

Planarization of CVD-interlevel-dielectric films can also be achieved using the *sacrificial-layer etchback technique* (Fig. 8-5). This method gained the most widespread acceptance in two-level-metal processes down to ~1-μm device tech-

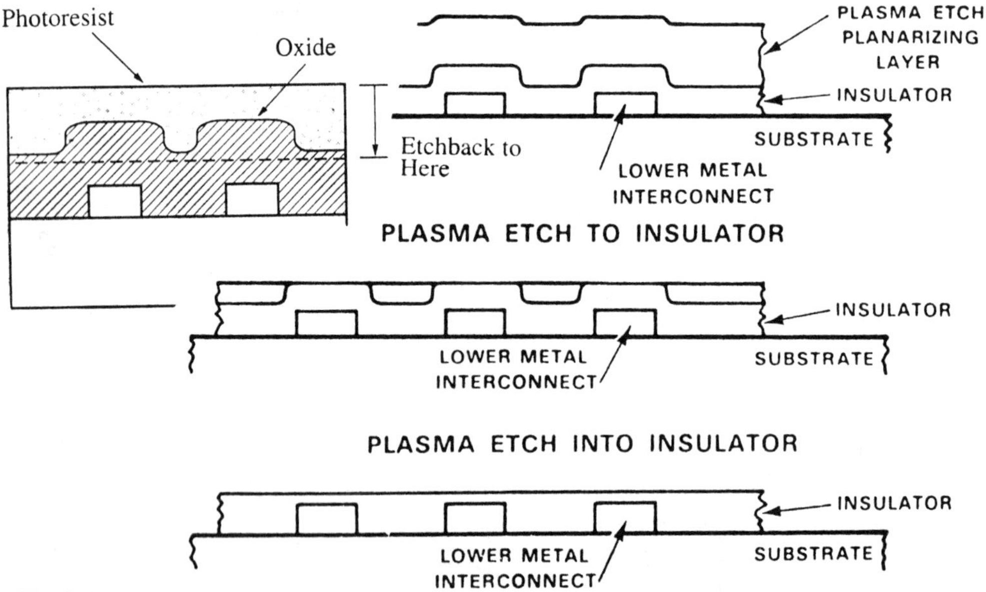

Fig. 8-5 Sequence of steps used in a sacrificial-layer etchback process for planarization.

nologies (and even in some reported three-level-metal bipolar processes). Using this approach, it is possible to achieve a high degree of planarization between steps that are ~2-10-μm apart. With more closely-spaced features, the technique runs into problems. For more widely spaced steps, planarization is less complete. The process is carried out by first depositing the CVD film that will serve as the interlevel dielectric (ILD) or intermetal dielectric (IMD). This layer is then coated with a film that will later be etched off (sacrificed). In most cases, such sacrificial-layers are photoresists, spun-on in liquid form. Upon being baked, these liquids become solid thin films that are thick enough so they provide a film surface that is close to being flat, as shown in Fig. 8-6. As long as the feature-size is smaller than about 10-μm, the degree of planarization can exceed 70% with most photoresists.

In the next step, the sacrificial-layer is first rapidly etched back in a plasma (typically, O_2, or O_2 mixed with CF_4) until the topmost regions of the dielectric layer are just exposed (Fig. 8-5b). The etch chemistry is then modified so the sacrificial-layer material and the dielectric are etched at approximately the same

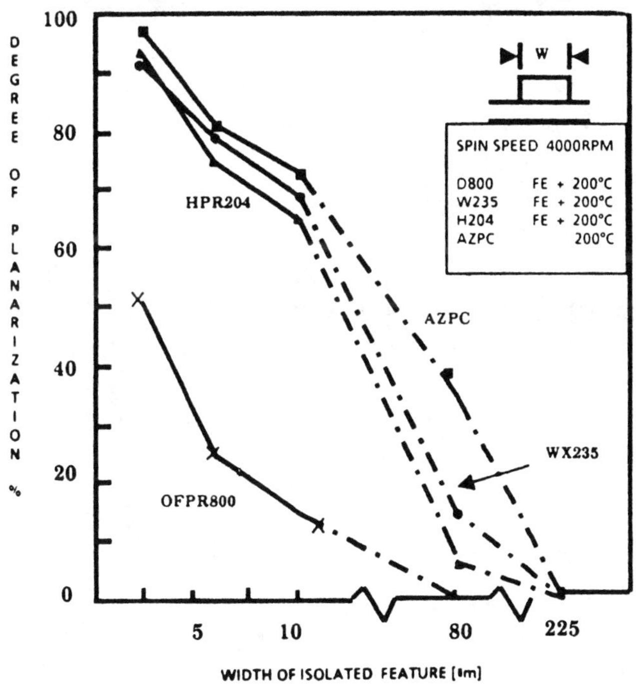

Fig. 8-6 Degree of planarization vs. various feature size for different polymers.[2] (© 1987 IEEE)

rate. The etch is continued under these conditions until all of the sacrificial layer has been etched away. At this point, the surface of the dielectric film is highly planarized since the profile of the sacrificial layer is transferred to the dielectric layer by the etchback procedure. The thickness of the dielectric film over underlying features (such as metal lines), may be thinner than desired after the etchback is completed. In fact, in some processes the etchback step is allowed to proceed until the Metal-1 lines are exposed. In either case, an additional layer of CVD dielectric is generally deposited in order to establish a minimum adequate thickness everywhere on the wafer surface.

8.1.5.1 Sacrificial-Etchback-Process Problems: To achieve the desired degree of planarity and final dielectric-film-thickness, very tight process control is necessary for such parameters as the magnitude of the etch-rate, the etch-rate uniformity across a wafer, end-point-detection, and oxide-thickness-uniformity under the resist layer. The resist etch-rate is sensitive to the cure cycle (especially if a high-temperature bake step [> 150°C] is used to improve surface planarity), and to chamber preconditioning. In batch processes, the etch-rate is also sensitive to the number of wafers present in the chamber. Such tight process-control of a complex process is hard to maintain on a production line.

The process also requires a multitude of steps, which implies a lengthy process, and thus low-throughput. Etchback is a difficult process to successfully implement in a batch-etch mode, resulting in an even lower-throughput. Finally, if the aspect-ratio between adjacent Metal-1 lines with vertical sidewalls exceeds 0.4, voids will form when most CVD-SiO_2 layers are deposited (Fig. 8-7). These voids can trap resist and moisture, which can subsequently be released if the voids are opened during the etchback process. Opened voids will also trap metal during the Metal-2 deposition-step. It will remain in the form of metal stringers following Metal-2 patterning. Thus, this process is not feasible for Metal-1 spacings of less than ~1.25-μm (unless a process is available for deposition of SiO_2 layers at low temperatures without formation of voids in high-aspect ratio spaces [e.g., PECVD TEOS films, O_3:TEOS films, or HDP CVD films]).

8.2 INTRODUCTION TO CHEMICAL MECHANICAL POLISHING (CMP)

As the number of interconnect layers in ICs has increased, the planarization of dielectric and metal layers has become more critical. In general, planarizing techniques such as the thermal flow of oxide layers, sacrificial-resist-etchback, and spin-on-glass, are inadequate for interconnect systems with more than three layers of metal. All of these techniques ultimately provide only a limited degree of smoothing and local planarization (i.e., over distances smaller than 10-μm).

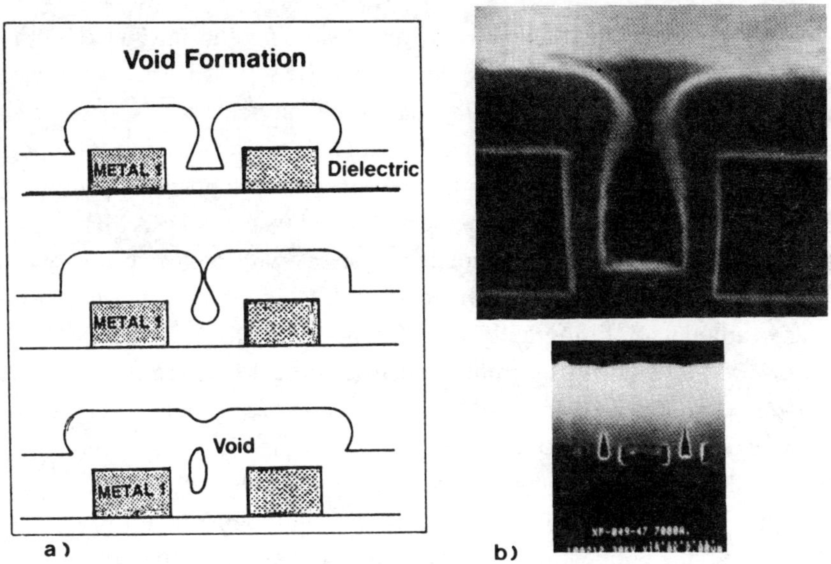

Fig. 8-7 (a) Diagram of void-formation due to cusping of the dielectric. (b) SEM of voids left between 1.5-μm-spaced Metal-1 lines in a 3.0-μm-thick PMD PECVD oxide.

They are not able to provide global planarization.[3] Global planarization of both dielectric and metal layers, however, is possible with chemical-mechanical-polishing (CMP), a process that was pioneered at IBM in the 1980s. In this technique, high elevation features are selectively removed (i.e., material from high elevation features is removed more rapidly than material at lower elevations), resulting in topography with improved planarity (see Fig. 8-10 [in which $h_{final} << h_{initial}$] and Fig. 8-10b). The degree of planarization is expressed by Eqs. 8-1 and 8-3. A step-height reduction of 90–95% is achievable using CMP, yielding values of β in the 0.90–0.95 range. CMP is also capable of planarization of distances up to several millimeters, far in excess of that possible with any other technique thus far developed.[4]

Mechanical polishing, assisted by chemical action, produces such selective material removal. The process is performed by mounting the wafer face down on a carrier. The carrier is then pressed against a moving platen containing a polishing pad. The carrier itself is rotated. An abrasive-containing aqueous slurry is dripped onto the table, and centrifugal force distributes the slurry across the pad. This forms a thin colloidal layer of slurry which saturates the pad (Fig. 8-8). The combination of mechanical effects and chemical reactions results in material removal from the surface of the wafer.

CMP offers three key advantages for fabricating sub-micron ICs. First, for logic devices (such as microprocessors and ASICs), CMP makes it possible to achieve high manufacturing yields and high device speeds with multilevel interconnects having up to eight levels of metal. By planarizing IC topography, CMP avoids the problem of metal thinning over steep topographies in multi-level interconnect structures. Second, memory devices (such as DRAMs) benefit from CMP even though they use interconnect systems with only two levels of metal. For memories, the flat wafer surface after CMP increases the depth-of-focus budget available for lithography (see Chap. 13, Vol. 1, 2nd Ed.). This allows memory-circuit designers to employ smaller critical dimensions, thus reducing chip sizes without decreasing yield. Smaller chip sizes provide a significant cost advantage in the competitive realm of DRAM manufacture. Third, CMP is paradoxically capable of reducing (rather than increasing) defect-density.

Because CMP has become an enabling technology for allowing ICs to be further scaled, the applications of CMP have become pervasive. For example, by 2000, stacked-capacitor DRAMs have upwards of 9 polishing steps, and microprocessors may incorporate as many as 16 polishing steps. As a result, the manufacturability aspects of CMP (namely, the overall cost per CMP step - expressed in terms of $/wafer/polish step) become as important as CMP yield.

In the remainder of this chapter the following topics will be discussed: **a**) the terminology used to characterize CMP processes; **b**) the history of CMP; **c**) the mechanisms involved in CMP of oxides and metals; **d**) CMP equipment and consumables; and **e**) the details of CMP for various materials; and **f**) the challenges facing CMP processing. The latter include: inadequate understanding of the polishing mechanisms, low throughput, inadequate reliability of the polishing tools, degradation of removal-rate with usage-time of a polishing pad,

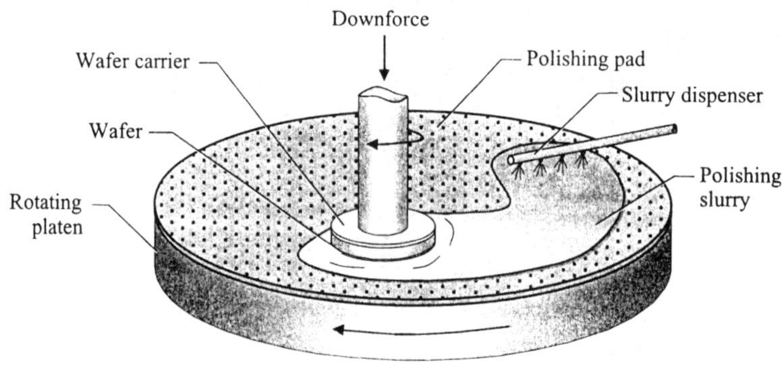

Fig. 8-8 Schematic of a rotary-table chemical-mechanical-polishing system.

non-uniformity of film-thickness-removal across the wafer, high cost of ownership, and high defect density.

8.3 THE TERMINOLOGY USED TO CHARACTERIZE CMP PROCESSES

Before beginning the discussion of the details of the CMP process, it is useful to define the metrics used to characterize the CMP process itself. This discussion is begun by listing the goals desired in a CMP process. They are:

1. The CMP process should have the ability to selectively reduce the step heights of features in the film being polished (i.e., a high degree-of-planarization must be possible);
2. Good planarity across a die (highest spot to lowest spot should be less than 150-nm);
3. Good film thickness uniformity within the wafer (±2–5%, 1σ);
4. High removal rates (100–140-nm/min), for high throughput and low cost-of-ownership;
5. Uniform removal rates from wafer-to-wafer;
6. Good selectivity and/or good endpoint technology to avoid under-polishing or over-polishing;
7. The CMP process should not add excessive defects (e.g., for 0.18-μm CMOS, fewer than 0.125-particles/cm^2 with a size larger than 0.06-μm are allowed at each mask level);
8. The overall cost of a CMP process should be as low as possible (expressed in $/wafer for each CMP process which is conducted).

While it is useful to list these goals in the qualitative manner as given above, the metrics for characterizing a CMP process can also be expressed in quantifiable form. These quantifiable forms are presented next.

8.3.1 CMP Removal-Rate (RR)

The *removal-rate* (RR) describes how rapidly a CMP process removes material from a non-patterned wafer (Fig. 8-9). This quantity is measured by performing a CMP process on a planarized, unpatterned wafer for a specific time, and determining how much material has been polished away. Generally speaking, the RR should be high so that the CMP process can be completed quickly enough to allow a satisfactory throughput. The RR is quantitatively expressed as follows:

$$RR = \Delta T/t \tag{8.3}$$

$$RR = \frac{\Delta T}{t}$$

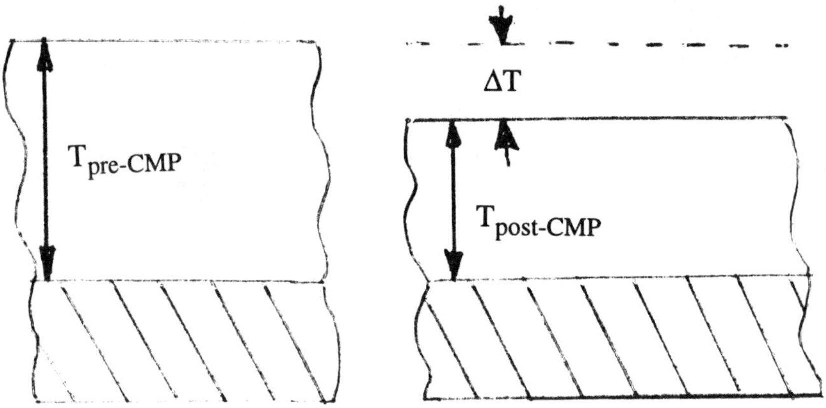

Fig. 8-9 Definition of *removal-rate* (RR).

where ΔT is the thickness of material removed in time t. RR is most frequently expressed in units of Å/min.

8.3.2 The Degree of Planarization (DOP)

In the ideal case, after CMP the surface should be completely planarized (see Fig. 8-1). The quantity that describes how well the CMP process meets this ideal is the *degree of planarization* (%DOP). The %DOP (see Fig. 8-10a) is quantitatively expressed as:

$$\%DOP = [(H_d^{\,i} - H_d^{\,f})/H_d^{\,i}] \times 100\% \tag{8.4}$$

where $H_d^{\,i}$ is the height of feature being polished prior to carrying out the CMP step, and $H_d^{\,f}$ is the height of the feature after the completion of the polishing step. Note that when the value of the %DOP is 100% this indicates that the feature has been completely planarized. Figure 8-11b shows a wafer surface on which the %DOP is 100% (complete global-planarization). Figure 8-11a shows a wafer surface on which the CMP process has not produced complete planarization (%DOP < 100%).

8.3.3 Within-Die Non-Uniformity (WIDNU)

The rate at which material is removed by CMP within a die may not be uniform. Such *within-die non-uniformity* (WIDNU) is quantitatively defined as the *standard deviation of the film thickness within a die*, σ_{WID}, divided by the *average value* of all "*n*" film-thickness measurements within the die, $\bar{x}_{WID}$:

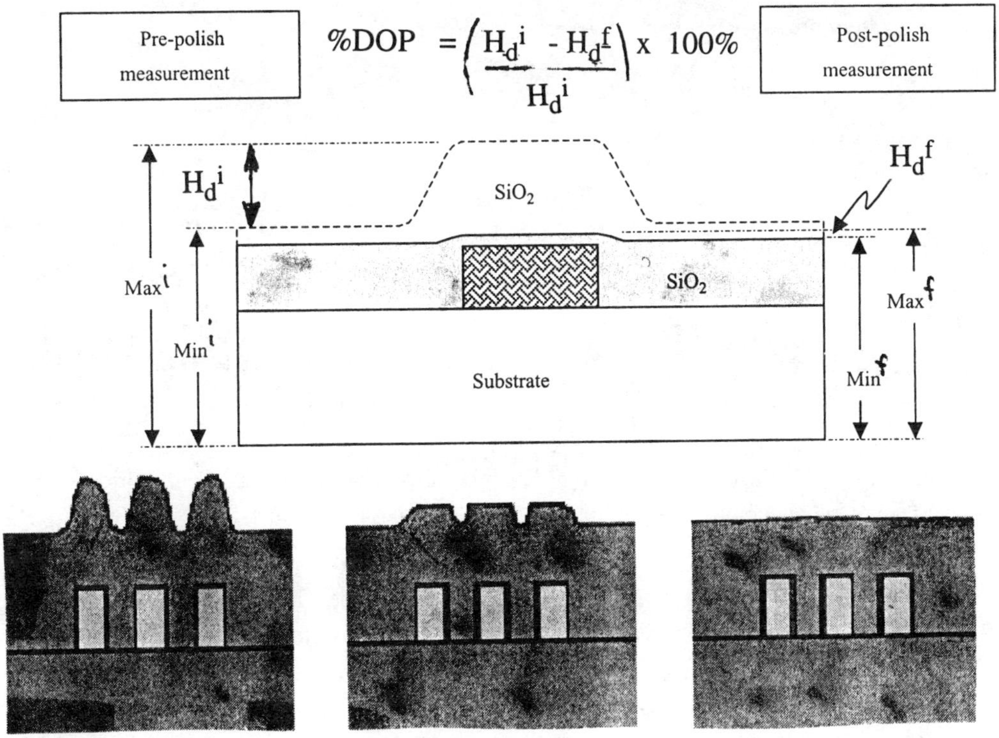

Fig. 8-10 Definition of *degree of planarization (*DOP).

$$\text{WIDNU} = \sigma_{\text{WID}} / \bar{x}_{\text{WID}} \qquad \textbf{(8.5)}$$

Note that σ_{WID} is the standard deviation of measurements within the die of some specific thickness, depending on the type of thickness non-uniformity being specified (e.g., the thickness variation of the oxide in the field between metal lines in an otherwise planarized oxide surface, see Fig. 8-11), and it is expressed quantitatively by:

$$\sigma_{\text{WID}} = \sqrt{[(x_1 - x_{\text{WID}})^2 + \ \ldots + (x_n - x_{\text{WID}})^2]/(n-1)} \qquad \textbf{(8.6)}$$

where $x_1 \ldots x_n$ are particular measurements on a die. Two examples of effects that cause WIDNU are shown in Figs. 8-12 and 8-13, respectively.

In Fig. 8-12c the step heights after CMP have not been reduced by the same amount within a die. Specifically, narrow, isolated-features tend to be planarized to a greater degree than wide, isolated-features, or narrow, densely-spaced-features. This form of WIDNU is termed *pattern-dependent-planarization.* The causes of such pattern-dependent-planarization are described in Sect. 8.5.3.2.

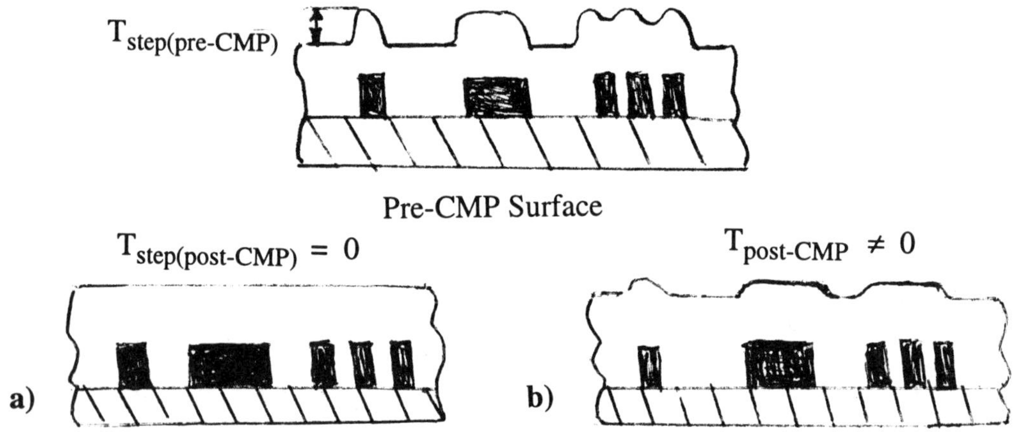

Fig. 8-11 a) Surface on which complete global-planarization has been achieved; b) Surface on which complete global-planarization has not been achieved.

In Fig. 8-13c, although the step heights of the oxide within a die (as shown in Fig. 8-13a) have been completely planarized, some film has been eroded in regions between the steps to a degree that the oxide surface is not planar throughout the die. This causes the distance from the top of the oxide to the planarized substrate surface below the oxide to differ within the die. This form of WIDNU is known as *dishing,* and the effects that cause such dishing are described later.

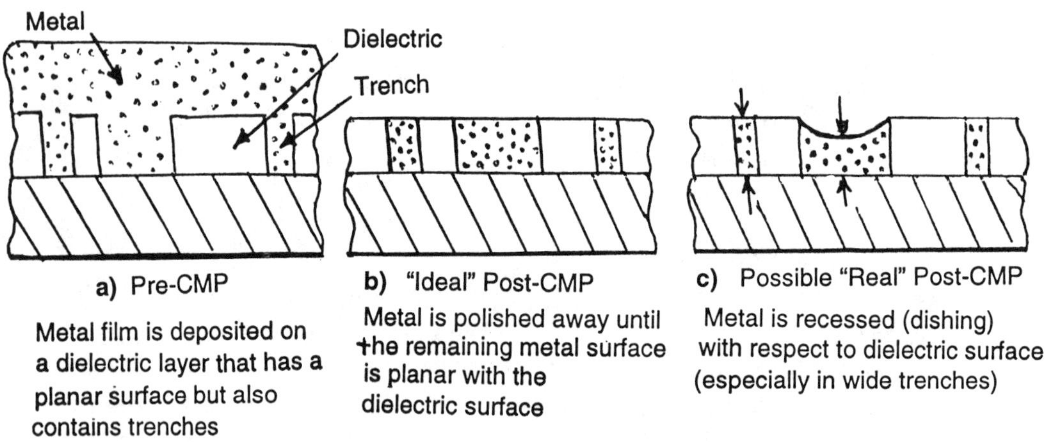

Fig. 8-12 Effect #1 that causes *within-die non-uniformity* (WIDNU).

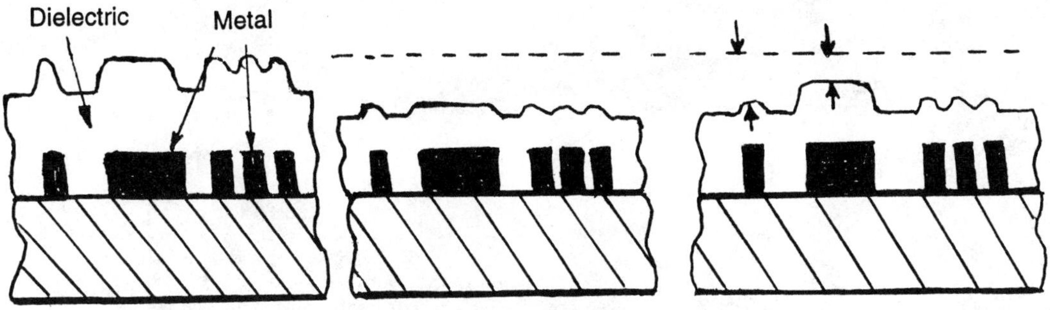

a) Pre-CMP

b) "Ideal" Post-CMP

Uniform planarization within the die, regardless of feature width or spacing between features

c) Possible "Real" Post-CMP

Planarization of all features is not the same (non-uniform planarization within a die)

$$\text{WIWNU} = \sigma_{WIW} / \bar{x}_{WIW}$$

Fig. 8-13 Effect #2 that causes *within-die non-uniformity,* WIDNU (dishing).

8.3.4 Within-Wafer Non-Uniformity (WIWNU)

The *within-wafer non-uniformity* (WIWNU) is defined as the *standard deviation of the film thickness within a wafer*, σ_{WIW}, divided by the *average value* of all "*n*" film thickness measurements on that same wafer, $\bar{x}_{WIW}$.

$$\text{WIWNU} = \sigma_{WIW} / \bar{x}_{WIW} \tag{8.7}$$

Note that σ_{WIW} is the standard deviation of measurements within the wafer of some specific thickness, depending on the type of thickness non-uniformity being specified (e.g., the thickness variation of the distance between the otherwise planarized surface of the oxide and the surface of the planarized substrate below, see Fig. 8-14), and it is expressed quantitatively by:

$$\sigma_{WIW} = \sqrt{[(x_1 - x_{WIW})^2 + \ \dots + (x_n - x_{WIW})^2]/(n - 1)} \tag{8.8}$$

where $x_1 \ \dots \ x_n$ are particular measurements on a wafer. Two examples of WIWNU are shown in Fig, 8-15.

In Fig. 8-15a the surface of the oxide after CMP is shown to be globally planarized across the entire wafer. However, the distance from the surface of the oxide to the planarized surface of the substrate below is non-uniform across the wafer. An example of two effects that can lead to such WIWNU are: **1**) If the thickness of the substrate (wafer) itself is non-uniform, and the CMP process is performed with a completely rigid pad, this will cause WIWNU (see Sect.

$$\bar{x}_{WIW} = (x_1 + x_2 + x_3 + x_4)/4 \qquad WIWNU = \frac{\sigma_{WIW}}{\bar{x}_{WIW}}$$

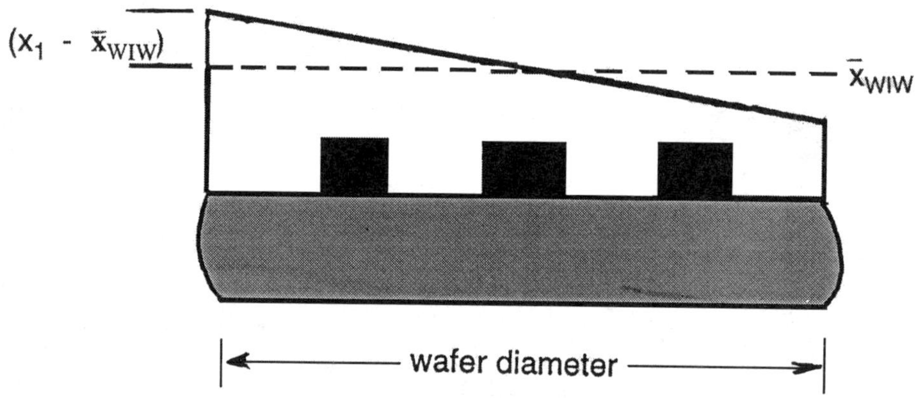

Fig. 8-14 Definition of *within-wafer non-uniformity* (WIWNU).

8.12.2); and **2**) In rotary CMP polishing tools, WIWNU can be caused by the relationship between the angular rotation velocities of the wafer carrier and the polishing pad platen. This topic is discussed in Sect. 8.5.

Another effect that gives rise to a special type of WIWNU is shown in Fig. 8-15b. This figure shows that the wafer surface after CMP has an edge bead of material, caused by a smaller removal rate of material by CMP near the edge of the wafer. This topic is described further in Sect. 8.6.2.1.

8.3.5 Wafer-to-Wafer Non-Uniformity (WTWNU)

The *wafer-to-wafer non-uniformity* (WTWNU)) is defined as the *standard deviation of the film thickness from wafer-to-wafer*, σ_{WTW}, divided by the

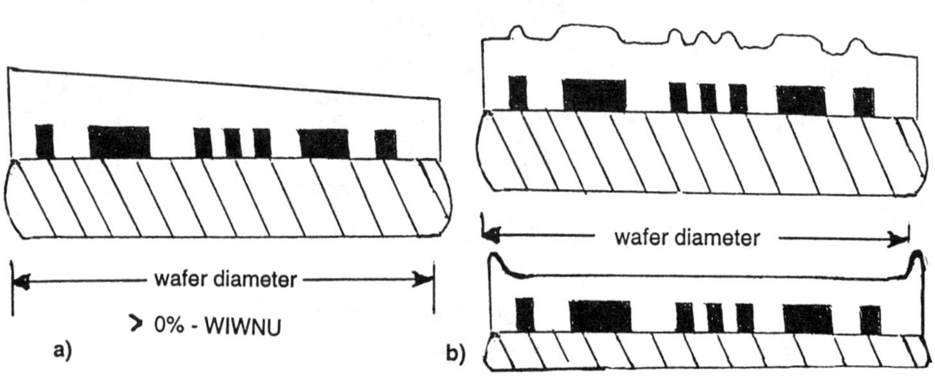

Fig. 8-15 a) Example of one type of WIWNU; b) Example of a second type of WIWNU caused by the *edge-bead-effect*.

average value of all the measurements of the film thickness measurements taken over all *n* wafers, $\bar{\bar{x}}_{WTW}$.

$$WTWNU = \sigma_{WTW} / \bar{\bar{x}}_{WTW} \qquad \textbf{(8.9)}$$

Note that σ_{WTW} is the standard deviation of measurements from wafer-to-wafer for "*n*" wafers of some specific thickness, depending on the type of thickness non-uniformity being specified (e.g., the thickness variation of the distance between the otherwise planarized surface of the oxide and the surface of the planarized substrate below, see Fig. 8-16), and it is expressed quantitatively by:

$$\sigma_{WTW} = \sqrt{[(x_1 - x_{WTW})^2 + \ldots + (x_n - x_{WTW})^2]/(n-1)} \qquad \textbf{(8.10)}$$

where $x_1 \ldots x_n$ are particular measurements on a wafer. An example of WTWNU is shown in Fig. 8-16. Two of the many causes of WTWNU include pad-aging (see Sect. 8.6.7), and slurry inconsistency from batch-to-batch (see Sect 8.6.3).

8.3.6 Efficiency of Planarization (EOP)

When an oxide is deposited over a metal line, enough oxide over the metal line must exist after the CMP step to minimize electrical cross-talk and parasitic capacitance between metal layers. This oxide thickness is determined by the IC designer. The CMP process-engineer must set up the CMP process so that after CMP is completed the oxide minimally meets the designers' requirements. The *efficiency of planarization* of a CMP process has an impact on the process integration choices that must be made to ensure that the minimum oxide thick-

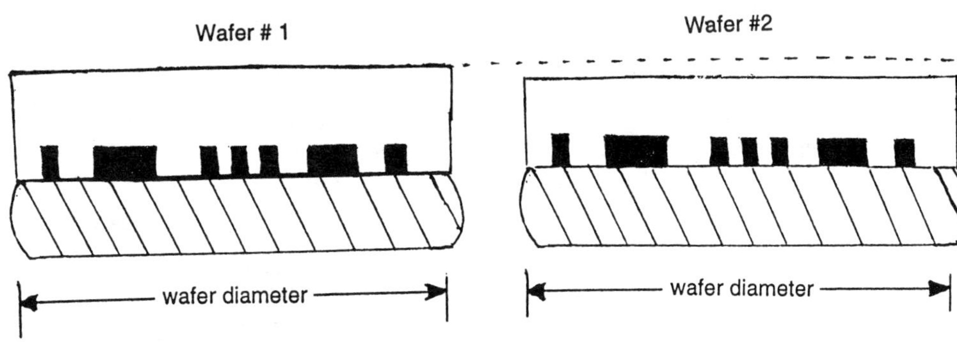

Fig. 8-16 Definition of *wafer-to-wafer non-uniformity* (WTWNU).

ness remains after CMP. The *efficiency of planarization* (%EOP) is expressed as:

$$\%EOP = [(T_r - T_d)/ T_r] \times 100\% \quad \textbf{(8.11a)}$$

where T_r is the dielectric thickness removed by the CMP process on top of a feature (e.g., a metal line) and T_d is the oxide-thickness-loss at the center of a wide space between metal lines (see Fig. 8-17). The ideal case would be a CMP process with an %EOP = 100%. If this occurs, that means that complete planarization is achieved without the erosion of any oxide in the center of the wide space between metal lines. However, the %EOP is generally not 100%, and thus some erosion of the oxide between metal lines usually occurs.

It is the task of the CMP process engineer to determine how much total oxide must be deposited prior to the CMP step to ensure that the oxide on the metal lines is at least as thick as that specified by the IC designer. This thickness quantity is normally determined during tool setup, but it is useful to have an expression which allows the CMP engineer a way to estimate the correct value prior to performing the experimental task itself.

To allow the CMP engineer to calculate this thickness value more easily, the %EOP expression given by Eq. 8-11a is rewritten using different parameters. That is, Eq. 8-11a provides an expression for %EOP that makes it easier to visualize the concept of %EOP, but using it makes it cumbersome to calculate the required oxide deposition-thickness. Therefore, another expression is used instead of Eq. 8.11a for the deposited-oxide-thickness calculation. This alternate relationship is given as Eq. 8-11b (see Figs. 8-17c and 8-17d)

$$\%EOP = [H_{metal} /(T_{total} - T_{cov})] \times 100\% \quad \textbf{(8.11b)}$$

where H_{metal} is the height of the metal line being covered by the oxide, T_{total} is the total amount of the oxide film deposited over the metal line prior to CMP, and T_{cov} is the target amount of oxide that should remain over the metal line (as specified by the IC designer). Equation 8-11b is then rearranged to allow T_{total} to be determined:

$$T_{total} = (H_{metal}/EOP) + T_{cov} \quad \textbf{(8.12)}$$

As an example, if the %EOP is 50%, and H_{metal} = 0.5-μm, and T_{cov} is specified as 0.5-μm, then using Eq. 8-12 (with EOP = %EOP/100% = 50%/100% = 0.5), then the amount of oxide that would have to be deposited prior to the CMP process that would result in 0.5-μm of oxide over the metal lines after CMP would be T_{total} = 1.5-μm.

The least-expensive process involves depositing the smallest thickness of oxide (and this would occur if %EOP was 100%, corresponding to a T_{total} = 1.0-

$$\%EOP = [H_{metal}/(T_{total} - T_{cov})] \times 100\%$$

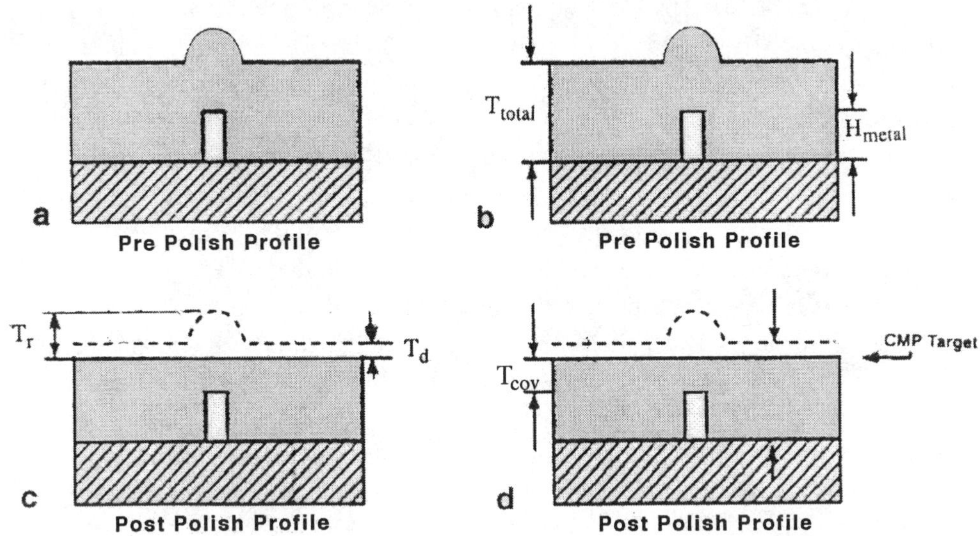

Fig. 8-17 The basic CMP process is one of reducing film-thickness selectivity over the wafer, ending with improved planarity. Here that metric is defined as *efficiency of planarization* (EOP).

μm). Thus, it can be seen that if the %EOP can be increased, the overall IC fabrication cost associated with forming the planarized oxide film over the metal steps will be reduced. For example, if %EOP could be increased to 80% (from 50%), Eq. 8-12 would predict that the $T_{total} = 1.1$-μm.

8.4 THE HISTORY OF CMP

The early history of CMP was chronicled by Fury.[5] He recounts that the process of chemical-mechanical-polishing originated at IBM in 1983. Research and development on CMP was exclusively performed within IBM until 1988. During those five years IBM closely guarded all information about the process. By the late-1980s several factors changed their secretive attitude. First, Intel microprocessors were selected for IBM's PC product line, and IBM decided to share some CMP data to help improve Intel's interconnect technology. Second, at that time the DRAM market was dominated by Japanese manufacturing companies. IBM was wary of having to rely entirely on foreign suppliers for such a critical component of its computer systems. Hence, they elected to help strengthen a domestic DRAM supplier (Micron Technology, Inc.) by teaching them about the CMP process. Information about CMP was included in the process for the 4-MB DRAM that IBM sold to Micron Technology in the mid-1980s. In 1988, IBM agreed to disclose to SEMATECH its process scheme for multi-level interconnects, although they did not give all the details on CMP. They did, however, provide one of their CMP polishing tools (a Si-wafer polishing-tool sold by Westech, and marketed later as the IPEC Model 372, shown in Fig. 8-18). Motorola was introduced to CMP through the Power-PC microprocessor project, a joint venture between IBM, Motorola, and Apple Computer. In 1989 IBM published the first paper which publicly discussed the CMP of oxide films.[6] These events initiated the rapid proliferation of CMP throughout the IC industry.

The spread of CMP was hastened by other events, chiefly the downsizing of IBM from 1990–94. Many engineers with expertise in CMP left IBM and went to work for other IC companies who were seeking to implement the technology. The effort at SEMATECH involved improving the CMP tool-performance and reliability. Their work with Westech tools led to the development of the early popular CMP-tools, sold later as the IPEC 372 and 472 models. (Westech was acquired by IPEC.)

Although early development of CMP was centered in the U.S., worldwide efforts soon got underway. Polishing tools were initially offered by companies that had been building machines for polishing silicon wafers (Speedfam and Strasbaugh, as well as IPEC [Westech]). Overseas suppliers soon joined them (e.g., Ebara and Cybeq from Japan, and Wolters from Europe). Eventually the

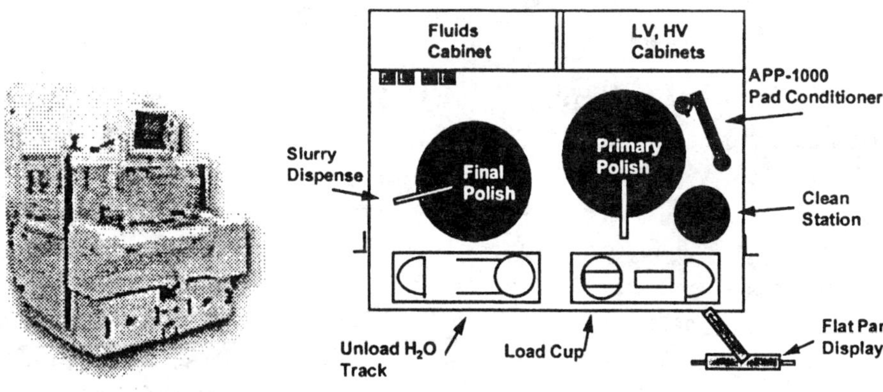

Fig. 8-18 The Speedfam 372 CMP tool.

larger semiconductor equipment vendors entered the market. Applied Materials introduced its tool in 1996, and Lam Research acquired On-Trak Systems, another commercial CMP vendor. The large number of vendors gave rise to some consolidation as well (an example of which was the merger of Speedfam and IPEC in 1999).

CMP was the fastest growing IC process technology of the 1990s, and it was also the most rapidly growing segment of the semiconductor process equipment market in that time frame. From 1995 to 1997 sales tripled, reaching $520 million in 1997. Sales were expected to exceed $1 billion by 2000.[7] The number of CMP polishing steps per year was also estimated to reach 125 million by that date. The first application of CMP was the planarization of dielectric films (mainly CVD oxide), but the number of applications has increased. Currently it is being used to planarize metal films (including tungsten, aluminum, and copper), as well as oxides and low-*k* dielectric films. By the end of the century, CMP had evolved from being a cleanroom-engineers' nightmare, to a critical process technology without which the production of ULSI chips with multilevel-interconnects might not be possible.

8.5 MODELING THE MECHANISMS OF CMP

The mechanisms of CMP are quite complex, and differ for each type of film being polished. Many variables also impact the removal rate, including: film type, pressure (force applied to the polishing pad), velocity, temperature, slurry feed rate, type of polishing motion, slurry particle size, pad hardness, and pad conditioning. The pressing need for production-worthy CMP capabilities forced early researchers to ignore the complexity of the process and go straight to an empirical trial-and-error approach. But, to logically optimize a CMP process, a more fundamental understanding of the mechanical and chemical forces at work is needed. While a complete understanding of CMP mechanisms has not yet been achieved, some insight has been gained. A brief introduction as to what is known about CMP mechanisms is presented here, divided into three categories: **1**) the mechanical aspects of material removal by CMP (Preston's Equation); **2**) the chemical and mechanical effects that cause planarization of oxides by CMP; and **3**) the chemical and mechanical effects that cause planarization of metals by CMP.

8.5.1 The Mechanical Aspects of Silicon Dioxide Removal by CMP: (Preston's Law)

The CMP process for planarizing films in ULSI devices evolved from the technology of chemical-mechanical-polishing used to produce silicon wafers. That is, CMP in the Si wafering process is used to: **1**) remove residual sawing and grinding damage that occurs when single-crystal silicon ingots are sliced into wafers; and **2**) enhance the flatness (planarity) of the bare, starting-wafer surface.

However, the CMP process used in the Si wafering procedure differs significantly in two ways from the oxide-CMP process used to planarize oxide-layers a ULSI IC:

1. The oxide-CMP process is much more exacting than the Si-wafering CMP process. That is, the thickness of material removed in the CMP step of wafering is on the order of 10-μm ± 0.1 μm. However, the thickness of material removed in the oxide-CMP processes is only on the order of 0.5-μm ±0.01 μm (i.e., about 5% of the thickness of the material removed in Si wafer polishing). Thus, the degree of control in oxide-CMP processes is much higher than that needed for Si wafer polishing.

2. When a *bare Si wafer is being polished*, the process involves oxidizing the Si surface with a chemical in the slurry solution, and then the slurry particles abrade away this oxide. That is, the CMP of Si wafers is carried out with a slurry containing fine slurry particles dispersed in a solution that *oxidizes the bare-Si wafer*. This is the *chemical component* of the CMP process. The abrasion of this oxide by the slurry-particles is the *mechanical component* of the process. The abrasive process exposes the Si beneath the oxide to the chemical in the slurry, which re-oxidizes the exposed Si, and so on. In fact, the abrasion of the oxide performs two functions: **a**) The polishing process is continued long enough so that the damaged region near the Si surface from sawing and grinding is eventually consumed by the oxidation and abrasion sequence; and **b**) Since the abrasion occurs more rapidly on "high" spots on the wafer surface, these "high" spots are eroded more rapidly than the "low" areas. Thus, the wafer-surface becomes flatter (more planar).

However, when an oxide is being polished in a ULSI IC (a process that is carried out in such applications as the CMP of the trench-oxide in shallow-trench-isolation structures (STI), and the CMP-step used to planarize the inter-level dielectric [ILD] and inter-metal dielectric [IMD] layers), the material to be polished *is already an oxide*. Hence, the mechanism employed to polish the Si wafer (i.e., chemically oxidizing the Si surface, and mechanically abrading it away), is not an operative mechanism in oxide-CMP. A discussion of what are believed to be the phenomena that govern the removal of oxide films in oxide-CMP follows.

In 1997 Tomozawa stated that, like optical-glass polishing, much of the oxide CMP process was still an art rather than a technology. To qualify as a technology, the basic mechanisms of a process must be understood, allowing it to be logically optimized. However. at the time Tomozawa made his comment, the basic mechanisms of oxide CMP were not yet well understood.[8] The descriptions provided in this section briefly summarize what is known about the phenomena associated with oxide-CMP. Since silicon oxides used in IC fabrication are a form of glass, oxide-CMP is a specialized form of the glass polishing technology used to produce optical lenses. Thus, some aspects of oxide-CMP can be understood from the long history of glass polishing (which generally have been derived from trial-and-error, rather than from a fundamental understanding of the process). Optical-glass and SiO_2-films-on-Si-wafers are polished with a slurry consisting of abrasive silica particles (whose hardness is about equal to that of the oxide) suspended in an aqueous alkaline solution (typically KOH or NH_4OH, diluted in DI-water).

An empirical relationship that models the optical-glass-polishing removal-rate, RR, of flat-glass regions (defined as the thickness of glass removed per unit time), was developed by Preston, and is known as *Preston's equation* or *Preston's Law*.[9] This law states that the RR is proportional to the applied-pressure P and linear-velocity v at which a polishing pad moves relative to the optical-glass specimen, or:

$$RR = K_p P v \tag{8.13}$$

where K_p (*Preston's constant*) is a proportionality constant that is independent of pressure and velocity. The same law appears to apply to the rate of removal of oxide by CMP. (Typical values of P and v in CMP processes are 7-psi, and 30-40-ft/min, respectively.)

Note that Preston's equation describes the relationship between the RR of oxide and *mechanical effects*. (That is, mechanical effects are those caused the application of physical forces on a substance, namely in Preston's equation, the pressure exerted on the oxide and the relative velocity between the glass and the polishing-pad.) The Preston equation predicts that oxide-RR is linearly proportional to P and v, but gives no insight into other phenomena (i.e., such physical and chemical factors as the glass-temperature, the chemical nature of the polishing medium, the hardness of the glass, etc). These effects are all put into the Preston constant, K_p, which contains the influence of such phenomena as glass hardness, Young's modulus, the polishing slurry, and the composition and hardness of the polishing pads. Despite the fact that Preston's equation sheds no light *per se*, on how these other phenomena impact RR, it is still a model that provides useful predictability concerning oxide-CMP. Hence, it is informative to discuss some further issues related to this equation.

The most important criticism of the Preston equation is that it was formulated empirically rather from first principles (i.e., it was not mathematically derived from a physical model of the glass-polishing process). Several authors have proposed equations from first principles that predict the dependency of oxide RR on pressure and velocity, including Brown and Cook (1984),[10] Liu *et al.* (1995),[11] Runnels and Eyman (1994),[12] Tseng (1997),[13] and Zhao and Shi (1999).[14] A critical review of the evolution of these models is found in Shinn *et al.*[15]

Many of the models (and supporting experimental evidence) indicate that the dependence of oxide-RR in CMP on pressure and velocity is not a simple linear relationship. Recent work has proposed multiple variations of Preston's equation of the form:

$$RR = K_p \, v^{\beta} \, P^{\alpha} \tag{8.14}$$

Where α and β are adjustable parameters dependent on the system to be polished. The values of these coefficients must be experimentally determined.

In the model of oxide RR of Zhao and Shi (which is the first model that also includes the effect of pad and wafer hardness), the RR demonstrates a sub-linear dependence on pressure given by:

$$RR = K_p \, v \, P^{2/3} \tag{8.15}$$

Figure 8-19 shows experimental data demonstrating the dependence of RR for oxide films.[16] Zhao and Shi also found that the pressure dependence of RR significantly influences planarization efficiency and pattern-density dependence. For instance, not all slurry particles in contact with a wafer necessarily remove material efficiently. A particle rolling against the surface removes little material, yet a particle held firmly by the pad slides against the wafer surface, changing the surface structurally and chemically. Sliding and removing requires a minimum local pressure, giving a revised polishing rate of:

$$RR = K_p \, v \, (P^{2/3} - P_{th}^{2/3}) \quad \text{when } (P \geq P_{th}) \tag{8.16}$$

and

$$RR = 0 \quad \text{when } (P \leq P_{th}) \tag{8.17}$$

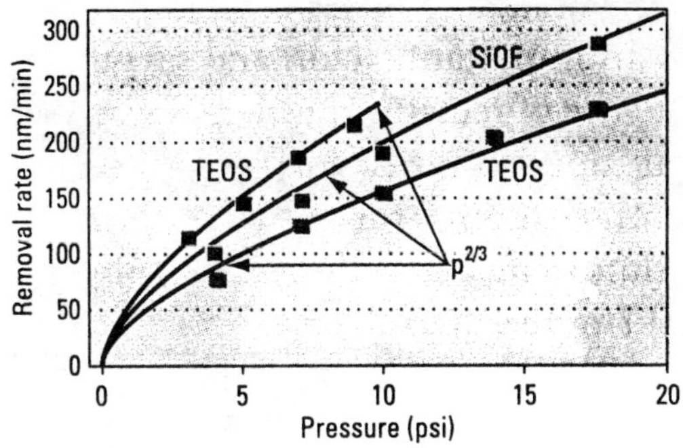

Fig. 8-19 Experimental data correlates well with the $P^{2/3}$ dependence of removal-rate on applied pressure.[16]

where P_{th} is the threshold pressure. In addition, they found that planarization efficiency also improves with P_{th} (Fig. 8-20), requiring less material removal to achieve a given planarization result.[16]

8.5.1.1 Model that Treats WIWNU Effects in Rotary-CMP Tools Based on Preston's Law: One of the limitations of CMP is that within-wafer-non-uniformities (WIWNU) occur, as defined in Sect. 8.3.4. The Preston equation states that the oxide-RR is linearly proportional to the *linear velocity* of the pad with respect to the oxide-surface being polished. Thus, one way to reduce such WIWNU is to ensure that the same linear velocity is experienced by all areas of the wafer. In rotary-CMP tools, the wafer carrier rotates, as does the platen that holds the polishing pad. If the pad did not rotate, and only the carrier rotated about its axis, the oxide-RR at center of the wafer would drop to zero, since there would be no relative motion between the wafer and the pad at that point. In addition, the RR would increase radially from the center to the edge of the wafer.

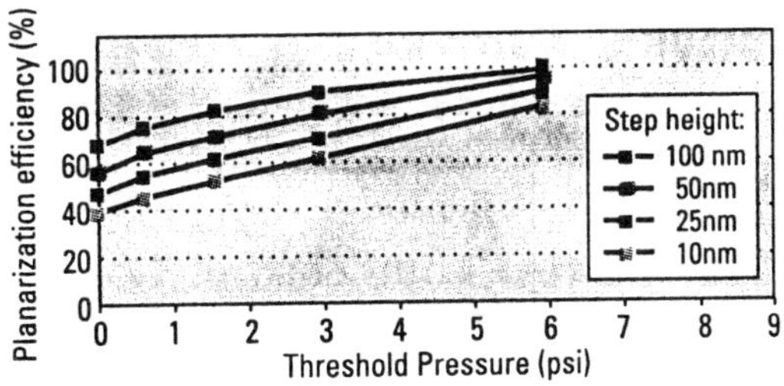

Fig. 8-20 Increasing the threshold pressure improves planarization efficiency for various final step heights (500-nm initial step height).[16]

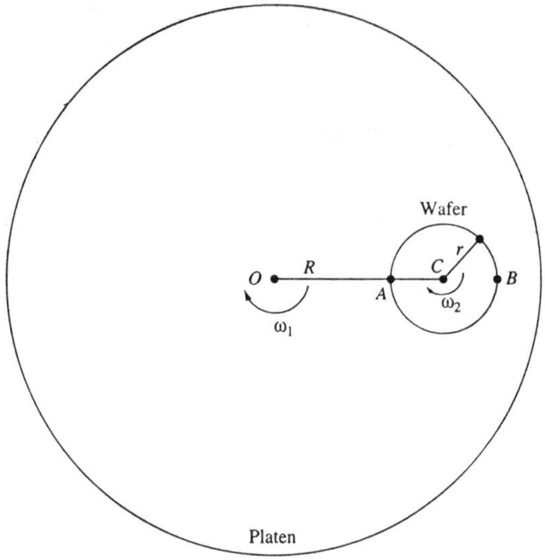

Fig. 8-21 Plan view of a single-carrier single-platen rotary-polisher, showing relative rotary motions of both the carrier and the platen.

Now assume that the platen and the wafer carrier both rotate. In this case (referring to Fig. 8-21), it can be noted that the centers-of-rotation of the wafer-carrier and the platen are offset from one another. If the wafer-carrier and the platen have the same rotational frequency, and are rotating in the same direction, point B of the outside edge of the wafer in that figure experiences zero relative velocity, and the point A on the inside edge of the wafer experiences the greatest relative speed of the surface of the wafer to the pad. Thus, the instantaneous removal rate at the edge of the wafer is greatest when the edge is directly between the two centers (point A), and zero when in the extreme position (point B). However, since all points on the edge of the wafer pass through points A and B at a constant frequency, the time-average RR at the edges will be the same as at the center of the wafer. Thus, it is the average-radius from the center of the platen, r_{av}, that defines the distance the wafer travels and the average speed of travel, S_t, for purposes of the total film removed and RR calculations:

$$S_t = 2\pi r_{av} f\,(\text{rpm}) \quad \textbf{(8.18)}$$

In practice, within-wafer-non-uniformity of RR is observed in rotary-CMP tools (e.g., WIWNU). One form of such WIWNU is a spoke-like pattern due to vibration harmonics. Driving the carrier in rotational synchronization with the platen usually causes the worst vibrations. Hence, the wafer-carrier is usually driven at a constant difference in frequency to the platen. Since the correct wafer-carrier rotational-frequency for perfect circular symmetry is not known, it must be experimentally determined.

8.5.2 Models Which Describe Factors that Impact Oxide-RR in CMP Through the Preston Constant

There are many factors that impact the oxide-RR in CMP besides the applied pressure and velocity. These are expressed in Preston's Law by changes in the Preston constant. Here two models are given that have been proposed to explain the dependence of oxide-RR on other phenomena besides pressure and velocity.

8.5.2.1 Impact on the Oxide-RR from the Electrochemical Interactions Between the Slurry Particles and the Oxide Surface: A model was proposed by Cook in which he postulated that a mechanism for oxide-removal by CMP involved contact-mediated reactions with the solvent (water).[17] That is, local events in CMP may be considered to be the approach to contact of two opposing asperities: **1**) pad plus adhering particle; and **2**) the oxide surface. Asperity-contact causes reactions in the vicinity of the contact, and these result in mass removal from the oxide. The depth and width of the swath removed is determined by the size of the slurry particle and the chemical-reaction-rate, which is largely controlled by the interaction force between the contacting surfaces. The length of the swath gives an indication of the duration of the asperity contact.

For the case where the slurry particle and the surface are of the same material (as occurs in oxide-CMP with a silica-based slurry), the forces due the Zeta-potential act to repel the silica particles from the oxide surface, while the Van-der-Waals forces cause the oxide surface to attract the slurry particles (see Sect. 8.7.1 for a discussion of the Zeta-potential and the Van-der-Waals forces). The curve that graphs the force-versus-distance for such a case of opposing forces between the particles and the wafer-surface exhibits a maximum repulsion at some characteristic separation distance. This may be perceived as an activation-energy-barrier for closest approach. The particle must thus have a kinetic-energy greater than this activation-energy-barrier in order for direct contact to occur between the slurry particle and the oxide-surface. A reduction in this activation-energy-barrier should result in a higher probability of contact, and therefore cause an increase in oxide-RR.

Based on this model, Cook suggested that an increase in RR could be achieved by using a slurry whose particles were different than that of the oxide (e.g., use CeO_2 slurry particles instead of silica), and in which the Zeta-potential had a value that caused the slurry particle and the oxide to attract one another. This would strongly increase the probability of contact between the slurry particle and the oxide-surface and thus increase RR.

8.5.2.2 Dependence of Oxide-RR in CMP on Dielectric Hardness: Dielectric materials vary in hardness from the very-hard and inelastic silicon-nitride, to softer dielectrics such as doped-oxides and spin-on-dielectrics (SODs). The RR of dielectrics has been found to increase with a decrease in their hardness. Figure 8-22 shows that as the hardness of fluorinated-silicate-glass (FSG) is decreased, the RR increases.[18] (The hardness of the FSG in this study was decreased by increasing the concentration of fluorine that is incorporated in the oxide.) In another work, the RR of PSG and BPSG was studied as a function of the P-concentration in a P-doped oxide-film, and as a function of the B- and P-doping in a BPSG oxide film (Fig. 8-23).[19] Higher doping is seen to cause the RR to increase. This is correlated with oxide hardness, since the incorporation of P_2O_3 and B_2O_5 into an SiO_2-network causes a softening of these oxide-films.

8.5.3 Models Which Describe Chemical-Mechanical Phenomena that Give Rise to Planarization Phenomena Associated With CMP of Oxides

Various models have been proposed to explain how chemical phenomena act in concert with the mechanical forces affecting oxide-RR (as expressed by Preston's Law). These models also provide insight as to why oxide-CMP can selectively remove the "up-regions" of an oxide film on wafers faster than the "down-

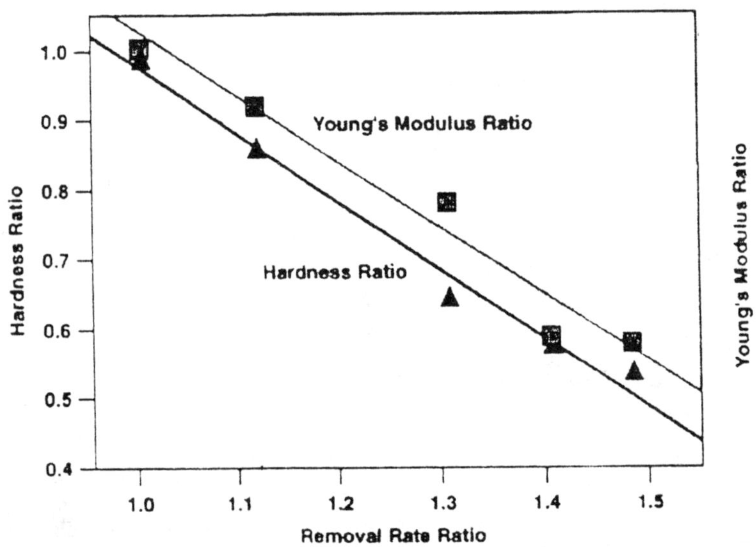

Fig. 8-22 A linear relationship exists between the CMP-removal-rate and the hardness and elastic modulus of F-SiO_2. Removal-rate, hardness, and elastic modulus are all normalized with respect to those parameters of undoped TEOS-oxides.[18]

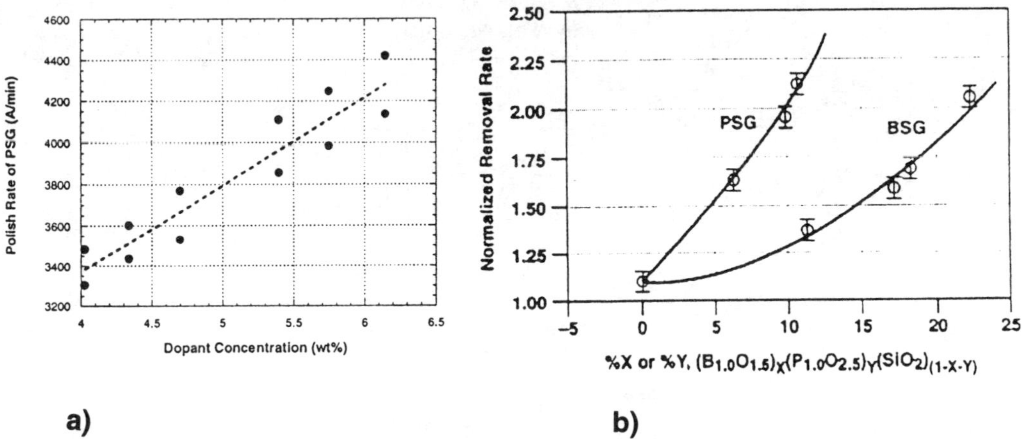

Fig. 8-23 a) Removal-rate of PSG as a function of phosphorus-concentration: pressure 4-psi, table speed: 40-rpm, carrier speed: 643-rpm; b) Removal rates of P-doped oxides and B-doped oxides as a function of P- and B-doping-concentration in the oxide.[19]

regions" (thereby allowing planarization of the film to occur). Some ideas from these models are discussed in this section.

Izumitani proposed that optical-glass-polishing "proceeds by the formation of a hydrated-layer by means of a chemical-reaction between the glass-surface and water and then the removal of this hydrated-layer by the abrasive particles."[20] This proposed mechanism is consistent with the observation that the polishing-rate is extremely low without the presence of water to cause the hydration. (Polishing with a slurry of silica-particles in a non-aqueous solution [e.g., in oil] exhibits substantially lower polishing rates.) The hydrated-layer that thus forms on the oxide-surface generally has a much lower hardness and mechanical strength than the SiO_2 beneath it. Therefore, plastic-deformation results in mechanical removal of the hydrated-layer (by a plowing effect) when it is rubbed by the abrasive particles.

The same type of mechanism is expected for oxide-CMP. Removal of the oxide takes place by a combination of mechanical and chemical action, using an abrasive slurry dispersed in alkaline solution (Fig. 8-24). Since silica-particles are used in the slurry of oxide CMP, they have about the same hardness as the SiO_2. Thus, the unreacted SiO_2 is not appreciably removed (nor damaged) by the plowing effect when it is rubbed by the silica particles in the slurry. However, the silica-slurry-particles still cause the oxide to plastically deform. This deformation

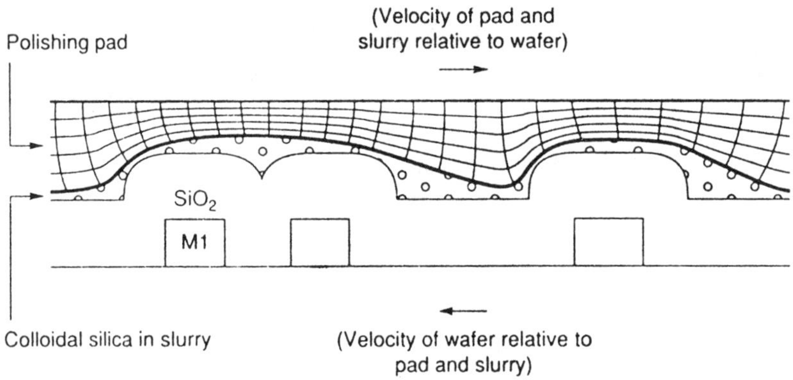

Fig. 8-24 Depiction of polishing-pad compression and localized pressures applied to topographic features. Pressures are greatest at the top-leading-edges of steps, and are lowest in the wakes of the steps.

is likely to be assisted by frictional heating. It has been determined that oxide readily undergoes hydration (water entry) if it is plastically deformed by a diamond indenter. When entered by water, the hardness and mechanical strength of the oxide is generally decreased. Thus, during CMP, it seems that hydration of the oxide film surface is caused by the plastic deformation of the oxide by the abrasive particles. The degree of deformation is increased as the pressure of these particles on the surface is increased, and by frictional heating. The softer hydrated-layer is then removed by the plowing action of the abrasive particles, which increases as the linear-velocity of the polishing pad is increased. These conclusions are substantiated by Preston's Law. The properties of the pad also appear to play an important role in CMP. It is probable that the polishing pad has to be soft enough to grip the abrasive particles, but hard enough to transmit the pressure to the oxide. It is generally agreed that some sort of chemical mechanism (such as SiO_2-dissolution) is a cofactor in causing oxide-CMP. Various chemical processes have been proposed.[21] The larger polishing rate in higher-pH slurry is often cited as evidence for silica-dissolution.

8.5.3.1 Model of the Mechanism that Produces Planarization of Steps in Oxide CMP: While the above model gives some insight into how oxide is polished with a silica-based slurry, it does not address the issue of how planarization of the oxide by such polishing is achieved. To obtain planarization of a film with topography, the surfaces of elevated features ("up-areas") must be eroded faster than those at lower elevations ("down-areas"). The selective removal of material in "up-areas" is empirically observed to depend on many factors, including the feature dimensions, the structure (and surface condition) of the polishing pad, and the height of the step being removed

It is important to note that the case being discussed here involves the CMP of an oxide layer that proceeds only until the oxide surface is planarized, and that the surface of the wafer after the CMP step is *homogeneous* (i.e., only oxide is exposed over the entire wafer surface at the conclusion of the CMP step, as shown in Fig. 8-25a). This differs from the case when a film is polished until it is removed from the surface in some regions, but is allowed to remain in recessed regions of the wafer surface after the CMP step. In that case (which is described in a later paragraph), the wafer surface is *heterogeneous*, insofar as some exposed areas of the wafer contain the polished film residing in recesses, and others contain material that underlaid the film that was polished (Fig. 8-25b). Note that in the second case, the CMP process is essentially a *blanket-removal-process of a planar film* because the film being removed is already roughly planar when the CMP process is started.

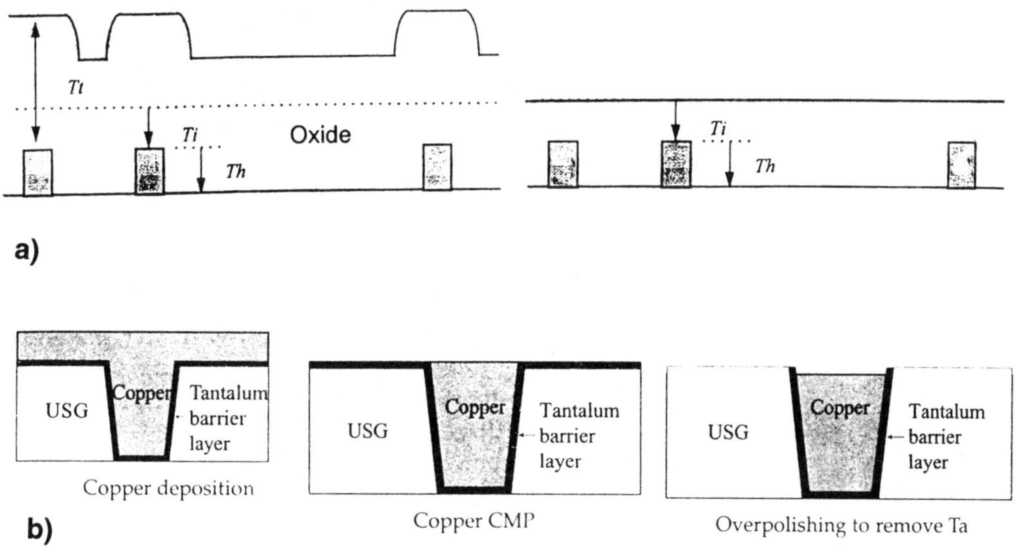

Fig. 8-25 The process of removing material by CMP falls into two categories: a) Removing homogeneous material and stopping while the removal of the surface film is incomplete (e.g., planarization of an ILD oxide); b) Removing deposited material, but continuing until all material is entirely removed on some regions of the wafer surface, while in other regions it remains recessed below the surface (e.g., Cu-damascene feature formation by CMP).

Selective removal of material that protrudes above the surrounding topography (as shown for the case of Fig. 8-25a) is thought to arise in part from the compliant nature of the polishing pads.[22] Figure 8-24 depicts the localized compression occurring within a pad as features on the wafer surface are driven across it. The pad applies pressure to the wafer surface through the slurry, with the largest pressures being exerted where the pad is most compressed. Higher pressure is applied to elevated regions ("up-areas" - Fig. 8-26) and the leading edges of the features that are moving against the compressed pad, compared to lower-lying regions ("down-areas") and trailing edges.[23] The difference in local applied pressure causes the material removal rate to be fastest at the upper corner of the leading edge of a step. Figure 8-27 shows how a feature would be planarized as the leading edge is eroded away as a function of time during a CMP process. (In the next section it will demonstrated how this phenomenon - in concert with wafer rotation during polishing - causes smaller features to be rounded-off and polished-aster than wider features.)

As the steps erode, the "down-areas" are exposed to increasing pressure, and the removal-rate there increases (approaching that of the "up-areas"). As this happens, the rate of step-height-reduction decreases toward zero. In the latter stages of the polishing process the selectivity is so low that small residual steps remain, and in practice cannot be fully removed by continued polishing.

In the second case described above (Fig. 8-25b), the material of the blanket-deposited layer is removed non-selectively by CMP. This non-selective removal continues until all of the film is polished away from the areas of the substrate, and only film material remains in the trenches. However, if the CMP process is continued beyond that point, material within the trenches will continue to be removed. This removal will be greater in wide trench areas than in narrower trenches. The result of such removal from wide trenches gives rise to a CMP problem called *dishing*. The phenomenon of dishing and methods developed to mitigate it are also discussed in other parts of the book. In Sect. 8.12.1, the general aspects of dishing are covered, and the impact of dishing effects on the fabrication shallow-trench-isolation structures is found in Chap. 9. Chapters 15 and 16 provide additional information on the impact of dishing on Cu-CMP and on the CMP processes used to polish inlaid metal lines in damascene processes.

8.5.3.2 Pattern-Dependence of the Planarization Rate in CMP Processes: The degree of planarization (DOP) of features for the CMP processes depicted by Fig. 8-17a has been found to depend on the feature sizes and the density of features within a region of a die. This DOP dependency on feature size and pattern density is one cause of WIDNU. The reasons for such WIDNU are described here.

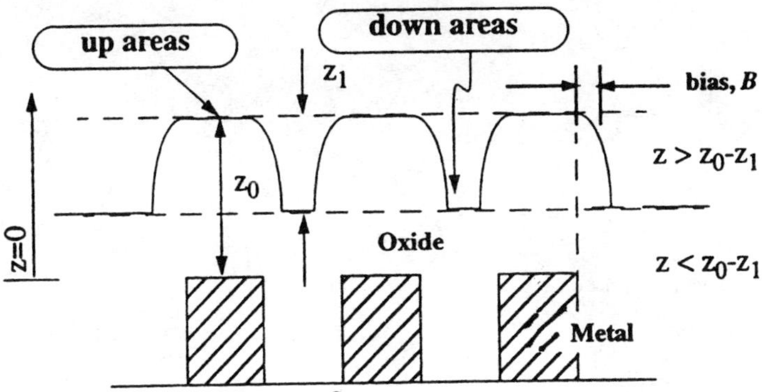

Fig. 8-26 Cross section of conformal dielectric on patterned metal, showing the "up-regions" and "down-regions" of the dielectric film.

The first question to answer is why narrow, isolated-features tend to planarize faster than wide isolated features. From Fig. 8-27 we see that the leading edge of a step will be removed as a function of time in an oxide CMP process. However, in a process in which the wafer carrier is rotating, all top edges of the feature being polished become the "leading edge" as the feature is rotated by the carrier. As a result, the entire feature is planarized according to the sequence shown in Fig. 8-28a. If the feature is wide, the leading edges will be polished away, as shown in Fig. 8-28b, but the material in the center of the wide feature will not be polished as quickly as material at the feature edges. As a result, wider, isolated-features will planarize more slowly than narrow, isolated features. This will cause such WIDNU as shown in Fig. 8-13.

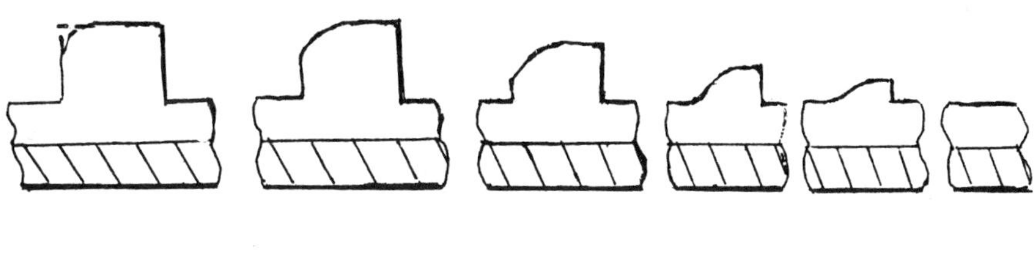

Fig. 8-27 A depiction of how the leading edge of a feature of an oxide step (i.e., the edge that is pressed against the pad by the combined movements of the carrier and platen), is eroded by CMP.

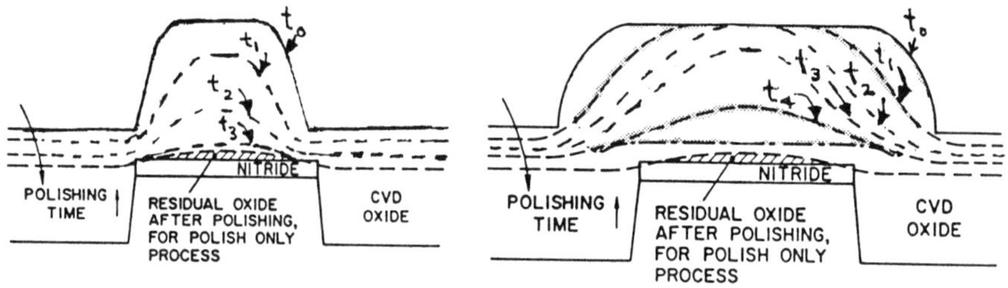

Fig. 8-28 Schematic cross-section of a narrow and wide feature of oxide being polished by CMP, as a function of time. Since the leading edges are polished most rapidly (and the feature continually rotates, so that all edges become the leading edge for the same amount of time), the narrow feature is planarized more rapidly than the wide feature.

The next question is: Why are densely-spaced narrow features planarized more slowly than isolated, narrow features? As first discussed by Burke,[24] the time required to planarize an "up-area" structure as shown in Fig. 8-26 (or to reduce its height to some fraction of its initial value) is dependent on the "down-area" adjacent to it. Specifically, the time to planarize a step depends on the width of the "up-area," the width of the "down-areas" between "up-areas," and the ability of the polishing-pad to rigidly span that distance between "up-areas." The stiffer the pad, the more rigid the span - and the slower the rate of material removal from "down-areas." This gives good planarization. The softer the pad, the less able it is to bridge larger distances. Thus, more material is removed by softer pads in "down-areas" between adjacent "up-areas." This gives poorer planarization. When features to be polished are separated by distances greater than the leveling-distance of the pad, there is no selectivity between "up-areas" and "down-areas," and planarization does not occur. In practice, some selectivity in RR occurs between "up-" and "down-areas," but this selectivity is not complete. Ideally, the leveling-distance of a pad should be large enough to planarize the widest "down-areas" on a chip. On the other hand, the greater the leveling-distance, the more a pad will respond to wafer-scale non-flatness parameters, such as TLV and bow. This results in non-uniformity of the polished film thickness across the wafer (a subject discussed in more detail in Sect. 8.12.2).

In summary, planarization will occur most slowly for step heights in an oxide over the largest or densest interconnect structures adjacent to wide unpatterned field areas, scribe lines, and wafer edges. It will occur most rapidly for step heights in the oxide over smaller arrays and narrow, isolated features. It is also necessary to specify an acceptable degree of planarization, because in the latter stages of the CMP process, material is removed nearly as fast from "down-areas" as from "up areas." Step heights no longer decrease at appreciable rates.

Like metal-CMP processes, oxide-CMP in a production environment is often a multi-step process. When in-line endpoint-detection is employed (as will be discussed in Sect. 8.6.9), two or more polishing steps may be needed to obtain the correct oxide thickness. In addition, a separate final-polish (or buff) procedure is frequently used to provide light polishing and to remove slurry particles and small wafer-defects. A small amount of oxide can also be buffed away with this buffing procedure (typically 30–40-nm), which aids in removing defects (especially residual slurry particles and the small scratches created by the primary polish). Use of a buff procedure is one of the reasons why CMP can actually *reduce* defect-densities (including eliminating some types of defects created in other processes), rather than increase them.[25]

8.5.4 Metal-CMP Mechanisms

The mechanisms involved in the CMP of metals which cause planarization of the metal film surface, follow a relatively simple three-step sequence proposed by Kaufman.[26] This three-step sequence is as follows (Fig. 8-29):

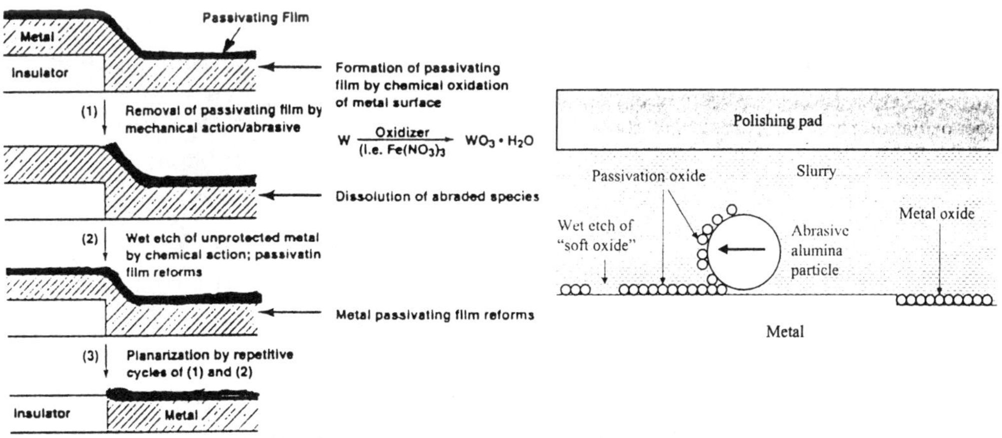

Fig. 8-29 Schematic representation of the model of CMP of metal films.[26]

1. A passivation layer (e.g., an oxide of the metal) is chemically formed on both the "up-areas" and "down-areas" of the metal film, due to the oxidation potential of the solution. (For effective polishing, it must be possible to both oxidize the metal and closely match the hardness of the slurry abrasive to the hardness of the material being removed.)

2. This freshly formed oxide film is softer than the metal itself and is mechanically removed from the "up-areas" by the pad/abrasive. (While the "down-areas" are not contacted due to the partial flexibility of the pad.)

3. The corrosive nature of the slurry re-grows the passive film on the "up-ares," while the corrosive elements are blocked from the metal in the "down-areas" by the passivation film.

The "up-areas" are continually corroded and polished until the pad flexibility allows it to contact the "down-areas." At that point, the planarization process again depends on the hardness of the pad and the down-force applied to the wafer. Thus, the corrosion properties of the slurry are chosen to convert the metal surface into a film that is soft enough to be abraded by the pad-plus-abrasive-material, yet hard and impermeable enough to prevent outright etching of the metal.

The goal is to find slurries with just the right combination of chemical and abrasive aggressiveness to produce high removal-rates, good-planarity, local-uniformity, and high-selectivity. Although the metal-CMP-model is simple to understand, successfully carrying out the above process for specific metals can be challenging. The slurry's chemical-composition tends to play a more important role in metal-CMP than in dielectric-CMP. As a result, it is possible to achieve fairly high metal-to-oxide selectivity (e.g., 120:1 for tungsten to silicon dioxide). However, the slurry solution may also etch the metal-oxide that is formed. A careful balance between etching and oxidation must therefore be maintained. If etching dominates, then the metal will etch isotropically and planarization will not be obtained. The metal-deposition-step may also have to be modified to prevent metal deposition near the wafer edge. Otherwise, metal residue may re-main on the wafer's beveled edge after CMP, causing yield and reliability problems.

Pourbaix diagrams are often used to identify which pH-range is most ideal for CMP of metals.[27] *Pourbaix diagrams* are graphical representations of the domains of stability for different oxidation states of metals in aqueous solutions with respect to pH and electrochemical potential (i.e., contour maps in a solution potential versus pH). For example, the Pourbaix diagram for W (shown in Fig. 8-30) indicates that in order to form an un-dissolved protective WO_3 layer over the W, the pH of the solution should be below 4, and have a slightly positive oxida-

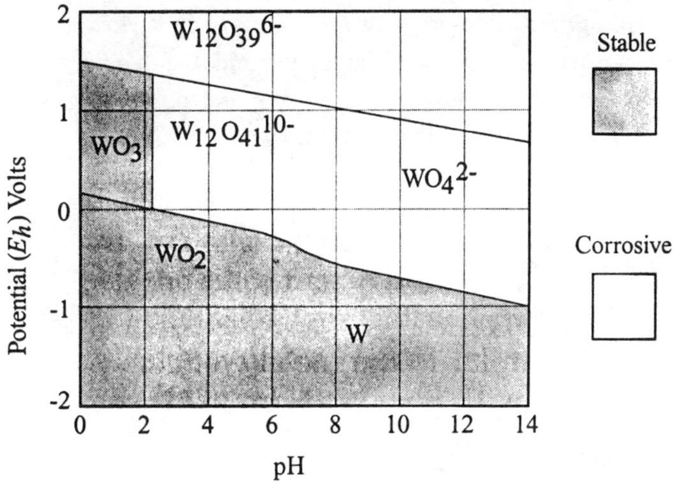

Fig. 8-30 Tungsten solubility versus solution-pH (Pourbaix) diagram.[27]

tion potential. For this reason, most tungsten slurries have a low pH, typically with pH-values between 2 and 3.

Unlike oxide CMP processes (whose purpose is to remove a specific amount of the dielectric layer), the metal-CMP processes require some degree of overpolish to ensure that metal is completely removed from field areas. The challenge here is to optimize such overpolish so that process time and oxide loss are minimized (i.e., to maximize throughput), while compensating for any non-uniformities from prior steps (and without causing any new uniformity problems from dishing or erosion).

It should also be mentioned that one aspect which makes metal-CMP processes easier to control than oxide-CMP processes, is that commercially available endpoint methods are much more sensitive for metal CMP. That is, metal-CMP processes are carried out by polishing-back blanket-metal-layers until all of the metal on the wafer surface is removed (but the metal in the via holes or within the trenches remains). Once the metal on the top surface is removed, the underlying oxide in the field is exposed. Consequently, this endpoint is easier to detect by monitoring a variety of phenomena associated with the clearing of the metal over the oxide, such as: **a**) a change in surface reflectivity; **b**) a change in surface temperature; or **c**) a change in the torque of the platens or wafer carrier (due to the different frictional coefficients of the materials being polished). In any case, the optimum metal-CMP process would have a clear endpoint-signal -followed by a short overpolish step (to remove any metal remaining in dished areas of the wafer, or in surface scratches). The result should be a surface in which the metal and the barrier-layer is completely removed, with a minimum of the underlying dielectric being eroded, and where a minimum amount of dishing is produced.

8.6 CMP EQUIPMENT

There are number of different tools and consumables used in a complete CMP process, including:

1. CMP polishing tools;
2. Slurries;
3. Pads;
4. Slurry distribution systems;
5. Endpoint detection systems;
6. Cleaning tools;
7. Metrology tools.

These items are described in the following sections.

8.6.1 CMP Polishing Tools

The basic polishing tool used for CMP evolved from the machines used to polish bare-Si-substrate-surfaces. When the idea for CMP was first conceived at IBM, they purchased silicon polishing tools (used by wafer manufacturers) from Westech (now SpeedFam-IPEC) and Strasbaugh. The basic configuration of the wafer-polisher was retained, but the tools were modified to make them suitable for developing CMP. As mentioned earlier, in such polishers the wafer is mounted upside down on a wafer-carrier, and the carrier is pressed downward against a polishing pad (Figs. 8-31 and 8-32). Both the wafer carrier and polishing pad are rotated. Slurry is dripped onto the polishing pad, and it becomes saturated with slurry. The pad thus drags slurry beneath the rotating wafer carrier, where it can polish the wafer surface (as described earlier in the section on mechanisms of CMP). The first polishers adapted for CMP had a single wafer-carrier.

The CMP-polisher differs from the wafer-polishing tool in a number of ways. In polishing silicon-wafers, several tens-of-microns of material are removed, while the CMP-process removes only about 0.5–1.0-μm of material. The uniformity of material removal across a wafer in CMP is also much more stringent than in silicon-wafer polishing (±0.5–0.1-nm for CMP). Furthermore, CMP-tools require a higher-level of automation, throughput, reliability, and critical process-parameter-control. Thus, despite more than 30 years of experience in silicon-wafer polishing, new processes and equipment had to be developed by the IC industry for carrying out CMP.

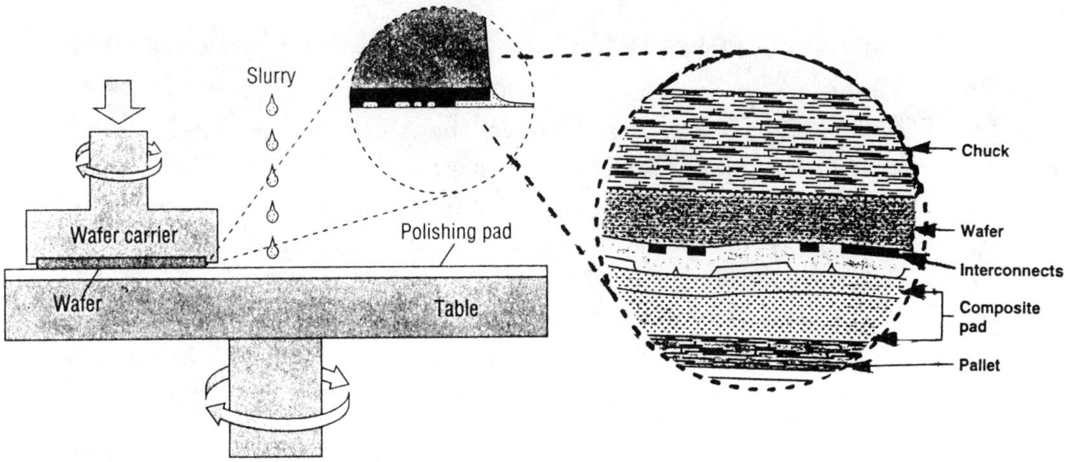

Fig. 8-31 Schematic diagram of a wafer being polished by CMP.

The Westech and Strasbaugh polishers first used by IBM were also the tools that found widest use in the early CMP processes. IBM provided to SEMATECH one of these silicon-wafer polishing tools from Westech. This tool was used by SEMATECH to launch a program that focused on improving the tool's ability to do CMP on the thin films encountered in VLSI structures, and to increase the tool reliability. The results of this program were used in designing the IPEC model 472®-CMP-tool, which achieved wide adoption for CMP applications during the mid-1990's. Figure 8-33 shows a plan view of an IPEC Avanti 776® polisher, a modified version of the 472 tool.

8.6.1.1 CMP Polishing-Tool Configurations: While early CMP tools used a single wafer-carrier, more recent designs contain multiple-carriers (and multiple-platens). This increases the throughput of the polisher and allows multi-step CMP processes to be carried out on a single tool. The classes of CMP polishing tools can be categorized according to the manner in which motion between the wafer-carrier and polishing table is produced. These categories are as follows:

1. *Rotary CMP Tools* (in which both the table [or platen] and the carrier rotate);
2. *Orbital CMP Tools* (in which the carrier rotates, but the table moves in an orbital path);
3. *Linear CMP Tools* (in which the carrier rotates and the table moves in linear path);
4. *Fixed Table CMP tools* (in which the table is stationary, but the carrier moves in an orbital path).

Table 8-1 lists some commercial CMP polishing tools and categories into which they fall.

8.6.1.2 Rotary CMP Tools: Figure 8-32a shows a schematic of a single-carrier/single-platen *rotary CMP tool.* The motion between the wafers and the pad is created by the rotation of the platen together with the rotation of the wafer-carrier. (The carrier also holds the wafer face down, and applies a downward force against the surface of the pad mounted on the rotating platen.) The pad is afixed to the flat table by a waterproof-adhesive. The following tools are examples of rotary-motion CMP tools with multiple wafer-carriers: Strasbaugh Symphony,® which has four wafer-carriers and three platens (Fig. 8-34); SpeedFam-IPEC Auriga® which has five wafer-carriers and two platens (for primary and secondary polishing, Fig. 8-35); Cybeq Nano-Tech-Isoplanar-8000® which has six-wafer carriers; and Applied Materials Mirra® with four wafer-carriers and three platens (each with an *in-situ* pad-conditioning arm, Fig. 8-36). In the Mirra,® a central spindle transports the wafer-carriers from one platen to the next. A three-platen-configuration (such as is available on the Symphony® and the Mirra®) permits three-step CMP-processes to be performed on a single tool.

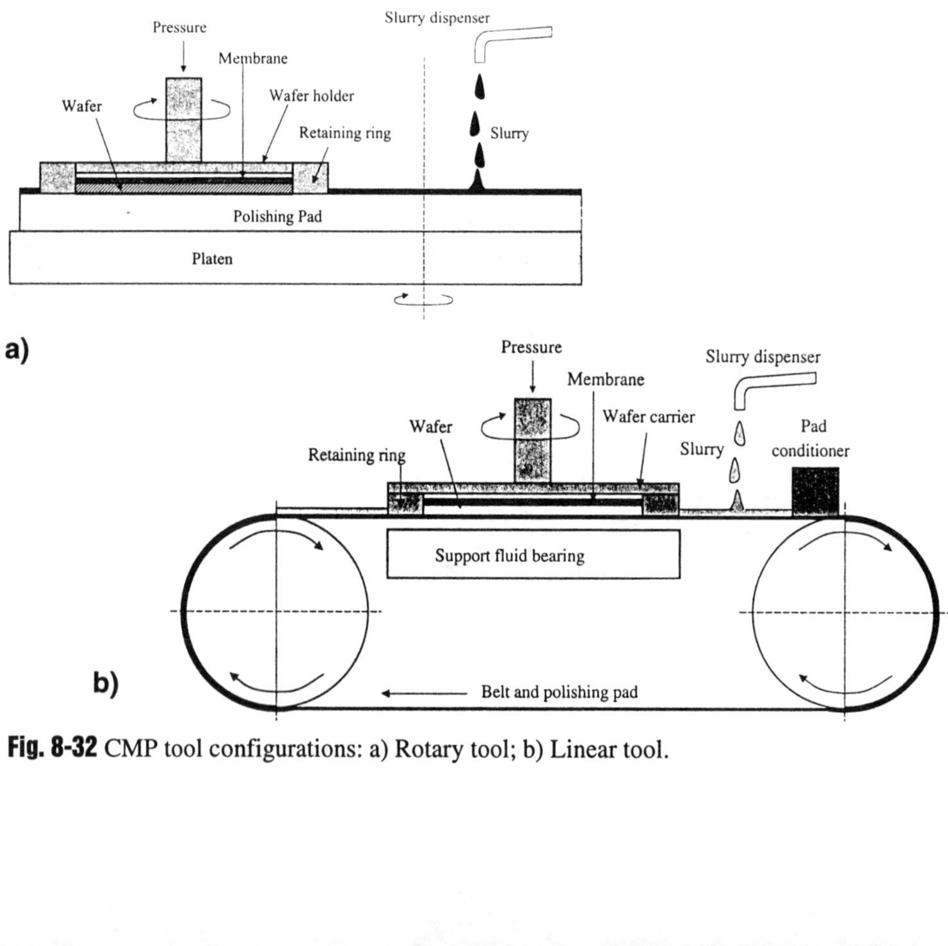

Fig. 8-32 CMP tool configurations: a) Rotary tool; b) Linear tool.

TABLE 8-1 Classification of Some CMP Polishers

Vendor & Model	Table motion–number	Carrier type-number	Built-in endpoint methods1	Special features
	Established	**Polishers**		
AMAT Mirra	Rotary–3	BB-CRRC–4	Red laser reflectometer	Sequential use of 3 tables
Cybeq 8000	Rotary–1	BB-CRRC–6		
IPEC 372/472	Rotary–2	FBC/NRRC–1	Absolute temperature	Sequential use of 2 tables
IPEC 676	Orbital–4	BB-CRRC–4		Parallel tool
IPEC 776	Orbital–4	BB-CRRC–4		Ditto, Integrated Ontrak scrubber
Speedfam/ Auriga	Rotary–1	FB-NRRC-5	Visible light reflectometer	
Strasbaugh/6DS-SP	Rotary–1 or 2	FB-NRRC-2		2 carrier size buffing platens, optional 2nd table
Strasbaugh/ Symphony	Rotary–3	FB-CRRC–4		
	Developing	**Polishers**		
Lam/Teres-	Linear-2	?–2		Fluid bearing platen
Ontrak/Synergy	Integrated Cleaner			
Obsidian	Fixed–1	Orbital, BBC?–1		Disposable 3-M slurryless pad

8.6.1.3 Orbital CMP Tools: In *orbital CMP tools*, the wafer-carrier rotates in the same manner as in the rotary CMP tools. The platen, however, executes an orbital path about the center of the rotating carrier (instead of rotating about its own central axis (Fig. 8-37). In this approach the removal-rate-non-uniformity problem encountered in rotary CMP tools is circumvented. This occurs because the travel distance of all loci on the wafer surface relative to loci on the platen surface are equivalent. Another advantage of this tool configuration is that it uses a platen with a smaller radius than that needed in a rotary polisher. As a result, the footprint of the polisher can be significantly reduced. Orbital CMP tools are sold by IPEC-Speedfam in their AvantGaard 676,® AvantGaard776,® and Momentum® models. Figure 8-38 shows the SpeedFam-IPEC Momentum,® which has four wafer carriers and four platens.

8.6.1.4 Linear CMP Tools: CMP tools that have a pad in the form of a continuous belt (that moves in the same fashion as a treadmill) are also available. The pad is thus moved past the rotating wafer carrier in a linear (rather than rotary) motion (*linear CMP tools*). These tools offer the potential benefits of higher linear

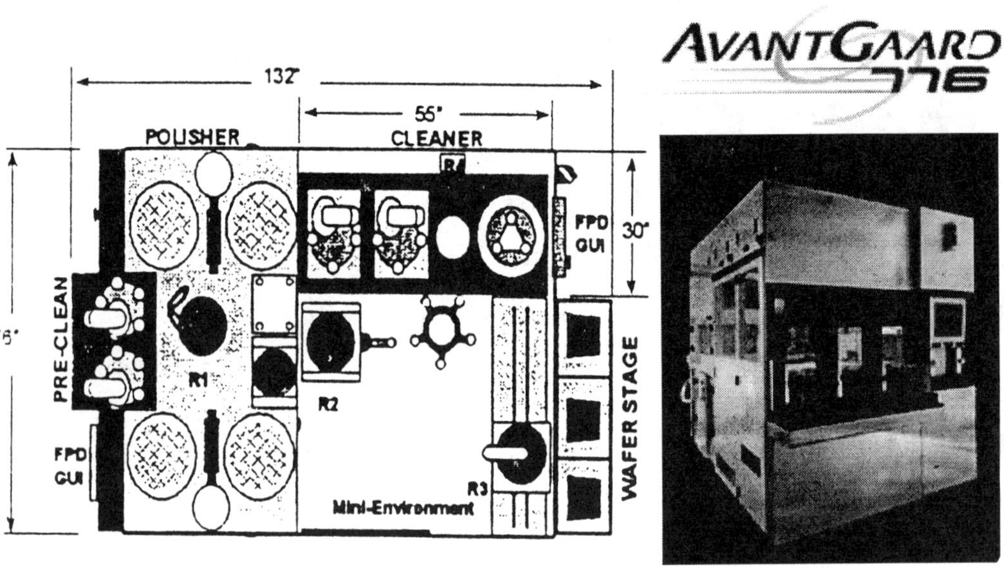

Fig. 8-33 Plan view of an IPEC *Avanti* 776® CMP tool. Courtesy of Speedfam/IPEC.

linear speeds 400–500-ft/min (versus 34-ft/min in rotary CMP tools), harder pad materials, and reduced polishing-pressure (i.e., because the polish rate can be maintained at a high value with reduced applied pressure, as a result of increased pad velocity). This type of tool may be able to produce better planarization of oxides in a shorter time and reduce the problem of dishing in CMP (see Sect. 8.12). Linear CMP tools are offered by Lam Research (Teres® Fig. 8-39), Aplex, and Doosan. The Teres® has two linear polishing belts, and one station with a rotating platen (for carrying out buffing or cleaning steps). The wafer carriers move independently from one station to another. The Teres® also employs an *air bearing* beneath the polishing belt. This allows the pressure distribution across the wafers to be profiled. At a given pressure and speed, adjustments to the pressure-distribution profile can be made to significantly reduce film-thickness non-uniformities (Sect. 8.12). The removal-rate non-uniformity problem encountered in rotary CMP tools is also circumvented in linear CMP tools. Improved planarization efficiencies (i.e., less oxide needs to be deposited to achieve complete planarity) have also been claimed for this configuration.

8.6.1.5 Fixed-Table CMP Tools: A CMP tool that employs a stationary platen and a wafer-carrier that moves in an orbital path over the platen (Fig. 8-40) was announced by Obsidian Corp. in 1999. (Another possibility would be to rotate the carrier while at the same time moving it linearly back and forth across the table,

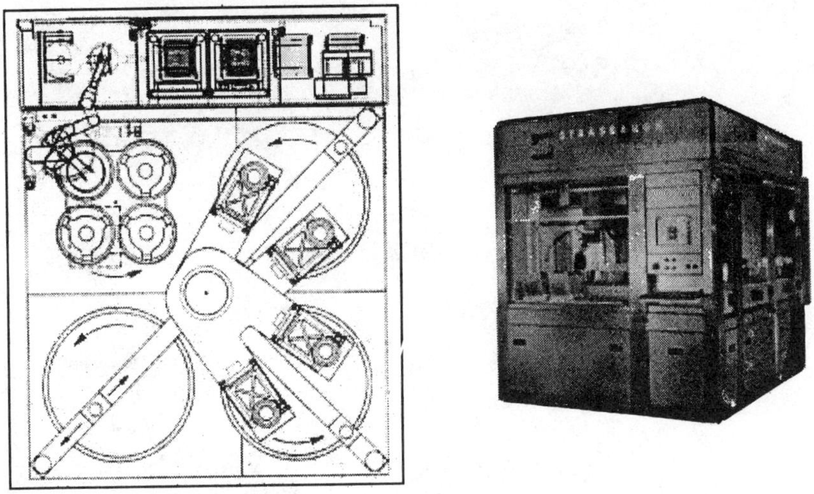

Fig. 8-34 Schematic drawing of the Symphony® CMP tool. Courtesy of Strasbaugh Corp.

see Fig. 8-41). This tool also employs a slurry-less pad supplied by the 3M Corp. The pad material is mounted on a roller (akin to a roll of paper towels). After each wafer is polished, the roll is advanced to remove the used pad-surface. This makes a new segment of pad material available to polish the next wafer (roll-to-roll transport of the slurry-free pad). Applied Materials acquired Obsidian in 2000, but by the end of 2000, the tool had not yet been commercially released.

8.6.1.6 300-mm CMP Tools: Many semiconductor equipment manufacturers are readying CMP polishing tools for 300-mm wafers. By mid-2000 more than a dozen CMP equipment manufacturers have exhibited or announced intentions to participate in the 300-mm market.[27] It is reported that most 300-mm CMP tools are capable of reaching the performance capabilities of the 200-mm tools. However, challenges remain with respect to the issues of consumables costs, integration of metrology, and full automation. In some cases a reduction in consumables cost is being achieved. For example, in the Speedfam-IPEC 300-mm polishing tool, slurry use is increased by only 20% over that needed in the 200-mm tool, whereas the area of the wafer being polished is increased by a factor of 2.25. This has a net positive effect on the CoO. Edge exclusion zones of 3-mm or less are also reportedly being achieved on these tools.

8.6.2 Wafer Carriers in CMP Polishing Tools (and the CMP Edge Effect)

In Fig. 8-42 the wafer is shown mounted on the carrier. The traditional wafer carrier consists of a circular plate (chuck) within a pocket (which holds the

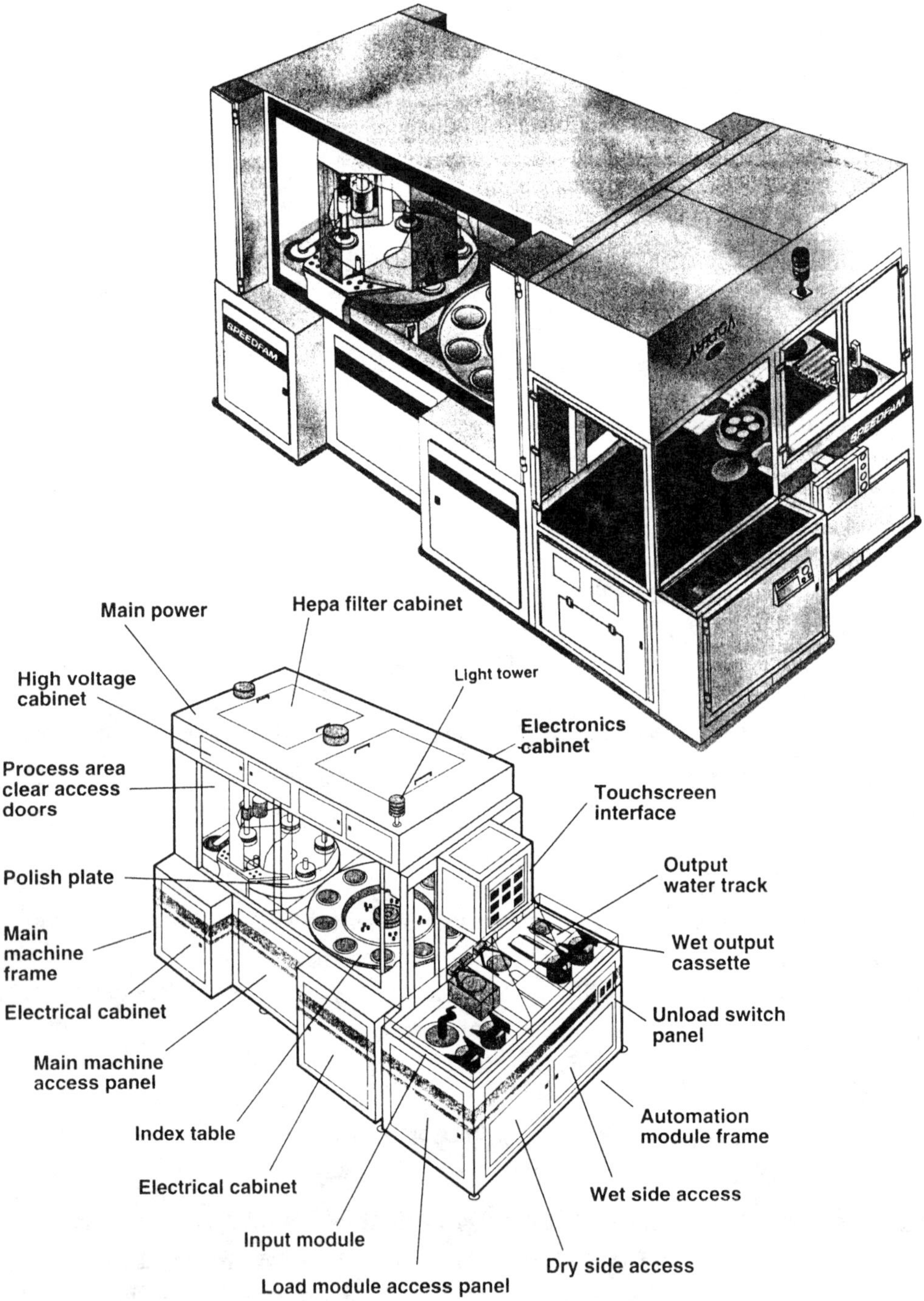

Fig. 8-35 Detailed view of the SpeedFam-IPEC Auriga® five-carrier polishing tool. Courtesy of SpeedFam-IPEC, Inc.

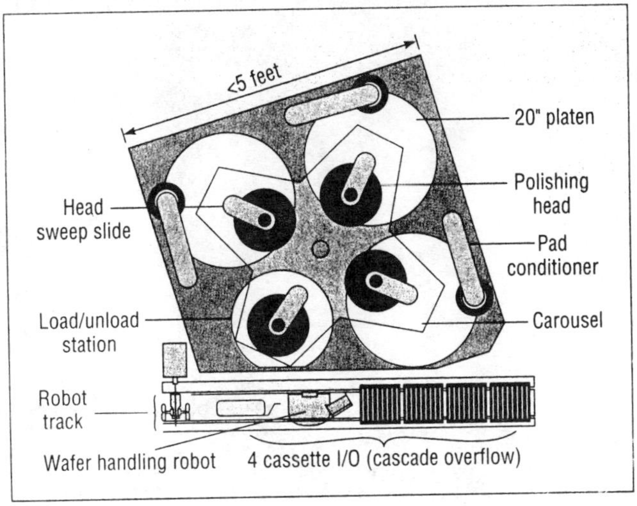

Fig. 8-36 Architecture of the Mirra® three-platen CMP tool. Courtesy of Applied Materials, Inc.

wafer), attached to a rod by a gimbal (Fig. 5-43)). Downward force is applied to the plate by the rod during polishing. The depth of the pocket in which the wafer resides is chosen to allow the top-surface of the wafer to extend beyond the pocket, permitting its surface to be polished. Instead of the wafer being directly in contact with the steel chuck, a backing film is sandwiched between the wafer and the chuck. This film provides elasticity. Without a flexible backing film, any

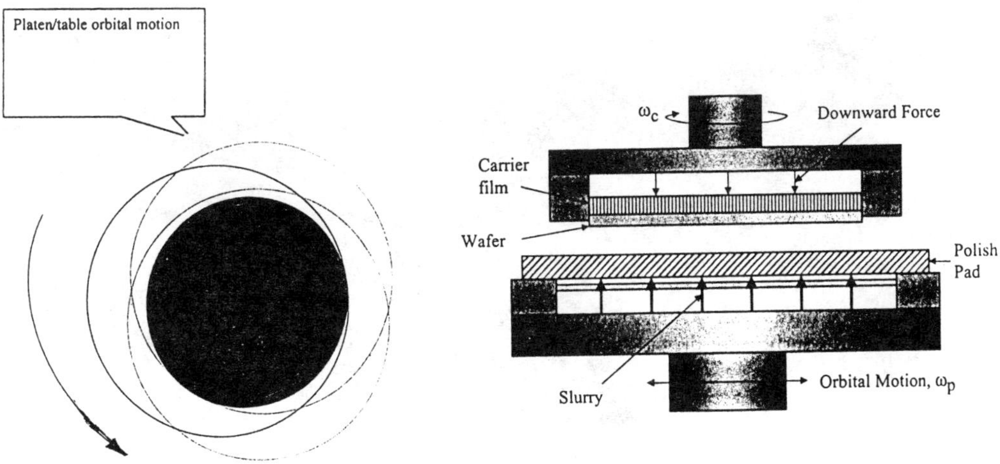

Fig. 8-37 Schematic drawing of orbital platen motion polisher.

Fig. 8-38 Photograph of the Momentum,[10] an orbital platen motion polisher. Courtesy of SpeedFam-IPEC, Inc.

defect or particle on the chuck (or the back of the wafer) will cause a thin spot in the film being polished. In addition, the possibility of wafer breakage is

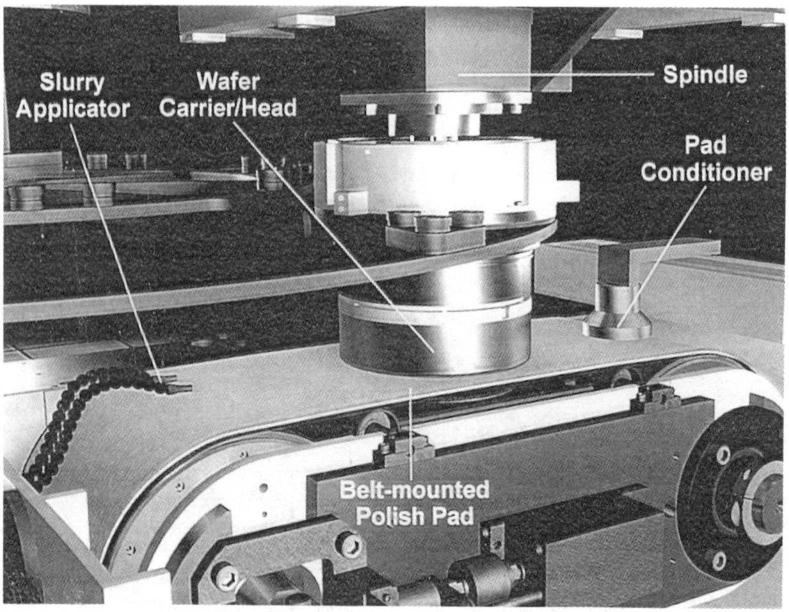

Fig. 8-39 Schematic drawing of the Teres® linear CMP tool. Courtesy Lam Research, Inc.

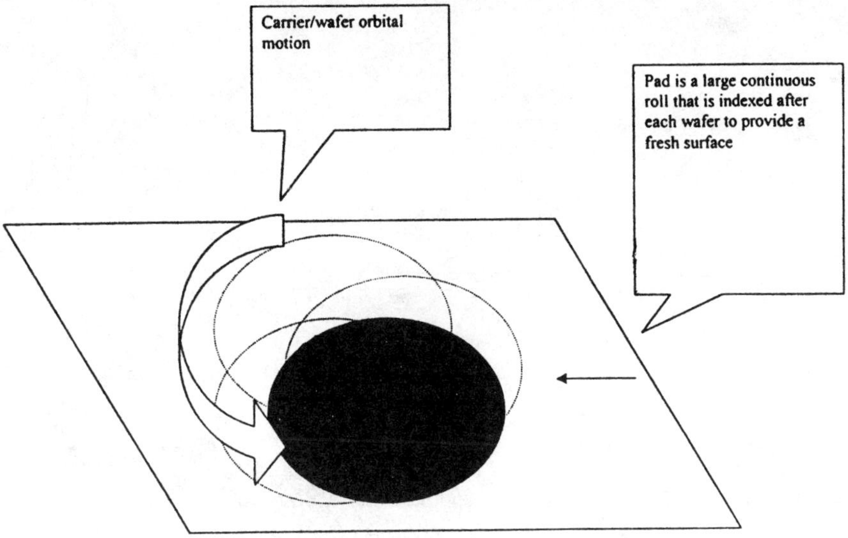

Fig. 8-40 Schematic drawing of orbital carrier-motion with indexing pad.

increased. Early backing films were merely thin foam layers. They did little to lessen the polish-performance-dependency on the variations of wafer-thickness or flatness. In order to more fully compensate for wafer-to-wafer variations in flatness and film-thickness, much development work has been done to improve the wafer-carrier design.

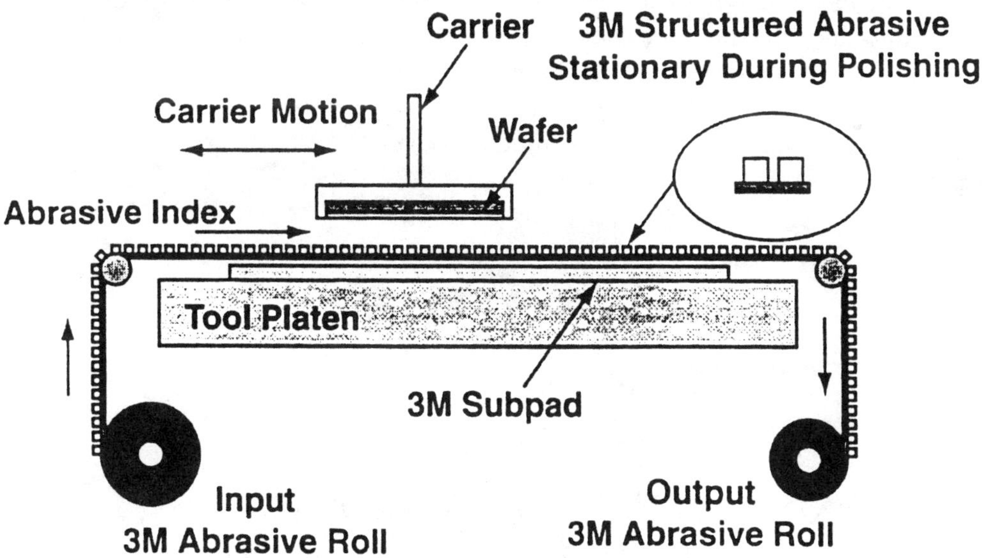

Fig. 8-41 Fixed-table with roll-to-roll pad. Carrier rotates and also executes linear motion.

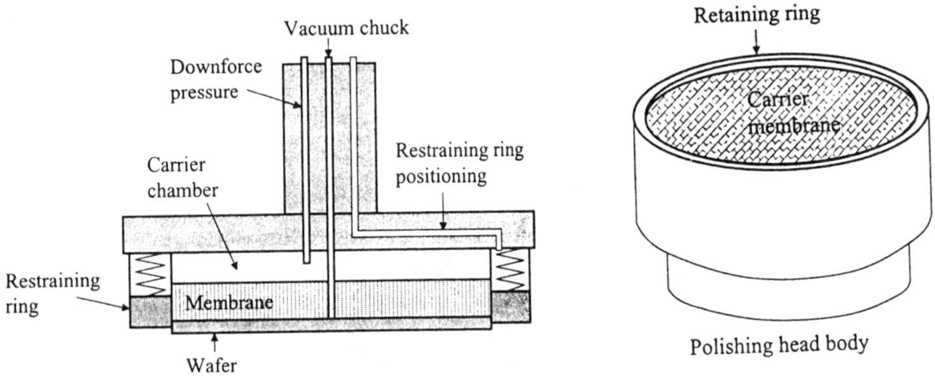

Fig. 8-42 Schematic drawing of the components of a CMP wafer-carrier.

Two key design elements in the improved wafer-carriers are:

1. The retaining-ring;
2. The wafer-leveling mechanism.

The primary function of the *retaining-ring* is to prevent the wafer from slipping out of the pocket when high down-force and high-speeds are used. Wafer retention can be maintained even with an average down-force of about 2-psi if the retaining-ring extends past the mid-thickness of the wafer. To keep the retaining-ring surface from being polished, some wafer-carriers may keep the retaining-ring about 100-microns off the pad surface. Wafer-carriers with such rings are termed *non-contact retaining-ring carriers* (NRRC). In NRRCs there is no pressure applied to the pad by this ring. The other type of retaining-ring

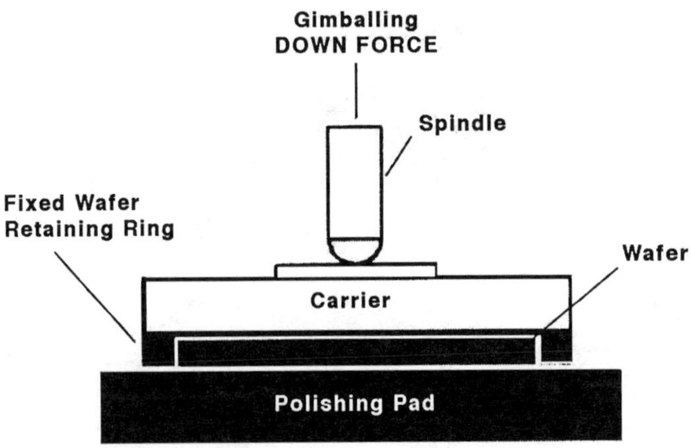

Fig. 8-43 Carrier being subjected to gimballing down force.

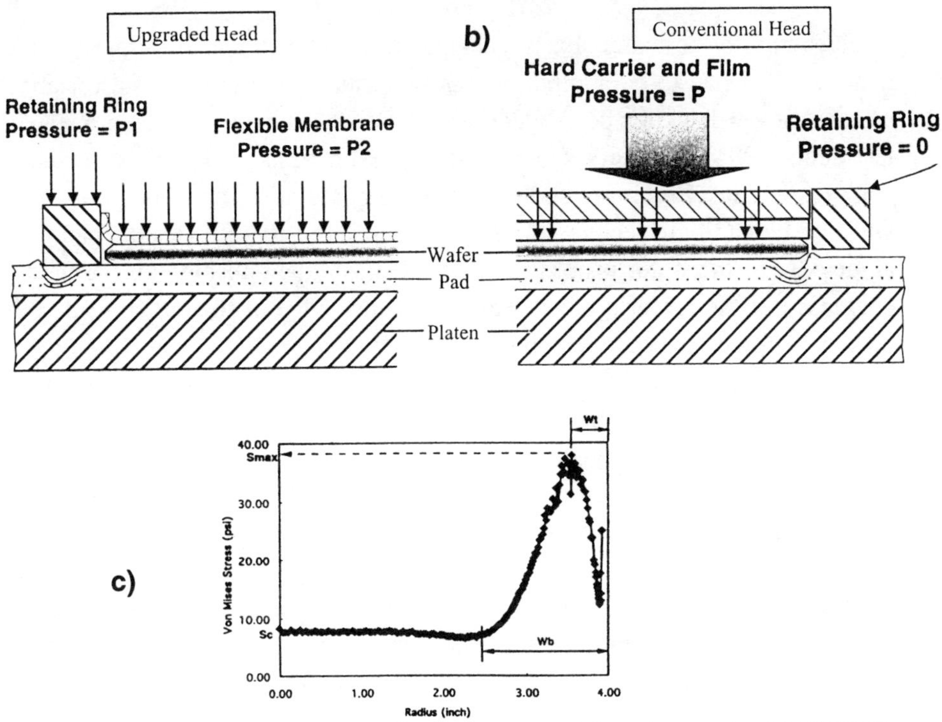

Fig. 8-44 Comparison of: a) film-backed/non-contact retaining-ring carrier (FB-NRRC); and b) bladder-backed/contact retaining-ring carrier (BB-CRRC); c) distribution of pressure profile at wafer edge for FB-NRRC.

design causes the retaining-ring to contact the pad. Hence, carriers with these rings are called *contact retaining-ring carriers* (CRRC). The CRRC also has the capability of varying the down-force on the wide (typically 10-25 mm-wide) bearing surface of the ring. Wafer-carriers used in some polishing tools (e.g., Applied Materials' Mirra® and Strasbaugh's Symphony®) are designed with a retaining-ring that can apply variable pressure to the pad.[28] This allows the pad profile at the wafers edge to be controlled. A smaller edge-exclusion zone on the wafer can therefore be used (2-3 mm).

Wafer leveling by the wafer-carrier is accomplished by backing the wafer with: **1**) a compressible film (i.e., *film back carrier,* FBC); **2**) a pneumatic bladder (*bladder-backed carrier*, BBC); or **3**) a lip-sealed air bearing *(air-backed carrier*, ABC).

The earliest wafer carriers were FBC-NRRC (Fig. 8-44a). Perforations in the backing film and the flat surface plate are used to vacuum-attach the wafer to the carrier during transport. The backing film acts as a cushion to distribute the down force evenly to the wafer. Early backing films were chemically-impregnated felt materials that had to be applied by hand, but they were only useful for about 25 wafers. Newer backing films (such as the DF200, the WB, and the R200 films) last about 200-800 wafers. (The reason backing films wear out is that they are compressed and de-compressed each time they are used. At some point, the film will lose its ability to recover and takes a "set.") In FBC-NRRCs the pad pressure at the edge of the wafer is not well controlled, and the polishing rate of the film is decreased at the wafer's edge. Hence, a wider edge-exclusion ring of 5-7-mm must be enforced for wafers with such carriers.

The second type of wafer-carrier is the BBC-design (Fig. 8-44b). In the BBC a rubber inner-tube-style bladder is substituted for the backing film. The chamber created by the bladder can be alternately vacuumed or pressurized. When the carrier brings the wafer into contact with the platen, first pressure is applied to the retaining-ring bladder (with a sufficient down force to clamp the wafer within the retaining-ring). Next, the wafer-bladder may be vented or pressurized. Usually the platen is stationary when the wafer-carrier retaining-ring makes contact and the wafer-bladder is vented until the wafer-carrier comes to speed. Then the wafer-bladder is pressurized. Since the wafer "floats" on the bladder bubble, global and local planarity of the wafer surface is accomplished without gimbaling the wafer-carrier to the head of the tool.

The air-backed wafer-carrier (ABC) functions much like the BBC. However, instead of a bladder to contain the air pressure (or vacuum) behind the wafer, it has a compliant lip-seal around the backside edge of the wafer. However, the seal is not easy to maintain, and slurry particles can enter and contaminate the backside of the wafer. As a result, ABC types have not yet been commercialized.

8.6.2.1 CMP Edge-Effect

The removal-rate of material by CMP should ideally be uniform across the wafer (i.e., the WIWNU should = 0). However, for various reasons, non-uniformity of material removal within a wafer is generally observed. One of the most common forms of WIWNU is the *CMP edge-effect*, in which the removal-rate is much smaller at the edge of the wafer, leaving an edge-bead of material at the conclusion of the CMP process. Figure 8-45 shows two typical thickness profiles of wafers polished by two different polishers, along a wafer-diameter. In both cases, an edge-bead, 100-200-nm high, exists after the polishing step. This edge-bead will generally render devices fabricated in that area inoperable. As a result, devices fabricated within a region near the edge of the wafer are considered to be excluded as candidates for working die, and this ring near the edge of the wafer

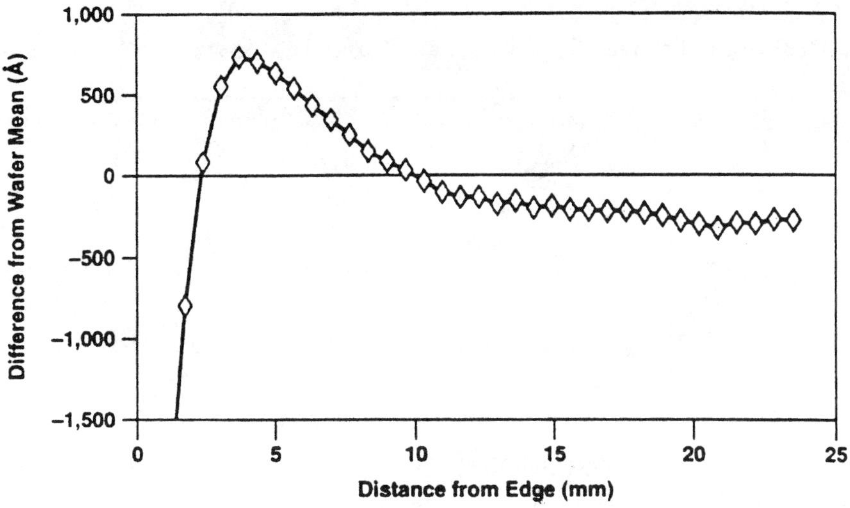

Fig. 8-45 Edge effect observed in CMP.

is termed the *exclusion ring* (Fig. 8-46). In current polishers, the exclusion ring is 3-mm wide. However, as wafer sizes are increased, the number of die at the wafer's edges keeps increasing (Table 8-2).[29] In order to obtain more yielding die per wafer, methods to further reduce the edge exclusion ring are being pursued.

The edge-bead-effect is attributed to the pad-deformation-profile and the pressure-distribution near the edge of the wafer. As a result, this effect can be significantly reduced by optimizing the pressure distribution at the edge of the wafer, primarily through the use of contact-retaining-ring-carriers (CRRC). The bulk physical properties of the polishing pad (such as its hardness, flexibility, compressibility, energy-absorption, and rebound) also all influence the width of

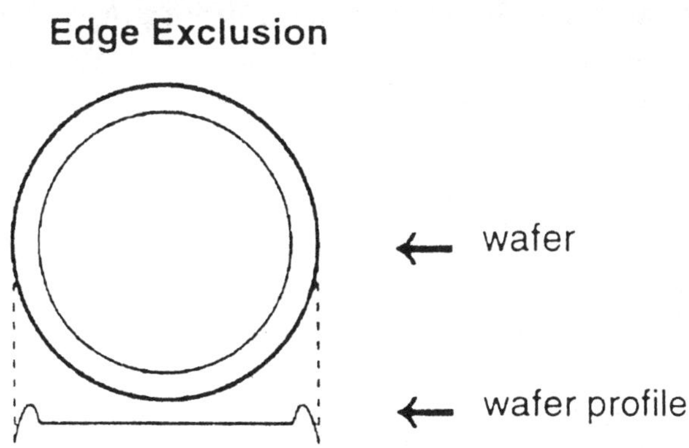

Fig. 8-46 Exclusion ring.

TABLE 8-2 Edge Die Count versus Wafer Diameter

Wafer diameter	Total wafer die count	Edge die count	Edge die percentage	Max edge revenue ($B)
300 mm	515	107	21%	2.76
200 mm	217	69	32%	1.78
150 mm	115	48	42%	1.24
127 mm	71	38	53%	1.00
100 mm	43	26	60%	0.68
76 mm	21	16	76%	0.43

the exclusion ring. Pad materials that provide the hardness necessary for planarization (but remain compliant enough to transfer dynamic contact-pressure more evenly), will deliver improved edge-exclusion. New materials for polishing pads and sub-pads have been designed to improve the uniformity of contact.[30]

8.6.3 CMP Consumables (Slurries)

Slurries consist of small, abrasive particles (generally inorganic oxides) of specific shape and size (typically 10–100-nm) suspended in an aqueous solution. The abrasive particles have roughly the same hardness as the film being polished. Reagent chemicals, such as acids, bases, surfactants, buffers, corrosion inhibitors, and complexing agents, are also added to the solution, depending upon the material to be polished. The slurry parameters that impact the polishing-rate include the chemical composition of the slurry solution (and its pH) and the concentration (in wt %) of the solid-particles in the slurry. Also important are the shape, size, and size distribution of the slurry-particles. An automatic slurry feeding system is used to ensure uniform wetting of the polishing pad and proper delivery of the slurry. Slurries are marketed by a number of suppliers, the largest of which is Cabot Corp. More slurry vendors are entering the market, including Rodel Inc., Fujimi Inc., Olin Microelectronic Materials, Praxair Surface Technologies, and EKC Technology, Inc.

The goal for each CMP process is to find a slurry that produces high removal-rates, good-planarity, local-film-uniformity, and high-selectivity. Also important are the ease with which the slurry can be cleaned from the wafer, the aggressiveness with which it attacks the polishing equipment, its storage stability, and the cost of disposing of the slurry waste. From a manufacturability perspective, consistency is a critical issue. That is, slurries have tended to exhibit quite a lot of variability from batch-to-batch. In the future, suppliers will be expected to deliver slurry that has the same performance consistently. If this can be done, the need for monitor wafers will be significantly reduced, allowing the cost of the CMP process to be decreased.

The shelf-lifc of slurry is also an important characteristic. This determines how long a slurry can be stored before use after manufacture. Some slurrics need to be kept within a specific temperature-range during shipping and storage to maintain their performance. The shorter the shelf-life of the slurry, the narrower the window in which it can be used, and inventory-control becomes important. An ideal shelf life for slurries is at least one year.

As CMP technology has evolved, slurry technology has therefore also had to progress to keep pace with the demanding requirements of the industry. Today's CMP slurries are increasingly tailored for specific process applications (like STI or Cu interconnects) as well as for specific polishing tools or other consumables (i.e., pads). Some of these issues have been discussed in Sect. 8.5.2, and others will be addressed in Sect. 8.11. The remaining slurry topics are considered here.

One other aspect of slurry management should also be pointed out. Slurries tend to dry rapidly on a wafer after leaving the wet environment of the polishing table. Once a slurry dries, it is very difficult (if not impossible), to remove all the slurry particles from a wafer surface. To prevent this from occurring, wafers are

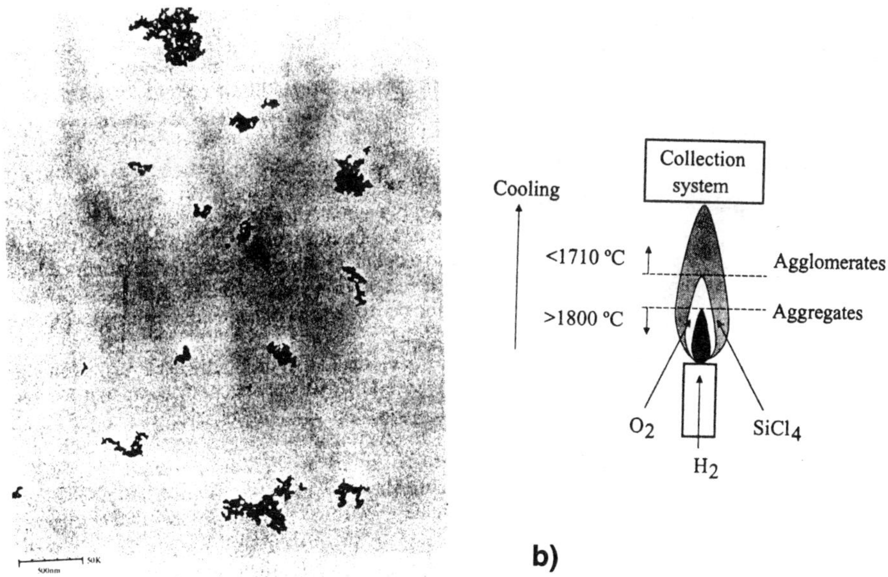

a) b)

Fig. 8-47 a) SEM of fumed-silica slurry-particles. Courtesy Rodel; b) Fumed-silica formation.

rinsed while still on the pad, and then quickly transported into wet-cleaning tanks to remove the slurry. In fabs that do not use polishing tools with integrated cleaning systems, logistical difficulties may delay the transport of wafers from the polisher to the cleaning system, inviting the above problem.

8.6.3.1 Slurries for Oxide-CMP: Slurries for oxide-CMP traditionally consisted of colloidal-silica or fumed-silica suspended in an alkaline solution (typically potassium-hydroxide [KOH] or ammonium-hydroxide [NH_4OH], diluted in DI-water). The first generation of slurries for oxide-CMP contained fumed-silica abrasives, and they still retain a large fraction of the market share. Fumed-silica is manufactured as a dry powder by the oxidation of silicon tetrachloride ($SiCl_4$) at high temperature in a flame (see Fig. 8-47). The powder must be dispersed and blended into the final slurry. The fact that the fumed-silica must be mixed is inconvenient, and adds cost. The average selling price of fumed-silica slurries in 1998 was $10-$15 per gallon. Figure 8-47 also shows a SEM of fumed-silica slurry particles.[31]

Colloidal-silica slurries have entered the market as an alternative to fumed-silica. Colloidal-silica is manufactured by the hydrolysis of sodium-silicate (or other silicates). This is followed by an ion-exchange to reduce the sodium levels for IC applications. (Early on, colloidal-silica was avoided due to the possibility of sodium contamination, but improvements in this aspect of colloidal-silica processing have opened the door for its adoption.) Colloidal-silica is prepared directly in solution, thus avoiding most of the handling problems of fumed-silica. Its microstructure is a distribution of spherical particles (see Fig. 8-48).[32] Colloidal-silica slurries also tend to be more stable and less prone to agglomeration than fumed-silica slurries. In addition, their more uniform size

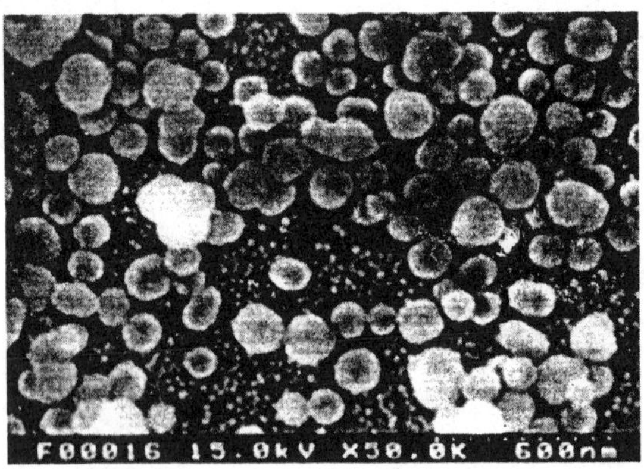

Fig. 8-48 SEM of colloidal-silica slurry particles. Courtesy of Fujimi Corporation

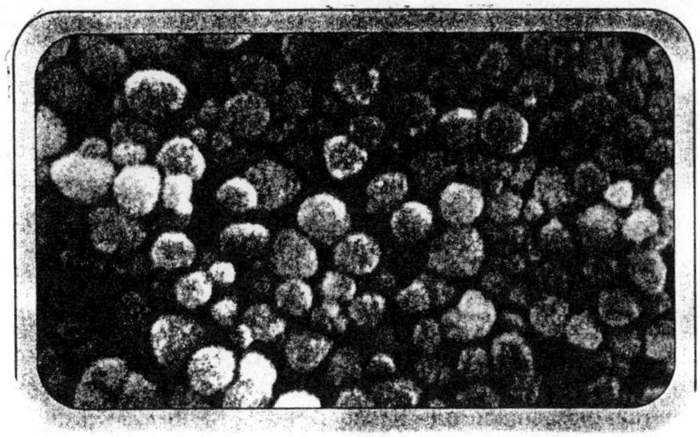

Fig. 8-49 SEM of cerium oxide slurry particles. Courtesy of Fujimi Corporation.

distribution makes it less likely that they will cause microscratches. The cost of colloidal-silica slurries is about 20% higher than that of fumed-silica slurries ($18 versus $15 per gallon). The use of colloidal-silica slurries has rapidly increased (replacing fumed-silica slurries), and they find use in STI, ILD, and polysilicon CMP processes.

The solids content of oxide-CMP slurries is 5-35-wt%, and the pH is between 9.0–11.0. By controlling pH within this range, agglomeration is minimized. CMP slurries generally contain few chemical additives. As mentioned above, these few typically consist of: **1**) an alkaline chemical to provide colloidal stability (i.e., to prevent slurry-particle-agglomeration) and to increase oxide removal rates; and **2**) a buffering agent (used to keep the pH constant throughout the slurry volume and over time). The polishing rate of oxide increases with: **1**) increasing pH,; **2**) increasing slurry particle concentration; and **3**) increasing particle size.

Slurries for oxide-CMP based on other abrasives than silica have also been introduced in recent years. Ceria-slurries (CeO_2-based) have been formulated for STI and ILD applications, although ceria is difficult to manufacture at small particle sizes. A self-stopping, dishing-free process using a ceria-slurry was reported by Nojo.[33] Ceria-slurries exhibit a higher removal rate at a much lower pH than silica-slurries (e.g., at a pH of 6 to 7). Ceria-slurries have been adopted for STI-CMP because of their high-selectivity to oxide over nitride (as high as 250:1). Ceria particles can also be blended with silica particles in a slurry. The addition of silica helps keep ceria particles in suspension. In 1998, the average selling price of ceria-slurries was $18 to $20 per gallon. Figure 8-49 shows cerium oxide (CeO_2) particles.[34]

Zirconia-based slurries (ZrO_2) have been investigated for use in low-*k*-dielectric-CMP, especially the CMP of non-fluorinated polyarylene ethers (e.g.,

FLARE™, from Honeywell, see Chap. 14). Zirconia is harder than silica but softer than alumina. Manganese-oxide slurries (MnO_2) have also been studied for the direct polishing of low-*k* dielectrics (and for the CMP of W). It was reported that the deep scratches formed with CMP using a conventional silica-based slurry were reduced from a 20-nm depth to a 2-nm depth by the use of a MnO_2-based slurry. Residual MnO_2 slurry particles can also be removed using H_2O_2 with a pH of 6, which does not attack the fluoro-carbon low-*k* material being polished (e.g., a poly-AE layer, FLARE®).[35]

8.6.3.2 Slurries for Metal-CMP: Slurries for metal-CMP applications typically have more constituents than do oxide slurries. Besides the *oxidizer* (e.g., ferric-nitrate and hydrogen-peroxide), and a buffer agent, they may contain: **1**) a *surfactant* (to assist in keeping the slurry particles dispersed, see Sect. 8.7.1.4); **2**) a *complexing agent* (used to facilitate the removal of metal from surfaces, especially in the case of copper); **3**) *corrosion inhibitors* (e.g., benzotriazole [BTA] in Cu slurries to permit the use of acidic slurries); and **4**) *additives to change the selectivity to other films*. The oxidizer is often separated from the abrasive for reasons of colloidal stability or shelf life. In the year 2000 time-frame, choosing and balancing all of these constituents still makes the formulation of metal-CMP slurries as much an art as a science.

Metal-CMP slurries typically use alumina particles suspended in various liquids (depending on the specific metal being polished). Alumina slurries are preferred because they exhibit higher metal-removal-rates and higher selectivity to the oxide ILD-layers (> 20:1, metal-to-oxide). The alumina particles may be fabricated either by a fumed process similar to silica, or by precipitation from aluminum hydroxide.

The pH range of metal-CMP slurries is low (e.g., pH = 1.0-5.0 for W-CMP alumina slurries, and pH = 3.0- 9.0 for Cu-CMP alumina slurries). As noted in the previous section, silica particles in oxide-slurries can be kept well dispersed for relatively long periods of time by controlling the pH of the slurry (giving them a long shelf-life). However, alumina-based slurries have traditionally exhibited the drawback of poor colloidal stability. This causes slurry particle agglomerates to form. Such clumps produce unacceptable microscratches in the films being polished. Recently, improvements have been made in the colloidal stability of alumina based slurries.[36] Nevertheless, to overcome this problem from another direction development of slurries based on other particle types (e.g., cerium oxide and zirconium oxide particles) for metal-CMP has been pursued. In 1998, the cost of alumina slurries was between \$25 and \$30 per gallon.

8.6.3.3 Multi-Step Slurries: When carrying out a metal-CMP process (e.g., W or Cu) two materials must be polished (i.e., the main metal, and the underlying barrier [or glue] layer). This CMP-step can be carried out with a process that uses one slurry, or two slurries. While the one-slurry process is simpler, it has not been possible to restrict the problems of dishing and erosion to the necessary degree with such a process. Hence, the two-slurry approach has been adopted. That is, for Cu-CMP, the first slurry removes the bulk of the Cu in the field areas, while the second removes the barrier (e.g., Ta or TaN) from the field areas. These two-step CMP processes significantly reduce the copper dishing and erosion effects. (Generally, the formulations of these slurries still remains proprietary). Use of a CMP-tool with multiple-platens also facilitates the manufacturability of such two-step CMP processes.

8.6.4 Slurry Distribution Systems

Slurry management (i.e., mixing, storage, distribution, and dispensing) is another key issue of a CMP process. Reliable *slurry distribution systems (SDS)* are an increasingly important component of the overall CMP-equipment-set. It is critical that the user of a CMP system view the SDS as an integral part of the CMP manufacturing process, and treat it with as much care as a polisher is given. This will greatly minimize process variations, ensure smooth functioning of manufacturing, and reduce unscheduled downtime.

Slurry management is challenging for a number of reasons.[37] Precise mixing and consistent batch blends are necessary to achieve the repeatability demanded by process engineers. The slurry consists of particles in a liquid, and they must be kept suspended in it in a homogeneous state. However, these particles will agglomerate into clumps unless steps are taken to suppress this tendency. (The main source of surface deformations in CMP is generally believed to be the presence of undetected coarser-size particles in the CMP slurry.)[38] The slurry-pH plays a role in the dispersion of silica particles. As is described in more detail in the section *CMP Cleaning*, at a pH above 7.5 the silica particles acquire sufficient surface charge to generate electrostatic repulsion, and this effectively keeps the slurry well dispersed.[39] (Above a pH of 10.7, however, the particles dissolve and form a silicate.) In addition, from the time the raw slurry is introduced into the SDS it is also critical that a blanket of humidified air be provided over the slurry to prevent agglomeration in the dispense system. Finally, slurries for Cu- and W-CMP require constant re-circulation to maintain particle suspension. (It is typically recommended that a minimum flow-rate of 1-m/sec be used for re-circ-

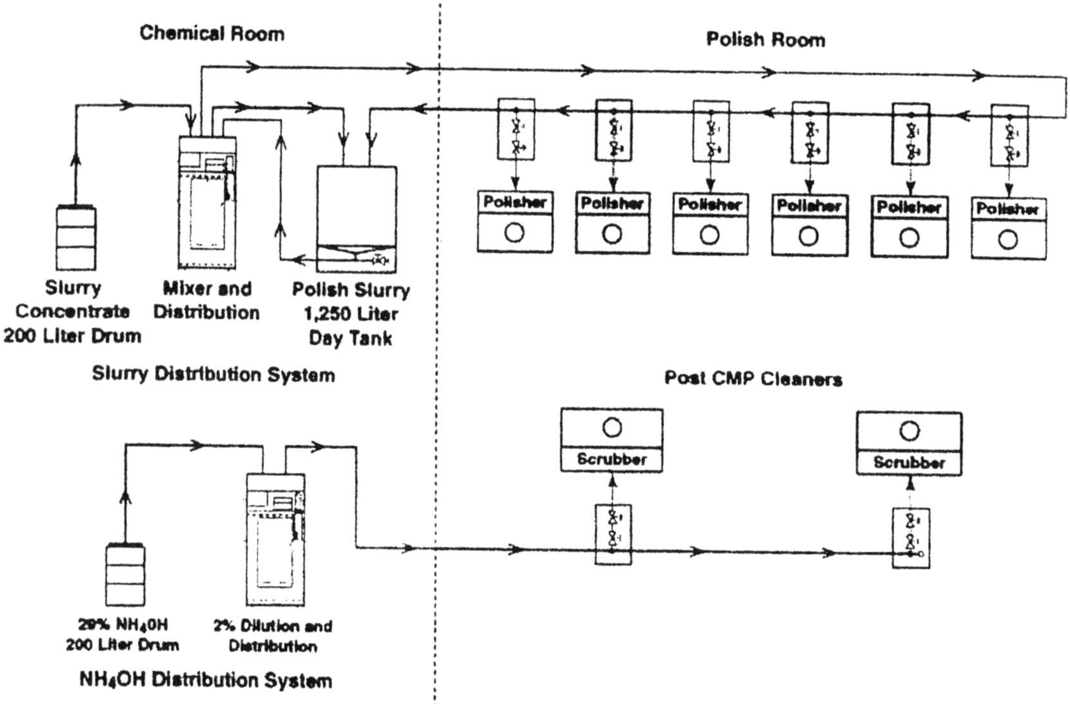

Fig. 8-50 Slurry distribution system.

ulation). Slurries have a finite shelf-life (after which the agglomerate-level becomes excessive). This shelf-life ranges from less than one week to about one month. Filtering is therefore necessary to prevent any agglomerates (which might have formed during that time), from reaching the wafer being polished.

Slurry handling in an SDS is a delicate task and care must be exercised to prevent "damaging" the slurry. That is, the slurry must not be subjected to excessive shear (which could cause the slurry to agglomerate), or to insufficient shear (which would permit slurry settling). The SDS must also be regularly maintained. Such maintenance involves periodic cleaning out and flushing of the system (as well as the polisher).

The SDS must continuously provide slurry to a number of polishers in a sustained manufacturing production environment (i.e., 24 hours a day, seven days a week, and in high volumes). The pH and solids-content of the slurry must be monitored in-line to ensure consistency. Furthermore, since slurries are abrasive particles contained in chemically aggressive liquids, they present a harsh environment to the components of the distribution system. That is, pumps, piping, valves, and measurement apparatus in contact with slurry, can be stressed to the point of failure in a short period of time. Nevertheless, bulk slurry distribution systems have been designed to handle these conditions. These sys-systems are sophisticated pieces of equipment, far more advanced than other bulk-delivery systems. Figure 8-50 shows the schematic layout of an SDS, and Fig. 8-51 shows the actual pieces of equipment in an SDS. Such SDS systems are available from vendors including: Mega Systems & Chemicals and FSI International. These vendors supply detailed recommendations regarding design and maintenance of SDS, piping, mixing, and dilution procedures, filtration of the slurry, filter changeouts, and re-circulation of the slurry.

8.6.5 Environment, Safety, and Handling Issues of CMP Slurries

As CMP further moves into large scale IC production, the disposal of spent slurry, CMP process by-products, and metal contaminants (i.e., material removed from the wafer surface) become important issues.

Slurry also always contains aggressive chemicals. In some cases these chemicals are highly toxic. Furthermore, the material removed by a CMP process ends up in the expended slurry and the rinse water. Some IC fabs estimate that CMP processes may come to consume 30-40% of the total water utilized in IC manufacturing. The issue of purification and disposal of this waste water (or

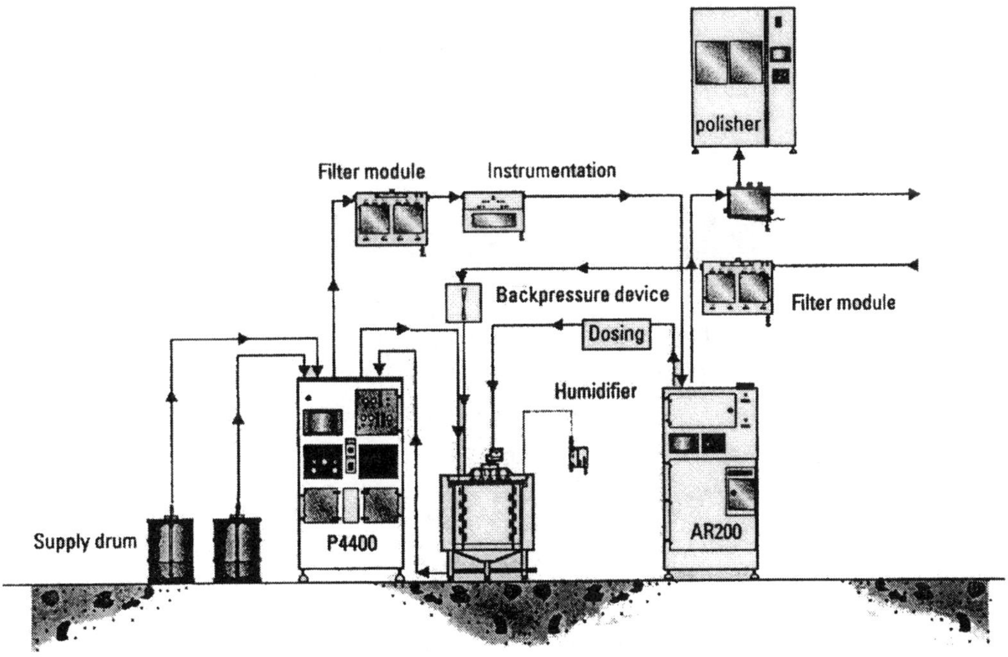

Fig. 8-51 Schematic of a slurry management-system with metrology, auto-dosing and filtration.

recycling), becomes more critical as new materials such as copper, gold, and tantalum become adopted in IC manufacturing. For example, copper is a known toxin in water. Therefore, fabs require waste-water treatment systems to recycle the drain water of CMP processes. The most common limits for waste streams are those for Total Suspended Solids (TSS), heavy metals (like copper, chromium, zinc, and nickel), pH, and total organic content (TOC).[40]

The slurry particles are suspended particles in a solution, and hence they can be removed from the waste water by a process of *flocculation* (i.e., a process by which the dispersed slurry particles are caused to clump together to form larger clusters of matter [i.e., to flock]) followed by filtration. Flocculation involves adding a flocculent and adjusting the pH to a level where the flocculent works best. The slurry particles then agglomerate and settle out of suspension. After the solids are separated, the next step involves ion removal. Either ion-exchange or reverse-osmosis may be used to do this. Note that as slurry volumes go up, they will generate huge amounts of solid waste, estimated at 100,000-400,000 pounds per year, per typical fab, and this solid waste must also be disposed of safely. A decrease in slurry consumption by the CMP-processes will be needed to reduce this waste-disposal burden.

Some companies are also examining the issue of reusing the water from a polisher. That is, while a polisher is in an idle (or pad-wet) mode, there is a continuous stream of DI-water that is used to keep all parts of the polisher wet and reduce the possibility of formation of dry particles. This water generally retains a high purity level and offers the potential of recycle and reuse, thus reducing the waste-treatment-costs for CMP.

One suggested method of reducing slurry consumption is to re-cycle the slurry. In one study of re-using slurry, KOH-based fumed-silica slurry was used to polish a PECVD-TEOS film. The slurry was then captured after being flushed from the CMP tool. By adding new slurry and adjusting the pH to compensate for the decreased solids content and pH, the slurry appeared to be capable of re-use up to six times, while maintaining roughly equal removal rates.[41] In a more comprehensive slurry-recycling process, the slurry-waste would be treated by the following processes: filtration, oxidizer-replenishment, removal of metals, and replenishment of buffers and other necessary additives. The task of slurry re-cycling (and eventual disposal) could be carried out by the slurry manufacturer. On the other hand, it has also been discussed that the slurry could be leased by the end user, and that it would be sent back to the slurry manufacturer after use for regeneration. The concept of slurry re-use is fraught with a number of difficult challenges, and is today still in its infancy. The issues of co-operation and development between the end-users and suppliers still need to be addressed before this process becomes an accepted procedure.

8.6.6 CMP Consumables (Polishing Pads)

Pads and slurries are the consumables of a CMP process. Historically, the polishing tools were considered to be the technology's major contributors. But consumables are now being recognized as being equally important. With a new group of consumables being made available, some customers select the slurries first, then review their compatibility with existing tools, and based on the results, select a polishing tool. The importance of CMP consumables is expected to increase with the advent of Cu-CMP and low-*k*-CMP. The general characteristics of polishing pads are described here.

Polishing pads are made of either: **1**) polyurethane-impregnated felts; or **2**) closed-cell cast and sliced polyurethane (with filler material to control hardness). Urethane is utilized because urethane chemistry allows the pad characteristics to be tailored to meet specific mechanical property needs. Table 8-3 lists the important properties of five polishing pads (all manufactured by Rodel). A thorough discussion of the mechanical properties and the surface roughness of CMP pads is found in pp. 68-83 of Ref. 42.

The pads made from impregnated felts are more porous and can carry a larger volume of water or slurry, due to the basic fiber structure of felt. The Rodel Suba

TABLE 8-3 Rodel Polyurethane Pad Properties

Pad	Specific Gravity	Compressibility	Hardness	Type
Politex Supreme	*Not available*	High	Soft	Micro porous felt
Suba IV	0.3	15%	61 (Shore A)	Urethane impregnated polyester felt
Suba 500	0.34	7%	72 (Shore A)	Urethane impregnated polyester felt
IC-1000	0.63–0.85	2.25%	57 (Shore D)	Rigid micro porous cast polyurethane
IC-1400	Top layer IC-1000, bottom layer foam	0.7%–7%	Flexible	Equal to stacked IC1000/Suba IV

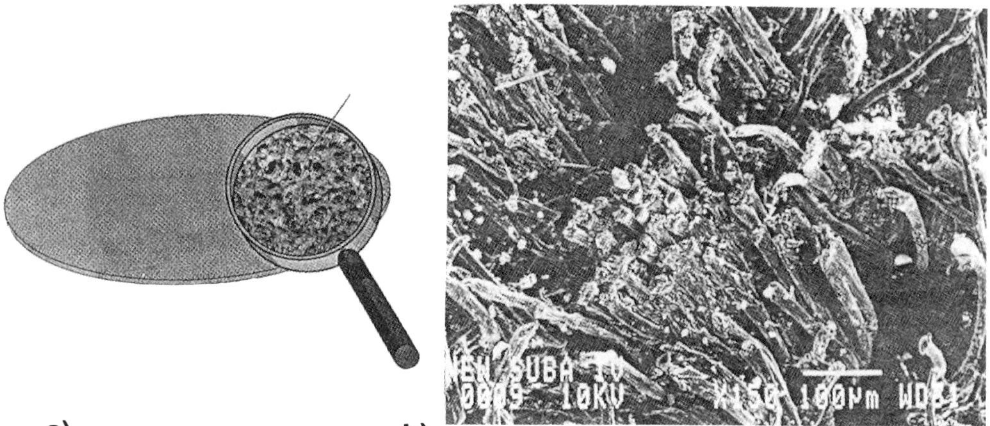

Fig. 8-52 a) Schematic of polishing pad; b) SEM of a new Rodel Suba IV® Pad. Courtesy of Rodel Inc.

IV pads are made of polyurethane impregnated polyethylene felt (Fig. 8-52), which shows a SEM of the surface of a new Suba IV pad.[43] They are classified as *soft-polish pads*, and exhibit greater compressibility than the closed-cell polyurethane types. The Rodel Politex pad is also made from a flexible felt-like material, and it is classified as a *finishing pad*, used to buff wafers with water and other chemicals to remove particles and reduce scratch defects.

The Rodel IC-1000 pads are micro-porous, filled, cast pads that have low compressibility (Fig. 8-53).[44] Pores make up about 30% of the volume of the pad, and the average diameter of a pore is about 30-nm. Such micro-porous, cast pads are classified as *hard-polish pads.* The IC-1000 pad has been the most widely used pad for oxide-CMP processes. The Rodel IC-1400 is a hybrid pad. While there is no consensus, it is generally agreed that the hardness and porosity

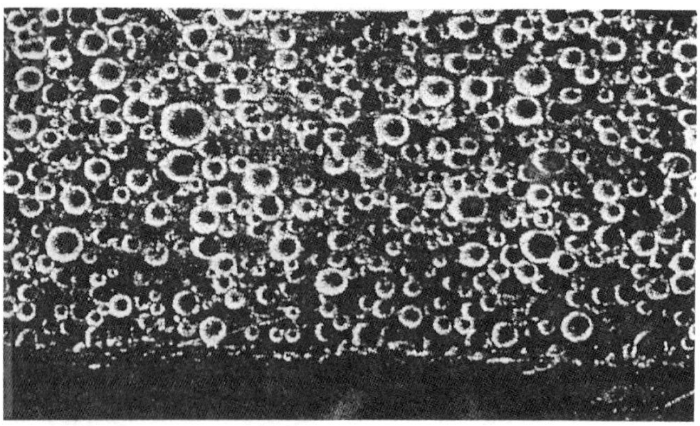

Fig. 8-52 SEM of a new Rodel 1000 IC Pad. Courtesy of Rodel Inc.

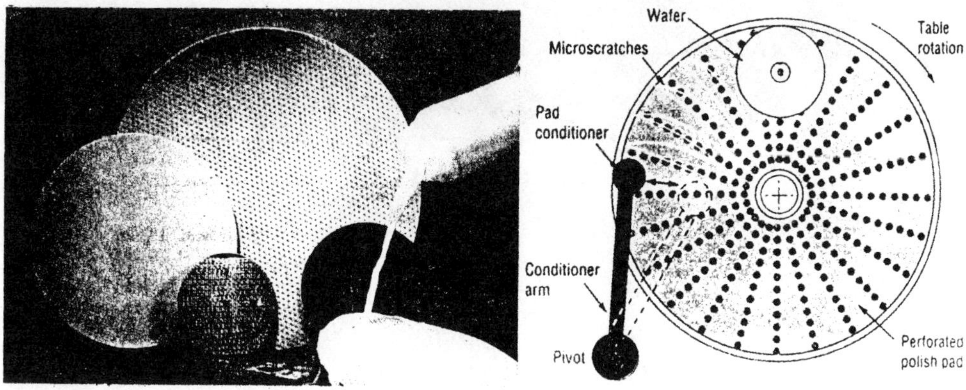

Fig. 8-54 Photograph of polishing pads, showing pad perforations. Courtesy of Rodel Inc.

of the pad are important parameters and that the pad plays a crucial (but still poorly understood) role in determining the final polish results. However, there is no single physical property of a polish pad that is known to lead to good quality for all CMP tools and consumable sets. Polishing pads are supplied by a number of vendors, with Rodel Corp. having the largest market share. Other suppliers of pads include Thomas West, Freudenberg, Fujimi, and Teijin.

In many CMP applications, two pads are simultaneously used, because a hard pad gives better local (within die) planarity, but a softer pad gives better uniformity of material removal across the entire wafer. Using a combination of two pads provides a compromise between these extremes. For example a hard IC-1000 pad is mounted onto a softer Suba-IV pad to form a stacked pad. On the other hand, the IC-1400 is a single molded pad. That is, it is an IC-1000 pad molded to a more porous backing pad that simulates the compressibility of the Suba-IV pad.

As shown in Figures 8-54 and 8-55, the surface of the pad is either cut with concentric-grooves or contains perforations 1-mm-wide and 250-μm-deep punched into the pad. These grooves (or perforations) form channels to help transport the slurry between the pad and wafer. Pores present on the pad surface also aid in slurry transport.

The desired goal of cost-of-ownership for CMP pads has been identified as less than \$0.05 per wafer. The cost of closed-cell polyurethane pads is high (\$0.06-\$0.10/cm^2). Thus, for a 200-mm CMP rotary tool (which requires a 300-mm diameter pad) the initial cost for the pad is about \$50. For a 300-mm CMP tool, the pad must have a diameter of 510-550-mm, and the cost of the pad is about \$125. To achieve a CoO for the pad of less than \$0.05 per wafer, the

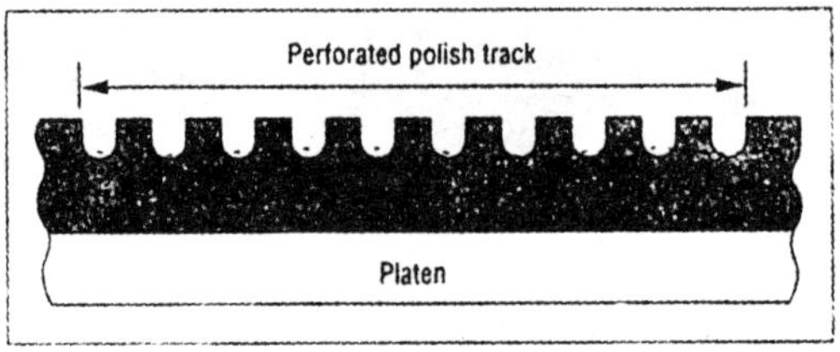

Fig. 8-55 Polishing pad cross-section, showing grooves in pad surface.

smaller pad (i.e., for 200-mm wafers) must polish 800 wafers, and the pad for the 300-mm CMP tool must polish 2500 wafers before being discarded.[45]

8.6.7 Pad Conditioners

One crucial aspect of CMP is the state of the surface of the pad. That is, surface roughness and porosity determine: **1**) slurry transport to the wafer surface; **2**) material transport away from the wafer surface; and **3**) the contact area of the pad to the wafer surface. As a result, unless the surface of a pad is continuously maintained in the optimal state of roughness and porosity, removal-rates can vary with time. In fact, a decrease in removal-rate has been most commonly observed (approximately logarithmic with time).[46] For example, starting with a new pad, the removal-rate can drop from 210-nm/min to 75-nm/min after only 50 wafers have been polished. This effect occurs because during the polishing process the surface of the pad rapidly undergoes plastic deformation so that the surface becomes smoother and the pores become filled with pad material (Fig. 8-56). This phenomenon has been termed *glazing*. Such glazing is undesirable, since as the pores of the pad become closed, slurry delivery is decreased to the wafer surface, and reduced polishing rates result. In addition, large waste-particles can become embedded in the pad and cause scratches.

Fig. 8-56 Surface of a glazed pad (Rodel's Suba IV® pad) after polishing.

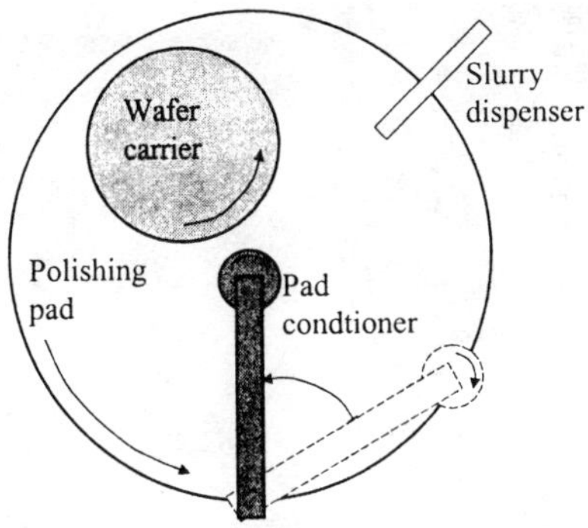

Fig. 8-57 Schematic drawing of pad conditioning. Reprinted with permission of Solid State Technology. Published by PennWell.

A consistent (and rapid) removal rate is necessary for production-worthy CMP processes. This can be achieved by a process that refreshes the surface of the pad, and this technique is called *pad conditioning*. Figures 8-57 and 8-58 show a typical conditioning process in which the pad-conditioner (a disk whose surface is embedded with a diamond grit) sweeps back-and-forth across the rotating pad to restore the pad to its required surface planarity and surface roughness. The pad-conditioning process also opens the pores of the pad by forming micro-scratches on the pad surface. An example of what the pad-conditioning process accomplishes is shown in Fig. 8-59, which contains SEM micrographs of the

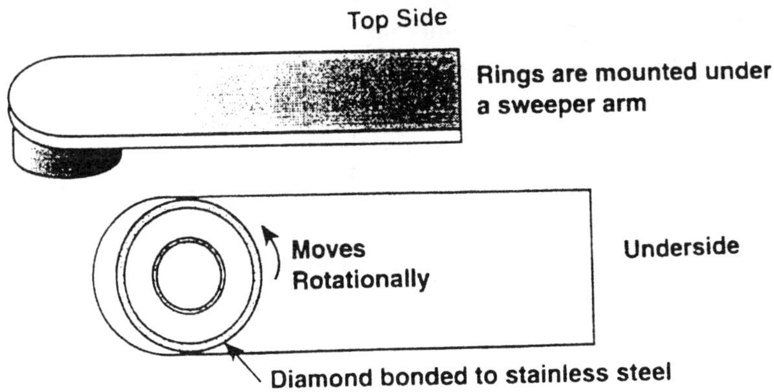

Fig. 8-58 Diamond embedded ring pad conditioner.

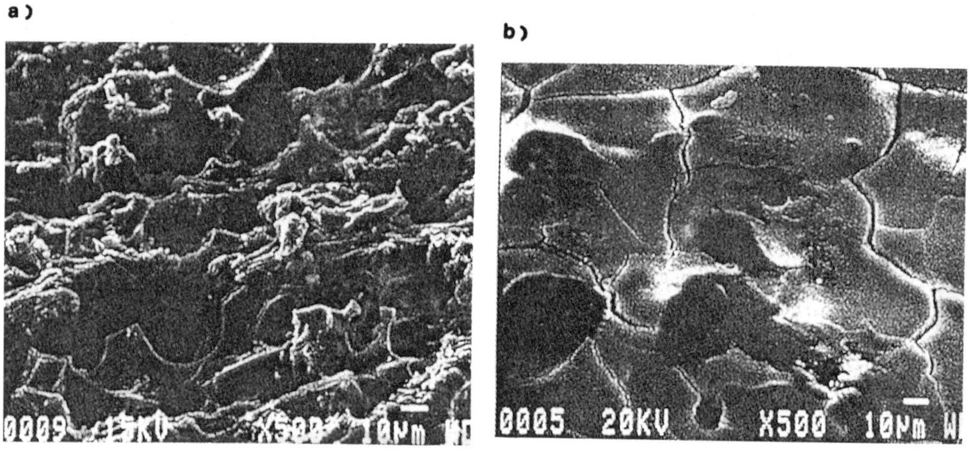

Fig. 8-59 (a) Polishing pad cross section. (b) Effect of pad conditioning.[47] Reprinted with permission of Solid State Technology. Published by PennWell.

pad surface of a Rodel IC-1000/Suba IV stacked pad, with and without conditioning. The SEM of the non-conditioned pad shows that its pores are completely clogged.[47]

The simplest form of pad-conditioning is done with a nylon-toothed brush. This is adequate for large particle removal or for conditioning soft, woven (or high-nap) pads. However, for the harder, molded closed-cell polyurethane pads (i.e., the Rodel IC-1000 or IC-1400 series), pad-conditioning is carried out by sweeping the pad surface with a disc containing 80-120-grit size diamonds embedded in a nickel plate (Fig. 8-60). Diamond, the hardest material available, is well suited for conditioning the hard-polyurethane-pad materials, but it is also brittle and can fracture. Alternative conditioning-abrasives have been pursued, since diamond particles may fracture and break away from the conditioning disk and become embedded in the pad. Since 80-grit diamonds are 100-200-microns in diameter, this can lead to severe scratching of the surfaces being polished.

Several approaches are used to mitigate the scratching problem. First, superior types of bonding mechanisms have been developed, which chemically bond the diamonds to the base of the conditioning-pad. Diamonds adhere better to such pads than to those that employ a plating technique for bonding. Second, periodic cleaning of the conditioning wheel is also carried out. Finally, some pads (such as the Rodel IC-1000) are fabricated with grooves created in the top surface of the pad. These grooves not only provide the pad with more slurry-carrying capacity, but dislodged 80-grit diamonds tend to be collected in the deep surface

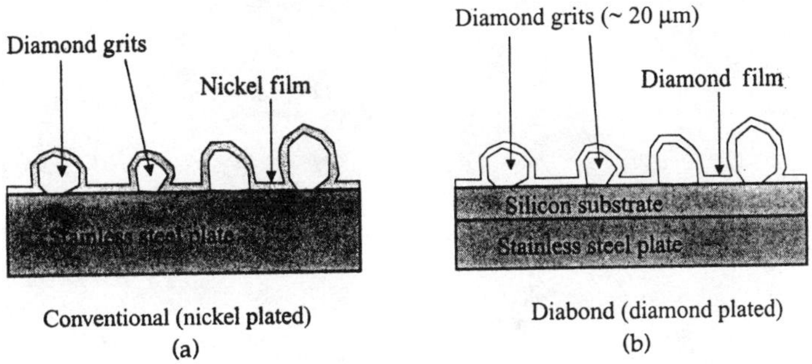

Fig. 8-60 Surface of pad conditioner showing bonded diamond grit.

grooves. Since the ridge tops of the pad between the grooves perform the polishing work (and not the pad surface within the grooves), the collection of the diamonds in the grooves minimizes scratches on the wafer surface. The pad is also cleaned with a high-pressure spray between wafers to dislodge and wash away the diamonds embedded in the grooves. (This high-pressure spray is more critical near the end of a pad's life, because at that point the grooves have become shallow. At that point, the diameter of the diamonds collected in the shallow grooves can exceed the groove depth and scratch the wafers). By using all of these techniques, diamond loss (and resultant microscratching), now represents much less of a problem.

Many polishers incorporate a continuous *in-situ* pad conditioning capability (a unit that moves across the pad during polishing, see Fig. 8-61 and Fig. 8-62).[48] This allows each wafer on the platen exposure to a pad with the same amount of conditioning, resulting in a more uniform removal-rate from wafer-to-wafer.

Fig. 8-61 In situ pad conditioner. Courtesy of Strasbaugh Inc.

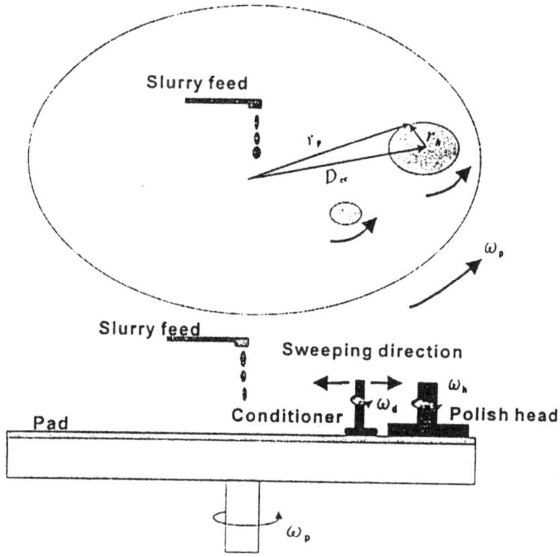

Fig. 8-62 *In situ* pad conditioner.

New pads also require a break-in conditioning-cycle to establish a constant removal-rate. In this procedure a set number of dummy wafers are polished for the purpose of stabilizing the pad condition and performance. Break-in conditioning-cycles can consume a significant fraction of the pad's useful life, and the polishing tool is not available for productive use while this is in-progress.

As a final note, it should be pointed out that conditioning of the pad is used to reduce the variation in removal-rate that occurs over a pad's lifetime. Yet the act of pad conditioning becomes yet another variable. As a result, elimination of pad conditioning would be an approach toward improving process stability. Use of a fixed-abrasive, one-use-pad, that is disposable or continuously replaced, is a potential path toward that goal.

8.6.8 Slurry-Free Pads

Slurry-free (or *fixed abrasive, FA*) CMP pads have also been developed. In these pads the traditional slurry-and-pad consumable-set is replaced with a pad in which the abrasive is attached to the pad. The main benefits of slurry-free pads are higher degrees of planarization and lower cost-of-ownership. The reduction in the cost-of-ownership is due to a number of factors. First, there is the reduced cost of consumables. Second, the cost of slurry handling is eliminated. Third, there is no longer a need for a slurry-distribution system. Finally, the problems of slurry-stability and solution-variation are eliminated. To date, most development work on slurry-free pads has centered on ILD and STI applications, but programs for slurry-free copper-CMP are also underway.

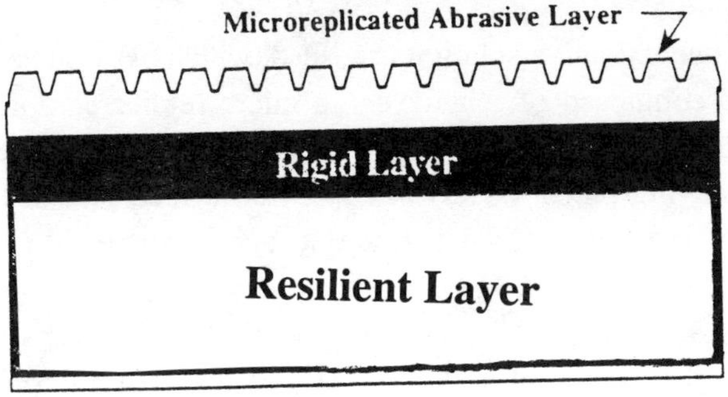

Fig. 8-63 Slurry-free (Fixed abrasive) pad structure.

A pad supplied by the 3M Co. embeds the abrasive material in the pad itself, in the form of a precise micro-replicated abrasive surface and a bi-component sub-pad system (Fig. 8-63). The abrasive surface can be constructed in one of several configurations chosen for specific performance criteria. Figure 8-64 shows two of the surface features currently in use.[49] For dielectric applications the slurry-free pad is used with a simple basic solution. For metal-CMP, a

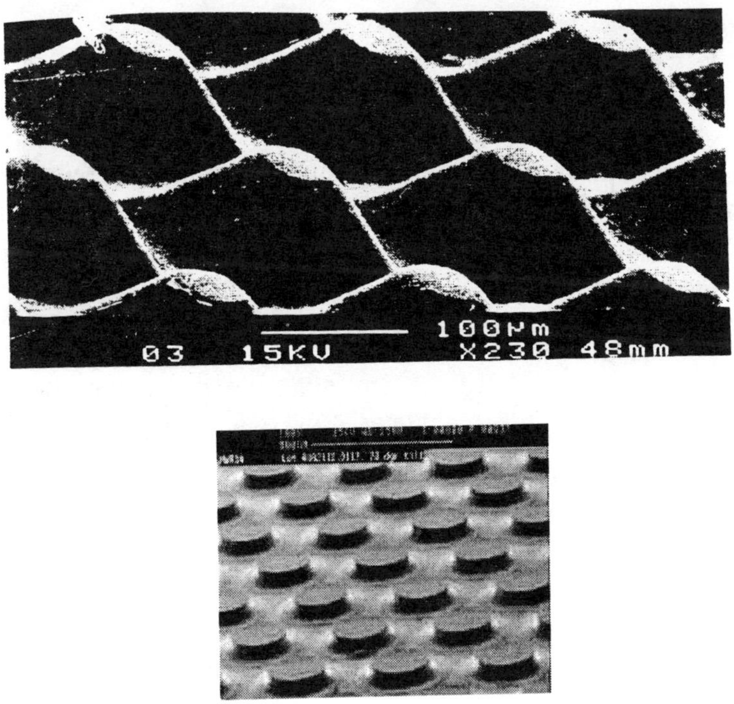

Fig. 8-64 Examples of slurry-free pad surface structures. Courtesy of 3M Co.

a specially formulated oxidative solution is needed, with a pH value of 2 to 7.

These pads are composed of four layers: a micro-replicated abrasive; a rigid layer; a resilient layer; and a self-adhesive backing. These slurry-free pads can be used either on a moving-platen (rotary or linear) or a fixed-platen. In early 2000 Applied Materials and 3M entered a strategic agreement to jointly develop a slurry-free pad for the Applied Mirra® multi-platen rotary CMP-tool. 3M also has a marketing alliance with pad supplier Rodel. SpeedFam-IPEC also announced an effort to develop a slurry-free pad for their rotary/orbital CMP tools. There is some concern about the consistency of slurry-free pads for rotary CMP tools throughout the pad life. As the pad wears, the surface area tends to change, which may cause a non-stable process.

These concerns are avoided in web-format pads for fixed table, roll-to-roll pads. The use of a slurry-free pad in a fixed-table (platen) CMP system was pioneered by Obsidian Corp. In Obsidian's web-based tool, a roll of abrasive pad material is spooled across a platen and incremented after each wafer. No slurry or pad-conditioning is required.[50] (Fig. 8-65). However, one concern about this slurry-free approach is the uniformity of polishing performance from roll-to-roll.

Slurry-free pads have demonstrated excellent planarization capability. This capability allowed a slurry-free STI CMP-process performed on an Obsidian 501 CMP tool to demonstrate exceptionally tight nitride ranges and minimal dishing (Fig. 8-66).[51] That is, the removal of nitride during the trench oxide CMP process is very uniform across the wafer, with only a 250-Å variation in the remaining nitride films after the polish step. The dishing in the field oxide regions was also very small, with the maximum dishing being less than 400-Å.

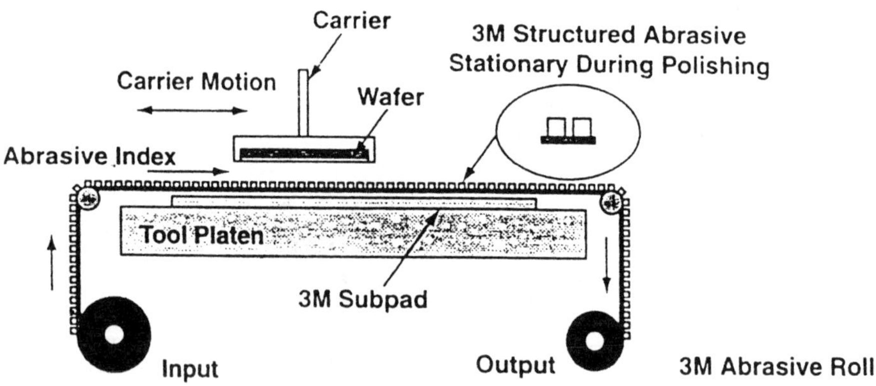

Fig. 8-65 Schematic of linear web format CMP tool, with indexing, slurry-free pad.

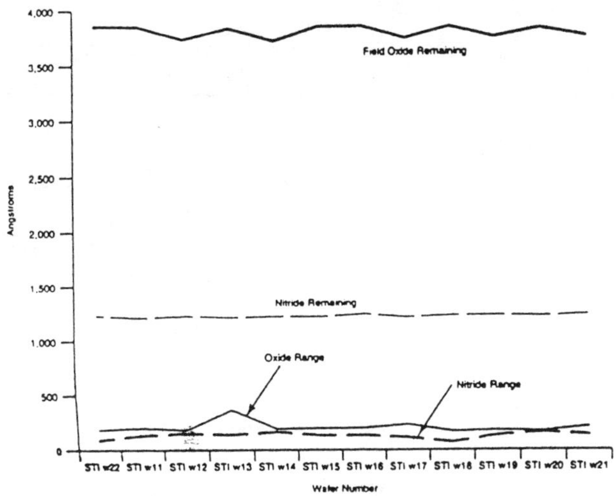

Fig. 8-66 Fixed abrasive polishing results on Linear CMP tool (roll-to-roll pad). Test performed on the Obsidian 501 CMP tool at the 3M Applications Lab. Courtesy 3M Co.

8.6.9 Endpoint-Detection in CMP

Since CMP is a process for reducing thickness at selected locations on a wafer, it is necessary to be able to determine when the process has reached its *endpoint*. There are two types of processes that need to be controlled: **1**) polishing a dielectric to a known final film thickness; and **2**) polishing an inlaid, usually multi-layered, metal film and stopping on the surrounding dielectric (after all of intra-inlaid metal is removed). The control of the first of these processes, that of polishing an oxide, can only be accomplished by making measurements on the film to be polished before the CMP process, and again after the wafer has been polished for some pre-determined time (based on the pre-polish film measurements). This is termed an *in-line endpoint-measuring* process. Another approach is to measure the film thickness as the polishing process is proceeding. This is called a *real-time, in-situ endpoint-detection* process. At the present time it is possible to control the second type of CMP process listed above (but not the first) by using *in situ endpoint-detection.* Nevertheless, both techniques will be described.

In the *in-line endpoint-measuring* method used for controlling dielectric-CMP processes, a pre-CMP measurement of the dielectric film to be polished is performed.[82] This measurement is made over an area containing dense device features to determine the thickness, T_0, shown in Fig. 8-17. Based on the

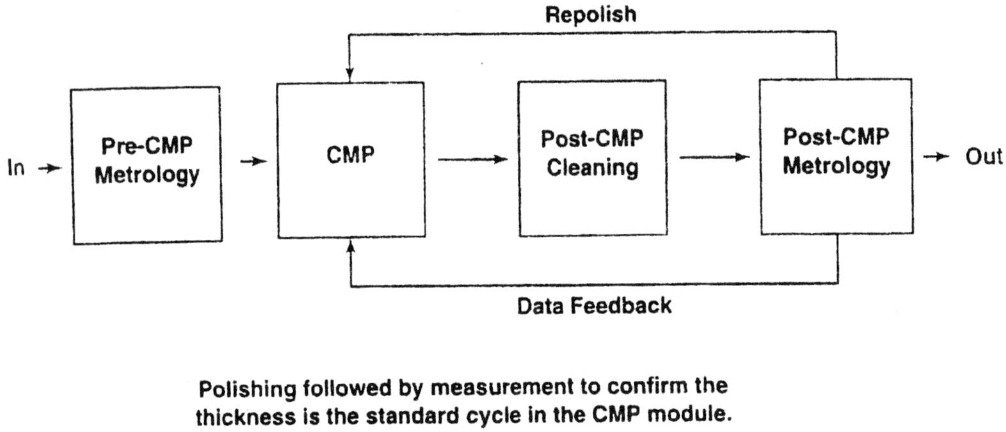

Fig. 8-67 *In-line* endpoint detection method used in CMP (i.e., polishing, followed by measurement to confirm thickness - and re-polishing if necessary - is the cycle used with *in-line* endpoint detection).

measured value of T_0, the polishing time (t_{final}) needed to reduce the film to a final required thickness, T_{final} (see Fig. 8-17) is calculated. The value of t_{final} is chosen from CMP calibration data based on applied pressure, rotation speed, and average pad-life-degradation. This produces a value of T_{final} centered within the allowed process window. Various factors may alter the actual value of T_{final} realized at t_{final} (e.g., by as much as 20%, 1-σ, during process development). If the material removed by t_{final} exceeds the allowed limit, the wafer must either be scrapped or more dielectric material must be deposited. If the remaining thickness is excessive, the wafer can be returned to the polisher for more removal. In fact, a typical manufacturing approach to oxide-CMP in which in-line endpoint-detection is employed, is to deliberately *underpolish* the SiO_2 layer. The film is then measured and polished a second time, using an adjusted polishing time. The cycle is repeated until the measured thickness is within specification limits. This procedure is often carried out on a look-ahead, or "pilot" wafer for each lot (Fig. 8-67), and the remainder of the lot is polished according to the time calculated when polishing the pilot wafer.[53]

This loop is the basic approach used in dielectric CMP (Fig. 8-68), but it has some drawbacks. The most significant are that the technique is time-consuming, measurement-intensive, and reduces the overall time the polisher can be kept in use. In addition, there is a risk of wafer breakage when being removed for measurement. Some thickness metrology tools are designed to be placed inside the polishing tool housing as another subsystem (e.g., the ITM® tool of Nova

Measuring Instruments). A central handling mechanism transports the wafer to the measurement position and then to a scrubber, then back to polish or to the output cassette. Multiple points on the wafer can be measured with a small optical head. This head (rather than the wafer stage) is moved, providing about 5 times faster movement.

The use of a *real-time, in situ endpoint-detection* method allows the problems of in-line endpoint-detection to be circumvented. *In situ* endpoint-detection is a way of identifying the completion of a polish step without having to move the wafer outside of the tool platform. *In-situ* endpoint-detection would permit the user to stop at the desired remaining thickness without having to re-polish. This would also improve throughput and increase the process window. Several methods have been developed to perform *in-situ* endpoint detection in CMP, including: **1**) front-side laser interferometry or spectrometry; **2**) optical-sensing of multiple-points on the backside of the wafer; and **3**) monitoring the current

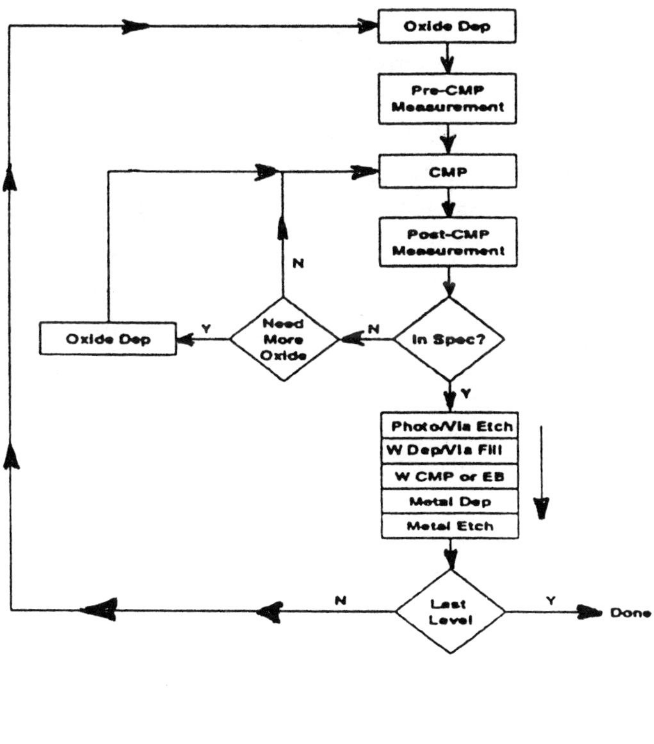

Fig. 8-68 Implementing the *in-line* endpoint-detection method used in CMP.

supplied to the motor of the wafer carrier. Unfortunately, as noted above, no satisfactory *in situ* methods yet exist for polishing homogeneous dielectric films to a known final thickness.

The *wafer-frontside laser-interferometry technique* is used by Applied Materials in their Mirra® tool, the *spectrometer reflectivity technique* on the Lam Research Teres® tool and in the Speedfam Auriga® tool. The name of the Applied Materials endpoint system is the *in-situ removal monitor* (ISRM).[54,55] Each polishing platen has a red laser and a detector mounted on the polishing table. A window in the polishing pad allows the laser to be aimed at the top surface of the wafer at 45° to the wafer surface. A detector is used to sense the reflected light (also collected at 45° from the surface). As the laser/detector passes under the wafer, the laser light passes through the clear window in the polishing pad, and reflections are sampled. The intensity of the reflected light is measured as a function of time (the light-intensity vs. time curve in Fig. 8-69).[56]

A typical reflectance curve for polishing Cu/Ta films on an oxide dielectric is shown in Fig. 8-70. After the maximum reflectance is reached, the process starts to break through the Cu layer to the semi-transparent Ta layer. At that point a sharp drop in reflectance occurs. Eventually, the copper filling the trenches continues to reflect a small constant amount of light.

The Applied Materials Mirra® CMP tool uses sophisticated signal processing/ detection algorithms to reduce noise from the slurry and pad surface condition (Fig. 8-71). In development tests, endpoint accuracies to within 20-nm were reportedly achieved. In W-CMP polishing, the ISRM system measures the reflectivity of the metal film and signals the transition from metal film to oxide. An over-polish time is specified to ensure complete removal of the liner film.

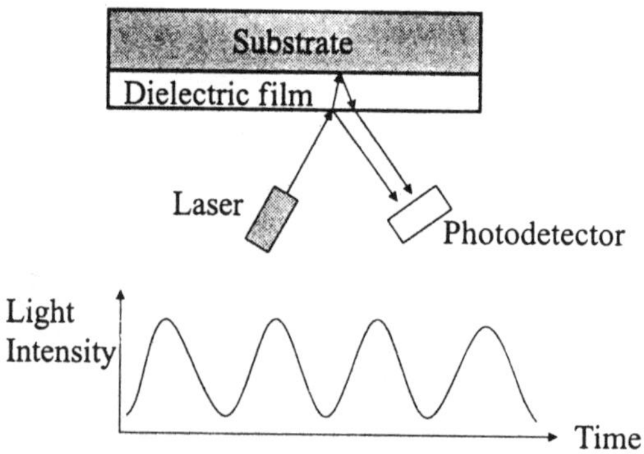

Fig. 8-69 *Wafer-frontside laser-interferometry* endpoint-detection for dielectric CMP.

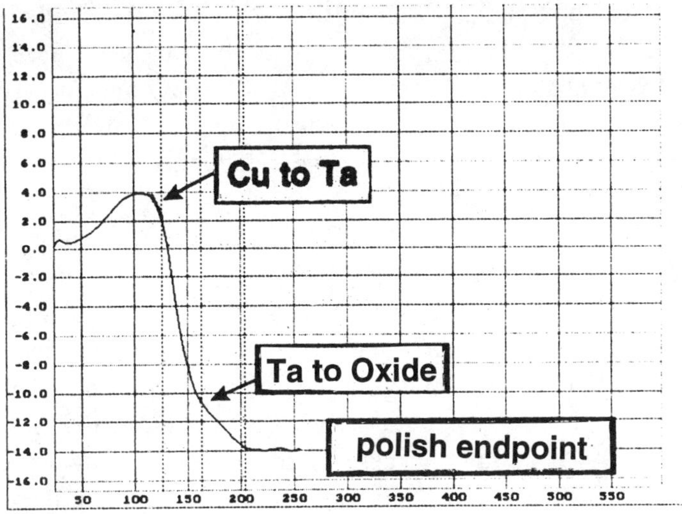

Fig. 8-70 Typical reflectance curve for polishing a Cu/Ta/oxide film.

In the Lam Research Teres® tool, a *multiple-wavelength, spectrometry-based endpoint-detection system* is employed. It can continuously monitor the thickness of transparent films as they are being polished. It can also measure the thickness of very-thin metal films. Thus, during a Cu CMP-process, the endpoint of the Cu polishing step can be sighted before all the Cu is removed. This allows a set of process conditions that rapidly removes Cu to be initially used. When the point of complete Cu removal is detected, the polish conditions can be switched to those that result in a slower Cu removal rate. This allows the Cu to be completely

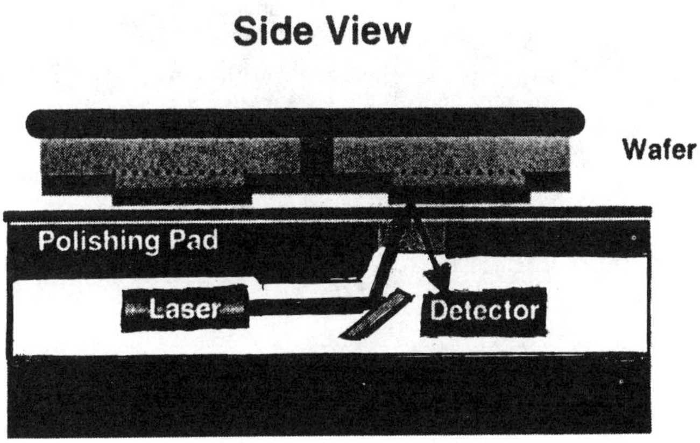

Fig. 8-71 Schematic of the Mirra® laser-reflectometer light path. Courtesy Applied Materials.

removed over the underlying Ta liner film without significantly eroding the exposed Cu in the trenches. In the Speedfam Auriga® multi-head polisher, a halogen lamp is used as the light source and diode arrays are employed to detect the light reflected from the surface of the wafer being polished. Endpoint in W-CMP is determined from an intensity change in the reflected-light as the exposed surface changes from tungsten to oxide.

The *optical monitoring of the wafer backside* as an endpoint-detection technique for monitoring the oxide-CMP process has been developed by Luxtron.[57] In this approach, the wafer's backside is illuminated with infrared radiation (to which silicon is substantially transparent). The radiation is therefore able to sense the film thickness on the wafer's front side as it is being polished.[58] By mounting several sensors onto the wafer carrier, multiple points on the wafer can be monitored. The film thickness is measured using the basic principles of optical interferometry as was described in earlier paragraphs. For wafers with no metal on the front surface, this technique is quite straightforward. However, if there is some metal on the front surface (e.g., in a multilevel-metal process), the situation becomes more complicated, because the metal will block some of the incident light. Fortunately, there is generally sufficient space between device features to allow light to pass. Figure 8-72 shows a schematic of a single sensor probe. Two optical fibers connected to an infrared emitter and detector are mounted in a modified wafer-carrier of a commercial CMP tool. Electrical communication from the sensors is accomplished via a wireless link. As many as four sensors can be mounted in a single carrier.

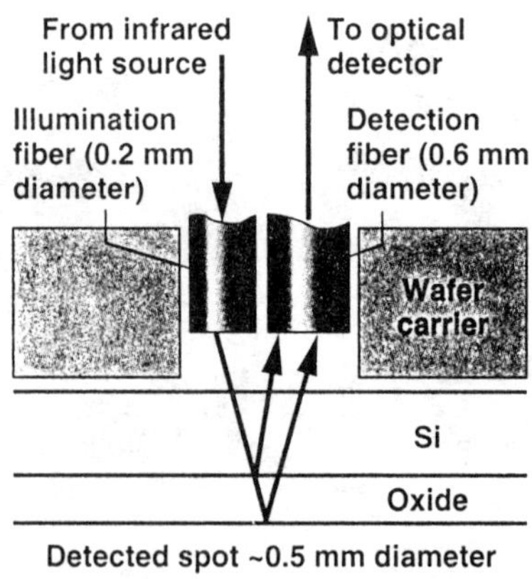

Fig. 8-72 Schematic of an optical sensor embedded in a wafer carrier.[57] Wafer is backlighted with IR light to measure films on the front surface. Reprinted with permission of Semiconductor International.

The *motor-current-monitoring technique*[57] is a production-proven endpoint-detection method that works well when polishing a material down to a stop layer (such as the polishing of a CVD-SiO_2 film on a silicon nitride stop layer in shallow-trench isolation,[58] and the polishing of W over oxide in W-plug CMP processes[57]). It is also relatively easy to implement. Circuitry such as a *current shunt* or a *Hall-effect probe* is used to monitor the current supplied to the motor that rotates the wafer-carrier (Fig. 8-73a). Since the wafer-carrier is driven at a constant rotational speed to maintain a constant polishing rate, the drive-current is varied to compensate for any load changes on the motor. This makes the motor-current sensitive to frictional changes at the wafer surface. The largest changes occur when one material has been polished away, exposing a layer which has different polishing characteristics (e.g., that could function as a CMP-stop layer, see also Fig. 8-73b). Since there is no layer breakthrough in oxide-CMP (as can be seen in Fig. 8-25a), the motor-current-monitoring technique is not appropriate for such applications. Luxtron and Aplex are vendors of such motor-current endpoint controllers (e.g., the Luxtron Xinix Model 2450®).

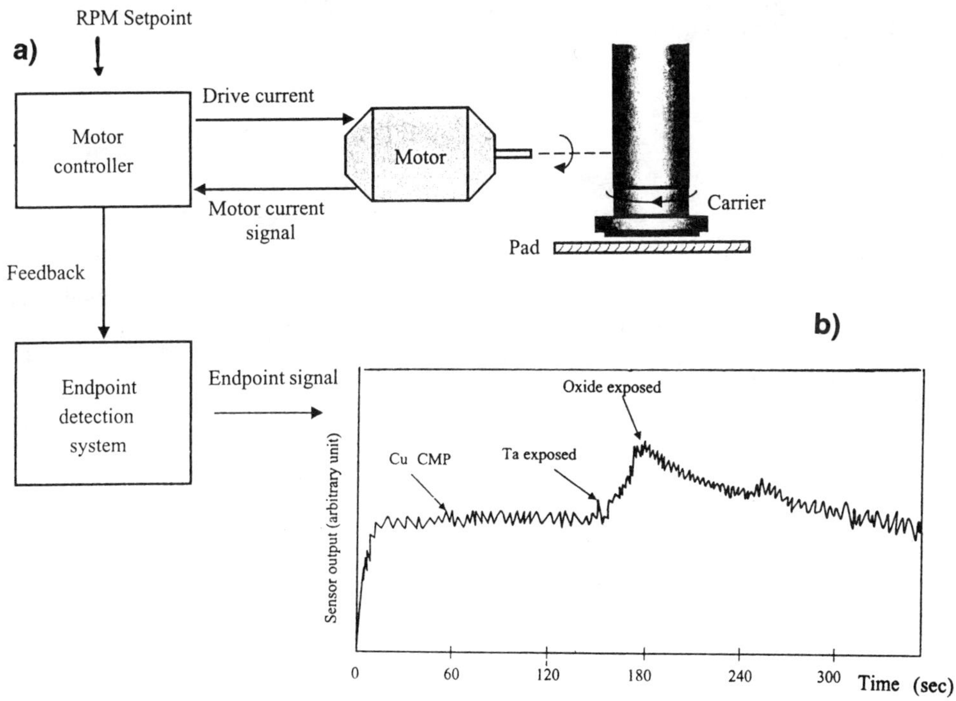

Fig. 8-73 a) Motor current endpoint detection. b) Motor current output during Cu/Ta/oxide CMP process. Courtesy of Aplex.

8.7 CLEANING ISSUES IN CMP

CMP is an inherently dirty process that can introduce yield-reducing defects to the wafer. That is, after a polishing step, the wafer-surface contains slurry particles, abraded pad-material, and debris that was removed from the wafer-surface. Residual slurry specifically introduces foreign particles, metal contaminants, and chemicals; all of which are undesirable for the remaining steps of chip fabrication. If the slurry particles alone were allowed to remain on the wafer, such contamination could obstruct electrical connections, create voids during subsequent oxide- or metal-deposition steps, or cause corrosion of exposed metal areas. Some early resistance to the adoption of CMP stemmed from such contamination concerns. As a result, significant effort has been expended to develop post-polish cleaning techniques so wafers returned to the cleanroom after CMP are not contaminated.[60]

To prevent CMP-based contamination, wafers are subjected to post-CMP cleaning steps that return the wafer-surface to an acceptable level of cleanliness. While the main objective is to provide a wafer-surface that is free of slurry particles, the cleaning process must also remove organic-residues and trace metal ions. The cleaning process must do this without introducing further defects. In the case where inlaid metal is exposed (e.g., in a W-plug CMP process or a in Cu-damascene CMP process), corrosion must also be controlled or avoided.

Since CMP is a wet process, CMP cleaning-technologies are mostly aqueous. The post-CMP wafers are kept wet as they are moved from the polishing tool to the cleaning station. The initial wafer-rinsing (after the polishing is completed), is done while the wafers are still on the polishing pad. Spray rinsing is also done as soon as the wafers are lifted from the pad. If wafers are exposed to air for even one or two minutes before such spraying (e.g., while waiting to be unloaded), subsequent cleaning becomes difficult (and sometimes impossible). Immediate removal of metal-polish slurries is also critical, because these can continue to react and remove material even after polishing has ceased.

Current models of CMP polishing tools either have an on-board (i.e., fully integrated) cleaning station, or they have the capability to interface with a stand-alone cleaner (in the form of a cassette-to-cassette track with stations for brush scrubbing [one or more], rinsing, and spin-drying). If a stand-alone cleaner is employed (as was common in early CMP processes), the wafers are transferred from the polisher to the cleaner by a human worker. If the cleaner is integrated with the polisher, (the most common method in the latest generation of CMP tools), automated wafer transfer is used (i.e., the wafer is moved from the polisher to the cleaner by a robot).

To prevent additional wafer and slurry contamination, the number of airborne particles should be minimized by filtering the air fed to the CMP tool with the same HEPA filtering techniques used for the cleanroom environment. The cleanliness of the DI water used to rinse or submerse the wafers during cleaning is also an issue. Tubs or baths in which wafers are kept wet must be continuously refreshed with new DI water. The CMP tool should automatically rinse all surfaces that come in contact with slurry, including the polishing table, drain tubs, carriers, and pad conditioner. These areas should also be cleaned periodically to prevent dried slurry buildup. As mentioned previously, slurry is composed of very small abrasive particles, and filtering by conventional methods is not easy. Nevertheless, such filtering is needed because large agglomerates can develop within the slurry and have serious device-defect consequences.

8.7.1 Mechanisms of Particle Adhesion and Removal from Surfaces

Particles are deposited on wafers as aerosols from the air or as particles present in the liquids in which wafers are immersed. A description of the techniques used to remove them (e.g., after a CMP polishing step) rests on an understanding of how such deposited particles adhere to wafer surfaces. Thus, a short description of this phenomenon follows.

If a solid particle wanders close enough to the vicinity of a solid surface, the *Van-der-Waals force* (*VdW*) can attract and capture it. The particle is then *physisorbed* on the surface. The Van-der-Waals force is universal, and is thus effective in trapping all types of solid particles (and gas molecules, as well). Other forces that physisorb particles include *static charge on the particle*, and *capillary adhesion* (the force exerted by a liquid film between the particle and the surface), but the VdW force is usually the strongest of these. In fact, it can be so strong that the particle may elastically (and sometimes even plastically) deform so that the area of contact between the particle and the surface is increased. This makes such binding even stronger. Nevertheless, particles held by VdW forces can be removed by the techniques described here.

There are three methods used to remove physisorbed particles: **1**) *electrical repulsion of the particle and the surface*; **2**) *chemically assisted removal* (particle dissolution, particle oxidation [and then dissolution], and lift-off by slight etching of the surface); and **3**) *mechanical dislodgement* (sliding and rotating the particle by the application of a shear force). These mechanisms are described in more detail in the next sections.

Note that cleaning by immersion-only is not effective for post-CMP cleaning, due to the particle adhesion forces and boundary layer effects. That is, merely rinsing in DI-water or immersing the wafers in a wet-chemical solution has not been found to adequately remove slurry particles from the wafer-surface.

8.7.1.1 Electrical Repulsion of the Particle and the Wafer Surface: Besides the particle adhesion forces discussed above, other inter-particle (or particle-to-surface) forces exist which can be either attractive or repulsive. The most important of these forces are those involving the charge near the particle surface. That is, the outer-most charge near the surface of a particle most influences the way a particle will interact with other particles or surfaces. Particles and surfaces immersed in a liquid can be charged by the adsorption of ions from the solution. The charge thus adsorbed on the surface is balanced by an equal but oppositely charged layer in the adjacent liquid, resulting a so-called *electric double layer.* (Fig. 8-74) The potential associated with this adsorbed charge on a particle or surface is termed the *Zeta potential* (ZP) and is also depicted in Fig. 8-74. Since the particle or surface adsorbs ions from the liquid, the ZP of the particle is also

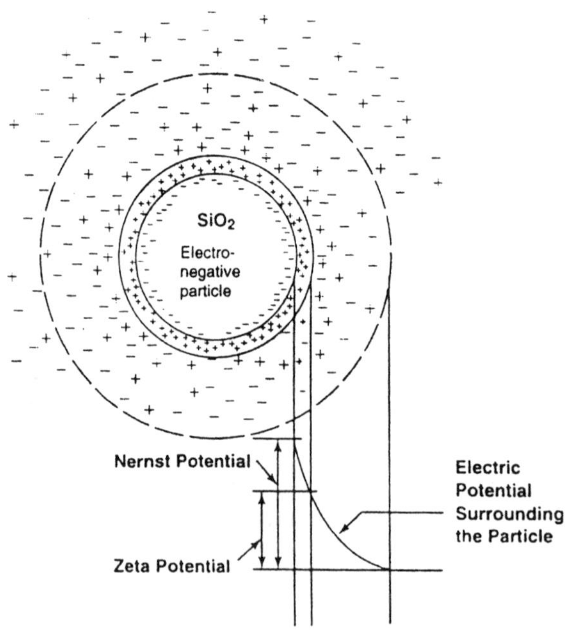

Fig. 8-74 Slurry chemistry electrostatic layers surrounding an SiO_2 particle (including a depiction of the electric dipole layer).

consequently dependent on the pH of this surrounding media. For example, Fig. 8-75 shows the ZP of a variety of particle types (SiO_2, Al_2O_3, W, poly-vinyl alcohol [PVA], and Si_3N_4), as a function of the pH of the liquid in which they are immersed. The silica particles have a negative potential of –30-mV if they are immersed in a liquid with neutral pH (i.e., pH = 7). Also note that their ZP (as well as those of the other particles shown in Fig. 8-75) becomes strongly negative as the pH is increased above pH = 7. (The ZP is also affected by the addition of a surfactant to a liquid, and thus the ZP of a solid-surface can likewise be manipulated by the addition of surfactant to the liquid.)

To prevent slurry particles from sticking together, it is best to immerse them in a solution that will stabilize their respective ZPs to either a positive or a negative value. This will ensure that the particles will repel each other (i.e., this repulsion - see Fig. 8-76 - will counteract such attractive forces as the Van-der-Waals forces). Consider the case of silica particles in a silica-based slurry. Since silica particles tend to have negative values of ZP, the net-negative electric charge near the particle-surface produces a force that tends to make the silica particles repel each other. However, to ensure that these repulsive forces are strong enough to overcome the opposing attractive forces (such as VdW), a stabilizing chemical that increases the pH of an aqueous solution (e.g., NH_4OH) is added to the slurry.

From the data shown in Fig. 8-75 it can be seen that to keep the silica particles in a slurry from agglomerating, they should be immersed in a liquid with pH values > 7 (basic). On the other hand, for slurries containing alumina particles, one might want to consider using a slurry solution with a low pH (i.e., < 7 [acidic]), because in such an acidic solution the alumina particles exhibit a positive ZP.*

To remove particles from a surface, consider the case of removing silica-slurry particles from a wafer-surface covered with oxide (i.e., a condition that exists at the completion of an oxide-CMP step). Since the ZP of the SiO_2 surface is also negative (at neutral pH), it is preferable to use a cleaning chemistry with a high pH (i.e., > 7). This causes the silica slurry particles adhering to the oxide

* We noted earlier that the CMP process also consumes large volumes of DI water. Work has thus been carried out to develop methods of treating the CMP wastewater so that it can be re-used. One phase of such reclaim treatments involves removing the slurry particles from the wastewater. For the case of W-CMP waste water, it was found that by a simple adjustment of the pH of this water, the slurry particles would agglomerate into large clumps, making their capture and removal quite easy.

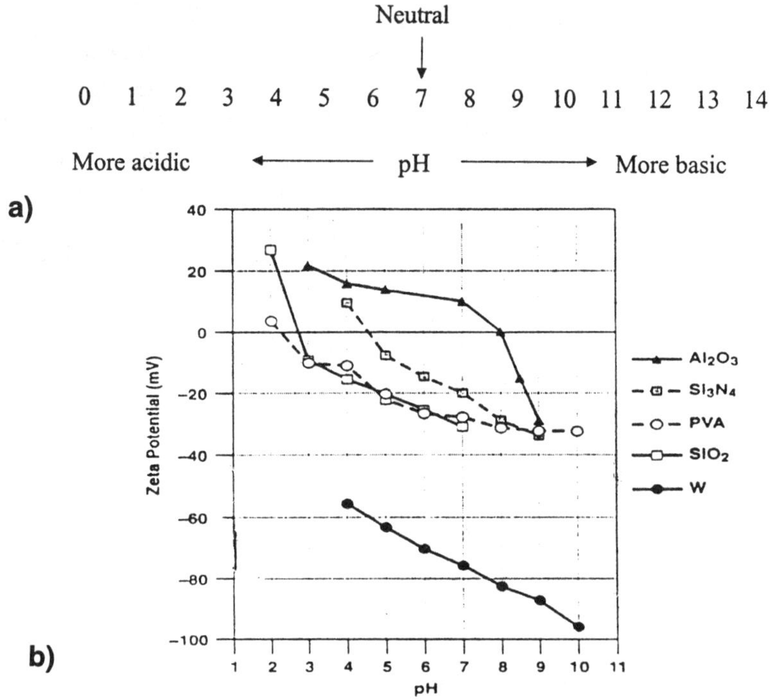

Fig. 8-75 a) Relationship of pH and acidity and alkalinity. b) Zeta potentials versus pH for particles of various types of slurry particles and materials in an aqueous solution.

surface to acquire additional negative charge, which increases the repulsive component of the force between particles and the surface, making any par-ticles adhering to the surface easier to remove. Note that in most cleaning solutions, chemistries are extremely dilute (i.e., 1:50, chemical:DI water).

It should also be mentioned that for many CMP processes, including W-CMP and Cu-CMP, the exposed wafer-surface at the conclusion of the polish-step contains more than one type of material (i.e., not just oxide). For example, when the CMP-process for polishing W-plugs (or Cu-interconnect-lines) is completed, more than 50% of the exposed wafer-surface area consists of oxide, and the remainder consists of metal. Nevertheless, since the majority of the exposed surface is still oxide, the most effective cleaning chemistries will still be those that remove slurry particles from oxide-surfaces. Hence, cleaning solutions used to remove silica-slurry- particles from these surfaces should have pH values > 8.

The same considerations apply to the cleaning of slurries that use alumina particles. That is, as shown in Fig. 8-75, alumina particles have a positive ZP if they are immersed in a solution with a low (acidic) pH (i.e., < 8). Thus, if an acidic cleaning solution were to be used, the alumina particles would be attracted to the electronegative oxide surface. To prevent this situation, the cleaning

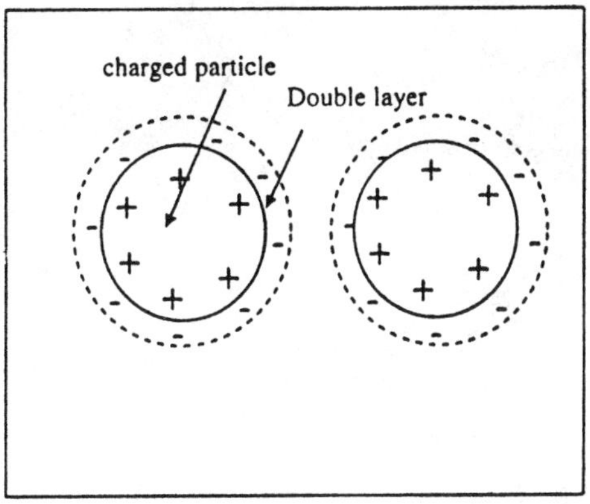

Fig. 8-76 Electrostatic repulsion keeps particles separated by repulsion of charged double layers.

solution must again have a high pH (> 8). This makes the ZP of the alumina particles negative, so they are repelled from the oxide surface on the wafer.

Figure 8-77a compares brush-cleaning results after oxide-CMP with and without NH_4OH in the cleaning solution. Figure 8-77b compares the results after W-CMP with and without NH_4OH. In both cases, the use of NH_4OH (which raises the pH of the solution) is seen to reduce defects by as much as 75%.

8.7.1.2 Chemically Assisted Particle Removal: The *chemically assisted* particle removal techniques are implemented in some wet-cleaning procedures (Fig. 8-78). For example, in the SC-1 step of the RCA clean (see Vol. 1, 2nd Ed. Chap. 5), several chemically-assisted particle removal mechanisms are believed to work in tandem. That is, the SC-1 solution removes some types of particles by dissolution (i.e., organic particles). For the case of cleaning bare silicon wafers, it removes other types of particles by slightly simultaneously oxidizing and etching the Si surface. That is, it undercuts the oxide under a particle, which reduces the contact area between the particle and the surface. This makes the particle more prone to being lifted off.

8.7.1.3 Mechanically Assisted Particle Removal: The *mechanical removal of particles* generally proceeds by momentum transfer. That is, shear forces impact a particle and dislodge it by either sliding it off the surface, or by rotating it until it detaches. As particle sizes decrease, they become more difficult to dislodge by mechanical force. Although the Van-der-Waals-adhesion force may get smaller, the electrostatic forces increase, and the efficiency of transferring translational momentum also rapidly declines. Mechanical particle removal techniques have traditionally involved: **1**) mechanical brush scrubbing; and **2**) vibrational

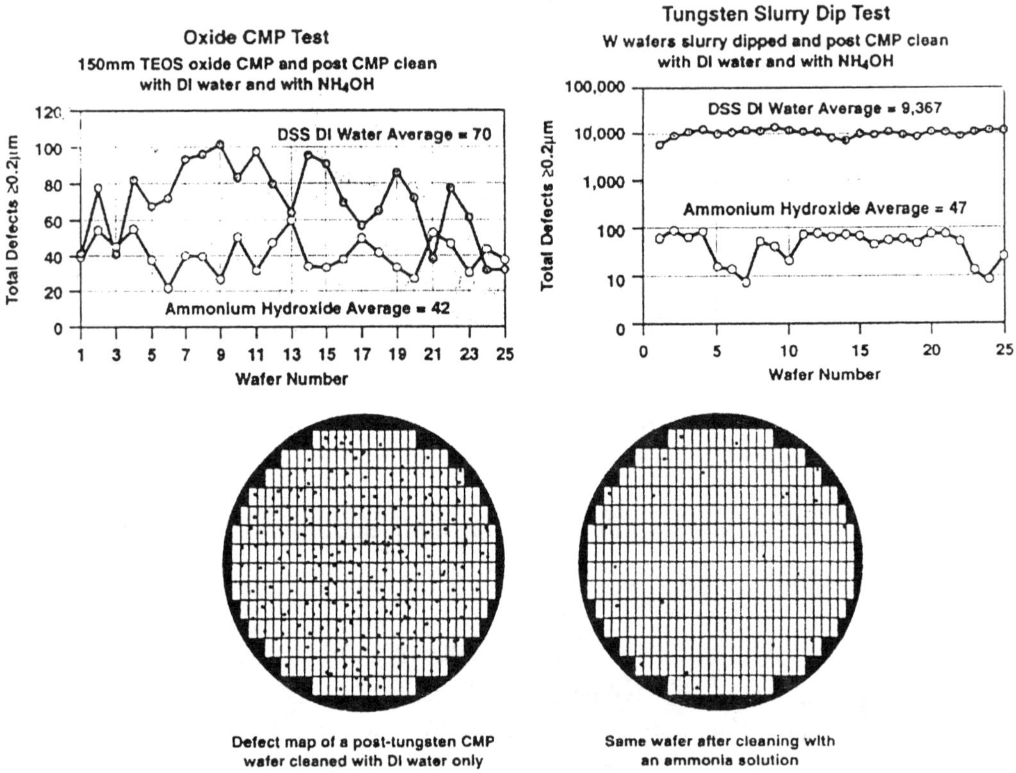

Fig. 8-77 Brush cleaning results comparing use of H_2O(DI) vs. NH_4OH: a) Oxide-CMP cleanup; b) W-CMP cleanup.

(*megasonic*) action. In fact, these particle removal mechanisms are employed in tandem in post-CMP cleaning processes. An explanation of how this is done will be provided in the upcoming sections.

8.7.1.4 Steric Stabilization for Suspending Particles in a Solution: Particles can be prevented from attaching to each other, or from settling on a surface, by the phenomenon of *steric stabilization.* That is, a molecular layer is attached to the particle surface (often a *dispersant* like a polymer or a surfactant is used for this role). This attached layer causes *entropic repulsion* as the particles approach each other (Fig. 8-79). This technique is a second method used to prevent settling and agglomeration in CMP slurries. If this mechanism is used to stabilize a slurry, care must be taken to ensure that the additive does not adversely affect any of the other slurry properties.[61]

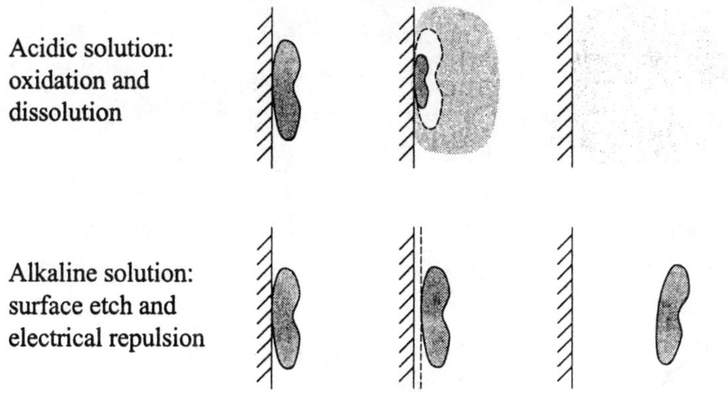

a)

Fig. 8-78 Wet-chemical mechanisms that remove particles from surfaces: a) in a powerful oxidizing agent; b) in an alkaline solution. Courtesy of Semiconductor International.

8.7.2 Practical Particle Removal Processes: Brush Scrubbing

The most widely used method of post-CMP-cleaning involves brush scrubbing the wafers while they are sprayed with appropriate chemical solutions (or while they are immersed in these solutions). Brush cleaning systems have long been used in many wafer-cleaning applications. Despite its name, it actually uses a hydrodynamic drag to exert a removal force on the surface-particles (mechanical dislodgment). Brush bristles do not contact the particle or the surface, but rather act as oars or paddles that push liquid across the wafer surface.

While the brush scrubbing technology is not new, some modifications had to be made to allow a stand-alone brush scrubber to be integrated with a CMP

b)

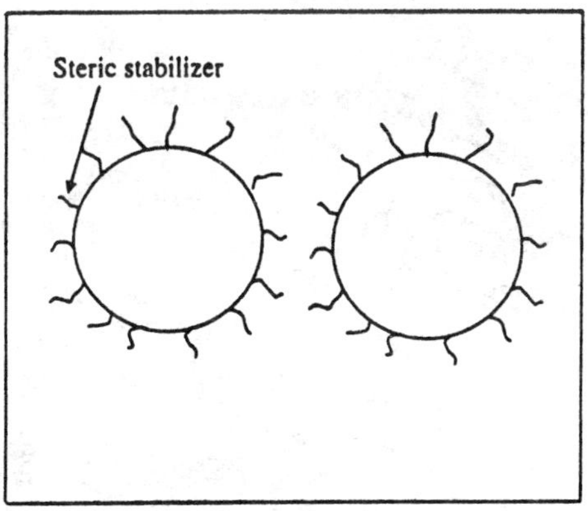

Fig. 8-79 Steric-stabilization of particles in aqueous solution. Particles are stabilized by attaching a molecule to the surface. The molecular layer causes an entropic repulsion as the particles approach each other.

polishing tool. For example, the current technology of choice for both oxide- and metal-CMP employs simultaneous double-sided brush scrubbing (see Fig. 8-80).[62] This simultaneously scrubs both the front and back of the wafer. Another change involves the fact that the scrubbing equipment must function reliably in the presence of both basic slurry chemistries (e.g., KOH and NH_4OH), and acidic cleaning chemistries (such as citric acid and HF), in order to optimally remove both metals and mobile ions. Alumina-bearing slurries used for tungsten-polishing more readily adhere to the wafer surface and are generally more difficult to remove. Particles as small as 0.12-μm have reportedly been effectively removed by brush scrubbing.

Particles are dislodged and they then stay suspended in the liquid until they are pushed over the edge of the wafer into the drain, as liquid is continuously delivered to the wafer surface. Dislodged particles, however, must be prevented from re-adhering to the wafer surface or the brush material. This is done by controlling the electrostatic forces between the particles and the wafer or brush surfaces. As described earlier, particle re-adhesion to both the wafer surface and the brush is suppressed by keeping the pH values of the cleaning liquid greater

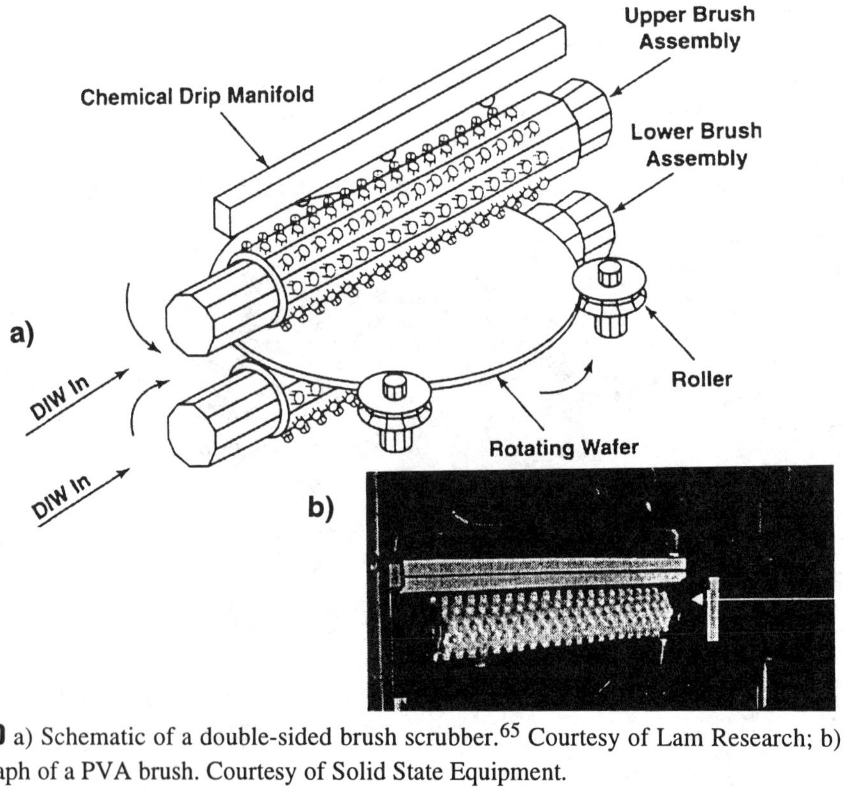

Fig. 8-80 a) Schematic of a double-sided brush scrubber.[65] Courtesy of Lam Research; b) Photograph of a PVA brush. Courtesy of Solid State Equipment.

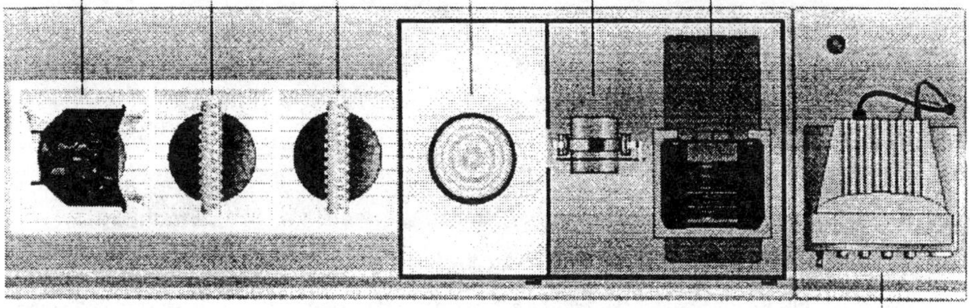

Fig. 8-81 Example of a post-CMP cleaning tool. Courtesy of Lam Research.

than 8. The post-CMP cleaning tool typically consists of a load station, a brush scrubbing module (or modules), a rinse-and-dry module, and an unload station (Fig. 8-81). Isopropyl alcohol and a heat lamp may be used to speed up drying.

8.7.2.1 Brush Scrubbing to Remove Particles in Post-Oxide-CMP Processes: When brush scrubbing wafers after an oxide-CMP process, a cylindrical polyvinyl-alcohol (PVA) brush (with or without nodules that protrude from its surface) is rotated across the surface of the wafer (Fig. 8-80), as the wafer is continuously flushed with the cleaning solution. The PVA brush-material is a soft, sponge-like material that is chemically resistant and highly compressible. This material allows the mechanical component of cleaning to occur without damaging the oxide surface. The brush is thus brought into direct contact with the wafer surface. Unlike rigid brushes such as nylon (commonly used in the 1970s), PVA does not cause scratches when brought in contact with the wafer surface.[63]

The PVA material is a sponge-like material that has an open structure consisting of interconnecting cells that allow the brush to be constantly flushed with DI water during wafer cleaning. These softer brushes clean more efficiently when they are compressed against the wafer. PVA brushes eventually wear out and need replacement. The average use rate is 20,000 wafers for post-oxide CMP cleaning, and 10,000 wafers for post-metal CMP cleaning.

It has been reported that higher brush speed can improve the particle removal efficiency without introducing scratches. For post-oxide CMP-cleaning (in which the CMP process uses silica-based slurries), the cleaning solutions use mixtures of DI-water and NH_4OH (and often H_2O_2, as well) to decrease silica-particle adhesion to the exposed oxide surface of the wafer.

One problem has been identified with respect to post-CMP cleaning, namely the difficulty of removing particles from non- or partially-planarized recesses. That is, as shown in the AFM image of Fig. 8-82,[64] a slurry particle resides inside a 0.7-μm contact hole with a 3:1 aspect-ratio after brush cleaning with dilute NH_4OH, and similar results were obtained with megasonic cleaning. It was observed that the ratio of the step-height to the horizontal-dimension of the feature is crucial for the cleaning capability of a specific cleaning technique.

8.7.2.2 Brush-Scrubbing to Remove Particles in Post-Metal-CMP Processes: In post-metal-CMP cleaning,, brush-scrubbing is also effective, but the process is modified from the one used for post-oxide-CMP brush-scrubbing. This is necessary because the post-metal-CMP process must: **1**) leave the surface of the metal without an oxide layer; **2**) leave the metal smooth; and **3**) not attack the metal. In addition, when cleaning wafers after a W-CMP-step, the wafers come from an acidic environment (2–4 pH) to a water-rinse and PVA-brush-clean. Some PVA materials cannot endure this chemistry well and must be frequently replaced. Other PVA materials (believed to be free of unreacted residues that hydrolyze in the acidic environment) appear to render them stable over an operating pH range of 2–12. These materials are reportedly compatible with post-W-CMP scrubbing.[66] It is reported that rinsing the wafers with dilute NH_4OH during post W-CMP scrubbing is also highly effective in removing alumina slurry-particles.[67] An additional benefit of using NH_4OH is that it prevents loading of the PVA brushes by alumina particles or tungsten oxides.

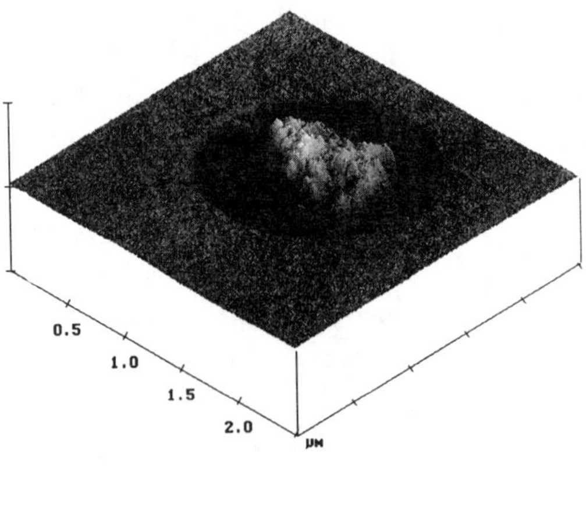

Fig. 8-82 AFM image of an agglomerated-silica slurry-particle in a contact hole.

Cleaning wafers after copper CMP presents serious challenges, including the following:

1. *The CMP-clean-step must not allow Cu-contamination of the fab cleanroom area*: That is, the Cu-CMP cleaning process must remove Cu from the front, back, and beveled edge of the wafer. However, the copper oxides formed and removed as part of the CMP process do not dissolve in DI water and may be easily transferred to the PVA brushes. If the brushes become "loaded" with these copper oxides, they may transfer them to subsequently processed wafers, thus producing severe Cu-contamination in the fab. Proprietary chemicals have been developed to dissolve the copper oxides and prevent such brush loading.[68] In the On-Trak systems brush-scrubbing tool, such a chemical is delivered to the wafer surface through the porous core of the brush itself (Fig. 8-83).
2. *The cleaning chemistry must not corrode the Cu*: If NH_4OH is used in the cleaning solution, great care must be exercised. This is because ammonia is known to complex with copper and cause surface roughness and corrosion products.[69]
3. *The Cu surface must be protected during the cleaning step*: Copper has a soft, porous oxide-layer through which chemicals can diffuse and cause further oxidation/corrosion. Protection of the copper-surface while cleaning the dielectric surface free of metals may be accomplished by using inhibitors.[70]

Note that more information on CMP of Cu is found in Chap. 16.

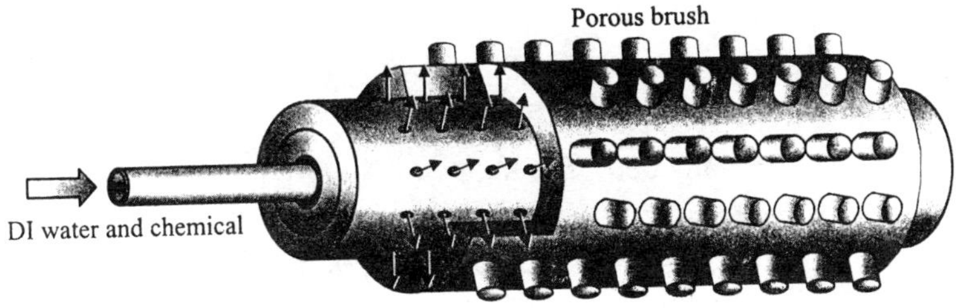

Fig. 8-83 *Through-the-brush* Chemical Delivery for post-CMP cleaning.

8.7.3 Vibrational Scrubbing to Remove Particles (Megasonic)

Brush scrubbing, while effective in removing particles, may cause scratching and brush shedding (which contributes additional particles). Vibrational scrubbing of wafers is an alternative non-contact and brushless technique for removing particles (Fig. 8-84a). Megasonic immersion cleaning systems for CMP using high-frequency sonic waves (0.7-1-MHz) are commercially available. The model of megasonic particle removal is shown in Fig. 8-84b. The wafers are immersed into an SC-1 solution contained in a megasonic tank. Sonic energy from acoustic transducers is directed parallel to the wafer surface, and this reduces the liquid boundary layer that surrounds a particle. The particle is thus subjected to larger fluid-acoustic-streaming velocities and energy from cavitation. These detach the particle from the surface.[71] The particles in the liquid must then be removed by passing the liquid through a continuous filtration system. This keeps them from remaining in the liquid where they could wander into the vicinity of the surface and be re-captured. Such filtration should also remove any particles generated by the megasonic system itself. (That is, particles shed by its own gaskets and transducer-bonding-material when the transducers are vibrating.)

The most common megasonic systems are batch-cleaning-systems. However, one recent report described a single-wafer megasonic-system that used a dilute basic-chemistry and cleaned each wager for 15-60 sec.[72] Cleaning with this system after an oxide-CMP-step yielded wafers with only 50 defects per wafer, well below the levels obtained when using brush scrubbing.

Megasonic cleaners may also offer cleaning improvement when topography such as alignment marks, scribe lines, or W-keyholes are encountered. That is, as discussed in the section dealing with *Brush Scrubbing for Oxide-CMP*, when particles are trapped within high-aspect-ratio recesses on the wafer surface, brush scrubbing is not effective in removing them. On the other hand, megasonic cleaning does a better job of dislodging particles in recessed features. Another advantage of megasonic cleaners involves cost. That is, cost is reduced since brush changes are not needed. The Applied Materials Mesa® combines brush scrubbing with a megasonic cleaner to remove particles in deep recesses. Other vendors, such as Verteq, also offer non-contact, non-scrubber CMP cleaners.[73]

8.7.4 Secondary (or Buff) Polish Procedure as a CMP Cleaning-Step

The polishing process itself can play a role in achieving low particulate levels at the conclusion of the CMP process. That is, by using a secondary buff-polish (i.e., polishing with a soft pad) and following this buff-polish with a high-pressure DI-water-rinse, fewer particles remain on the wafer after the subsequent cleaning step. As a result, a second platen for buff-polish has become a standard feature for commercial polishers. It is also reported that if DI-water is used in the slurry (i.e., instead of a basic solution), particles counts are further reduced as the pressure during the buff-polish is increased.[74]

8.7.5 Post-CMP-Cleaning of Metallic Contaminants

Besides the removal of particles, post-CMP cleaning processes must also ensure that no metallic contaminants remain on the wafers when they are returned to the cleanroom area of the wafer fab. Since some commercial slurries are KOH-based, an effective clean must be employed to remove surface metals, especially alkaline metals. It has been reported that at least several-hundred angstroms of the top-oxide must also be removed to ensure that alkaline metal levels are adequately removed from the wafer-surface. Consequently, an "HF-deglaze step" is usually employed to reduce metallic contamination (and also any defects embedded within, or close to, the surface).

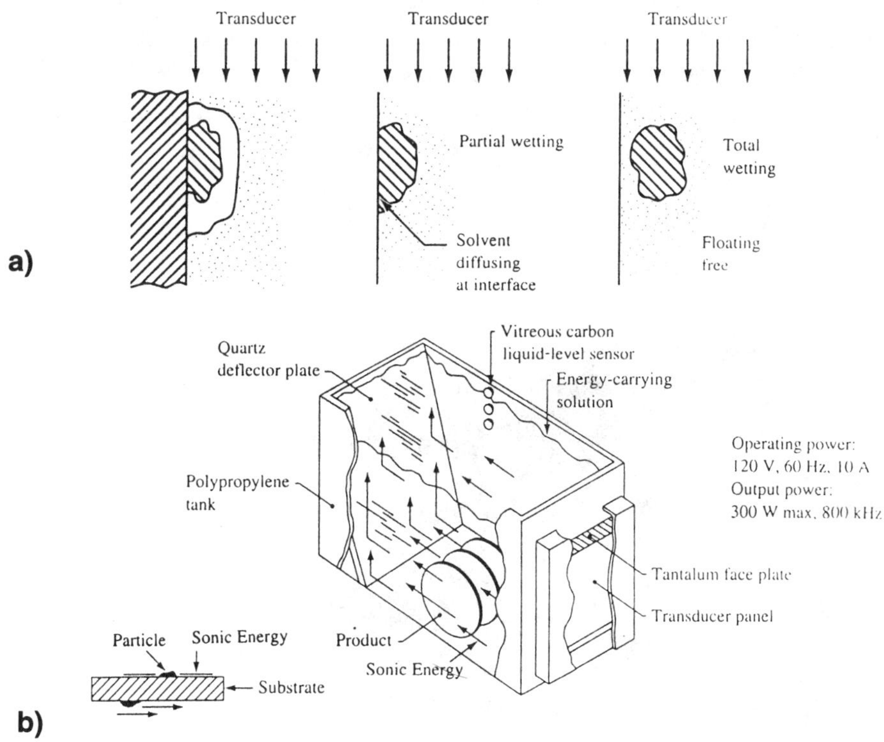

Fig. 8-84 a) Model for the mechanism of particle removal by megasonic scrubbing; b) Schematic of megasonic cleaning tank.

8.7.6 CMP Cleaning Equipment

On-Trak Systems (a division of Lam Research), supplies stand-alone double-sided scrubbing units for the post-CMP cleaning, as does Dainippon Screen. Other LAM post-CMP cleaning tools (e.g., the Synergy Integra® model, Fig. 8-85), are designed to be integrated with the polishing tools of other CMP vendors (including Applied Materials, SpeedFam, and Strasbaugh). Lam Research held about 70% of the post-CMP cleaning-tool market-share in 1999.

However, the Mirra® can also be integrated with an AMAT cleaning tool (the Mesa® Fig. 8-86).[75] It consists of four separate process modules: a megasonic bath; two-double-sided brush-scrub modules; and a spin-rinse-dry module. Each module cleans both sides of the wafer, and all handling is done by edge grip. Another fully integrated cleaner in a CMP tool is shown in Fig. 8-87.[76]

CMP processes are generally located adjacent to the cleanroom, connected through a bulkhead load, similar to etchers or CVD systems. In some cases, the CMP-tool-configuration consists of a polisher and stand-alone cleaning system with a robot that transfers wafers between them. A problem with this configuration being adjacent to the cleanroom is the potential for water and slurry spillage (and then leakage into the cleanroom) as transfers are made. A better solution is a *dry-in/dry-out CMP system*, in which the polisher and cleaning system are completely integrated. Wafers are planarized, cleaned, dried, measured, and returned back to the cassette located in the cleanroom.

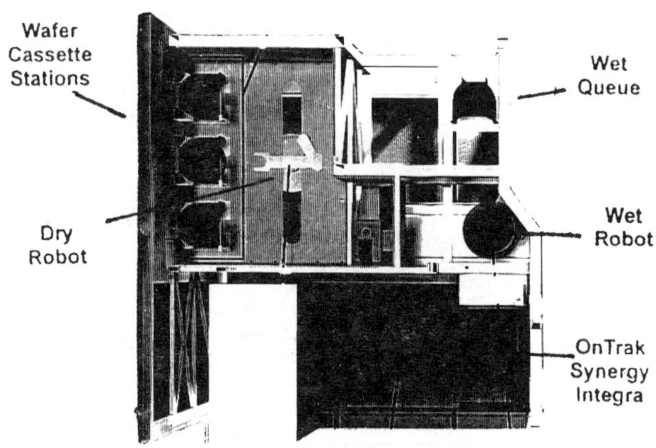

Fig. 8-85 Schematic of the Lam Research Synergy® post-CMP cleaning tool. Courtesy Lam Research.

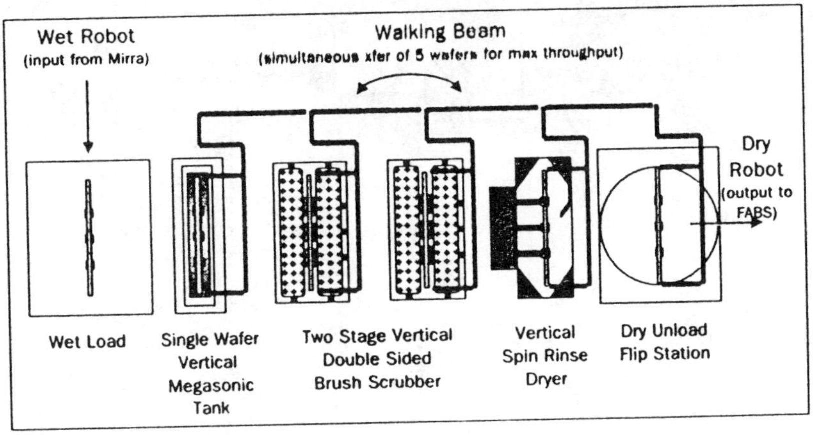

Fig. 8-86 Mirra Mesa® cleaning module layout. Courtesy of Applied Materials, Inc.

8.8 CMP METROLOGY

There are a number of metrology requirements in CMP. First, it must be determined when the CMP process is completed (*endpoint detection*, Sect. 8.5.3.4). Second, the film-thickness after CMP must be measured, including thickness-uniformity across a wafer. Third, the degree of film-planarity must be quantified after CMP. Finally, films subjected to CMP must be inspected for defects and damage. The latter two CMP metrology topics are discussed here.

Film thickness after CMP (and before CMP in in-line endpoint detection), is performed with a number of different techniques, including: spectrophotometry, ellipsometry, and beam-profile reflectometry. One problem associated with these techniques is that of *cycle-skipping* (or *order-skipping*). Since the signal obtained from the thickness-measurement is periodic, the same measured-data can fit more than one thickness-value. For oxide, the resulting errors are multiples of about 240-nm, enough to lead to incorrect conclusions about the post-CMP oxide thickness (which might cause properly processed wafers to be scrapped). Different strategies by each measurement technique are employed to avoid such problems. Ellipsometers use additional wavelengths, spectrophotometers rely on better signal-to-noise data-collection, and beam-profile-reflectometry is combined with spectrophotometry to avoid cycle-skipping errors.

One factor that impacts *thickness uniformity after CMP* is the uniformity of the wafer itself. It is observed that the *within-wafer-nonuniformity* (*WIWNU*) of the film removed by a CMP-process increases linearly with an increase in the

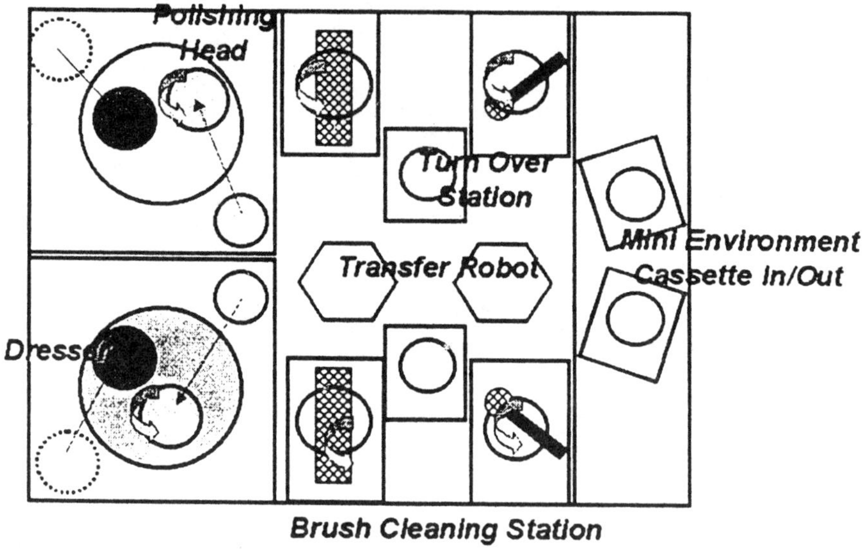

Fig. 8-87 Ebarra CMP cleaning module. Courtesy of Ebara.

total-thickness-variation, TTV, of the wafer.[77] In addition, wafers may not retain their initial flatness as they undergo the various steps of the process sequence. In particular, stress arising from thin-film-deposition processes can deform a wafer into a concave-shape (negative bow) or a convex-shape (positive bow). Although bow is reduced by the pressure applied to the wafer during CMP, some remains (as the degree of bow can vary from wafer-to-wafer). In addition some residual local-warpage persists after the wafer has been pressed against the wafer platen. Thus, the CMP-process may polish the edges faster than the center in a concave wafer. Wafers with lower pre-polish-bow reportedly had a smaller WIWNU than wafers with higher pre-polish-bow. As a result of the above issues, measurement of the wafer TTV as received from the wafer suppliers has been recommended.[77] Some CMP users have set an empirical TTV-cut-off-value of 5-μm for 200-mm-wafer pre-polish-sorting. Capacitive-gauging-tools are used to perform such TTV and bow measurements. More than 16,000 data points can be collected on a 200-mm-wafer in about one minute, and the results plotted to give a high-resolution map of the wafer geometry. Such gauging-tools are available from ADE Corp.

The *degree-of-planarity* achieved with a CMP-process is typically measured on a single-die or a few dice on a wafer with profilometry or atomic-force-microscopy. Such planarity can be location- and pattern-dependent. Within-die-planarity is influenced by the tendency of CMP to polish smaller, individual-features faster, and larger and densely-packed-features more slowly. Thus, CMP-thickness-measurements made with optical-thickness-tools commonly include several features at each selected site (from the smallest that the tool can conveniently measure, to a large [field] area where the polishing-rate reaches its lowest limit).

8.8.1 Detection of Defects Associated with the CMP Process

Detection of defects on post-CMP wafers is another important aspect of CMP metrology.[78] This importance arises from the fact that CMP is probably the most yield-impacting of all IC process steps given its intrinsically "dirty" processing environment (which includes chemically corrosive slurries, nano-particulate abrasives, and polish by-products). The defects associated with the CMP process can take the form of: **1**) *damage*; **2**) *contamination*; and **3**) *particles*.

The acceptable density of defects is becoming lower for each device generation. In 0.18-μm CMOS-technology the SIA Roadmap calls for a defect-density of 0.14-defects/cm^2, and the minimum-detectable-particle-size is decreased to 0.06-μm from 0.08-μm for the 0.25-μm generation. In oxide-CMP, defects can be separated into two categories: **1**) *foreign material* (such as residual-slurry, surface-particles, embedded-particles, and residual-metal); and **2**) *voids in the surface material* (such as microscratches and dishing). *Microscratches* can occur when a particle is dragged across the wafer surface, causing a shallow void in the oxide layer. Figure 8-88 shows the defects arising from oxide-CMP processes.[79]

In metal-CMP, defect types include: metal-loss, puddles (which can cause shorts if they remain between metal lines), surface-voids, residual-slurry, surface-particles, metal-filled-micro-scratches (which can cause shorts), and plug-holes from which metal has been removed. Figure 8-89 depicts a variety of defects associated with the W-CMP-process.[79]

The detection signals caused by such defects are obscured to some extent by noise sources associated with the CMP process. The three prevalent noise sources that impact defect inspection at post-CMP processing are: **1**) layer-thickness-variation; **2**) grain-noise; and **3**) pattern-noise. *SiO_2-layer-thickness-variation* is caused by nonuniform-polishing across the wafer. *Grain-noise* arises from the surface-roughness caused by the polycrystalline structure of materials beneath the oxide being polished (e.g., tungsten, aluminum, and polycrystalline-Si). The noise occurs from randomly-distributed pyramidal-peaks at the surface of the poly-crystalline-films. These scatter the detection-signal in a non-consistent manner so it is not completely subtracted when two adjacent-die are compared. *Pattern-noise* arises from metal-features beneath the oxide (which distort the incoming signal by causing background-scattering). Defect-detection-tools

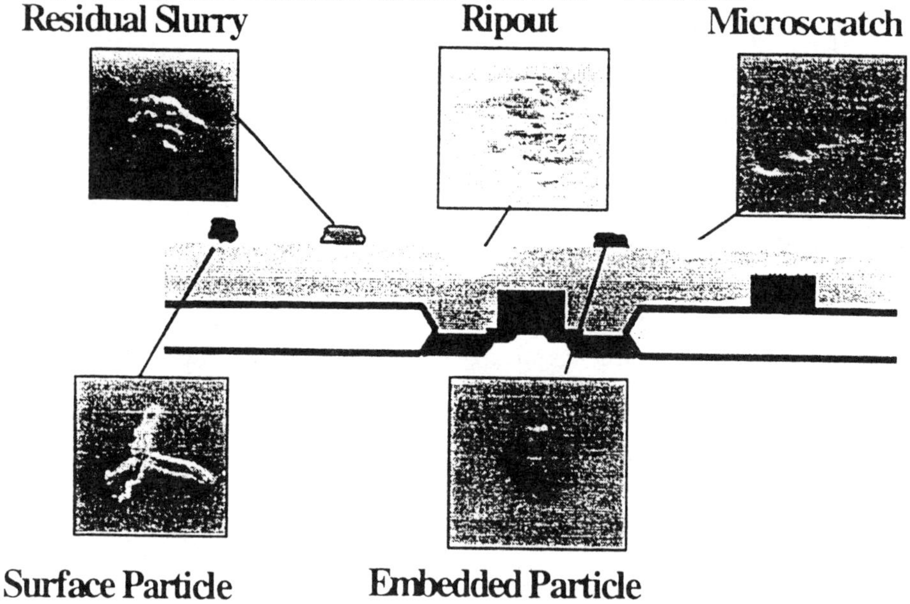

Fig. 8-88 Defects that arise during oxide CMP processes. Courtesy of KLA-Tencor.

employ image-processing algorithms to limit false alarms caused by such noise (i.e., that would indicate the presence of a defect where none actually exists).

Two illumination modes (*bright-field* and *dark-field*) are commonly used in the laser-scattering defect-counters employed for in-line automated-defect-inspection (such as the KLA-Tencor Surfscan AIT® model). Since reflected light is used in *bright-field inspection*, one of the main challenges for CMP-metrology is to reduce the interference between reflection at the oxide-surface and reflections that arise from underlying features. Some inspection tools use an ultra-broadband illumination-source (in place of a monochromatic-source) to suppress such interference effects. In *dark-field systems* only scattered light is collected, while the spectrally reflected light is rejected. Dark-field systems with a low-angle of illumination appear to work best for detecting slurry-particles on an oxide-layer over a grainy sub-layer (Fig. 8-90).[79] Dark-field systems with optical repetitive-pattern-filtering are reportedly more effective for detecting defects over dense repetitive arrays.[78]

Some 0.18-μm device-inspections for detecting CMP-defects may require use of SEM technology (such as the KLA-Tencor 2710). Because of their high resolution and large depth-of-focus, SEMs can locate ultra-small-defects hidden between dense geometries or in high-aspect-ratio areas. Small CuO particles ~0.1-μm from a dual-damascene process are easily detected using SEMs.

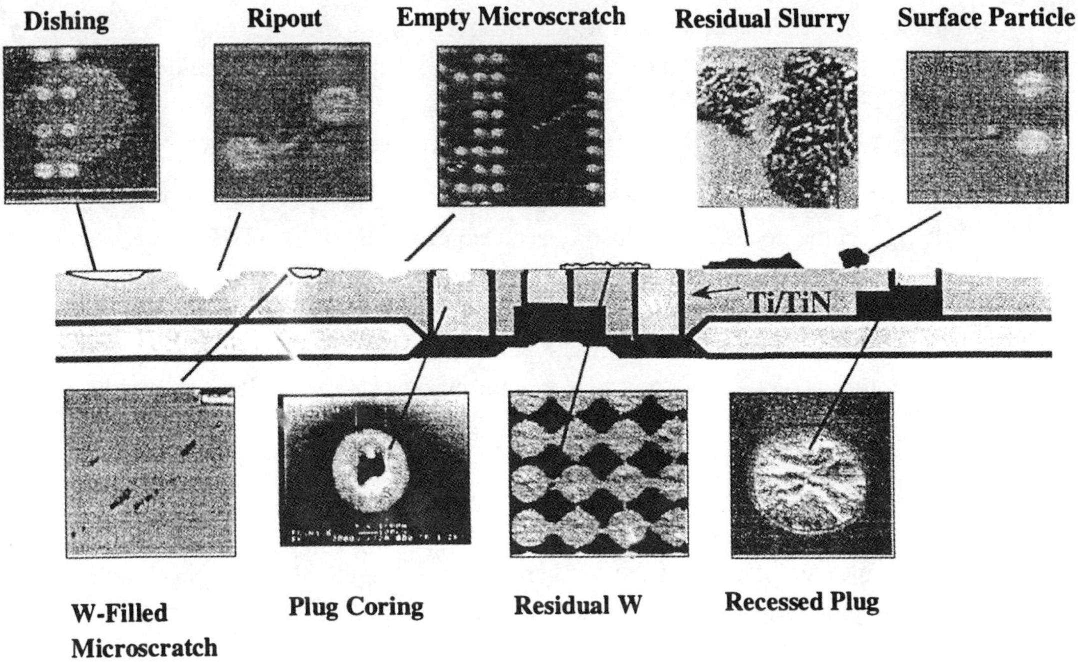

Fig. 8-89 Defects that arise during W-CMP processes. Courtesy of KLA-Tencor.

8.9 CMP POLISHER TOOL RELIABILITY

The CMP process creates a harsh set of working conditions for the polishers. Some have compared the operation of a CMP-polisher to "flying a helicopter in a sandstorm."[80] In the early days of CMP, production polishing-tools could have a mean-time-between-failure (MTBF) as small as a few days. Some of the problems of early CMP tools included the following:[39] insufficient vacuum for wafer handling; wafer sensor failure, misalignment of the load cup and sender and receiving cassettes, wafer release on polish pads (resulting in wafer breakage), equipment hangup (causing slurry to dry on the wafer-surface, and leakage of slurry and de-ionized water because of lack of quality control on valves and rotating unions. To make the systems more reliable, tools were designed to protect moving mechanisms from the abrasive and corrosive slurries (including sealing of parts and using air and water labyrinths).

Exposed materials were made with corrosion-resistant materials. Improved robotics for wafer loading, unloading, and handling were implemented. New system components were subjected to accelerated-testing to simulate one-year of production in three months. As a result, MTBF rates in CMP tools have been continuously improving. The goal of CMP tool suppliers is to build polishers with a MTBF of at least 400 hours (about 16 days).

8.10 CMP SYSTEMS AND PROCESS INTEGRATION

The CMP process requires a host of tools and consumables (Fig. 8-91). When CMP first came to the IC industry, each component of the CMP process was supplied by a separate company. The burden of integrating the various components fell to the user. As the market for CMP tools and consumables grew, equipment and materials suppliers realized the need for a better CMP-infrastructure. They understood that to gain market share, CMP users should be offered an off-the-shelf, "one-stop-shopping" package, perhaps with guaranteed process results. One way to accomplish this was to form strategic partnerships with other (non-competitive) suppliers. Most polisher suppliers included slurry distribution systems, cleaning tools, pads, slurries, endpoint detection, and film thickness measurement tools from other vendors as available options compatible with their polishing tools. Some companies (including Lam Research, Applied

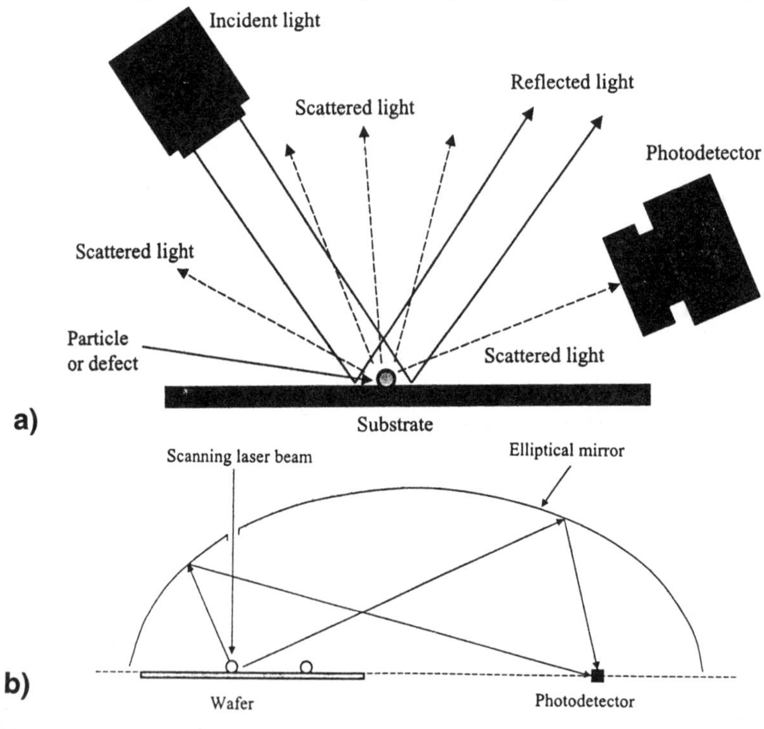

Fig. 8-90 a) Particle detection by Darkfield Light-Scattering; b) Light-scattering particle detector with an elliptical mirror.

Materials, SpeedFam-IPEC, Ebara, Strasbaugh, and Peter Wolters) designed CMP systems as cluster tools. Such systems integrate polishing, endpoint detection, cleaning (and perhaps inspection) into a single system. The Lam Research Teres® and the Applied Materials Mirra Mesa® are examples of such cluster CMP tools.

These integration trends have been accompanied by business consolidations as well. Speedfam and IPEC (two large polishing tool suppliers) merged in 1999. Lam Research acquired On-Track Systems (a maker of polishing tools and post-CMP-cleaning-systems). Applied Materials announced it would acquire Obsidian, Inc., a company that had developed a fixed-table CMP tool which uses a slurry-less pad technology. As CMP technology matures it is likely that such consolidations will continue.[82]

8.11 CMP OF VARIOUS MATERIALS

Here the issues of oxide-CMP and W-CMP are described in detail. The subject of Cu-CMP is briefly introduced in Sect. 8.11.3 here, but more details on Cu CMP are provided in Chap. 16 (that deals with the process technology of Cu in ULSI).

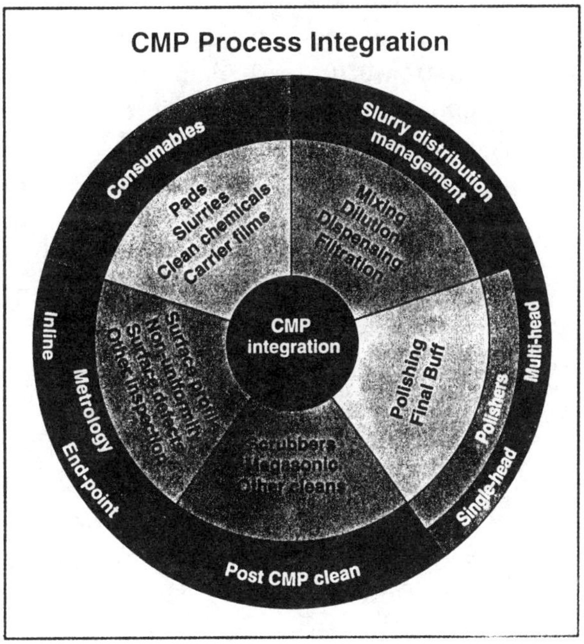

Fig. 8-91 Many components of CMP must be successfully integrated to develop a production-worthy CMP process. Reprinted with permission of Semiconductor International.

Note that the CMP issues associated with shallow-trench-isolation (STI) are mentioned briefly in this chapter in Sect. 8.12 (which deals with the WIDNU phenomenon of *dishing*). However, additional details concerning CMP-issues in STI are found in Chap. 9 (which is dedicated to the topic of STI).

8.11.1 CMP of Oxide Interlevel and Intermetal Dielectrics (ILDs and IMDs)

A blanket-dielectric-layer (typically an oxide-like film) is deposited (generally by CVD) atop polysilicon after gate-patterning. This layer is called an *interlevel dielectric* (ILD). Likewise, a blanket dielectric-layer (again, usually of silicon oxide) is CVD-deposited onto patterned metal-features, and such layers are called *intermetal dielectrics* (IMD). Topography is produced when oxide films are deposited by CVD onto etched polysilicon and metal patterns (see Fig. 8-1a). Many techniques have been attempted to planarize this topography (including: reflow for ILDs; and resist-etchback and spin-on-glass for IMDs). CMP has replaced these planarization techniques because they can only provide local-planarization (see Fig. 8-1d), whereas CMP can provide global-planarization (see Fig. 8-1e). However, unlike CMP for STI-applications, there is typically not a polish-stop-layer, and repeatable control of the polish-rate and uniformity is required. As noted previously, process variations (such as pad-age, slurry batch-changes, machine-calibration-drift, pad-conditioner-age, and chamber-to-chamber dielectric-film-property-variations) can all result in significant removal rate and process non-uniformity.

The industry standard pad for ILD and IMD CMP has been the IC-1000/Suba-IV stacked pad. This pad consists of a hard-cast polyurethane-pad (IC-1000) on top of a soft, compressive felt-pad (Suba-IV). The hard-top-pad is needed for planarization, and the soft-pad underneath improves the WIWNU (compared to polishing without the soft underpad). The drawback for including the soft underpad is a lower degree of planarization. Total stack compressibility can be modified by varying the thickness of the two pads.

Another issue of ILD-CMP is that doped silicon-oxides, such as PSG, and BPSG are softer glasses that undoped silicon-oxides (USG). As a result, they exhibit faster removal rates than USG. Furthermore, doped ILD-films exhibit higher potassium and calcium levels on the wafer-surface after scrubbing and cleaning than do USG films. This occurs because the phosphorus in the PSG and BPSG films acts as a gettering species for metal-impurities in the oxide-film. An HF-dip in the post-CMP-cleaning -equence can reduce the level of such metallic-contamination on the PSG surface (as well as slurry-residue). However, soft films, such as PSG tend to have high post-CMP defect-counts, and the HF etches microscratches isotropically, and enlarges their size. Thus, the post-CMP-cleaning-processes of PSG and BPSG ILDs need to be optimized to simultaneously reduce metallic-contamination and defectivity (e.g., by performing an oxide buff, which is very effective in reducing microscratching).[82]

8.11.2 CMP of Tungsten

Tungsten (W) CMP has been developed to replace the W-etchback step of the W-plug formation sequence (see Chap. 6, Vol. 1, 2nd Ed.). The W-CMP process, however, requires a surface that is planar prior to via-formation and metal-deposition (a condition not required in processes that form W-plugs by W-etchback). W-CMP has been described as a balance between the electrochemical oxidation of W to form a blanket passivation layer WO_3 on the entire W surface, and the mechanical removal of that reaction-product by abrasive slurry particles. Tungsten, being one of the refractory metals, tends to form strong oxygen bonds under oxidizing conditions. This oxide is so dense that the metal is protected from further chemical attack and the metal is "passivated."

Polishing of tungsten films raises three key issues. The first involves selection

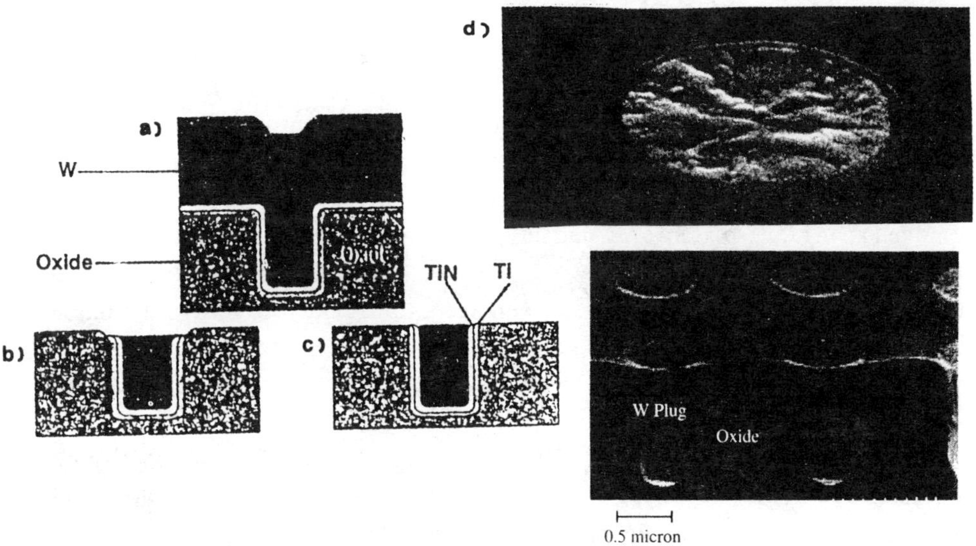

Fig. 8-92 W-plug formation using CMP: a) W deposition over TN/Ti barrier layers; b) CMP of W and TiN/Ti leaves recessed plugs; c) Oxide buff eliminates recess.[84] Reprinted with permission of Solid State Technology. Published by PennWell.; d) SEMs of W-plugs after W and TiN/Ti CMP steps and oxide buff step.[85] Reprinted with permission of Semiconductor International.

of an appropriate slurry. Tungsten polishing at IBM was originally done with a slurry-mixture of potassium ferricyanide and potassium phosphate (having a pH adjusted to about 5.0 to 6.5), together with colloidal-silica or dispersed-alumina, Al_2O_3.[83] (Alumina-based slurries have come to be the more commonly used.) However, potassium ferricyanide is highly toxic, and its disposal causes severe environmental impact. As a result, this chemical has been replaced, with the most widely used substitutes being ferric nitrate {$Fe(NO_3)_3$}, potassium iodate (KIO_3), and hydrogen peroxide (H_2O_2).[84] An example of commercially available slurries for W-CMP are Cabot's SEMI-SPERSE® WA-355 and SEMI-SPERSE® FE-10.

The second key issue of W-CMP arises because blanket-W-films are deposited over a liner-film of TiN/Ti. When removing the W-film, this liner-film must also be polished away. A high removal-rate of the TiN/Ti layer is essential to prevent erosion of the oxide underlying the metal-layers, and to also prevent plug-recessing (Fig. 8-92a). However, polish-rates of TiN/Ti are typically three to eight times slower than that of W (which is removed at rates exceeding 350-nm/min). It is also necessary to have a high selectivity to the underlying oxide to obtain local planarity of the plug-surface and to prevent excessive loss of the surrounding oxide by dishing. To remove the liner-film, *over-polishing* is typically used (i.e., so named because it causes some recessing of the W-plug as shown in Fig. 8-92b). During this step the field-oxide is also damaged to some degree.

To solve the problem of having to polish two-different metals, a multi-step polishing sequence is often used in W-CMP processes (typically a three-step CMP process is needed).[84,85] For example, in the first step, the W-layer is removed using one platen of a polishing tool. In the second step, the wafer is transported to a second platen where the liner-film (e.g., TiN/Ti) is polished away. Finally, a third oxide-buff-step (on yet a third platen) removes the oxide-damage by polishing away a small amount of the now-exposed underlying oxide (using a silica-based slurry that minimizes surface-roughness and facilitates the removal of alumina slurries). This also restores the level of the W-plug to that of the oxide-surface by eliminating the recess (Fig. 8-92c). Such CMP-W-plug-processes reportedly produce higher yields than processes which employ a W-etchback-process.[86,87] The higher yields with W-CMP are attributed to improved defectivity and superior final plug-morphology.

The third problem associated with CMP of tungsten involves the voids or seams that can occur when the blanket-W-film fills the contact or via holes. If such voids exist, they will be exposed during the CMP process. These voids become regions in which slurry can be trapped. Slurry chemistry can also attack a seam along a W-plug center and enlarge the seam to form a void or widen/expose

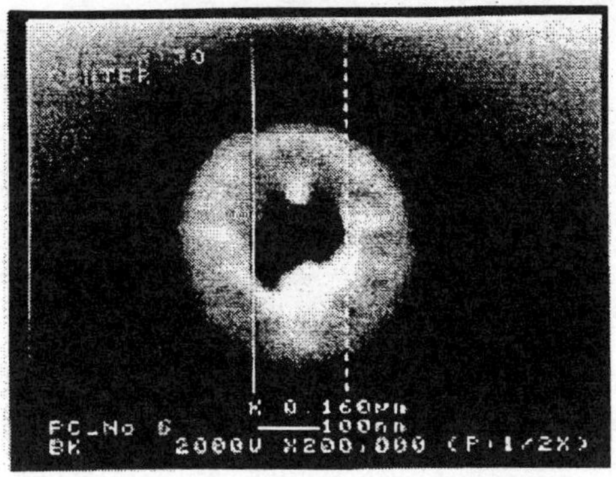

Fig. 8-93 Plug-coring.

an existing void, a phenomenon termed *plug-coring* (Fig. 8-93). Prevention measures include developing W-deposition-processes in which void-formation does not occur, and the use of less aggressive slurry chemistries (which will have less tendency to attack weak seams).

8.11.3 CMP of Copper

The CMP of copper (Cu) interconnect films involves a whole set of unique challenges,[88.89] and some of these issues are described here

The removal-rate of Cu by CMP is about 350-nm/min with a selectivity of 150:1 to the underlying oxide. However, Ta is commonly used as the liner-layer for Cu, and removing it by CMP represents some difficulties. Cu is relatively soft and oxidizes easily with standard chemistry (such as hydrogen peroxide or hydrogen persulfate). On the other hand, Ta is very hard and is also much more chemically inert. This typically leads to large differences in selectivity when simultaneously polishing Cu, Ta, and the underlying oxide (i.e., the Ta removal-rate is only about 30-nm/min). This leads to significant *dishing* of wide Cu and oxide features (the term *dishing* is defined in Sect. 8.12). Damascene CMP applications require low selectivity between the primary conductor-material and barrier-films (1:1) and a high selectivity between the primary conductor and insulator (> 100:1) to essentially stop the process. Since Ta is the harder and more inert metal, it is thought to be removed primarily by abrasion. An aggressive oxidizer could be used, but this would create the possibility of polishing through the Cu and removing device features. Work has been proceeding to find an oxidizer that selectively attacks the Ta without removing Cu.[90]

Another related problem in Cu/Ta-CMP is the electrochemical-interaction between Cu and barrier-materials during CMP.[91] Two dissimilar metals in simultaneous electrical-contact and contact with an electrolyte-solution constitute a galvanic-cell. This chemistry frequently leads to anodic-corrosion. Such a situation arises near the end of a polishing cycle of Cu and Ta (when the barrier-metal becomes exposed in the field-regions, and Cu is exposed in the trenches). Cu in the trench can be removed (sometimes entirely), even if the trench is initially well-filled with Cu. Obviously, this effect must be suppressed to enable CMP of Cu/Ta films to be successfully performed.

Use of multi-step polishing-processes has also been pursued for CMP of Cu. As in W-plug-CMP, a three-step damascene-Cu-CMP process uses three platens, each with its own slurry mixture. Copper is removed with the first platen, Ta layers with the second, and an oxide-buffing step is carried out on the third. It must be ensured that all of the Cu in the field-regions is removed during the first step, since the removal-rate of Cu in the second is deliberately kept very small. In another approach used by Speedfam, multiple slurries are simultaneously delivered through-the-pad to each of four autonomous polishing modules. In this way, a multi-step Cu CMP process can be carried out without transferring wafers from one platen to another (which increases throughput and yield).

For multi-step Cu-CMP processes one of 3 or 4 types of slurry might be best for the second polishing step. The selection of a particular slurry can depend on the liner-layer being used and the type of ILD material beneath the metal film. A large number of companies are developing slurries for Cu-CMP, hoping to capture a significant market share of this important emerging process.

8.11.4 CMP of Aluminum

CMP of aluminum is being examined for use in damascene-Al-interconnects. One of the challenges facing Al-CMP is that the metal is soft and susceptible to scratching, but its oxide Al_2O_3 is hard and adheres tightly to the metal. This makes Al_2O_3 difficult to remove. If a strong oxidizer is used, the entire Al-film (not only its elevated regions) tends to disappear. It has been shown that if a correct set of process conditions is established (e.g., an appropriate combination of pad, slurry type, and applied pressure), Al films can be planarized with CMP. While removal-rates of about 200-nm/min have been demonstrated, resolution of the issues concerning the Al-CMP process had not yet occurred by 1997.[92]

8.11.5 CMP of Polysilicon

Planarization of polysilicon layers by CMP is carried out in some IC fabrication applications. These include the following:

1. Removal of the polysilicon from the wafer surface when a blanket-polysilicon-film is deposited to fill the trenches of trench-capacitor based

DRAMs. (The issues of poly center-seam attack observed in some RIE etchback-processes used to remove the poly do not occur if a poly-CMP process is used. In addition, the polysilicon slurries are highly selective to oxide or nitride hard-mask materials and thus, are self-stopping);

2. Topography that exists when a polysilicon-film is deposited over an STI structure may need to be planarized. As shown in Fig. 8-94, the topography near the edge of an STI structure is often not perfectly planar. Since the polysilicon deposition process is inherently conformal, the non-planar surface of the STI-structure is replicated in the overlying poly film. In some applications it is advantageous to remove such undulations.

Polishing of polysilicon-films is done with a CMP-process similar to that used to planarize oxide.

8.11.6 CMP of Photoresist

There are some applications where the ability to CMP photoresist (PR) is beneficial. One example involves the fabrication of trench-based DRAM capacitors in embedded-DRAM technology. In the fabrication of the deep trenches used in such DRAM-capacitors, a sacrificial PR-layer is spun onto the wafer to control the height of the surface on the wafer and to facilitate alignment when building other device structures in the vicinity of the trench array. Both within-wafer and within-die non-uniformities of such sacrificial-PR-layers are important, as schematically shown in Fig. 8-95. These PR layers were found to

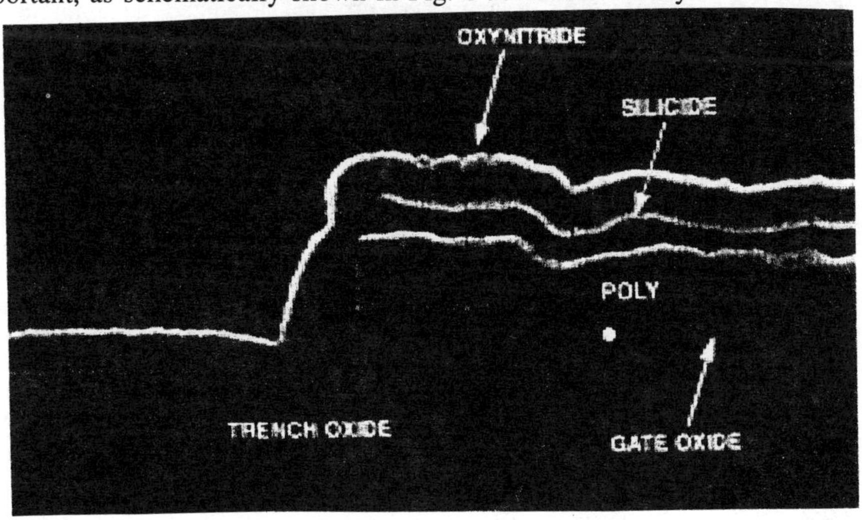

Fig. 8-94 CMP of polysilicon.

be successfully polished with a low-cost CMP process that utilized a slurry-free pH compensated solution of DI water (i.e., no slurry particles).[93]

8.11.7 CMP of Low-*k* Dielectrics

Other materials besides CVD-SiO_2 are being investigated for use as intermetal dielectric films (especially those with a lower dielectric constant - low-*k* dielectrics). These materials are discussed in detail in Chap. 14, but here the issues of low-*k* dielectric CMP are considered. That is, to be used in damascene interconnect structures, such low-*k* materials must be compatible with CMP.

Many of the CMP issues involving low-*k* dielectrics are common to both the subtractive-Al and damascene interconnect schemes. Most of the low-*k* stack options intend on keeping the low-*k* material encapsulated, never exposing it to the CMP process and post-CMP clean. In these cases, the investments in processes, materials, and equipment for silicon-oxide-based CMP systems will carry over and maintain their value. Film-adhesion is fundamental to all on-chip materials, particularly during CMP, with high friction and shear forces at work. Most proposals include capping the low-*k* dielectric with a traditional layer of silicon-oxide (or in some cases, with silicon nitride). The low-*k* materials must adhere well to both the cap-layer on top and to the underlying layer below (which is also likely to be silicon-oxide or silicon-nitride). In addition to adhesion, the compressibility of low-*k* films will make it possible for CMP processes to cause some local flexing within the interconnect stack. Such flexing could lead to cracking of the more rigid cap layers and produce localized delamination. In addition to unplanned exposure of the low-*k* film to slurry and post-CMP cleaning chemicals, the cap fragments could lead to scratching. Consequently,

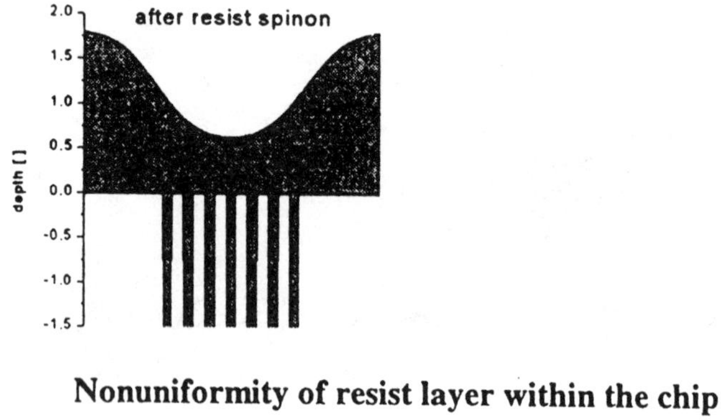

Fig. 8-95 CMP of photoresist.

polishing of cap layers on low-*k* is one of the earliest tests performed before a new low-*k* material can be considered integration-worthy. Yet another low-*k* CMP problem involves the low-thermal-conductivity of such films compared to silicon-oxide. Frictional heating in the CMP process may give rise to hot spots, possibly resulting in localized-stress and cap-delamination. High-linear-speed equipment being developed for greater planarization efficiencies would exacerbate this problem. In the extreme, local temperature excursions may further cure the low-*k* film, resulting in non-uniform dielectric properties.

Up to the end of 2000, there was little published information dealing with the issues of carrying out CMP of low-*k* materials in a volume-manufacturing environment. Part of the difficulty lies in the fact that process and consumables requirements may vary widely, depending on the specific low-*k* material chosen. One published report on low-*k* dielectric CMP involves the CMP of fluorinated oxide (F-SiO_2).[94] It was found that both mechanical-abrasion and chemical-erosion are intensified by fluorine-doping (relative to undoped TEOS oxides). This happens because of the reduced hardness and chemical reactivity of fluorinated oxides. One way to avoid these problems is to cap the F-SiO_2-layers with a CVD-SiO_2-layer. This cap-layer must be thick enough so that the dielectric-stack is planarized by CMP. In such cases the F-SiO_2-layer is not reached. Dielectric-erosion due to CMP-dishing must also be less than the thickness of the cap-layer. Furthermore, the cap-layer increases the overall dielectric-constant of the IMD-stack. In addition, depositing the stacked-film is a multi-step-deposition process. Another approach is to pursue the development of more chemically-stable low-*k*-materials, and such efforts are continuing.

A second study of direct polishing of low-*k*-dielectrics using a MnO_2-slurry reported that the deep scratches formed with CMP using a conventional silica-based slurry were reduced from a 20-nm depth to a 2-nm depth. Residual MnO_2 slurry particles can also be removed using H_2O_2 with a pH of 6, which does not attack the fluoro-carbon low-*k* material being polished (a poly-AE layer, FLARE®).[95]

8.12 MISCELLANEOUS ISSUES CMP

8.12.1 Dishing and Erosion

Dishing, another CMP issue, can represent a significant problem when a film of one type of material is being polished down until it remains only in recessed (or trench regions - Fig. 8-96). In such cases the layer under the film being polished

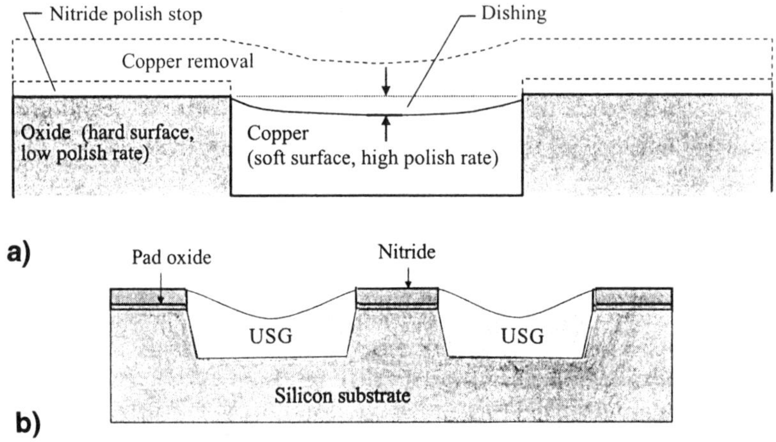

Fig. 8-96 Dishing-effect, shown here for: a) CMP of C in oxide-trenches; b) CMP of oxide in Si-trenches during formation of STI.

becomes exposed to the CMP process during the over-polish time (used to ensure that the film is completely removed over all regions of the wafer by the CMP process). Ideally, such underlayers should act as perfect polish-stop-layers. However, during the over-polish time, the film being planarized can continue to be polished away (especially in wide areas), causing it to be significantly thinned in such areas. One case where this occurs is in the CMP of a CVD-oxide-layer, deposited to fill trenches in the course of fabricating shallow-trench-isolation (STI) structures. A silicon-nitride-layer is typically selected to act as the polish-stop-layer. (See Chap. 9 for a discussion of STI technology.) During the over-polish time, the oxide in the trenches continues to be removed. The degree of thinning of the oxide in such areas (*dishing*) is known to be highly dependent on feature size, with the dishing-effect becoming larger as the width of the field-oxide is increased (Fig. 8-97). For example, in some processes dishing can cause the field-oxide thickness to decrease by more than 50-nm when the field-oxide width exceeds 100-μm.[96]

Dishing occurs because the polishing-pad is not perfectly rigid. If the pad were to exhibit perfect rigidity, CMP of the exposed oxide-regions would produce a perfectly planar surface (because the pressure on the exposed recessed regions would be zero). Since actual pads exhibit some flexibility, above-zero polishing-pressures are produced in recessed areas, especially in the middle of wide exposed-regions. Thus, oxide is removed from these regions, and a "dished" region exists after planarization. Dishing must be minimized to maintain planarity for subsequent wafer processing. It has remained a serious impediment to developing damascene processes (as is *erosion*, which is covered later).

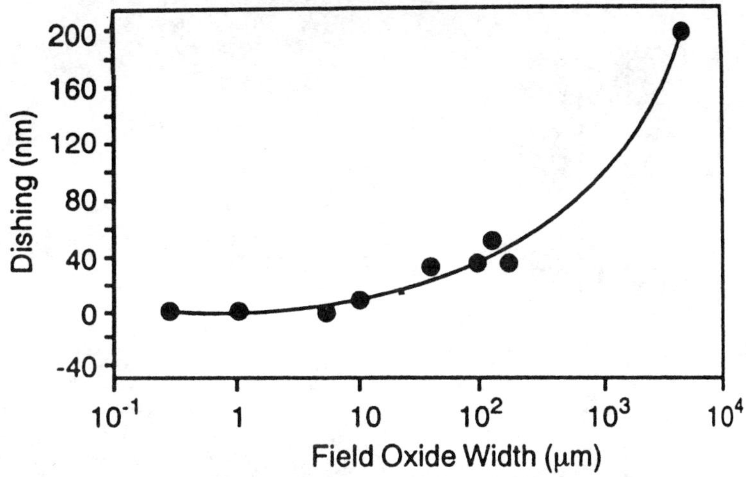

Fig. 8-97 Dishing versus field-oxide-region width for a specific oxide CMP process.

A number of approaches have been proposed to decrease dishing arising from the CMP-process used to fabricate STI, including: **1**) a combination CMP/RIE process; **2**) dummy structures added in the field-oxide regions of STI structures; **3**) adding an extra nitride-layer on the trench-dielectric after it is deposited;[103,104] **4**) adding an extra nitride-layer onto the dielectric after it has been deposited *and* using a reverse block-mask;[105] and **5**) implementing a direct-polish CMP-process using either a linear-polishing-tool, or a rotating-polishing-tool with a harder polishing-pad and an efficient endpoint-detection-system. Techniques #3 and #4 are described in Chap. 9 (STI), and the others here.

In the *CMP/RIE-approach*, the oxide is polished by CMP to a point where the nitride is not yet exposed and the surface is thus still completely planar (no dishing has yet occurred). Then RIE is used to etch the oxide-layer, with an endpoint-algorithm used to stop the etching just as the nitride is exposed to plasma.[97] The drawback of this method is that it increases the complexity of the process used to planarize the trench-oxide.

In the *dummy-active-region-approach* some areas of the trench are not etched. Instead, as shown in Fig. 8-98, regions of Si remain in the center of the trench. This essentially reduces the trench-width so that dishing effects are decreased to an acceptable degree. The CMP-removal-rate over regions that are densely packed with active-areas is also slower than over regions containing small isolated areas. Thus, to expose the nitride CMP-stop-layer over such densely packed regions, the nitride over small isolated active areas gets stripped, causing damage to Si in areas beneath the stripped nitride. Thus, dummy-active-regions

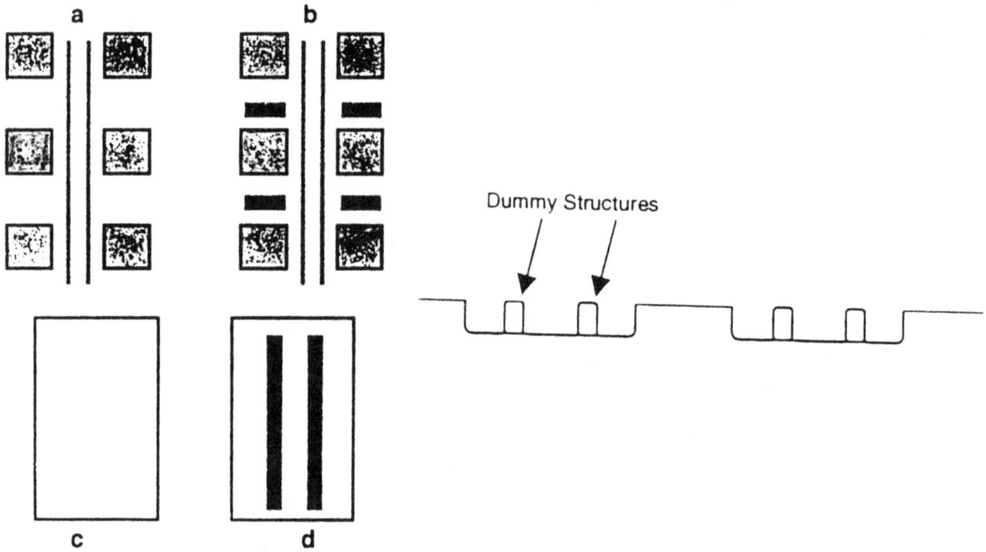

Fig. 8-98 Dummy (b) and dishing support features of small (a) and large wafer (c) features can prevent dishing.[89] e) Cross-section of dummy features in a trench.

must also be added to areas around isolated active regions. Finally, since dishing also occurs when wide areas of metal are polished (e.g., bond-pads, power-busses, and wide interconnect-lines), such dummy structures can also be added to these features. The technique, however, poses electrical, design, and space constraints, with the main issue being that several times as many structures must be added within a die over previous designs. This complicates the design and mask-making process, and significantly raises defect levels. Thus, some see this approach as merely a stop-gap measure.

The implementation of a *direct-polish CMP-technique* has also been pursued to avoid the drawbacks of the above methods. When a direct-polishing-process involves the use of a *linear-polishing-tool*, higher pad velocities (300–500-ft/min versus 34-ft/min for rotary-pad speeds) and lower polishing-pressures (3–5-psi versus 7-psi for rotary polishers) are used to suppress dishing. A highly-selective slurry (with cerium-oxide particles dispersed in an aqueous suspension of material that inhibits silicon-nitride removal) is also employed. This set of process conditions has been reported to significantly reduce dishing-effects (maximum oxide-removal across a die was 20-nm) and decrease the nitride-removal over isolated active-regions.[98] Decreased dishing using a direct-polish approach has also been described using *rotary-polishing-tools.* To achieve decreased dishing the correct set of process conditions and CMP consumables must be selected. That is, dishing effects are reduced by using a harder polishing pad, a relatively non-selective slurry, an efficient endpoint-detection method, and an optimized CMP process.[99] As also discussed in Sect. 8.6.8, the use of slurry-free pads holds the promise of also reducing dishing.

The dishing problem also occurs during other CMP processes as well. As noted in earlier, dishing can occur when polishing ILD and IMD layers. Furthermore, as will be discussed in Chap. 16 (which deals with Cu interconnect processing issues), dishing and erosion of the ILD layers can occur when the Cu (and Cu barrier layer) are being polished back. Refer to Chaps. 15 and 16 for more information on these topics.

Erosion is an effect that occurs during CMP of Cu or W which results in the thinning of the ILD layer. That is, erosion is defined as the difference in the ILD thickness before and after the metal-CMP-step. The Cu (or W) film is also thinned, but its final thickness is reduced by the sum of the dishing plus erosion that has occurred. Erosion occurs because there is also a non-zero removal rate during the over-polish of the metal.

The extent of erosion depends heavily on the pattern-density (but not on metal linewidth, as is the case with dishing). Erosion is worst in regions of dense metal pattern (Fig. 8-99). This is so because when the polish begins the metal (e.g., Cu) covers the entire wafer. In regions of high Cu-pattern-density there is a lower

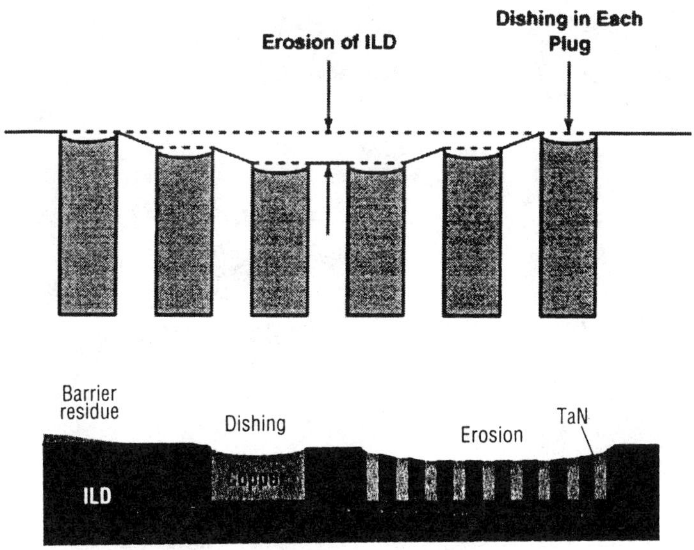

Fig. 8-99 Erosion of the ILD that occurs during over-polish of Cu during CMP.

aerial-density of Cu contacting the pad ("up-areas" of Cu). This is because the Cu in low areas (the "down-areas" of Cu) does not contact the pad. Thus, there is a higher pressure on the "up-areas" of Cu (and consequently higher polish-rate) than in regions where the Cu pattern-density is lower. Because of this difference in polish-rates, Cu is cleared in regions of high pattern-density sooner than in regions of lower pattern-density. Thus, the ILD in regions of high pattern-density receives a longer over-polish, adding to increase in erosion with pattern density.

8.12.2 Thickness Non-Uniformity Within a Wafer After CMP

Rigid polishing-pads provide good planarity within a die (chip), but poor thickness-uniformity across a wafer, for reasons explained here. If wafers were perfectly flat, this problem would not occur. A film on the surface of a flat wafer would be polished to a uniform thickness across the wafer (Fig. 8-100a). But not all wafers are uniformly flat. For example, if a wafer is tapered and the pad is rigid, the film on the thicker end of the wafer will be polished more than on the thinner end (Fig. 8-100b). If the wafer thickness undulates (with some regions thicker than others), a film covering regions where the wafer is thicker will be polished more (and will end up thinner) than a film on the thin parts (Fig. 8-100c). Softer (i.e., more compliant) pads produce improved uniformity across-the-wafer thickness, but at the price of degraded planarity. As discussed earlier,

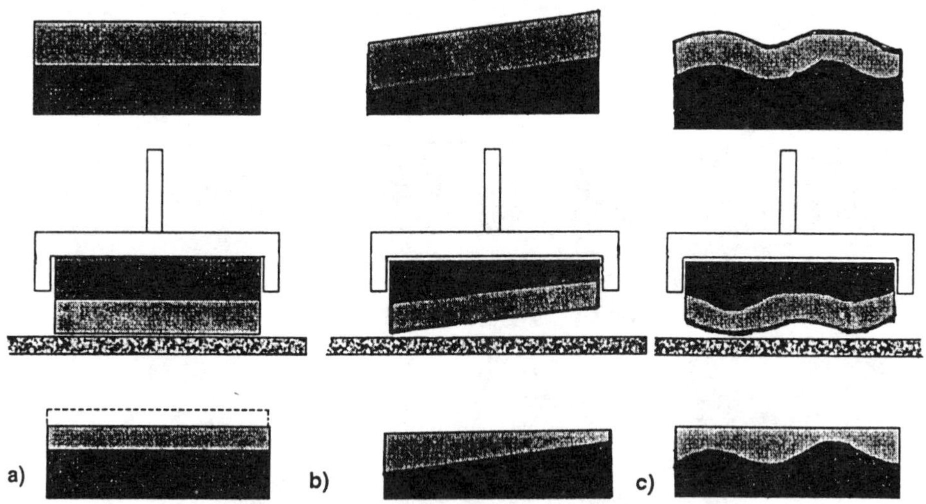

Fig. 8-100 If a rigid pad is used for CMP: a) A film on a flat-wafer is uniformly polished; b) A film on a tapered-wafer is polished thinner over the thicker end of the wafer; (c) For wafers with variations in wafer-thickness across the wafer, the film is polished thinner over the thicker wafer regions.

one solution is to use a composite pad (a rigid-pad backed by a softer one). Another approach is to use a linear polishing tool with a flexible (e.g., air) bearing beneath the polishing belt. This allows the belt to conform better to the wafer-surface, and reportedly produces improved across-the-wafer thickness-uniformity (and significantly less dishing).

8.12.3 Economic Considerations: Throughput and Cost of Ownership

While CMP is an enabling process for deep-submicron IC-fabrication, it is also a relatively expensive one. The cost of a single CMP-pass in 1999 (without a cleaning process) ranged from \$5 to \$8, depending on the material being polished. The post-CMP cleaning-step adds about another \$1 to \$2 per wafer. The main contributors to the cost of CMP include the polishing (and cleaning) tools, the slurry, the test- and monitor-wafers, and the need to do off-line measurements. The CMP-facilities, labor-costs, utilities, DI-water, pads, and waste-disposal are additional expenses. The 2002-target for the cost of a CMP process on 300-mm wafers is \$5.56/wafer/pass.[101]

The price of a fully-integrated polishing-tool in 1999 ranged from \$2–2.5 million, with the cleaning-module costing about \$500,000. Such tools are capable of polishing 50-60 200-mm wafers-per-hour. Slurries are also expensive. Oxide-slurries cost about \$10/gallon, tungsten-slurries about \$30/gallon, and Cu-slurries from \$40–50/gallon. One gallon is consumed for about every ten wafers.

Test- and monitor-wafers are needed not only during the set-up of the process, but also to check tool-performance for every lot. Typically, the removal-rate and the within-wafer-non-uniformity are monitored for each lot to ensure the tool is operating within spec. Some IC manufacturers have reported that up to two hours per shift are needed to process and measure the monitor-wafers and product test-wafers off-line, and that this reduces the CMP-equipment-capacity by nearly 50%. This is the reason that *in situ* measuring techniques are so attractive.

When the *cost-of-ownership* (CoO) of various CMP-tools are being compared, the three parameters that have the most significant impact on this cost are:

1. Throughput
2. Mean-Time-Between-Failure (MTBF)
3. Availability

The *throughput* of a CMP-tool is perhaps the most important parameter that contributes to the cost-of-ownership. It is defined as the quantity of wafers that can be completely polished within a given period of time (e.g., wafers-per-hour). A throughput of 50-60 wafers-per-hour is considered to be a good throughput range. This parameter is highly dependent on the number of polishing platens and carrier heads. The 2002 performance-target for CMP-tool-throughput is 75 wafers-per-hour-per-tool.[102]

The *mean-time-between-failure* (*MTBF*) is a measure of a tool's reliability. It is defined as the mean-value of the number of hours that a specific tool can run continuously, non-stop, without breaking down. Most CMP tool-manufacturers seek to design their tools with an MTBF of at least 400 hours.

The *availability* of a CMP tool is the actual time the tool can run product wafers (i.e., the time when a tool is being qualified or tested, or is shut down for preventative-maintenance is subtracted from the total-time a tool is in service). This parameter reflects the ease with which a tool can be serviced or maintained

It should be added as a final note that the test-time is not always taken into account when calculating CoO. However, the amount of time required to qualify a CMP-tool for processing and the *mean-time-to-test* (*MTTT*), can also have a noticeable impact on CoO. If one assumes that a CMP-tool needs to be qualified once during each shift (i.e., three times a day), and that each qualification test (i.e., off-line thickness-measurement) takes about 30 minutes, this means that the MTTT is 1.5 hours per day, or 10.5 hours per week. This 10.5 hours could be considered lost CMP-processing-time. The shorter is this time, the higher the CMP throughput-number, which would lower the CoO.

REFERENCES

1. L. Rothman, *J. Electrochem. Soc.*, **127**, 2216 (1980).

2. C. Jang *et al., Proc. 4th Intl. IEEE VMIC Conf.*, Santa Clara CA, p. 357, 1987.

3. S. Morimoto *et al.,* "Characterization of the Chemical-Mechanical Polishing of Intermetal Dielectric Films," *J. Electrochem. Soc.* 1993, p. 449.

4. J.M. Steigerwald, S.P. Murarka, and R.J. Gutmann, *Chemical Mechanical Planarization of Microelectronic Materials*, Wiley & Sons, New York, 1996.

5. M.A. Fury, "The Early Days of CMP," *Solid State Technol.,* May 1997, p. 81.

6 B. Davari *et al*, "A New Planarization Technique, Using a Combination of RIE and Chemical Mechanical Polishing (CMP)," *Tech. Dig. IEDM*, 1989, p. 61.

7. R. DeJule, "CMP Grows in Sophistication" *Semicond. Internatl,* Nov. 1998, p. 56.

8. M. Tomozawa, "Oxide CMP Mechanisms," *Solid State Tech.*, July 1997, p. 169.

9. F. Preston, "Theory & Design of Plate Glass Polishing Machines," *J. Soc. Glass Tech.,* **11**, 214 (1927).

10. Paper TuB-A4, *Tech. Dig., Topical Meeting on the Science of Polishing,* Optical Soc. of America, 17 April 1984.

11. C.W. Liu *et al.* "Characterization of CMP Process Based on Nanoindentation Measurement of Dielectric Films," *J. Electrochem. Soc.,* **142,** p. 3098 (1995).

12. S.R. Runnels and L.M. Eyman, "Tribology Analysis of CMP," *J. Electrochem. Soc.,* **144**, p. 1698 (1994).
13. W.T. Tseng, "Re-examination of Pressure & Speed Dependence of Removal rates during CMP processes," *J. Electrochem.. Soc.,* **144**, p. L15 (1997).
14. B. Zhao and F.G. Shi (1999). "CMP: Threshold Pressure and Mechanism," *Electrochemical Solid State Letters* **2**, p. 145 (1999).
15. G.B. Shinn *et al.,* "Chemical-Mechanical Polish," in *Handbook of Semiconductor Manufacturing Technology,* Y. Nishi and R. Doering, Eds., Marcel Dekker, New York, 2000, p. 415.
16. "New Model Could Vastly Improve CMP Processes," *Semiconductor International,* May 1999, p. 21.
17. L. Cook, "Chemical Processes in Glass Polishing," *J. Non-Crystalline Solids*, **126**, 160 (1990).
18. W.-T. Tseng *et al.,* "CMP of Fluorinated Silicon Dioxide: Is it Necessary and Feasible?" *Solid State Technology*, February 1997, p. 64.
19. W.J. Schaffer and H.W. Fry, "BPSG Improves CMP Performance for Deep Submicron ICs," *Semiconductor International*, July 1996, p. 205.
20. T. Izumitani and S. Harada, "Polishing Mechanism of Optical Glasses," *Glass Technology,* **12**, 131 (1971).
21. *ibid. Ref. 17.*
22. J. Olsen and F. Moghadam, "Planarization Techniques," in *Handbook of Multilevel Metallization for Integrated Circuits,* Eds. S.R. Wilson, C.J. Tracy, and J.L. Freeman, Noyes Pubs., Park Ridge, NJ, 1993, Chap. 6, pp. 426-438.
23. W.J. Patrick *et al.,* "Application of CMP to the Fabrication of VLSI Circuit Interconnections," *J. Electrochem. Soc.*, **138**, June 1991, p. 1778.
24. P.A. Burke, "Semi-Empirical Modeling of SiO_2 CMP Planarization," *Proc. VMIC* 1991, p. 379.
25. J. Schlueter, "Controlling Contamination During CMP Processes," *Solid State Technology,* May 1996, p. S8.
26. F. Kaufman., "Chemical Mechanical Polishing for Fabricating W Metal Features as Chip Interconnects," *J. Electrochemical Society,* **138**, Nov.1993, p. 3460.
27. M. Pourbaix, *Atlas of Electrochemical Equilibria in Aqueous Solutions,"* Nat. Assoc. of Corrosion Engr., Houston TX (1974).
28. J. Gisler, "Enhancing CMP Performance, *Semicond. Internatl,* July 2000,p. 150.
29. C.C. Yu., *Advanced Metallization for ULSI Applications Proceedings*, Oct., 1993.
30. A.E. Braun, "CMP Battles Low-k Fundamentals," *Semiconductor International,* October 2000, p. 72.
31. K. Witt and L. Cook, "Getting an 'Edge' with CMP," *Semiconductor International,* October 2000, p. 75.
32. Courtesy of Rodel Inc.

33. Courtesy of Rodel Inc.
34. H. Nojo, *et al.,* "Slurry Engineering for Self-Stopping & Dishing-Free SiO_2 CMP," *Tech. Dig. IEDM*, 1996, p. 349.
35. A.E. Braun, "Slurries and Pads Face 2001 Challenges," *Semiconductor International,* November 1998, p. 65.
36. T. Hara *et al.,* "Direct CMP of Low-*k* Organic by MnO_2 Slurry," *Mtg. Abs. Electrochem. Soc.,* Fall 1999, Abs. 1269.
37. J. McCann and C. Smith, "Bulk Slurry Distribution for Integrated Manufacture," *Semiconductor International,* November 1996, p. 94.
38. G. Basim *et al.,* "Effect of Particle Size of Chemical Mechanical Polishing Slurries for Enhanced Polishing with Minimal Defects," *J. Electrochem. Soc.,* September 2000, p. 3523.
39. I. Ali *et al.,* "Chemical-Mechanical Polishing of Interlevel Dielectric: A Review," *Solid State Technology,* October 1994, p. 63.
40. J.H. Golden *et al.,* "Evaluating and Treating CMP Wastewater," *Semiconductor International,* October 2000, p. 85.
41. H.-J. Kim *et al.*, "Physical and Chemical Characterization of Re-Used Oxide CMP Slurry," *Mtg. Abs. Electrochem.. Soc.,* Fall 2000, Abs. 477.
42. *ibid. Ref. 4.*
43. Courtesy of Rodel Inc.
44. Courtesy of Rodel Inc.
45. A.E. Braun, "Slurries and Pads Face 2001 Challenges," *Semiconductor International,* Nov. 1998, p. 65.
46. M.R. Oliver *et al.,* "CMP Pad Surface Roughness and CMP Removal Rate," *Mtg. Abs. Electrochem. Soc.,* Fall 2000, Abs. 487.
47. I. Ali and S.R. Roy, "Pad Conditioning in Interlayer Dielectric CMP," *Solid State Technology,* June 1997, p. 185
48. C.-Y. Chen *et al.*, "Operational Aspects of Chemical Mechanical Polishing: Polishing Pad Profile Optimization," *J. Electrochem. Soc.* October 2000, p. 3922.
49. Courtesy of 3M Company.
50. Courtesy of 3M Company.
51. Courtesy of 3M Company.
52. W.L. Smith *et al.,* "Film Thickness Measurements for Chemical Mechanical Planarization," *Solid State Technology*, January 1996, p. 77.
53. A.S. Lawing, "Pattern Dependent Polish Rate Behavior in Oxide Chemical Mechanical Polishing," *Mtg. Abs. Electrochem. Soc.,* Fall 2000, Abs. 493.
54. E. Worthington, "New CMP Architecture Addresses Key Process Issues," *Solid State Technology,* January 1996, p. 61.
55. R. DeJule, "Advances in CMP," *Semiconductor International,* November 1996, p. 88. **56.** Courtesy of Applied Materials

57. H.E. Litvak and H.-M. Tzeng, "Implementing Real-Time Endpoint Control in CMP," *Semiconductor International,* July 1996, p. 259.
58. H.E. Litvak, *U.S. Patent No. 5,499,733.* Issued March 19, 1996.
59. H. Litvak, *Proc of the SEMI Taiwan CMP Seminar* Sept. 1995, Taipei, p. 123.
60. *ibid.* ref. 25.
61. J. Skrovan, K. Robinson. "Planarization slurry including a dispersant and method of using same." *U.S. Patent 5,827, 781.* Oct. 27, 1998.
62. F. Preston, "Theory & Design of Plate Glass Polishing Machines" *J. Soc. Glass Tech.,* **11**, 214 (1927).
63. V.B. Menon, R.P. Donovan, in W. Kern Ed., *Handbook of Wafer Cleaning Technology*, Noyes Publications, Park Ridge, N.J., 1993.
64. K. Mikhaylich *et al.,* "Evaluation of Contact and Non-Contact Cleaning for CMP Applications," *Mtg. Abs. Electrochem. Soc.,* Fall 1999, Abs. 1268.
65 D. Hymes *et al.,* "Brush Scrubbing Emerges as a Future Wafer-Cleaning Technology," *Solid State Technology*, July 1997, p. 209.
66. T.L. Myers, "Post-Tungsten CMP Cleaning," *Solid State Tech.* Oct. 1995, p. 59.
67. J.M. de Larios *et al.,* "Post-CMP Cleaning of Oxide and Tungsten Applications," *Semiconductor International,* May 1996, p. 121.
68. D. Hymes *et al.,* "The Challenges of the Copper CMP Clean," *Semiconductor International,* June 1998, p. 117.
69. K. Osseo-Asare, K.K. Mishra, "Solution Chemical Constraints in the CMP of Copper: Aqueous Stability Diagrams for the Cu-H_2O and Cu-NH_3-H_2O Systems," *J. Electron. Materials,* **25**, p. 1599 (1997).
70.Y. Feng, *et al.,* "Corrosion Protection of Copper by a Self-Assembled Monolayer of Alkenathiol," *J. Electrochem. Soc.*, **144**, p. 55 (1997).
71. A.A. Busnaina and F. Dai, "Megasonic Cleaning," *Semiconductor International.* August 1997, p. 85.
72. A. Busnaina *et al.,* "Single Wafer Non-Contact Post-CMP cleaning Using DI Water and Dilute Chemistry," *Mtg. Abs. Electrochem. Soc.,* Fall 1999, Abs. 1267.
73. M. Elissa *et al.,* "Post Copper CMP Cleaning: A Non-Contact Method," *Mtg. Abs. Electrochem. Soc.,* Fall 1999, Abs. 1270.
74. I. Ali *et al.,* "Investigating the Effect of Secondary Platen Pressure on Post-CMP Cleaning," *Microcontamination* **12**(10), p. 45, 1994.
75. R. Davenport and B. Brown, "Integrated CMP Clean Methods," *European Semiconductor,* July 1999, p. 36.
76. Courtesy of Ebarra.
77. Y. Zhang *et al.,* "Wafer Dimensional Analysis for Chemical Mechanical Planarization," *Solid State Technology*, July 1997, p. 179.
78. J. Reynolds *et al.,* "Defect Inspection Technologies to Meet the Challenges of Advanced CMP Processes," *Solid State Technology,* March 1998, p. 39.

79. Courtesy of KLA-Tencor Inc.
80. *ibid., Ref. 55.*
81. *ibid., Ref. 39.*
82. W. O'Mara, "The Inevitable Shakeout in the CMP Equipment Market," *Solid State Technology*, May 1998, p. 84.
83. F. Kaufman *et al.*, "Chemical Mechanical Polishing for Fabricating Patterned W Metal Features as Chip Interconnects," *J. Electrochemical Society,* Vol. **138**, November 1993, p. 3460.
84. K. Wijekoon *et al.*, "Tungsten CMP Process Developed," *Solid State Technology,* April 1998, p. 53.
85. I. Kim *et al.*, "Optimized Process Developed for Tungsten CMP," *Semiconductor International,* November 1996, p. 119.
86. S. Roy, "Advanced Metal & Intercon. Syst. for ULSI Applications," 1995, p. 733.
87. *ibid. Ref. 29.*
88. J.-F. Wang "CMP of Cu Multilevel Interconnects," *Procs. VMIC* 1995, p. 505.
89. R. DeJule, "CMP Challenges Below Quarter Micron," *Semiconductor International,* Nov. 1997, p. 54.
90. D.R. Evans, "Electrochemical Interaction Between Copper and Barrier Materials During CMP," *Mtg. Abs. Electrochem. Soc. Mtg.,* Fall 1996, Abs. No. 504,
91. A. Beverina *et al.*, "Copper Photocrrosion Phenomenon during Post-CMP Cleaning," *Electrochem. Soc. Letts.,* March 2000, p. 156.
92. A.R. Seturaman *et al.*, "Copper vs Aluminum: A Planarization Perspective," *Semiconductor International,* June 1996, p. 177.
93. J. Nuetzel *et al.*, "Conventional and Slurry Free CMP of Polymer Layers for Gbit DRAM," *Mtg. Abs. Electrochem. Soc.,* Fall 1999, Abs. 1232.
94. W.-T. Tseng, "CMP of Fluorinated Silicon Dioxide: Is it Necessary and Feasible?" *Solid State Technology*, February, 1997, p. 61.
95. *ibid. Ref. 36.*
100. S.J. Fang *et al.*, "Optimization of the CMP Process for Pre-Metal Dielectrics," *J. Electrochem. Soc.,* February 2000, p. 682.
101. T. Marbeiter, *et al.*, "An Update: Transition to 300-mm CMP," *Semiconductor International,* November 1998, p. 76.
102. *ibid. Ref. 101.*
103. M. Nandakumar *et al.*, "Shallow Trench Isolation for Advanced ULSI CMOS Technologies," *Tech. Dig. IEDM,* 1998, p. 134.
104. L. Peters, "Choices and Challenges for Shallow Trench Isolation." *Semiconductor International,* April 1999, p. 71.
105. G. Badenes *et al.*, "A New Dummy-Free Shallow Trench Isolation Concept for Mixed-Signal Applications," *J. Electrochem. Soc.,* **147**, Oct. 2000, p. 3827.

Chapter 9

SHALLOW TRENCH ISOLATION (STI)

Despite advances made to decrease bird's beak and channel-stop encroachment in LOCOS isolation (by poly-buffered LOCOS and other approaches), such techniques eventually become inadequate for deep-submicron IC technologies. In particular, the scalability of LOCOS for sub-half-micron CMOS technologies comes into question. The limitations of LOCOS include: the penetration of the LOCOS bird's beak into the device active regions; non-planarity; LOCOS field-oxide-thinning at narrow isolation spacings; and stress-induced silicon defects. (LOCOS-based isolation technologies are described in more detail in Vols. 1 [2nd Ed], 2, and 3 of *Silicon Processing for the VLSI Era.*) The successor to LOCOS in CMOS is *shallow trench isolation* (STI). In STI a relatively shallow trench is first etched into the Si substrate (0.3-0.5-μm deep), which is then refilled with an insulator material (following a short thermal-oxidation growth step that forms a thin film of SiO_2 on the trench walls to control the Si-SiO_2 interface quality). The surface is planarized by CMP after refill to complete the isolation structure. Figure 9-1a depicts a shallow-trench isolation (STI) structure in CMOS.

Since the bird's beak is entirely eliminated in STI and the redistribution of channel-stop-dopant is reduced, smaller isolation spacing is also possible than with LOCOS. In addition, the field oxide in STI is fully recessed, offering the potential of a completely planar surface after the isolation structure formation. (Note that while CMP can also be used to reduce topography resulting from the LOCOS process, the oxide encroachment with LOCOS is not eliminated by such CMP.) Furthermore, the problem of field-oxide-thinning in narrow isolation spaces does not occur, and the threshold voltage is constant as a function of channel width (because of the elimination of the narrow-channel effect found in LOCOS - see Chap. 9, *Silicon Processing for the VLSI Era*, Vol. 3). Also, MOSFETs made with fully-recessed oxide isolation exhibit larger g_m and smaller S_t values than those using semi-recessed LOCOS. Finally, the fully-recessed STI structures have trenches whose lower corners, though rounded, have

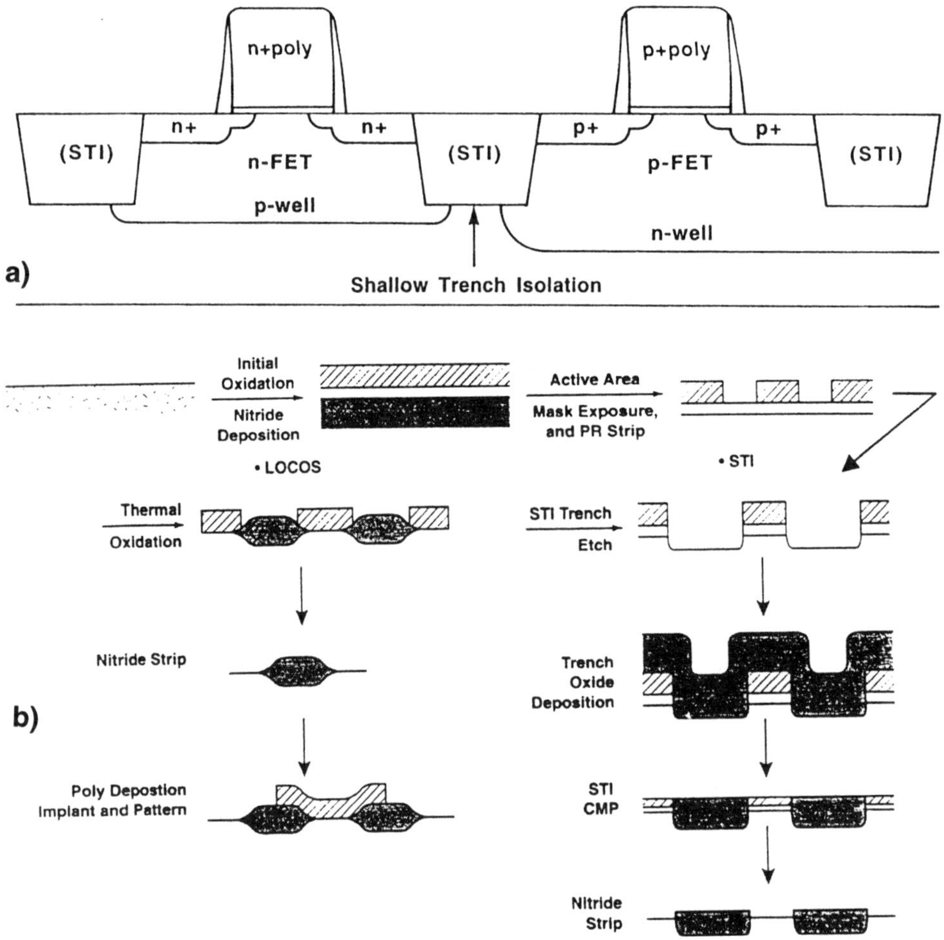

Fig. 9-1 a) Schematic of STI in a CMOS structure. b) LOCOS vs. STI process flow, indicating that STI requires more steps.

a smaller curvature than exists with semi-recessed LOCOS field oxides. These trench corners provide improved isolation efficiency due to the so-called "corner effect."[1] That is, the field lines at the bottom corners of the trench spread out at the vicinity of the corner, leading to a higher threshold voltage at the corners. Until the silicon surface at these corners is inverted, isolation is maintained.

Despite the impressive advantages of shallow trench isolation listed above, a variety of new difficulties associated with the fabrication and process integration of STI structures arise. For one thing, the STI process involves a larger number of steps than does LOCOS isolation process (Fig. 9-1b). Yet, the overall benefits of STI over LOCOS provided the impetus for the industry to find ways to overcome these challenges. STI began to find use in mainstream IC technology when it was integrated into 166 MHz Pentium microprocessors (which used a 0.35-μm process) and the PowerPC 603e chip (which was fabricated with a 0.45-μm process). In this chapter we will describe the details of the STI process fabrication sequence, the process integration challenges that have been overcome, and those that it faces going forward.

9.1 EARLY SHALLOW-TRENCH-ISOLATION STRUCTURES

The earliest example of shallow-trench-and-refill is the so-called *buried-oxide* (or BOX) isolation technology. The original BOX technique (BOX-I) was introduced in 1981,[2] and an improved version was presented in 1983 (BOX-II).[3] In the latter, shallow trenches (0.5-0.8 μm deep) were etched with an orientation-dependent silicon-etch to provide an inclination of the trench sidewalls of ~60°. Next a channel-stop-implant step was performed. (The sloped sidewalls of the trenches ensure that they receive sufficient doping during the channel-stop implant step to prevent weak inversion of the silicon-sidewall surface. Thus, subthreshold leakage from source-to-drain along this parasitic conduction path is suppressed.) A thin thermal oxide (20-50-nm-thick) was then grown to passivate the trench sidewalls. Next, CVD oxide was deposited to fill the trenches. This layer was etched back, leaving oxide only in the recesses. Ideally, the top of the oxide should be level with the original silicon surface. Etchback was performed by eroding photoresist and SiO_2 at the same rate in a plasma (see Chap. 4, Vol. 2). In the improved process, two layers of resist were applied (to try to produce a globally planarized surface prior to etchback). Active regions are those protected from etch when the trenches are formed.

These early BOX processes, however, exhibited several drawbacks, including:

1. Void formation occurred if the CVD oxide was deposited into 0.5-μm-deep trenches that are narrower than about 2-μm.

2. Inversion of the silicon at the sidewalls of *p*-type active regions could still occur, especially if vertical trench sidewalls were used to reduce the isolation spacing. This problem was made worse by boron segregation into the thin, sidewall-passivating thermal oxide. This effect lowers the boron concentration at the sidewall surface, as will be discussed in following section.

3. Despite complex planarization schemes, the early BOX processes were unable to yield adequately planar surfaces. (In particular, global planarity was poor.)

4. The BOX isolation structures could negatively impact the device operating behavior.[4,5,6,7] That is, in a conventional LOCOS-isolated MOSFET, a polysilicon gate that passes across the device will curve upward and away from the device once it reaches the isolation oxide, as shown in Fig. 9-2a. (This figure represents the cross section of the region under the gate of the MOSFET, where the source and drain of the device are situated above and behind Fig. 9-2, which lies in the plane of the paper.) For this particular configuration of the polysilicon gate, the portion of the gate poly over the LOCOS region will have little effect upon the device behavior. Consequently, a simple 2-D simulation will give an adequate description of the behavior of the device, where the 2-D region is taken in the plane perpendicular to Fig. 9-2 and along the axis of the device, so that it includes the source and drain regions.

In the ideal case of a shallow-trench-isolated MOSFET, the gate poly runs flat across the isolation, as shown in Fig. 9-2b. Consequently, the part of the gate poly over the field region creates more of a two-dimensional fringing electric-field on the corner and sidewall of the silicon mesa than in the LOCOS case of Fig. 9-2a. This establishes a *corner parasitic-MOSFET* in parallel with the main device.[8] The parasitic device turns-on at gate voltages lower than those required by the main channel (because the same voltage occurring at the corner produces a larger electric field at that point, inducing a higher concentration of

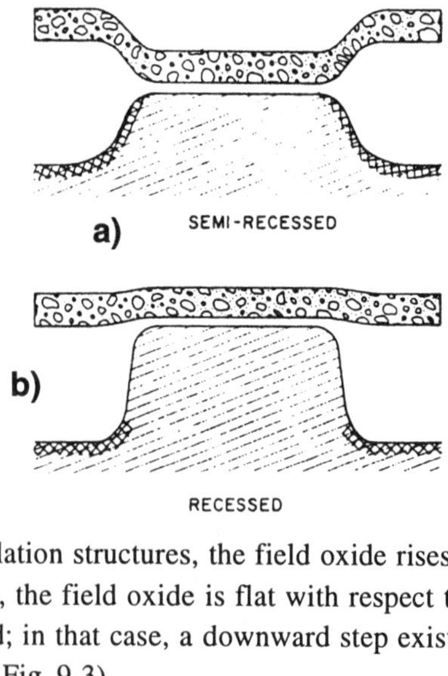

Fig. 9-2 a) In LOCOS isolation structures, the field oxide rises above the substrate. b) In trench isolation structures, the field oxide is flat with respect to the substrate (unless the trench oxide is overetched; in that case, a downward step exists at the edge of the active regions – see, for example Fig. 9-3).

inversion-layer charge than the same voltage induces in the nominal channel region). This causes a "hump" or "kink" in the drain-current versus gate-voltage subthreshold curves (see Fig. 9-3). Thus, the parasitic device increases the subthreshold leakage-current of the active device.

This effect becomes even stronger if the isolation region is partly etched below the top surface of the silicon mesa (i.e., the surface of the active region of a MOSFET IC). In that case, the polysilicon gate wraps around the corner of the mesa as shown in Fig. 9-3. Now the regions of the channel near the edges of the device will behave much differently than the region at the center of the mesa. Consequently, 3-D modeling is required to accurately simulate the device characteristics. Such simulations verify that the corner regions have a lower threshold voltage than do the center regions of the device.

5. Thinning of the gate oxide at the trench edge is another concern. Thinned gate oxide not only degrades the gate-oxide-reliability, but also worsens the parasitic device problem (since a thinner gate oxide increases the electric-field-strength at the trench corner).

6. Problems arose as result of the etchback step. That is, to ensure complete removal of the CVD SiO_2 from the active areas, the field oxide must be overetched by about 100 nm. The field oxide surface is then lower than active-area surface, causing a portion of the active-area sidewall to become exposed. This exacerbates the parasitic sidewall conduction described above and also enhances the inverse narrow-width effect (see Chap. 6, *Silicon Processing for the VLSI Era* - Vol. 3).

The combination of the above problems (especially the complex processing required to produce adequately planar surfaces), kept BOX-isolation from displacing LOCOS-type isolation-structures. However, the benefits offered by shallow-trench isolation are so attractive that work continued in an effort to overcome the above problems. Several enabling technologies were developed to overcome the drawbacks of the early shallow-trench isolation processes. These included: *dry-etch processes* that allowed trenches with controlled sidewall slopes to be produced; *deposition techniques* that permitted the trenches to be filled without the formation of voids; and *chemical-mechanical polishing* (CMP) methods that could provide global planarization of the wafer surface at the conclusion of the STI process. We will describe these technologies in more detail in the next section, emphasizing how their use enabled the implementation of STI. Figure 9-4 shows an SEM of an STI structure.

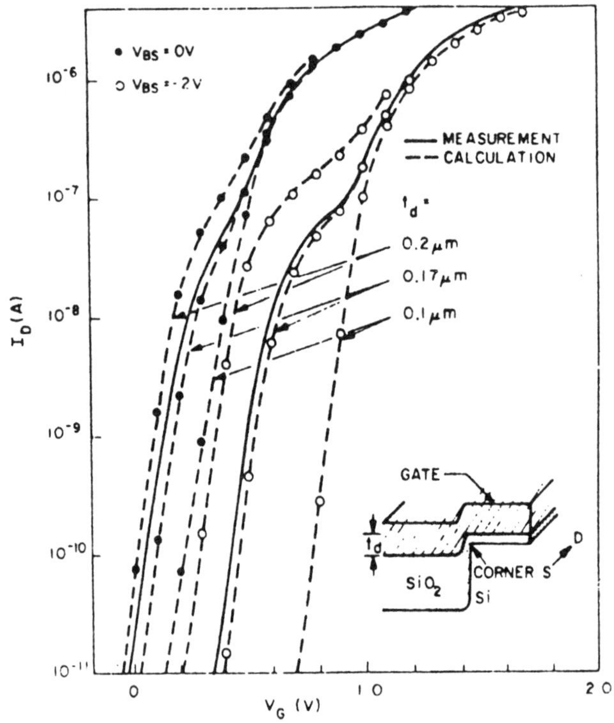

Fig. 9-3 Measured and simulated drain current versus gate potential for devices having a downward step in the field oxide (t_d).[9] (© 1981 IEEE)

9.2 STI-ENABLING PROCESSES

Progress in overcoming the four major problems of early-STI isolation structures is summarized as follows:

1. The problem of void formation when refilling narrow trenches with CVD-oxide was ameliorated in two ways: **1**) dry-etch processes which could etch the trenches with a controlled slope (and rounded-bottom trench corners) were developed to replace the wet-etch techniques used to etch the trenches in early-STI structures; and **2**) various so-called "gap-filling" CVD techniques were developed that could fill the trenches with CVD-oxides without creation of voids. Such gap-filling-CVD processes include: *LPCVD TEOS*; *atmospheric-pressure CVD of ozone-TEOS*; and *high-density-plasma CVD (HDP-CVD)*. Note that the film stress of the trench-fill dielectrics must also be smaller (<200 MPa) than in intermetal dielectric films, due to the use of thicker films.

2. The introduction of *chemical-mechanical polishing* (*CMP*) allowed the oxide that is deposited into the trench to be polished back, leaving a planar surface over the wafer. The CMP process was developed in the late 1980s at IBM, and was introduced to the rest of the IC industry in the early 1990's. Details of the CMP process are provided in Chap. 8.

3. As described above, the subthreshold characteristics of narrow MOSFETs (e.g., ≤ 0.5-μm-wide) isolated by refilled shallow-trench structures are sensitive to the shape of the upper corners of the trenches.[10] That is, the edges of the silicon mesa at the upper corners are turned-on by the portion of the gate that overlaps such corner regions (see also Sect. 5.2.2 in *Silicon Processing for the VLSI Era*, Vol. 3). This edge-effect becomes even stronger if the isolation region is partly eroded away, in which case the gate wraps around the corner of the device, as shown in Fig. 9-3. The optimum condition for the shallow-trench-isolation structure is thus a gate that runs flat across the isolation, requiring a planar surface prior to poly deposition. A process utilizing RIE-etchback followed by a chemical-mechanical-polishing step offers the ability to produce a highly planar surface after etchback. The top surface exhibits virtually no step. However, single-wafer processing with *in-situ* control of the CMP process is likely to be needed to obtain the exact surface required.

9.3 SHALLOW TRENCH ISOLATION FOR CMOS

Trench isolation has also been explored for twin-tub CMOS as a means of implementing isolation structures, primarily as a means for achieving smaller interwell (n^+-to-p^+) isolation spacing. (Other advantages of trench isolation over LOCOS were described earlier.) A trench at the n-well-to-p-well interface prevents dopant interdiffusion where the two wells join. Thus, dopant compensa-

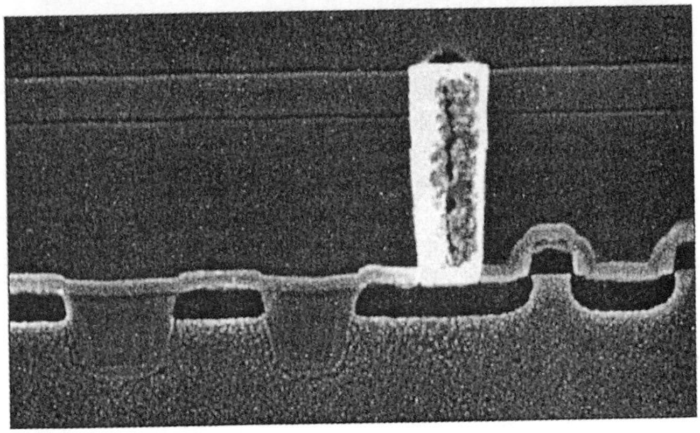

Fig. 9-4 SEM of a shallow-trench isolation structure.

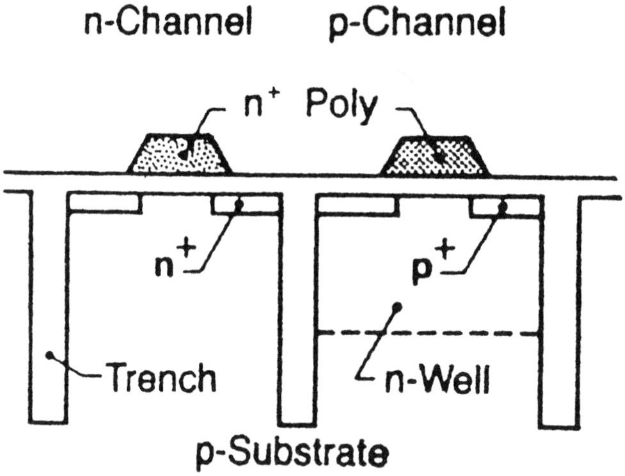

Fig. 9-5 Schematic of an *n*-well CMOS with deep trench isolation.[11] (© 1986 IEEE).

tion (which lowers the surface dopant concentration at the well borders) is eliminated. This permits larger parasitic-MOSFET-threshold-voltage values (V_{TFiso}) to be maintained at smaller isolation spacings. Shallow and deep trenches of submicron widths have been used to reduce p^+-to-n^+ spacings.

When trenches deeper than the well-depth are used, they replace the reverse-biased *pn*-junction as the isolation structure at the well sidewalls (Fig. 9-5). However, since fabrication of deep-trench-isolation for CMOS is substantially more complex than shallow-trench-isolation, and latch-up measurements have not fulfilled expectations, this approach was not widely implemented for such applications. Shallow-trench-isolation structures, however, became the standard isolation structure for CMOS at about the 0.25-μm node.

It should be noted, however, that while their use permits smaller n^+-to-p^+ spacing, the presence of trench structures also gives rise to additional leakage paths not present in LOCOS-CMOS-isolation structures. These new paths can degrade the isolation between the active devices and the well. In some cases these paths can also increase the subthreshold leakage-current within *active* devices. Generally, these new leakage phenomena involve the formation of an inversion channel along the vertical trench-sidewall (see Fig. 9-6a).

The sidewall inversion is caused by the *horizontal* parasitic MOS device, in which the positively biased *n*-well acts as the gate electrode, and the trench dielectric serves as the MOS gate oxide (with a thickness equal to the trench width). If the *n*-well is biased at $+V_{DD}$, and the trench is narrow (e.g., ≤ 1-μm), inversion along the sidewall of the trench between the source and drain of the

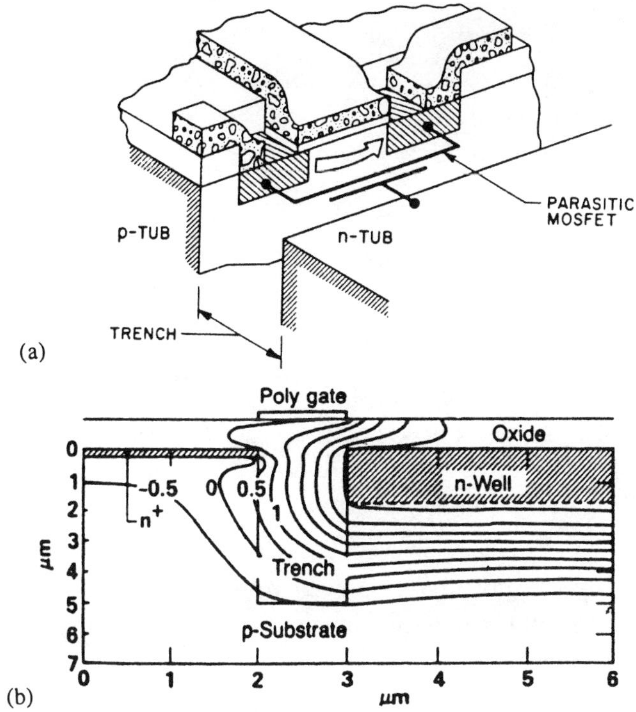

Fig. 9-6 a) A parasitic-sidewall-MOSFET with the *n*-well as the gate and the trench serving as the gate oxide.[12] From S.M. Sze Ed., *High-Speed Semiconductor Devices*, Copyright 1990 by John Wiley & Sons. Reprinted with permission. b) Cross-sectional view of a simulation of the 2-D equipotential contours in an *n*-well CMOS with trench isolation. The simulation is done at Q_{fc} = 5x10^{10} cm^{-2}, N_A = 9x10^{14} cm^{-3}, V_{well} = 3V, V_{GS} = 0V, V_{sub} = -1V.[13] (© 1983 IEEE)

MOSFET can easily occur. As seen in Fig. 9-6a, this is the trench sidewall that is outside of, but facing the *n*-well. In Fig. 9-6b, the simulated 2-D potential contours in a trench-isolated CMOS structure are shown. Note that the potential on the trench sidewall on the *p*-substrate side is > 0.5V, indicating that it is inverted.

It would appear that to take maximum advantage of such trench structures for reducing isolation spacing, the active devices should be placed right at the well edge (as shown in Fig. 9-6a). However, if an active NMOS-device abuts an inverted trench sidewall, a conducting path will exist between the source and drain of the active NMOS device. This path will exist regardless of the bias on the "real" gate of the active NMOS device, giving rise to an increase in the subthreshold leakage-current of the active device. In practice, to avoid this problem, active devices are usually moved back somewhat from the well edge.

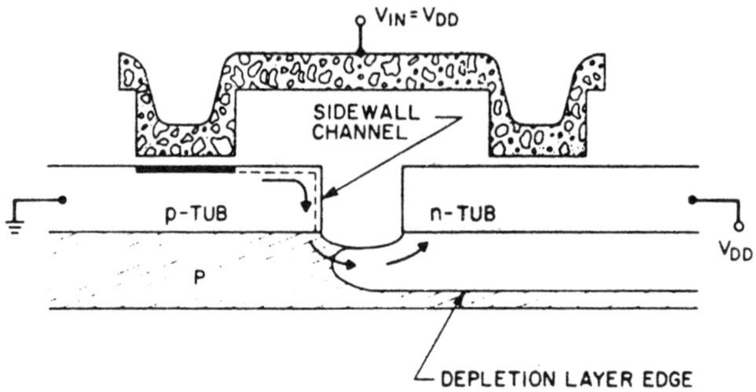

Fig. 9-7 A trench-isolated *p*-well and *n*-well with a parasitic vertical-sidewall-MOSFET (with the *n*-well as the gate) and a parasitic-horizontal-MOSFET (with the V_{in} poly-line as the gate).[14] From S.M. Sze Ed., *High-Speed Semiconductor Devices*, Copyright 1990 by John Wiley & Sons. Reprinted with permission.

Even if the NMOS device is moved back from the trench edge, the trench sidewall can still become inverted by a positively-biased *n*-well. If this happens and the poly over the field-region also inverts the top surface of the *p*-well, it is possible for a leakage path from the NMOS-source to the *n*-well to be established, as shown in Fig. 9-7.

Assuming that the leakage path of Fig. 9-7 is somehow suppressed (e.g., by implanting channel-stop-dopants into the bottom of the trench), the inverted trench sidewall can still produce other, unexpected leakage paths. One such dc leakage path between V_{DD} (e.g., 5V) and V_{SS} (e.g., 0V) is illustrated in Fig. 9-8 for the case of two series-connected CMOS inverters which suffer from inadequate CMOS isolation.[15] This path arises from leakage along both the inverted surface of the field-regions and the trench-sidewall (which, in this example, is offset from the active device edge). Here V_{in1} to the first inverter is at logic 1 (i.e., V_{in1} = 5V), causing the NMOS transistor of the first inverter to be turned *ON*. The output of the first inverter V_{out1} (and thus the input to the second inverter V_{in2}) is at logic 0. This causes the output of the second inverter V_{out2} to be at logic 1, and causes the PMOS device of the second inverter to be turned *ON*. The black highly conductive paths in Fig. 9-8 from V_{DD} to the edge of the n^+-region in the second inverter and from V_{SS} to the p^+-region in the first inverter are connected by a less-well-conducting parasitic conductive path. This path consists of an inverted region that runs under the poly of gate 1 in the *p*-well, then along the inverted trench sidewall, then under the poly of gate 2 (note that for clarity, in Fig. 9-8 the path under the poly of gates 1 and 2 is drawn as running adjacent to the poly, whereas the path actually exists beneath the poly).

More severe trench-sidewall-inversion problems are exhibited in p-wells than in n-wells. This is due to two interrelated phenomena. The first involves the dielectric (insulator) material used to fill the trench (typically SiO_2), and the resultant Si/SiO_2 interface at the trench-sidewall. The interface fixed or trapped charge that exists at the Si/SiO_2 interface is positive, which tends to lower the inversion-threshold of p-channel regions (in this case the p-well). To minimize this interface charge, a thin, good-quality thermal-oxide-layer must be grown inside the trench. This oxide growth process, however, gives rise to the second problem. That is, during thermal oxidation, some of the boron dopant in the p-well segregates into the oxide. Thus, the B surface-concentration at the interface is reduced, further decreasing the inversion-threshold. As a result, inversion of p-well trench-sidewalls occurs at lower threshold-voltages than do those of an n-well for comparable well-doping concentrations. In 0.25-μm (and smaller) CMOS technologies, this last problem is circumvented by forming the well-regions with an implant after the STI-structure is completed. A high-energy-implant process is used to implant well-dopants through the STI-oxide (and in

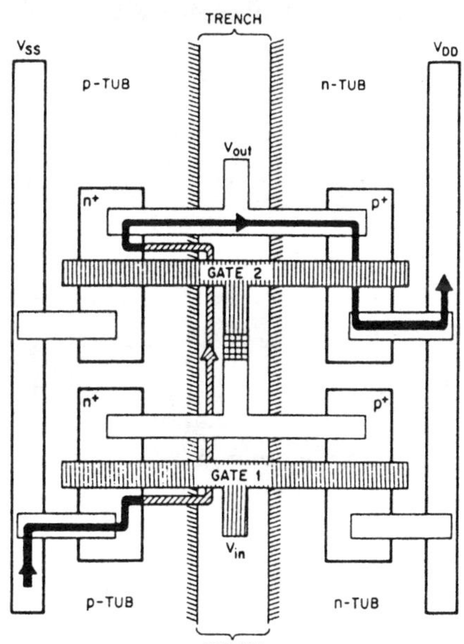

Fig. 9-8 A dc leakage path through two inverter stages in series. The solid black path is a good conducting route and the shaded path is a parasitic route. The parasitic consists of a field oxide corner under Gate-1 (cf. Fig. 9-7), a sidewall path along the trench (cf. Fig. 9-6a), and a corner path under the V_{out} strap.[15] From S.M. Sze Ed., *High-Speed Semiconductor Devices*, Copyright 1990 John Wiley & Sons. Reprinted by permission.

the active regions). Thus, the boron-segregation problem that would occur during the trench-liner-oxidation process is avoided.

To counteract intra-well and inter-well leakage in trench-isolated-CMOS several steps are necessary. First, the surface-leakage-path must be suppressed by use of sufficient channel-stop-doping between the active-device and well-edge (as is also done in LOCOS-isolated-CMOS). Second, to avoid subthreshold leakage in single active devices, the devices are moved back somewhat from trench edge. Third, inversion of the trench sidewall should be prevented. One way to do this is to increase the trench width. However, this will increase the isolation spacing (as will moving the active devices away from the active device to the trench edge).

Another way to prevent inversion of the trench sidewalls is to increase the doping at the sidewall surface. Use of a direct, low-tilt-angle implant (e.g., boron, at an angle of 8° off the perpendicular to the top wafer surface, or at an angle of 82° off the perpendicular to the trench walls) is one reported way to do this (Fig. 9-9a).[16] However, since only one of the walls is doped by one implantation, the implant must be done four times, with the wafer rotated 90° after each step to ensure that all four walls of the trench are doped. In addition, this procedure is only effective for intra-well isolation (i.e., to isolate like-types of devices).

An alternative method for doping the sidewall of the p-well without doping the n-well with Si (thus making it suitable for CMOS inter-well isolation), was proposed by Davari *et al.*[17] In this approach, the p-well-sidewall-dopant is provided by boron diffusion from p^+-poly or borosilicate glass. The n-well does not require additional sidewall dopant, since phosphorus piles up during the brief thermal oxidation of the trench prior to CVD-oxide-refill.

In this selective-sidewall-doping process, a nitride/pad-oxide mask-stack is first formed and patterned. Next, after shallow trenches are etched by RIE to a depth of 0.3-0.5-μm, a boron diffusion source (*in-situ*-doped p^+-poly or borosilicate glass, BSG) is deposited. A lithography step is then used to etch the poly or BSG away from the n-well-sidewall and trench-bottom (see Fig. 9-9b). A thermal cycle is then carried out to drive the boron into the sidewall and bottom of the p-well trench. No boron is diffused into the active device areas since they are covered with nitride. The boron source is then etched away, and the remainder of the trench-refill process is conducted. The boron concentration at the sidewall-surface is controlled by the segregation during the brief thermal 50-nm-thick oxide-growth step. This approach was used for trench widths as small as 0.8-μm (i.e., used, for example in a 0.35-μm process). For the smaller-width trenches used in 0.25-μm and smaller technology nodes, the problem is handled

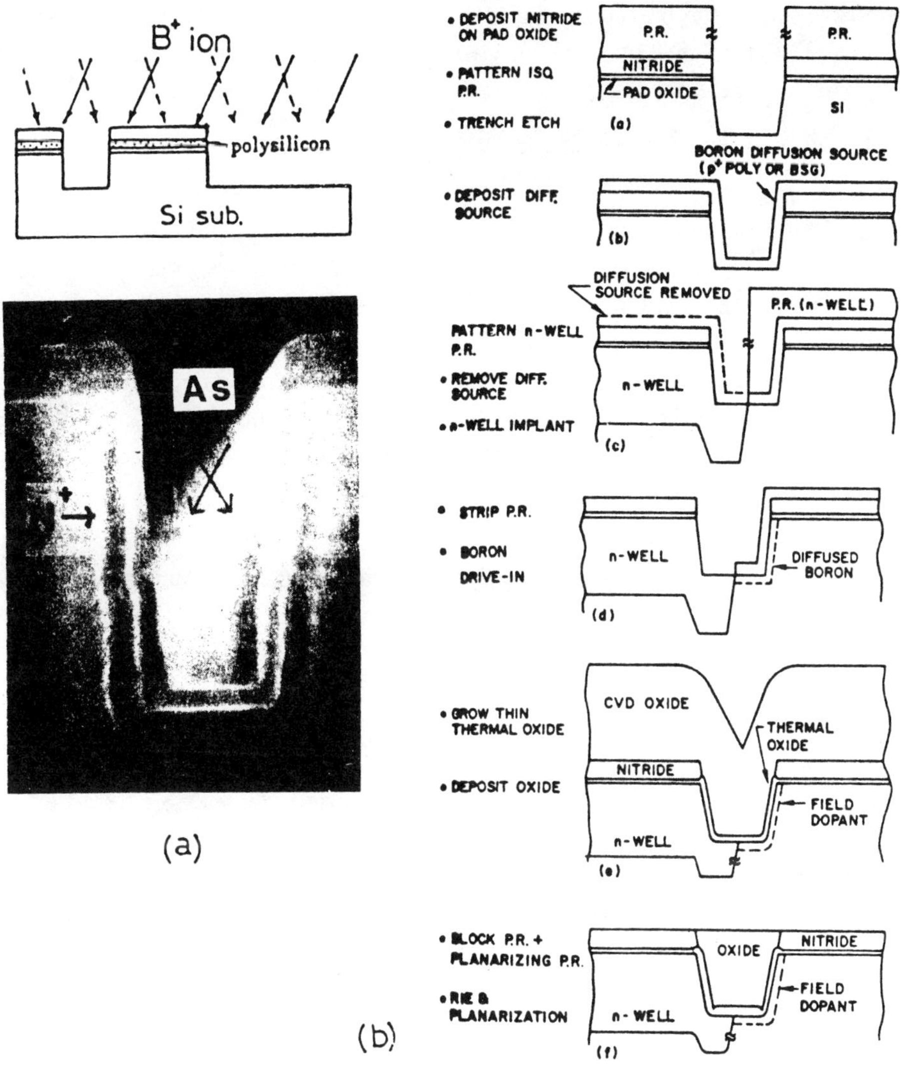

Fig. 9-9 a) Schematic of oblique ion implantation and SEM cross section result of this procedure.[16] (© 1987 IEEE) b) Shallow trench isolation process flow using a doped oxide as diffusion source of trench sidewall doping.[17] (© 1988 IEEE)

differently (e.g., the typical minimum trench width in a 0.18-μm process is about 0.4-μm with an active-area width of about 0.6-μm). That is, the p-well-doping implant-step (carried out after the STI structure is completed, using a high-energy-implant process) also places boron channel-stop-dopants at the bottom of the trench in sufficient concentration to prevent an inversion path from occurring there. In addition, the boron-doping at the silicon surface in the channel-region of the NMOSFETs required to prevent DIBL in deep-submicron MOSFETs is high enough to keep the sidewalls of the trench from inverting.

9.4 DETAILS OF THE PROCESS FLOW TO FORM A BASELINE SHALLOW-TRENCH-ISOLATION (STI) STRUCTURE

In this section we present a process for fabricating STI structures. Note that the process sequence we describe is that of a baseline process, and one that is likely carried out for 0.25-μm CMOS generations. Some modifications can be made to this baseline process (even at the 0.25-μm technology node), and we will describe these variations in a later section. As ICs are scaled further, other modifications to the STI process will have to be made, and we will also discuss the need for such changes in the sections that follow this one that cover those process integration challenges in more detail.

The baseline sequence of steps for forming STI structures is given as follows:

1. The first several steps performed in fabricating an STI structure are the same as those used in the LOCOS isolation process sequence, with the exception being that the trenches for shallow trench isolation are the first structures formed on the bare silicon starting wafer. (On the other hand, in the conventional twin-well-CMOS process that used LOCOS, the wells were the first structures created on the bare-Si starting wafer.) In the STI process sequence, the bare, starting Si wafer is first cleaned (e.g., with an RCA clean), and then a pad oxide is thermally grown on the Si surface (typically to a thickness of 10-20-nm). Note that the role of the pad oxide layer is to prevent stresses that would exist if the silicon nitride layer (which is deposited next) were deposited directly on the bare-Si wafer. Such stresses could cause defects in the Si-substrate to occur. Note, however, that in the STI process the stresses associated with the formation of the STI structure are smaller than those involved in forming a LOCOS isolation structure. That is, in the LOCOS process, a long, high-temperature operation is performed to grow the field-oxide. During this field-oxide-growth step, the stresses between the nitride and the Si-substrate can be severe, and the pad oxide must therefore be sufficiently thick to keep these stresses from getting large enough to cause defects in the Si-substrate. In the STI process, there is no high-temperature field-oxide-growth process, and thus the stresses between the nitride film and the Si surface will not be as large as those in the LOCOS process. Hence,

the pad oxide in the STI process can be significantly thinner than the ones used in the LOCOS process (10-nm-thick vs. 25-40-nm-thick).

2. A silicon-nitride layer is then deposited on the pad-oxide layer (Fig. 9-10a). The silicon nitride plays two roles in the fabrication of the STI structure: **i**) it serves as a hard mask for etching the trench; and **ii**) it acts a CMP polish-stop-layer when the trench-fill dielectric material is planarized with a CMP process. The nitride-layer-deposition step is typically performed with an LPCVD process that deposits a silicon-nitride layer that is 100-200-nm thick. Note that the LPCVD of silicon nitride is a batch process that is simultaneously carried out carried out in a furnace tube on many wafers at once. As a result, the backsides of the wafers are also exposed, and silicon nitride is deposited on the back of the wafers (as well as on the front). Hence, this backside-nitride must somehow later be removed.

3. Then photoresist is applied and *Mask #1* is used to pattern the regions on the wafer surface where STI trenches will be opened.

4. The nitride and pad-oxide are then anisotropically etched, using a plasma-etching process (Fig. 9-10b). Usually this etch is carried out in a dry etcher designed for etching dielectric materials (*Etcher #1*).

5. The trenches are then etched (Fig. 9-10c). This silicon-etch step is also carried using a plasma-etching process. The goals of this etch process include: **a**) etching the trenches to the correct depth; **b**) controllably producing trenches with the sidewalls having the desired slope (e.g., from 70-85°, depending on the technology node, the trench depth, and the type of circuit being fabricated); **c**) producing trenches with rounded bottom corners; and **d**) having the etch process produce smooth sidewalls and minimum Si damage.

In general, it is easier to fill trenches without forming voids in the trench-fill material if the trench openings are sloped. Obviously, this requires more Si area for the trenches, and thus as ICs are scaled, the sidewall profiles will have to become more vertical. Gasses used for etching the trenches are mixtures of HBr and other gases (e.g., $HBr/Cl_2/O_2$, or $HBr/NF_3/He/O_2$).[18]

The trench-etch process is usually carried out in a second etcher (*Etcher #2*), different from the one used to etch the nitride and oxide layers (*Etcher #1*). That is, *Etcher #2* is usually a modified version of a gate-etch tool (i.e., one in which polysilicon is etched).

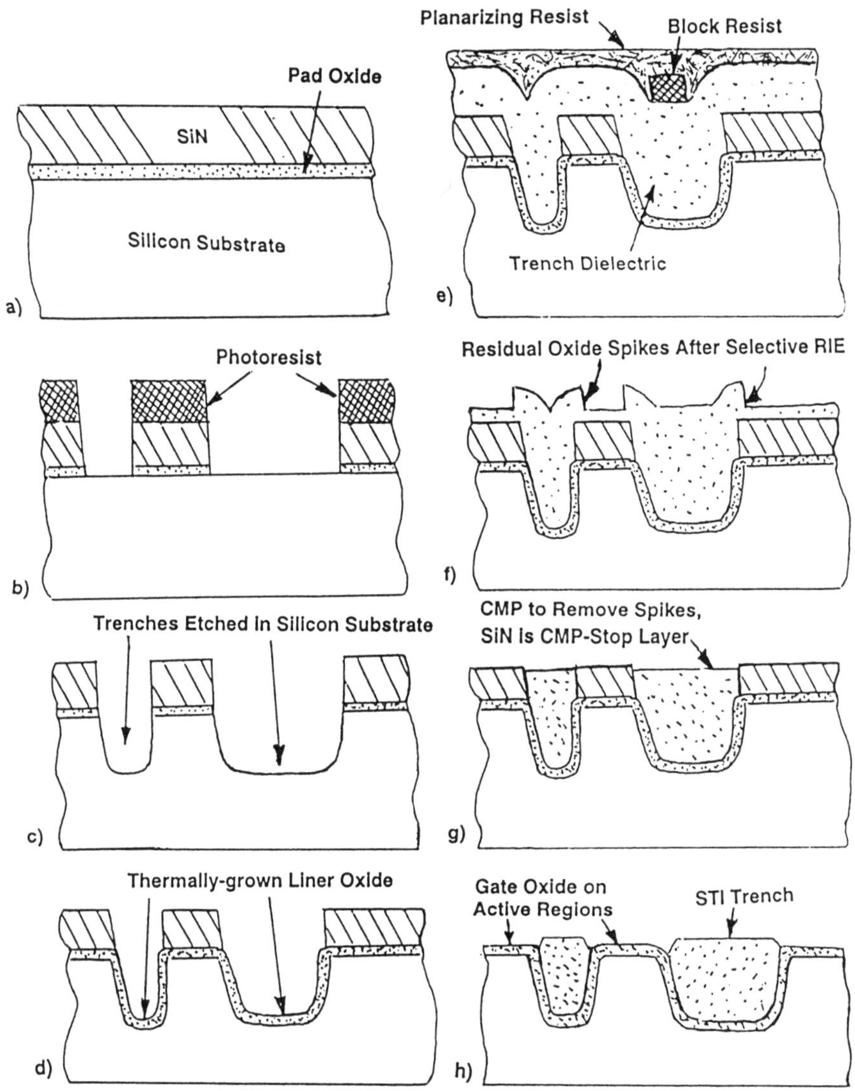

Fig. 9-10 Baseline process flow used to form an STI structure: a) Form pad oxide by thermal oxidation, then deposit a silicon-nitride layer by LPCVD; b) Pattern the nitride and pad-oxide layers; c) Etch trenches, strip resist; d) Grow liner-oxide on the trench walls, using thermal oxidation; e) Deposit the trench dielectric & densify it, then deposit a block resist layer & pattern it. Finally deposit the planarizing resist layer; f) Use a selective-RIE-etch process to simultaneously remove the resist and the trench-oxide-layer, leaving residual oxide spikes; g) CMP the CVD-oxide-layer to abrade spikes and planarize its surface. The silicon nitride acts as a CMP-stop-layer; h) Strip the nitride and then grow the gate-oxide on active regions. The STI structure is now completed.

Note, however, that these two etch steps could also be carried out in a single etcher. Significant enhancements to process productivity (in the form of higher throughput and lower cost) could be obtained if the opening of the silicon-nitride hard-mask-layer and the etching of the trenches were carried out *in-situ* in a single etcher (and such a process has been termed *in-situ hard-mask-open*). Etch tools capable of *in-situ* hard-mask-open are available from Applied Materials, Lam Research, and TEL America, and some such processes have reportedly been put into production by 2000.[19] The main challenge of successfully implementing an *in-situ* hard-mask-open process is selectivity to the photoresist.[20]

The trench etching process is difficult for a number of the following reasons:

a. The trench depths across the wafer must have uniform depth to ensure correct circuit operation (the dependence of circuit performance on trench depth will be discussed later, in the section that deals with CMP-dishing problems). However, some dry-etch processes exhibit *aspect-ratio-dependent-etching* (ARDE) characteristics, in which material from wider-spaced trenches is etched more rapidly than from narrower trenches. Thus, etch-process development must be employed to find etch-process conditions in which ARDE effects are limited.

b. The trench etch process is a 'blind' process, in that there is no etch-stop layer at the bottom of the trench, nor is there any material present at the trench bottom that will create an end-point-signal. Hence, an etch process must be available that is so well-controlled

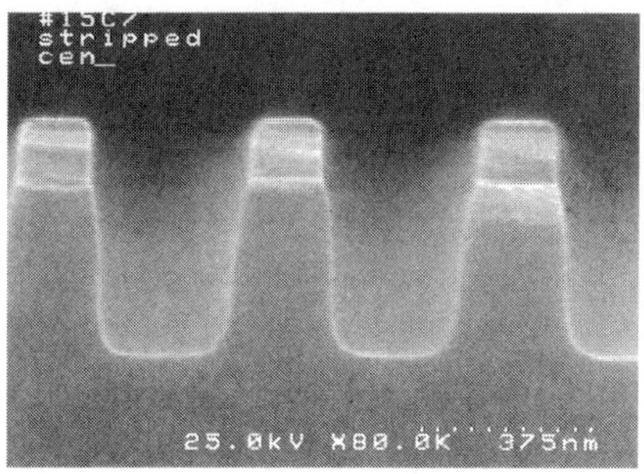

Fig. 9-11 SEM of etched trenches of an STI structure with trench walls having an 87° slope and bottom corner rounding. Courtesy of Lam Research.

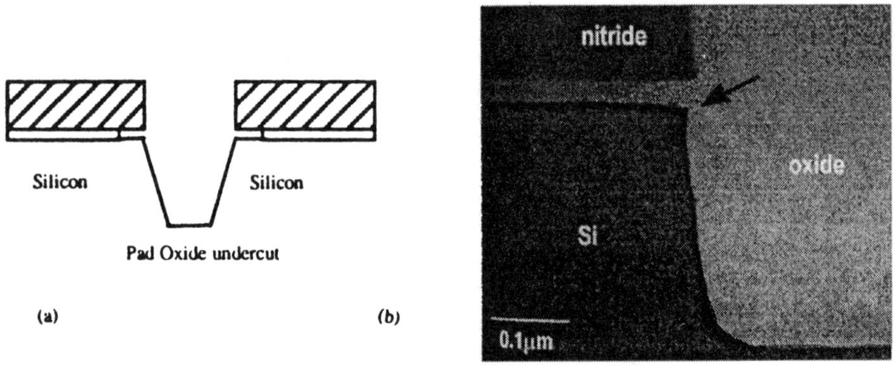

Fig. 9-12 a) Pad-oxide trench-undercut in the STI-baseline process. b) TEM of an as-etched trench profile showing the sharp corner of the active Si region at the top of the trench.

that a timed etch can be relied upon to produce trenches with the correct depth, from wafer-to-wafer, and across a wafer.

c. The etch process should yield smooth trench sidewalls with angles of 70–85°, rounded bottom-corners (to minimize stress), and a residue-free silicon-surface after etch (see Fig. 9-11). Trenches with rounded-bottom corners minimizes stress to the silicon lattice encountered during the trench-liner-oxidation process of Step **8** and the trench-filling-process of Step **9**. It was reported that using gas etch mixtures consisting of $HBr/NF_3/He/O_2$ gave improved corner rounding results than mixtures containing HBr/Cl_2.[21]

6. After the trenches are etched, the resist is stripped.

7. The pad oxide is next undercut with a short HF dip (Fig. 9-12a). The reason this step is carried out has to do with the fact that upon its completion, a silicon dioxide layer is thermally grown on the exposed trench surfaces. This so-called *liner-oxide* layer serves several important roles. First, the process of growing this oxide consumes any residual damage caused by the plasma-trench-etch process. Second, the thermal oxide passivates the Si surface of the trench, providing as stable interface between the Si and the trench fill material. (The threshold voltage of the parasitic field MOSFETs must be kept high enough so that leakage paths at the Si-surfaces of the trenches are not electrically activated during normal device operation. To ensure that this requirement is met, the trench-Si/filler-material interface must be electrically stable. A thermal oxide in contact with the Si surface provides the most electrically stable of interfaces.) Third, at the top corner of the active-region (i.e., where the active-region and the isolation-region meet, see Fig. 9-12b) the Si normally has a sharp corner after the trench etching has been completed. Such a sharp corner will enhance the electric field lines at this corner and degrade the MOSFET turn-off characteristics (as will be discussed in more detail later). To suppress this effect, the top-sharp-corner of the silicon active-region must therefore be rounded. Device simulations have demonstrated that the parasitic double-hump in the I_D-V_{GS} curves can be eliminated by having a trench corner with a radius of 50-nm if a trench with no more than an 80° sidewall slope is specified.[22] This can be done in a number of ways, but here, we will assume that such a "corner-rounding" process is also performed by the thermal-oxidation step that forms the liner-oxide.

It has been demonstrated that the top-sharp-corner will be more effectively rounded by the thermal-oxidation step that forms the liner if the pad-oxide is undercut prior to performing the thermal-oxidation liner-step. Hence, as shown in Fig. 9-12a, this pad-oxide-undercutting step is performed next, i.e., by dipping the wafer in HF to etch-back the pad-oxide to a controlled extent.

8. The thermal-oxidation step that forms the liner oxide and rounds the sharp-top-corner of the trench is performed next. The oxide grown in this procedure is typically 40-50-nm thick, and it is grown in dry-O_2. It has also been reported that maximum corner-rounding is achieved if the liner-oxide is grown with a high-temperature (1050-1100°C) rapid-thermal-oxidation (RTO) process (Fig. 9-13).

Note that a thin layer of the nitride is converted to oxide during this liner

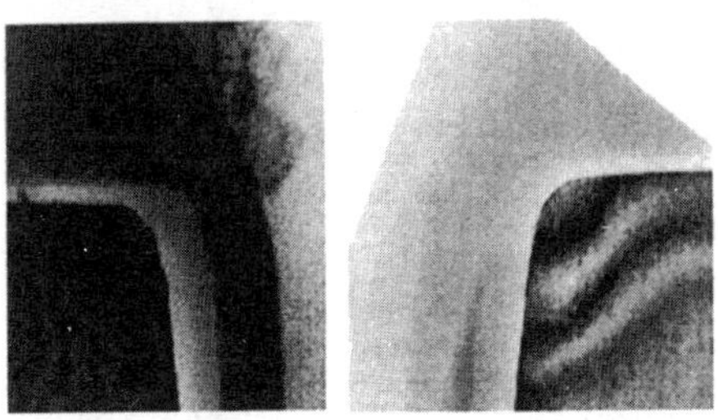

(a) Lower temp+Hcl (b) High temp RTO

Fig. 9-13 Cross-sectional TEM of the top trench corner after pad oxide undercut & liner oxidation: HCl in the ambient (a); and higher-temp RTP-oxidation (b) improve corner rounding.

oxidation step, an effect that must be taken into account when the nitride is later stripped. Also note that in DRAMs fabricated with STI, a thin silicon-nitride liner may also be deposited by CVD prior to filling the trench with a CVD oxide. This nitride liner helps decrease leakage current, a critical parameter in DRAM operation.

9. Next, a CVD-dielectric film is used to fill the trenches. Since the deposition is a blanket process, some of the material is deposited onto the active regions that are still covered with nitride (i.e., 8000-9000-Å of gap-fill dielectric are deposited when filling a 5000-Å-deep [0.5-μm] trench). As a result, the material must be removed from the active regions after the deposition process is completed (Fig. 9-10d). As mentioned earlier, a number of CVD processes have been pursued for filling the trenches without void formation, including: LPCVD TEOS; (sub)atmospheric-pressure CVD of ozone/TEOS; and high-density plasma CVD (HDP-CVD).

10. The dielectric material is densified at about 900°C. This step is carried out so that the trench-fill dielectric will exhibit etch rates comparable to that of thermal SiO_2. This characteristic is necessary so that excessive loss of the trench-fill dielectric is not incurred during later HF deglazing steps (i.e., steps **15**, **17**, and **20**). This densification process reduces the wet-etch rate by about 20%.

11. A photoresist layer is next spun-onto the wafer. This resist is then patterned to form a reverse-mask of the active regions and a planarizing-resist layer is deposited over the patterned reverse-masked wafer (Fig. 9-10e). The oxide and the resist are etched simultaneously in an RIE-etch-back step that etches resist and oxide at the same rate. This reduces oxide thickness in the field areas. The result of partially planarizing the surface with this REI step is that the time needed to planarize the wafer surface during the subsequent CMP step is decreased. A shorter-duration CMP step decreases the problem of dishing (as discussed further in Sec. 9.5.3).

12. After the resist is patterned and the planarizing-resist layer is deposited, the wafer is etched back with an RIE process described in Step **11**, and as shown in Fig. 9-10f.

13. Any remaining photoresist is then stripped. Small spikes of oxide, however, remain near the edges of the active/field regions (see also Fig. 9-25). These must be removed to produce a globally planarized STI surface.

14. A CMP-step is used to polish back the dielectric layer to remove these spikes, which is continued until the nitride is reached (the nitride acts as a CMP-stop layer, Fig. 9-10g). A major issue involved with this CMP-step is dishing of the trench-fill-dielectric in wide trench regions. The reverse-mask procedure of Steps **11,12,** and **13** is one method used to mitigate the dishing problem. Other dishing-suppression techniques are discussed in Section 9.5.4.

15. The nitride is then stripped, using a hot-phosphoric acid, wet-cleaning process. But, it should be noted that the stripping process is actually done using two wet-etching steps. That is, the thin layer of nitride that was converted to an oxynitride layer during the trench-liner-formation thermal-oxidation-step must be removed prior to carrying out the nitride-stripping-step itself. The oxidized layer on the nitride will not etch in hot phosphoric acid, and unless it is removed, the phosphoric acid will thus not strip the nitride. Thus, a short HF-etch step is first performed to remove the oxide layer on the nitride. Note, also that the CVD-oxide that has filled the trench is also attacked by this HF-etch-step. Since the CVD-trench-oxide etches more rapidly than does the oxide layer on the nitride, this HF-etch-step must be carefully optimized. Another factor that should be considered when determining this HF-etch is the possibility of there being insufficient CVD-oxide removal from the nitride by the CMP process. If the oxide on the nitride remaining after the CMP step is too thick, the HF-etch of the process in this step may not remove all of it, making the nitride-layer-etching process ineffective, leading to potentially significant yield degradation.

16. After removing the oxide layer on the nitride, the nitride can be stripped by the hot phosphoric acid. There are several reasons why a wet process is used to strip the nitride, rather than a dry process. First, the wet process is inexpensive and selective (i.e., the hot phosphoric acid will not attack the underlying pad oxide, or Si substrate). Second, a wet process will remove the nitride deposited on the back of the wafer earlier by LPCVD, as well as the nitride on the front wafer surface.

17. Normally the pad oxide is stripped in HF at this point. Again, the oxide in the trench is also attacked by this HF etch step. This etching plays an important role in forming the topography at the edge of the active region, where the polysilicon gate is deposited on both the active area and the trench region (see Fig. 9-14). This topic will be considered again later.

18. Grow sacrificial oxide, about 20 nm thick, to protect the exposed Si surface of the active areas from contamination during the implant process.

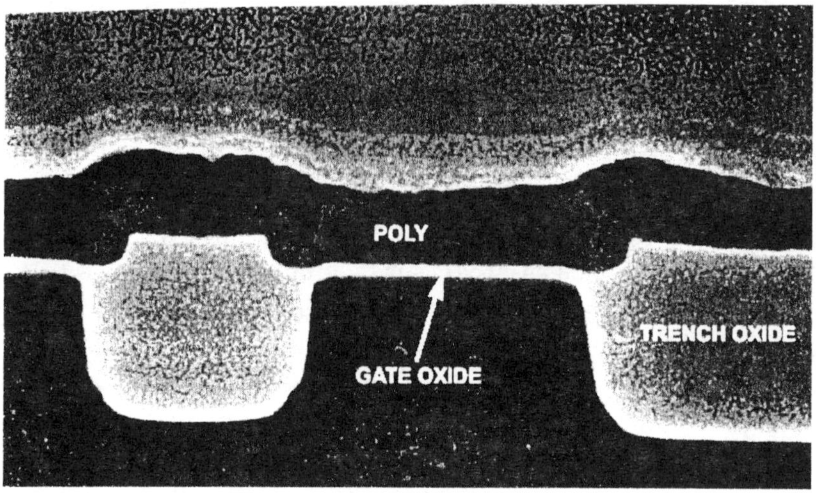

Fig. 9-14 SEM showing polysilicon crossing the region at the edge of the active device region and the trench in an STI structure.

Note that another reason for stripping the pad oxide and growing a sacrificial oxide is that an effect similar to the Kooi-effect (that occurs during LOCOS isolation - see Vols. 2 and 3 for more details on the Kooi effect) has also been reported to possibly occur during STI processes. That is, if the trench-oxide is densified in an ambient containing O_2, this O_2 can react with residual H_2 in the trench oxide film or in the SiN layers present on the chip to form H_2O. This H_2O in turn can react with the SiN to form NH_3 which would lead to nitridation of the Si as in the Kooi effect. The net result of such a phenomenon would be that the gate-oxide grown where such a nitrided region existed could be thinner there, as shown in Fig. 9-15.[23] To avoid the possibility of such gate oxide thinning (and a resultant degradation of the gate oxide integrity [GOI]), a sacrificial oxide is grown which will consume any

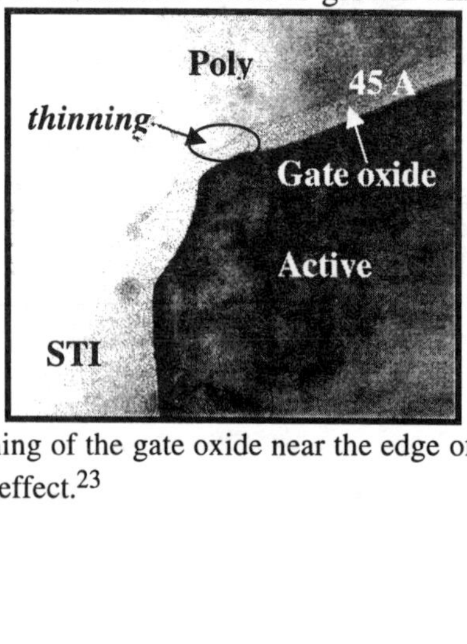

Fig. 9-15 TEM showing thinning of the gate oxide near the edge of an active region in an STI structure due to the Kooi effect.[23]

nitrided regions that formed on the active Si surface regions.

19. The wells are now created by performing a high-energy-implant to place dopants within the Si substrate in the active regions, and beneath the STI structures to produce a channel-stop layer of dopants. In addition, the V_T-adjust- and punchthrough-implants are carried out at this time.

20. The sacrificial oxide is then etched, again using a wet (HF-etch) process. Once again, the oxide in the trench is subjected to an HF etchant, further reducing the thickness of the oxide in the trench (and also further modifying the trench edge corner region - Fig. 9-16). By the way, it should not necessarily be construed that the HF etches will remove enough oxide in the trench regions to cause the trench oxide to be recessed. While this might be a possibility, not much oxide thickness loss occurs as a result of the HF etches. The more significant effect is on the region of the trench oxide near the trench top corner.

21. Finally, the gate-oxide growth process is carried out (Fig. 9-10h).

9.5 ISSUES INVOLVED WITH STI PROCESS INTEGRATION

9.5.1 Issues Involved With Etching Trenches for STI

Some of the issues involving trench etching for a baseline-STI structure were discussed in Section 9.4. Here, we elaborate on some other issues that arise as more advanced STI structures are needed. In the baseline-STI structures, the trench sidewalls were typically sloped, with slopes ranging from 70-85°. As the trench widths are scaled to ever decreasing dimensions, the trench sidewall profile requirements are becoming steeper, with slopes as high as 88° being called for in some applications.[24]

One consequence of this change is that the *in-situ* hard-mask-open etch process may become more feasible. That is, in the baseline process of Sect. 9.4,

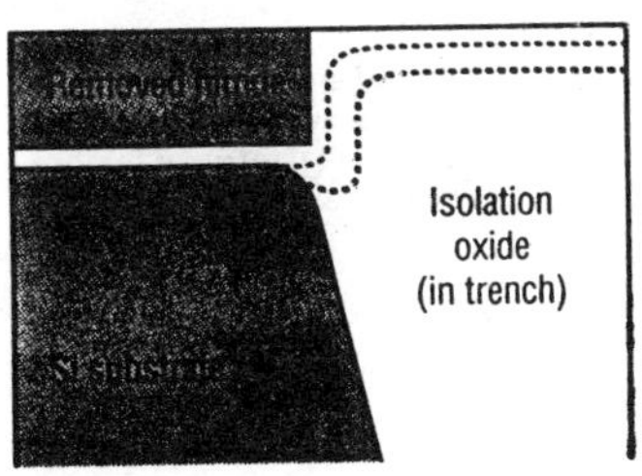

Fig. 9-16 Topography of the top corner of the Si active region and the oxide surface in the trench. The dotted lines represent the removal of oxide that occurs during the wet HF dips before the sacrificial oxide and gate oxide growth steps.

STI structures had sidewall profiles with smaller slopes (e.g., 75-80°). If the trenches were etched while the resist remained on the wafer, the polymer deposition on the sidewalls during the trench etch would lead to increased ARDE effects, and this would have contributed to trench-depth and profile-variations between trenches of different widths. Thus, the two-etcher trench-etch-processes (with the resist being stripped after the 1st trench-etch step) were more desirable. To produce steeper trench profiles, lower-pressure etch-processes with higher ion-bombardment-fluxes are used. In these high-ion-bombardment-flux processes, there is much less polymer deposition needed on the trench sidewalls to protect them from lateral etching. As a result, the presence of the resist during the trench-etching stage can be more tolerated, opening the door for the *in-situ* hard-mask-open single-etcher solution. Note that the steep trench-sidewall profiles are also controlled by using low-pressure, high-density-plasma etchers. The gases employed seek to exploit a non-depositing etch process, with gas mixtures for example consisting of Cl_2 and Ar.

Another trend as devices are scaled, is toward using shallower trench depths. The baseline-STI structures used trenches with depths of ~0.5 μm. As devices approach the 0.1-μm node, the trench depths are expected to be decreased to < 0.2-μm, at least for some circuit applications at that technology node.[25] The trench depth is set by two main considerations:

1. The oxide thickness in the trench (field-oxide), together with the channel-stop-doping under the field-oxide determine the threshold voltage of the parasitic-isolation-region-MOSFETs. To keep this threshold-voltage high enough to prevent significant isolation leakage, a minimum field-oxide - thickness (i.e., trench depth) is needed. As the operating voltages are scaled for future device generations, smaller threshold voltages of the parasitic-isolation-region-MOSFET will be acceptable. Thus, a shallower STI-trench will be feasible.

2. The trench-oxide-depth impacts the capacitance value of the polysilicon and 1st-level-interconnect-metal to the substrate. The thinner the trench depth, the higher the capacitance. However, use of smaller operating voltages will also allow thinner STI-trench-depths to be used, with the exact depth value to be determined by the circuit application. Note that the severity of dishing problem, to be discussed in Sect. 9.5.4, will also play a role in determining how much the trench depth will be permitted to be reduced.

The contamination caused by the trench-etch-process is also an issue. Since an isolation structure is being fabricated, concerns regarding metallic contamination increase with ever-decreasing device dimensions. To address this problem, Lam Research reportedly replaced all ceramic chamber components with quartz (as the ceramic material may contain trace levels of metal that could be released into the etch-gas ambient during the etch process, and then these could redeposit on the exposed Si of the trench walls). Applied Materials attacked this question by implementing a Clean Chemistry™ using fluorine throughout the etch, which eliminates the need for frequent dry-cleans of the chamber.

As already mentioned in the baseline-STI process sequence, maintaining *trench-depth-uniformity* across a die (and from wafer-to-wafer), is a challenging problem. However, merely controlling the *trench-depth* is also particularly important, because the uniformity of fill and also uniformity of CMP, in turn, depend on trench-depth-uniformity. The trench etch is essentially a blind etch, with no etch-stop-layer, nor any material differences in the Si substrate regions available to provide an end-point-detection signal. Thus, in the 2000-timeframe, the etch-depth-control must rely only on a timed-etch process.

Depth- and profile-control-nonuniformity are functions of reactor design and etch-gas chemistry. The stability of the chamber conditions also affect the trench-etch-rate-uniformity. The etch-rate will vary for different patterns, structures, open areas, and different microloading conditions. Thus, the specific etch-rate may need to be re-calibrated, depending on a particular set of conditions being faced. This indicates that there is currently a great opportunity for a production-worthy endpoint-detection-system for the STI-trench-etching process.[26]

9.5.2 Corner-Rounding Techniques in STI Structures

In Steps **7** and **8** of the baseline-STI process we described how the top of corner of the STI trench must be rounded so that adequate leakage performance charac-

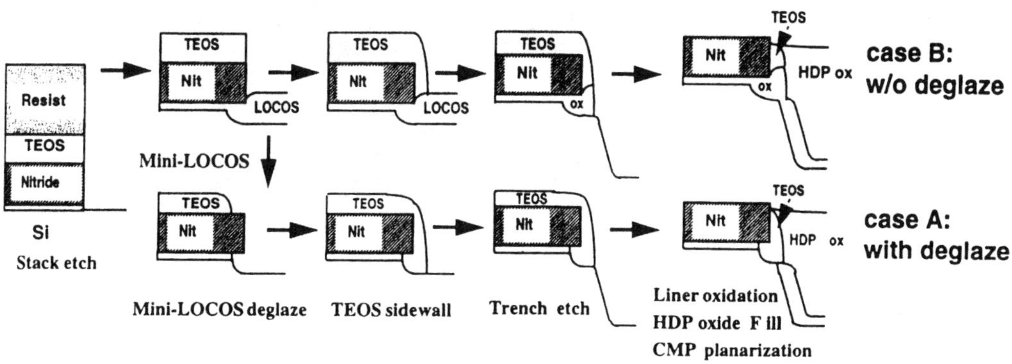

Fig. 9-17 Process sequence of STI using mini-LOCOS edge.[29] (© IEEE 1997)

teristics will be exhibited by CMOS ICs that employ STI. The corner-rounding process described in the baseline-STI flow involved undercutting the pad-oxide just before performing the rounding (aka *liner*) oxide procedure. The best corner-rounding results are obtained if an RTO step in dry-O_2 at 1050-1100°C is used.[27] In the paper that described this step, several other corner-rounding procedures were also studied. These included: **a**) adding HCl to the dry-O_2-RTO process;[28] **b**) carrying out a post-CMP-oxidation in dry-O_2 at 1050-1100°C;[29] and **c**) using a "mini-LOCOS" process (see Fig. 9-17).[30] Good rounding is observed in the post-CMP-oxidation-process, but it comes at the expense of reducing the active-device area to some degree. The mini-LOCOS process ads considerable process complexity. A hydrogen-annealing procedure was also reported as an alternative method of achieving corner rounding. In this approach, the wafer is annealed at 1100°C after the trenches are etched (but before being filled). It is proposed that at this temperature, Si-surface atoms migrate and establish positions that represent the structure with the lowest free energy (namely a sharp corner becomes a rounded corner, see Fig. 9-18).[31,32] A novel approach to rounding the trench corner during the trench process itself is described in Sect. 9.5.6.

9.5.3 Trench-Fill Dielectrics for STI

We know that the trenches of the STI structure are filled with a dielectric, and that deposition of oxide-like films by CVD has been the chief technique for

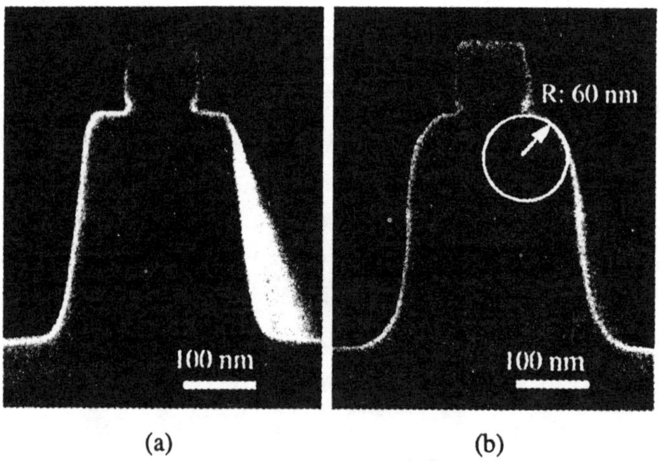

Fig. 9-18 Cross-sectional SEM of before (a) and after (b) H_2-annealing. Annealing condition was 950°C, 380 torr, 60 sec. Silicon migration phenomenon can round out STI's top-corner-rounding-radius up to 60-nm.[32] (© IEEE 1998)

filling the trenches. A report published by Nag *et al.*,[33] compared several such dielectrics deposited by such methods for their suitability as the trench insulating material. The criteria they used to evaluate these materials were as follows:

1. The deposited material must be void-free, to avoid opening seams during HF-deglazing steps (i.e., such voids could later lead to poly filaments).

2. Post-deposition HF-deglazing also reduces the trench-fill oxide thickness. This can expose silicon at trench corners and cause leakage. Therefore a comparison of the etch rates of the trench-fill-oxides relative to thermal oxide must be established.

3. Film stress, shrinkage, and composition after the high-temperature anneal (densification) affect the integration.

4. The CMP-polish-rate relative to nitride and polish uniformity of the oxide impacts the trench-fill-oxide thickness remaining after CMP that influences the "corner effect" problem.

5. Isolation characteristics of PMOS and NMOS diodes have to be tested to verify the electrical performance of trench-fill dielectrics.

In this study four films types were compared: LPCVD TEOS; HDP-CVD oxide; SACVD O_3-TEOS; and an HSQ-SOG (see Vol. 1, 2nd Ed for more information on HSQ-SOGs, as well as Chap. 14 of this text). In general, they concluded that the HDP-CVD oxides were the most suitable of these four films for STI-trench filling. Table 9-1 summarizes the results of their study. Trenches 0.5-μm-deep and widths as small as 0.22-μm could be filled without forming voids.

Table 9-1 TRENCH-FILL DIELECTRIC EVALUATION CRITERIA

Criteria	SACVD	LPTEOS	HDP	SOG
Gap-fill	OK	No	OK	OK
Hi-T Anneal required	Yes	Yes	No	Yes
As-dep'd HF-etch rate vs. thermal ox.	5	2.1	1.1	
Densified (850C) HF etch	1.3	1.1	1.1	1.1
CMP rate vs. SiN	5.1:1	5.1:1	5.1:1	5.2:1
Shrinkage at 850C	5-6%	3-4%	0-0.8%	22-28%
Moisture Absorption	Yes (cure)	No	No	Yes (cure)
Process	Seed/Cap	1 step	1 step	Seed/Cap

Some more recent reports, however, have indicated that some of the other films might still be good candidates as trench-fill materials. For example, a study by Xia *et al.,* demonstrated that a two-step deposition of a BPSG film using a sub-atmospheric (SA)CVD-process of O_3-TEOS is also capable of void-free gap filling of trenches whose widths are as small as 0.06-μm and whose depths are 0.35-μm deep.[34,35] Figure 9-19 shows a SACVD chamber used to deposit such films.

Nevertheless, since the HDP-CVD method appears to have emerged as the technology of choice for depositing STI trench oxides, some comments should be made about this process (and the potential impact that the HDP-CVD method can have on the STI structure). Note that Novellus (SPEED® HDP-CVD model) and Applied Materials are vendors of HDP-CVD tools for dielectric deposition.

The basis for the improved gap-filling capability by films deposited with HDP-CVD is that HDP sources provide a high-density of low-energy ions to the wafer surface. These ions simultaneously sputter the film being deposited by CVD. The sputtering removes film material from some regions on the surface more rapidly than from other regions. This effect is exploited to yield films that fill narrow gaps without voids. Another benefit of HDP-CVD is that films can be deposited at lower temperatures. This mechanism is described in detail here.

The film being deposited by CVD initially covers all the surfaces of the wafer conformally (including the sidewalls and bottom of any trenches or contact holes). As the film grows thicker, an overhang develops at the top of the trench. This will close-off the opening of the trench before the dielectric material completely fills it, leaving a void. If the overhang can be prevented from

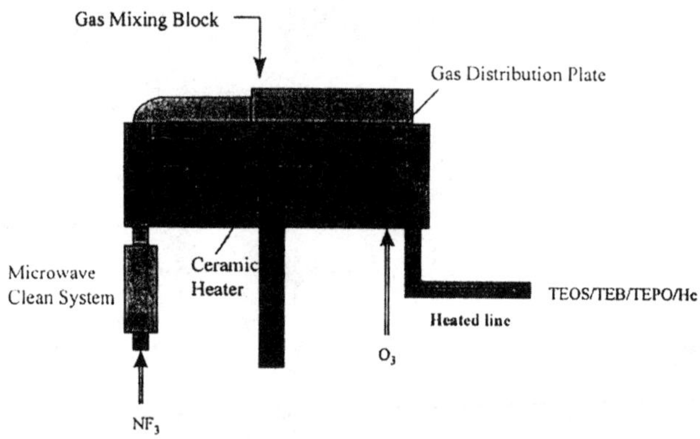

Fig. 9-19 Schematic of Applied Materials SACVD Giga-Fill BPSG chamber, equipped with a high-temperature heater and remote microwave system for predeposition cleaning.[34]

forming, it may be possible to fill the trench without void formation. This is accomplished in an HDP source by the bombardment of the wafer surface with a high density of ions. For depositing oxides by HDP-CVD, a mixture of SiH_4, O_2, and Ar (or He) gases is most commonly used. The HDP source excites the oxygen and argon gas molecules. A bias applied to the wafer pulls the energetic ions out of the plasma and directs them toward the wafer surface. The oxygen reacts with silane to form SiO_2 while the Ar ions simultaneously sputter deposited material away. A high-density-plasma source is needed to maintain a high-ion-density at the relatively low pressures (a few mTorr) needed to obtain the directionality required for good gap-fill. The good gap-fill is achieved as described below.

The sputter yield is higher for surfaces that face the incoming ion beams at an angle, than at those surfaces perpendicular to the incoming ion directions. The maximum sputter-etching-rate occurs on surfaces at about 45° to the normal. Thus, a deposition-to-etch-ratio exists that varies with location on the wafer surface. The net deposition-rate is higher on flat surfaces (the field-regions of the wafer and the trench bottoms), than it is on the trench corners. (These angular surface regions are the 45° facets formed by depositing dielectric film at the top of the trenches.) By adjusting the deposition-to-etch-ratio, the trench openings can be kept from closing up while the dielectric film fills the trenches from the bottom up (Fig. 9-20). Sputtering from the via bottoms also causes re-deposition of some material onto the sidewalls. This helps re-distribute the film so the gap is

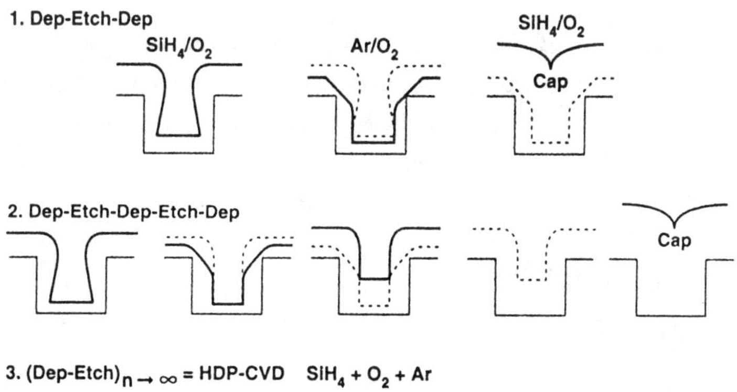

Fig. 9-20 High-density plasma CVD achieves good gap filling through simultaneous deposition and etching. The deposition-to-etch-ratio is usually kept between 0.14 and 0.33.[43] Reprinted with permission of Semiconductor International.

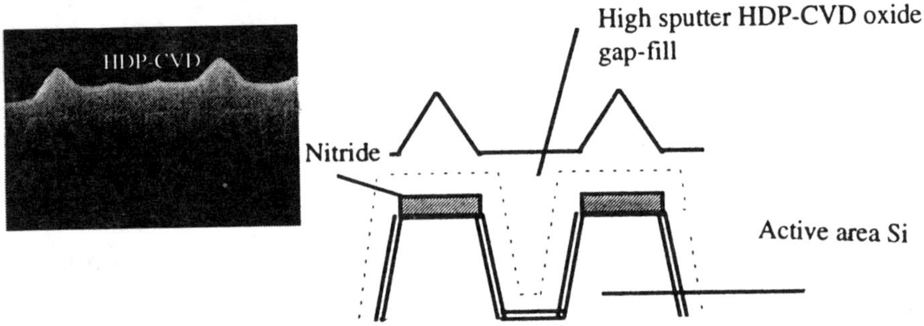

Fig. 9-21 The surface topography of a dielectric film deposited by HDP-CVD showing the sharp spikes that remain as a result of the dep-etch nature of the HDP-CVD process.

filled without forming a void. To achieve good filling, the ratio of deposition to removal rate should be no larger than 3.0. The wafers are also heated by the ion bombardment. Thus, the deposition-rate is limited by the highest sputter-etch-rate that can be tolerated. The process is run at low pressures (< 10 mtorr), which allows ions striking the wafer surface to cross the sheath with no collisions. They arrive with the full energy they receive from the voltage drop across the sheath.

At the conclusion of the deposition process, the surface of the dielectric will generally still contain sharp ridges over the underlying metal lines (Fig. 9-21). CMP is needed to restore the wafer surface to a planar condition (although the amount of CMP required is reduced from that needed for a dielectric film deposited with a non-HDP-CVD process).

One issue that arises with HDP-CVD is that the combination of deposition and etching can sputter the top of the nitride corners, in effect "clipping them." One technique that suppresses this effect is to use He instead of Ar as the sputtering species in the HDP-CVD process. Another approach is to decrease the deposition-to-(sputter)-etching ratio of the process. Since this reduces the gap-filling capability, a trade-off must be made to optimize the overall deposition process.

9.5.4 Dishing Problem Associated with the CMP of STI Trench Dielectrics

Another CMP problem that occurs when fabricating STI structures is *dishing*. The effect involves the thinning of trench oxide regions that are exposed during the overpolish time that occurs after the CMP-stop layer is reached (Fig. 9-22). A silicon nitride layer is used to act as a CMP-stop layer. The thinning of the exposed oxide areas between the nitride-CMP-stop features is known to be highly dependent on feature size, with the dishing effect becoming larger as the

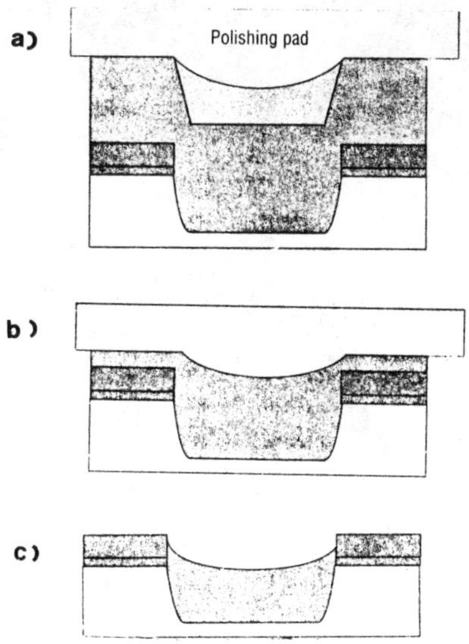

Fig. 9-23 Origin of the dishing effect in CMP: (a) polishing pad flexes into the trench opening; (b) formation of a concave surface; (c) trench area remains "dished" after planarization. Reprinted with permission of Solid State Technology. Published by PennWell.

width of the field oxide is increased (Fig. 9-24). For example, in the CMP process of Fig. 9-24, dishing causes the field oxide thickness to decrease by more than 50-nm when the field oxide width exceeds 100-μm.[36] Dishing occurs because the polishing pad is not perfectly rigid, and because oxide removal is faster than nitride removal. If a pad exhibited perfect rigidity, CMP of the exposed oxide regions would produce a perfectly planar surface (because the exposed recessed regions would not be subjected to any polishing pressure). However, since actual pads exhibit some flexibility, above-zero polishing pressures are produced in recessed areas (especially in the middle of wide exposed regions) when polishing continues after the nitride has been exposed. Thus, some oxide is removed from these regions, and a "dished" region is observed after the polish step is completed. To minimize dishing, the silicon dioxide polish should be stopped as soon as the oxide over the nitride clears. Accordingly, the quality of an STI-CMP-process is measured by the control of trench-oxide and nitride thickness. This requires tight control of the step height between the trench-oxide surface and silicon surface after nitride strip.

Dishing must be minimized to maintain planarity for subsequent wafer processing. It still represents a significant limitation of the STI process. A number of

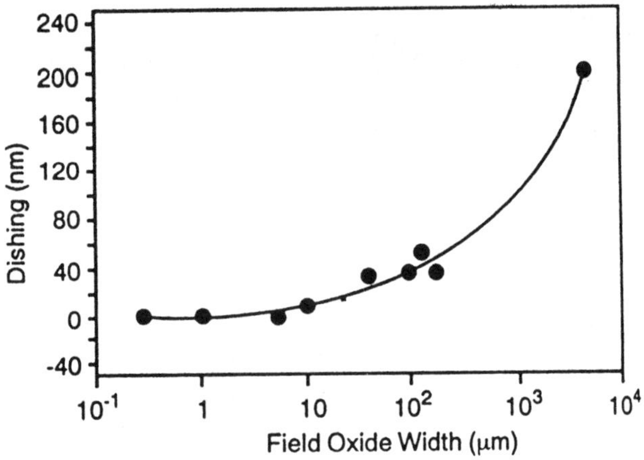

Fig. 9-24 Dishing versus field-oxide-region width for a specific oxide CMP process[36] (©1993 IEEE).

approaches have been proposed to decrease dishing, including: **1**) a combination RIE/CMP process (as described in Sect 9.4); **2**) dummy active-regions; **3**) adding an extra nitride layer onto the trench-dielectric after it has been deposited; [37,38] **4**) adding an extra nitride layer onto the dielectric after it has been deposited, and using a reverse block-mask;[39] **5**) implementing a direct-polish CMP-process using either a linear-polishing tool, or a rotating-polishing tool with a harder polishing pad and an efficient endpoint-detection-system.

As described earlier, in the *RIE/CMP approach* the goal is to reduce the oxide thickness over active areas prior to CMP, so that a shorter CMP-oxide-removal step is needed. Since CMP is performed for a shorter time, the degree of dishing in wide trench regions is reduced. In the first step of the RIE/CMP process, a mask that contains a pattern that is the reverse of the trench-opening mask is used (but, with the exception that active-regions below a specified nar-row width are also not exposed - Fig. 9-24).[40] This mask is aligned so that its opaque patterns cover the trench regions on the chip surface. That is, this mask exposes the active regions of the chip, leaving the trench regions covered with resist (Fig. 9-10e). After the subsequent deposition of another planarizing resist layer (see Fig. 9-24), an etchback step is carried out using RIE. This process re-moves resist and dielectric material at the same rate, and so thins the oxide covering the active regions. Note that oxide in the trenches is not etched while oxide over the active regions is being removed, because the trench oxide is still covered with resist.

After this RIE step the resist is stripped, leaving the oxide over the active areas partially removed, Now, when the CMP step is performed, it only has to remove a thinner oxide layer, reducing the CMP time This not only reduces

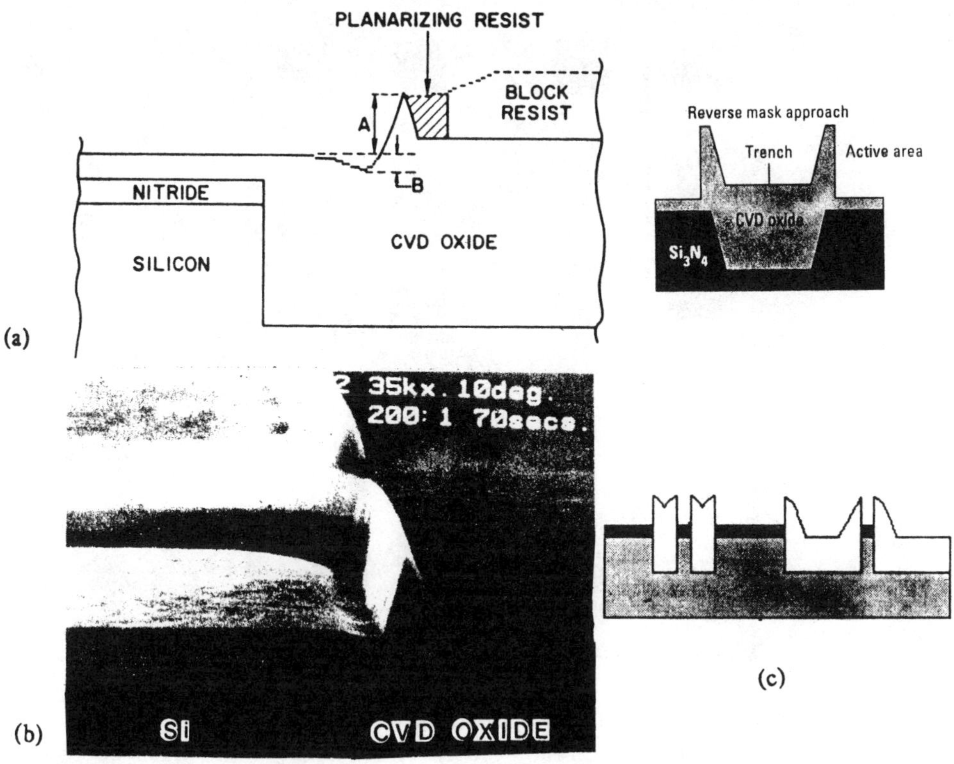

(a) (b) (c)

Fig. 9-25 Spikes in the dielectric layer after the RIE step of the RIE/CMP process: a) Simulated cross-section after RIE; b) SEM cross-section after RIE and resist strip, showing dielectric spikes; c) Schematics of the dielectric spikes after RIE and resist strip.

dishing, but reduces CMP cost (since a shortened CMP step allows the throughput of the CMP procedure to be increased). Note also that dimensions on the reverse-trench mask are actually slightly smaller than the dimensions of the active-regions themselves. Thus, even when the mask is aligned to the active-regions, a small border of the active-region oxide remains covered by the resist. As a result, at the conclusion of the RIE step, small "spikes" of oxide remain on the wafer surface (Fig. 9-25).[41,42] But, these spikes can be effectively removed by the CMP step, allowing a planarized oxide surface over the chip to be obtained. The main drawback of the RIE/CMP method is that it increases the complexity (and thus also the cost) of the process used to planarize the trench oxide.

In the *dummy-active-region approach*, some areas of the trench-region are not etched. Instead, as shown in Fig. 9-26,[42] regions of Si remain in the center of trench. This has the same effect as if the trench width was reduced, and the degree to which dishing occurs decreases with trench width. Consequently, the

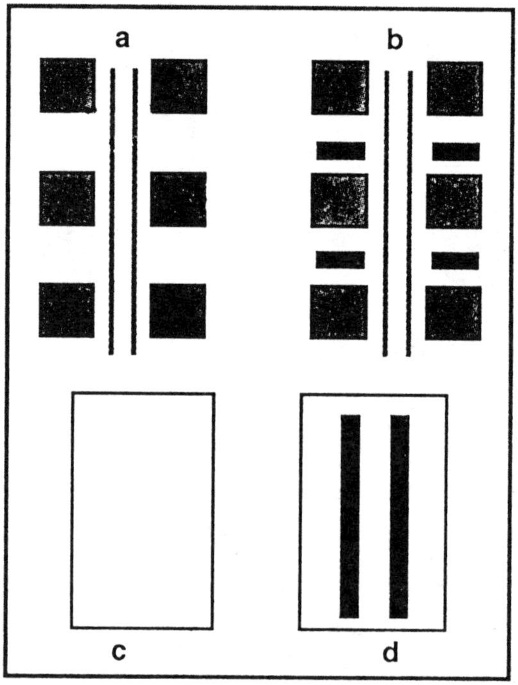

Fig. 9-26 Dummy (b) and dishing support features of small (a) and large wafer (c) features can prevent dishing.[50] Reprinted by permission of Semiconductor International.

dummy structure method is also effective in reducing dishing effects to an acceptable degree. The CMP-removal rate over regions that are densely packed with active areas is also slower than over regions containing small isolated areas. Thus, to expose the nitride-CMP-stop-layer over such densely packed regions, the nitride over small isolated active-areas gets stripped, causing damage to Si in areas beneath the stripped nitride. As a result, dummy-active-regions must also be added to areas around isolated active regions. The dummy-structure technique, however, poses several difficulties. First, it imposes electrical, design, and space constraints, with the main problem being that several times as many structures must be added within a die over previous designs. This complicates the design and mask making process, and significantly raises defect levels. In addition, such dummy structures in trenches cannot be used for analog- and mixed-signal chips because the coupling capacitance over trenches becomes variable, and this disrupts the correct operation of analog circuitry.

In Procedure **3** used to reduce dishing during trench-CMP, a thin layer of silicon nitride is blanket-deposited onto the trench-oxide before it is polished-back with CMP (Fig. 9-27). A slurry with low selectivity between nitride and

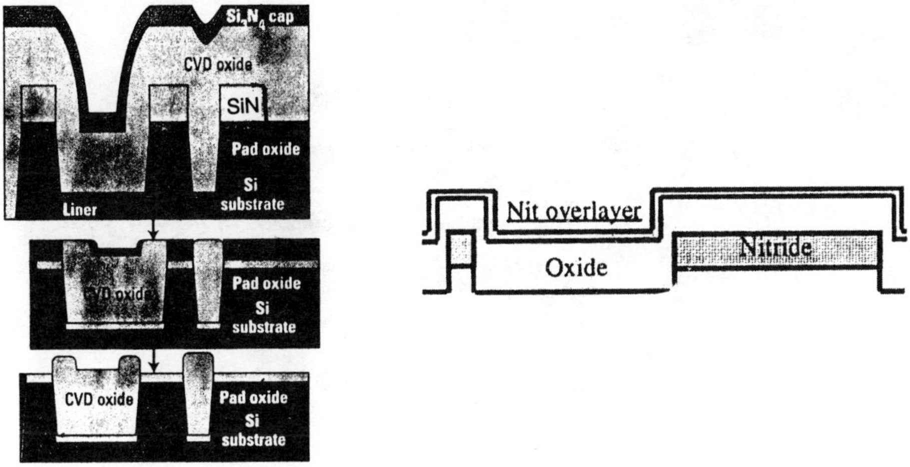

Fig. 9-27 *Nitride cap* approach to minimize dishing during CMP of the trench-fill oxide.

oxide is used for the CMP process, and thus the thin-nitride-layer on the top of the oxide covering the active areas (which are "up" regions) is polished away. This exposes the oxide under the nitride in the active-areas to the CMP mechanism. However, the surface of oxide that fills the trenches is recessed below the surface of the oxide on the active-regions (as these are "down" regions). Hence, the SiN on these recessed-regions is not polished away when the areas of nitride on the active-regions are removed. This SiN thus remains on the wafer to act as a polish-stop-layer when the trench-oxide-thickness is reduced by the CMP process to the degree that it is essentially planar with the surface of the active-regions. In this way dishing over wide trench regions is decreased.

In Procedure **4**, the thin nitride layer is again deposited on top of the trench oxide surface before CMP, as is done in Procedure **3**.[44] However, in Procedure **4**, this thin nitride is also patterned with a mask such that after this lithography step is completed, the nitride on the active area regions is exposed (Fig. 9-28). This nitride is then etched, so that it is no longer present on the active-areas of the chip. However, the thin nitride does remain covering the trench regions, where it acts as a polish-stop-layer during the oxide-planarization CMP-step. Thus, dishing of the oxide in the trenches is again suppressed.

Use of the *direct-CMP method* avoids the added complexity of the previous methods used to combat the trench-dishing problem. That is, in the direct-CMP-approach the trench dielectric is etched directly after deposition and densification without applying any of the above procedures. To permit the trench-oxide to be directly polished, the CMP process itself must exhibit reduced tendency to dishing (putting the burden the CMP tool, the slurry, and the overall CMP process).

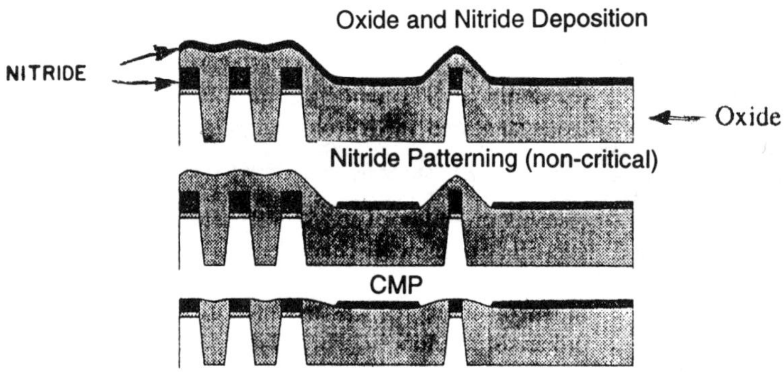

Fig. 9-28 *Patterned second-nitride layer* approach to minimize dishing during CMP of trench-fill oxide.

The three methods being pursued to allow direct-CMP are: **1**) the use of *linear polishing tools*; **2**) the use of *rotary polishing tools with a slurry that exhibits relatively low selectivity between oxide and nitride*; and **3**) the use of *rotary polishing tools with a fixed abrasive pad.* Linear polishers employ higher pad velocities (300–500-ft/min versus 34 ft/min for rotary pad speeds) and lower polishing pressures (3–5 psi versus 7 psi for rotary polishers). In the linear polishing process, a *highly selective slurry* (with cerium oxide particles dispersed in an aqueous suspension of material that inhibits silicon-nitride removal) is also employed.[45] This set of process conditions significantly reduces dishing effects (maximum oxide removal across a die was 20-nm) and decreases the nitride removal over isolated active regions.[46,56] On the other hand, dishing is reportedly decreased when *rotary polishing tools* are used by selecting a different set of process conditions and CMP consumables. That is, decreased dishing effects are obtained by using a harder polishing pad, a relatively non-selective slurry, efficient endpoint detection method, and an optimized CMP process.[47,48] The use of a fixed-abrasive pad for the direct-CMP method of STI trenches has also reportedly shown good results.[49]

The degree to which a CMP process used to polish STI trenches causes dishing can be evaluated with the help of a special STI density- and pitch-pattern test-wafer developed by MIT.[55] These test-wafers have 4800-Å-deep trenches, 8000-Å-TEOS oxide deposited over 1960-Å of silicon nitride and 150-Å of thermal liner-oxide. Process engineers can use this test wafer to measure the effects of trench dishing and nitride erosion.

9.5.5 Corner Engineering and Its Effects on MOSFET TURN-OFF Characteristics of MOSFETs Fabricated with STI

As we described in Section 9.1, the relationship of the elevation of the isolation dielectric surface to that of the active region of the MOSFET can impact the TURN-OFF behavior of the MOSFET. Specifically, if there is a recess of the isolation-region dielectric with respect to the active-region-edge at the top corner of the trench-edge, the polysilicon gate will wrap around the corner as shown in Figs. 9-3 and 9-29.[51] This can cause the leakage current of the MOSFET in the subthreshold region of operation to increase above the value it would exhibit if the polysilicon and the isolation were at the same elevation (and there then would be no poly wrap-around).

In the sequence that described the baseline-STI structure we indicated that there are three HF-deglaze steps that typically occur in such a process flow. The effect of the latter two (see Steps **17** and **20**) is to cause the dielectric in the trench to be etched at the same time the pad oxide and sacrificial oxides are etched (in Steps **16** and **19**, respectively). The effect on the topography of the dielectric at the top corner of the trench is shown in Fig. 9-30, in which it is seen that a *divot* is formed in the trench oxide at the active-area edge.[52] This divot can have four related and detrimental effects on the active MOSFET. First, it causes poly wrap-around, that may lead to increased leakage of the MOSFET operating in the subthreshold region. As described in the baseline-STI process, however, if sufficient rounding of the top corner is implemented, the enhanced subthreshold leakage (as evidenced by the "double-hump" or "kink" in the MOSFET TURN-OFF characteristics) can be suppressed. Second, the wrap-around causes the MOSFET to behave like a MOSFET which has an isolation region recessed

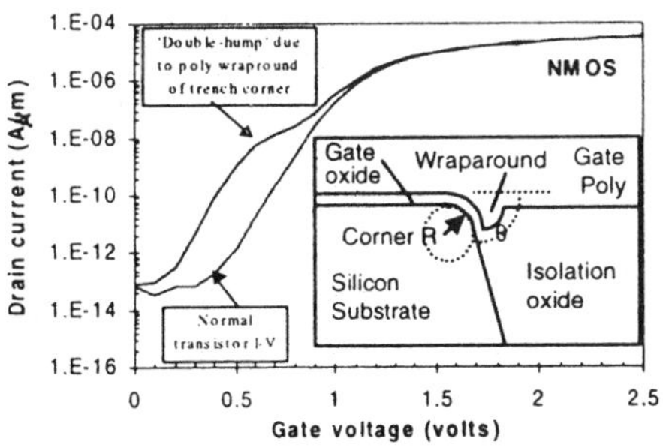

Fig. 9-29 "Kink" in MOSFET I_D-V_{GS} characteristic due to parasitic corner conduction caused by poly wraparound.

below the silicon surface. This enhances the so-called reverse-narrow-width effect, in which the threshold voltage decreases with MOSFET channel width, (see Fig. 9-31).[53] Note that this figure also shows that corner rounding reduces this effect, even if some poly wraparound occurs. Third, the threshold voltage of the MOSFET decreases as the depth of the divot is increased (Fig. 9-30). These effects produce unacceptably large device threshold voltage variations (± 200 mV). Fourth, the rounded edge encroaches into the active region of the device (i.e., under the nitride edge). This encroachment reduces the electrical width of the MOSFET. As the 0.1-μm technology node is approached, this width-reduction can represent a high percentage of the width of a minimum-sized MOSFET.

One reported technique that is used to combat this problem involves implementing a new corner rounding process that is accomplished after the nitride hard mask is opened (but before the silicon trench etch is performed). That is, a new lithographic technique from Clariant-AZ termed RELACS,[54] causes the trench width to be slightly reduced at the time the silicon nitride layer is opened. This is accomplished by applying a material to the resist pattern after it has been exposed and developed, and this material remains attached as a thin layer to the sidewalls of the resist pattern. This makes the openings of the resist pattern that define the silicon-nitride patterns slightly smaller (Fig. 9-32a). After the nitride layer has been opened, a short etch of the silicon trench surface undercuts the nitride edge, leaving a rounded top corner on the Si trench corner surface. The remainder of the trench is etched with an anisotropic etch. The overall trench profile after these etches is shown in Figs. 9-32b and 9-32c (here with the nitride left in place to show the location of the rounding with respect to the active area defined by the silicon nitride edge). The top corner rounding is clearly visible.

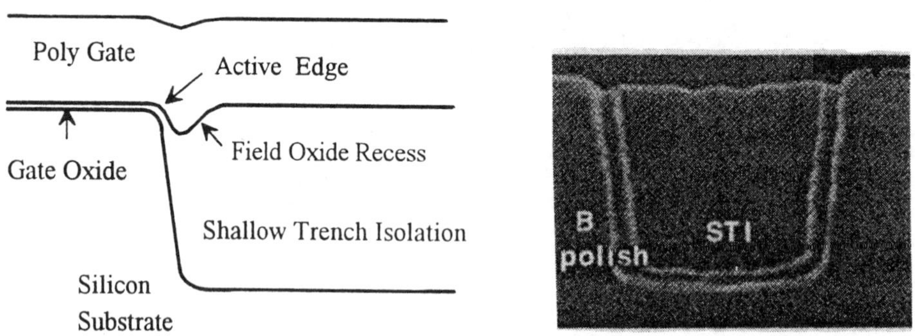

Fig. 9-30 Field-oxide recess (or *divot*) formed at the active edge of an STI structure.

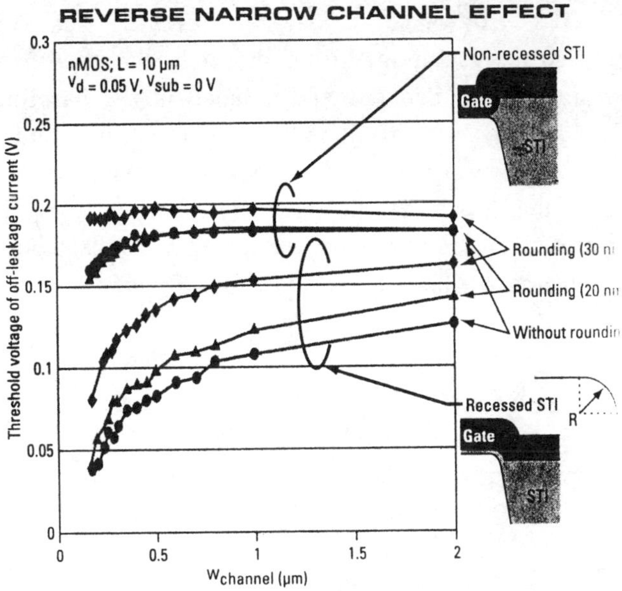

Fig. 9-31 Corner rounding helps eliminate the reverse narrow channel effect that causes subthreshold leakage. Recess of the isolation region below the silicon surface exacerbates the problem.[26] Reprinted with permission of Semiconductor International.

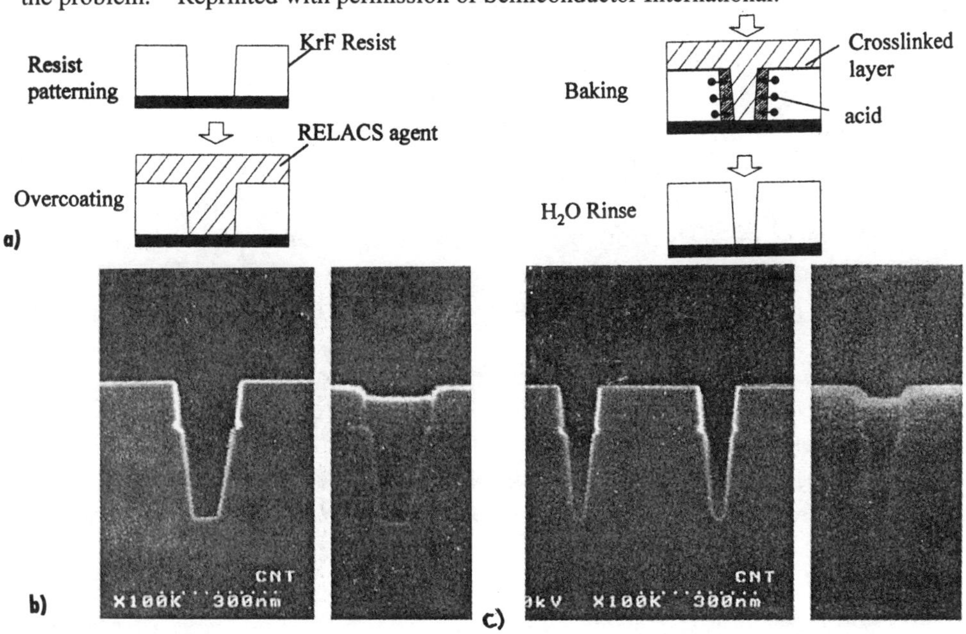

Fig. 9-32 a) Process flow of RELACS. (b & c) Top corner rounding provided during etch in a TCP system. An *in-situ* hard mask/rounding/trench etch scheme was used. The top corner is rounded to ~15 nm radius. b) 180-nm-wide trenches as etched and after fill and CMP; c) 100-nmwide trenches as etched and after fill and CMP.[54]

By providing the top-corner-rounding during the etch, the encroachment under the nitride edge by the rounded corner produced in the baseline-STI sequence is eliminated, and the severity of the poly wrap-around phenomenon is also reduced. Furthermore, the corner is rounded without encroachment. Another benefit of this technique is that less variability in the threshold-voltage value is observed from-wafer-to-wafer. The reverse narrow-channel effect becomes more severe if the trench-oxide encroaches beneath the silicon-nitride edge.

REFERENCES

1. E. DuBois *et al.,* "A Study of the Electrical Performances of Isolation Structures," *IEEE Trans. Electron Dev,*. **ED-37**(6), p. 1477, June 1990.

2. K. Kurosawa, T. Shibata, and H. Iizuka, "A New Bird's-Beak-Free Field Isolation Technology for VLSI devices," *IEDM Tech. Dig.* 1981, p. 384.

3. T. Shibata *et al, IEDM Tech. Dig.* 1983, p. 27.

4. D. C. Cole *et al.*, "The Use of Simulation in Semiconductor Technology Development," *Solid State Electronics*, **33**, No. 6, p. 591, 1990.

5. D. Foty, J. Mandelman, and T. Furukawa, "Behavior of an NMOS Trench-Isolated Corner Parasitic Device at Low Temperature," *Electrochemical Soc. Mtg. Abs.*, Fall, 1989, p. 295.

6. T. Furukawa, J. Mandelman, "Process and Device Simulation of Trench Isolation Corner Parasitic Device," *Electrochem. Soc. Mtg. Abs.,* Fall 1988, p. 329.

7. N. Shigyo *et al.*, "Three-Dimensional Analysis of Subthreshold Swing and Transconductance for Fully Recessed Oxide (Trench) Isolated 0.25-μm-Width MOSFETs," *IEEE Trans. Electron Dev.*, **ED-35**, July 1988, p. 944.

8. A. Bryant *et al.*, "The Current-Carrying Corner Inherent to Trench Isolation," *IEEE Electron Dev. Letts.,* **EDL-14**, August 1993, p. 412.

9. T. Iizuka *et al.,* "Double Threshold MOSFETs in Bird's-Beak Free Structures," *Tech. Dig. IEDM* 1981, p. 384.

10. K. Ohe *et al.,* "Narrow-Width Effects of Shallow-Trench-Isolated CMOS with n^+ Polysi Gate," *IEEE Trans. Electron Dev,*. **ED-36**, June 1989, p. 1110.

11. J.Y. Chen, "CMOS - The Emerging Technology," *IEEE Circuits and Devices Magazine,* March 1986, p. 16.

12. J.R. Brews, "The Submicron MOSFET," in *High-Speed Semiconductor Devices,* Ed. S.M. Sze, John Wiley & Sons, New York, 1990, p. 183.

13. K.M. Cham *et al.*, "Characterization and Modeling of the Trench Surface Inversion Problem for the Trench Isolated CMOS technology," *Tech. Dig. IEDM* 1983, p. 23.

14. *ibid. Ref. 12.*

15. *ibid. Ref. 12, p. 185.*

16. G. Fuse, "A New Isolation Method with Boron-Implanted Sidewalls for Controlling Narrow-Width Effect," *IEEE Trans. Electron Dev.*, **ED-34**(2), p. 356, Feb. 1987.

17. B. Davari *et al.*, "A Variable-Size Shallow Trench Isolation Technology with Diffused Sidewall Doping for Submicron CMOS," *IEDM Tech. Dig.*, 1988 p. 92.

18. S. Ambadi *et al.,* "Impacting Device Performance and Yield Through Sacrificial Oxidation Improvements," *Mtg. Abs. Electrochem. Soc.,* Fall 2000, Abs. 713.

19. S. Lassig *et al.,* "An Integrated etch Approach as STI Evolves for the 100-nm Regime," *Solid State Technology,* July 2000, p. 157.

20. L. Peters, "Gaining Control Over STI Processes," *Semiconductor International,* March 2000, p. 74.

21. A. Ambadi *et al.,* "Development of Shallow Trenches with Round Bottoms and Factors Affecting Etch Profile and Uniformity," *Mtg. Abs. Electrochem.. Soc.,* Fall 1999, Abs. 1312.

22. N. Nandakumar *et al.,* "Shallow Trench Isolation for Advanced ULSI CMOS Technologies," *Tech.. Dig. IEDM,* 1998, p. 133.

23. N. Balasubramanian, "Gate Oxide Thinning in MOS Structures with Shallow Trench Isolation," *Mtg. Abs. Electrochem. Soc.,* Spring 2000, Abs. 454.

24. *ibid., Ref. 20.*

25. *ibid., Ref. 19.*

26. L. Peters, "Choices and Challenges for Shallow Trench Isolation." *Semiconductor International,* April 1999, p. 69.

27. *ibid., Ref. 22.*

28. A. Chatterjee, *J. Vac. Sci. Technol.,* 1997, p. 1936.

29. C.P. Chang *et al.*, "A Highly Manufacturable Corner Rounding Solution for 0.18-μm Shallow Trench Isolation," *Tech. Dig. IEDM,* 1997, p. 661.

30. A. Chatterjee *et al.,* "Shallow Trench Isolation Using LOCOS Edge for Preventing Corner Effects for 0.25/0.18-μm CMOS Technology & Beyond," *Tech. Dig. IEDM*, 1996 p. 829.

31. T. Sato *et al.,* "Trench Transformation Technology using Hydrogen annealing for Realizing Highly reliable Device Structure with Thin Dielectric Films," *Dig. Symp. VLSI Tech.,* 1998, p. 206.

32. T. Sato, "Novel Corner-Rounding Process for Shallow Trench Isolation Utilizing Micro-Structure Transformation of Silicon," *Tech. Dig. IEDM,* 1998, p. 137.

33. S. Nag *et al.,* "Comparative Evaluation of Gap-Fill Dielectrics in Shallow Trench Isolation for Sub-0.25 μm Technologies," *Tech. Dig. IEDM,* 1996, p. 841.

34. L.-Q. Xia *et al.*, "High Aspect Ratio Trench filling Using Two-Step Subatmospheric Chemical Vapor Deposited Borophosphosilicate Glass for <0.18-µm *Device Application,"* *J. Electrochem. Soc.,* May 1999 p. 1884,

35. L.-Q. Xia *et al.*, "High Temperature Subatmospheric CVD Deposited Undoped Silicate Glass," *J. Electrochem. Soc.,* March 1999 p. 1181.

36. S.S. Roth *et al.*, "PELOX Integrated PBL," *IEEE Trans. Semi. Manf.,* **6**, August, 1993, p. 246.

37. *ibid., Ref. 22.*

38. *ibid. Ref. 26.*

39. G. Badenes *et al*, "A New Dummy-free Shallow Trench Isolation Concept for Mixed-Signal Applications," *J. Electrochem. Soc.,* October 2000, p. 3827.

40. *ibid., Ref. 22.*

41. *ibid., Ref. 26.*

42. P. Tsui *et al.,* "A Manufacturable and Modular 0.25-µm CMOS Platform Technology," *Dig. Symp. VLSI Technology,* 1998, p. 152.

43. P. Singer, "The Future of Dielectric CVD: High-Density Plasmas?" *Semiconductor International*, July 1997, p. 127.

44. *ibid. Ref. 39.*

45. J.Y. Kim. "Novel STI CMP Process Using High Selective Ceria Slurry with Thin Nitride stopper," *Mtg. Abs. Electrochem. Soc.,* Spring 2000, Abs. 379.

46. B. Withers *et al.,* "Wide Margin CMP for STI," *Solid State Technology*, July 1998, p. 173.

47. *ibid. Ref. 26.*

48. *ibid. Ref. 26.*

49. J. Schlueter, "Trench Warfare: CMP and Shallow Trench Isolation," *Semiconductor International,* October 1999, p. 123.

50. R. DeJule, "CMP Challenges Below a Quarter Micron," *Semiconductor International,* November 1997, p. 54.

51. *ibid. Ref. 49.*

52. *ibid., Ref. 19.*

53. *ibid., Ref. 48.*

54. T. Toyoshima *et al.,* "0.1-µm-Level Contact-Hole Pattern Formation with KrF Lithography by Resolution Enhancement Lithography Assisted by Chemical Shrink (RELACS)," *Tech. Dig. IEDM,* 1998, p. 333.

55. Massachusetts Institute of Technology.

56. E. Zhao and C. Shan Xu, "Direct CMP for STI," *Semiconductor International,* June 2001, p. 145.

Chapter 10

SILICON-GERMANIUM (Si:Ge) TECHNOLOGY FOR HIGH-PERFORMANCE TELECOMMUNICATIONS ICs

Silicon-Germanium (Si:Ge) BiCMOS technology has become an important technology with many new and exciting product applications. Si:Ge-based heterojunction-bipolar-transistors (HBTs) are now found in a wide variety of technology offerings, and are incorporated in the product roadmaps of many telecommunications companies. In this chapter we describe the process technology issues of fabricating Si:Ge-HBTs and Si:Ge-BiCMOS ICs.

10.1 INTRODUCTION TO TELECOMMUNICATIONS IC APPLICATIONS

There is a rapidly growing market for wide-bandwidth and wireless telecommunications. That is, consumer demand for data from the Internet, for digital audio and video, and for video-on-demand is on the rise. In general, data is delivered to the consumer in two ways: *guided-wave propagation* and *free-space transmission*. Guided-wave propagation occurs along optical-fiber links and coaxial cables. Free-space transmission is the territory of radio, television broadcasting, satellite links, local multi-point distribution systems, and wireless local-area networks. (For example, such wireless applications as cellular phones currently operate in the 2.5-10-GHz range.) For free-space propagation, expensive installations are not necessarily required, but governments do regulate the use of bandwidth and frequency with a firm hand.

Since most of the *wireless* part of the frequency spectrum has been allocated, new or large bandwidth applications will be forced to resort to unallocated higher carrier frequencies. For example, a new wireless local-area network may have to operate at 60 GHz. Besides possible economy, a second attraction of higher frequencies is that they use smaller antennas. Because antenna efficiency is proportional to carrier frequency, receivers of higher carrier frequencies need only compact antennas. For instance, an antenna just 10-cm in length suffices for collision-avoidance automotive radar at 77-GHz.

As for *guided-wave systems*, cable has been deployed, and once it is in place, bandwidth is relatively unregulated. In this kind of system (such as an optical-fiber data-link), the development of high-frequency components is mainly a response to the call for higher data rates. As optical-fiber trunk lines are more fully exploited, the next generation of communications channels will come to operate at 40, 80, or 100 Gb/s, well above the current 2.5-10 Gb/s. In fact, 100 Gb/s links could be in production as soon as 2010.

These data rates raise another obstacle. The signals must be received and processed at high speed and then eventually slowed down to match silicon CMOS processing speeds of, at best, 1.5-Gb/s. Such a receiving circuit requires complexity of at least the LSI level.

There is also a trend to higher operation frequencies in mobile communications and high-speed data-links. Future optical communications systems will also operate at over 10-Gb/sec and microwave/millimeter-wave systems will be used for mobile communication and intelligent traffic-control. The diverse range of applications in the broadband space includes *local multi-point distribution systems* (LMDS) that distribute information within major metropolitan centers via microwave transmission; *global-positioning systems* (GPS); *wide-* and *local-area networks*; *cellular base stations* and *handsets*,

Table 10-1 Performance Characteristic Comparison of GaAs and Si:Ge-HBTs

Characteristic	GaAs HBT	Si:Ge HBT
GAIN		
f_t, GHz	50	30
f_{max}, GHz	70	62
G_{max} @ 2 GHz, dB	20	23
NOISE		
F_{min} @ 2 GHz	1.5	0.5
G_{assoc}	> 20	> 0.5
1/f corner	1-10	0.1-1
POWER		
V_{be}, V	1.4	0.85
V_{knee}, V	0.5	0.25
FEATURE SIZE (μm)	2.0	0.5

including *cellular, 3rd-generation* (3G) *personal-communications systems* (PCS) for wireless telephony; *asynchronous transfer mode* (ATM) *switching and multiplexing for information transfer*; and the *synchronous optical network* (SONET) protocols and systems for optical communication.

In 1990, conventional wisdom predicted that devices built using silicon-based technology could operate only at high frequencies below 2-GHz. As noted above, there are many applications in telecommunications, radar, and data transmission that need higher frequencies. It was felt that all of these applications would require devices built using the more expensive III-V-based materials. In fact, since that time such wireless and high-performance (wide bandwidth) applications have turned increasingly to gallium arsenide (GaAs) and other III-V compound-semiconductor technologies to fabricate the high-frequency devices that such applications require. However, in the wireless personal communications systems that use GaAs devices, designers have had to use many technologies for one system (such as GaAs technology for the power amplifier [PA], low-noise amplifier [LNA] and other rf components), and CMOS-silicon-VLSI-technology for the baseband operations. To overcome this issue, in 1992 IBM decided to develop a competitive, commercially-acceptable analog technology for such applications based entirely on silicon (that is to say, strictly speaking, on silicon devices fabricated with some added *silicon-germanium* [Si:Ge] layers).[36]

The premise for this technical and commercial breakthrough was to transform standard Si into a heterojunction semiconductor material by doping the Si lattice with germanium (Ge). Such a film can be created by epitaxially depositing such a Si:Ge layer onto a Si substrate. This epitaxial Si:Ge layer can be used as the base region of a *heterojunction bipolar transistor* (HBT). Table 10-1 compares the performance of such Si:Ge-HBTs to those made with GaAs. It can be seen that the maximum operating frequency of operation of Si:Ge-HBTs is close to those made from GaAs.

When IBM decided to pursue this path, they realized that there were many process integration challenges that would have to be met. However, if the venture was successful, it would become possible for silicon-based devices to compete with III-V semiconductor performance while simultaneously maintaining the low-cost advantages traditionally associated with Si. Furthermore, the intent was to make the ultra-high-speed Si:Ge-HBTs fully compatible with existing CMOS technologies. This would enable ICs to be designed with a BiCMOS process that integrated these ultra-high-frequency (bipolar) HBTs with CMOS on the same chip. In this fashion, a technology would be created that combines the radio-frequency (rf) front-end circuits with CMOS signal-processing circuitry. Since such a BiCMOS technology would offer both high-speed HBTs and CMOS on the same chip, it would be ideally suited for providing the high-performance and low-power characteristics required by wireless applications. In addition, it would be possible to manufacture the products on 200-mm wafers, with low defect densities, and in mature silicon-IC production facilities.

The overall approach to establishing such a technology was to develop a process in which the fabrication sequence of the Si:Ge-HBT became a module that could be integrated into a state-of-the-art CMOS process with no change to the CMOS flow or parameters. If this could be possible, it would represent an attractive route to cost-effective, single-chip solutions for the wireless market. That is, the cost of manufacturing Si:Ge-ICs is only marginally higher than for ICs in Si alone, while the fabrication of ICs made of GaAs or InP is considerably higher. In addition, such an approach would also pave the way for achieving high levels of integration not possible with GaAs technologies. One of the pioneers of Si:Ge technology, Bernard Meyerson of IBM, was asked in 1999 about the future of Si:Ge technology. He replied, "In the next 3-5 years, the entire world will move from global systems for mobile communications to code division multiple access (CDMA) with added focus on high-speed data-links."[1]

10.1.1 Basic Research Underpins IBM's Si:Ge Bid

The decision by IBM to launch an IC product-line for wide-bandwidth and telecommunications applications based on Si:Ge-technology was bolstered by more than a decade of research that was performed at the IBM Watson Research Center on Si:Ge. The original intention when this research was launched in the mid-1980's was to extend bipolar technology for mainframe computing applications. While IBM eventually abandoned bipolar technology in favor of CMOS for this application, it nevertheless became apparent that due to their high-frequency performance and low-noise characteristics, Si:Ge *heterojunction bipolar transistors* (Si:Ge-HBTs) were also ideally suited for telecommunications applications. Thus, IBM decided to continue to fund the Si:Ge technology program, with the new application in mind.

IBM published the first papers on Si:Ge devices in 1989. The devices described in these papers demonstrated much better performance than conventional silicon devices. Maximum clock speeds were 75-GHz compared to 25 GHz for the fastest silicon devices. This high speed could be traded off for improved linearity and lower power. That is, a Si:Ge device running at 20-GHz requires only 10-20% of the power needed to run a conventional silicon device at the same speed. This reduction in power is critical to the operation of battery-operated systems like cell phones. Operating at lower power improves the linearity, lowers the amount of noise generated, and increases reliability.

By 2000 Si:Ge-based HBTs were available with a below-10-ps ECL-gate-delay, and a cutoff frequency of over 100 GHz. In fact, with these operating characteristics, it is expected that Si:Ge-HBTs will gradually replace GaAs devices in such key antenna interface circuits as low-noise amplifiers, switches, and power amplifiers. Indeed, the development of Si:Ge-technology has opened the entire frequency range to silicon-based devices. That is, it is now possible to build complex high-speed systems on a chip which combine complex logic arrays, high-speed bipolar devices, and passive components. These ICs offer unmatched performance yet can be built as routine products in conventional commercial production fabs. Furthermore, essentially any new process step developed for silicon can be incorporated into Si:Ge-technology.

The first Si:Ge commercial process was qualified for production at IBM in 1996. This BiCMOS process revolved around an HBT with a maximum available power-gain frequency of 65-GHz, together with various active and passive support elements that allowed designers to produce a wide array of high-performance analog ICs.[2] The BiCMOS process used a pre-existing ASIC CMOS process to merge large CMOS digital logic-functions with a high-performance HBT. A variety of products have since been introduced using ICs built with this process. For example, in the second half of 1999, Alcatel announced a 10-Gb/sec SONET transceiver built with IBM's Si:Ge-technology. By the year 2000, IBM had developed its third generation of Si:Ge-technology, which includes 0.4-μm and 0.25-μm BiCMOS.[3,4] A cross-sectional SEM of a 0.2-μm Si:Ge HBT is shown in Fig. 10-1.[5]

By 2000 a large number of companies had joined the rush to carry out research, development, or production on Si:Ge-HBTs. While the adoption of Si:Ge ICs by 2001 is still at an early stage, the overall market for such products was rapidly growing due to the lower-cost, high-performance opportunity created at the boundary of Si and GaAs.[6] At the beginning of 1999, total sales of Si:Ge ICs was believed to be only $3.7 million. But, it is predicted that by 2005 the market for Si:Ge chips alone will exceed $1 billion, and the systems into which they will be inserted will have sales of between $5-$10 billion. Increasingly, participants in the analog segment of the IC industry are developing proprietary process recipes to compete for market share in this lucrative business.

However, there are some barriers that have limited the growth of the market for this technology (although, as we shall note, this situation is changing):

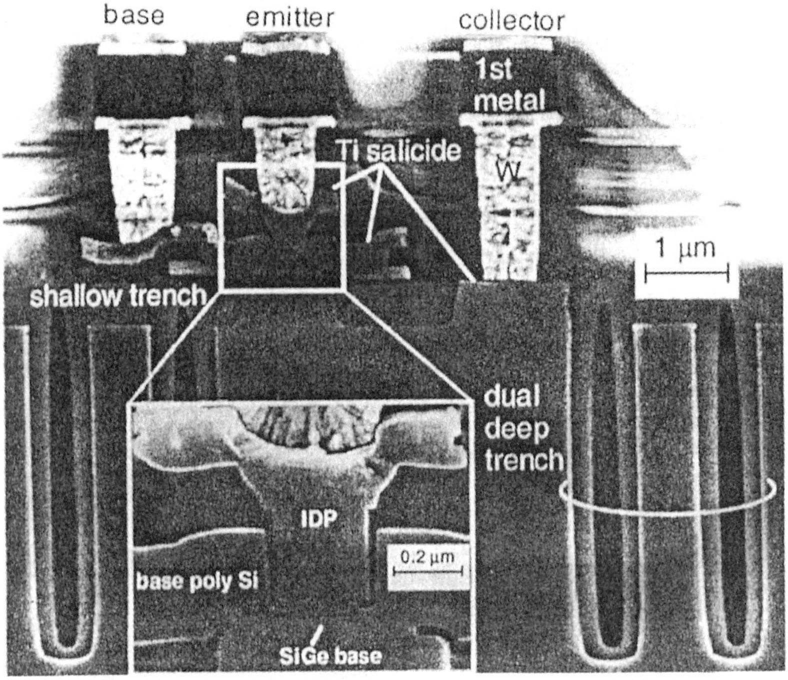

Fig. 10-1 SEM cross-sectional view of a 0.2-μm Si:Ge-HBT.[5]

1. The intellectual property is narrowly held by only a few organizations.

2. Until recently, relatively few manufacturers or foundries had developed the process-technology expertise for manufacturing such devices. However, by late-2000, most state-of-the-art BiCMOS foundries had announced Si/Si:Ge-HBTs as their baseline bipolar technologies.[37] For example, the Taiwanese foundry, TSMC, announced in early 2001 that they had licensed Si:Ge technology from Conexant.

3. The devices being made with Si:Ge are still primarily bipolar HBTs. The lack of commercially-available Si:Ge CMOS structures is therefore also an impediment. However, development of MOS-based Si:Ge devices has been described.[7]

10.2 INTRODUCTION TO HETEROJUNCTION BIPOLAR TRANSISTORS

As the vertical dimensions of the bipolar transistor were being scaled, it was predicted that some serious device-operational limits would be encountered, even if such innovations as the use polysilicon emitters were implemented (see Vol. 2, *Silicon Processing for the VLSI Era; Process Integration - Chapter 7 - Bipolar and BiCMOS Process Integration* for a detailed discussion on bipolar transistor device physics and the evolution of bipolar transistor technology). The following scenario illustrates this point: When the base-width of the bipolar transistor is decreased, it's doping level must be increased in order to control punchthrough, and to maintain a sufficiently low-value of base resistance. Since the doping level in the emitter must remain high in order for a sufficient emitter-injection efficiency to be maintained, the emitter-base capacitance C_{EB} will also increase. The current density will thus need to be increased so that the propagation delay associated with this extra capacitance can be overcome. To handle the increased current density without suffering the Kirk effect, it will also be necessary to raise the doping in the collector. Consequently, even though such a device will have a higher f_T, it will have a narrower operating-current range. Furthermore, the increased doping levels will reduce the values of the emitter-base and collector base breakdown-voltages (BV_{EBO} and BV_{CEO}, respectively), to such a degree that it will not be possible to use the device in many practical applications.

One technique that can be used to overcome the above limits is to build bipolar heterojunction transistors (HBTs). An HBT is a bipolar transistor in which the emitter, base, or collector use materials with different energy bandgaps. Bipolar transistor speed is determined mostly by the time a charge-carrier (electron or hole) takes to pass through the base of the device and the time the junction capacitances take to charge up. Thinner base-region layers, higher charge-carrier velocities, and higher current densities all increase the frequency of operation, as do lower parasitic capacitances and lower device resistances.

Recognized metrics for the speed of transistors are the following:

1. *Current-gain cutoff frequency* (f_T): This is the frequency at which the current gain of the active-device falls to 1. The highest clock rate at which the technology based on these active devices can operate is related to f_T. The value of f_T is strongly influenced by minimum device dimensions in the direction of current flow and by device capacitances.

2. *Maximum frequency of oscillation* (f_{max}): This is the maximum frequency at which it is possible to make an oscillator out of a given device. It is also referred to as the *unity power gain cutoff frequency.* Unlike f_T (which is influenced by minimum device dimensions in the direction of current flow and device capacitances), the value of f_{max} is instead quite sensitive to parasitic elements external to the core of the device, such as the transistor contacts.

High values of both f_T and f_{max} are desirable for both digital and analog designs. While f_T and f_{max} values can be traded off against each other, balanced values are ideal for a general-purpose IC. Normally, small *gate lengths* in

MOSFETs and small *base widths* in bipolar transistors lead to high cutoff frequencies. If NMOSFETs are compared to Si:Ge bipolar transistors, it is found that NMOS devices with gate lengths of 0.18-μm have an f_T of about 70-GHz and an f_{max} of less than 40-GHz, while the most advanced Si:Ge-HBTs have cutoff frequencies in the 80-100-GHz range. It is predicted that Si:Ge-technology will be scaled to provide Si:Ge-HBTs with 150-GHz f_T-values by about 2005.[8] (Note that indium-phosphide [InP]-HBTs though not discussed here, have been fabricated with an f_T of 250-GHz, and an f_{max} over 1-THz [1000-GHz].)

10.3 HETEROJUNCTION BIPOLAR TRANSISTORS WITH LARGE-BANDGAP EMITTERS

One approach to fabricating HBTs is to build transistors with emitter materials whose band gaps are larger than that of the material used in the base. This technique appears promising because the emitter-injection-efficiency, γ, depends on the band-gap difference according to the equation:[9]

$$\gamma = (n_e\, v_e / p_b\, v_h) \exp(\Delta E_g / kT) \qquad \textbf{(10.1)}$$

where n_e is the emitter electron-concentration, v_e is the effective velocity of electrons injected into the base, p_b is the base hole-concentration, v_h is the corresponding hole velocity injected into the emitter, and ΔE_g is the difference in the band gaps of the emitter and base. From Fig. 10-2, it can be seen that the larger band gap in the emitter gives rise to a larger potential-energy barrier for holes going from the base to the emitter than for electrons going from the emitter to the base. Thus, if the band gap of the emitter in silicon devices is larger than that of the base, when ΔE_g is 0.2 eV or more, $\exp(\Delta E_g/kT)$ will exceed 2000. As a result, a sufficiently-high emitter-injection efficiency will be assured, no matter what doping levels exist in the base and emitter.

For example, doping levels in the base can be increased to above 10^{19} cm^{-3}, and in the emitter they can be dropped to $5x10^{17}$ cm^{-3} at the emitter-base interface (Fig. 10-3a1). Under such conditions, many other device characteristics are also improved (among the improvements are a lower base resistance, a higher punchthrough voltage, a higher Early voltage, and a decreased E-B junction capacitance). However, for such devices to be successful in silicon technology, the technology must be compatible with advanced silicon bipolar processing.

Most approaches to heterojunction fabrication are based on growing an epitaxial layer on the Si substrate, such that the epi material has a different band gap than the Si, but also has a lattice constant that is at least approximately

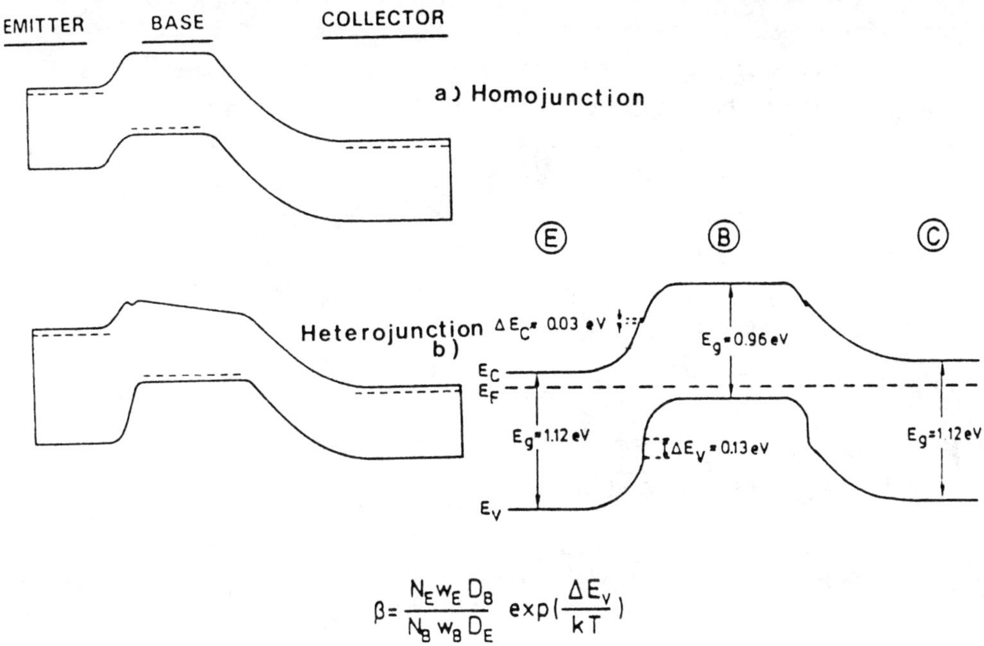

Fig. 10-2 Band diagrams of representative *npn:* a) homojunction, and; b) heterojunction bipolar transistors.[9] (© 1989 IEEE)

matched to the lattice constants of the Si substrate. If such deposited layers are sufficiently thin, the requirement for exact lattice-match is relaxed. That is, thin layers with an almost-matched lattice constant will deform so that they can accommodate to the lattice constant of the substrate. In this deformed condition they are either "coherently strained" or "pseudomorphic" layers. The latter are thermodynamically stable for thicknesses below a strain-dependent "critical" value. However, it is important that subsequent processing temperatures be restricted to prevent the stress of the layers from being relaxed through the formation of an array of misfit dislocations.

From the point of view of process integration, the most attractive approach to building devices with hetero-E-B junctions is to increase the band gap of the emitter (since this represents the only change from a conventional bipolar process). Several methods for implementing this approach have been attempted, including the use of the following materials as the wider band-gap emitters: GaP;[10] semi-insulating polycrystalline-silicon (SIPOS, E_g ~1.7 eV);[11] oxygen-doped Si-epitaxial films (OXSEF, E_g ~1.7 eV);[12] β-SiC (E_g ~2.2 eV);[13] and phosphorus-doped hydrogenated micro-crystalline silicon (μc-Si, E_g ~1.7 eV).[14]

Despite an intense effort to develop such processes, no completely successful fabrication procedure using an emitter with a larger bandgap was demonstrated. Although numerous high-gain transistors have been reported, these devices exhibited other problems, including high-temperature requirements, high bulk and contact resistivities, high emitter resistances, or poor stability during the multilevel-interconnect fabrication steps. The most promising candidate of this group was the μc-Si emitter. Fujioka *et al.* reported achieving reduced emitter resistances with such structures and built ECL gate arrays with them using a non-self-aligned emitter structure.[14] Basic gate delays of 295-ps at 6.5-mW/gate were obtained.

10.4 HETEROJUNCTION BIPOLAR TRANSISTORS WITH SMALLER BANDGAP BASE REGIONS (Si:Ge BASE)

An alternative approach to using a *larger bandgap* material for the emitter is to use a material in the base of the transistor that has a *smaller bandgap* than that of

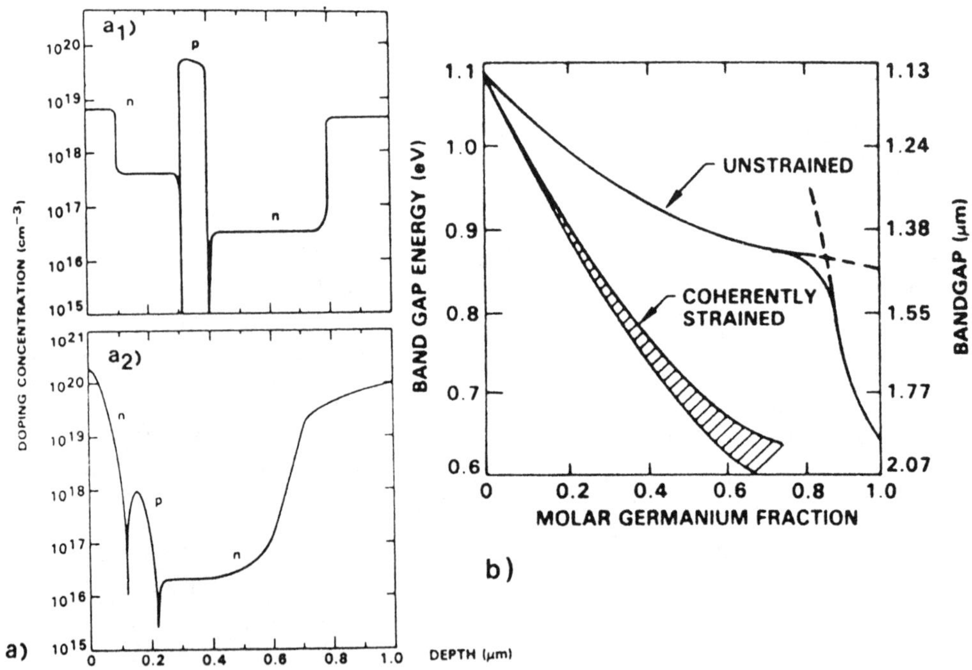

Fig. 10-3 a) Representative doping profiles for npn heterojunction bipolar transistors.[9] (© 1989 IEEE); b) Band-gap energy versus alloy composition for Si:Ge alloys.[19] (after People and Bean Copyright 1986 AIOP)

silicon. Layers of Si to which Ge has been added exhibit such reduced bandgaps (Fig. 10-3b). For example, pseudomorphic epitaxial Si_xGe_{1-x}-alloy layers of up to 100-nm in thickness can be formed with a Ge content of 20%, and the ΔE_g between the emitter and base in Si transistors fabricated with such base layers (i.e., *Si_xGe_{1-x}-base-HBTs*) can be as large as 0.1 eV.[15]

Pseudomorphic epitaxial Si_xGe_{1-x} layers can be deposited by means of molecular beam epitaxy (MBE) or limited-reaction-processing (LRP) CVD.[17] The most successful commercial CVD methods have involved ultra-high-vacuum-CVD (UHV-CVD),[16] or low-pressure CVD (LP-CVD). Note that appropriate levels of boron doping must also be incorporated into these films during the epi-growth process of the base-region in order to form the *p*-base of an *npn*-HBT. Only *npn-HBTs* are being pursued at present because *pnp*-HBT devices do not benefit substantially from the use of a Si:Ge base (due to a less favorable band-offset). We will describe the details of the UHV-CVD process used to form the Si:Ge-epitaxial-layers in the next section.

It should also be noted that to add a Si:Ge layer to a conventional Si device requires a special piece of semiconductor equipment (i.e., an MBE machine, or a special CVD reactor), which in 2000 cost approximately $2.5 million. However, the only other way to make a major performance improvement in a family of semiconductor devices is to shrink the geometry of the devices. Implementing a device shrink would typically require installing a whole new generation of process equipment and would cost more than $750 million!

HBTs with narrow-band-gap Si_xGe_{1-x} bases are able to achieve very high performances (e.g., f_T's as high as 100-GHz have been built with a 20%-Ge-concentration in a 100-nm base, a 0.2-μm-wide Si emitter, and a collector-current-density of $1x10^5$ A/cm^2).[18] Conversely, Si_xGe_{1-x}-base-HBTs offer the potential benefits of the same performance characteristics as those of conventional homojunction transistors, but with a relaxed profile (i.e., the base can be doped more heavily for a lower base-sheet resistance, or it can be made thicker to achieve lower doping levels with higher breakdown voltages).

The chief drawback of the reduced-band-gap-base approach is that it is more difficult to integrate into existing advanced bipolar processes. The metastable nature of the Si_xGe_{1-x} films places constraints on the Ge content, layer thickness, subsequent annealing cycles, and thermal oxidation. On the other hand, this method does not suffer from the other problems exhibited when wide-band-gap-emitter materials are used. In addition, to provide the ultrathin base profiles for future bipolar devices, the conventional ion-implantation approaches are becoming increasingly difficult to implement successfully, while the low-temperature epitaxial processes show great promise. We will discuss the process technology and process integration issues of Si:Ge ICs in the following sections.

10.5 PROCESS TECHNOLOGY FOR FABRICATING Si:Ge-HBTS

As noted in the previous section, Si:Ge heterojunction bipolar transistors have the same general structure as homojunction bipolar transistors, with the exception that they contain a Si:Ge layer in place of a Si layer in the intrinsic-base-region of the transistor. This Si:Ge layer is formed by epitaxial deposition onto a Si substrate, and a polysilicon layer is deposited onto the Si:Ge base-layer to serve as the polysilicon emitter. Since only *npn*-HBTs are fabricated (i.e., as noted earlier, *pnp* HBTs do not exhibit substantial performance improvement over their homojunction *pnp* counterparts, so there is no advantage to manufacturing them), the *p*-type intrinsic base is fabricated by also incorporating boron into the Si:Ge layer as it is being grown.

The use of an epitaxial layer as a specific structure of a bipolar transistor (i.e., the base region), is a significant departure from past uses of epitaxial layers in the fabrication of Si ICs. That is, in conventional (i.e., non-Si:Ge) IC fabrication, epitaxial layers have been used to form a lightly-doped Si layer on a heavily-doped Si substrate. That is, the epitaxial layer becomes a part of the *substrate* in which the device is formed, but *does not act* as a *specific device structure* itself. The fact that the epitaxial Si:Ge layer in an HBT is an active-device-structure-layer allows one of the fundamental limitations of bipolar device fabrication to be overcome, namely the fact that the base region of modern homojunction bipolar transistors must be formed using an ion-implantation process.

The ion implantation process produces a base-doping-profile with a skewed Gaussian distribution (due to the Gaussian-distribution of the as-implanted dopants, plus channeling effects associated with the ion-implant process). By using an epitaxial process to form the base region, it is theoretically possible to grow a *uniformly-doped* layer with atomically abrupt boundaries, and all of this doping can be confined within the Si:Ge layer. In practice, such atomically abrupt boundaries are not possible, but one can create a boron-doping profile within the Si:Ge layer that is much more uniformly doped than are the base regions of bipolar transistors created with an ion-implantation process. Note that the Si:Ge epitaxial layers must be grown at temperatures low enough that the Si:Ge layers retain their pseudomorphic structure (500-550°C). Furthermore, by performing the epitaxial-deposition process at such low temperatures the outdiffusion of the dopant from the layer during the epitaxial growth process is essentially suppressed, keeping the boron well-confined within the Si:Ge layer.

A low-temperature-epitaxy (LTE) technique using ultra-high-vacuum chemical vapor deposition (UHV/CVD) in a multi-wafer reactor is the most mature process used to grow epitaxial layers of Si:Ge-alloys for use in HBTs.[21] It is reported that device-quality epitaxy can be achieved at a growth temperature of 550°C. This LTE technique originated in the early 1980's with Meyerson's work at IBM on a novel UHV/CVD reactor. The Si:Ge epitaxy process has two stages: **1**) preparation of the Si substrate surface so that an epitaxial layer can be grown on it at a low temperatures; and **2**) growth of the layer itself.

To enable growth of an epitaxial layer on a Si surface, the surface must be free of native oxide and other contaminants. To eliminate the need to use high temperatures it is necessary to ensure that the surface of the Si is free of such species prior to initiating the epitaxial-growth-stage of the process. A simple wet chemical procedure is carried out to remove any such native oxide from the Si surface and at the same time to passivate the wafer surface with hydrogen. This process entails etching the wafer for 10-15-sec in dilute (10:1) H_2O/HF. The layer of hydrogen adsorbed on the Si surface during this wet etch acts to retard the growth of native oxide on the Si surface. That is, after the 10-sec HF dip, there is a passivation layer consisting of hydrogen-terminated bonds across the silicon surface. The presence of the hydrogen reduces the reactivity of the growth interface by approximately 13 orders of magnitude from that of a bare-silicon surface with respect to its oxidation rate in ambient air. Since the wafer surfaces are oxide free and passivated as-pulled from the HF dip, the wafers may be loaded directly into the growth chamber (Fig. 10-4) after this HF-last step. This

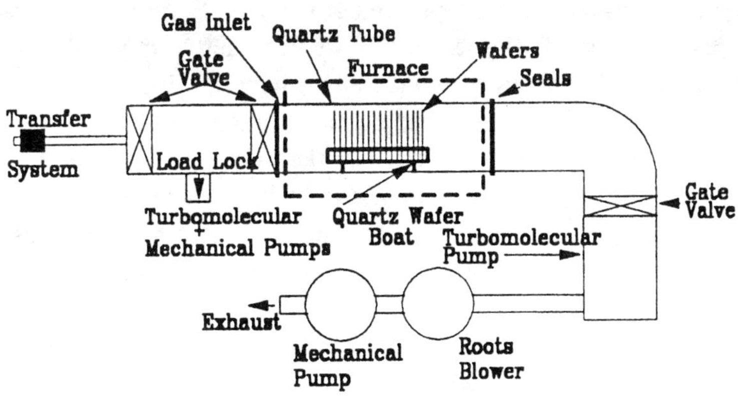

Fig. 10-4 Cross-section schematic of a UHV/CVD reactor.[15]

surface-preparation step has been shown to reliably keep the Si surface free of native oxide until the growth process can be begun (provided the temperature/time history of the load-sequence prevents hydrogen desorption until growth is initiated). Note that bare-silicon wafers removed from an HF bath are hydrophobic (i.e., they do not wet). Thus, the surface readily de-wets after being pulled from the HF. This makes it easy to eliminate residual HF from such a surface, allowing this removal process to be automated.

The batch of wafers (typically 25) is introduced into the loadlock of the UHV-CVD tool. After a pumpdown to 10^{-6} torr, the wafers are transferred into the growth chamber under an H_2 flow and growth is commenced immediately. The growth chamber is pumped down to a base pressure of 10^{-9} torr (UHV), and the precursor gases (silane, germane, and diborane) plus hydrogen are introduced. Films are deposited at temperatures in the range of 500-600°C. During deposition the pressure is maintained at 10^{-3} torr. The growth rate during this process is only about 0.4-4.0-nm/min. This very-low growth rate provides precise control of the layer thickness (on the order of 1-2 atomic layers in this instance), doping concentration, and formation of a pseudomorphic strained SiGe layer. Versions of this system are available for processing 200-mm wafers (based on IBM designs), from Unaxis (formerly Balzers Process Systems) and CVD Equipment Corp (Fig. 10-5). To ensure these tools are robust and productive in a manufacturing environment, good vacuum practice and contamination-prevention methods must be incorporated when designing them. Note that these reactors are batch systems that process up to 25 wafers at a time. The industry, however, is moving toward single-wafer processing, and single-wafer Si:Ge deposition tools are thus also available.

High-quality Si- and SiGe-epitaxial-films can be formed at such low temperatures for the following reason.[36] The hydrogen-terminated silicon wafer surface desorbs its hydrogen at temperatures between 500-600°C. The silane (SiH_4) that is fed into the reactor chamber and that adsorbs on the wafer surface also decomposes at these temperatures. Of the products of this decomposition, silicon incorporates into the crystal, and the hydrogen remains adsorbed on the silicon surface, replenishing the hydrogen that was initially there. Thus, at these temperatures the silicon-epitaxial-layer grows while maintaining a hydrogen-passivated surface. Thus, the surface is protected from ambient oxygen contamination, allowing high quality epitaxial layers to form. (Note that silicon can undergo solid-phase-recrystallization at reasonable rates in the range of 500-700°C). However, at higher temperatures (> 650°C), the hydrogen passivating film is removed instantaneously, allowing even low residual oxygen content in

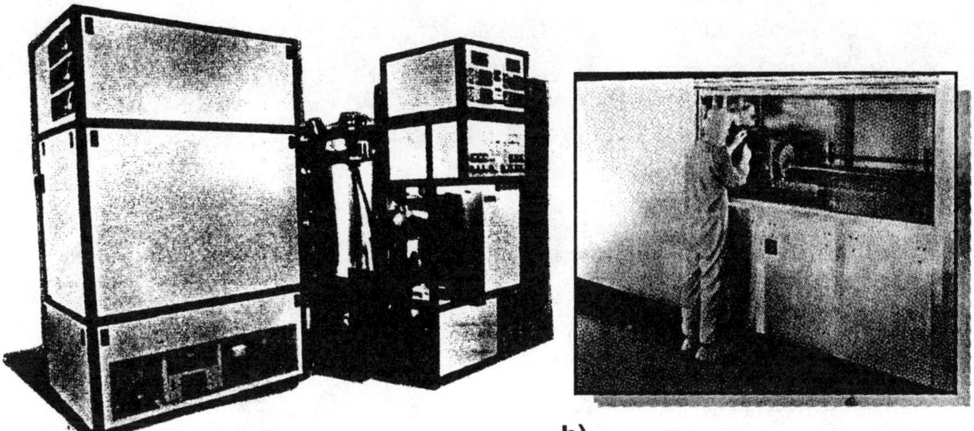

a) b)

Fig. 10-5 a) Commercial UHV/CVD Si"Ge deposition system. Courtesy of CVD Systems; b) Commercial Low-energy plasma-enhanced CVD system for Si:Ge deposition. Courtesy of Unaxis Ltd.

the growth ambient to cause the formation of native oxide on the silicon surface. This causes high defect densities in the epitaxial film. As a result, to obtain high-quality films when silicon wafers are heated to temperatures greater than 650°C (initially hydrogen-passivated or not), a high-temperature pre-treatment step just prior to epitaxial growth is needed to remove the now-present native oxide.

Single-wafer, ultra-high-vacuum, rapid-thermal-CVD (UHV/RTCVD) has been reported as a means for achieving LTE.[22] An advantage of this single-wafer technology is that the process can be initiated and terminated by rapid changes in the substrate temperature, rather than by changing gases. Figure 10-6 shows a schematic of such a system.[37]

Conventional single-wafer epi reactors can also be operated with an LP-CVD process at low temperatures (about 600°C) to produce strained-layer Si:Ge-alloys at a deposition-rate of about 10-nm/min. Note that in single-wafer, load-locked, RTP tools, the surface-cleaning of the Si-substrate prior to the Si:Ge-epitaxial-growth stage is carried out by a high-temperature RTP step which sublimates SiO present on the wafer surface (instead of relying on an HF-etch as is done in the batch-UHV/CVD process). It must be rigorously ensured that the ambient during both the cleaning stage and the growth stage of the epitaxial-growth process in the single-wafer reactor is kept completely free of oxygen and moisture. This is accomplished by using a loadlock, point-of-use gas-purification, and ultra-clean gas handling. Two vendors that supply single-wafer reactors in which Si:Ge can be deposited by LP-CVD are ASM International[23,24] and Applied Materials.[39,40]

A substantial number of these reactors are reportedly being sold to the industry.[25] ASMI claims they have 75% of the installed Si:Ge-reactors in 2000 (ASMI Epsilon E2000®).

Etching the Si:Ge layers also presents some challenges. While Si:Ge can be etched with plasma processes that use chemistries similar to those used to etch Si, the fine features and profiles required to maintain high-frequency performance, place increased demands on the processes. Novel chemistries can be an aid, and endpoint-detection plays a vital role. Figure 10-7 shows a super-self-aligned selectively-grown-epitaxial Si:Ge-base transistor.

The use of *selective-epitaxial-growth* (SEG) to grow the Si:Ge-base of the HBT has been pursued to achieve higher f_T and f_{max} values. For example, a double-polysilicon self-aligned-HBT with a low-resistivity Si:Ge-base was fabricated having a 88-GHz f_T-value and a 78-GHz f_{max}-value. Si:Ge-HBTs with such performance characteristics would make then suitable for wireless communications up to 10-GHz and high-speed data-links at 40-Gbit/sec and above.[26] Another report described how such a Si:Ge-SEG-process was used to build a 0.2-μm, self-aligned Si:Ge-HBT that was then integrated into an SOI BiCMOS technology. The base Si:Ge layer in this HBT was deposited by UHV/CVD. The HBT in this report had an f_{max} of 180-GHz.[27]

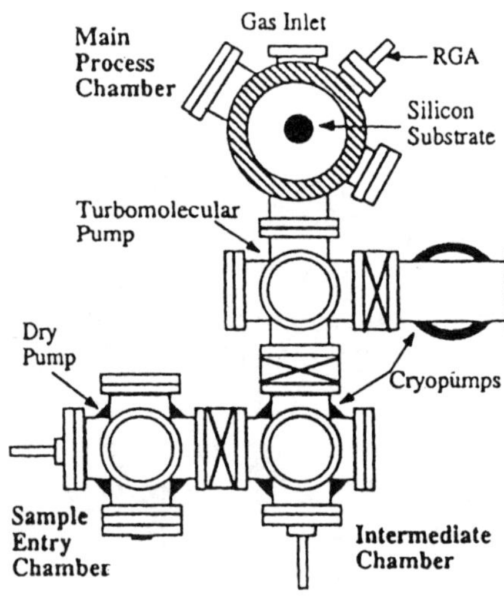

Fig. 10-6 Schematic of a UHV/RTCVD epitaxial system.[22] Reprinted with permission of the American Physical Society.

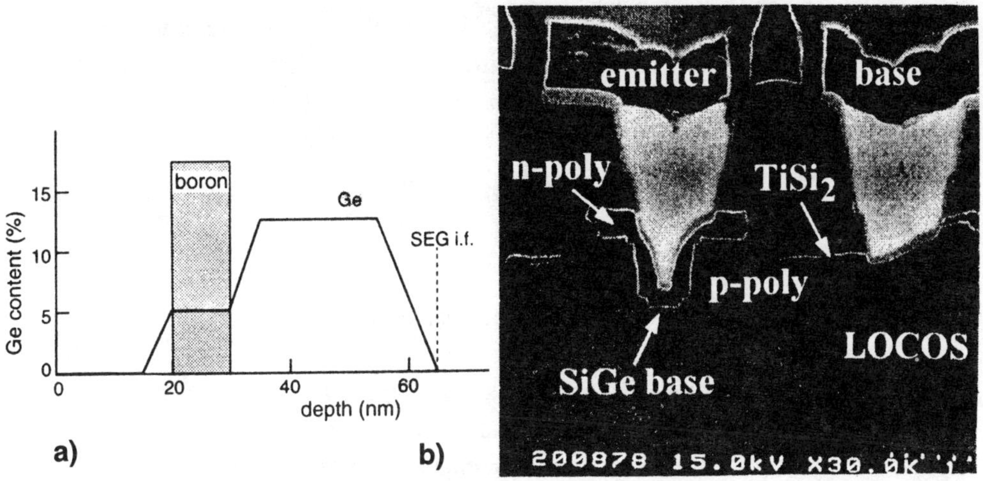

Fig. 10-7 a) Schematic of the Ge profile in the Si/Si:Ge layer of a Si:Ge-HBT;[27] b) SEM cross-section of the emitter-base complex of a 0.2-μm Si:Ge-HBT.[26]

10.6 PROCESS INTEGRATION ISSUES FOR BiCMOS TECHNOLOGY WITH Si:Ge-HBTs

The previous section dealt with the process technology of forming epitaxial layers of Si:Ge on Si-substrates. Here we consider the issues involved with integrating such an epitaxial-growth-step into a CMOS process so that HBT structures can be fabricated that are integrated on the same chip as CMOS circuitry. That is, a BiCMOS process is implemented instead of a CMOS process. The goal of creating such a BiCMOS technology is to end up with a process that results in no change (or minimal change) to the standard-CMOS flow, and that the electrical characteristics of the CMOS in a BiCMOS structure matches the baseline-CMOS as closely as possible (and conversely, that the CMOS process steps do not degrade the performance-characteristics of the HBT).

As noted earlier, basic physics explains that when a Si:Ge-lattice is formed, the atomic size difference between Si and Ge atoms necessarily results in the formation of a layer that will create a lattice-mismatch if the layer is formed on a Si substrate acting as a growth template. This lattice-mismatch must be dealt with when integrating this layer with the overall fabrication sequence. While a certain level of strain is required for the bandgap-reduction of the Si:Ge-layer (and this bandgap-reduction is a driving force behind HBT performance improvement), the strain must be kept low enough to avoid a generation of lattice defects and unstable Si:Ge films. This limitation has been extensively studied and the Matthews and Blakeslee curves relate film-thickness to strain in a Si/Si:Ge system.[34] To allow the Si:Ge-layer to be integrated with the remainder of the processing steps used to fabricate standard CMOS, this implies that the Si:Ge-strained-layer must be able to tolerate total emitter-thermal-cycles such as an 850°C furnace-anneal followed by a short 900-1000°C rapid-thermal-anneal.[28]

10.7 HIGH-QUALITY Si:Ge EPITAXIAL FILM GROWTH ON PATTERNED WAFERS

One of the most significant issues facing the integration of an HBT device into a CMOS process is that of being able to deposit high-quality epitaxial films onto wafers containing pre-existing oxide-isolation-patterns. The use of wafers patterned prior to epitaxial deposition is the process factor that separates this approach from the approach that is used to fabricate discrete HBTs or mesa bipolar transistors that contain a Si:Ge-epitaxial- base-region. The problem is that the parts of the wafer covered by oxide are hydrophyillic (these surfaces *do* wet). The wetted surface makes it difficult to remove residual-HF without a DI-water rinse-and-dry step. But, such a water rinse would degrade the passivation quality of the hydrogen-termination layer present on the regions of the wafer where silicon is exposed (through hydrolysis of less stable F-terminated surface and/or kink sites which cannot be ideally passivated).

A pragmatic two-step solution when depositing Si:Ge-films in the UHV/CVD reactor is to first assure that the entire exposed wafer surfaces (front and back) are composed of silicon. This is done by depositing a by a polysilicon "protect" layer on the wafer. Then windows are cut into this layer where epitaxial growth is required. The polysilicon layer also serves to protect the field isolation and/or critical oxides from HF attack during the HF-passivation dip, and to reduce the overall moisture load in the UHV/CVD chamber. As a result, the basic approach is to provide a substrate which resembles as closely as possible the ideal of an all-silicon, hydrogen-passivated surface of a bare Si-wafer.

10.8 THERMAL STABILITY PROCESS INTEGRATION ISSUES

It was noted previously that it must be possible to maintain a nearly defect-free Si:Ge-strained-layer on Si throughout the entire BiCMOS process. Although unstrained Si:Ge layers also exhibit significant bandgap-reduction, the presence of an extensive misfit dislocation network in the material would preclude the fabrication of high quality bipolar devices. However, by adopting effective epitaxial growth techniques and optimizing the Si:Ge-strained-layer characteristics, it is possible to minimize strain-induced-defects. It turns out that Si:Ge films with doping levels of Ge up to 13% are able to maintain their mechanical stability under annealing conditions of up to 900°C for 30 min, making it feasible to use such Si:Ge layers in BiCMOS processes (provided the layers are not subjected to temperatures and times above this limit).

10.9 COMPATIBILITY WITH ION-IMPLANTATION PROCESSES

There have been some questions concerning the compatibility of a Si:Ge layer with subsequent ion-implantation steps. That is, one might wonder if introducing significant amounts of lattice damage to a strained layer through implantation would degrade its integrity. Such implantation damage could provide a source of defect nucleation sites, which in turn could lead to deleterious formation of an extensive misfit dislocation network upon annealing. It turns out, however, that Si:Ge layers which are mechanically stable are apparently able to undergo implantation and anneal cycles with no additional defect-generation or movement beyond that expected in single-crystal silicon.

The two processes likely to cause problems are the moderate-energy, low-dose implants used to selectively implant dopants for the sub-collector of the HBT (phosphorus, at 100-keV, $1x10^{13}/cm^2$), and the high-dose, low-energy implants used to form the extrinsic base of the HBT and the source/drain regions of the PMOSFETs (boron, 5-keV, $1x10^{15}/cm^2$). The effects of these implants on the Si:Ge-layers was examined, and it was found that no significant yield impact was observed. A examination of the Si:Ge-strained-layer after ion implantation was studied using transmission electron microscopy (TEM). There was no evidence of misfit-dislocation formation within the device regions.

10.10 COMPATIBILITY WITH THERMAL OXIDATION PROCESSES

It must also be possible to oxidize exposed regions of Si to create an insulator in the integrated BiCMOS process. Studies indicate that the presence of Ge in the Si dramatically enhances the oxidation rate, and that Ge segregates to the growth-interface during oxidation. Since local-oxidation is generally used to provide electrical junction isolation, the Ge "pile-up" effect is likely to lead to high levels of leakage and potential reliability concerns under electrical stress. A generally accepted way to control both of these effects is to use an all-Si capping-layer for oxide growth and isolation purposes.

A low-thermal-budget CMOS process is required for thermal-cycle compatibility with a narrow-base *npn* process. After the base-epitaxial-layer is grown, the thermal-cycle should be kept to a minimum, ideally consisting of only the bipolar *npn*-emitter-cycle, which entails a furnace anneal and an RTA anneal. The two largest post-gate-etch thermal-cycles in a CMOS process are the polysilicon reoxidation step, and the activation anneal performed after the source/drain implantation step. Both of these steps in conventional CMOS are typically furnace anneals at temperatures > 900°C, which causes significant diffusion of the base. However, the poly reoxidation step has been recently carried out at somewhat lower temperatures (850°C), and short RTP anneals have generally been replacing furnace S/D anneal steps. As the Si:Ge processes have been evolving through several generations, the thermal budget issues have been handled differently. We will discuss this subject again in Sects. 10.11 and 10.12.

10.11 VERTICAL SCALING OF Si:Ge HBTS

Using Si:Ge and bandgap-engineering, bipolar transistors will continue to provide circuit designers with the performance necessary to meet the stringent circuit and systems requirements of the next decade. In demonstration of this trend, companies are leveraging the learning from 50-GHz-f_T Si:Ge-devices to develop 75 to 90-GHz-f_T Si:Ge-devices. Continued progress in process and device design will certainly push performance of the Si:Ge-HBT to higher levels in the future.

Higher f_T performance is principally achieved through the vertical scaling of dopants and dimensions in the HBT device. Such scaling primarily involves: **1**) a reduction in transit time through the base and collector by reduction of the vertical-layer dimensions; and **2**) a commensurate increase in Ge-grading across the (now thinner) neutral base.

Key to vertical dimension reduction is the minimization of thermal cycles in post-base-deposition processing. The modular integration trend exemplified by the base-after-gate process sequence (described in Sect. 10.12) provides one important aspect to obtaining vertically scaled HBTs. In addition, movement towards RTP processing is favorable to achieving low overall thermal budgets.

Another technique for reducing the base-width-dimension is to incorporate carbon into the base epitaxy. Carbon incorporation at concentrations lower than $10^{20}/cm^3$ has been shown to significantly reduce B diffusion (by lowering the Si self-interstitial concentration within the C-rich regions).[37]

The choice of emitter dopant is closely linked to the device thermal-cycles. Various emitter dopants have been described in the literature, including implanted arsenic or phosphorus emitters, and *in situ*-doped phosphorus emitters. Because thermal-cycles required to diffuse and activate arsenic emitters will adversely impact base-widths as thinner bases are used, *in-situ*-doped phosphorus emitters will become prevalent in future Si:Ge-HBT device designs targeting greater than 100-GHz-f_T.[38]

10.12 BASE-DURING-GATE AND BASE-AFTER-GATE BiCMOS PROCESS FLOWS

As noted earlier, the goal of producing a BiCMOS process that integrates HBTs into a CMOS process flow is to ensure that the electrical characteristics of the CMOS devices remains as close to those obtained in the CMOS process alone. Therefore, the process must maintain the thermal excursions needed to produce the CMOS structures. In the first two generations of BiCMOS technologies developed at IBM this was accomplished by sharing layers and thermalcycles. That is, the HBT was fabricated *during* the fabrication of the CMOS layers, and this sequence was termed *base-during-gate* (BDGate). A flow diagram of the base-during-gate process is shown in Fig. 10-8.[29] To avoid significant widening of the base during the post-base-deposition thermal-processes of poly-reoxidation and NMOSFET-S/D-annealing, a low-temperature poly-reox-step was used, together with a low-thermal-budget CMOS-process technology. The BDGate flow worked well down to the 0.5-μm-BiCMOS technology node, but it was difficult to extend it into the next, scaled generations. This was because the thermal-cycles needed in the next generations became too high for the HBT-device.

As a result, a new flow, termed *base-after-gate* (BAGate) was developed. In the base-after-gate sequence the thermal-cycles for the CMOS-devices and the HBTs were decoupled. This is achieved by building the HBT after the CMOS is fabricated. That is, the CMOS is largely fabricated in one complete block before the Si:Ge-base is deposited. Therefore, the CMOS-technology steps can be completely copied without modification. The bipolar-transistor is fabricated in a complete block after the CMOS without sharing any thermal-cycles (except for a final emitter drive-in RTA). Since the HBT-processing is performed at low temperatures, thermal effects on the FETs are minimal. A flow diagram of the base-after-gate process for the 0.25-μm-BiCMOS process is given in Fig. 10-9. By decoupling the fabrication of the CMOS- and HBT-structures, the migration from generation-to-generation becomes easier.[30,35,38] The modularity of the BAGate process supports rapidly developing derivatives and low-cost alternative BiCMOS technologies. The baseline-CMOS processes are also incorporating more analog elements (capacitors, inductors, and resistors) because of the expanding use of CMOS for analog products. This trend is favorable for BiCMOS technology because fewer processing steps need to be added to the CMOS to have an analog-Si:Ge-BiCMOS process.

Both processes employ a planarized deep-trench-plus-shallow-trench isolation-structure fabricated with the aid of CMP technology. The Si:Ge is blanket deposited so that it is single-crystal over the exposed silicon-regions and polysilicon over the field-oxide isolation-regions. Standard CMOS processing is

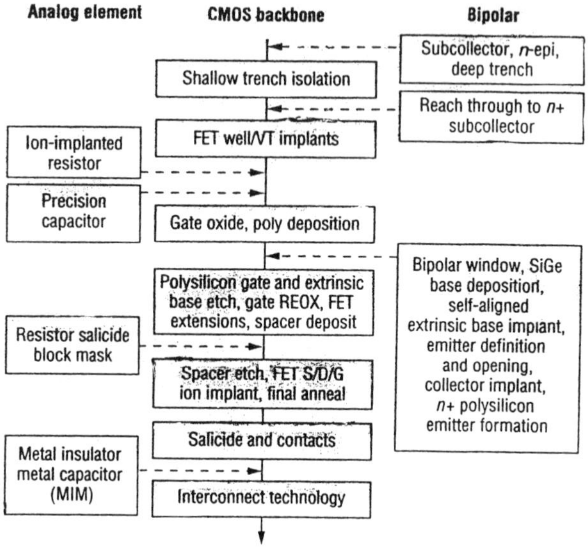

Fig. 10-8 CMOS process flow for fabricating *Base-During-Gate* Si:Ge BiCMOS structure.[29]

used, including the use of titanium salicide. The metallization is planarized Al-Cu metal, patterned by RIE, with tungsten interlevel plugs. Fig. 10-10 shows a schematic cross-section of this BiCMOS structure.

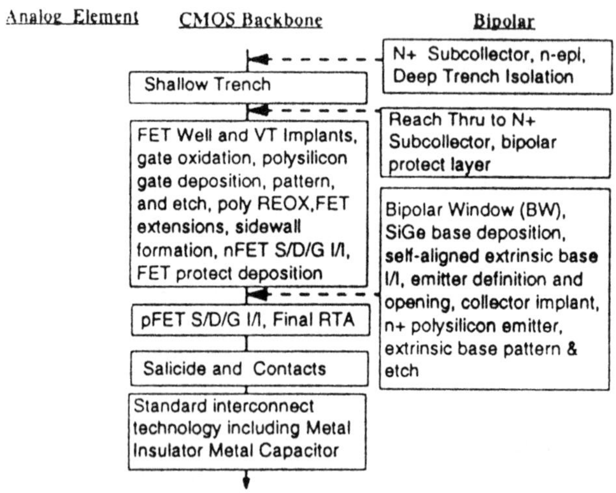

Fig. 10-9 CMOS process flow for fabricating *base-after-gate* Si:Ge BiCMOS structures.[29]

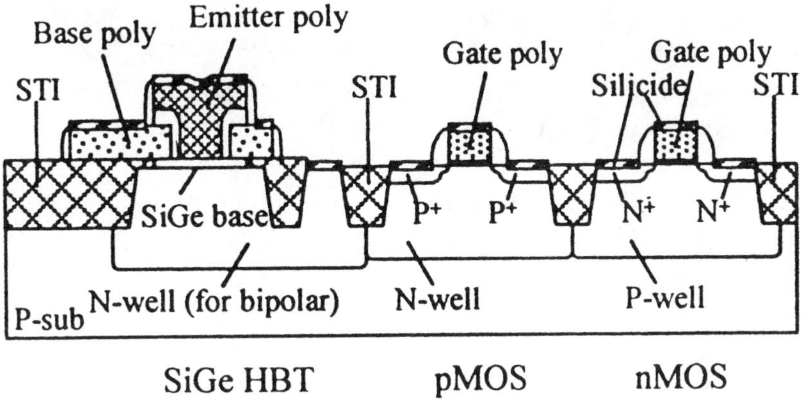

Fig. 10-10 Schematic cross-sectional view of the Si:Ge BiCMOS. The Si:Ge HBT has a self-aligned structure.[16]

10.13 EXAMPLES OF THE MODULAR INTEGRATION Si:Ge-HBTS INTO A BiCMOS PROCESS FLOW

The first Si:Ge-IC that IBM offered commercially was built using 0.25-μm geometries and combined 1200 Si:Ge-bipolar-transistors with over million CMOS-transistors. This IC was the highest-performance, lowest-noise *data-channel-circuit* in the world. It is used to read data from high-speed hard-drives. The IBM fab at Burlington VT was processing 75 wafers per day of this device (starting in 1999), with yields above 75%.[20]

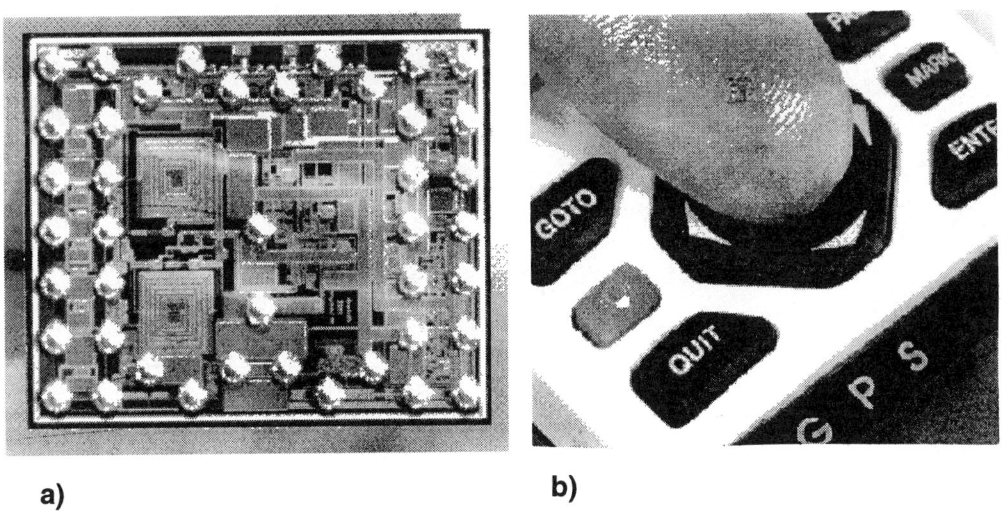

Fig. 10-11 a) Single-chip GPS Receiver; b) Relative size of GPS Receiver.[8]

In 1998 IBM described the characteristics of a 1.8-million-transistor ASIC chip, fabricated using a Si:Ge-BiCMOS technology.[31] Another description of the modular integration of an HBT that used a Si:Ge-base-layer (in which carbon was also incorporated) into a standard 0.25-μm-CMOS process to yield a BiCMOS technology is given in Ref. 32. Neither the HBT nor the CMOS parameters were adversely impacted by the additional thermal budget of the complete BiCMOS process. The technology is low-cost, has high-performance, and allows re-use of existing CMOS designs. The Si/Si:Ge-HBTs in this technology are able to perform at very high speeds; f_{max} = 65-GHz. Because such high device-switching-speeds are not necessary for most wireless circuits operating at 900-MHz and 2.4-GHz, the Si:Ge leverage comes from being able to trade the excess speed for improvement in other device attributes; most notably, low-power operation (Fig. 10-11). Another example of integrating Si:Ge-HBTs into a CMOS process is given in Ref. 33. In this process four additional masking steps were needed to create this BiCMOS process. The result is a BiCMOS technology that includes a super self-aligned (SSA) Si:Ge-HBT with a peak f_T of 52-GHz and a peak f_{max} of 70-GHz. A 0.18-μm rf-BiCMOS technology with a 73-GHz f_T S i:Ge-HBT was described in late 2000.[18] A two-step annealing technique is used to solve the thermal budget tradeoff between Si:Ge HBTs and CMOS. IBM has also devoted significant effort to develop high-quality capacitors and inductors, which can be integrated into the Si:Ge-process. These passive components allow IBM to build the world's only single chip Global Positioning Satellite (GPS) receiver (Fig. 10-11).[8]

A recent report described the effects of scaling on the performance characteristic of deep-submicron Si:Ge HBTs (Fig. 10-12).[41]

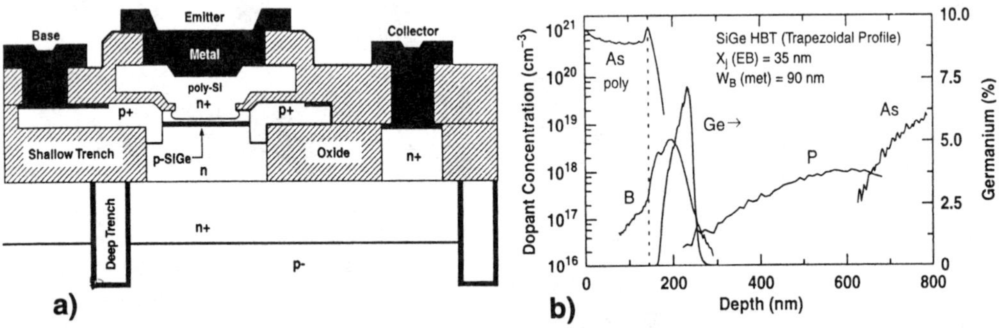

Fig. 10-12 a) Schematic cross-section of deep-submicron Si:Ge HBT; b) SIMS doping profiles of the dee-submicron Si:Ge HBT.[41] (© 2002 IEEE)

REFERENCES

1. M. Helms, Chip: Semiconductor Industry Source Book, 1999, pp. 26-27, www.bps-IT.com.

2. D.C. Ahlgren *et al.*, "Manufacturability of an Integrated SiGe HBT Technology for the Analog and Wireless Marketplace," *Tech Dig. IEDM* 1996, p. 859.

3. S. St. Onge *et al.*, "A 0.24-μm SiGe BiCMOS Mixed Signal RF Technology Featuring a 47 GHz f_T HBT and 0.18 μm-L_{eff} CMOS," BCTM Proc. 1999.

4. G. Niu, *et al.*, "SiGe Profile Design Tradeoffs for RF Circuit applications," *Tech. Dig. IEDM* 1999 p. 573.

5. K. Washio *et al.*, "A 0.2 μm Self-Aligned SiGe HBT Featuring 107-GHz f_{max} and 6.7-ps ECL," *Tech. Dig. IEDM* 1999 p. 558.

6. A. Catalano, *Solid State Technology,* September 2000, p. 59.

7. Y.-C. Chao *et al.,* "Enhanced Performance of Sub-100 nm CMOSFETs Using Strained Epitaxial Si-Ge," *Tech. Dig. IEDM*, 2000 p. 753.

8. L.E. Larson, "Device and Technology Requirements for Next Generation Communications Systems," *Tech. Dig. IEDM*, 2000 p. 737.

9. J. Woo and J. Plummer, *IEEE Trans. Electron Dev.,* August 1988, p. 1311.

10. S.L Wright *et al.,* "Molecular Beam Epitaxy Growth of GaP on Si," *J. Appl. Physics*, Vol. 55, April 1984, p. 2916.

11. Y.H. Kwark and R.M. Swanson, "*n*-Type SIPOS and Poly-Silicon Emitters," *Solid-State Electronics,* Vol. 30, November 1987, p. 1121.

12. M. Takahashi *et al., IEEE Electron Dev. Letts.*, October 1987, p. 475.

13. T. Sugii *et al., IEEE Electron Dev. Letts.*, February 1988, p. 87.

14. H. Fujioka *et al., tech. Dig. IEDM*, 1988, p. 574.

15. D. L. Harame *et al.,* "Si/Si-Ge Epitaxial-Base Transistors - Part I: Materials, Physics and Circuits," *IEEE Trans. Electron Dev,.* Vol. 42 March 1995, p. 455.

16. B.S. Meyerson, "Low-Temperature Silicon Epitaxy by Ultrahigh Vacuum Chemical Vapor Deposition," *Appl. Phys. Letts.,* Vol. 47, p. 797 (1985).

17. J.F. Gibbons *et al.,* "Limited Reaction Processing: Silicon Epitaxy," *Appl. Phys. Letts.,* Vol. 10, p. 721 (1985).

18. T. Hashimoto *et al.*, "A 73 GHz 0.18 μm f_T RF-Si:Ge BiCMOS Technology Considering Thermal Budget Trade-Off and With Reduced Boron-Spike Effect on HBT Characteristics," *Tech Dig. IEDM*, 2000, p. 149.

19. R. People and J.C. Bean, *Appl. Phys. Letts.,* **48**, p. 538 (1986).

20. B.S. Meyerson, "Silicon Germanium Technology: A 'Material of the Future' Gone Mainstream," *IEEE/SEMI Advanced Semiconductor Manufacturing Conference,* Fall 2000, Boston MA.

21. B.S. Meyerson, "Low-Temperature Si & SiGe Epitaxy by Ultrahigh Vacuum/ CVD: Process Fundamentals," *IBM J. Res. Dev.,* Vol. 34, p. 806 (1990).

22. K.E. Violette *et al*, "Low-Temperature Selective Silicon Epitaxy by Ultrahigh Vacuum Rapid Thermal Chemical Vapor Deposition Using Si_2H_6, H_2 and HCl," *Appl. Phys. Letts.*, Vol. 68, p. 66 (1996).
23. "Motorola's Next Si:Ge Wireless Technology Will Use the ASM Epsilon E2000 Reactor." *Press Release: ASMl*, March 1, 2001, Bilthoven, Netherlands.
24. D.J. Meyer and J. Italiano, "Productive Single-Wafer Si:Ge Processing," *Meeting Abstracts of the Fall Mtg. 1999, Electrochem. Society*, Abs. No. 1314.
25. ibid. Ref. 6.
26. J. Bock *et al.*, "Si-Ge Bipolar Technology for Mixed Digital and Analogue RF Applications," *Tech. Dig. IEDM* 2000 p. 745.
27. K. Washio *et al.*, "A 0.2 μm 180-GHz - 6.7 ps-ECL SOI/HRS Self-Aligned SEG Si-Ge HBT/CMOS Technology for Microwave and High-Speed Digital Applications," *Tech. Dig. IEDM* 2000 p. 741.
28. D. L. Harame *et al.*, "Si/Si-Ge Epitaxial-Base Transistors - Part II: Process Integration and Analog Applications," *IEEE Trans. Electron Dev.* **42** Mar 1995, p. 469.
29. D.C. Ahlgren and B. Jaganathan, "SiGe for Mainstream Semiconductor Manufacturing," *Solid State Technology,* January 2000, p. 53.
30. D. Harame *et al.*, "Trends in SiGe BiCMOS Process Integration," *Meeting Abstracts of the Fall 1999 Mtg. Electrochemical Society,* Abs. No. 1333.
31. R.A. Johnson *et al.*, "1.8 Million Transistor CMOS ASIC Fabricated in a Si-Ge BiCMOS Technology," *Tech. Dig. IEDM* 1998, p. 217.
32. K.E. Ewald *et al.*, "Modular Integration of High-Performance SiGe:C HBTs in a Deep Submicron, Epi-Free CMOS Process," *Tech Dig. IEDM* 1999 p. 561.
33. C.A. King, "Very-Low Cost Graded SiGe-Base Bipolar Transistors for a High Performance Modular BiCMOS Process," *Tech Dig. IEDM* 1999 p. 565.
34. S.R. Stiffler *et al.*, "The Thermal Stability of Si-Ge Films Deposited by UHV/CVD," *Appl. Phys. Lett.*, **70**, p. 1416, 1991.
35. D.C. Ahlgren *et al.*, "Si-Ge Fuels High-Speed Communications," *Semiconductor International,* November 2000, p. 62.
36. D.L. Harame and. B.S, Meyerson, "The Early History of IBM's SiGe Mixed Signal Technology," *IEEE Trans. Electron Dev.,* **48**, November 2001, p. 2555.
37. H. Rucker *et al.*, "Dopant Diffusion in C-doped Si and SiGe: Physical Model and Experimental Verification," *Tech. Dig. IEDM*, 1999, p. 345.
38. D.L. Harame *et al.*, "Current Status and Future Trends in SiGe BiCMOS Technology," *IEEE Trans. Electron Dev.,* **48**, November 2001, p. 2575.
39. A.V. Samilov *et al.*, "Dedicated Pre-Clean Reduces Epi's Thermal Budget," *Semiconductor International,* November 2000, p. 73.
40. V. Loup *et al.*, "Fine Control of Low-Temperature Epitaxial Growth," *Solid State Technology,* July 2001, p. 91.
41. S.Zhang *et al.*, "Effects of Geometrical Scaling on Frequency Response and Noise Performance of SiGe HBTs," *IEEE Trans. Electron Dev.* Mar. 2001, p. 429.

Chapter 11

SILICON-ON-INSULATOR (SOI) TECHNOLOGY

In this chapter the characteristics of silicon-on-insulator (SOI) technologies fabricated with deep-submicron MOSFETs are described. The focus is on the SOI-devices themselves as well as on the process technologies used to fabricate both the starting SOI-wafers and the SOI ICs. An argument is also presented as to why SOI technology may provide a path for continuing to advance CMOS ICs for several more generations. That is, implementation of SOI may make it possible to exceed the limits facing bulk-silicon-CMOS in the near term.

11.1 WHAT IS SILICON-ON-INSULATOR (SOI)?

Silicon-on-insulator (*SOI*) circuits consist of ICs made from single-device islands dielectrically isolated from the substrate and each other. Each such island contains a single transistor, with the isolated transistor being either a MOSFET or a bipolar transistor. However, since about 90% of the ICs manufactured in 2002 are based on CMOS-technology, the discussion in this chapter will be restricted to SOI technology involving MOSFETs. Figure 11-1 shows a cross-section of a MOSFET fabricated on an SOI wafer that also uses STI as the isolation structure to provide lateral isolation. Note that in the case of a MOSFET fabricated on an

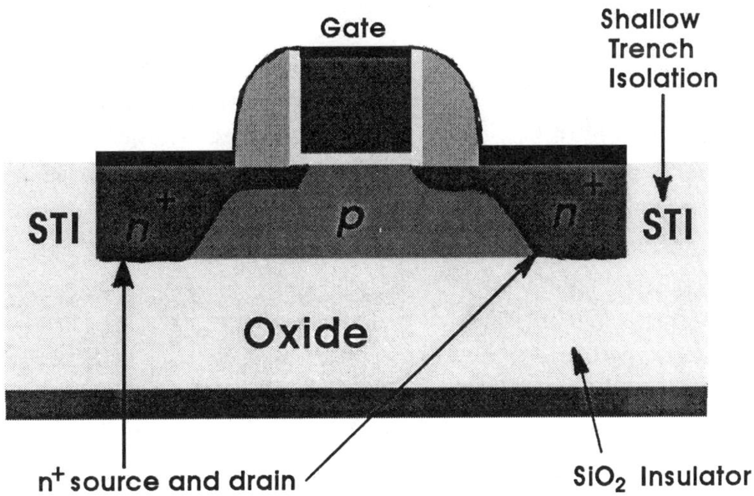

Fig. 11-1 Cross-section of a MOSFET fabricated on an SOI wafer.

SOI wafer, a *buried oxide* (*BOX*) underneath the device provides vertical isolation, and a trench oxide (in this example), provides the lateral isolation. Since the trench oxide extends all the way through the silicon device layer atop the BOX (and a thermal oxide exists on the wafer top surface), dielectric material continuously surrounds each MOSFET on all sides. Thus, complete electrical isolation is provided for the MOSFET with respect to all other devices, and the substrate, on the wafer.

11.1.1 Isolation Issues in CMOS Fabricated in Bulk-Silicon Wafers

The electrical isolation requirements of CMOS ICs built in SOI are quite different from those encountered when CMOS ICs are fabricated in bulk-silicon. To see why, the circuit strategies and structures needed for isolation purposes in CMOS circuits on bulk-silicon must be examined. For reference purposes a single-MOSFET built in a bulk-silicon-substrate is shown in Fig. 11-2a (with various device structures and regions defined). Figure 11-2b shows two NMOSFETs built on a *p*-doped substrate, together with the parasitic-transistor that exists in the field-region separating the active devices. Semirecessed-LOCOS is used to produce the field-oxide (which is part of the device structure employed to maintain electrical isolation among such MOSFETs). Figure 11-2c shows an *n*-well CMOS structure fabricated in a *p*-type bulk-silicon wafer. This figure also shows the parasitic MOSFETs that are associated with that structure (as well as the *pn*-junction that separates the *n*-wells from the *p*-substrate). Again, a semirecessed-LOCOS field-oxide plays a role in maintaining electrical isolation among the MOSFETs of the CMOS structure. (It should be noted that shallow-trench isolation [STI] is used in place of LOCOS to form the field-oxide in CMOS technologies below 0.35-μm. Nevertheless, the isolation issues described here for CMOS on bulk-silicon remain the same, regardless how the field-oxide is formed). Finally, in Fig. 11-2d a cross-section of a CMOS IC structure is depicted, together with a schematic depiction of the parasitic-bipolar-transistors and associated lateral currents.

Using the IC device structures described above, the isolation issues involved in CMOS ICs can be undertaken. The first item to be noted is that the *pn*-junction separating a well from the substrate must be kept under reverse-bias to prevent current from flowing from the power-supply-line to ground through forward-bias of this junction. Second, inter-device isolation for devices within a well (*intra-well isolation)* or within the substrate regions, entails the use of a field-oxide to suppress spurious surface-leakage effects between active devices within a well. Third, a field-oxide and channel-stop implants are also used to prevent leakage

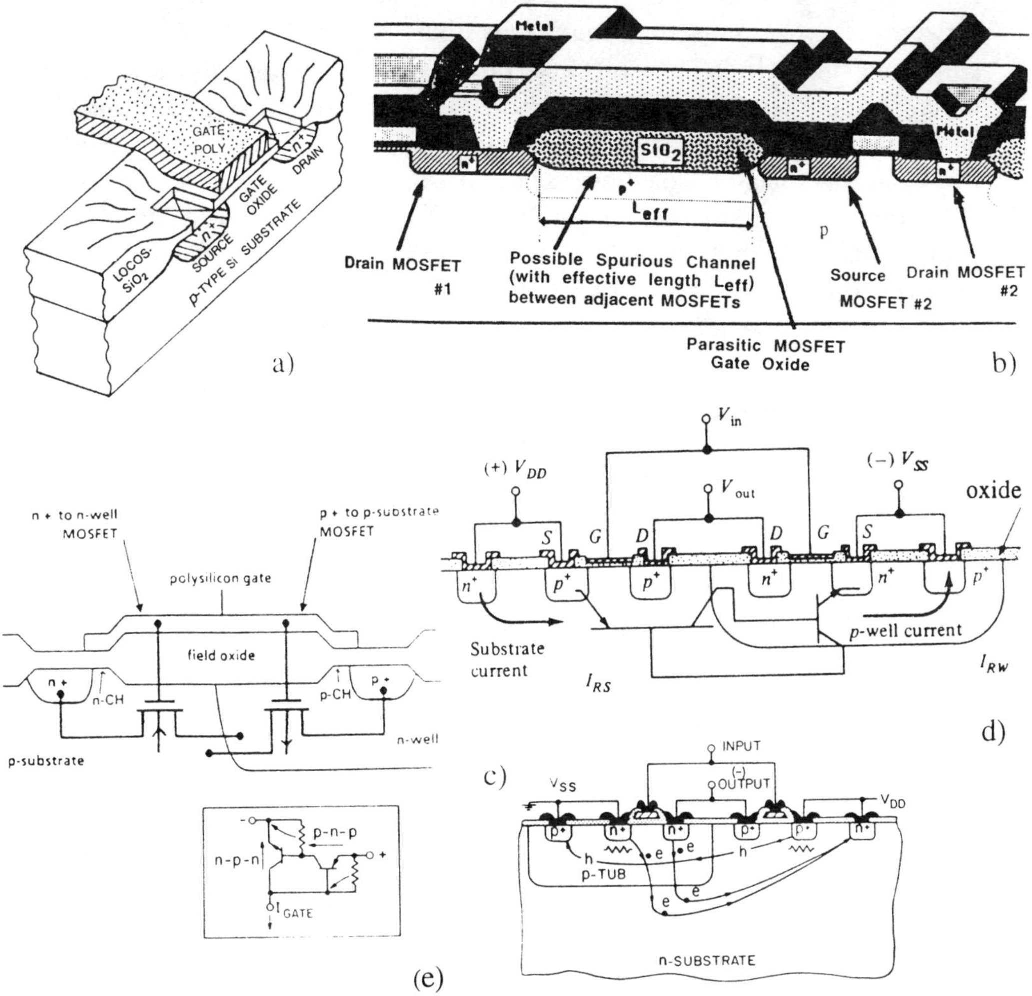

Fig. 11-2 a) A single MOSFET built in a bulk-silicon substrate, with various device structures and regions defined; b) Two MOSFETs (e.g., both NMOSFETs) built in a substrate having one type of doping (i.e., on a *p*-doped substrate), together with the parasitic transistor that exists in the field-region separating these MOSFETs. Semirecessed-LOCOS is used to produce the field-oxide (which is part of the device structure employed to maintain electrical isolation among such MOSFETs); c) An *n*-well CMOS structure fabricated in a *p*-type bulk-silicon wafer. Also shown in the figure are the parasitic MOSFETs that are associated with that structure, as well as the *pn* junction the separates the *n*-wells from the *p*-substrate; d) A cross-section of a CMOS-IC-structure is depicted, together with a schematic depiction of the parasitic-bipolar-transistors and associated lateral currents; e) The parasitic bipolar transistors that exist in CMOS structures can be *turned on*, which can cause the CMOS IC to go into a latchup state.

between devices among different wells (*inter-well isolation*). Yet, despite employing these isolation structures, intra-well- and inter-well-leakage still occurs to some degree in CMOS ICs made in bulk-silicon (and the problem gets worse as the devices are brought closer together). Note that Volume 3 can be consulted for a more detailed discussion of isolation issues in CMOS built in bulk-silicon wafers. In addition, the parasitic-bipolar-transistors that exist in CMOS structures can be *turned-on*, and this can cause the CMOS IC to go into a *latchup state* (Fig. 11-2e). That is, latchup can occur because two parasitic-bipolar-transistors are always simultaneously formed in the course of fabricating a CMOS structure. In bulk-CMOS, these two transistors are connected in such a way that if the current gain of the loop formed by these two devices is larger than unity, latchup can occur. Latchup can be triggered by different mechanisms, such as node- voltage overshoots, displacement-current, junction-avalanching, and pho-tocurrents. It is true that latchup is a well-understood phenomenon which can be dealt with in bulk-CMOS using proper processing and design methodologies. However, with the constant reduction in feature spacing, latchup immunity in bulk CMOS ICs remains a continuing challenge.

11.1.2 Isolation in SOI CMOS

The isolation issues of CMOS fabricated in SOI wafers are much less complex than those involving CMOS ICs in bulk-silicon. This is because each MOSFET in SOI is built in an island of silicon on an insulating substrate. Hence, all of the previously-mentioned possible isolation causes are eliminated. Let us explain this assertion in more detail.

First, the requirement of keeping the *pn*-junction that separates the CMOS well regions from the substrate under reverse-bias is eliminated. That is, this requirement no longer exists because a continuous oxide barrier exists in SOI between the well-regions and the substrate. Consequently, current cannot flow between the power and ground lines, even if the junction in question becomes forward-biased. Second, the possibility of leakage current between devices within a well (intra-well-leakage) and from well-to-well (inter-well-leakage) is eliminated. This is because the parasitic-field-region-MOSFET does not exist (as a result of the field oxide extending through the silicon film and joining the buried oxide). In other words, the channel between the source and drain of any parasitic-field-region-MOSFET is electrically interrupted by the oxide moat surrounding each MOSFET within an SOI IC. Consequently, leakage current due to: 1) field-inversion of the channel of a parasitic-field-region-MOSFET; or

Table 11-1 ITRS Roadmap of the Required MOSFET Junction Characteristics

Year	2000	2001	2002	2003	2004	2005	2008	2010	2014
Gate size (nm)	120	100	85	80	70	65	45	32	22
S/D junction depths (nm)									
Contact	65–125	55–105	45–90	43–5	38–75	35–70	30–55	20–40	15–35
Extension/channel	36–60	30–50	25–43	24–40	20–35	20–33	16–26	11–19	8–13
SSRD channel	18–30	15–25	12–22	12–20	10–18	10–17	8–13	5–10	4–8
Extension doping									
Sheet resistance (Ω/sq.)	310–760	280–730	250–700	240–675	220–650	200–625	150–525	120–450	100–400
Resistivity (10^{-3}Ω-cm)	1.1–4.6	0.84–3.6	0.62–3.0	0.57–2.7	0.44–2.3	0.40–2.1	0.24–1.4	0.13–0.86	0.08–0.52
Lateral extension junction abruptness (nm/decade)									
For V_t-roll-off control	12	10	8.5	8	7	6.5	4.5	3.2	2.2
For spreading resistance control	4.8	4.1	3.4	2.9	2.7	2.4	1.25	0.8	0.5

2) parasitic-MOSFET subsurface-punchthrough, do not exist in SOI-technologies. Finally, the existence of vertical- (as well as lateral-) isolation within SOI ICs eliminates the possibility of latchup. That is, the current paths from the collectors of the parasitic-bipolar-transistors to the bases of the complimentary bipolar-transistors (which set up feedback paths in CMOS in bulk-silicon that cause the circuits to enter latchup), are interrupted by the vertical oxide of the SOI. Hence, latchup can no longer occur in SOI ICs.

Despite the fact that intra-well and inter-well leakage paths are interrupted in SOI CMOS structures, new *intra-device* parasitic transistor paths inherent in SOI CMOS structures can still produce leakage current, as is discussed in Sect. 11.7.

11.2 THE "END OF THE ROAD FOR BULK CMOS" SCENARIO PREDICTED BY THE 2001 INTERNATIONAL TECHNOLOGY ROADMAP FOR SEMICONDUCTORS

The *2001 International Technology Roadmap for Semiconductors* (ITRS-2001) indicates that planar CMOS transistors on bulk-silicon are expected to reach their limits at gate lengths of about 50-nm in the 2005-2006 time frame. That is, the ITRS-2001 analysis documents a clear "end-of-roadmap" event-horizon at the onset of IC device scaling to 50-nm gate-lengths.[1] The premise of the ITRS study was to predict, as far as possible, the characteristics of an extension of the industry's "standard technology," namely, planar-CMOS-transistors fabricated on bulk-silicon wafers. The study concludes that many process and materials constraints combine to force this change in the technology path. That is, the conclusions drawn in this study are that sometime between 2005 and 2008 nearly all of the 30 or so transistor characteristics required for the extension of Moore's Law are listed as "no known solutions." This is a clear signal that IC transistor technology requires fundamental changes in the next half-decade. In this section the principal issues limiting *junction-formation* in sub-100-nm CMOS ICs on bulk-silicon will be detailed. As part of the discussion to be presented in the next section (Section 11.3), it will be described how some of these constraints can be removed or relaxed through the use of SOI. That is, by implementing CMOS-on-SOI a path may be provided to allow CMOS ICs to be fabricated beyond the 50-nm gate-length-limit faced by CMOS-on-bulk-silicon. An extension of as much as a decade beyond the "end-of-the roa-map" scenario posited by the ITRS-2001 may be possible by fabricating CMOS ICs on SOI-wafers.

The three principal challenges for forming shallow-junctions in CMOS on bulk-silicon wafers are as follows:

1. Limiting dopant-diffusion in the source/drain regions to shallow depths;
2. Activating a sufficient percentage of the dopants implanted into the source/drain regions to provide source/drain regions that exhibit sufficiently low contact-resistance and sheet-resistivities:
3. Obtaining source/drain junctions that have highly-abrupt vertical and lateral doping-profiles.

The ITRS-2001 values for these key transistor parameters, as a function of the device generation (which correlates to the gate-length of that generation and the expected time of its introduction) are given in Table 11-1.

11.2.1 Source/Drain Junction Scaling

As MOSFET gate lengths (L_L) are scaled, the other device structures (i.e., the gate-oxide-thickness and the junction-depth of the source/drain extension) are also scaled proportionally. To maintain such "proportional scaling" for CMOS when the gate-length decreases below 100-nm, the design rule used in the ITRS-2001 study was that the junction depth of the source/drain extension $X_{j(extension)}$ would scale according to: $X_{j(extension)}$ ~0.55 x L_L within a range of ±25% (see Fig. 11-2a for a graphic that depicts the region referred to as the source/drain extension). Controlling junction depths to less than 55-nm for sub-100-nm MOSFETs will push the limits of sub-keV boron implantation and "spike" anneals.[2]

11.2.2 Dopant Activation in Shallow Source/Drain Junctions

In the ITRS-2001 model, it is expected that ways will be found to produce highly doped source/drain extension-regions having a sufficiently low resistivity (to reduce the sheet-resistance of these regions) at the same time the depth of the junctions of these regions is reduced. If this can be done (i.e., if the resistivity of the source/drain extension regions could somehow be reduced from the value of 10^{-3} Ω-cm [that is produced in the source/drain extension-regions of 0.15-μm NMOS transistors by using low-energy boron implantation {0.2-2.0-keV} and "spike" anneals] to say 10^{-4} Ω-cm) a sheet-resistance versus junction-depth curve (as plotted in Fig. 11-3 for the value of $\rho = 10^{-4}$ Ω-cm) could be obtained.[3] Overlaid on these sheet-resistance versus junction-depth curves (with ρ as the parameter), are the industry expectations of the value of ρ for: **1**) high-performance ICs; and **2**) cost-focus (i.e., lower-cost) ICs, as taken from the ITRS-2001. That is, the values of ρ that the ITRS-2001 roadmap expects in order to be able to fabricate CMOS ICs on bulk-silicon for smaller gate-length-devices, decreases from about 10^{-3} Ω-cm for NMOSFETs with gate lengths of 0.14-μm, down to

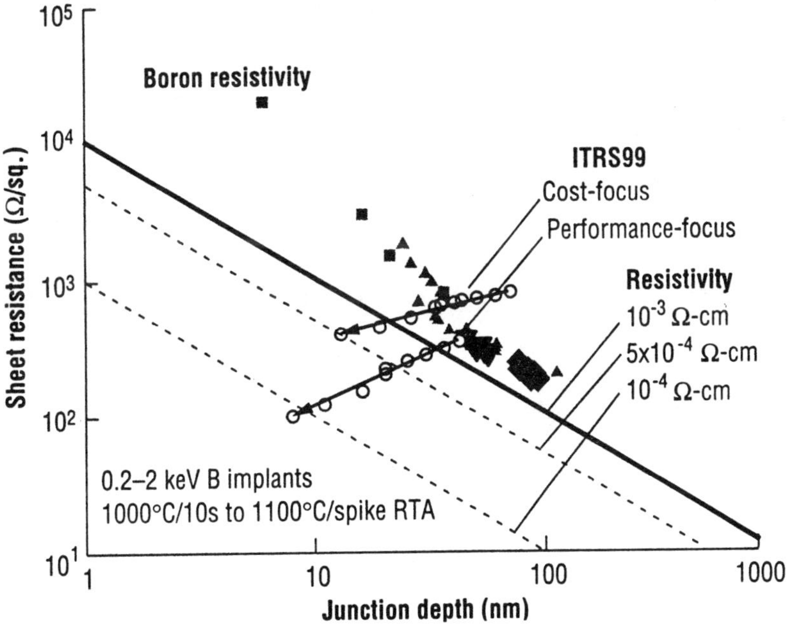

Fig. 11-3 Sheet-resistance and junction-depths for low-energy boron implants and spike-implant anneals, and industry expectations for S/D extension junctions (starting with 140-nm gates to 22-nm gates at the head of the arrows) for cost-focus and performance focus devices. The dopant resistivity for sub-keV boron implant and spike-RTP process (in CZ-Si) is limited by a "wall" at about 10^{-3} Ω-cm.

about 10^{-4} Ω-cm for NMOSFETs with 22-nm gate lengths (corresponding to a junction-depth of 8-nm and a sheet resistance of 100 Ω/sq). Some alternative doping technology may be required to achieve such low resistivity. Possible alternative doping techniques include: **a**) low-temperature (or selective) epitaxy; **b**) diffusion from a doped-glass source; **c**) diffusion from a low-temperature Si-Ge alloy; **d**) direct gas-source-diffusion in an RTP-system; **e**) gas-immersion-laser-diffusion (GILD); and **f**) planar-doping-layers using atomic-layer-epitaxy (ALE). Of these many possible processes, low-temperature-epitaxy may hold some key advantages, as it can also be done in a selective manner (which may possibly minimize masking steps). However, with selective epitaxy, separate hard-masks are required for *n*- and *p*-S/D-regions. On the other hand, GILD with projection masking (PGILD) does not require resists or hard-masking and may become an important doping technology.

At this time, however, it is not clear whether any of the above methods of reducing the value of ρ in the source/drain extension regions of NMOSFETs will be able to reduce ρ down to 10^{-4} Ω-cm. In fact, the resistivities of junctions formed by the present day "best methods" of sub-keV boron implants and spike anneals appears to be limited by the "activation-wall" to a value no smaller than 10^{-3} Ω-cm.[4]

11.2.3 Doping-Profile Abruptness in Shallow Source/Drain Junctions

To permit sub-100-nm MOSFETs to exhibit the performance characteristics expected in the ITRS-2001 roadmap, the total source-to-drain resistance of such MOSFETs must be kept sufficiently small.[5] Controlling this total source-to-drain resistance-value in deep-submicron MOSFETs requires that special attention be paid to the contact-resistance and spreading-resistance at the junction of the source extension and the channel. As shown in Fig. 11-4, parasitic-resistances exist in the path between the metal-to-Si substrate interface and the region in the device where the actual transistor action begins (see also Sect. 5.4, Chap. 5).

In the planar MOSFET, the device-region in which transistor-action occurs is the channel-region under the gate. The current flowing from the metal-interconnect lines to the channel-region must be transported from the metal-Si contact to the edge of the channel through the source or drain regions (Fig. 11-4). In the MOSFET, current enters the contact perpendicular to the wafer surface, and then travels parallel to the surface in order to reach the channel. The total parasitic-series-resistance, R_S, of the current path from the contact to the edge of the channel can be modeled by the sum of four components,

$$R_S = R_{co} + R_{sh} + R_{sp} + R_{ac} \tag{11.1}$$

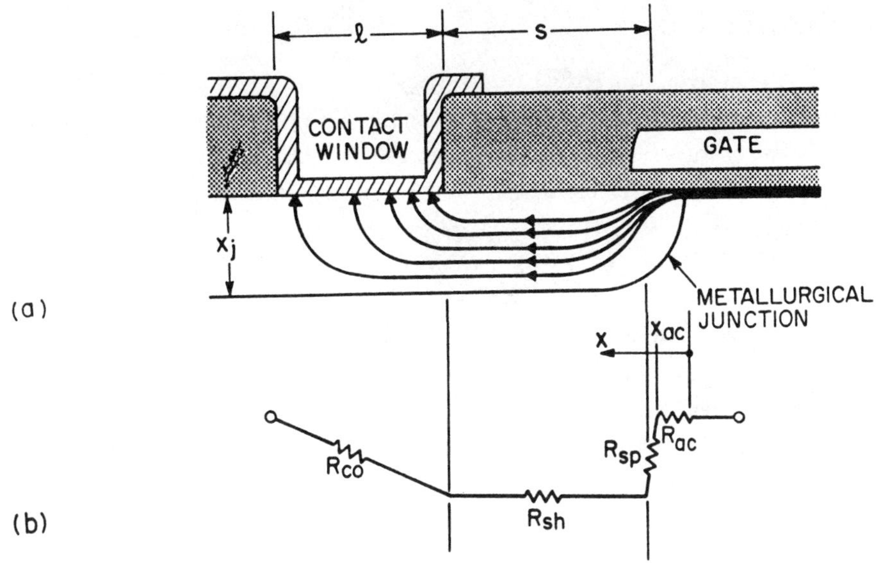

Fig. 11-4 Schematic drawing showing: a) current paths within the source and drain regions, of a planar MOSFET and; b) their corresponding representative series resistance components.[5] (© 1979 IEEE)

where R_{co} is the *contact-resistance* between the metal and the source/drain region, R_{sh} is the *sheet-resistance* of the bulk-region of the source and drain, R_{sp} describes the *resistance of the current lines crowding near the channel end of the source*, and R_{ac} is the *accumulation-layer resistance.* Qualitatively, for a non-abrupt junction, R_{sp} and R_{ac} arise because, after leaving the channel, the current does not immediately spread into the bulk of the source-drain regions (the parasitic-resistance-effect in these bulk regions is R_{sh}). Thus, R_{sp} and R_{ac} exist because the accumulation-layer conductance is much higher than the semiconductor conductance in the vicinity of the junction (see again Sect 5.3).

The key issue for sub-100-nm MOSFETs is the abruptness of the doping-profile at the lateral edge of the source extension and its effect on R_{sp} and R_{ac}. The ITRS-2001 study calls for lateral-doping-profiles that decrease into the channel region at a slope of ~3 nm/decade in concentration, for 100-nm gates. Obtaining a lateral-doping-profile abruptness of less than 3-nm/decade presents severe challenges for bulk silicon doping by implantation and rapid thermal annealing. As-implanted profiles for 0.2-keV boron into pre-amorphized silicon have a vertical profile steepness of 2.5-nm/decade.[6] The key S/D-extension profile in the lateral direction, still remains to be characterized, and will be further broadened by thermal annealing in actual transistors.

11.3 WHY SOI?

Silicon-on-insulator (SOI) technology has long been used in many specialized applications, such as radiation-hardened or high-voltage integrated circuits. It is only in recent years (since about the mid-1990s), however, that SOI has emerged as serious contender for low-power, high-performance applications.[7] The power consumption, P, in CMOS ICs is given approximately by:

$$P \sim C\,V^2 f + V\,I_{leakage} \tag{11.2}$$

where C is the total capacitance, V is the supply voltage, f is the frequency, and $I_{leakage}$ is the standby-leakage-current. Thus, in low-power applications the supply voltage must be reduced to decrease both terms of this equation. This power-supply reduction has had to follow the evolution in battery-voltage and capacity. The SEMATECH roadmap target has been to reach 0.9V by the end of the century. However, at such a low voltage, the IC performance is also reduced due to a smaller value of I_D in MOSFETs. As will be seen, however, the use of SOI allows the drain current to be significantly increased, even at such low voltages. SOI therefore also becomes appropriate for emerging system-on-chip ICs used for smart power, as well as for micro-electromechanical systems (MEMS), mixed-signal, and integrated-optics applications.

SOI also provides a path to permit the scaling of ICs to go forward. That is, to continue to increase performance and keep up with Moore's Law, several cumulative approaches have been adopted. The evolution of IC technology in the near future which uses this strategy is summarized in Fig. 11-5. This diagram indicates that integrating copper wiring and SOI will represent the next step toward the continued improvement in IC performance. Here we list the advant-

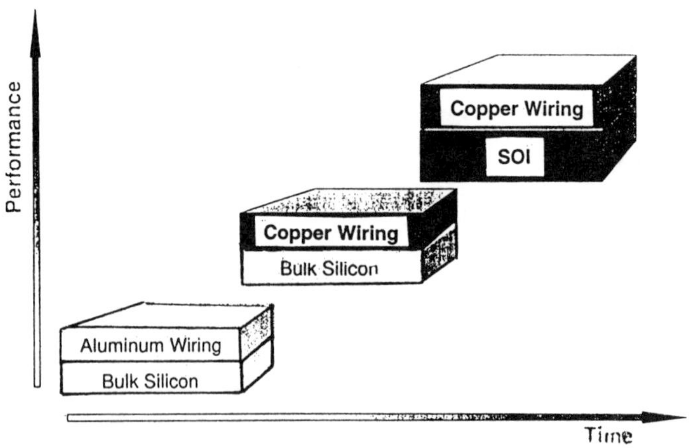

Fig. 11-5 Technology migration for future IC progress.

ages offered by CMOS-onSOI which contribute to such performance improvement.

1. **Improved Speed-Performance.** The speed performance of SOI ICs is enhanced by 20-35% compared to the same circuits fabricated in CMOS on bulk-silicon if they are operated at the same voltages. (The reasons for this increase in speed performance is due to a number of factors that are listed later as other advantages of SOI.) Thus, SOI circuits built using the design rules of "generation-n" of some CMOS technology in bulk silicon will operate as well as the ICs of "generation-$n + 1$" of CMOS-on-bulk-silicon. This also implies that if a roadblock occurs in some upcoming generation of CMOS on bulk silicon, by having the capability to produce these ICs in SOI, it will be possible to fabricate ICs that allow that roadblock to be exceeded for at least one generation (if not more).

2. **Reduced Power-Consumption.** If CMOS-ICs in bulk-silicon and CMOS ICs in SOI are operated to give the same speed-performance, there is a 35-70% reduction in power-consumption in the CMOS-on-SOI ICs. This advantage provides CMOS-on-SOI a huge competitive edge for low-voltage/low-power applications operated with a one-battery supply. Figure 11-6 depicts this performance-versus-power tradeoff issue.

3. **Reduced Junction-Capacitance.** One reason for the increased performance of CMOS-on-SOI is that the junction-capacitance of the MOSFETs is significantly reduced. For example, a depletion-capacitance exists at the drain-substrate junction, and its value is a function of the

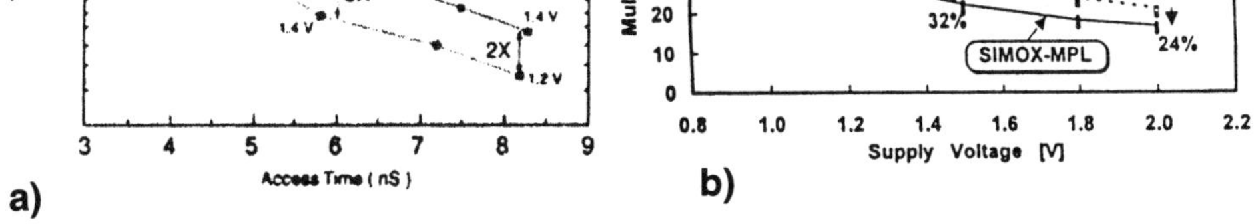

Fig. 11-6 a) Power drain and access time for bulk and SOI CMOS for drive voltages ranging from 1.2 V; b) Comparison of 48-bit-multiplier performance between bulk and SOI using a 0.25-μm CMOS technology.

substrate doping-concentration. The higher the substrate doping-concentration, the higher the capacitance. Modern submicron circuits increase the substrate doping-concentrations with each new technology generation, which contributes to an increased junction-capacitance. However, in SOI circuits, the maximum capacitance between the junctions and the substrate is the capacitance of the buried-insulator. This capacitance is proportional to the dielectric constant of the capacitance material. Silicon dioxide (which is used as the buried insulator), has a dielectric constant k_{SiO_2} = 3.9 which is three times smaller than that of silicon k_{Si} = 11.7 Therefore, a source/drain junction located on a buried-oxide gives a parasitic-capacitance approximately three times smaller than that of a bulk-junction. Since the SOI buried-insulator-thickness does not scale with device scaling, parasitic-capacitances do not necessarily increase as SOI technology progresses.

As a result of the above phenomena, and depending on the circuit layout, the total capacitance of a CMOS SOI-IC can be reduced by 15-50% compared with the same IC fabricated in bulk-silicon. In fully-depleted SOI (FD-SOI) MOSFETs (described in Sect. 11.5.1), the source/drain-regions butt-up against buried oxide (BOX), and thus no *pn*-junction exists at that device-region boundary (and hence there is no depletion-region that needs to charged and discharged there). Even in partially-depleted SOI devices (PD-SOI), the depletion-region of the junction of the source/drain-regions extends to the BOX, and does not increase in volume beyond that point once this depth is reached. Thus, even in PD-SOIs the maximum junction-capacitance of the source-to-body, C_{jSB}, and drain-to-body, C_{jDB}, is limited to a smaller value than that existing a comparably-sized MOSFET in bulk-silicon. The wiring-capacitance (i.e., associated with interconnect lines running over the field regions of an IC) is also slightly reduced by the presence of the BOX in SOI wafers.

4. **Latchup Elimination.** As noted in Sect. 11.1.2, the phenomenon of latchup is eliminated in CMOS-on-SOI ICs and the techniques needed to suppress latchup in bulk CMOS are unnecessary. This also means that the use of epitaxial-wafers with a heavily-doped-substrate to control latchup in some applications is no longer needed.

5. **Reduction of Interwell-Isolation-Spacing.** Due to the elimination of the possibility of latchup, and the ability to retain adequately low intra-well and inter-well leakage-currents due to the dielectric isolation, it is possible to reduce the isolation-spacing design rules, compared to those needed in

CMOS ICs in bulk-silicon (especially the n^+-p^+ spacing). In addition, the contact-overlap-spacing can be smaller. Thus, the packing density of the CMOS SOI-IC is increased, allowing circuits with greater functional-density to be fabricated on the same area of silicon. There is also no need for high-energy implantation of well-dopants.

6. **Reduced Soft-Error-Sensitivity.** SOI reduces the substrate-collection-volume for alpha-particle- and cosmic-ray-generated electron-hole tracks, hence lowering soft-error-sensitivity. In particular, SOI- technology for ECL-bipolar/CMOS ICs is preferred because of the high sensitivity of the ECL peripheral-circuitry to radiation.[8]

7. **Improved MOSFET Turn-Off Characteristic (Steeper Subthreshold-Slope).** The subthreshold slope, S_t, of MOSFETs built on SOI-wafers is much steeper than that of MOSFETs in bulk-Si-wafers. Subthreshold-slopes as small as 62-mV/decade have been observed in SOI MOSFETs.[9] The small value of S_t implies that an SOI-MOSFET can be turned-off with a smaller gate-voltage swing. This allows faster switching in MOSFETs, along with the use of lower threshold-voltages. A lower threshold-voltage allows lower power-supply-voltages to be used, and an increase in the drive-currents of SOI-MOSFETs compared to MOSFETs built in bulk-silicon.

8. **Reduced Leakage-Current Arising from Carrier-Generation in Junction Depletion Regions.** The ability to make the source/drain junction-areas of SOI-MOSFETs smaller by approximately two decades, also decreases the volume of the depletion-regions of these junctions. Thus, less leakage - current arising from the generation of carriers in these regions is produced, which reduces the standby leakage-current of SOI-ICs to a comparable degree. The negligible leakage current makes SOI ideal for low-power ICs.

9. **Compatibility with Bulk-Si CMOS Process Tools & Processes.** The SOI process-technology is compatible with the conventional process-technology used to fabricate CMOS ICs in bulk-silicon. Thus, the tools and facilities used to fabricate CMOS-on-bulk-silicon can also be used to fabricate CMOS-on-SOI. The SOI process-sequence is even somewhat simpler than that used to fabricate CMOS on bulk-silicon. That is, there is no need for the well-formation process in SOI-ICs, nor any need for channel-stop-implants to suppress lateral-surface-leakage between devices due to field-inversion.

10. **Advantages for System-On-Chip (SOC).** A high-resistivity substrate under the BOX can be used, which can significantly reduce the device

capacitance and the loss of passive components, such as inductors and transmission-lines, commonly used in analog- and rf-devices.[10] Therefore, SOI offers advantages over bulk-Si to realize a system-on-a-chip by integrating high-performance digital and analog/rf circuits on the same chip.

11. **Increased Drain-Current.** The drain current of SOI-MOSFETs, I_D, is also increased due to the significantly lower doping that can be used in the channel. The ability to tolerate lower-channel-doping occurs because of the natural punchthrough suppression arising from the backgating effects of the insulator interface.

12. **Improved Circuit Reliability Due to Elimination of Junction-Spiking.** When contact is made to a MOSFET in bulk-CMOS, unwanted reactions between the metal or the silicide can take place, leading to the problem of "junction-spiking" (and the attendant rise in uncontrolled leakage currents). While barrier materials normally control this problem, the potential for its occurrence is still there in bulk-CMOS. If MOSFETs are fabricated in thin SOI material, the source and drain junctions extend to the buried insulator. In that case, no leakage will be produced if some uncontrolled metal-silicon reaction occurs.

11.3.1 The Path of Using SOI May Allow Extension of the ITRS Roadmap

Many of the challenges for CMOS transistors in bulk-silicon discussed in Sect. 11.2 are relaxed or removed when SOI-wafers are used instead of bulk-silicon wafers, and SOI technologies offer new options for transistor design. With the 20-50% increase in transistor speed that can be achieved by switching to SOI, the performance of ICs can be increased without the cost of a shift to a higher lithography resolution. SOI "thin-body" CMOS transistors provide excellent control for the off-current (leakage) characteristics that limit the endurance of battery-operated computer and communication devices. Furthermore, soft-error-immunity and radiation-hardness of SOI devices is seen as a key requirement for expansion of high-data-rate network-servers that form the core of Internet and other global data-transfer-links.

In addition, SOI-wafers are opening a path for the extension of planar CMOS transistor design by perhaps a decade beyond the "end-of-roadmap" limits foreseen in the ITRS-2001 study for 50-nm gates on bulk-silicon. Operational SOI transistors have been studied with gate lengths as small as 25-nm and Si channel thicknesses of 1-nm.

TABLE 11-2 COMPARISON OF KEY ISSUES FOR SUB-100-nm CMOS FOR BULK-SILICON AND FULLY-DEPLETED SOI MOSFETS

Critical Issue	Bulk silicon	SOI
Junction depth	RTP/(LTP)	Set by Si-SOI film
Dopant activation	Thermal budget on vertical and lateral diffusion	Lateral diffusion limit
Junction abruptness	Critical vertical and lateral abruptness limits	Carrier injection still requires abrupt lateral junction in thin-body
Contacts	Extended S/D	Extended S/D
Off-current	Requires halo	Dual gate
Threshold control	SSRD (super-steep retrograde doped channels) or SiGe gate	SiGe gate
Well doping	High-energy implant	No well

To begin, it is easier to control the junction-depths of the source/drain extensions in SOI devices. In bulk-silicon, obtaining shallow-junctions requires careful control of thermal budgets, with the use of rapid-thermal-annealing-"spikes" or laser-annealing to limit both the vertical- and lateral-dopant-diffusion, while obtaining high degrees of dopant-activation and adequate defect annealing. With fully-depleted, thin SOI (FD-SOI) MOSFETs (discussed in Sect. 11.5.2), the source/drain-extension junction-depth is determined by the Si-SOI-layer thickness. This means that the thermal-budget-constraints on vertical-junction-depth are essentially removed for fully-depleted-SOI CMOS devices. Second, the ability to activate dopants in the source/drain-regions of CMOS in bulk-silicon is limited by the thermal-budget on vertical- and lateral-diffusion effects, while in FD-SOI MOSFETs only the lateral-diffusion sets a limit on this thermal-budget (a potentially less difficult problem to solve). Third, in CMOS-on bulk-silicon both the vertical- and lateral-abruptness of the source/drain-extension doping-profile needs to be controlled, whereas in FD-SOI, only the abruptness of the lateral-doping-profile of the source/drain extension-region needs to be controlled. Fourth, the well-doping in CMOS-on-bulk-silicon needs to be carried out with high-energy ion-implanters, but in CMOS-on-SOI, no wells even need to be formed. Finally, SOI-technologies can be developed to include new device structures, such as dual-gate and high-channel-mobility (with

Si:Ge-CMOS), that provide new tools for transistor designers to increase signal speed while decreasing off-current leakage-levels. The transistor characteristics and process tradeoffs for bulk-silicon- and SOI-wafers are summarized in Table 11-2. The process challenges shift to controlling lateral-diffusion and obtaining abrupt-junction-interfaces.

One of the critical processes that will be needed to allow thin-SOI-wafers to serve as the substrates for sub-50-nm CMOS is the ability to fabricate thin-SOI wafers with high-levels of control over the SOI-Si device-layer-thickness, electrical-properties, and interface-characteristics. The Smart-Cut® and Nanocleave® processes described in Sects. 11.6.3.2 and 11.6.3.3, respectively, are two approaches that may allow these goals to be met.

The timeline for the use of the various types of "thin" SOI MOSFETs can be linked to the projections made by the ITRS-2001 (Fig. 11-7). The thickness of the Si-to-SOI surface layer (T_{Si}) will shrink according to the approximate design rules of $T_{Si} \sim 2 \times X_{j(S/Dcontact\ region)}$ for partially depleted SOI MOSFETs (see Fig. 11-2a, which shows what is meant by the terms "*S/D contact-region*" and "*S/D extension region*"), and $T_{Si} \sim 1 \times X_{j(S/Dextension)}$. Up to the year 2000 timeframe, production SOI-ICs were being manufactured using PD-SOI.

For the time between 2001 to 2005 (i.e., for MOSFETs with gate lengths of 0.15-μm down to 0.07-μm), the Si-SOI layer thicknesses being used is expected to cause the type of SOI FETs to be a mix of partially-depleted (PD) CMOS for mainstream logic and ASIC devices, and fully-depleted (FD) SOI for high-performance processor and server/router devices.

For CMOS devices with gate-lengths in the range of 50-nm (beginning in about 2006), the multitude of process and performance issues described in the ITRS study will preclude the use of planar-CMOS on bulk-Si as well as in partially-depleted SOI. By migrating mainstream transistor structures to fully-depleted SOI, it may also be possible to extend CMOS circuit technologies a full decade beyond the "end-of-roadmap" for planar-CMOS on bulk-silicon. "Thin-body" SOI, with gates sizes at 20-nm on Si-SOI layers of ~10-nm are now at the threshold of research devices. If scaling continues at the historical rate of a 0.7 shrink every three years, these devices would be in production by about 2015.

11.4 HISTORY OF SILICON-ON-INSULATOR (SOI) TECHNOLOGY

Silicon-on-insulator technology has been available since the late 1960's, and has been used since that time for military integrated-circuit applications. However, it has taken more than 30 years for the first mainstream commercial SOI devices

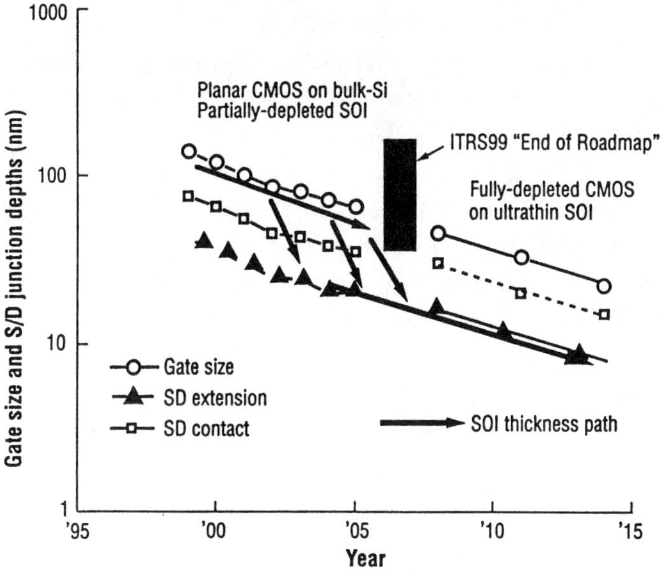

Fig. 11-7 Timeline of Si-SOI layer thickness linked to junction depth and gate size predictions from ITRS-2001.

to be offered. As described in Sect. 11.3, the advantages of SOI technology are well known. First, circuits fabricated in SOI wafers have reduced parasitic-capacitance when compared to bulk or epi wafers. Less capacitance translates to lower power or higher speed. Second, SOI devices have improved radiation-induced single-event-upset (SEU) immunity, and thus they are useful for space applications. Even large memory arrays that are used in terrestrial applications are subject to SEU and may require SOI circuits. Third, SOI devices are completely free of latchup. Finally, the fabrication process in SOI wafers can be simplified by reducing the number of masks by as much as 30%. However, SOI wafers are more expensive to fabricate and may contain more defects than bulk or epi-silicon. As the requirements for improved speed, reduced power, and less expensive processing have become more important, a keen interest in SOI technology has been re-kindled among mainstream device suppliers.[41]

Currently there are a number of useful approaches for making SOI wafers. The first group includes the "legacy" SOI technologies of *dielectric isolation* (see Vol. 2, Chap. 2, for more information on the dielectric-isolation-technology) and *silicon-on-sapphire* (*SOS*) that were developed in the 1960's, and that even today, still find limited application. (As we will note, an improved SOS technology is currently being used to fabricate high-performance rf-devices.) The more modern SOI-technologies are based on either a buried-insulator-layer created by ion-implantation technology (e.g., *Separation by IMplanted OXygen* [*SIMOX*]), or by a *wafer bonding* (*WB*) technique. (The insulating substrate for SOS wafers is sapphire [i.e., very pure single-crystal, aluminum oxide, Al_2O_3]. On the other hand, the insulator for the SIMOX and wafer-bonding technologies is silicon dioxide.) Various (and to-date, unsuccessful) attempts have been made to grow single-crystal silicon (or large-grain polycrystalline-silicon) directly on amorphous insulators.[42] When devices were first built on SOI-wafers, small islands of silicon were formed (typically by etch techniques) on top of the insulator. In modern SOI ICs, either LOCOS or STI is used to complete the dielectric isolation. Individual devices are then fashioned in the islands and these devices are then interconnected in a conventional manner. SOI-technology has been the subject of intense research and development in the late 1990s. A number of symposia that address SOI technology are held on a regular basis.[44,45] For more information on the evolution of SOI circuits, see Sect. 11.8.

11.5 SILICON-ON-INSULATOR DEVICES

The MOSFETs built using SOI-substrates are categorized in a number of ways. The first such grouping differentiates between so-called *thick-SOI-MOSFETs* (built on SOI-substrates whose Si layer is thicker than 1 micron - i.e., > 1.0-μm), and *thin-SOI-MOSFETs* (in which this layer is less than 1 micron thick – i.e., < 1.0-μm). The applications for "thick" SOI include power-IC switches, rf bipolar-signal-processors, and a wide variety of MEMS and optoelectronic devices. Thick-SOI devices were also historically used for a variety of military applications (e.g., radiation-hardened ICs) and for such commercial-applications as high-voltage and high-power devices. The principal SOI-material being presently used for VLSI and ULSI CMOS signal, logic, and memory applications is "thin" SOI (50-1000-nm thick [0.005-1-μm]. The "thick" SOI market segment, at only $50 million/year is a very small piece of the $110 billion/year commercial, mainstream CMOS IC market (and is not likely to grow significantly in the future). Hence, we will not further deal with such devices, but instead will focus on "thin" SOI devices in the rest of this section.

Within the category of "thin" SOI devices, there are a number of additional subcategories, namely *partially-depleted SOI-MOSFETs* and *fully-depleted SOI-MOSFETs*. The thickness of the silicon-device-layer of the SOI-structure can be chosen so that the SOI-MOSFETs will operate in one or the other of these modes. Each of these SOI device types will be discussed separately in more detail. Note once again that one of the ways that MOSFETs fabricated in SOI differ from the MOSFETs fabricated in a bulk (i.e., non-SOI) substrate is that the channel region (or body) of the SOI-MOSFET is not automatically connected to any external terminal (i.e., intrinsically the body of the SOI MOSFET is electrically floating). This aspect of SOI-MOSFETs (a floating body), leads to so-called *floating-body-effects,* which can be detrimental to device performance. Floating-body effects in SOI-MOSFETs will be described in more detail later.

11.5.1 Partially-Depleted Thin-SOI-MOSFETs (PD-SOI)

Partially-depleted, thin-SOI-MOSFETs (PD-SOIs) are given this name because when the maximum gate-bias is applied to the MOSFET, the depth of the resultant depletion-region in the channel of such devices is shallower than the thickness of the silicon layer (Fig. 11-8). Note that "thick" SOI MOSFETs are also operated as partially-depleted MOSFETs.

As mentioned earlier, the body of SOI-MOSFETs is intrinsically floating. That is, unless a contact to some circuit potential is explicitly provided, the potential of the SOI-body is floating (Fig. 11-9a). However, if the SOI-MOSFET body is grounded (i.e., by independent body contacts or by body-source ties), partially depleted SOI MOSFETs behave very much like bulk-Si MOSFETs, and most of the standard $I_D(V_{GS}, V_{DS})$ equations and design concepts apply. If body-contacts are not supplied, so-called *floating-body effects* arise, leading to potentially detrimental consequences (including the introduction of a *kink* in the dc I-V characteristics, a reduction in breakdown-voltage, and the causing of hysteresis and instability during dynamic operations). The issue of floating-body

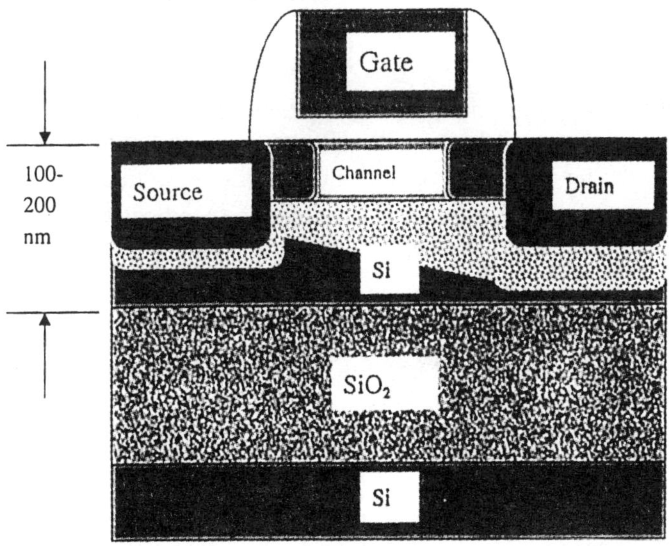

Fig. 11-8 Partially-depleted transistor in a Si layer over an isolation layer.

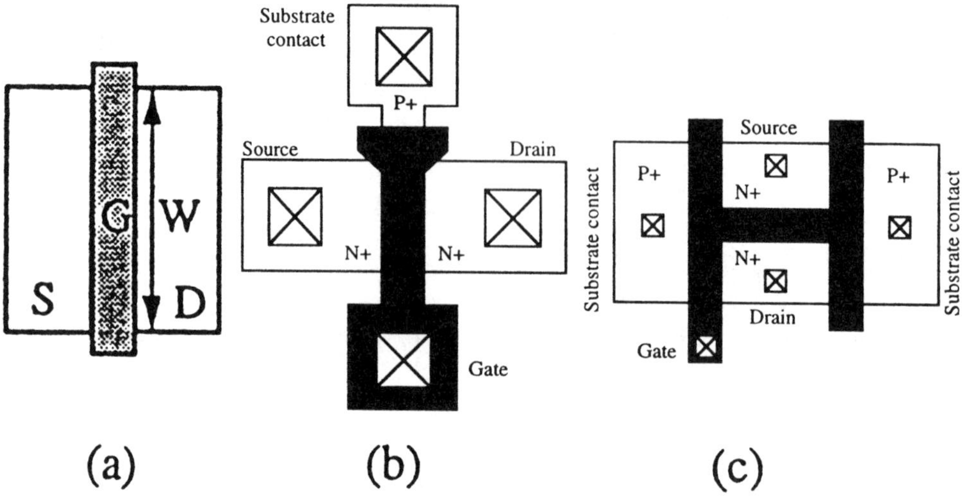

Fig. 11-9 a) Top-down view of an SOI-MOSFET in which no provision is made for contacting the body. Unless a contact to some circuit potential is explicitly provided, the potential of the SOI-body is floating; b) MOSFET with body-contact made on one side of the MOSFET channel; c) An "H-gate" SOI-MOSFET (i.e., a body-contact is made on each side of the MOSFET channel).

effects in partially depleted SOI MOSFETs is described in further in Sect. 11.9.

Note that body- and substrate-contacts were routinely used when "thick" SOI devices were fabricated, as shown in Fig. 11-9b. To avoid the possible device-performance-degradation that floating-body-effects could cause, early partially-depleted, thin-SOI-MOSFETs also used either separate body-contacts or source-body ties, again as shown in Fig. 11-9b. However, since the body-region of the MOSFET beneath the channel can exhibit a high resistance, a single body-contact tied to ground may not be adequate. In some applications a body contact is made on both sides of the MOSFET channel, forming a so-called "H-gate" (Fig. 11-9c).[11] Figure 11-10 shows how a body-contact is made to an SOI-NMOSFET both from a top view, and from a cross-sectional view. Figure 11-11 shows the layout of a DRAM sense-amplifier-circuit fabricated in SOI, in which a body-to-source-tie is employed.

While body-contacts (and body-source-ties) can be used to avoid the problems associated with the floating-body-effect, these structures consume valuable silicon-circuit-area. That is, the global use of body-contacts on every SOI MOSFET in a circuit may severely degrade the density (and hence also the

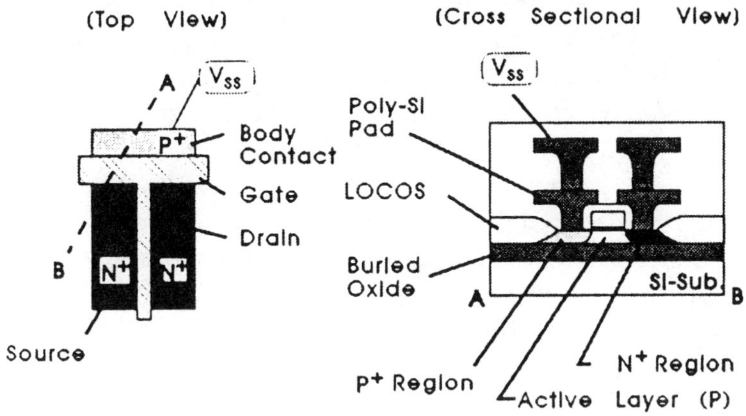

Fig. 11-10 A top-down and diagonal cross-sectional view of an SOI-NMOSFET with a body-contact made on one side of the MOSFET channel.

performance), of a large-scale IC design. As a result, in more recent SOI ICs built using partially-depleted thin SOI MOSFETs, attempts were made to find ways to eliminate the need to use such body-contacts. As shall be seen in Sect. 11.9, it has been demonstrated that in some circuit applications, such body-contacts may not need to used. That is, the device-performance-degradation caused by floating-body-effects may be small enough to allow the elimination of body-contacts from such SOI circuits.

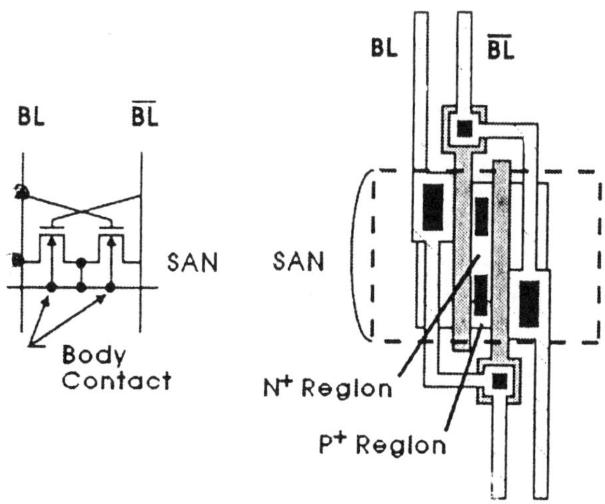

Fig. 11-11 Layout of a SOI-NMOSFET DRAM sense amplifier incorporating a body contact.

When CMOS circuits are built with partially-depleted thin-SOI-MOSFETs (and are powered with the same voltages), switching-speeds are observed to increase by 20-30% compared to equivalent circuits in bulk-silicon devices. Alternatively, partially-depleted SOI devices can be operated at lower voltages and lower battery-power-drains with a significantly-smaller speed-penalty than bulk-silicon devices. As noted earlier, since partially-depleted SOI-MOSFETs operate electrically in essentially the same manner as bulk MOSFETs, an immediately useful advantage is that circuit designs from bulk-silicon devices can be imported into the SOI environment with only a modest investment in circuit redesign. As of 2000, the SOI circuits being offered commercially were largely based on partially-depleted SOI-MOSFETs.

11.5.2 Fully-Depleted Thin SOI MOSFETs (FD-SOIs)

In *fully-depleted, thin-SOI-MOSFETs* (FD-SOIs) the depth of the depletion-region in the channel of the transistor extends entirely through the thickness of the surface-silicon-layer of the SOI-structure (Fig. 11-13b), even when there is no-bias applied to the gate. (In addition, the junction-depth formed in the source/drain-regions in fully-depleted SOI-structures is also as deep as the Si-surface-layer-thickness. Hence, there is no *pn*-junction at the [source/drain]/BOX interface, and thus there is no depletion-region at that interface to contribute to the source/body- and drain/body-capacitance.) This also implies that there is less overall junction-surface area than in a CMOS-structure formed in bulk-silicon (which results in reduced leakage-currents and junction-capacitances). In the FD-SOI case, the depletion-charge is thus constant and cannot extend further into the body as the gate-bias is increased. As a result, the application of a gate-bias results in more inversion-layer-charge for the same gate-bias as in a bulk-Si MOSFET. This has several beneficial effects, including: **1**) drain current is increased; and **2**) subthreshold-slope becomes steeper (Fig. 11-12). However, the $I_D(V_{GS}, V_{DS})$ equations used to model bulk MOSFET and PD-SOI MOSFETS are no longer valid, and new I-V relationships need to be used for FD-SOIs. Although circuit-designs need to take full account of the special conditions of operation in an SOI layer, the highest gains in circuit-speed, reduced-power requirements, and the highest level of soft-error-immunity are realized with fully-depleted SOI-devices. That is, FD-SOI devices exhibit the following advantages over bulk and partially-depleted SOI-MOSFETs: **1**) sharper subthreshold slope;[12] **2**) higher drive-current; **3**) higher-transconductance; **4**) reduced parasitic-capacitance; and **5**) reduced short-channel-effects.[13] The latter benefit is gained because some of the charge that would otherwise be needed to support the

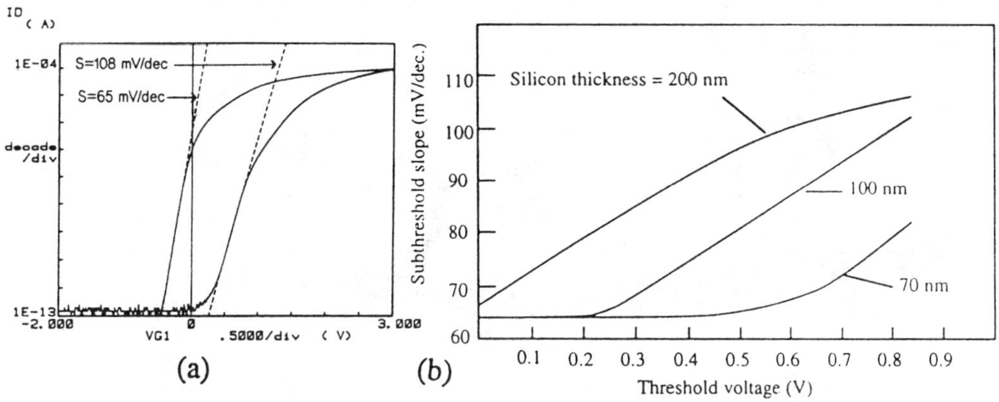

Fig. 11-12 a) Log-plot of drain-current I_D versus gate-voltage in SOI-MOSFETs made with two different silicon film thicknesses: 90-nm (left curve) and 450-nm (right curve). The inverse-subthreshold-slope, S_t (in millivolts per decade) is indicated. V_{DS} = 1 V; b) S_t as a function of threshold voltage for various SOI film thicknesses (*n*-channel device).

channel depletion-region-charge is instead produced by charge in the substrate under the BOX. However, as of 2001, fully-depleted SOI MOSFETs were not yet being widely implemented in commercially available ICs.

The reasons fully-depleted SOI MOSFETs were not being used in commercial production by 2001 is that these devices must meet the requirement that the silicon-film-thickness must always remain well below the depletion-region-width in the channel, during all modes of the MOSFET operation (i.e., from when the device is OFF through inversion). This dictates the use of a low threshold-voltage-value, and the ability to keep the Si-SOI surface-layer-thickness well below the depletion-region-width under conditions of thickness variation. That is, it must be possible to control the Si-surface-layer-thickness to within 5-10-nm.[14] This imposes severe limitations on the manufacturability of the device. (Use of partially-depleted SOI MOSFETs in an SOI-circuit, on the other hand, alleviates the constraint on V_T and its sensitivity, allowing the channel-doping-profile to be tailored to meet any desired V_T value, thus easing the manufacturing problem.) Nevertheless, by the end of 2001 this situation was starting to change. Intel announced a 50-nm FD-SOI structure, and said that they would use FD-SOI MOSFETs to build ICs from 2005 and beyond.[66] In addition to FD-SOI, the Intel MOSFETs incorporate elevated source/drains (see Sect. 5.3.2).

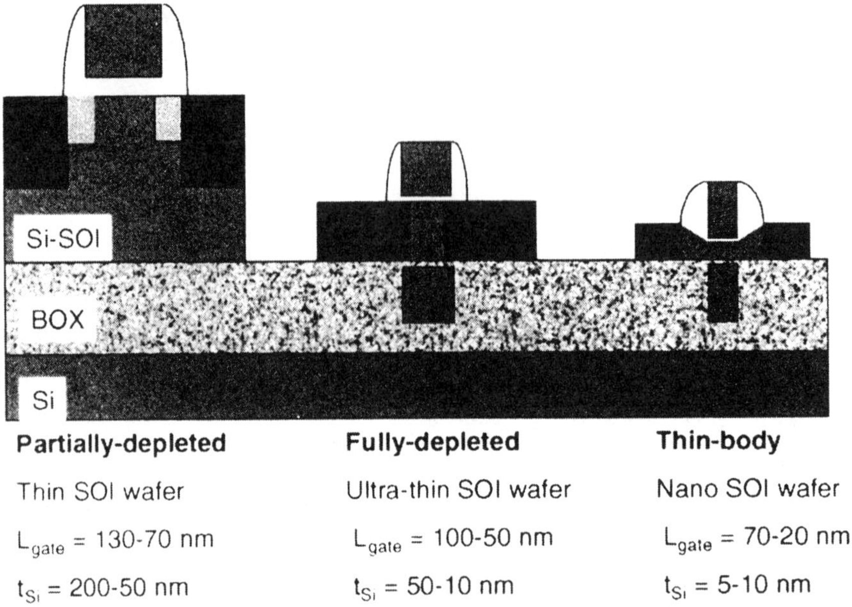

Fig. 11-13 Schematics of partially-, fully-depleted, and "thin-body" (double gate) SOI MOSFETs.

In FD-SOIs some of the gate-charge is balanced by charge in the depletion-region of the channel, but additional charge will arise in the substrate beneath the BOX layer. This induced charge in the substrate will have the effect of trying to establish an inversion layer *along the interface between the Si surface layer and the BOX* (i.e., the substrate beneath the BOX acts as a *back-gate*). As a result, a coupling exists between the potentials on the top-gate and the back-gate, and the electrical characteristics of the channel along the back-gate vary with bias applied to the top-gate. For example, leakage along the back-gate-channel will be caused by the application of a bias on the top-gate. In fact, when the Si-surface-layer of an FD-SOI becomes too thin, the top-gate eventually can no adequately longer turn-off the FD-SOI MOSFET. In such cases it may be necessary to fabricate these FD-SOI MOSFETs with a dual-gate structure and a thinned body-region (as discussed in Sect. 11.5.3.2 and as shown schematically in Fig. 11-13c). Such dual-gate FD-SOI-devices offer the prospect of excellent control of the off-current of small channel-length SOI-MOSFETs.

A dual-gate FD-SOI has both a top-gate and bottom-gate (Fig. 11-13c). The channel is sufficiently thin so as to be fully-depleted and an inversion-layer is established at both the top and bottom gates. To minimize short-channel-effects the channel-thickness must be considerably smaller than the gate-length.[48] For gate lengths smaller than 50-nm, this will require thin silicon layers of 20-nm or less. The potential transconductance of a dual-gate device is approximately twice that of a bulk-device because of the presence of two gates. Also, the subthreshold slope, S_t, is potentially much improved over a conventional MOS device. Several theoretical studies have shown that the silicon thickness for a single-gate FD-SOI device must be about 1/2 that of a dual-gate device for equivalent short-channel effects.[48]

However, by 2000, dual-gate FD-SOIs were still in their infancy, and were far from being ready for consideration as devices in commercial ICs. That is, there are major problems in the practical manufacturing of dual-gate FD-SOIs. In a horizontal configuration, a gate must be fabricated below a thin silicon layer and accurate alignment of the top-gate over the bottom-gate must be achieved. All of these steps are very difficult.

11.5.3 Process Technologies Used to Fabricate Fully-Depleted SOI Devices

As noted earlier, one reason it is difficult to manufacture FD-SOI is that it must be possible to fabricate very thin Si-SOI surface-layers and to control their thickness to within 5-10-nm. Here some of the process sequences that have been developed to try to meet these requirements are listed.

11.5.3.1 FD-SOI Fabrication by Thinning the Si-SOI Surface Layer by Thermal Oxidation: An early process for fabricating FD-SOI devices was reported in 1992.[15] In this work a 220-nm-thick Si-SOI layer formed with a SIMOX process was the starting substrate. This layer was thinned to a 40-nm-thickness by carrying out a blanket thermal-oxidation of the Si-SOI layer and then etching off the oxide in an HF etch step. A LOCOS isolation-structure was used to provide lateral isolation. After forming the gate, a titanium salicide process was used to form $TiSi_2$ on the source-drain regions and on the polysilicon gate. The silicon in the source-drain regions was completely converted to $TiSi_2$ so that the silicide layer lies on top of the buried-oxide. A silicidation that fully consumes the SOI layer in the source/drain regions was chosen for ease of control. Figure 11-14 shows a cross-section of the finished device.

11.5.3.2 FD-SOI Fabrication by a LOCOS-Recess of the MOSFET Channel-Region: As the thickness of the Si-SOI layer (T_{Si}) has to become even thinner than the 40-nm-thick Si layer used in the FD-SOI device of the previous section, the parasitic resistance of the source/drain regions of such devices becomes too large to obtain the required device performance characteristics (even if these regions

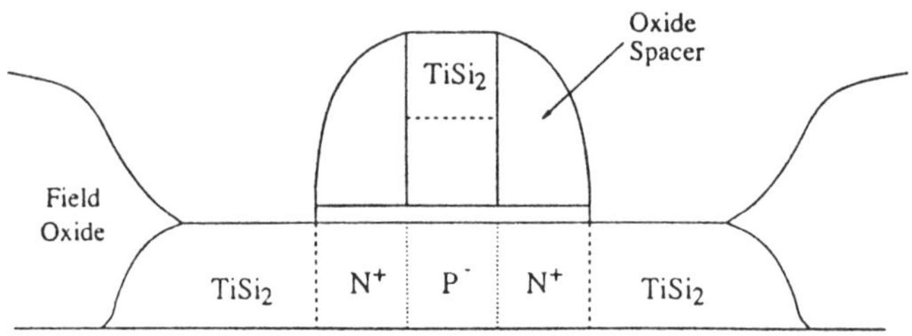

Fig. 11-14 Cross-section of the ultrathin SOI transistor with silicided source/drain and gate.

are silicided). An approach which allows the S/D regions to remain thicker while reducing the thickness of the body has thus been developed. In this method, the body region of the FD-SOI device is thinned by using LOCOS process, as shown in Fig. 11-15.[16] In one example, a SOI wafer with a T_{Si} of 195-nm was the starting material. A LOCOS process was used to thin the body region of the FD-SOI device to a 110-nm thickness. MESA active islands were formed by selectively etching back the Si-SOI to the BOX layer. These MESA islands were then oxidized to provide the lateral isolation.[17] Figure 11-16 shows an SEM of a completely fabricated FD-SOI device.

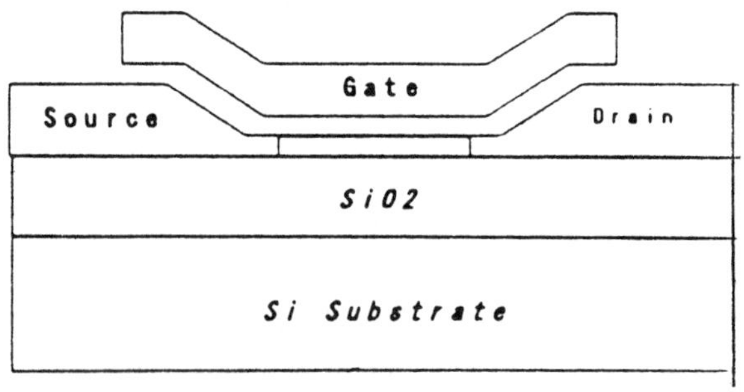

Fig. 11-15 Cross-section of a LOCOS-thinned "thin-body" SOI MOSFET.

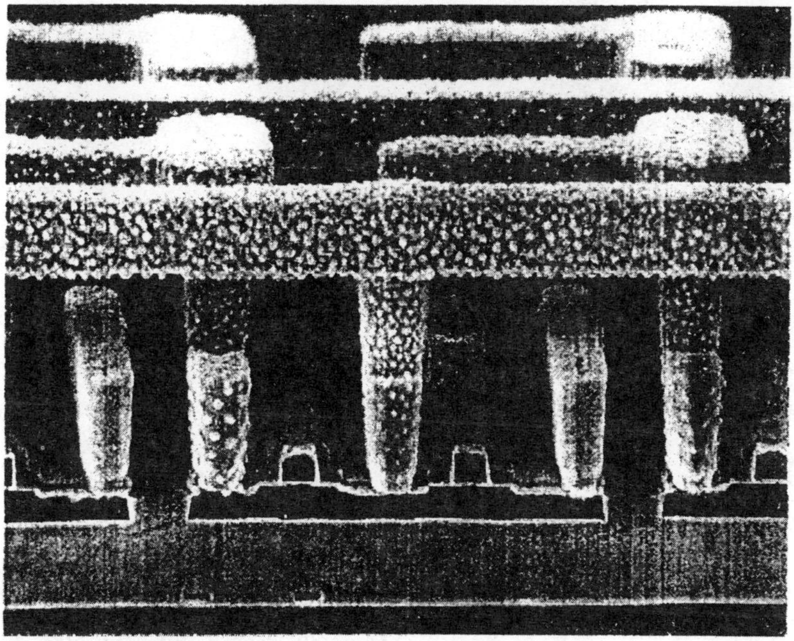

Fig. 11-16 SEM of completely fabricated SOI-MOSFET.

11.6 FABRICATION TECHNOLOGIES USED TO PRODUCE STARTING SOI-WAFERS

There a number of methods for fabricating starting SOI-wafers. Here those being currently used to make SOI wafers will be discussed. The annual production of SOI wafers in 1998 was only 10-million square-inches of silicon wafer. However, it is predicted that the total SOI-wafer-demand will grow by a compound annual growth rate (CAGR) of 55% between 1999 and 2005. In 2005, this should make the production of SOI wafers reach 187-million square-inches. The SOI merchant-wafer market registered $100 in revenue in 1999 (this did not include the captive SOI wafer production by such companies as IBM).

11.6.1 Silicon-on-Sapphire (SOS)

The work on silicon-on-sapphire (SOS) dates back to the 1960s when there was interest in this technology for its well-known immunity to radiation damage. Sapphire is insulating, and so circuits built in SOS substrates have high speed and excellent device-isolation. That is, unlike in other SOI approaches (where a buried oxide-layer is used to provide vertical isolation), the entire sapphire substrate is an insulating material. This reduces microwave losses and makes SOS an ideal candidate for the integration of rf-circuits for wireless communications. Furthermore, the high thermal-conductivity of sapphire facilitates thermal dissipation through the substrate, which alleviates the self-heating effects of the other SOI technologies as described in Sect. 11.9. Device isolation can be easily achieved through simply removing the conducting silicon between the transistors by etching down to the insulating sapphire-substrate. SOS circuits were historically used for radiation-hardened applications. There are several excellent review articles on this subject and the reader is referred to them for additional details.[18,19]

Since SOS technology is limited to fabrication of MOS devices, a (100) orientation of the crystalline-silicon film is desirable. This orientation has been hetero-epitaxially grown on several different sapphire (i.e., crystalline alumina) orientations, specifically (0112), (1012), and (1102). The lattice-parameter mismatch between the Si-epi layer and the sapphire-substrate is sufficiently large to cause the generation of misfit dislocations, stacking faults, and edge dislocations in the Si-epi layer. The defect-concentration is heaviest near the substrate/epi interface and decreases as the film grows.[20] Figure 11-17 shows the stacking fault density as a function of epitaxial-film-thickness in a thick SOS film. However, current requirements for fabrication employ so-called *fully-depleted MOSFETs* (which are defined in Sect. 11.5) and these must be built in a Si-epi films only 0.1–0.3-μm thick. These films can be highly faulted, and require additional processing to improve their crystalline quality.

The accepted way to improve the film quality is the technique of amorphization (by implanting a high dose of Si into the deposited silicon

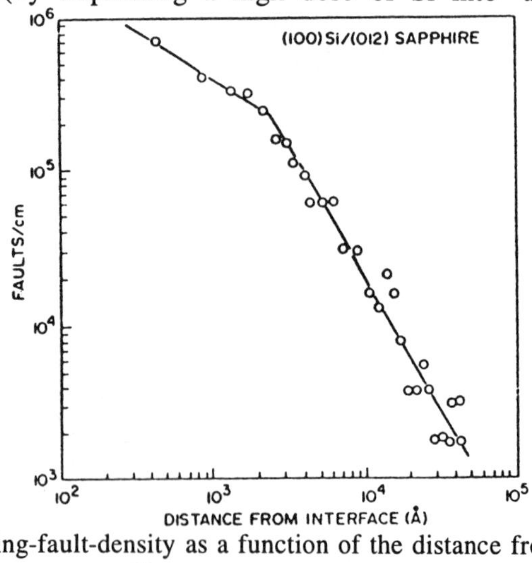

Fig. 11-17 Stacking-fault-density as a function of the distance from the silicon/sapphire interface in an MOS structure.[21] Reprinted by permission of the American Physical Soc.

epitaxial layer) followed by *solid-phase-epitaxial regrowth* (*SPE*).[22] In *thick SOS films* only a single amorphization is required. That is, an ion implantation step causes the region near the interface to become amorphized. Then the undamaged single-crystal-Si near the surface is used as seed material for the epitaxial regrowth of the amorphous layer. Since the regrowth is primarily along the <100> direction, stacking faults and twins do not propagate, leaving a controllable number of dislocations in the film.

For *thinner SOS films* a *double SPE* (*DSPE*) process may be required. In this approach, the Si-region near the sapphire-interface is first amorphized by a high-energy ion-implantation step, and the surface-regions are used as a seed for regrowth.[23] This is followed by a lower-energy ion-implantation step that then amorphizes the surface-region of the Si. The previously re-crystallized bottom layer is now utilized as the seed to re-grow the surface-amorphous-layer. Devices fabricated in such material show higher surface mobilities and lower junction leakage than either untreated SOS, or material obtained from a single SPE-process.[24] Peregrine Semiconductor, which markets high-performance SOS-rf devices for the commercial market, builds these devices using standard ULSI processing techniques on 200-mm SOS wafers. SOS films may require additional characterization over that used for conventional epi-films. These methods are comprehensively reviewed by Cullen *et al*,[25] and Munteanu *et al.*[26]

11.6.2 SOI by Separation by Implanted Oxygen (SIMOX)

A widely used approach to manufacture SOI wafers is that of *Separation by IMplanted Oxygen*, or SIMOX. This process was first proposed by Izumi of NTT in 1978 (and who also coined the acronym SIMOX).[27] The SIMOX technology

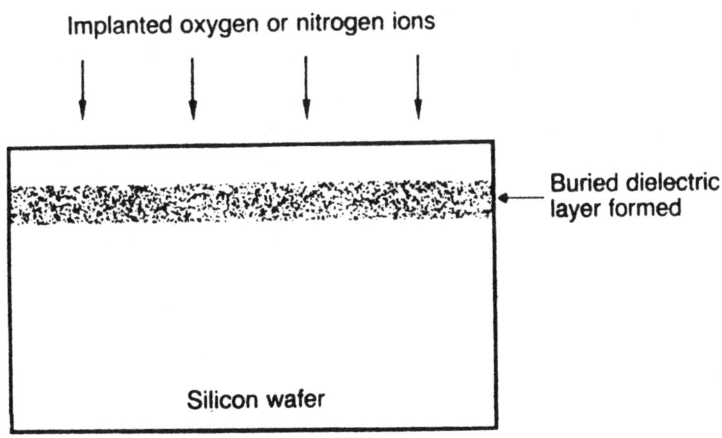

Fig. 11-18 Implanted oxygen ion process for the formation of a SIMOX wafer.

is based on the ion-implanting a high-dose of oxygen ions (O^+) into a silicon wafer (~$1.4x10^{18}$ O^+ ions/cm^2), such that the peak concentration lies about a half micron beneath the silicon surface. After implantation these wafers are subjected to a high-temperature-anneal which forms a continuous stoichiometric subsurface-layer of silicon-dioxide (Fig. 11-18). This *buried-oxide* (or BOX) layer electrically separates the top and bottom portions of the wafer, and is typically 350-nm thick. The single-crystal-region of silicon above the BOX is typically 250-nm thick. The result is a wafer with a thin surface-silicon-film of high-crystalline-quality, which also has an abrupt interface to a buried layer of stoichiometric-SiO_2.

The SIMOX process has undergone significant evolution since it was first introduced. In early-SIMOX the material was of very-poor quality. That is, the defect-density (dislocations, oxide-precipitates, and polycrystalline-silicon-inclusions) was so high that epitaxial-silicon had to be grown on top of the silicon overlayer in order to obtain a silicon film good enough for device fabrication. In addition, at that time, the implantation of sufficient oxygen to form a buried-oxide-layer in a 2x2 cm^2 area on a silicon-wafer required 24 hours (using a conventional implanter). Thus, most SOI researchers viewed SIMOX as just a laboratory curiosity. Nevertheless, when NTT produced a 1-k SRAM using SIMOX in 1982, the SOI community began to consider SIMOX from a less skeptical perspective.

In 1985, Eaton (now Axcelis) produced the first high-current oxygen-implanter, the NV-200, according to specifications provided by NTT. This machine could implant O^+ ions at a beam-current of 100-mA and at energies up to 200-keV. The availability of such an implanter was a key step toward allowing SIMOX development to go forward.

Before 1985, SIMOX wafers were annealed at 1150°C (which is the highest temperature available in conventional furnaces equipped with quartz tubes). In 1985, it was demonstrated that annealing at 1300°C in a furnace equipped with either a polysilicon or a silicon-carbide tube (and even at higher temperatures - up to 1405°C - in a lamp-annealing system) could dramatically improve the quality of the material, and produce atomically sharp silicon/silicon-dioxide interfaces. Other improvements (to be discussed in more detail later), included the use of multiple-implants, and a lower oxygen-dose. Eventually, by the late 1980s, SIMOX-wafers began to be produced commercially by publicly-traded companies (e.g., Ibis Technology).

As a result of all of this progress, three forms of SIMOX have been developed. The first two forms (*high-dose SIMOX* and *low-dose SIMOX*) use conventional ion-implantation technology to implant a high-concentration of oxygen. A third version of the SIMOX process, which uses a plasma immersion implanter for the oxygen implant, is called *Separation by Plasma IMplantation of OXygen* (SPIMOX). The latter technique is a potentially low-cost process for fabricating SOI substrates, because the oxygen atoms are implanted by *plasma-immersion ion-implantation* (PIII, see Fig. 11-26). The whole wafer is implanted at once and implantation is thus independent of wafer size. This results in high wafer throughput. A possible drawback of this technique is the lack of ion-beam-selection and the fact that this process may cause more contamination in the SOI wafer than would a conventional ion-implanter.[28] High-dose SIMOX was the first technique to be developed, and it was followed more recently by the low-dose SIMOX-process. The high-dose SIMOX and low-dose SIMOX forms are described in more detail here.

Although SIMOX wafers can be obtained from several commercial suppliers, the high-dose SIMOX wafers are still significantly more costly than bulk-Si or even epi-Si wafers.[29] However, the development of the low-dose SIMOX process appears to significantly cut the premium that must be paid for SIMOX wafers. In fact, it is estimated that SIMOX wafers in the future may not cost any more than epitaxial wafers (whereas in 2000 they cost 2-3x the cost of epi wafers). Future cost-reductions in SIMOX wafers will come from such advances as increased beam-current (e.g., 2x that of present implanter-technology). Note that the SIMOX process has only two major steps: **1**) oxygen implantation; and **2**) annealing. Since SIMOX wafers are commercially available from various suppliers, introducing SIMOX wafers into production does not require any major infrastructure changes in a fab. Furthermore, after years of development work, SIMOX-SOI-wafer and device-fabrication processes have achieved sufficiently high-quality and low-cost to be applied commercially. In August 1998, IBM announced it would apply its leading-edge CMOS technology to commercialize SOI-CMOS products based on low-dose-SIMOX SOI.

11.6.2.1 High-Dose SIMOX

Figure 11-18 shows a schematic of the high-dose SIMOX process. The sequence of steps involved in the high-dose SIMOX process is as follows:

1. The starting material is typically a (100) device-quality Si-wafer. The wafer is first subjected to a high dose ($1\text{-}2 \times 10^{18}/cm^2$) oxygen ($O^+$) ion-implantation step at high enough energy (150–300 keV) so that the peak (projected range) of the implant is deep within the silicon (0.3–0.5-μm).

This process is carried out using a dedicated high-current oxygen-ion-implanter, typically with a beam current of 100-mA. For reasons described in the next paragraph, the wafer is maintained at a temperature of ~600°C during the oxygen-implant step.

Note that stoichiometric-silicon-dioxide contains 4.4×10^{22} oxygen atoms/cm^3. Therefore, from such considerations alone, an implantation dose of $4.4 \times 10^{17}/cm^2$ should be sufficient to produce a 100-nm-thick buried-oxide-layer. Unfortunately, due to the statistical nature of ion implantation, the oxygen-profile in silicon does not have a box shape, but rather a Gaussian profile. Thus, these implanted-oxygen atoms will spread out over a wider range than 100-nm, and silicon-dioxide-stoichiometry is not reached. If the wafer was annealed after such a $1\text{-}2 \times 10^{18}/cm^2$ oxygen-dose, oxygen precipitates would be formed at a depth equal to the depth of the maximum oxygen-concentration, but no continuous layer of SiO_2 would be formed. Experiments have shown that a dose of 1.4×10^{18} oxygen atoms/cm^2 must be implanted in order to create a continuous-buried-oxide-layer. The standard dose (which is the most common) is $1.8 \times 10^{18}/cm^2$, and produces a 400-nm-thick buried-oxide-layer upon annealing. As the dose is increased above $1.8 \times 10^{18}/cm^2$, the overall oxygen-implant-profile no longer shows a higher peak value, but instead becomes broadened. This is due to the fact that at the elevated temperature of the implant, oxygen can readily diffuse to the Si/SiO_2 interface, where oxidation occurs. The dose of oxygen at which the continuous-buried-oxide-layer begins to form ($1.4 \times 10^{18}/cm^2$) is called the "*critical dose.*"

2. The wafers are then given a post-implant anneal for sufficient time (3–5 -hours) at an extremely high temperature (1300-1350°C [in furnaces equipped with poly-crystalline-silicon or silicon-carbide tubes], are now considered standard annealing conditions) in N_2 (with about 2% oxygen added, to protect the silicon from the pitting phenomenon which occurs when silicon is annealed in pure-nitrogen at high temperatures), or Ar. It is generally agreed that the higher the annealing temperature, the better the quality of the SIMOX material. An anneal in pure-nitrogen can also be performed if a cap-layer of oxide (usually deposited by CVD) is formed on the wafer after the implant, and prior to the anneal.

This extremely-high-temperature annealing-step has three functions: **1**) It drives the chemical reaction which forms a continuous-buried-amorphous-oxide (BOX) layer of silicon-dioxide near the peak of the implantation. (The choice of a sufficiently-high O^+-implant-dose and energy provide the required oxygen concentration to enable stoichiometric SiO_2 to be formed); **2**) It removes many of the defects (dislocations) formed during the ion-implantation step; and **3**) It dissolves any SiO_X-precipitates in the silicon device-surface-layer and denudes this layer of oxygen as well (see Chap. 2 for a discussion of denuded-layer formation in CZ-silicon-wafers). To allow integrated-circuit devices to be fabricated on the wafer after the formation of the BOX, a thin layer of crystalline-Si must remain above the BOX-layer. The BOX-layer depth is defined by the depth of the ion-implantation and it's thickness is limited to about 0.5-μm.

The evolution of the SIMOX structure during the high-temperature-anneal (as mentioned above) is quite complicated, but it is useful to summarize the phenomena that occur from a thermodynamic perspective. During annealing, both dissolution of the oxide-precipitates and precipitation of the dissolved-oxide takes place in the oxygen-rich silicon-layers. Oxygen diffuses out of the top silicon-surface, initially denuding the zone near the silicon-surface. Within the bulk of the silicon, small precipitates dissolve into silicon and large precipitates grow from the dissolved oxygen. At any given temperature (and for a given concentration of oxygen in the silicon), there exists a *critical precipitate radius*, r_C, below which a precipitate will disappear, and above which it is stable. The critical precipitate radius increases with temperature, and becomes essentially infinite at very-high temperatures. At temperatures above 1300°C the only stable precipitate is the buried-oxide-layer itself, which has an essentially infinite radius-of-curvature. The complete dissolution of all precipitates at annealing temperatures greater than 1300°C, and the diffusion of dissolved oxygen towards the buried-oxide-layer explain the thickness increase of the BOX when high-temperature annealing is performed.

The fact that a thin-layer of high-quality crystalline-silicon must exist above the BOX-layer implies that the lattice structure of the silicon in this region must maintain a high-degree of crystallinity in the face of the high-dose oxygen-implant step. The high oxygen-implantation-dose does amorphize the layer of silicon that lies above the projected range. However, it would be unacceptable if complete amorphization occurred, and a polycrystalline layer was formed during further annealing. This would happen if the temperature of the wafer was kept too low during the oxygen implant step. On the other hand, by keeping wafers at a temperature greater than 500°C during the implant (by a combination of beam power and external heating), the amorphization damage anneals out during the implantation process ("self-annealing"), and a high-quality crystal-structure is maintained in the silicon layer at the wafer surface. By maintaining such an elevated temperature during the implant, the implanted oxygen also tends to diffuse away from the vicinity of the projected range to form a broad, flat concentration plateau. Note that the most commonly used implantation temperatures range between 600°C and 650°C. This is because SiO_2 precipitates form in the silicon overlayer at higher temperatures (700-800°C), mostly near the bottom interface. However, large precipitates are also observed near the surface region of the silicon film (and this therefore sets an upper limit of 700°C on the implantation temperature).

3. If the device requires thicker silicon layer than can be achieved with the depth of ion-implanted oxygen, an epitaxial-Si layer can be deposited to provide this additional thickness.

The are several shortcomings associated with the high-dose SIMOX technology, including the following:

1. The formation of SIMOX wafers requires extremely high doses of oxygen, and this necessitates very-long implantation times (which, as we shall see, contributes to the high-cost of SIMOX-wafers). The need to anneal the wafers at the extreme temperatures exceeding 1300°C may also be a limitation for 300-mm wafers.

2. Transition metals may be sputtered from the beam-line or end-station components during the implantation. These metals can deposit onto the SIMOX wafer surface and can then be driven into the silicon by the wafer-heating and subsequent thermal processing. Their presence in the silicon will enhance junction-leakage and reduce the minority-carrier-lifetime of the devices. Metallic contamination is present in all implant processes, but since high-dose SIMOX implants are nearly a thousand times greater than those of conventional high-dose source/drain implants, the metal contamination problem is potentially increased by approximately the same degree.

This issue was especially troublesome during the early days of SIMOX, when implanters were not yet built to limit transition-metal-sputtering. Since that time, however, great efforts were made to reduce the metal contamination sources, and silicon or silicon-dioxide shields have been placed in implanters such that the beam never "sees" any metallic part. The design of high-temperature furnaces used to anneal the SIMOX material has also been optimized in order to minimize the diffusion of metallic contaminants from the furnace heating-elements into the furnace tubes.

3. Pinholes in the buried-oxide-layer will occur if a particle is present on the wafer during the oxygen-implantation-process. Such particles screen the oxygen implant at that point. The result is a breach in the buried-oxide-layer that represents a flaw in the insulating barrier at that point. Careful wafer cleaning and attention to particulate control in the implant process can control this problem.

4. Despite the high-temperature anneal, the surface silicon may retain a dislocation density on the order of $10^3/cm^2$. The presence of such defects can cause reduced oxide-integrity and increased junction-leakage. However, devices fabricated in the best SIMOX material have been shown to be able to build excellent quality CMOS and bipolar ULSI circuits.[30]

5. SIMOX-wafers are more costly than bulk-Si-wafers, although the cost is expected to sharply decrease as the popularity of this technique expands. The cost reduction will be partly achieved by improvement in the implant equipment used to fabricate SIMOX-wafers. For example, SIMOX-dedicated ion-implanters are now available from IBIS Technology. Such implanters can furnish adequate throughput and acceptably-low levels of metallic-contamination. These production tools have an output of 100-mA of O^+ at an acceleration voltage 210-keV. The total transition-metal contamination on the wafers (from Ti to Zn) has been measured to be less than $5x10^{10}$ atoms/cm^2.

11.6.2.2 Low-Dose SIMOX

A recent trend with SIMOX fabrication is to use a lower oxygen-implant-dose than is used in the high-dose SIMOX approach. With the low-dose technique, improved, lower-cost SOI substrates are obtained. For example, a typical modern SIMOX implanter with a nominal 50-mA of O^+ beam current has a maximum output of only about 25 wafers of 200-mm diameter per day, round the clock operation. The low-dose SIMOX process also drastically increases the crystal quality of the silicon layer above the stoichiometric BOX layer. The oxygen implant dose is reduced to about $4x10^{17}$-$1.0x10^{18}$ ions/cm^2, and at an energy around 120-keV. Note that only a very-narrow process window close to the dose of $4.0x10^{17}$ ions/cm^2 will produce a continuous, precipitate-free BOX layer.

Table 11-3 compares the process conditions used to produce SIMOX with high-, medium-, and low-doses of oxygen.

A silicon-layer between 150-200-nm deep is produced, but the lower oxygen-implant-dose also reduces the BOX layer to thicknesses of only 90-150-nm (depending on the selected oxygen dose).[31] A high-temperature anneal in Ar (i.e., at 1350°C) is then carried out, just as is done in the high-dose SIMOX process. However, this high-temperature-anneal in an inert ambient is followed by a subsequent high-temperature anneal in a gas mixture of inert gas and 30-60% oxygen (at ~1350°C).

The oxygen in the furnace-ambient at this high-temperature both oxidizes some of the superficial silicon layer, but also diffuses through the surface silicon layer and reacts at the buried-oxide/Si-interface to form new oxide within the Si (Fig. 11-19). This has the effect of increasing the thickness of the BOX-layer.[31] This process of growing the BOX-layer to a greater thickness by annealing the low-dose SIMOX wafers in oxygen is called an *internal thermal oxidation process* (*ITOX*). Note that as long as the oxide grown on the surface of the silicon during ITOX is thinner than 500-nm, there exists a linear-relationship between the thickness of this oxide layer (t_{ox}) and the thickness increase in the buried

TABLE 11-3 Physical and Electrical Properties of Commercial SIMOX

Parameter	Units	Standard Dose SIMOX	Medium Dose SIMOX	Low Dose SIMOX
Silicon Thickness	A°	1500 - 2000	1500 - 2200	1500 - 1800
BOX Thickness	A°	370 - 400	170 - 190	90 - 150
HF-Defects Density	cm^{-2}	< 1	< 1	< 1
Dislocation Density	cm^{-2}	$< 10^7$	$< 10^6$	$< 10^5$
Stacking Fault Tetrahedara	cm^{-2}	$< 10^5$	$10^5 - 10^6$	$10^5 - 10^6$
Surface Roughness (10um x 10um)	A°	7 - 10	5 - 7	3 - 5
Metallics (TXRF)	cm^{-2}	$< 10^{11}$	$< 10^{11}$	$< 10^{11}$
Short Density	cm^{-2}	< 0.1	< 2	< 2
Electron Mobility	$cm^2/V\text{-}s$	850 - 1000	850 - 1000	850 - 1000
Hole Mobility	$cm^2/V\text{-}s$	200 - 325	200 - 325	200 - 325
Oxide Charge	C/cm^{-2}	$2 - 3 \times 10^{11}$	$1.5 - 3 \times 10^{11}$	$1 - 2.5 \times 10^{11}$
Interface State Density	$cm^{-2}eV^{-1}$	$2\text{-}3 \times 10^{11}$	$1 - 3 \times 10^{11}$	$1 - 2.5 \times 10^{11}$
Transconductance (Gmsat)	mS	400 - 440	900 - 1200	1000 - 1300

oxide Δt_{BOX}): $t_{ox} = 0.06\ \Delta t_{BOX}$ (Fig. 11-19c). The internal-oxide also grows at the expense of the bottom of the silicon overlayer.

In the low-dose SIMOX process, the BOX formation consists of oxygen precipitation, growth of the precipitates, and their coalescence during the high-temperature-anneal process. This sequence of events occurs because the oxygen implantation dose is less than the SiO_2 stoichiometry, even at the depth of the peak-oxygen-concentration, and a stoichiometric-SiO_2-buried-layer is required to be formed to produce effective dielectric isolation. By carrying out the ITOX process, the BOX in the low-dose-SIMOX-process is made thicker and exhibits better isolation characteristics. The upper part of the BOX after the ITOX process is essentially a thermal-oxide-layer. Since the upper segment of an ITOX-SIMOX layer is essentially a thermal oxide, such important BOX properties as the breakdown-field, show a marked improvement (i.e., the breakdown-field is increased from 3-4-MV/cm for the case of conventional low-dose SIMOX to more than 5-MV/cm for ITOX-SIMOX, see Fig. 11-19d).[49] This improvement can be attributed to the BOX-thickness incremental-increase due to ITOX. In addition, the ITOX reduces the size and density of silicon pipes in the BOX, which is the main cause of BOX breakdown.[50] In 2001 IBIS Technology licensed a low-dose-SIMOX-technology from IBM called Advantox® MLD.

In 2000, IBM said recent advances in the production of low-dose-SIMOX-wafers will cut incremental processing costs of SOI to only 3% over the costs of producing ICs on bulk-Si-wafers. Low-dose-SIMOX-wafers are commercially available from such vendors as Ibis, Komatsu, and Nippon Steel. For example, the Ibis-1000 model is currently available. The next generation oxygen implanter, the Ibis-2000 (which can accommodate both 200-mm and 300-mm wafers) was under development in 2001, and is forecast to be ready for delivery in the first half of 2002. Figure 11-20 is a photograph of a Power4™ processor made by IBM using a SIMOX-SOI-based process.

11.6.3 Wafer Bonding

Another popular technology for producing SOI wafers is *wafer bonding*.[32] Wafer bonding is a simple concept. Basically, wafer bonding refers to the phenomenon that mirror-polished, flat and clean wafers of almost any material, when brought into contact at room temperature, are locally attracted to each other by van der Waals forces and adhere, or *bond* to each other. The bonding process is started by locally applying a slight pressure to the wafer pair, and the bonded area spreads laterally over the whole wafer area in a few seconds. Because the bonding achieved at room temperature is typically relatively weak, wafer pairs bonded

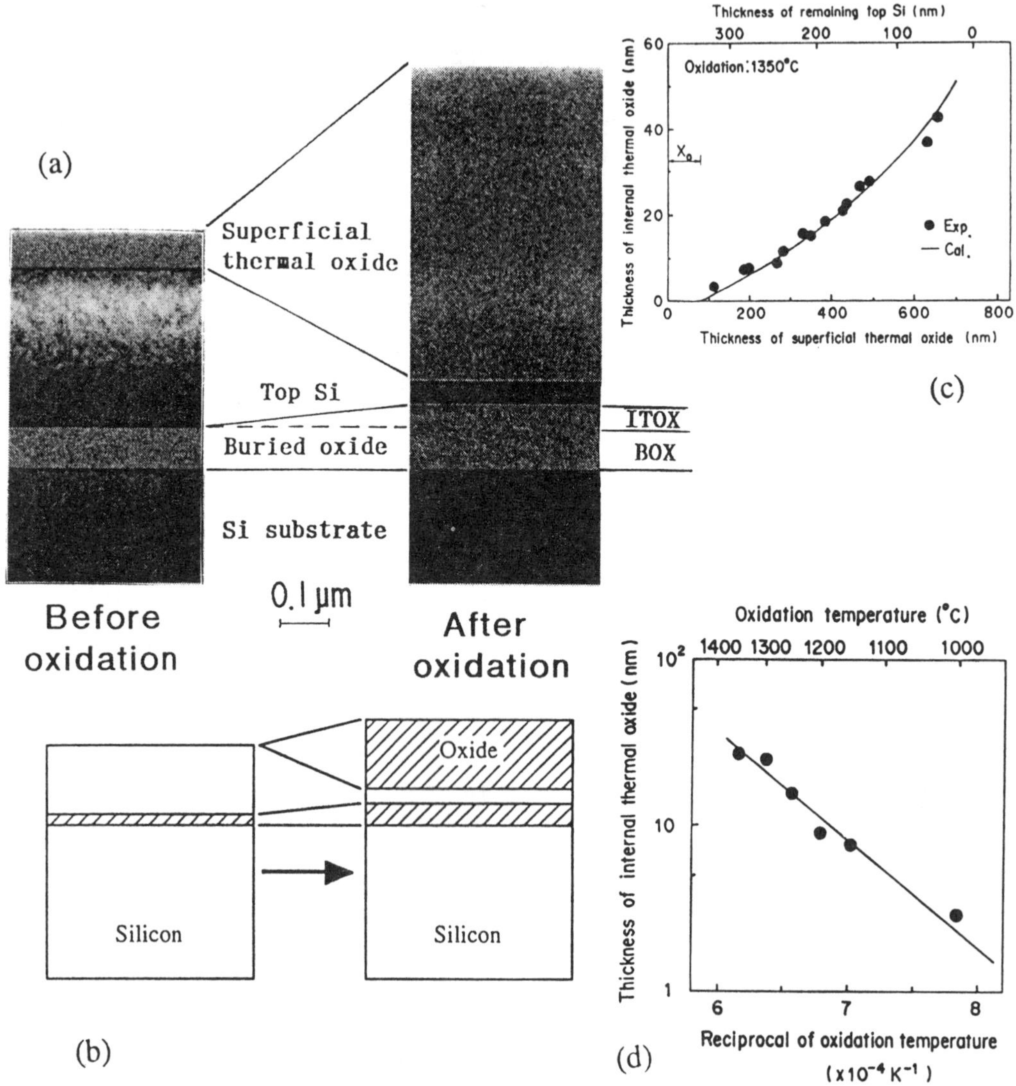

Fig. 11-19 a) An XTEM of SIMOX wafers before and after high-temperature-oxidation at 1350°C; b) A schematic drawing of SIMOX wafers before and after high-temperature oxidation at 1350°C; c) Relationship between the thickness of the ITOX layer and the superficial thermal-oxide-layer formed by high-temperature-oxidation at 1350°C; d) Relationship between the thickness of the ITOX-layer and the reciprocal of the oxidation-temperature.

at room temperature are usually subjected to a heat treatment to strengthen the bonding interface.

In the application of the bonding-process to form SOI-substrates, two thermally-oxidized-Si wafers (or one oxidized-wafer and one bare-Si-wafer) are first

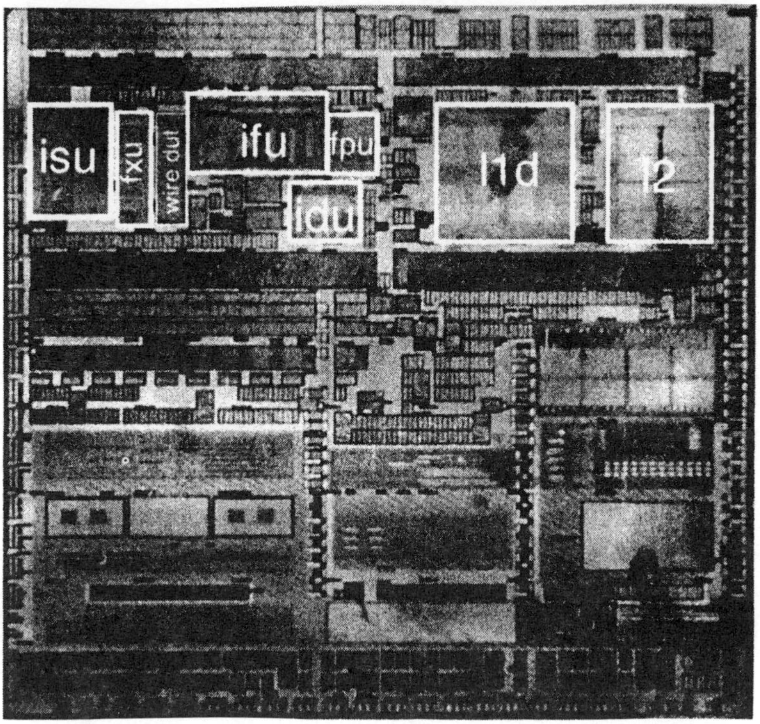

Fig. 11-20 Photograph of an IBM Power4^{TM} processor using a SIMOX-SOI-based process.

first bonded by applying pressure to the wafers at room temperature (Fig. 11-21). A subsequent furnace anneal is used to increase the bonding strength. After such bonding, most of the silicon from one of the wafers is removed, leaving a thinned

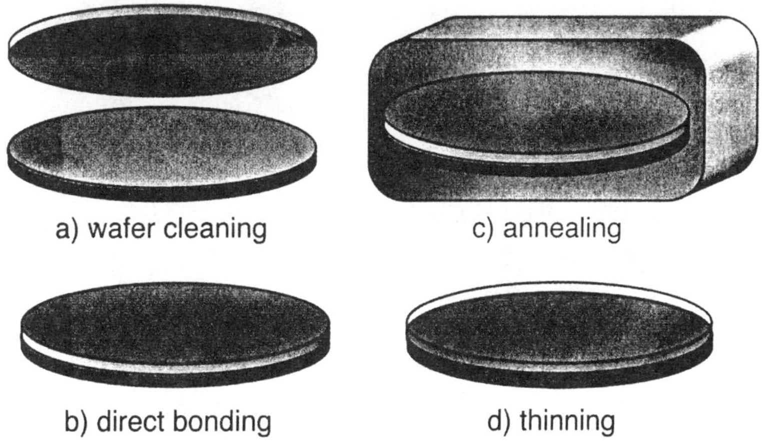

Fig. 11-21 Generic direct-wafer-bonding process flow.

layer of silicon from one of the wafers. The IC devices are built in this remaining thinned-silicon-layer, which now exists over the oxide present at the interface of the bonded wafers (i.e., a buried-oxide). The wide-availability of chemical-mechanical polishing (CMP) in IC processing, as well as a variety of other thinning approaches has led to wide and diverse use of wafer bonding. Unlike SIMOX, the wafer-bonding technique can form BOX-layers up to 50-μm thick. A book has recently become available dealing with the topic of wafer bonding.[33]

There are several approaches for achieving bonding. In one method, two oxidized-silicon-wafers are held together (face-to-face) by applying external pressure to the wafers while they are exposed to an oxidizing-ambient at about 700°C.[34] Bonding is believed to take place by the conversion of oxygen trapped between the wafers, into SiO_2. The conversion consumes the oxygen and creates a partial-vacuum that forces the wafers into intimate contact, allowing the chemical-bonding-reaction to proceed. The chemical-reaction is believed to be the polymerization of silanol bonds to form siloxane networks.

A second approach applies a moderate voltage across two oxidized-wafers placed face-to-face at temperatures in the range of 1100–1200°C.[35] Since no mechanical force is required when the wafers are first placed in the furnace, this method is less likely to introduce crystalline-defects. A third approach uses one oxidized-wafer and one bare-wafer. Both are treated with an acid solution that forms O-H groups on the surface of the wafers. After drying the wafers are placed face-to-face at room-temperature in clean air. After this treatment a self-adhesive contact is formed which holds the wafers together during the bonding process. The bonding is completed by annealing the two wafers at 1100°C in a nitrogen-ambient for 4 hours.[36]

Recently it has been demonstrated that bonding can occur at temperatures as low as 150°C, if the process is performed in a moderate vacuum (e.g., several torr).[37] It is believed that the vacuum significantly reduces the amount of trapped nitrogen at the interface. Trapped nitrogen can prevent an intimate contact of the bonding-surfaces during annealing, and hinders formation of the required covalent-bonds. It is also possible to achieve full-bonding-energy directly at room-temperature under ultra-high vacuum conditions if the wafer surfaces are free of adsorbates.[38]

It should be mentioned that particles or protrusions on either surface being bonded, do prevent bonding, (at least locally) around these disturbances. These lead to locally-unbonded-areas that are frequently termed *interface bubbles* or *voids.* Such voids are, of course, undesirable, since they can lead to delamination of devices located in imperfectly-bonded-areas during chip-assembly and packaging. Voids with a diameter of several millimeters are readily created by particles 1-μm or less in size (and these are easily revealed by infrared or ultrasonic imaging). Therefore particles on wafers have to be avoided by proper cleaning and by carrying out all bonding procedures in an ultra-clean environment (Fig. 11-22).

11.6.3.1 Bond-and-Etch-back SOI (BESOI): Once the wafers are bonded it is necessary to remove most of the silicon from one of the wafers to produce the SOI structure. There are several proposals being evaluated for this removal process. The *bond and etch-back SOI* (BESOI) technique is used to produce relatively thick-films of both oxide and silicon. The BESOI wafers typically have thicker, yet better-quality silicon and buried-oxide-layers than do SIMOX wafers.

The silicon-layer on BESOI substrates can be thinned in one of several ways. Mechanical grinding and polishing can be used to thin the Si to a thickness down to ~1-μm, but only to within 10-30% uniformity. Better uniformity can be achieved by using an etchstop. Several etch-stoplayers have been developed, including: a boron-doped-layer, a silicon-germanium-layer, a carbon-implanted-layer, and a porous-silicon-layer.

In the first of these etchstop methods, a lightly-doped B-region can act as an etchstop-layer when it is formed on a heavily boron-doped-substrate (Fig. 11-23). That is, the wafer that will ultimately contain the device consists of a lightly-B-doped epitaxial-layer on a heavily-B-doped-wafer. After oxidation and bonding of the lightly-doped surface of this wafer to another wafer, the heavily doped

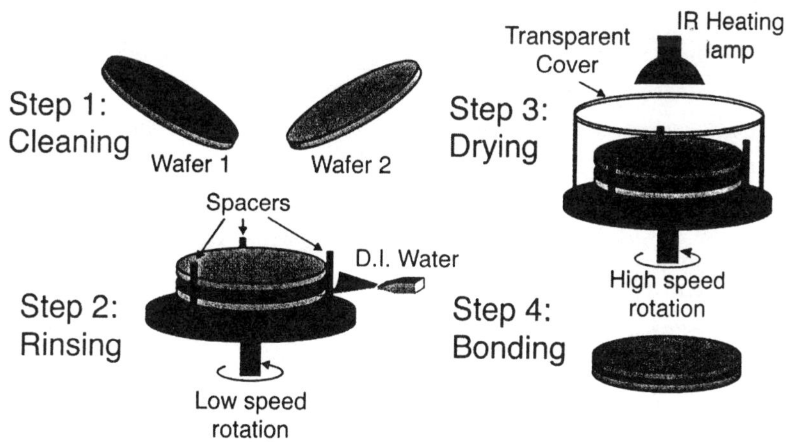

Fig. 11-22 Schematic of the microcleanroom setup and procedure.

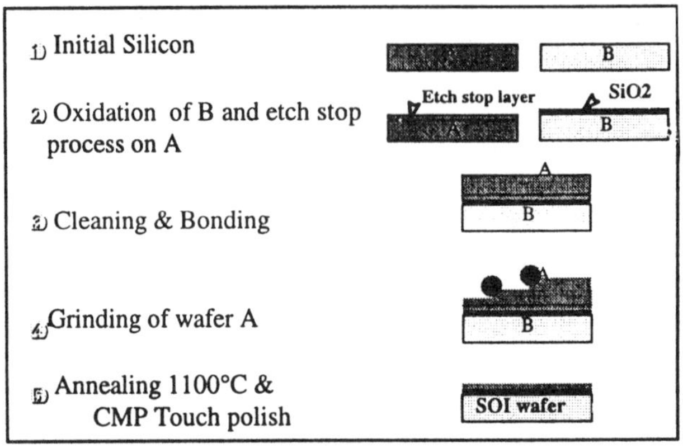

Fig. 11-23 Schematic of the Bonded-and-Etched-Back SOI technology (*BESOI*).

substrate is removed by an etch process that preferentially etches such heavily doped silicon. The epi-layer that remains bonded to the second wafer determines the thickness of the active-device-layer and the sum of the thickness of the oxides on the two wafers determines the thickness of the buried-oxide. One of the issues with this approach is that two silicon-wafers are required to make one device-SOI- wafer. Another is that the uniformity of the thickness of the silicon-layer depends on the etch-stop-selectivity (which can vary, depending on the etch-stop method used, from 10 to 10^5).

Such BESOI methods are widely used for "thick" SOI applications (i.e., > 1-μm), such as power devices (high-voltage and high-temperature devices) and micromechanical machines (MEMS). The power-devices are used for military, industrial, and aerospace applications. The share of the thick-film segment of the SOI market dropped from 80% in 1995 to 53% of the market in 1999 and was taken over by the thin-film segment. Future thick-film SOI applications will be driven by industrial and automotive applications.[54]

11.6.3.2 Hydrogen-Implantation-Induced Layer Splitting to Create SOI Substrates (SMART-CUT®): A recent (and elegant) wafer-bonding-based method has been developed for forming SOI substrates using a technique to obtain very-uniform and thin regions of single-crystal-silicon. This technique employs a process called *hydrogen-implantation-induced layer splitting.*[39] The process is also referred to by its inventor M. Bruel (and by the company that sells such SOI substrates [SOITEC SA, France]), as *Smart-Cut.*® The Smart-Cut process also allows the sacrificial-wafer to be reused and only one wafer is therefore required to make a bonded SOI substrate. (The SOI material made with the Smart-Cut process is termed UNIBOND.®) There are two reasons why this process may be significantly less-expensive than the process used to fabricate SIMOX-wafers. First, the sacrificial wafer can be re-used. Second, because the implant-dose is an order of magnitude lower than even one in the low-dose SIMOX-process (and the implant is done at room temperature in a conventional implanter), the cost of the implant step is significantly lower, compared to those used in SIMOX.

Figure 11-24 shows how the process works. Two device-quality-wafers (labeled A and B) start the process. Wafer-A is oxidized (to a thickness of about 0.5-μm) to form what will become the buried-oxide of the SOI-structure. Then this wafer is ion-implanted with a high-dose of hydrogen ions ($\approx 5 \times 10^{16}$ H^+ ions/cm^2) through the oxide, using a conventional high-current-implanter. The peak-concentration of this implanted hydrogen-dose is at a projected range R_p of about 0.2-μm below the oxide-surface. At this point, only micro-cavities or micro-bubbles are formed at the R_p of the implantation. For reasons explained next, the depth of the H^+ ion implantation determines the thickness of silicon above the BOX.

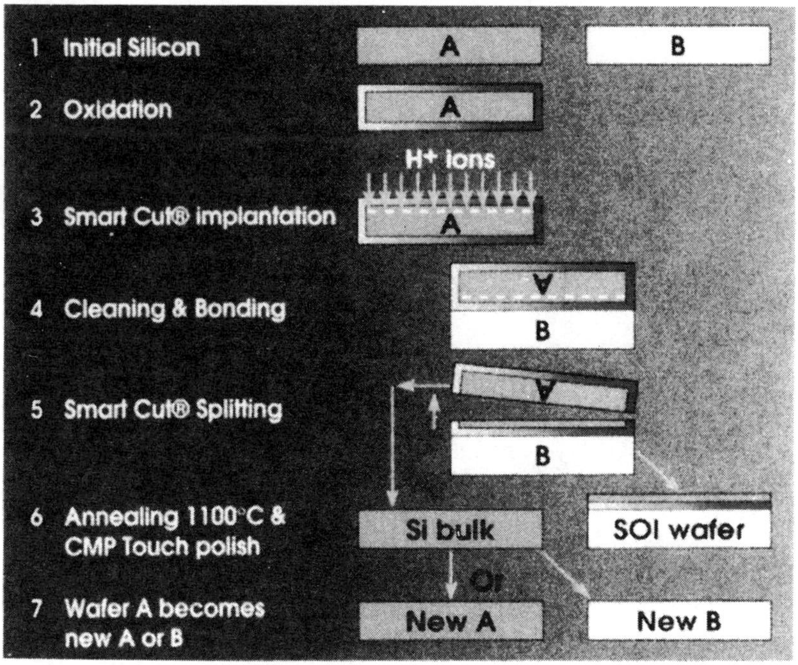

Fig. 11-24 Process flow for fabricating Smart-Cut® wafers.

After implanting the hydrogen into Wafer-A, both Wafers A and B are cleaned and low-temperature bonded face-to-face. The bonded wafers are next annealed (at 400-600°C). During this anneal the implanted-hydrogen precipitates, and this precipitation-process merges the buried network of bubbles and voids at the vicinity of the depth of R_p. As these hydrogen-filled microcavities continue to grow and merge, they eventually form a continuous microcrack along the line corresponding to the peak of the implanted-hydrogen. At that point Wafer-A delaminates from along that line, and Wafer-B (which now has a BOX-layer covered by the silicon layer from Wafer-A) becomes the SOI wafer. Meanwhile, Wafer-A remains available to be used as the sacrificial wafer of another UNIBOND substrate. Both wafers are then annealed at 1100°C to remove defects. This is followed by a "light" CMP (or "touch-polish") to erase the surface-roughness resulting from the layer-splitting process. Wafer-A, which is now only a fraction of a micron thinner than it was before the start of the process is used as Wafer-B when the process starts over. The use of ion implantation for layer separation improves the layer-thickness-uniformity over that of bond-and-etch-back SOI-wafers. The method is used to prepare commercially available 200-mm SOI wafers with a silicon thickness above the BOX in the 100-nm range and a uniformity of better than 5%. The dislocation density is $< 10^2/cm^3$, the surface roughness is < 2 Å (rms), and metal contamination is $< 5x10^{10}/cm^3$.[40]

11.6.3.3 Nanocleave® Wafer-Bonding SOI: The fabrication of starting wafers for SOI can also be carried out in other ways. One of these processes, named *Nanocleave®* was introduced by Silicon Genesis Inc. in 1998.[43] The process combines growing a SiGe/Si epitaxial layer, low-temperature wafer-bonding, and a room-temperature cleaving-process to produce sub-100-nm-thick layers of single-crystal-silicon over a buried-oxide substrate. One of the key advantages of the Nanocleave process is that it can produce such thin SOI-layers with a uniformity of several percent over an entire wafer. This type of uniformity is necessary for the successful fabrication of fully-depleted CMOS-SOI-FETs. Other thinning techniques (such as etch-back or CMP have increasing difficulty maintaining such SOI-layer thickness-control when silicon layers of sub-100-nm thickness are being produced. The Nanocleave process uses an atomic-scale cleaving-process that produces cleaved surfaces with an as-cleaved surface roughness less than 5-Å (as measured by AFM). In addition, the Nanocleave process potentially offers a method that is lower in cost than any of the other fabrication processes used to fabricate SOI-substrates. The Nanocleave technique also offers another path toward the capability of fabricating both "ultrathin" Si-

SOI-layers (10-to-50-nm-thick Si) and "nanothin" Si-SOI (5-to-10-nm-thick Si) required for the MOSFETs with gate-lengths smaller than 50-nm that are scheduled to begin production by 2010.

The Nanocleave process begins with the formation of a *cleave-plane*, the growth of a thermal-oxide-layer and Si-device layer on a high-quality "donor" wafer (Fig. 11-25a).[44] This "donor" wafer is then bonded to an inexpensive oxidized "handle" wafer (on which the device layer will ultimately reside).

The cleave-plane is created by first epitaxially growing a layer of Si-Ge on the "handle wafer, and then continuing the growth of this layer with an epitaxial-Si device layer (Fig. 11-26a).[63] The buried Si-Ge-layer is a compressive-strain-layer of a non-dislocated Si-Ge alloy, which contains no porous-regions, inclusions, or voids, as shown in Fig. 11-26a. When device wafer is separated, the resulting separated surface has a surface roughness, as measured by AFM profiling of 0.2-nm, or less.

After the cleave plane has been formed and a thermal-oxide is grown on the surface of the wafer with the cleave-plane (which will become part of the buried-oxide-layer in the finished SOI-device), the "donor" wafer is bonded to the

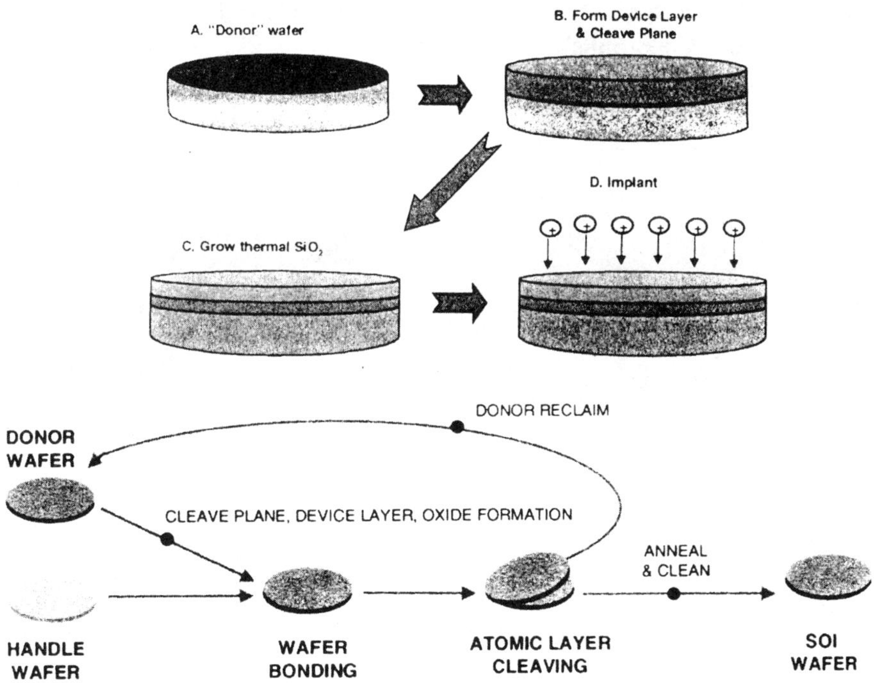

Fig. 11-25 Process flow of the Nanocleave® SOI technology.

"handle" wafer (which also contains a surface-oxide layer). Plasma-bonding techniques are employed to increase the initial bond-strength to sufficient levels for the room-temperature process used to transfer the silicon-layer from the "donor" to the "handle" wafer. This transfer process (by cleaving the device layer from the "donor" wafer from the "handle" wafer) is carried out next (Fig. 11-25b). Since the bond-strength is significantly greater than the interface-strength along the cleave plane, a novel cleaving process is used to separate the Si device layer from the "donor" wafer (which results in the creation of an SOI structure on the "handle" wafer). In the Nanocleave process, the donor wafer cleaving is done at room temperature (unlike in the Smart-Cut-process that uses an elevated temperature anneal for the cleaving step). In the Nanocleave process, the drive force is a static pressure of N_2 gas. A mechanical fixture secures the bonded wafer pair as a high-pressure nitrogen-gas is introduced along the strained Si-Ge layer. That is, cleaving is controlled by fracture along a cleave-front within the Ge-rich Si-layer. A thin-film of material from the "donor" wafer remains bonded to the substrate "handle" wafer, producing an SOI-structure. The result is an SOI-wafer with a silicon-layer whose thickness uniformity is within 1% across a 200-mm wafer. Furthermore, the atomic-layer cleaving-process is claimed to produce an as-cleaved surface roughness of less than 1-nm (typically 2-5 Å).[63] This is an order-of-magnitude smoother than the 80 Å of hydrogen-induced thermal cleaving obtained with the Smart-Cut technique. The smoother surface in the Nanocleave process is acceptable for many IC applications with no additional

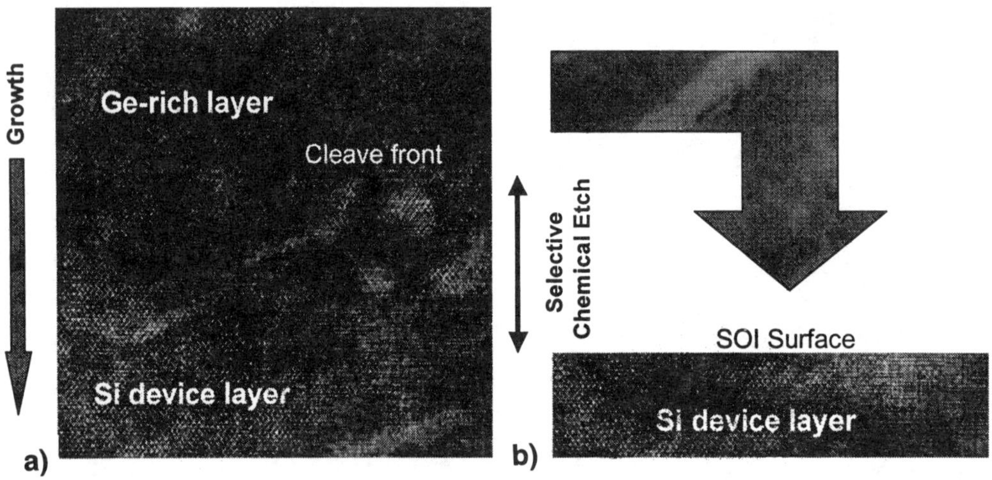

Fig. 11-26 a) HR-TEM image of the boundary area of a Ge-rich, epitaxial strain-layer and a crystalline-Si-device-layer in a state prior to atomic-layer-cleaving; b) The residual Ge-rich layer on the as-cleaved-wafers are subsequently removed through a selective etch step.

surface polishing. The edge of the SOI-layer has a smooth and regular character without the need for edge polishing. The residual Ge-rich layer on the as-cleaved wafers is subsequently removed through a selective etch step (11-26b).

A fully automated tool that separates bonded donor-handle wafer-pairs is available for this purpose. The high-quality wafer can be used to produce many thin-film transfers, reducing the process cost. Another factor that contributes to further cost-reduction arises from the fact that the "handle"-wafer is separated from the device-layer by the buried-oxide-layer. Consequently, a "handle" wafer with considerably relaxed electrical and chemical specifications (and therefore lower cost) can be used in this process.

Post-cleave-processes for layer-thickness-adjustment include: **1**) Si layer thinning using etching processes; and **2**) layer-thickening by growth of additional Si or Si-Ge multilayers. The layer-thickness-control for thickened wafers is better than 0.5% with a within-wafer thickness-uniformity of 2% or better (Fig. 11-27).

11.6.3.4 ELTRAN® Wafer Bonded SOI

A relatively new SOI structure (originated and developed by CANON Inc.), which is also based on wafer bonding, is called *ELTRAN.®* The name ELTRAN comes from *Epitaxial LAyer TRANsfer by BESOI* (Bond and Etch-back SOI method) with selective-etching of porous-silicon (Fig. 11-28).[45,46] ELTRAN wafers became commercially available in 1997 for 200-mm substrates.

The porous-silicon surface on an otherwise bare-silicon-wafer is formed by an anodizing treatment of the silicon surface. The ELTRAN method is characterized as BESOI implemented by growing an epitaxial-layer on a layer of porous-silicon. (It has been demonstrated that high-quality epitaxial-layers can be grown on porous-silicon.) The surface of the epitaxial-layer is then thermally oxidized, and this layer will eventually become the buried-oxide-layer of the SOI-structure. This wafer (with a thermally-oxidized epitaxial-film grown on porous-silicon) is then bonded to a "handle" wafer as the "donor" wafer. In early ELTRAN wafer fabrication process, the "donor" wafer material was removed after bonding by a grinding process (down to the porous-silicon-layer). The exposed porous-silicon that remains after the grinding-process is then removed by a wet-chemical etch. A mixture of HF and H_2O_2 is used for this etching step, since it provides an etch selectivity as high as 105:1 between the porous-silicon and the epitaxial-layer. At the end of the process, only the epi-layer remains on the top of the oxide layer and handle-wafer. The ELTRAN process has been shown to produce 470-nm-thick SOI-layers with a 14-nm thickness uniformity.

In the more recent ELTRAN process (Fig. 11-29), a double porous-silicon

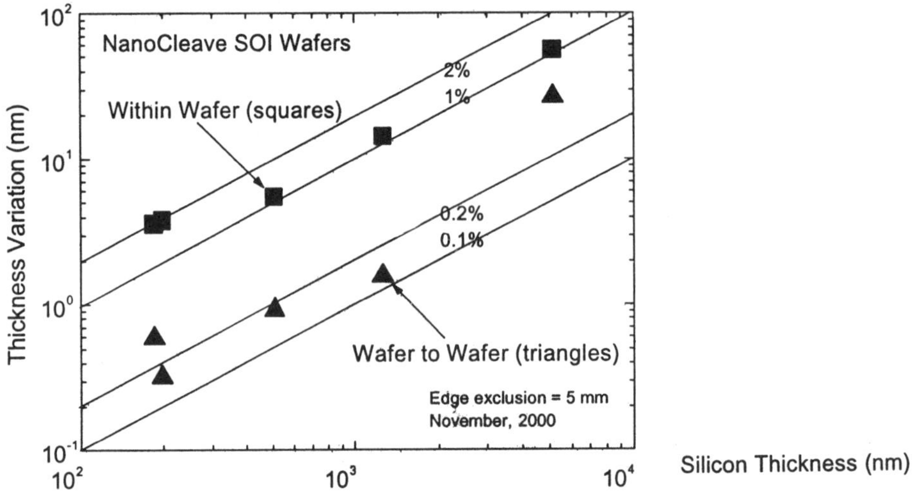

Fig. 11-27 Si-SOI layer-thickness-uniformity and average layer-thickness-control for post-cleave thickening of as-cleaved SOI-wafers with cleaved 200-nm Si-SOI-layers.[63]

layer is grown on the "donor" wafer, with each of these porous-silicon-layers having a different porosity. The porosity of the porous-silicon-films on the wafer surface can be adjusted by changing the value of the anodization current. After bonding it becomes possible to physically separate the "donor"-wafer from the

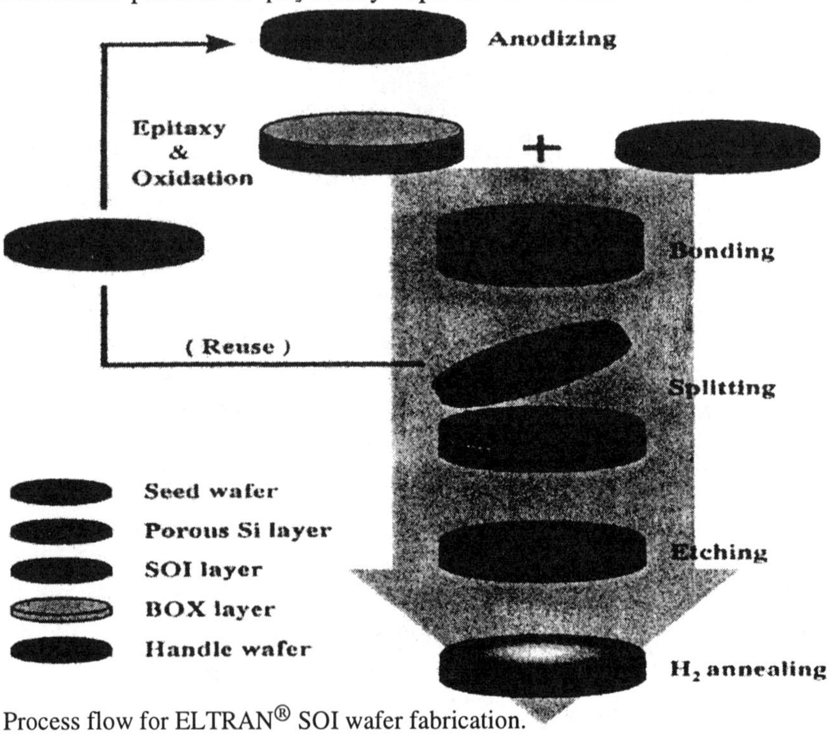

Fig. 11-28 Process flow for ELTRAN® SOI wafer fabrication.

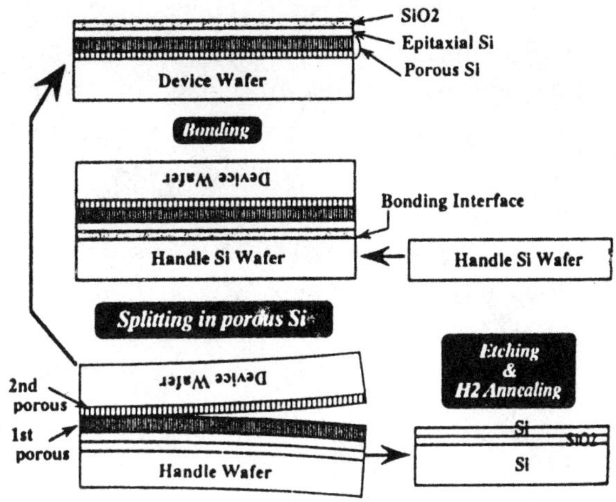

Fig. 11-29 The ELTRAN process flow for device wafer re-usage. Bonded pairs are split within the porous-Si layer. The device wafer sides are used as a starting device wafer. The handle sides result in ELTRAN wafers after etching the porous-Si and H_2-annealing.

"handle" wafer using a physical splitting process (i.e., a water-jet slices the SOI-layer from the seed layer).[62] The splitting occurs along the interface between the two porous-silicon layers. That is, since the porous-layers are a double-stacked structure with different porosities, the strain-energy accumulated at the interfacial area limits splitting to an extremely narrow range (because the interface between the first and second porous layers is mechanically the weakest point in the structure). Therefore, the stress-configuration induced by the double porous-layers is responsible for the splitting mechanism.

For splitting a "water-jet" technology is used. The water jet – a small collimated water stream – is injected at the edge of the bonded-pair and splits the two wafers at the thin porous-silicon layer. The double-layered porous silicon structure contains the stress concentration, with the result that both of the split wafers (SOI and seed wafer) are covered by the porous silicon protective layers. The water-jet drives high-pressure water into narrow gaps by fluid wedging and applies uniform pressure, reducing the chance of damage. The splitting yield is reported to be over 90%, and automatic water-jet splitting-machines are in production use.

The device layer over a buried-oxide is then formed by the etchback of the porous-silicon. After the porous-silicon-film is removed, the surface of the device silicon-layer is flattened by hydrogen-annealing (see Chap. 2, for a discussion of hydrogen-annealing of Si wafers).

There is another economic benefit of the more recent, double-porous-silicon-layer process used to form ELTRAN wafers. In this process (and like in the Smart-Cut and Nanocleave processes discussed in the previous sections), after the splitting process, the "donor"-wafer becomes available to be used as another a "handle"-wafer (instead of being ground away, as in the early ELTRAN process). Canon announced that it has expanded the production of ELTRAN wafers from 2000-per-month to 10,000-per-month in November 2000. This production expansion is being supported by a greater marketing of ELTRAN wafers. To this effect Canon opened a marketing office in the U.S. to achieve this goal.[53]

11.7 LATERAL ISOLATION IN SOI

SOI-CMOS-structures eliminate all inter-device leakage-currents by building *n*- and *p*-type islands on an insulating substrate for *p*- and *n*-channel transistors, respectively. Field-inversion and device-to-device-punchthrough do not exist in SOI-technology, but new parasitic transistors in an SOI-structure can still give rise to leakage-currents. SOI-CMOS-technology depends on high-quality isolation. The methods used to laterally isolate SOI structures will be described here, as well as a description of the leakage paths that arise as a result of each of them. The three techniques for providing lateral isolation in SOI are: **1**) MESA isolation; **2**) LOCOS isolation; and **3**) shallow-trench isolation (STI).

11.7.1 MESA Isolation

MESA isolation was the early method used to create islands of silicon on an insulating substrate (Fig. 11-30). The technique is attractive because of its simplicity. That is, in MESA isolation the active-device-regions (the silicon islands) are formed into a "mesa" by using a mask-step and a silicon-etch step to remove the silicon in the field-areas down to the insulating-substrate. The SOI oxide helps as the etch-stop-layer, while anisotropic-etching allows for minimization of the distance between adjacent mesas. Passivation of the islands (mesas) is performed at the gate-oxidation step, where the gate-dielectric is grown not only on top of the silicon islands, but on their edges as well.

Although the MESA isolation method is simple, it also has some inherent problems. The first involves the etching of the mesas and the thermal-growth of the gate-oxide that is subsequently grown over the entire mesa. That is, the etching of the mesas produces silicon islands with sharp upper-corners. When the gate oxide is grown on these islands, the oxide at the top corner is thinner than on the flat areas of the mesa (typically 30-50% thinner than on the top surface - see

Vols. 1 and 2 for a discussion of oxide-thinning at corners during thermal-oxidation processes.) When the gate-material is run over the mesa and voltage is applied to turn-on the MOSFETs built in the mesa, the thin-oxide at the corner can cause premature gate-oxide-breakdown. The gate-oxide-breakdown-voltage may also be reduced on the sidewalls, since the crystal orientation may be different from that of the top surface of the silicon wafer. The growth-rate of SiO_2 is dependent on the silicon-crystal-orientation of the surface on which it is being grown. Furthermore, crystal-orientations other than (100)-Si generally have larger fixed-oxide-charge-densities and surface-state-charge-densities than (100)-Si, and this can decrease the threshold-voltage of MOSFETs built on such mesas.

The second problem relates to the steep sidewalls produced by an anisotropic-etch of the silicon-mesas. When a polysilicon-layer (used to form the MOSFET gate) is deposited on the mesa (typically to a thickness of 400-nm), an anisotropic-etching of the poly will leave stringers along the mesas, and this may short the gates of devices within the same mesa. Overetching is therefore needed to remove such stringers, but this can lead to poor dimensional-control of the poly gate-lengths.

The third problem involves early inversion along the top of the mesa sidewalls, which behave like lateral (edge) transistors in parallel with the top (main) device. This is due to charge-sharing of the depletion-regions caused by the poly running over the top and sides of the mesa-sidewalls. Thus, the corners show a lower threshold-voltage than the main channel of the MOSFET. This manifests itself as a kink in the subthreshold-characteristic of the drain-current. Such a deterioration of the subthreshold behavior is most pronounced in NMOSFETs in SOI-mesas, and can be alleviated by performing a boron-implant into the field-regions of the silicon wafer before etching the mesas. This boron-implant is then driven-in laterally into the active-area. The mesas are then etched, leaving the sidewalls more highly doped than the active-channel. Figure 11-31 shows a perspective drawing of a laterally-isolated MESA SOI-NMOSFET with a T-gate and a body-substrate contact.

11.7.2 LOCOS-Isolation for SOI

LOCOS-isolation is the second technique that has been used to produce lateral-isolation among active-silicon-islands in SOI. LOCOS is more complicated than MESA isolation, but it has the advantage of already being in wide use in bulk-CMOS technology. LOCOS-isolation in SOI is much the same as LOCOS in bulk-CMOS (see Vols. 1, 2, or 3 of *Silicon Processing for the VLSI Era* for more details on the LOCOS process.) However, there are some issues that do arise

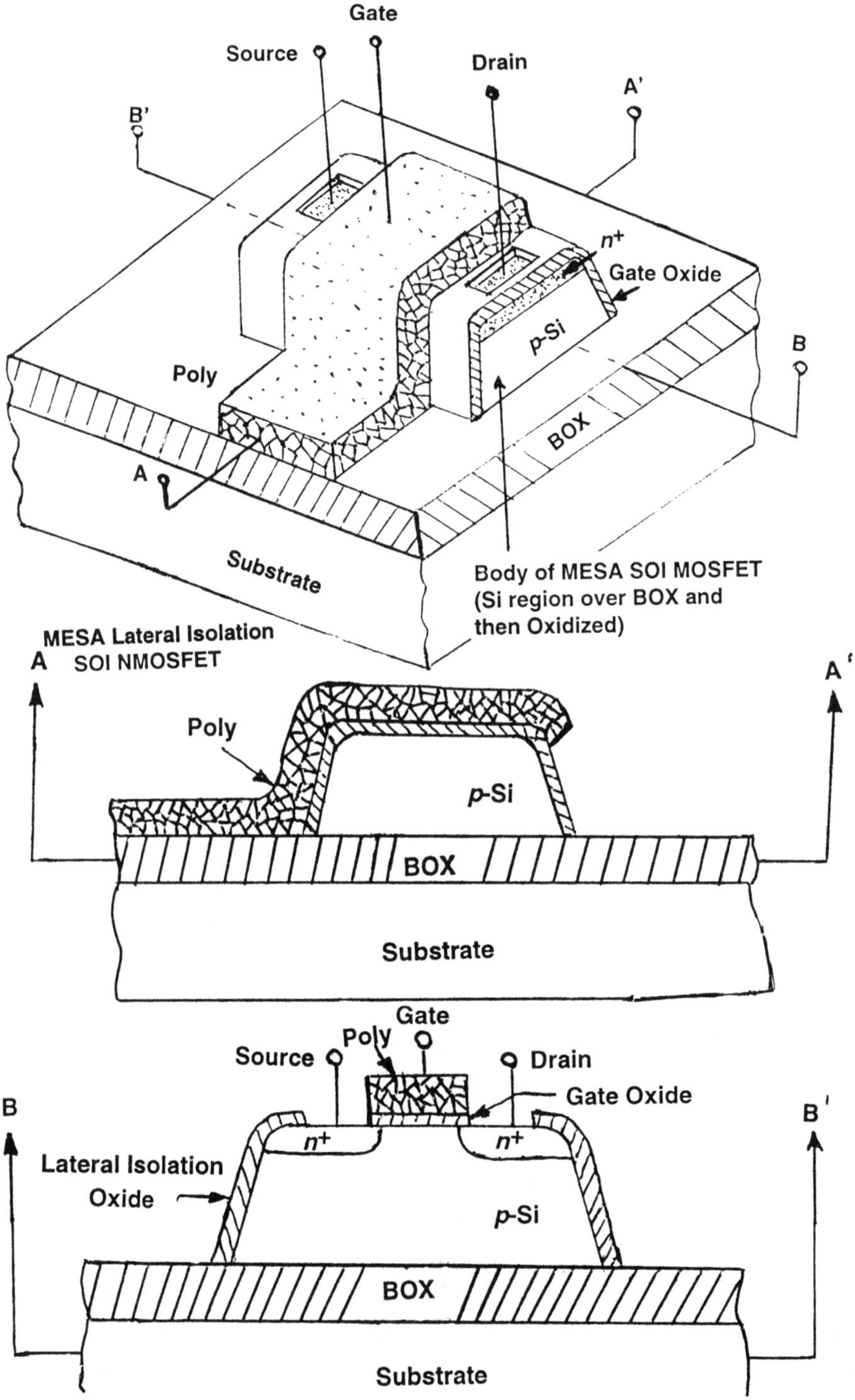

Fig. 11-30 MESA lateral isolation SOI NMOSFET.

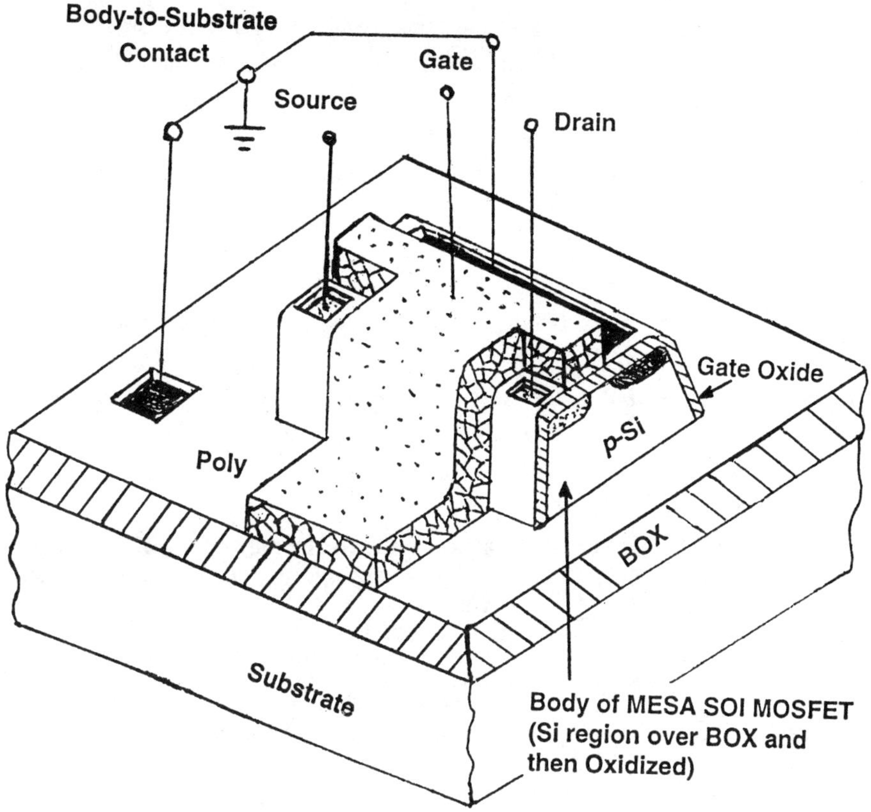

11-31 Mesa lateral-isolation SOI NMOSFET with a T-gate and body-to-substrate contact.

when LOCOS is applied as the technique for producing lateral-isolation in SOI. These issues are discussed in this section.

The LOCOS lateral-isolation-technique in SOI consists of the following steps (for producing NMOSFETs): a thin pad-oxide is grown, and a silicon-nitride-layer is then deposited by LPCVD onto the pad-oxide. Using a masking step, the nitride is patterned to define the active-silicon-areas. Boron is then implanted around the islands of *p*-type silicon that will contain the NMOSFETs (i.e., a channel-stop-implant). Next, the field-oxide is thermally grown. No oxide growth occurs over the active-areas which are protected by the silicon-nitride. A nitride and pad-oxide stripping step completes the process. In SOI the field-oxide-growth is carried out so that the thermally-grown field-oxide completely consumes the silicon in the regions not covered by the nitride, with the thermally-grown field-oxide reaching through to the buried oxide (see Fig. 11-32).

The lateral-encroachment of the field-oxide (called the "bird's beak") is proportional to the thickness of the grown field-oxide. This encroachment has typical values of a fraction of a micron, and this hampers the use of the LOCOS technique for submicron bulk-CMOS processes (where field-oxide-thicknesses of 0.3-05 μm are grown). The problem of "bird's-beak"-encroachment in SOI-CMOS is even worse than in bulk-CMOS processes. This is because the oxidation-kinetics in SOI are somewhat different, particularly when the growing field-oxide reaches the buried-oxide. That is, consumption of silicon during oxide-growth on a bulk-silicon-wafer is 0.45-t_{ox}, where t_{ox} is the oxide-thickness. That means that a thermal-field-oxidation process that produces a silicon-dioxide-layer about twice as thick as the consumed silicon-layer must be used in a bulk-CMOS. For example, a silicon-layer of about 125-nm is consumed when a 250-nm SiO_2 field-oxide-layer is grown.

By comparison, a 125-nm silicon-layer above the buried-oxide of an SOI-wafer is not fully oxidized if only such a 250-nm silicon-oxide growth-process is used. That is, unoxidized-filaments of silicon remain in SiO_2-layers grown by the 250-nm SiO_2-process on an SOI-wafer with a 125-nm silicon-layer above the buried-oxide. To ensure that such silicon-filaments no longer remain in the silicon-dioxide-layer, an oxidation-process equivalent to one that would grow a 350-nm-thick SiO_2-field-oxide on a bulk-wafer must be used. In other words, a process that grows an oxide 3-times-as-thick as the silicon-layer consumed has to be used in SOI to completely laterally-isolate the active-regions with a LOCOS process. The longer field-oxidation-process produces even more "bird's-beak" encroachment (and more lateral-boron-diffusion from the channel-stop-implant into the active-region) than a LOCOS-process in a bulk-CMOS process.

Another problem of the LOCOS method of lateral-isolation is that source-to-drain leakage-current in the subthreshold-region of operation can also be increased by the parasitic-edge-transistor that exists along the sidewalls of the silicon-active-regions in parallel with the main-channel of the SOI-MOSFETs. The problem is most severe in LOCOS-isolated SOI-NMOSFETs for the following reason: A p^+ channel-stop-implant is carried out before the field-oxidation to prevent such edge-transistors from being turned-on prematurely and causing excess subthreshold-leakage-current. However, during the field-oxidation step, such implanted-boron segregates into the growing-oxide on the top of the silicon-surface - and into the buried-oxide as well. As a result, the boron doping at the tip of the bottom corners of the active-device islands ends up being

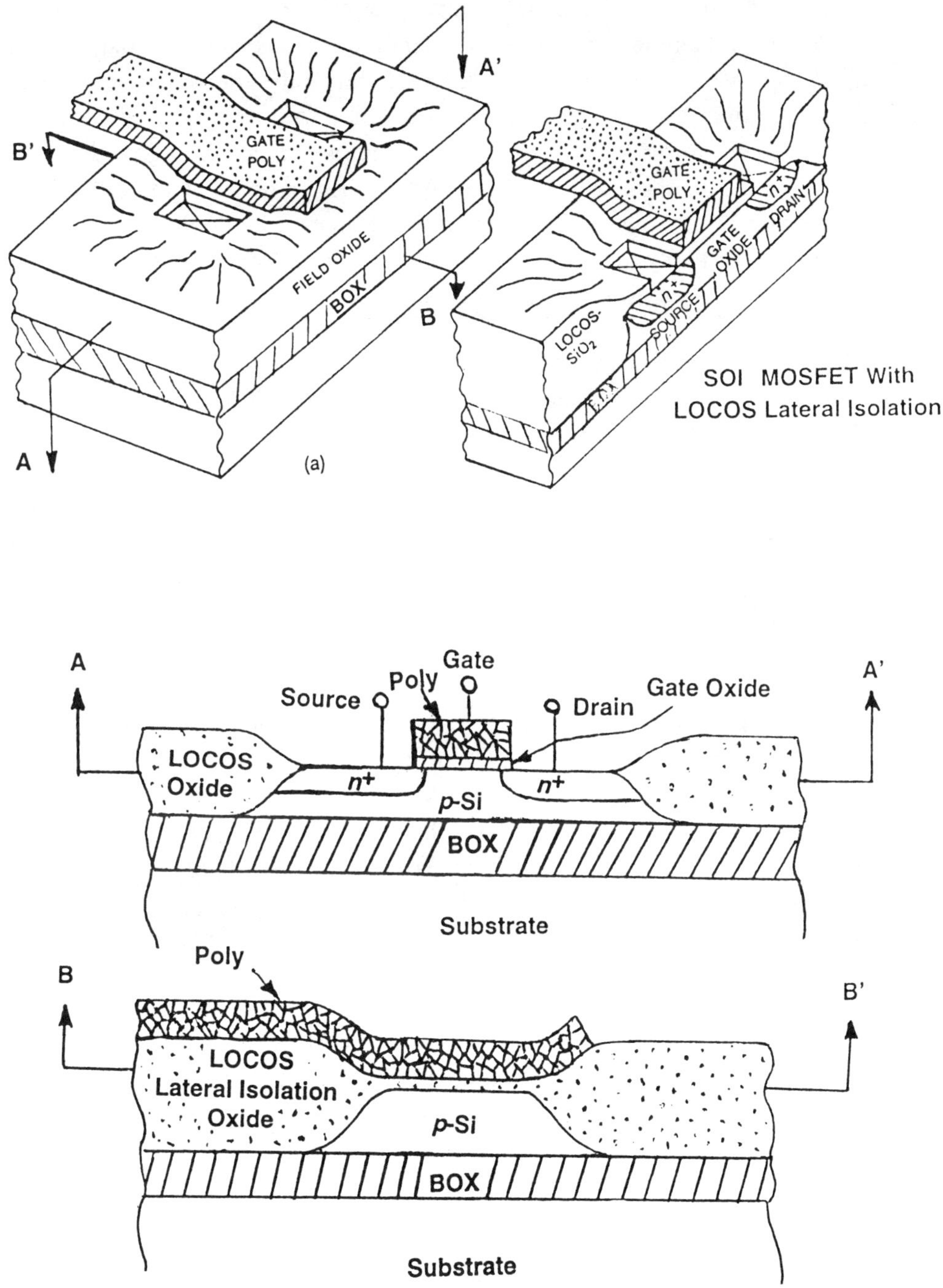

Fig. 11-32 SOI-MOSFET that uses LOCOS lateral isolation.

quite low, allowing edge-transistor-leakage to arise. One method suggested to alleviate this problem is to perform a second boron channel-stop-implant after the LOCOS field-oxide-step has been completed. Since the implant step is carried out after the field-oxide step, such B will be implanted into both active-regions and field-regions of the wafer. The energy of the implant can be adjusted to place the peak of the B-implant-dose close to the meeting point of the grown- and buried-oxides. Thus, this implant step will not impact the doping-concentration of the main MOSFET channels near the surface of the active-regions.

11.7.3 Shallow-Trench Isolation (STI) for SOI

Because of the problems associated with the MESA and LOCOS isolation techniques for providing lateral isolation in SOI technology, the more advanced SOI technologies have adopted shallow-trench-isolation as the lateral-isolation method of preference (Figs. 11-1 and 11-33). That is, the active-regions of the silicon are formed as in the MESA isolation method, but in STI the regions between the mesas are filled with a CVD-dielectric-material and then the surface is planarized by CMP to yield a flat wafer surface. Oxide remains in the trenches between the mesas, and its top surface lies in the same plane as that of the silicon-islands. STI eliminates the problems created by the sidewalls of the mesas, and the issues associated with the LOCOS-isolation method for SOI. (That is, the problem of "bird's-beak" encroachment does not occur when STI is used. Also, anomalies in the oxidation kinetics associated with field-oxide growth in SOI do not have to be addressed since the field-regions between the active devices are filled using CVD rather than a thermal-oxidation process). However, the same general issues involved with shallow-trench-isolation methods in bulk-CMOS still have to be addressed when applying an STI approach to SOI. Chapter 9 of this text provides a detailed discussion of STI technology.

11.8 SOI PRODUCT EVOLUTION

Table 11-4 presents a list of the SOI circuits that have been introduced during the International Electron Devices Meeting (IEDM) from 1989 to 1998.[64] It provides an indicator of SOI R&D efforts. The initial publication of the gain in the speed of a relatively small IC fabricated in SOI has been followed by a series of studies of performance-enhancements of larger ICs, mostly using SIMOX wafers. In general, these ICs were fabricated using partially-depleted SOI-MOSFET technologies, and silicon- and BOX-thicknesses were all comparable (and fairly thick). In 1998, IBM announced it would apply its leading-edge SOI-CMOS technology (including copper interconnects), to commercialize SOI-CMOS prod-

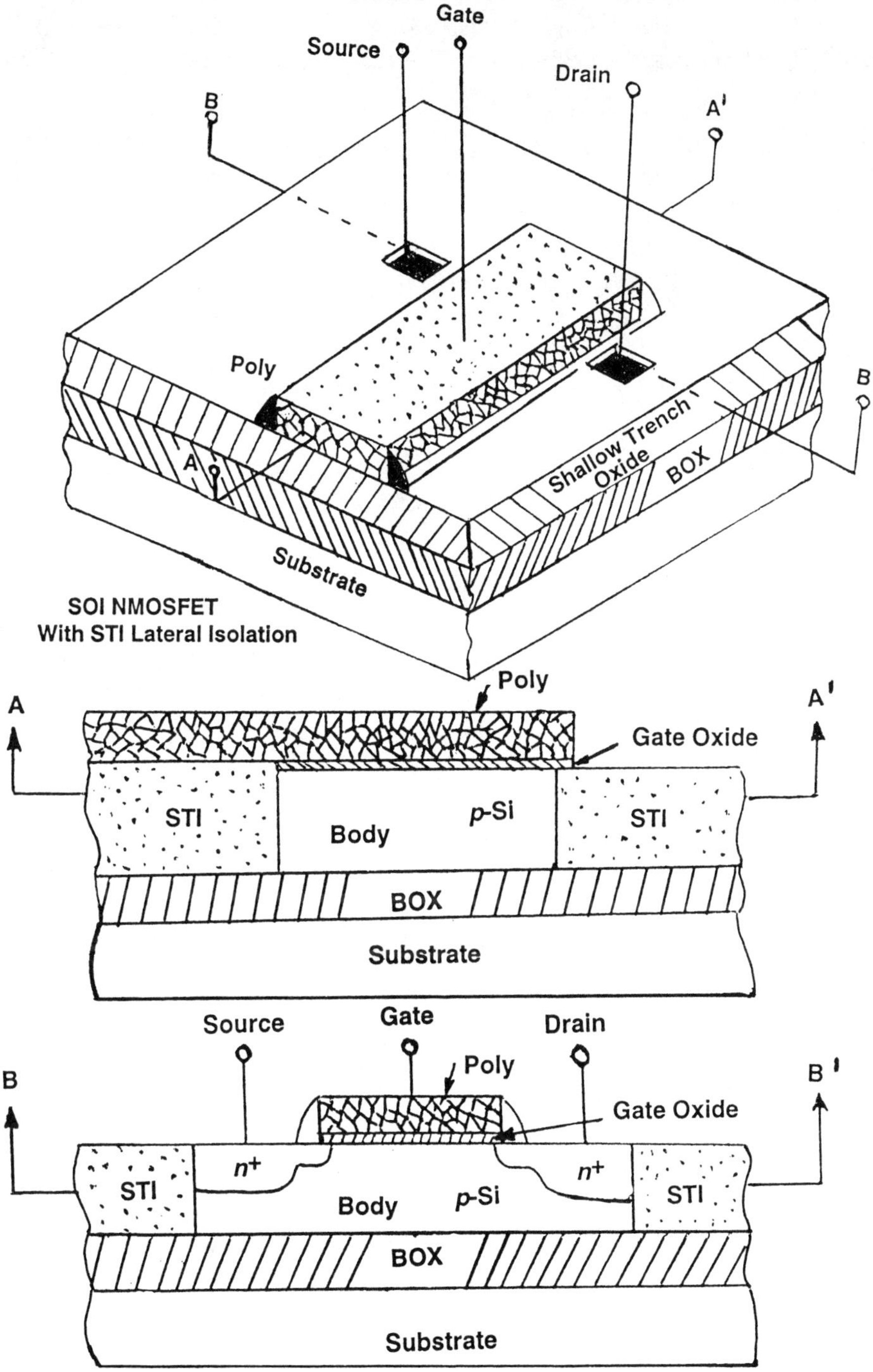

Fig. 11-33 Shallow-trench isolation (STI) lateral isolation SOI MOSFET.

Table 11-4 Summary of SOI Circuits Presented Annually at IEDM[64]

	Bell Labs 1989	IBM 1993	Motorola 1995	Mitsubishi 1996	Hyundai 1997	DEC 1997	IBM 1997	Intel 1997
Circuit	Prescaler	0.5Mb SRAM	CPU	16M DRAM	1G DRAM	1M SRAM	4 Mb SRAM	4 Mb SRAM
Technology (μm)	0.50	0.50	0.50	0.50	0.17	0.35	0.25	Sub-0.25
V_{DD}		1.0		1.0		2.0	1.8	1.8
Silicon thickness	134		100	100	200	200	200?	150
BOX (nm)	235		400	400	250	200	400	400
Lateral isolation	LOCOS			MESA	LOCOS	STI	STI	STI
Silicon layer characteristics	PD/FD	Non-depleted	PD		Body-cont., thin Si film (75nm) beneath FOX	PD	Non-depleted SOI	PD
Metal levels	1		2			4	6	5
Performance improvement	50%						25–35%	
Power reduction						50%	170–300%	18%
Yield (if known)	Low		Same as bulk			Approaching bulk		Low
Conclusions and claims			Bulk CMOS design, performance 2 × bulk	Both body-tied and floating body MOSFETs	Not clear if circuit is functional!	20% performance improvement	Highest-performance 0.25μm CMOS technology reported. FB effects negligible	No speed advantage over bulk for sub-0.25μm logic applications

ucts based on SIMOX. Figure 11-16 shows a SEM of a 0.18-μm IBM IC structure built using SOI. The company believes SOI processors are 30% faster than bulk-CMOS versions, and that they provide comparable circuit yields and add less than 10% to wafer-processing costs. Other companies have also announced the transition to SOI for high-speed microprocessors or low-power logic circuits. This is a positive development that supports IBM's claim that CMOS could move more towards SOI in the next few years (see Fig. 11-5).

In 1999, the presentation of new SOI devices occurred primarily at the *International Solid-State Circuits Conference* (instead of at the IEDM). Announcements were made concerning the successful fabrication of a number of SOI circuits in this ISSCC 1999 Conference. These circuits included 32- and 64-bit microprocessors, and 16-Mbit DRAMS on SOI. Depending on the technology and circuit layout performance, 10-40% enhancement was reported. A 0.18-μm SOI-CMOS technology for high-performance logic applications, however, was described by IBM in the 1999 IEDM. This SOI technology used a low-dose SIMOX substrate, shallow-trench-isolation, dual gate-oxide-thickness, dual threshold-voltage (V_T), dual-doped-polysilicon, cobalt-silicide, low-k dielectric in their interconnect scheme, and seven layers of copper metallization. The SOI circuit devices were partially-depleted MOSFETs, and the test chip built using this technology contained a microprocessor-circuit that demonstrated performance above 1-GHz.[58,59] These reports indicate a new-degree of maturity in SOI technology.

The 0.18-μm SOI process of IBM was qualified by that firm for production in the summer of 2000, and work was continuing to qualify a 0.13-μm SOI process in 2001. The 0.13-μm SOI technology would allow production of high-performance devices with 44-million transistors. IBM also said that it had begun shipping its AS400 servers with 0.22-μm SOI devices early in 2000, and that their newer "Turbo" Unix machines would employ 0.18-μm SOI-processors running at well over 800-MHz. The 0.18-μm designs utilize shallow-trench-isolation, dual-work-function-poly, local-interconnect, and borderless tungsten-interlayer interconnect.

At the end of 2001, IBM reported on their 2nd generation 0.13-μm SOI process. The power supply voltage was 1.2 V, and a thinner SOI-layer and a gate-dielectric material with a higher concentration of nitrogen were used. The interconnect structure can have as many as 8 levels of metal. A completely low-k dielectric stack (i.e., with SiLK for the main dielectric material and SiC as the hardmask material and the post-CMP cap-dielectric). The Cu wires are formed using a dual-damascene process.

11.9 PROBLEMS ASSOCIATED WITH SOI

11.9.1 Floating Body Effects in Partially-Depleted SOI MOSFETs

In partially-depleted SOI MOSFETs, the depletion-region under the gate when the MOSFET is operated in inversion does not extend all the way to the buried-oxide interface. Instead, a neutral-region still exists. When the body is grounded (say, with an independent body-contact), a partially-depleted SOI-MOSFET behaves very much like a bulk-MOSFET, and most standard $I_D(V_{GS},V_{DS})$ equations and design concepts still apply. However, if a body-contact (connected by a conducting-line to ground) is not made to the SOI-MOSFET, so called *floating-body-effects* arise, some of which have a detrimental impact on the SOI MOSFET behavior. These are discussed in more detail in the following sections.

11.9.1.1 Kink Effect in Partially Depleted SOI NMOSFETs: The *kink effect* manifests itself as a "kink" in the output characteristics of partially-depleted SOI MOSFETs (Fig. 11-34). The kink appears in the graph of the drain current of

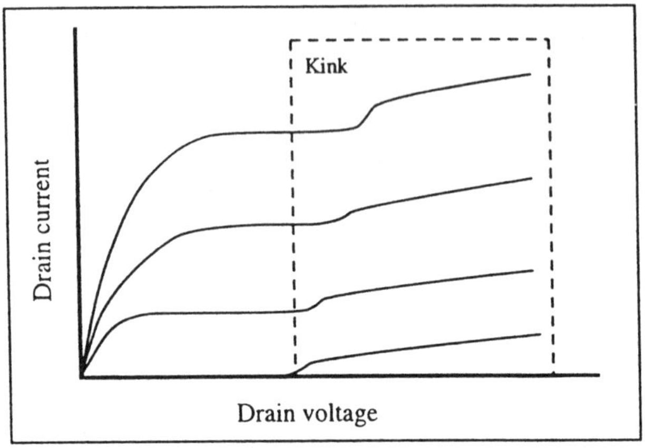

Fig. 11-34 Illustration of the kink-effect in the output characteristics of a partially-depleted SOI-MOSFET.

such devices above some value of the drain voltage. It is often quite pronounced in SOI-NMOSFETs, but is usually absent in SOI-PMOSFETs. The kink-effect is also not observed in any bulk-MOSFETs operated at room temperature if a substrate or well contact exists. It may be observed in bulk-NMOSFETs if they are operated a low temperatures or in NMOSFETs fabricated in a floating *p*-well.

The kink-effect occurs for the following reason. When the drain voltage is large enough in an NMOSFET, the electrons near the drain are accelerated across the channel-drain depletion-region by an electric-field that is strong enough to cause electron-hole pair generation by impact-ionization (in PMOSFETs the coefficient of electron-hole pair generation by energetic-holes is much lower than that of energetic-electrons - which are the carriers in NMOSFETs). Drain voltages typically get this high when the MOSFET is operating in the saturation region (Fig. 11-34).

In a bulk-NMOSFET, electrons generated by this effect rapidly move into the drain, adding to the drain-current. The generated holes exit the device through the body-contact where they become substrate-current. In partially-depleted SOI-NMOSFETs, the generated electrons still move out of the drain and add to the drain-current. The generated holes, however, collect in the NMOSFET body, where they raise the potential of the body, with respect to the source (which is grounded). The amount by which the potential of the body is raised can be calculated, but this requires iterative solving of a set of complex equations, and will not be described here. However, when the body-potential rises about 700-800 mV, it starts to significantly impact the output-characteristics of the NMOSFET.

That is, an increase in the body-potential has the same effect as biasing the substrate (body-effect). In other words, when the body of an NMOSFET is positively biased, it will give rise to a decrease in the threshold-voltage, V_T, of the NMOSFET (this is termed the *body-effect*). The decrease in V_T causes an increase in the drain-current, and this increase shows up as a "kink" in the drain-current curve of the SOI-NMOSFET. The kink represents an annoyance when analog circuits are designed with SOI NMOSFETs, but can actually lead to the fabrication of faster digital-circuits.

The situation regarding floating-body-effects in fully-depleted SOI-MOSFETs is somewhat different than in partially-depleted devices for several reasons. First, in the fully-depleted SOI-MOSFET, the body of the MOSFET at threshold has the same potential as the surface of the silicon at the top Si/SiO_2 interface. Hence, any holes generated by impact ionization at the drain will migrate out of the source and not collect in the body (which, in a partially-depleted SOI MOSFET, will have a lower potential than the source, and will thus attract the generated holes). Hence, there is no collection of holes in the body region, and the body-effect is also insignificant. In other words, the generation of holes by impact-ionization in fully-depleted SOI-MOSFETs does not decrease V_T, and hence no kink-effect is observed in fully-depleted SOI-MOSFETs, even in the absence of a body contact (i.e., body contacts are thus no longer needed). Second, since the potential of the body-region is higher, the electric-field at the drain will be smaller, reducing the number of electron-hole-pairs generated by impact-ionization.

There is, however, some effect on the subthreshold-slope of both partially-depleted and fully-depleted SOI-MOSFETs that is produced by the electron-hole pairs generated by impact-ionization at the drain. That is, as the MOSFET is being turned-on (but is still in the subthreshold-regime), the generated-holes will collect in the body (which has a lower potential than the source in both partially- and fully-depleted SOI-NMOSFETs if it is being operated in the subthreshold-regime). These injected-holes charge the body and produce a forward-bias on the body-source junction that reduces the threshold-voltage. As a result, the subthreshold-current increases, and the subthreshold-slope of the $I_D(V_{GS})$ curve is not as steep (i.e., the turn-off behavior of the device is degraded).

11.9.1.2 Transient I_D Effects: The floating-body may also produce transient-effects in the drain-current of a partially-depleted SOI-MOSFET. For example, an *overshoot* in the drain-current occurs when the MOSFET is turned-on (e.g., by increasing the gate-voltage from 0 to V_{DD}). What is happening is that majority-carriers are being expelled from the volume of the silicon of the floating-body under the gate that is becoming a depletion-region, and these carriers collect in the remaining neutral-region of the floating-body. This causes the potential of the floating-body to rise, and (as is described in the section on the kink-effect), this causes an increase in the drain- current. However, these excess-majority-carriers recombine and eventually equilibrium is re-established. At that point, the potential of the floating-body is returned to the value it would have without the presence of the excess injected- majority-carriers. Thus, the drain-current gradually subsides with time to a lower value. A reciprocal transient *undershoot* in the drain-current is observed when the gate voltage is switched off (e.g., if the gate-voltage decreases from V_{DD} to 0).

Such short-duration transient spikes in the drain-current can impact such SOI-circuits as SRAMs and DRAMs, which use pass-gate configurations. The amount of charge that is passed by the transient floating-body-effects must be accounted for in the design of such circuits.

11.9.1.3 Effect of Building Partially Depleted SOI-MOSFETs Without Body-Contacts: As discussed in the previous sections, the floating-body of partially-depleted SOI- MOSFETs has an impact on various MOSFET performance characteristics, most of which appear to be undesirable. However, Shahidi made a radical proposal in 1994. He suggested that in many important commercial circuit applications, partially-depleted SOI-MOSFETs could be used without contacts to the floating-body, and that this could be accomplished by making some modifications to the circuit design. Furthermore, by leaving the body floating, not only would the density be increased, but circuit-speed would be improved as well.

Early applications of SOI were limited to space and military programs where radiation-hardness was more important than performance and cost. These products were built with partially-depleted SOI, and fixed potentials to the body were provided by well- and substrate-contacts, just as is done in bulk-CMOS technology. In commercial applications, the considerations of performance and cost become paramount, and radiation-hardness is almost never an issue. Thus, Shahidi suggested that the floating-body-effects would not significantly degrade the performance of many SOI-ICs tailored for specific applications. If that was the case, the floating-body-contacts could be left out. In fact, several bulk-SRAM circuits and microprocessors were implemented in a partially-depleted SOI-technology that did not employ floating-body-contacts, and these SOI-ICs did indeed exhibit high-performance. In addition, they met all other commercial product specifications (such as stress, reliability, burn-in, and yield). Although some circuit-modifications to the bulk-CMOS circuit design had to be made when implementing them in SOI, these modifications resulted in a higher performance gain than would have been obtained by merely re-mapping the bulk-CMOS design into SOI.[51] The floating-body-effects on the change in drain-current were compensated for in the circuit designs by a number of solutions, and these allowed the SOI-ICs to be successfully designed and fabricated using partially-depleted SOI-MOSFETs without contacts to the floating-body.[52]

Another floating-body issue arises when designing 0.13-μm SOI-CMOS circuits (and beyond), as a result of the very thin gate-oxide-thicknesses needed for these generations of devices. That is, such thin-gate-oxides give rise to a significant amount of gate-to-body tunneling-current. The charge introduced into the floating-body modulates the body-voltage and therefore causes a so-called *history-effect*, which can become excessive on 0.1-μm SOI-CMOS circuits.[60]

This history-effect can be explained as follows: The gate-current in these circuits can change the device "ON" V_{bs}, but does not usually change the "OFF" V_{bs}. Therefore, the gate current will not impact the delay when the input signal switches from "0" to "1" for the first time, but it will impact the delay when the input signal then falls from "1" to "0" (1st switch V_{bs}) - and the impact of this delay will persist when the input signal switches the second time from "0" to "1" (2nd switch V_{bs} – Fig. 11-35). This history-effect must be minimized in 0.1-μm SOI-CMOS by appropriate device-design.

Minimizing the history-effect can accomplished by making two modifications to the SOI-MOSFET device-design: **1**) the silicon-dioxide gate-dielectric is replaced with a dielectric having a higher *k*-value (which reduces the gate-current for an equivalent-oxide-thickness); and **2**) the value of gate-side junction capacitance C_j is adjusted.[60] (That is, C_j controls the drain-capacitance coupling to the body, and hence determines the 2nd switch V_{bs}.) By adjusting C_j at a given level of gate leakage-current, the 1st and 2nd switch delays can be balanced (thereby eliminating the history effect in the switching delay). From a performance point of view, a low-value of C_j is beneficial. C_j can be reduced effectively by scaling the SOI silicon-layer thickness.

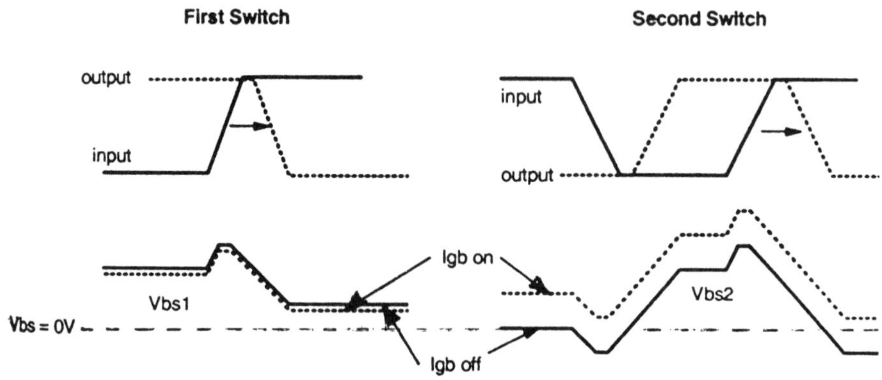

Fig. 11-35 Definition of 1st and 2nd switch rise-to-fall delay for the NMOSFET in an SOI-CMOS inverter. (© IEEE 2000).[60]

11.9.2 Short-Channel-Effects in SOI-MOSFETs

SOI-MOSFETs offer good short-channel behavior, especially for devices with channel lengths smaller than 0.1-μm. Fully-depleted SOI-devices are being studied for low-power/low-voltage applications because they appear to exhibit the best short-channel behavior. However, the silicon-thickness and the buried-oxide-thickness have to be made ultrathin in single-gate fully-depleted SOI-MOSFETs to achieve the best short-channel behavior (i.e., the silicon-thickness must be considerably smaller than the junction-depth of a comparable bulk technology). Even in such ultra-thin SOI-MOSFETs, back-channel-potential control-issues can exacerbate such short-channel-effects as threshold-voltage-rolloff and this issue must be addressed.[54,55]

The problem of threshold-voltage-rolloff arises in deep-submicron single-gated fully-depleted SOI-devices because the depletion-region-charge of the source/drain-regions cannot terminate in the buried-oxide region, and thus must terminate in the substrate below the buried-oxide or in the channel-region (Fig. 11-36).[55] Because the buried-oxide is thick, a portion of the depletion-charge will terminate in the channel. Thus, the source and drain behave like fringing back-gates. If the channel is short, the fringing-fields from these parasitic back-gates can induce back-channel-conduction, which enhances the problem of threshold-voltage-rolloff with decreasing channel-length. This phenomenon has been

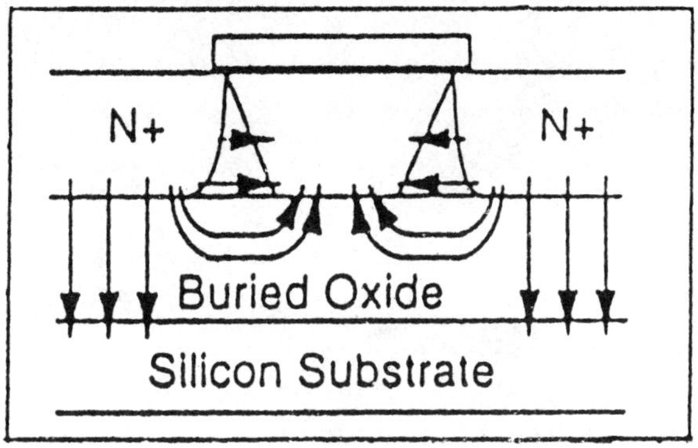

Fig. 11-36 Electric-field lines in the source/drain and body *pn*-junctions in a short-channel SOI-MOSFET. Some of these lines terminate in the substrate beneath the buried-oxide (BOX), while others terminate in the body, producing fringing-electric-fields.[55]

studied by simulation and empirically modeled.[55] Reducing the buried-oxide thickness can also help alleviate the 2-D charge-sharing through the buried-oxide, because the bottom substrate is brought closer to the source and drain regions, provided there is no significant depletion in the substrate. Simulation studies have been presented that predict that single-gate SOI-MOSFETs, even with ultra-thin silicon channels, do not provide adequate immunity against short-channel-effects in 25-nm channel-length devices.[57]

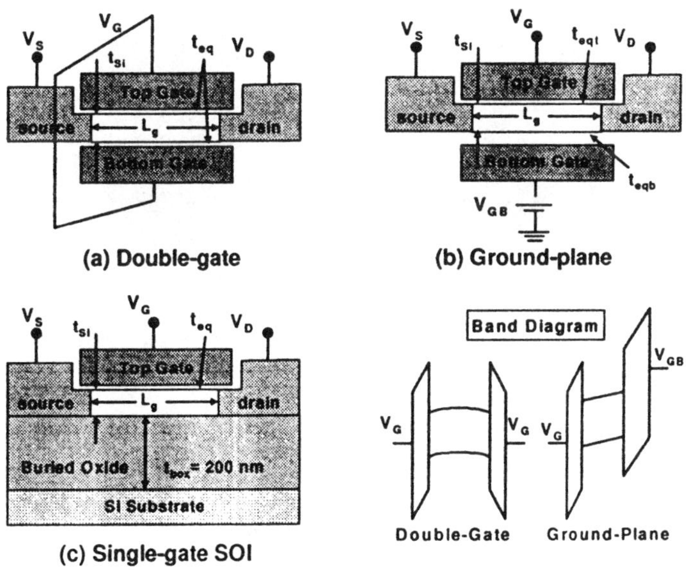

Fig. 11-37 a) Double-gate; b) Ground-plane; c) Single-gated SOI-MOSFET. Gate material work-function is set at the mid-gap of the silicon band gap. On-chip biasing of the ground plane is assumed. The source/drain junction is abrupt. The channel length (L_g) is the metallurgical channel length. (© 1998 IEEE).[57]

As a result, two alternative approaches have been proposed to reduce the threshold-voltage-rolloff problem described above. That is, it has been suggested that either a *double-gate SOI-device*, or a *ground-plane SOI-device*, be used instead of a single-gate SOI-device (Fig. 11-37). Simulation studies show that double-gate SOI-MOSFETs give the best control over the threshold voltage rolloff problem, but the difficulties in fabricating double-gate SOI MOSFETs are daunting. Instead, a compromise structure (namely an SOI device with a low-resistance ground plane) has been proposed.[56,57] In the ground-plane SOI MOSFET, the low resistance ground plane can be biased with a power supply separate from that used to control the gate. This offers the flexibility of tuning the threshold voltage by means of the bottom gate bias.[57] Both double-gate and ground-plane SOI-MOSFETs are shown to able to meet the electrostatic design requirements.

11.9.3 Self-Heating Effects in SOI MOSFETs

SOI MOSFETs are thermally insulated from the substrate by the poorly heat-conducting buried-oxide layer. Consequently, removal of excess heat generated by Joule heating within the transistors is less efficient than in bulk-CMOS technology. As a result, the device-temperature in SOI technologies can rise substantially above what it would be in a comparable bulk technology. Although the heat transfer rate is lower in an SOI device, heat is nevertheless still transported out through a number of paths, namely: **1**) through the buried oxide layer; and **2**) laterally through the silicon-island into the contacts and metallization. As we shall see below, however, such heating effects have not proved to be a significant problem in SOI CMOS technology.

The increased operating temperature within an SOI-MOSFET can have an impact on the device-performance. The most significant effect is that the mobility of the carriers in the channel is reduced. That is, if a MOSFET is being operated in the saturation-region, and the drain voltage is increased, there will be a larger voltage drop at the drain-junction. This will increase the power dissipated in the device and its temperature will rise. Once the device temperature is increased by this effect by 50°C (to 150°C), a significant reduction in carrier-mobility will occur. This effect will manifest itself in the output characteristics of the device in the form of a drop in the drain-current with increasing drain voltage (Fig. 11-38).

However, in CMOS circuits, all of the MOSFETs normally operate in a standby-mode. Large drain-currents in any device occur only during switching, and these cease after a very brief period of time (< 1-ns). Since the thermal time constants of SOI MOSFETs are on the order of several tens of nanoseconds, it

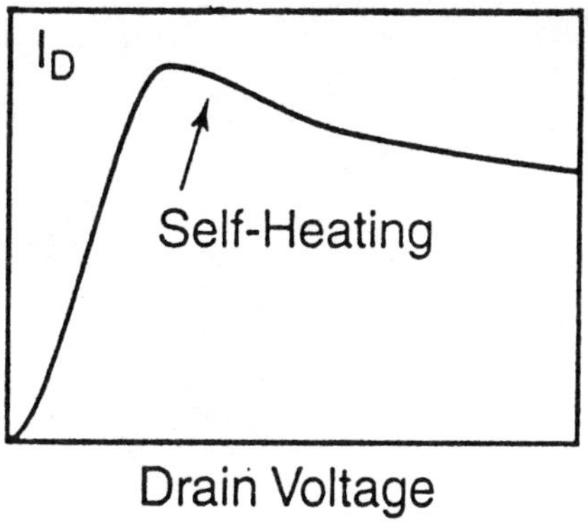

Fig. 11-38 The effect of self-heating in SOI-MOSFETs on output characteristics of the device.

seems that this self- heating effect generally will cause no problems in CMOS digital circuits, with a few exceptions. That is, in logic-circuits that exhibit short transition times, temperature rises of only 2-3 degrees are observed compared to bulk-CMOS. However, in such circuits as off-chip drivers (that can be ON for a significant portion of the switching cycle), the temperature rise is observed to be higher (e.g., ~10 degrees).

11.9.4 Reliability Issues for SOI

There are also some reliability issues unique to SOI-technology that should be addressed when considering SOI for ULSI applications.[61] Such an analysis indicates that SOI-circuits do not exhibit significantly worse reliability behavior than comparable bulk-CMOS-circuits. Some of the reliability issues associated with silicon-on-insulator ICs are reviewed here.

First, bulk- and SOI-CMOS circuits appear to react comparably with respect to *off-state current degradation* when subjected to bias-temperature stress-testing. Hence, this apparently does not represent a special reliability-problem for SOI. Second, channel hot-electron-degradation (CHE) in SOI-devices is more complex than in bulk-MOSFETs because charge-trapping and interface-state-generation can occur at both the gate-oxide-interface and at the buried-oxide-interface. Third, the defects responsible for gate-oxide yield and reliability have been minimized by 2000 in state–of-the-art SOI-materials to acceptable levels. This claim is supported by TDDB tests on gate-oxides on both bulk- and SOI-materials (i.e., the oxides grown on both the bulk- and SOI- materials show

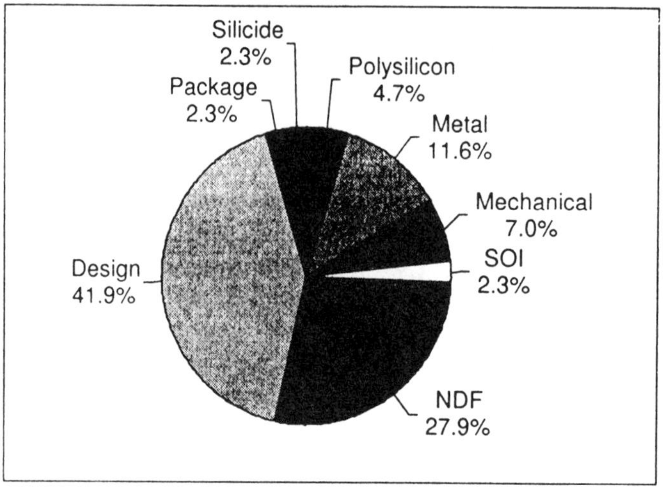

Fig. 11-39 SOI unique defect fails (induced by burn-in stress), are negligible part of the overall failure population.[61] (© 2000 IEEE)

similar failure distributions). Fourth, SOI-devices appear to be less susceptible to charging damage than bulk-CMOS devices. Finally, when failure mechanisms in SOI-microprocessor-circuits are catalogued, the percentage of fails due to SOI-unique defects represent a negligible portion of the failure distribution (Fig. 11-39). In summary, these reports indicate that by using state-of-the-art SOI material, highly reliable products can be produced.

REFERENCES

1. International Technology Roadmap for Semiconductors, 1999 Edition, Semiconductor Industry Association, San Jose, CA.

2. "Proc. 5th Inter. Workshop on Measurement, Characterization, and Modeling of Ultrashallow Doping Profiles in Semiconductors," *J. Vac. Sci. Technol.,* **B-18**, 2000, p. 337-604.

3. G. Shahidi, *et al.*, "Mainstreaming of the SOI Technology," *Proc. IEEE Inter. SOI Conf.*, p. 1-4, 1999.

4. A. Agarwal, "Ultra-Shallow Junction Formation by Spike Annealing in a Lamp-Based or Hot-Walled Rapid Thermal Annealing System: Effect of Ramp-Up Rate," *Materials Sci. in Semiconductor Processing*, 1, p. 237, 1998.

5. K.K. Ng and J. Lynch, "The Impact of Intrinsic Series Resistance on MOSFET Scaling," *IEEE Trans. Electron Dev.,* **ED-34**, p. 503, 1987.

6. M.I. Current *et. al.*, “Process Integration Issues for Doping of Ultrashallow Junctions,” *J. Vac. Sci. Technol.* **B-18**, p. 468, 2000.
7. C.-T. Chaung *et al.*, “SOI for Digital CMOS VLSI: Design Considerations and Advances,” *Proc. IEEE,* **86**, April 1998, p. 689.
8. S.T. Liu *et al.*, “Radiation Response of SOI Materials,” *Mtg. Abs. Electrochem. Soc,.* Spring 1999, Abs. 414.
9. J.-P. Colinge, “Subthreshold Slope of Thin-Film SOI MOSFETs,” *IEEE Electron Dev. Letts.,* **EDL-7**, pp. 244, 1986.
10. C.L Chen *et al.*, “High-Frequency Characterization of Sub-0.25-μm Fully-Depleted Silicon-on-Insulator MOSFETs,” *IEEE Electron Dev. Letts.,* **EDL-21,** October 2000, p. 497.
11. *ibid.* Ref. 7.
12. *ibid. Ref. 9.*
13. J.-P. Colinge, “Fully-Depleted SOI CMOS for Analog Applications,” *IEEE Trans. Electron Dev.*, **45**, May 1998, p. 1010.
14. G.G. Shahidi *et al.*, “A Room-Temperature 0.1-μm CMOS on SOI,” *IEEE Trans. Electron Dev.,* **41,** December 1994, p. 2405.
15. N. Kistler *et al.*, “Sub-Quarter Micron CMOS on Ultrathin SOI, *IEEE Electron Dev. Lett.*, **EDL-12**, May 1992, p. 235.
16. M. Chan *et al.*, “Recessed-Channel Structure for Fabricating Ultrathin SOI MOSFET with Low Series Resistance,” *IEEE Electron Dev. Lett,* **EDL-14**, January 1994, p. 22.
17. A. Toriumi *et al.*, “Performance and Reliability Concerns of Ultra-Thin SOI and Ultra-Thin Gate Oxide MOSFETs,” *Tech. Dig. IEDM,* 1995, p. 847.
18. G.W. Cullen and C.C. Wang, Eds., *Heteroepitaxial Semiconductors for Electronic Devices,* Springer-Verlag, 1978.
19. G.W. Cullen *et al.*, *J. Crystal Growth*, **56**, 287 (1982).
20. M.S. Abrahams and C.J. Buiocchi, "Cross-Sectional Electron Microscopy of Silicon on Sapphire", *Appl. Phys. Lett.*, **27**, 325 (1975).
21. K.E. Violette, P.A. O’Neill, M.C. Ozturk, K. Christensen, and D. Maher, “Low- Temperature Selective Silicon Epitaxy by Ultra High Vacuum Rapid Thermal Chemical Vapor Deposition Using Si_2H_6, H_2 and HCl,” *Appl. Phys. Lett.,* **68**, 66 (1996).
22. A. Gupta and P.K. Vasudev, "Recent Advances in Hetero-Epitaxial Silicon-on-Insulator Technology, Part II," *Solid State Technology,* **26**, 129 (1983).
23. R. Reedy, T.W. Sigmon, and L.A. Christer, “Suppressing Al Outdiffusion in Ion Implantation Amorphized and Recrystallized Silicon-on-Sapphire Films,” *Appl. Phys. Lett.*, **22**, 537 (1983).
24. G.A. Garcia, and R.E. Reedy, “Electron Mobility Within 100 nm of the Si/Sapphire Interface in Double-Solid Phase Epitaxially Regrown SOS,” *Electron. Lett.,* **22**, 537 (1986).

25. G.W. Cullen, M.T. Duffy, and R.K. Smeltzer, "Recent Advances in Heteroepitaxial Si on Sapphire Technology," *VLSI Science and Technology/1984*, Electrochemical Society, N.J., 1984, p. 230.
26. D. Munteanu *et al.*, "Silicon-on-Sapphire Material and Device Properties at High and Low Temperature," *Mtg. Abs. Electrochem. Soc.,* Spring 1999, Abs. 435.
27. K. Izumi *et al., Electron Lett.*, **14**, (1978) 593.
28. J. Liu *et al.*, "Formation of Buried Oxide in Silicon Using Separation by Plasma Implantation of Oxygen," *Appl. Phys. Letts.,* **67**, p. 2361, 1995.
29. One supplier of SIMOX wafers is Ibis Technology Corp, Danvers, MA.
30. C.E. Chen *et al.*, *1984 IEDM Tech. Digest*, 701 (1984).
31. S. Nakashima *et al.,* "Investigations on High-Temperature Thermal Oxidation Process at Top and Bottom Interfaces of Top Si of SIMOX Wafers," *J. Electrochem. Soc.,* **143**, p. 244, January 1996.
32. *Wafer Bonding and SOI Technologies*, Abstracts of the Fall 1997 Meeting of the Electrochemical Society, Paris, France.
33. Q.-Y. Tong and U. Gosele, *Semiconductor Wafer Bonding*, VCH-Wiley, New York (1999).
34. J.B. Lasky *et al.,* "SOI by Bonding and Etchback," *Tech. Dig. IEDM*, 1985 p. 684.
35. R.C. Frye *et al*, "A Field-Assisted Bonding Process for Silicon Dielectric Isolation," *J. Electrochem. Soc*, **133**, 1673 (1986).
36. H. Ohashi, "Improved Dielectrically Isolated Device Integration by Silicon-Wafer Direct Bonding Technique," *Tech. Dig. IEDM*, 1986 p. 210.
37. Q.-Y. Tong, W.J. Kim, T.-H. Lee, and U. Gosele, "Low-Vacuum Wafer Bonding," *Electrochemical and Solid State Letter*, **1**, 52 (1998).
38. A. Plossl and G. Krauter, *Mater. Sci. and Engr Reports,* **R25**, 1 (1999).
39. A.J. Auberton-Herve, "SOI: Materials to systems," *Tech. Dig. IEDM* 1996, p. 3.
40. Unibond Wafer Specification Sheet. SOITEC USA Inc. Peabody, MA, (1998).
41. IBM Microelectronics
42. H.W. Lam, A.F. Tasch, Jr., and R.F. Pinzzotto, "Silicon-on-Insulator for VLSI and VHSIC," in *VLSI Electronics*, Ed. N.G. Einspruch, Vol. 4, p. 1, Academic Press, NY (1982).
43. S. Farrens *et al.,* "High-Quality SOI Structures by the Genesis Process," *Mtg. Abs. Electrochem. Soc.,* Spring 1999, Abs 396.
44. M.I. Current *et al.*, "Ultrashallow Junctions or Ultrathin SOI?" *Solid State Technology*, September 2000, p. 66.
45. K. Sakaguchi *et al.,* "ELTRAN by Splitting Porous Si Wafers," *Mtg. Abs. Electrochem. Soc ,* Spring 1999, Abs. 395.

46. T. Yonehara *et al.*, "ELTRAN: Epitaxial Layer Transfer," *Mtg. Abs. Electrochem. Soc.*, Spring 1999, Abs. 394.
47. K. Suzuki *et al.,* "Scaling Theory for Double-Gate SOI MOSFETs," *IEEE Trans. Electron Dev.*, **40**, 2326 (1993).
48. C.H. Wann *et al.,* "A Comparative Study of Advanced MOSFET Concepts," *IEEE Trans. Electron Dev.*," **43**, 1742 (1996).
49. M. Tachimori *et al., Proc. Electroche,m. Soc. on SOI Tech & Dev.* Vol. 96-3, p. 53 (1996).
50. A. Matsamura *et al.,* "Recent Progress in Low-dose SIMOX Wafers Fabricated with Internal-Thermal-Oxidation (ITOX) Process," *Mtg. Abs. Electrochem. Soc.,* Spring 1999, Abs. 391.
51. D.H. Allen *et al.*, ISSCC, 1999.
52. F. Assaderaghi and G. Shahidi, "Advanced SOI Circuits for the Microelectronics Market," *Mtg. Abs. Electrochem. Soc.,* Spring 1999, Abs. 383.
53. "Canon Brings Big Company Muscle to the SOI Market With Yet Another Technology," *Solid State Technology*, June 2001, p. 50 (and 167, an Ad which announces the expanded production of ELTRAN wafers).
54. T. Ogawa, "At Long Last, SOI Market is On the Move," *Solid State Technology*, February 2001, p. 62.
55. L. Su *et al.,* "Deep-Submicrometer Channel Design in Silicon-On-Insulator (SOI) MOSFETs," *IEEE Electron Dev. Letts.*, Vol. 15, September 1994, p. 366.
56. T. Ernst and S. Cristoloveanu, "Ground Plane Concept for Reduction of Short-Channel Effects in Fully-Depleted SOI Devices," *Mtg. Abs. Electrochem. Soc.,* Spring 1999, Abs. 431.
57. H.-S. P. Wong, D.J. Frank, and P.M. Solomon, "Device Design Considerations for Double-Gate, Ground-Plane, and Single-Gated Ultra-Thin SOI MOSFETs at the 25 nm Channel Length Generation," *Tech. Dig. IEDM* 1998, p. 407.
58. E. Leobandung, *et al.*, "Scalability of SOI Technology into 0.13 μm 1.2 V CMOS Generation," *Tech. Dig. IEDM* 1998, p. 403.
59. E. Leobandung *et al.,* "High-Performance 0.18 μm SOI CMOS Technology," *Tech. Dig. IEDM* 1999, p. 679.
60. S.K.H. Fung *et al.*, "Controlling Floating-Body Effects for 0.13 μm and 0.1 μm SOI CMOS," *Tech. Dig. IEDM* 2000, p. 231.
61. R. Bolam *et al.,* "Reliability Issues for Silicon-on-Insulator," *Tech. Dig. IEDM* 2000, p. 131.
62. K. Sakaguchi and T. Yonehara, "SOI Wafers Based on Epitaxial Technology," *Solid State Technology,* June 2000, p. 88.
63. M. Current *et al.,* "Atomic-Layer Cleaving and Non-Contact Thinning and Thickening for Fabrication of Laminated Electronic and Photonic Materials," *Proceedings 2001 Materials Research Society Spring Meeting*, Paper 18.3.

64. A. Kamgar and J.T. Clemens, "Isolation's Path to SOI Technology," *Solid State Technology*, October 1999, p. 109.
65. J.W. Sleight *et al.,* "A High Performance 0.13 μm SOI CMOS Technology with a 70 nm Silicon film and with a Second Generation Low-*k* Cu BEOL," *Tech. Dig. IEDM*, 2001, p. 621.
66. R. Chau *et al.,* "A 50-nm Depleted Substrate CMOS Transistor (DST)," *Tech. Dig. IEDM,* 2001, p. 621.

Chapter 12

MULTILEVEL INTERCONNECTS FOR ULSI

The semiconductor industry continues to pursue the development of ever more powerful ICs. This effort basically involves a two-pronged approach. The first approach seeks to increase the speed of individual transistors through a continual reduction in the minimum size of device features. Such scaling also provides an increase in the device density on the chip. The second approach entails developing increasingly complex interconnect systems, which now use multilayer structures of metal wiring separated by interlayer dielectrics (ILDs). This enables the higher-speed devices to be interconnected (resulting in ICs with enhanced performance and system functionality). Efforts in both of these development arenas are supported by continually more sophisticated device and circuit designs.

The tradeoffs between individual device and interconnect performance depend on the details of the circuit architecture. When the IC industry was in its early stages (i.e., from the mid-1970s to about 1990), circuit design rules were in the 1-5-μm range. During that time, interconnect delays could generally be ignored because they were typically much smaller than the device switching times (which were ~50-Mz). However, by the mid-1990s, microprocessors and other large area ULSI devices with millions of circuit elements and speeds greater than 1-GHz were envisioned. Designers then began to realize that interconnect delays would become comparable to (and could eventually exceed) the intrinsic delays of the transistors. It was generally agreed that beyond the 0.25-μm technology node, the overall circuit performance would be dominated by the efficiency with which devices are connected, rather than by the speed of individual devices (Fig. 12-1).[1]

In addition, by the mid-1990s, two different technological domains had also started to collide – namely circuits designed for purely for radio-frequency (rf) applications and others designed merely for digital applications. That is, rf circuits were typically low-density and high-frequency (GHz) designs used in avionics and wireless communication, whereas digital circuits strove for high-density and covered a wide performance spectrum (1-100 MHz). The merger of performance requirements (or *mixed-signal products*), reflects a growing trend,

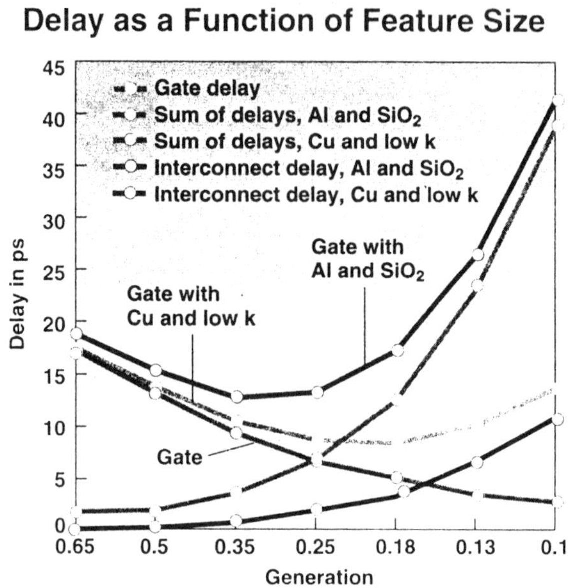

Fig. 12-1 RC-delay as a function of feature size.[1] (© IEEE 1997)

and places additional demands on interconnect systems as performance requirements continue to escalate.

In this chapter, some of the challenges related to the performance and density issues of ULSI interconnects will be discussed. The discussion will be divided into two parts:

1. The performance requirements of ULSI interconnect systems;
2. The evolution of the materials and architecture of interconnects for ICs.

The exact feature size at which the interconnect RC delay equals and then surpasses the intrinsic transistor delay depends on the details of the circuit design. Furthermore, the total overall interconnect delay t_{int} is composed of four components and can be approximated by:[2]

$$t_{int} = R_0 C + 0.4\,[(RC)^{1.6} + (tof)^{1.6}]^{1/1.6} + 0.7\,RC_{in} \tag{12.1}$$

where R_0 is the output resistance of the driver device, *tof* is the time of flight of the electromagnetic wave in the interconnect, and C_{in} is the input capacitance of the device in the next gate. These four terms contribute differently to the overall delay, depending on the gate-length and gate-width, the interconnect line-length, and other geometrical parameters. However, most signs point to the feature size of the order of 0.5-μm at which the interconnect RC-delay starts to become a significant part of the overall signal delay.

12.1 EVOLUTION OF THE STRUCTURE AND MATERIALS USED IN IC INTERCONNECTS

Interconnects for ICs have evolved from the single-level-metal system on the original IC patented by Noyce, to the multi-level-metal interconnect systems used in today's ULSI. In this section we will describe the evolution of these interconnect systems. This will point out the limitations that existed for previous generations of IC interconnect technologies, and how they were overcome by learning how to build more advanced interconnect systems. By describing the state-of-the-art of current interconnects, we will also be able to see the limitations they face, and what kinds of techniques are being explored to overcome them.

12.1.1 Single-Level-Metal Interconnects in Early ICs

In 1959 Robert N. Noyce invented the concept of the monolithic-integrated-circuit (IC) while at Fairchild Semiconductor. His patent application described how planar silicon-bipolar-transistors and resistors could be interconnected with thin and narrow aluminum lines adherent to an underlying silicon-dioxide film (SiO_2) that covered the surface of the silicon chip. Contact holes were etched in the oxide film, and these allowed the aluminum lines to make contact to the circuit devices fabricated within the silicon substrate. After the aluminum lines were patterned, a passivation layer (e.g., a CVD P-doped SiO_2 film) would be deposited over the entire wafer. Figure 12-2 shows a schematic of how such a single-level-metal interconnect system was fabricated. The materials used in these early ICs (Al and SiO_2) were selected because they had many unique advantages. These advantages proved to be so compelling that this same combination of materials was used for many generations of IC interconnect

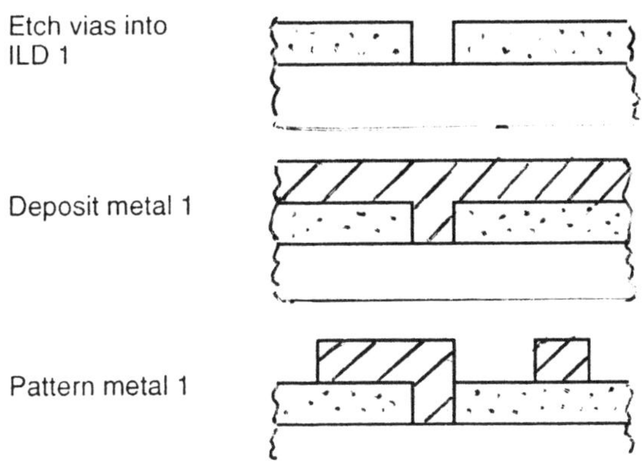

Fig. 12-2 Schematic of process steps used to produce a single-level-metal interconnect.

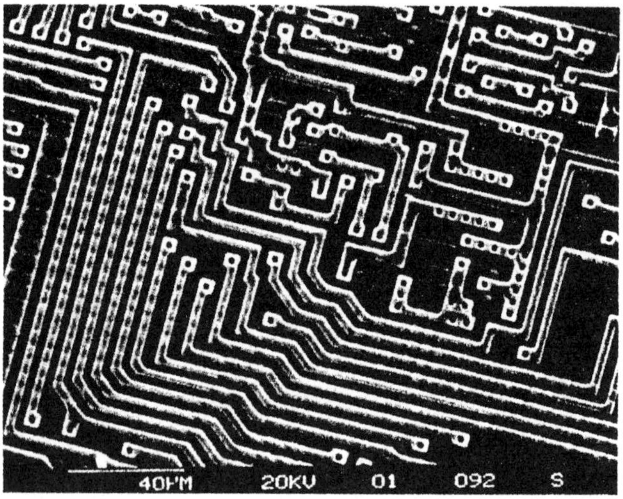

Fig. 12-3 SEM showing the patterned metal layer on a circuit with only one layer of metal.

technology. As a point of reference for materials parameters, the resistivity of pure Al is $\rho_{Al} = 2.7\text{-}\mu\Omega\text{-cm}$, and the dielectric constant of SiO_2 is $k_{SiO_2} = 4.2$ (for the type of SiO_2 used in dielectric films under the aluminum). Later versions of this interconnect system used Al films alloyed with Si to improve contact stability. Figure 12-3 shows an SEM of a circuit using only one-level of metal.

Note that the metal wires made in this traditional way of creating the IC interconnect structure are formed by: **1**) depositing a blanket layer of the metal over the entire wafer; and **2**) selectively removing the metal from regions on the wafer where the metal is not needed (i.e., with a *subtractive* etching process).

12.1.2 Double-Level-Metal Interconnects in Early Bipolar ICs

As the density of ICs increased with decreasing feature size, it became necessary to have more than one level of interconnect. In NMOS ICs, the polysilicon-gate material could serve as a local-interconnect level (eventually covered with a silicide to form a polycide), and this level could be covered with a single-level of metal (Al-Si) to form a two-level interconnect system.

In bipolar technology there is no MOSFET poly-gate structure, and thus to get two-levels of interconnect, it is necessary to use two-levels of metal (Fig. 12-4). Thus, the first two-level-metal interconnect systems for ICs were used in bipolar technology. (Figure 12-5 shows the steps used to fabricate such a 2-level-metal

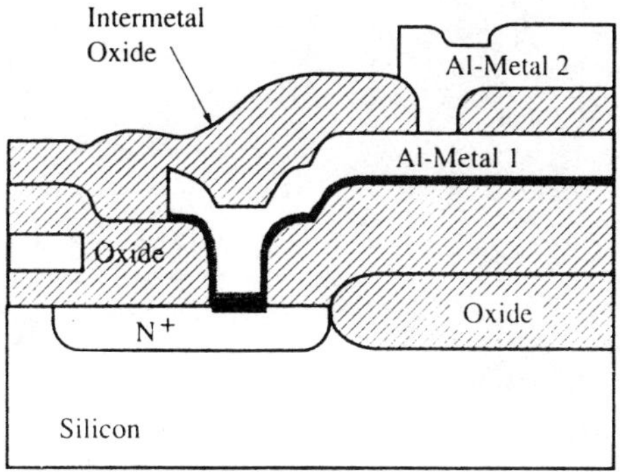

Fig. 12-4 Schematic of a double-level-metal structure used in early bipolar-ICs.

interconnect structure.) In these 2-level-metal bipolar-ICs there was no attempt made to planarize the dielectric layer that separated the 2-levels of metal (i.e., the *interlevel dielectric*, or *ILD*). The holes opened in the ILD to allow Metal-2 to contact Metal-1 were termed *vias*. While the two-level-metal process could be successfully implemented on bipolar-ICs, the step coverage of Metal-2 as it ran over Metal-1 lines was marginally acceptable. Thus, it would not have been possible to continue to use this process to form a three-level-metal interconnect, even in bipolar-ICs.

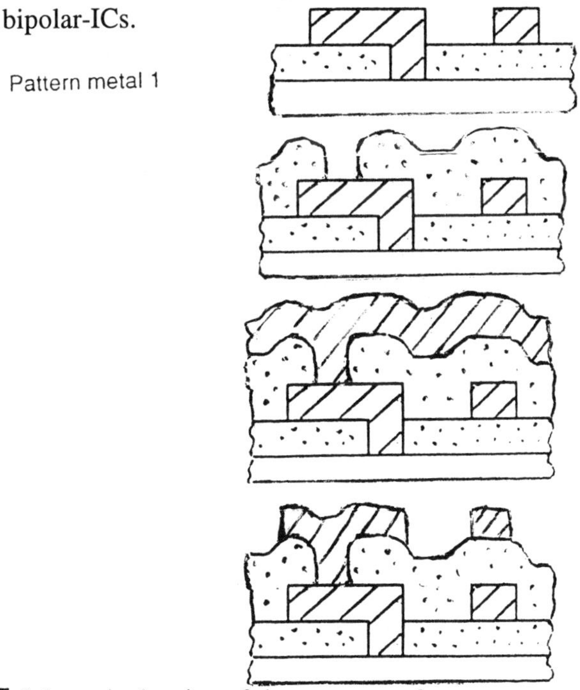

Fig. 12-5 Schematic drawing of the sequence of process steps used to produce a double-level-metal interconnect structure as shown in Fig. 12-4.

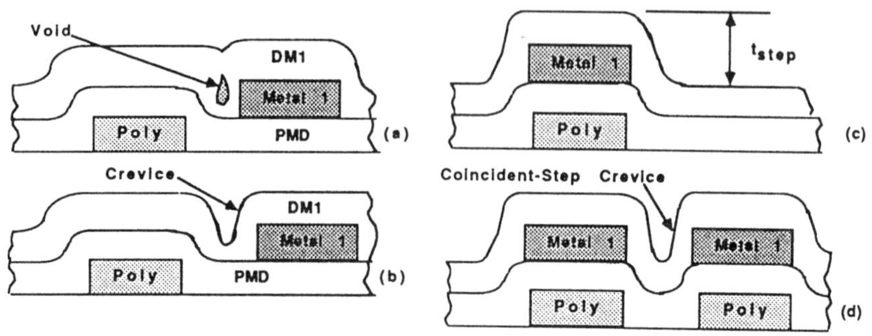

Fig. 12-6 Topography conditions that arise in MOSFETs that make it difficult to fabricate a double-level-metal structure without some kind of planarization method.

Likewise, when CMOS ICs came to replace NMOS ICs (in the mid-1980's), it was no longer possible to use the polysilicon as a local interconnect layer in the same manner that it had been used in NMOS. This was because polysilicon in early CMOS technologies was *n*-doped, and therefore it could not be used to make contact to PMOS source/drain regions. Instead, it also became necessary to use a double-level-metal interconnect structure in these CMOS ICs. Unfortunately, the same approach to fabricating a double-level-metal structure used in the early bipolar ICs could not be used in CMOS ICs. That is, the steps created by the polysilicon layer (together with the steps caused by Metal-1) were too high to allow adequate step coverage when Metal-2 lines had to run over such (poly + Metal-1 hills), as is shown in Fig. 12-6.

12.1.3 Double-Level-Metal Interconnects in CMOS ICs (Partial Planarization of the ILD)

To enable double-level-metal interconnect structures to be fabricated in CMOS ICs, the fabrication sequence used for making double-level-metal interconnects in bipolar ICs had to be modified. That is, it was necessary to be able to partially planarize the ILD prior to depositing Metal-2. The partial-planarization technologies used in the CMOS ICs of the late 1980's included *resist etchback* and the use of *spin-on glass* (SOG). Details concerning these partial-planarization methods can be found in Vol. 2, Chap. 4. Furthermore, the contact holes and vias were opened with a two-step etch process. The first half (which opened the top half of the via) used an *isotropic* etch process, while the second etch step (that opened the remainder of the via) used an *anisotropic* process. This

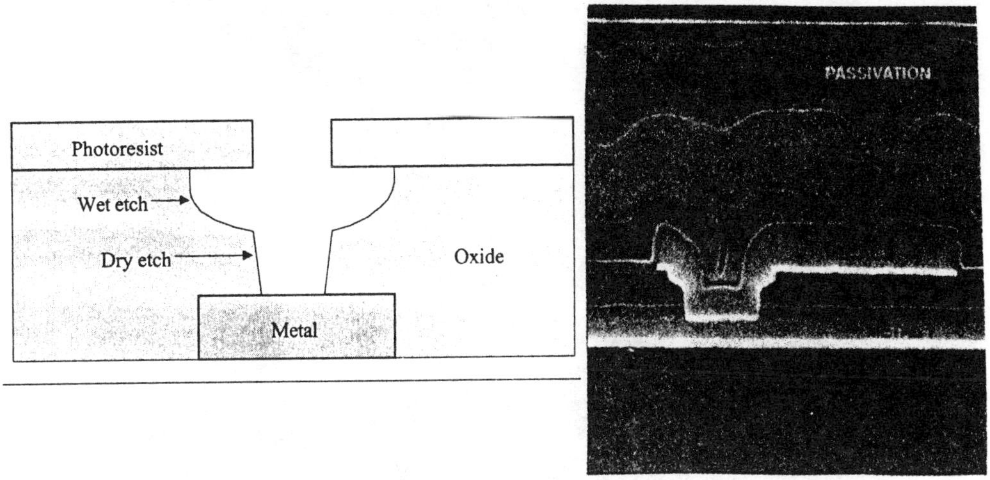

Fig. 12-7 a) Schematic of a via with a "champagne-glass" shape; b) SEM of actual via.

two-step via-opening process produced via profiles that had a "champagne-glass" shape (Fig. 12-7). Such vias were easier to adequately fill with Metal-2. Note that Al-alloy films were typically used (e.g., Al:2%Cu) because they exhibited better electromigration characteristics than did pure-Al films. Figure 12-8 depicts a schematic of a double-level-metal structure for CMOS technology, and Fig. 12-9 shows a SEM of an actual CMOS IC with such a two-level-metal interconnect-structure. Three-level-metal structures in bipolar-ICs could also be fabricated with this approach.

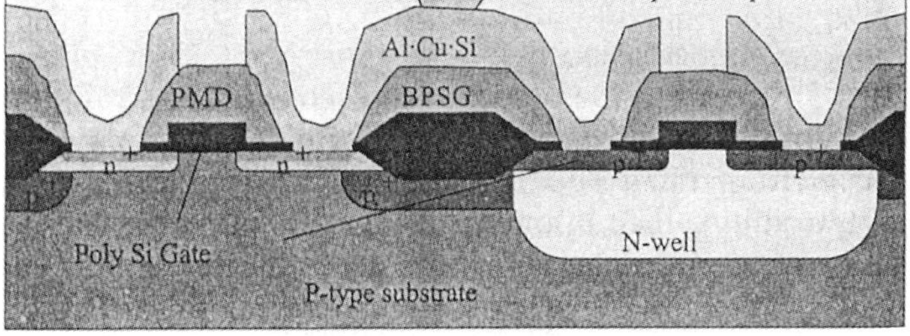

Fig. 12-8 Schematic of a double-level-metal structure in CMOS with partial-planarization.

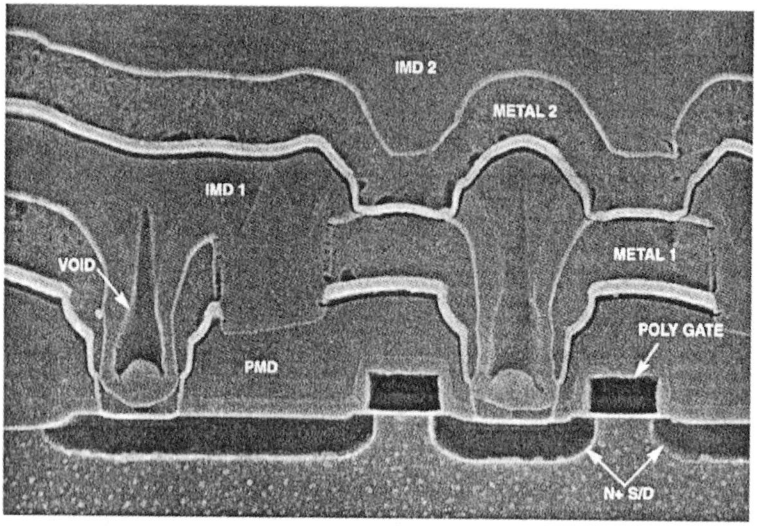

Fig. 12-9 SEM of an actual double-level-metal structure in CMOS with partial planarization.

The partial-planarization of the ILD allowed the step-coverage of Metal-2 to be sufficiently improved, and the "champagne-glass" via profile allowed Metal-2 to adequately fill the vias. Nevertheless, the topography of the IC remained too rough to allow this technology to be extended for producing three-level-metal interconnects in CMOS-ICs.

12.1.4 Three-Level Metal Interconnects in CMOS ICs: (CMP of the ILD & W-Plugs)

To enable three-level-metal interconnects (or more) to be fabricated in CMOS-ICs, a technique for globally-planarizing the ILD after it was deposited was needed. The technology developed for this purpose was *chemical-mechanical-polishing (CMP)* of the CVD-SiO_2 ILD. With this technique, the next level of metal would not have to cover any steps, but could be deposited on a completely planar surface (Fig. 12-10).

While CMP of the ILD solved the problem of metal lines having to cross high steps, it created a new difficulty. That is, as shown in Fig. 12-10, the vias that have to be opened in the planarized ILD are of varying depths (with some being quite deep). Such deep vias could no longer be adequately filled by the traditional PVD process used to deposit metal layers (sputtering), even if a two-step via-etch process was used. To address this issue, a CVD (instead of PVD) process was

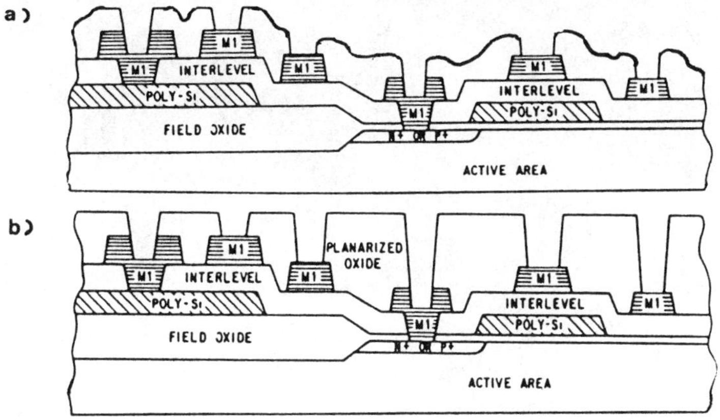

Fig. 12-10 a) If no planarization is used, the via and contact holes between adjacent conductor levels all have approximately the same depth; b) If planarization of the intermetal dielectric layer is achieved, the vias will vary widely in depth.

used to deposit the metal to fill the vias. The metal chosen to fill the vias in this manner was tungsten (CVD-W). A blanket CVD-W process was developed that could deposit a layer of W over the entire wafer covered with a planarized ILD (Fig. 12-11), such that variable-depth-vias across the wafer could also be completely filled. This blanket-W film would then removed from the planar wafer surface (by etching or CMP), leaving W only in the vias (i.e., *W-plugs* were created). The process of filling recesses in a dielectric layer with a metal film and removing the overburden with etchback or CMP (rather than depositing, patterning, and etching the metal on top of a dielectric layer) is called a

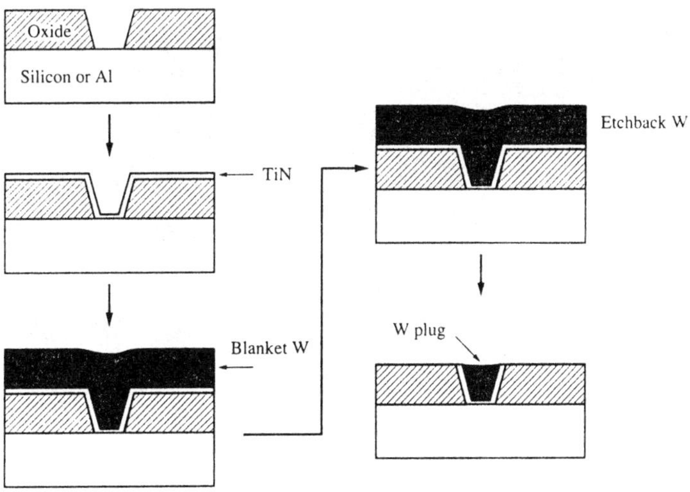

Fig. 12-11 Schematic of the process used to form W-plugs with CVD-W and etchback.

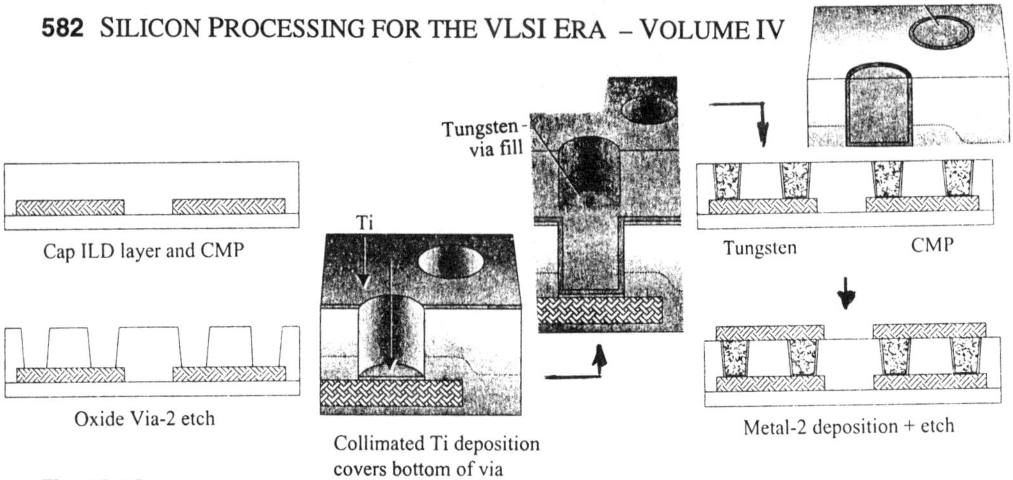

Fig. 12-12 Schematic of the process used to form the "hybrid" W-plug/Al-alloy structure.

"*damascene*" method. Its name comes from the ancient decorative practice of inlaying metal in wood or ceramics that originated in Damascus.

The layer of aluminum alloy (e.g., Al:2%Cu) from which the interconnect lines are patterned is then deposited and etched by the usual subtractive process. Fig. 12-12 shows a schematic of the steps used to form one-level of this type of interconnect structure. Figure 12-13 shows a schematic drawing and a SEM of such an interconnect structure with stacked vias.

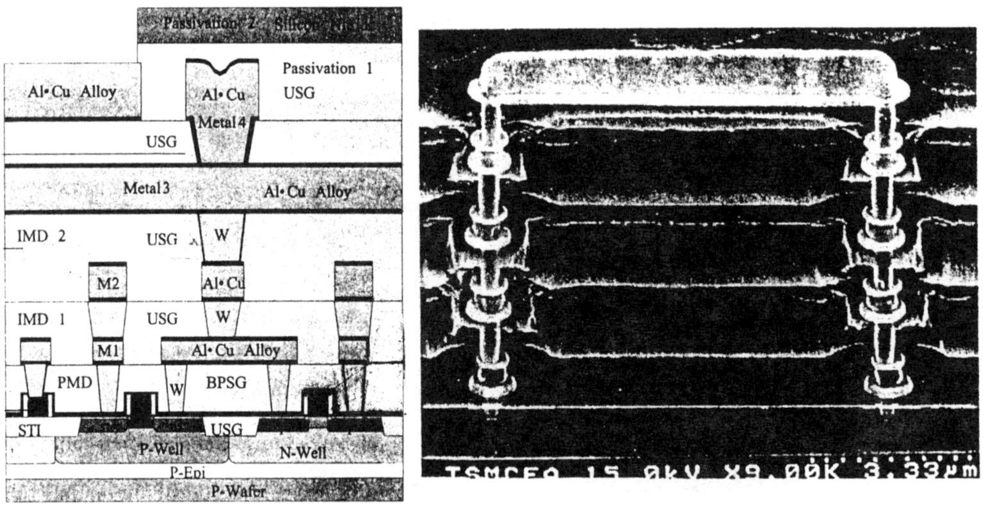

Fig. 12-13 a) Schematic of a W-plug/Al-alloy interconnect structure with stacked vias; b) SEM of an interconnect structure with stacked vias and five-levels of metal.

Note that in several aspects this sequence involves a hybrid process-technology approach. First, two-types of metal are used, one for the plug that fills the via (W), and another to form the metal lines (Al:Cu). Second, a single-damascene process is used to remove unwanted W from atop the entire surface of the chip, while a subtractive etch-process is used to selectively remove Al:Cu from some regions of the chip surface, and thus form the interconnect lines.

The ILD-CMP/W-plug/Al:Cu-line interconnect-fabrication-approach solved some significant problems, allowing it to be used to form interconnect structures with an indefinite number of levels of metal. These solutions are: **1**) the metal lines on every level have no steps to cover; **2**) vias of varying depth can be completely filled using the CVD-W step; and **3**) the vias can be stacked on top of one another if a CMP process is used to planarize the surface of the W-plugs (which allows the density of the interconnect structure to be increased).

12.1.4.1 Problems of W-plug/Al-line Interconnect Structures: While this approach to fabricating multi-level interconnects represented a significant advance, it also introduced some drawbacks. First, it is a complex process, in that it uses two-types of metal, and two-different metal-deposition-processes (each with its own set of process tools). This fabrication complexity is compounded by the fact that two-CMP steps, one ILD-etch-step, and one metal-etch-step are required. Second, the overall resistance of the Al-line/W-plug-structure can grow to be too large for small sized interconnect paths (whereas this resistance value can be reduced by using a lower-resistivity conductor-material - e.g., Cu). Finally, there are some problems associated with W-contacts and vias. That is, electromigration problems in Al often occur in the vicinity W-contacts and vias. This may be partly due to mechanical stresses that can develop at W/Al interfaces (which may enhance electromigration, or promote void and hillock formation). But it is also because the different material in the contact or via serves as a point for flux-divergence. Al also has a lower electrical resistivity than W. This can be detrimental with respect to electromigration effects, especially in circuits where current flow between different materials is significant.

12.1.5 Dual Damascene Cu/Low-*k* Interconnects

An alternative approach to forming interconnects on ULSI ICs is the *dual-damascene approach.* The Cu/low-*k* dual-damascene interconnect scheme has been pursued because it has several advantages over the subtractive W-plug/Al-alloy interconnect approach. First, an all-Cu interconnect structure has a lower resistivity and better electromigration resistance characteristics than the W-plug/Al-alloy system. Second, by incorporating low-*k* dielectrics, the parasitic capacitances of the interconnect structure are reduced, also improving the RC-delay. Third, only one CMP step is needed (i.e., CMP of the Cu/diffusion barrier film stack), compared to the two-CMP steps needed for the subtractive interconnect (i.e., CMP of the W-plug and the ILD dielectric layer). Fourth, the damascene approach avoids having to etch Cu lines, a difficult process to implement in manufacturing. Fifth, only a single interface exists for the via (i.e., the barrier metal in contact with the Cu film in the metal level below, versus the two interfaces at the top and bottom of the W-plug).

The Cu/low-*k* dual-damascene interconnect also has some difficult challenges, some of which are listed here. First, Cu is a fast-diffusing and device-failure-inducing contaminant, even when present in trace concentrations in the silicon. Therefore, it must be surrounded by a diffusion barrier on all sides to prevent it from entering the silicon substrate. Deposition of diffusion-barrier-materials into high-aspect-ratio trenches and vias with sufficient thickness everywhere becomes harder as the via and trench dimensions are scaled. Second, CMP of Cu (and the underlying diffusion-barrier-metal-layer) is also a complex process. Third, the deposition of Cu-layers by electroplating introduces another new, and potentially contaminant-producing process technology into the wafer fab. Fourth, developing low-*k* materials to replace CVD-SiO_2 as the ILD layer for the dual-damascene structure is not simple, and new materials will need to be found for several generations of low-*k* dielectric applications. These issues are discussed in detail in other chapters of this book, namely: low-*k* dielectrics (Chap. 14); issues of dual-damascene interconnects (Chap. 15); and copper deposition and CMP (Chap. 16).

The following sequence of steps is used to create a dual-damascene interconnect structure (see Fig. 15-3 in Chap. 15):

1. A low-*k* interlevel dielectric (ILD) film is first deposited;

2. Vias and trenches are next patterned and etched in the ILD (two masking and etch steps are needed to accomplish this result);

3. A multi-layer film (i.e., barrier-metal/Cu) is deposited into the vias and trenches of the ILD (the barrier-layer and Cu-seed-layers are deposited by PVD, and the main Cu-layer by electroplating);

4. The barrier-layer/Cu film is removed from regions above the surface of the ILD by CMP, leaving the barrier-layer/Cu recessed in trenches and vias;

5. The entire wafer surface is covered with a non-conductive diffusion-barrier-layer material (e.g., SiN) to complete the encapsulation of the Cu.

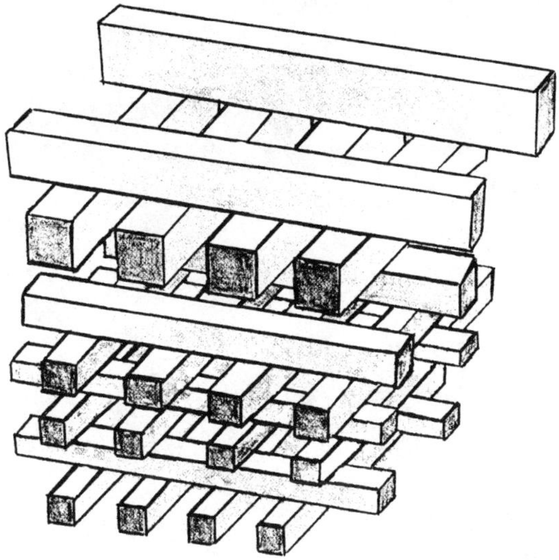

Fig. 12-14 Schematic showing the orthogonal architecture of multilevel interconnect lines.

12.1.6 Other Geometrical Aspects of Multilevel ULSI Interconnects

There are two other geometrical aspects of ULSI multi-level interconnects that should be discussed: **1**) orthogonal directions of alternating metal levels; and **2**) tiers of metal lines containing the same pitch.

In a multi-level-metal interconnect structure, alternating levels are run orthogonal to the levels above and below (Fig. 12-14). This serves three purposes. First, maximum flexibility of routing signals is made possible by using such an orthogonal multilevel structure. Second, the intralevel-capacitance between levels is minimized. Third, crosstalk can be reduced. Figures 12-15 and 12-16 show examples of such an orthogonal structure in a W-plug/Al-line interconnect structure, and in a Cu dual-damascene interconnect structure, respectively. Figure 12-17 shows a cross-sectional SEM of a six-level-metal interconnect structure that reveals the use of orthogonal metal-level layout as well.

Multilevel interconnect structures are also typically arranged in *tiers*. Tiers are pairs of orthogonal metal levels with equal pitch. In some interconnect systems, only two tiers are used (i.e., a *local interconnect-tier* and a *global interconnect-tier,* Fig. 12-18a). In a *local interconnect-tier,* the wires are only used within a cell or a block. In such local tiers the finest metal-pitch allowed by the design rules is employed. In a *global interconnect-tier,* the wires are used for inter-block communication. Thus, the lines may be quite long, and are thicker and wider than those in the local interconnect tier. In interconnect structures that contain six or more levels of metal, there may be a more optimized three-tier network (i.e., one

Fig. 12-15 SEM of orthogonal architecture of metal lines in a W-plug/Al interconnect.

that contains a third, *semi-global tier*, as shown in Fig. 12-18b).

12.2 PERFORMANCE REQUIREMENTS OF ULSI INTERCONNECT SYSTEMS

By 2001, microprocessors for consumer PCs were operating at speeds greater than 1-GHz (and even faster in the most advanced workstations). Data in Table

Fig. 12-16 SEM of orthogonal architecture of metal lines in a Cu interconnect.

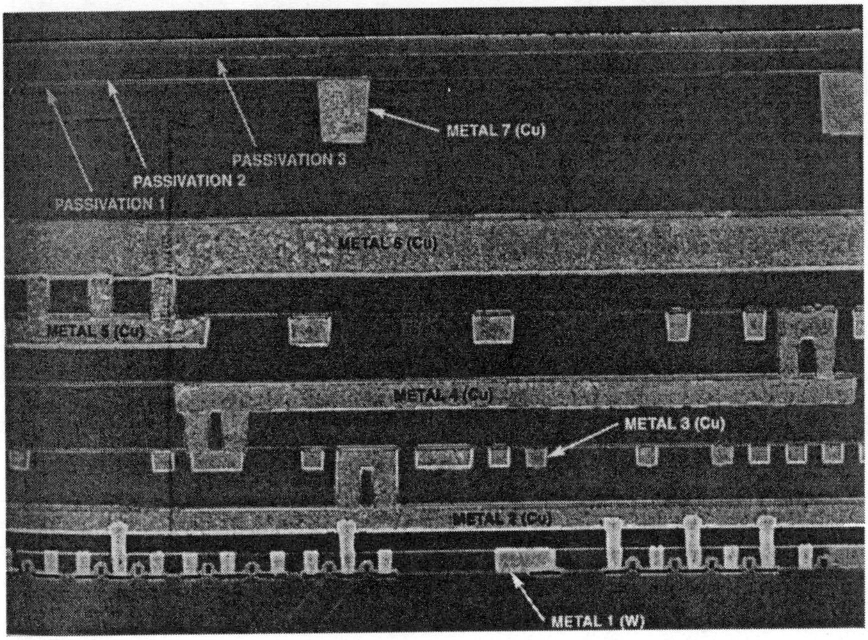

Fig. 12-17 SEM of a cross-section of a 6-metal-level interconnect.

12-1 taken from the 1999 ITRS,[3] summarizes the projected interconnect needs of microprocessors out to 2014, with respect to performance, density, feature size, and other critical technology parameters. Here we discuss those parameters and issues involved in allowing multilevel interconnect structures to achieve the

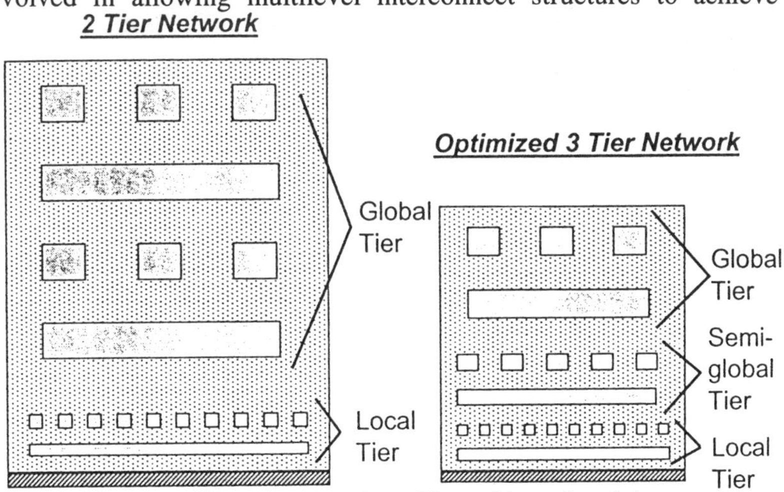

Fig. 12-18 The architecture of interconnects also utilizes a hierarchy of *tiers*.

TABLE 12-1 Selected Overall Technology Characteristics based on the 1999 International Technology Roadmap for Semiconductors.[3]

Year Technology node	1999 180 nm	2002 130 nm	2005 100 nm	2008 70 nm	2011 50 nm	2014 35 nm
MPU ½ pitch	230	160	115	80	55	40
MPU gate length (nm)	140	85–90	65	45	30–32	20–22
DRAM ½ Pitch (nm)	180	130	100	70	50	35
MPU chip frequency on-chip across chip clock–MHz (Cost-performance)	600	800	1100	1400	1800	2200
MPU chip frequency on-chip local clock–MHz (High performance)	1250	2100	3500	6000	10000	13000
High volume-MPU (Cost-performance) chip size at ramp (mm^2)	170	214	235	269	308	354
High volume-MPU (Cost-performance) Mtransistors/cm^2 at ramp (including SRAM)	7	18	41	100	247	609
Memory chip size at production (mm^2)	132	152	174	199	229	262
Memory chip Gbits/cm^2*	0.20	0.52	1.23	3.05	7.51	18.5
Maximum power w/high performance heat sink (Watts)	90	130	160	170	174	183
Maximum power (Watts) –Battery (hand-held)	1.4	2.4	2.0	2.0	2.2	2.4
Minimum logic Vdd–(Maximum performance) Volts	1.8	1.5	1.2	0.9	0.60	0.60
Minimum logic Vdd–(Lowest power) Volts (battery power)	1.5	1.2	0.90	0.60	0.50	0.30

performance goals listed in Table 12-1. The performance parameters that will be addressed are the following:

1. Propagation-delays (*RC*-delays) in digital circuits;
2. Power dissipation issues in ULSI interconnects;
3. Crosstalk in ULSI interconnects;
4. Contact resistance;
5. ESD protection.

12.2.1 Propagation-Delays in ULSI Multilevel Interconnects

As noted in the introduction of this chapter, for technology nodes beyond 0.25-μm the overall performance of ICs will generally become limited by the performance of the interconnect system on the chip. The most widely used figure of merit of an interconnect system is its $R_{int}C_{int}$-*delay*, which represents a measure of the time-delay for signal propagation. R_{int} denotes the resistance of the interconnect metal-line and C_{int} denotes the effective capacitance between the line and its surroundings. For the multi-level interconnect systems found in ULSI chips at the 0.25-μm technology nodes and beyond, the estimated effect of various geometrical and materials parameters on the $R_{int}C_{int}$ delay can be modeled using the test-structure shown in Fig. 12-19.[11] This test-structure consists of a set of metal lines between two grounded metal planes, all embedded in the interlayer-dielectric medium (ILD). Let us assign the following symbols to those aspects of the interconnect system depicted by this test-structure.

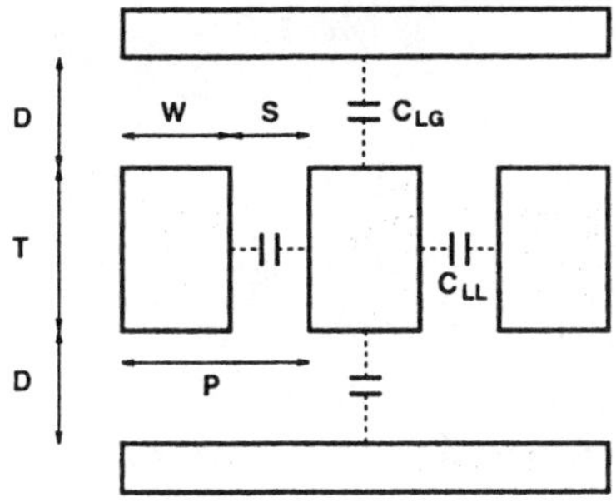

Fig. 12-19 Simplified *RC*-delay test-structure.[11] (© IEEE 1995)

That is, the metal-line *width* is w, the metal-line *thickness* is t, the *spacing* between the lines is s, the metal-line *pitch* is p ($p = w + s$), and the *thickness of the ILD between the top of the metal lines and the grounded planes* is d. Also let the *length* of the metal lines be l, and the *resistivity* of the metal be ρ Some equations can now be derived that allow the contributions made to R_{int} and C_{int} from: **1**) the materials properties; and **2**) the interconnect architecture, to be expressed. In these derivations the pitch p and the metal thickness t will serve serve as the parameters in the equations. In that way it will be possible to set $w = ap$, and $d = bt$, where a and b are constants appropriate for the given geometry.

Using this approach, the resistance of a metal line is calculated from:

$$R_{int} = \rho l/wt = \rho l/apt \qquad \textbf{(12.2)}$$

Next, the capacitance between the metal lines and the grounded planes C_{LG} (separated by an ILD whose relative dielectric constant is k), is calculated from:

$$C_{LG} = k\varepsilon_o wl/d = k\varepsilon_o apl/bt \qquad \textbf{(12.3)}$$

and the intralevel capacitance C_{LL} (assuming the same ILD) is found from:

$$C_{LL} = k\varepsilon_o tl/s = k\varepsilon_o tl/(p - w) = k\varepsilon_o tl/(1 - a)p \qquad \textbf{(12.4)}$$

where ε_o is the permittivity of free space. In Eqs. 12-3 and 12-4, C_{LG} represents the "vertical" capacitance between a metal line and a ground plane above (or below), and C_{LL} represents the "horizontal" capacitance between two neighboring lines. The total capacitance C_{int} of a metal line relative to its surroundings is thus written as:

$$C_{int} = 2(C_{LG} + C_{LL}) = 2k\varepsilon_o l\{[ap/bt] + [t/(1 - a)p]\} \qquad \textbf{(12.5)}$$

For the $R_{int}C_{int}$ time delay constant, it follows that:

$$R_{int}C_{int} = 2\, R_{int}\,(C_{LG} + C_{LL}) \qquad \textbf{(12.6)}$$

By combining Eqs. 12-2, 12-3, 12-4, and 12-6, we get:

$$R_{int}C_{int} = 2\rho k\varepsilon_o l^2\{[1/bt^2] + [1/a(1 - a)p^2]\} \quad \textbf{(12.7)}$$

This equation can also be rewritten as:

$$R_{int}C_{int} = 2\rho k\varepsilon_o(l^2/p^2)\{[p^2/bt^2] + [1/a(1 - a)]\} \quad \textbf{(12.8)}$$

The right-hand side of Eq. 12-7 has a natural interpretation, in that the factor $2\rho k\varepsilon_o$ represents the materials contributions, and the factor $l^2\{[1/bt^2] + [1/a(1 - a)p^2]\}$ represents the geometrical contributions, to the interconnect time delay.

Equation 12-8 can also be examined from a slightly different perspective if one defines the aspect ratio A of the metal lines as the ratio of line-thickness to line-width; i.e., $A = t/w$, or equivalently $A = t/ap$. In terms of A, Eq. 12-8 can be expressed as:

$$R_{int}C_{int} = 2\rho k\varepsilon_o(l^2/p^2)\{[1/a^2bA^2] + [1/a(1 - a)]\} \quad \textbf{(12.9)}$$

It can be observed that the term $1/a^2bA^2$ in Eq. 12-9 represents the "vertical" contribution to the total capacitance, and $1/a(1 - a)$ represents the "horizontal" contribution to the total capacitance.

Several conclusions about the performance of a multilevel interconnect system can be drawn by further examination of Eqs. 12-7 to 12-9. Two separate cases can be considered: **1**) where the same ILD is used in the interlevel and intralevel spaces (i.e., the value of k is constant throughout); and; **2**) where the spaces between metal lines are assumed to be filled with a material of a lower k.

12.2.1.1 Case When the *k*-Value of the ILD is Constant Throughout the Interconnect System: In this case, from Eqs. 12-8 and 12-9, it can be seen that for a fixed length l and a fixed aspect-ratio A, the RC-delay increases quadratically with decreasing feature size (i.e., decreasing pitch p). These two equations also indicate that for $A << 1$, the interlayer "vertical" capacitance is more important to the RC-delay than the intralayer (or "horizontal") capacitance., whereas the opposite is true for $A >> 1$. Also note that for the simplest geometry (where $a = 0.5$ and $b = 1$ (i.e., $d = t$), Eq. 12-9 reduces to:

$$R_{int}C_{int} = 8\rho k\varepsilon_o(l^2/p^2)\{[1/A^2] + [1]\} \quad \textbf{(12.10)}$$

12.2.1.2 Case When the *k*-Values of the Interlevel-ILD and the Intralevel-ILD are Different: In this case, one ILD is used between the metal lines and ground planes (whose "vertical" [between levels] k-value is given by k_{LG}), and another ILD is used between the metal lines themselves (whose "horizontal" [intralevel] k-value is given by k_{LL}). In that case, Eq. 12-9 becomes:

$$R_{int}C_{int} = 2\rho\varepsilon_o(l^2/p^2)\{[\,k_{LG}/a^2bA^2] + k_{LL}[1/a(1 - a)]\} \quad \textbf{(12.11)}$$

Again, for the simplest geometry where $a = 0.5$ and $b = 1$ (i.e., $d = t$), Eq. 12-11 reduces to:

$$R_{int}C_{int} = 8\rho\varepsilon_o(l^2/p^2)\{[\, k_{LG}/A^2] + [k_{LL}]\} \qquad \textbf{(12.12)}$$

Equation 12-12 demonstrates several important conclusions. First, if a high aspect ratio interconnect structure is used on a given level (i.e., A is large), the large value of A will mitigate the influence of k_{LG} (i.e., it doesn't matter as much if an ILD with a high or low value of k_{LG} is used between interconnect levels). However, the value of k_{LL} plays significant in determining $R_{int}C_{int}$ for this case. Second, Eqs. 12-11 and 12-12 indicate that the contribution of the intralayer capacitance to the total capacitance (namely, that due to the term 1/{a[1-a]}), is minimized when $a = 0.5$; i.e., when w is one half the pitch p.

While Eqs. 12-11 to 12-12 offer some general insight for minimizing the $R_{int}C_{int}$-delay, the interconnect structure can also be treated as consisting of two tiers, termed: **1**) the *local interconnect tier*; and **2**) the *global interconnect tier* (see Sect. 12.1.6). There are some differences in the ways in which the minimum *RC*-delay can be achieved in each of these tiers. The issues of are considered next.

12.2.1.3 Minimizing the *RC*-Delay in Local-Interconnect Tiers: The local interconnect tiers are those used to route interconnect lines *within* a cell or block, and in which the finest metal pitch allowed by the design rules are employed. In these tiers, the routing lines are so short that it can be assumed that $R_{int} << R_{trans}$ and $C_{int} >> C_{trans}$ (i.e., the wire resistance of lines shorter than about 1-mm is much less than the transistor on-resistance R_{trans}). In addition, the lines are so close that the largest component of C_{int} is intralevel capacitance, C_{LL}. In this case, the *RC*-delay in local interconnects is not found from Eqs. 12-2 through 12-4 (which assume the *RC* is controlled by R_{int} and C_{int}) but instead is described by the following equation:

$$R_{trans}C_{int} \sim R_{trans}\, k\varepsilon_o tl/s \qquad \textbf{(12.13)}$$

Thus, unlike in the global interconnect tiers where the *RC*-delay is controlled by $R_{int}C_{int}$ (and where the delay is thus dependent on the *square* of the interconnect length l), in the local interconnect tiers the delay is only *linearly-dependent* on the interconnect length. In addition, in this region C_{int} is dominated by the intralevel coupling between adjacent lines. Furthermore, in the local interconnect regions it is also clear that changes in the metal resistance will not play a big role in altering the *RC* delay, since $R_{trans} >> R_{int}$. This means that in this line-length regime, the resistance and capacitance are decoupled. Consequently, factors that reduce C_{int} (such as lowering the value of k between metal lines, and reducing

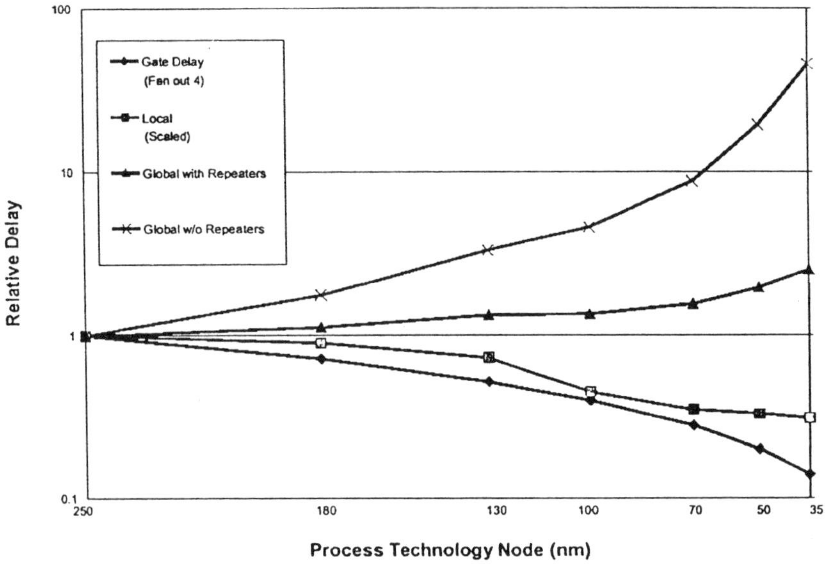

Fig. 12-20 Interconnect versus delay, from ITRS99.[3]

the height of the metal lines), will help decrease the *RC*-delay (provided that the interconnect resistance of the metal line does not exceed that of R_{trans} due to the reduction in the height of the metal lines). Figure 12-20 shows it is predicted that the relative delay of local interconnect tiers will remain roughly comparable to the gate delay as devices are scaled down toward 50-nm gate-lengths.

12.2.1.4 Minimizing the *RC*-Delay in the Global-Interconnect Tiers: The factors that minimize the *RC*-delay in the global tiers of the interconnect structure are somewhat different than the factors that govern minimizing *RC*-delay in the local interconnect tiers. (The *global interconnect tiers* are those used for inter-block communication employing the upper metal levels.) First, there are two coupling regimes that are encountered in global interconnect tiers – the *intralevel regime* and the *interlevel regime*. Figure 12-21a depicts the conditions that exist in the intralevel regime (which is typical of the denser interconnect lines found in the lower levels of global interconnect tiers). In this regime the intralevel capacitance dominates ($C_{LL} >> C_{LG}$). In the interlevel regime (which tends to exist in the upper levels of the global interconnect region, where wide wires are used to lower the resistance of long runs – Fig. 12-21b), the interlevel capacitance dominates ($C_{LG} >> C_{LL}$).

In the *intralevel global interconnect-regime*, $R_{int}C_{int}$ is calculated by combining Eq. 12-2 and 12-4, to get:

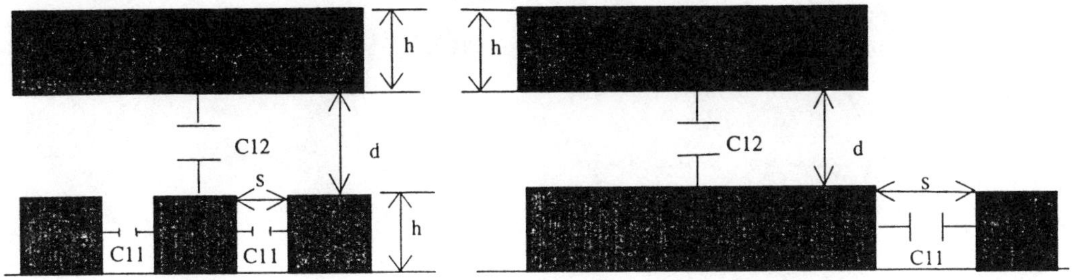

Fig. 12-21 Capacitance coupling contributions for interconnect systems: a) Dense interconnect case where major coupling is between minimum-spaced lines; and b) Wide interconnect lines where the major coupling is between metal-levels.

$$R_{int}C_{int} = 2R_{int}C_{LL} = 2\ k\varepsilon_o\rho l^2/ws \quad \textbf{(12.14)}$$

In the *interlevel global interconnect-regime* the interlevel-capacitance dominates, and hence $R_{int}C_{int}$ is calculated by combining Eqs. 12-2 and 12-3:

$$R_{int}C_{int} = 2\ k\varepsilon_o\rho l^2/td \quad \textbf{(12.15)}$$

We see that in both cases the *RC* delay is strongly impacted by the length of the interconnect line (i.e., *RC* increases as the square of the line-length).*

In many circuits it is undesirable to operate under these conditions (i.e., in which the interconnect limits the propagation delay of the circuit). To overcome this limitation, some circuit designs include *repeaters* or *buffers* along the interconnect lines that exceed a critical length. The use of repeaters (while not desirable from a circuit-density standpoint), is nevertheless an essential technique for optimizing signal propagation. Implementing repeaters can even extend the usefulness of a given process technology. Their use will optimally halve the delay of a long line, with a small penalty of the repeater-delay and some device area.[13] Figure 12-22 shows how the addition of repeaters to an interconnect system can reduce the number of levels of metal as the functional density of ICs is increased.[16] In another example, Figure 12-23 depicts the interconnect delay contribution vs. feature size for varying repeater insertion lengths of 3000-μm

* Note that the intermediate situation (i.e., where both intralevel- and interlevel-capacitance effects are comparable in magnitude) is treated by Eqs. 12-8 and 12-9, since this situation lies between the two extremes and includes capacitive contributions both in- and between-metals-levels.

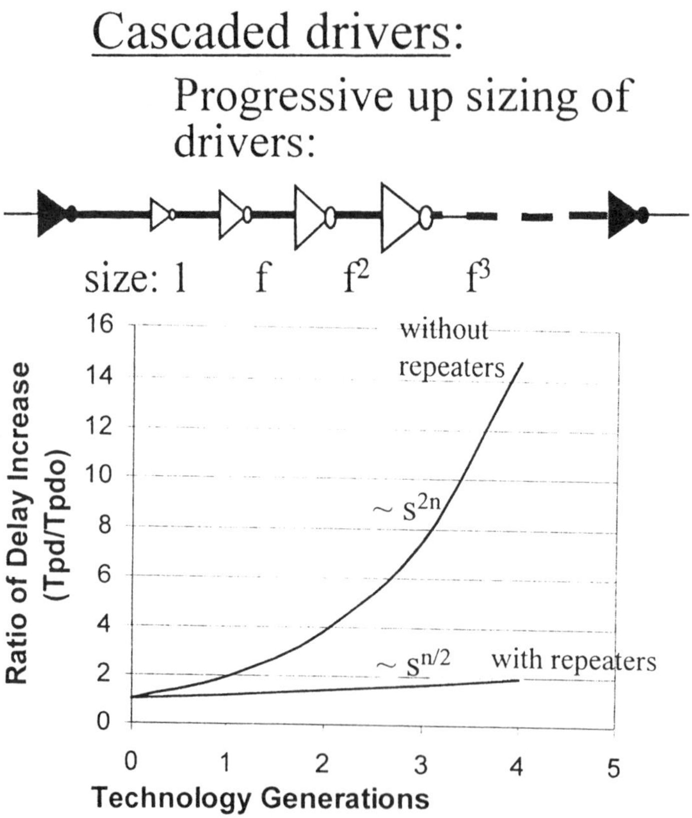

Fig. 12-22 Impact of repeaters on global propagation delay.[5]

and 5000-μm.[14] Note that reducing the spacing at which repeaters are placed can compensate for delays due to *RC* parasitics and circumvent the need for new materials (e.g., Cu and low-*k* dielectrics).

On the other hand, where long interconnect lines must be used, large pitches *p* must be employed, and the lines should be made wider and thicker. This implies that long interconnect lines should only be implemented in the upper interconnect metal layers where there is this room for increased pitch (wider lines and spaces), and where the lines can be made thicker. This is the approach that is commonly used today.

Thus, a multi-level interconnect system requires a hierarchy of scales, with the smallest lines in the first metal-level, and progressively wider (and perhaps thicker) and longer lines in the upper levels. One disadvantage of using thicker and wider lines in the upper levels is that fewer metal lines can be fabricated on any level. Consequently, to achieve the same functionality in the face of continued feature size scaling, the number of metal levels used in interconnect systems will also continue to grow.

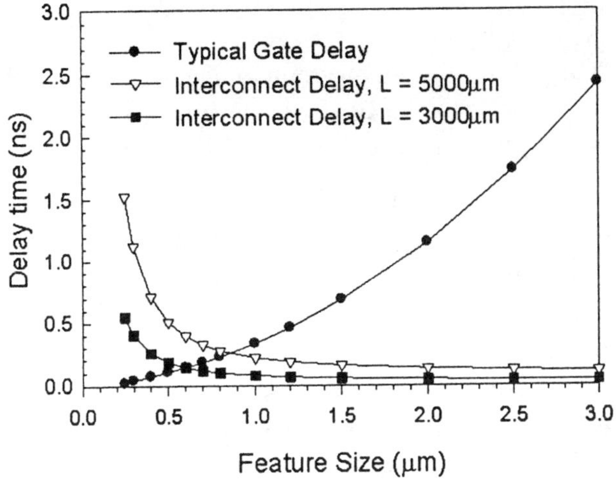

Fig. 12-23 Gate and interconnect delay versus feature size. Interconnect delay is shown for repeater spacings (L) of 3000-μm and 5000-μm.[14]

On the materials side of this issue, it is evident that the lowest $R_{int}C_{in}$-delay in global interconnect-layers will be achieved if a metal with the lowest possible ρ is used together with an ILD having the lowest possible k-value. With respect to the metal, this means using copper. (The details of fabricating metal interconnects with copper are described in Chap. 16). With regard to the ILD, it is especially beneficial to selectively reduce k_{LL} between the metal lines. We will explain how this can be accomplished in Chap. 14.

12.2.2 Power-Dissipation Issues in ULSI Interconnects

As the requirements for performance in ULSI continue to increase, the power demands will also arise unless the overall system capacitance, resistance and/or operating voltages can be reduced. The power-dissipation associated with a functioning IC can be calculated using either of the following two relationships:

$$P \sim f\, C_{total} V^2 \quad \textbf{(12.16)}$$

or

$$P \sim I_{total}^2\, R_{total} \quad \textbf{(12.17)}$$

where:
- P = power dissipated (watts)
- f = operating frequency (Hz)
- C_{total} = total IC capacitance = $C_{trans} + C_{int}$ (farads)
- R_{total} = total IC resistance = $R_{trans} + R_{int}$ (ohms)

V = IC operating voltage (volts)
I_{total} = IC operating current (amperes)
C_{trans}, C_{int} = transistor & interconnect capacitance contributions
R_{trans}, R_{int} = transistor & interconnect resistance contributions

Although Eqs. 12-16 and 12-17 indicate that the IC operating voltage plays the greatest role in determining the power dissipation, its selection is based on circuit and system design issues. On the other hand, the system resistance and capacitance are the chief parameters that can be controlled by the fabrication technology. Currently, the transistor resistive losses are responsible for most of the chip power dissipation (and these are discussed in Chaps. 1 and 5). However, as the ICs are being operated at higher and higher frequencies, an appreciable fraction of the power dissipation and voltage drops are associated with the resistive component of the interconnect. Due to the decreased size of interconnects on ULSI (~200-nm minimum feature sizes in 2000), interconnect resistance is increasing and losses can become significant if the wiring is not designed properly. The overall resistance of the interconnect includes both the line resistance and the resistance of one or more contacts and/or vias along the wiring path. Joule-heating caused by current flowing through the resistive interconnect enhances electromigration and represents a growing reliability concern. The power dissipated at the higher-resistance transition-regions of the contact or via interface is of special concern, especially for large pulsed currents. Since power-dissipation in interconnects is a growing problem, it is necessary to ensure interconnects are properly optimized to reduce such power-dissipation.

12.2.3 Crosstalk in ULSI Interconnects

Another issue that impacts the performance of ULSI interconnects is *crosstalk.* Interconnect lines of substantial length act as "antennae" which experience unwanted voltage pulses induced by the rapid voltage swings in neighboring wires of close proximity (which arise from the propagation of digital pulse waveforms along the driven lines). Induced voltages due to intralevel-coupling (the largest crosstalk component between dense wires in the same level) can easily achieve 20% of the supply voltage for very rapidly changing clock waveforms.[4] Figure 12-24 shows a simulation of the amount of crosstalk experienced by an undriven wire that is 1-mm long, separated by 100-nm from an adjacent interconnect line of the same length and driven with a very fast clock edge.[5] V1 is the driving waveform and V2 is the crosstalk voltage induced in the undriven line. The rate of rise of the clock edge imposes a large induced voltage

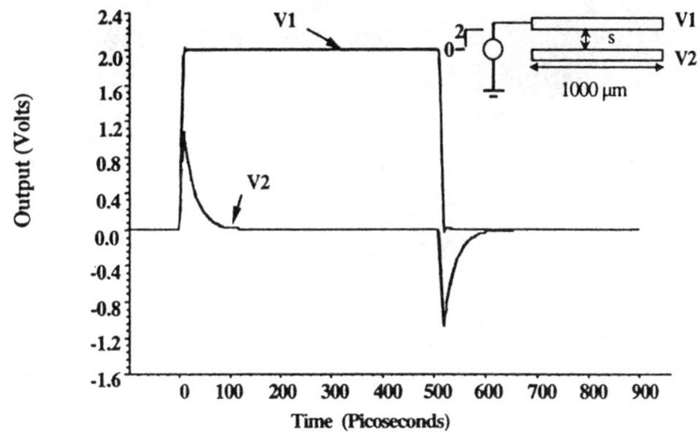

Fig. 12-24 Electrical modeling result showing the crosstalk voltage induced from a driven interconnect-line (V1) into an undriven line (V2) seeing a 20-psec waveform-edge. The lines were spaced at s = 1000-nm with a common length of 1-mm.[10]

which is related to the parasitics of the interconnect. In this example, where the clock-edge rise time is 20-psec, the magnitude of the imposed crosstalk voltage can be as much as 60% of the voltage waveform V1 (2 x 0.6 = 1.2V). As the intralevel-spacing between interconnects decreases, the magnitude of the crosstalk noise can become substantial, as is evident from the simulation data shown in Fig. 12-25.

It is more difficult to establish the level of crosstalk in dynamic systems than in the example given above. More sophisticated circuit simulators are required to

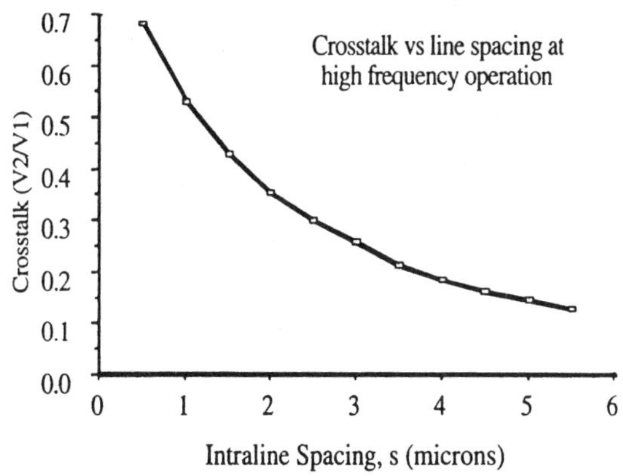

Fig. 12-25 Simulation results showing that the magnitude of the crosstalk between 2 interconnects separated by different line spacings, s.[10] Conditions described in Fig. 12-24.

estimate the impact of such crosstalk on circuit device performance. The potential of surrounding wires, wire resistance, wire length, and the capacitive load all affect the amplitude of crosstalk signals. Nevertheless, it is possible to reduce intralevel crosstalk by changing the relative dielectric environment. That is, by inserting an embedded low-k dielectric between dense intralevel wires, electric-field lines can be diverted to the short orthogonal lengths of underlying interlevel wiring (lowering capacitive-coupling between intralevel interconnects).

Crosstalk will become a significantly greater problem as ULSI mixed-signal devices begin to emerge in the marketplace. It is predicted that noisy digital wires and circuit elements will cause a number of problems for analog elements that are in close-proximity to such sources of noise. New techniques involving such novel electrical-isolation-methods as shielding and signal conditioning will be needed. This is because, in some circuits, analog devices must be immune to such "noise" down to levels that are three to six orders of magnitude smaller than the power-supply voltage-levels.

12.2.4 Contact Resistance in ULSI Interconnects

The magnitude and uniformity of contact and via resistance is another forthcoming challenge (see Vol. 2, Chap. 3 for more background information on contact and via resistance). Contact and via resistances can be substantial and have a significant impact on the overall wiring resistance of the interconnect system. These resistance contributions act in the same way as adding length to an interconnect line, and thus can factor heavily into voltage drops and delays in the wiring. For 0.25-μm contacts with an interfacial-contact-resistivity ρ_c to heavily doped silicon of 5×10^{-8}-1×10^{-7}-Ω-cm^2, such contact-resistance values are approximately 100-200-ohms/contact.[6] Vias for systems using Al alloys with W-plugs have practical metal-to-metal contact resistivities of approximately 2×10^{-9} Ω-cm^2 and a resistance of about 5-Ω/via at the same geometry.[7] Vias which use copper metallization, however, are expected to show substantially lower contact-resistance because they have only one critical barrier interface per via (instead of two interfaces when W-plugs and Al alloy interconnect lines are used), and because the interfacial-contact-resistivity should be lower than 1×10^{-9}-Ω-cm^2.[8]

Cleaning prior to the deposition of the contact- and via-filling conductor-films and control of dimensional tolerances are critical to the reproducibility of via or contact resistance. It is predicted that production variability will increase with the migration to smaller feature sizes and have a negative impact on resistance. This resistance must be accounted for in the delay analysis when multiple levels with a large number of contacts and vias are used.

12.2.5 Electrostatic Discharge Protection Issues of Interconnects

Electrostatic discharge events (ESD) can cause large transient currents to flow in an IC. This can cause "fusing" in upper wiring levels, thus jeopardizing the long-term reliability of the IC. ESD robustness has been shown to be a severe reliability concern by several research groups, for interconnect generations with metal features smaller than 0.25-μm.[9,10] Since short term discharges can exceed 2000 volts for durations of hundreds of nanoseconds, it has been shown interconnects are becoming more prone to failure due to this type of electrical stress. Copper interconnects exhibit a substantially better resistance to this type of failure, when compared to aluminum interconnects, due to the higher melting point and thermal conductivity of copper.

12.3 SUMMARY OF USLI INTERCONNECT ISSUES AND FUTURE TRENDS

Based on the discussion of this chapter, the following trends are likely to be encountered in ULSI interconnect systems:

1. If the materials of an interconnect system are not changed (i.e., the metal resistivity ρ and ILD-k values remain the same), the RC-delay will continue to increase as device dimensions (and therefore interconnect pitch p) are reduced. There are two main reasons why this increase in RC occurs:

    a. Increasing the aspect ratio A of metal lines (to keep R constant) beyond about 2 produces diminished improvements in RC - even for a constant pitch p (because C_{LL} rises while R remains constant). However, if the pitch is reduced, this makes C_{LL} grow even larger, which, in turn causes an additional rise in RC.

    b. The metal line lengths tend to scale less strongly than does the metal pitch p. Hence, even as devices keep shrinking, the ICs will continue to contain some rather long lines (which will have larger C_{total}/unit length, for lines with the same length).

We noted earlier, however, that in short- and intermediate-length local interconnects, changes in the metal resistance will typically not play a major role in the RC-delay (since in this regime $R_{trans} >> R_{int}$). For such cases, the resistance and capacitance are decoupled. That is, factors that reduce interconnect capacitance (such as a reduction in the dielectric constant of the ILD, and/or a reduction in the height of the metal lines), will help improve *RC* delay (as long as the interconnect resistance increase due to reducing the metal film thickness does not approach R_{trans}).

2. If ρ and/or k can be decreased (especially a reduction in the value of k of the ILD used in intralevel metal spaces k_{LL}) this will cause a decrease in *RC*. Hence, there will be continual effort expended to integrate Cu and low-*k* dielectrics into ULSI interconnect systems. The issues involved with incorporating low-*k* ILDs are covered in Chap. 14, and the challenges encountered when attempting to implement Cu-metal-interconnects are described in Chap. 16. Figure 12-26 shows a possible migration path for materials use in IC interconnect structures.[15]

3. More levels of metal will need to be used as the feature sizes are decreased.

4. Optimizing interconnect length is a powerful method for reducing *RC*-delay. This can be influenced by design philosophy and device architecture. Use of repeaters, while not desirable from a density standpoint, is another key approach for optimizing signal-propagation, and can extend the use of a given process technology. Since incorporation of repeaters into an interconnect system is a circuit-design issue, this approach not considered further in this text.

5. Increasing the metal line aspect-ratios and adding more levels of interconnect are two methods for improving interconnect performance and density. However, these approaches will reach practical limits in a few

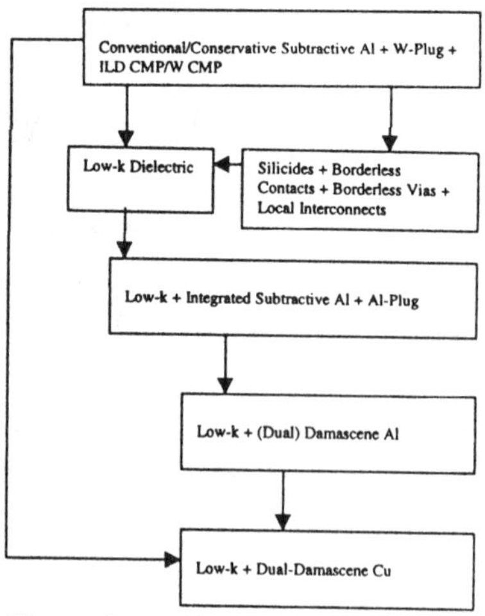

Fig. 12-26 Schematic illustrating an interconnect migration.[15]

generations.[11] (For example, continuing to increase the aspect ratio A above $A = 2$ will bring only diminished improvements in RC because the reduction in the metal resistance will be offset by an increase in intralevel capacitance.) New interconnect materials, improved circuit-design techniques, and possibly novel interconnect schemes will be needed to address the growing limitation that interconnects pose to ULSI performance. Circuit design needs to more accurately model interconnect performance to avoid overdesign or unexpected performance limitations, and automatic layout tools need to make more efficient use of existing interconnects. The novel interconnect approaches include RF-transmission and optical-interconnects. More information about these novel techniques can be found in Ref. 12.

REFERENCES

1. S.C. Sun, "Process Technologies for Advanced Metallization and Interconnect Systems," *Tech. Dig. IEDM*, 1997, p. 765.

2. R. Liu *et al., Proc. Int. Interconnect Technol. Conf. (*IEEE), 1999, p. 21.

3. International Technology Roadmap for Semiconductors, 1999 Edition, Semiconductor Industry Association, San Jose, CA.

4. C. Cregut *et al.,* "Low-*k* Dielectrics Influence on Crosstalk: EM Analysis and Characterization," *Proc. of the International Interconnect Technol. Conf.,* San Francisco CA, June 1-3, 1998, p. 59.

5. H.P. Bakoglu, *Circuits, Interconnections, and Packaging for VLSI,* Addison Wesley Publ., Reading MA, 1990.

6. I.S. Park, *et al.,* "Fabrication of 256 Mbit DRAM with a Fully Planarized Double Level Interconnection," *Proc. 12th VLSI Multilevel Interconnect Conf. (VMIC),* Santa Clara CA, 1995, p. 45.

7. K. Kikuta *et al.,* "Comparitive Study of W-plug, Al-plug, and Al-Dual-Damascene for 0.18 μm ULSI Multilevel Interconnect Technologies," *Proc. of the Internat. Interconnect Technol. Conf.,* San Francisco CA, June 1-3, 1998, p. 140.

8. J. Zhang *et al.,* "CVD Cu Process Integration for sub-0.25 μm Technologies," *Proc. of the Internat. Interconnect Technol. Conf.,* San Francisco CA, June 1-3, 1998, p. 163.

9. S.H. Voldman, "ESD Robustness and Scaling Implications of Aluminum and Copper Interconnects in Advanced Semiconductor Technology," *IEEE EOS/ESD Symposium , 1997, p. 316.*

10. K. Banerjee *et al.*, "Characterization of VLSI Circuit Interconnect Heating and Failure under ESD Conditions," *1996 IEEE 34th International Reliability Physics Proceedings*, April 30-May 2, 1996, Dallas, TX, p. 237.
11. M.T. Bohr, "Interconnect Scaling – The Real Limiter to High Performance ULSI," *Tech. Dig. IEDM*, 1995, p. 241.
12. M.E. Thomas and R.H. Havemann, "Overview of Interconnect," Ch.12, in *Handbook of Semiconductor Manufacturing Technology*, Y. Nishi and R. Doering, Eds. Marcel Dekker, New York, 2000, pp. 303-4.
13. H.P. Bakoglu and J.D. Meindl, " Optimal Interconnection Circuits for VLSI," *IEEE Trans. Electron Dev.,* May 1985, p. 903.
14. J.C. Eble *et al*, "Minimum Repeater Count, Size, and Energy Dissipation for Gigascale Integration (GSI) Interconnects," *Proceedings of the International Interconnect Conf.*, San Francisco CA, 1998, p. 56.
15. J. Givens, "Engineering CMP for Manufacturability and Yield," *Abs. Mtg. Electrochem. Soc.,* Fall 1999, Abs. 1275.
16. S. Wills, "Interconnect Architecture," in *IEDM* 2000 *Short Course*: *Advanced Interconnects: Design, Process, Integration*, p. 27.

Chapter 13

POLYCIDES AND SALICIDES OF $TiSi_2$, $CoSi_2$, AND NiSi

Silicon forms many stable metallic and semiconducting compounds with metals, and such compounds are termed *silicides.* Several metal-silicides exhibit low-resistivity and high thermal-stability, making them suitable materials for several ULSI applications. The formation kinetics, film properties, and potential applications such silicides have been extensively examined by Murarka[1] and by Nicolet and Lau.[2] In this chapter the use of silicides for the purposes of fabricating polycides and salicides in ULSI applications is discussed.

13.1 POLYCIDES AND SALICIDES

The term *polycide* is a condensed form of the words polysilicon-silicide. Polycides are widely used for gate structures in MOSFETs that contain a non-self-aligned silicide. In this structure, a silicide is typically blanket-deposited over a doped polysilicon film and the composite film is patterned together (Fig. 13-1). Polycide gates can achieve a gate sheet-resistance smaller than 10-Ω/sq, compared to that of a polysilicon gate without a silicide of more than 40-Ω/sq. Even with its considerably lower-resistivity, the polysilicon gate with a silicide can only be used for local-routing over short-distances in deep-submicron ICs.

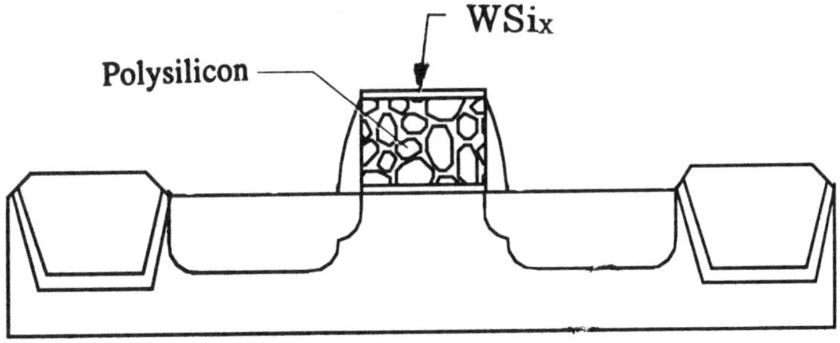

Fig. 13-1 Schematic cross-section of a polycide gate on a MOSFET.

Still, such a local-interconnect can be beneficial when only one or two levels of metal are used in a circuit - since the poly routing can help to reduce die size by as much as 20%. In fact, the incorporation of a polycide allowed the conductivity of poly lines serving as local interconnects to be so significantly reduced that the invention of the polycide-gate-structure was responsible for delaying implementation of multilevel-metallization in CMOS by several years.

Several types of polycides have been employed in IC technology, including WSi_2, $TaSi_2$, and $MoSi_2$. All are refractory-metal-based, and exhibit adequate thermal stability and resistance to processing chemicals. Since such polycides are formed as part of the gate fabrication sequence, they must be able to withstand the temperatures of any processes carried out after gate formation (typically 800-1000°C). Consequently, as WSi_2, $TaSi_2$, and $MoSi_2$ films remain stable at these elevated temperatures, they can be used for such gate polycides.

However, the silicides of Ti and Co (i.e., $TiSi_2$ and $CoSi_2$) are unstable above about 900°C (i.e., they agglomerate above ~900°C). Thus, it is not possible to use these materials as polycides if the gate formation sequence is followed by a thermal step that exceeds 900°C. Even if a process is used in which this temperature is not exceeded, neither $TiSi_2$ or $CoSi_2$ is a good candidate for a polycide that is patterned at the same time as the polysilicon gate. This is primarily due to an issue that arises during the patterning of the poly gate. That is, silicides composed of $TiSi_2$ are attacked by the Cl-based chemistry used for polysilicon dry-etching, and this makes linewidth control difficult if $TiSi_2$ must be etched during the patterning of the critical gate level. $CoSi_2$, on the other hand, lacks a volatile reaction by-product and is thus is difficult to pattern by dry etching. In addition, both $TiSi_2$ and $CoSi_2$ are vulnerable to processing chemicals, especially to HF solutions.

However, $TiSi_2$ and $CoSi_2$ are ideal candidates for *self-aligned-silicide* (also referred to as *salicide*) applications. As will be seen, such salicide processes are performed *after* the polysilicon lines for the gate structure have been patterned, and the source/drain regions of the MOSFET have been fabricated. Since the formation of the $TiSi_2$ or $CoSi_2$ in the salicide process are self-aligned techniques that require no etching, the etching difficulties mentioned above are avoided (as is exposure to temperatures > 900°C), permitting these materials to be used.

As the minimum dimensions of an IC approached 1-μm, the conventional contact structures used up to that point (see Chap. 3, *Silicon Processing for the VLSI Era*, Vol. 2 for details on contact formation in VLSI ICs) began to limit device performance in several ways. First, it was not possible to minimize the contact resistance if the contact hole was also of the minimum feature size, and problems with cleaning the small contact holes became another concern. In addition, the area of the source/drain-regions could not be minimized because the contact-hole had to be aligned to these regions with a separate masking step, and extra area thus had to be allocated for misalignment. (The larger area that had to be used for such source/drain-regions also resulted in increased source/drain-to-substrate junction-capacitance, which slowed-down device speed).

Finally, when non-minimum-width MOSFETs were fabricated with conventional contacts, several small, uniform-sized contact holes were usually used rather than one wider contact hole. (The reason for this is that if all of the contact holes on an entire chip are of the same size, they are more likely to clear simultaneously during the etching process.) A problem with using either one wide contact hole (as well as if several contact holes of equal size are used) is that a metal film must be used to contact each contact hole. The result is that valuable routing area is thus removed from the chip surface (because it must be used for running metal connections to the contacts), and this results in an increase in the die size. The main approach used to alleviate this problem has been to employ a self-aligned silicide on the source/drain-regions (where these silicides are formed at the same time as the polycide structure is created). The process of simultaneously fabricating such self-aligned-silicides and polycides on the gate is referred to as a *salicide* (self-aligned silicide) process.

13.2 POLYCIDES FABRICATED BY CVD OF TUNGSTEN SILICIDE (WSi_x)

As mentioned earlier, various refractory metal silicides including WSi_x, $TaSi_2$, and $MoSi_2$ were explored in the 1980s as possible shunt materials in polycide structures. Tungsten silicide (WSi_x) films ended up being the most widely adopted for this application, with CVD being the chosen method of forming the WSi_x thin-films. Polycides with WSi_x films now find widespread use as word-line and bit-line interconnects in IC memory chips. Note that unlike CVD-W, which is used for plug/interconnect applications (and is thus never exposed to temperatures above about 500°C), these WSi_x films are deposited on polysilicon during the gate formation process. This means that such WSi_x films will have to have to be able to endure temperatures up to 800-1000°C during BPSG reflow, source/drain implant-activation, and poly reoxidation. WSi_x has also found some application as a stand-alone adhesion layer for blanket CVD-tungsten films.

A process sequence for producing the WSi_x polycide gate-structure is depicted in Fig. 13-2. It shows that WSi_x is blanket-deposited onto the doped polysilicon film, and then this WSi_x/poly multilayer is etched to form the polycide gate-structure. The CVD deposition of WSi_x won out over other candidate processes

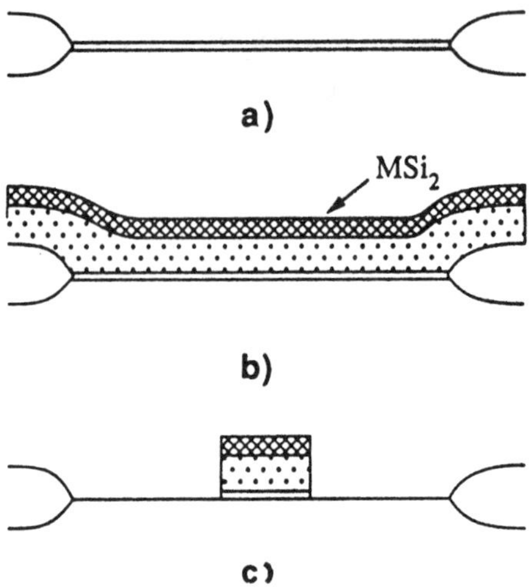

Fig. 13-2 Process sequence for producing a WSi_x-polycide-structure: a) Gate oxide growth; b) Polysilicon and WSi_x deposition by CVD; c) Patterning the polycide.

(i.e., evaporation and sputtering of WSi_x) for the following reasons: **a**) this process can produce high-purity WSi_x films without the need for high-vacuum deposition equipment; **b**) its throughput is acceptable; **c**) better step-coverage is obtained than with PVD; and **d**) good wafer-to-wafer and run-to-run uniformity is attainable. The first chemical reaction used to deposit WSi_x was the following, and it is capable of producing films with the benefits listed above:

$$WF_6\ (gas) + 2\ SiH_4\ (gas) \rightarrow WSi_2\ (solid) + 6\ HF\ (gas) + H_2\ (gas) \qquad \textbf{(13-1)}$$

Tungsten-silicide film deposition using this reaction is carried out at pressures between 50-300 mtorr (LPCVD) and at deposition temperatures between 300-400°C.[3] This reaction is similar to the silane reduction of WF_6 used to produce CVD-W films. As mentioned earlier, if it is desired that WSi_x be deposited, a higher flow rate of SiH_4 is needed. This ensures that the deposited film is WSi_x and not W. During film deposition, WSi_2 forms with excess Si collecting in the grain boundaries. Since an excess of silicon exists in such films, the chemical formula to denote them is given as WSi_x. However, unless $x \geq 2.0$, WSi_x films formed with this reaction are prone to cracking and peeling away from the underlying poly during later high-temperature steps. That is, the excess Si in the as-deposited silicide film avoids the possibility of consuming some of the underlying poly material, which could cause the above problems. Thus, in practice a SiH_4/WF_6 flow-ratio of more than 10:1 is used to ensure a deposition

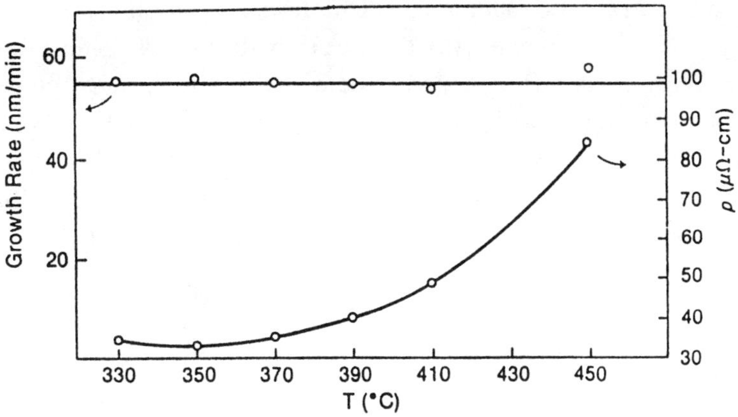

Fig. 13-3 Resistivity of CVD-WSi_x versus x, where x varies from 2.2 to 2.6.

with x ~2.2–2.6. As-deposited WSi_x-films exhibit high resistivity (~500-$\mu\Omega$-cm), but this drops to about 50-$\mu\Omega$-cm after an RTP anneal at 900°C. Note that the resistivity of CVD-WSi_x also depends on its stoichiometry as is shown in Fig. 13-3, where x varies from 2.2 to 2.6. The resistivity of the film increases as it becomes progressively richer in Si.

It was found that CVD-WSi_x films formed by the reaction of Eq. 13-1 had high fluorine-concentrations (~$10^{20}/cm^3$). This creates problems when such films are used with gate oxides thinner than about 20-nm, because low-field oxide breakdown and threshold-voltage shifts are observed.[4] Such problems arise because some of the fluorine gets incorporated into the gate oxide during anneal of the polycide stack. As a result, an alternative chemistry using dichlorosilane (DCS - SiH_2Cl_2) instead of SiH_4 was developed for CVD of WSi_x:[4]

$$WF_6(gas) + 3.5SiH_2Cl_2(gas) \rightarrow WSi_2(sol) + 1.5SiCl_4(gas) + 6HF(gas) + HCl(gas) \quad \textbf{(13-2)}$$

This DCS process is also carried at low pressure (LPCVD) in the temperature range of 570–600°C. The fluorine content is much lower than in films formed with the silane reduction of Eq. 13-1, and the chlorine content is also quite low. The resistivity is comparable in both types of films. However, the step-coverage is better in DCS films, and its deposition-rate is three to five times faster. Peeling and cracking is also less severe in WSi_x films formed from the reaction of WF_6 and DCS. DCS has thus largely replaced silane for CVD-WSi_x processes.

Exposure of the WSi_x films to high-temperature O_2-ambients (900°C) will cause a dense adherent layer of SiO_2 to grow on the silicide while the polycide structure underneath remains intact. The oxide forms by reaction of oxygen with the excess-silicon dissolved in the silicide film, and continues after this silicon is consumed by diffusion of additional silicon from the polysilicon to the surface of the WSi_x. The oxide-formation reaction occurs at the oxide/silicide interface.

CVD-WSi_x films are deposited in cold-wall reactors. Batch processes were initially used but single-wafer systems have also been introduced. A process that integrates polysilicon deposition, polysilicon doping, and WSi_x deposition in a single cluster tool is available.[6] Advantages of such a configuration include cleaner processing and less cleanroom floor space.

13.3 SALICIDE (SELF-ALIGNED-SILICIDE) STRUCTURES

In the salicide process, the entire surface area of the source and drain regions of a MOSFET are covered with a conductor film (i.e., a metal silicide). This approach becomes even more attractive if such a film can be formed using a self-aligned process, in that a self-aligned process does not entail any additional masking steps. The process sequence performed to fabricate such a self-aligned conducting silicide on the source/drain-regions (and to simultaneously form a conducting silicide on gate-polysilicon structure [polycide]) is as follows:

1. After the source- and drain-regions have been implanted and annealed to form the source/drain-junctions (Fig. 13-4a), CVD-SiO_2 sidewall-spacers on the polysilicon gate are fabricated (i.e., by CVD deposition of an oxide, followed by an etchback step which leaves spacers on the poly sidewalls - Fig. 13-4b).

2. The metal film (of either Ti or Co) used to form the silicide is deposited (typically by sputtering).

3. Next, the wafer is heated, which causes the silicide reaction to occur wherever the metal is in contact with the silicon. Everywhere else, the metal remains unreacted (Fig. 13-4c). Note that RTP is extensively used in this first silicidation step (1st RTP-step). At the conclusion of this step, the silicide is typically in a high-resistance phase. A second RTP-step (carried out after the unreacted metal is removed in step **4**), is performed at a higher temperature than the 1st RTP step. This transforms the silicide into the desired low-resistivity phase.

4. The unreacted-metal is selectively removed through the use of a wet-etchant that does not attack the silicide, the silicon-substrate, or the SiO_2 (Fig. 13-4d). As a result, each exposed source and drain region, and each exposed polysilicon gate region is now completely covered by a silicide film, but no metal or silicide film remains elsewhere.

5. As noted in step 3, after the unreacted metal is stripped, a 2nd RTP step is performed at a higher temperature than the 1st RTP step. This higher temperature process transforms the high-resistance phase of the silicide remaining on the source/drain- and poly-regions to the desired low-resistance silicide-phase.

6. A dielectric layer is deposited onto the silicide, and contact holes are opened in it down to the silicide layer.

7. Metal is deposited into the contact holes to make contact with the silicide (Fig. 13-4e).

The goals of the above salicide process are as follows:

1. The silicide should meet the required low sheet-resistivity specifications called for by the particular technology node being fabricated.

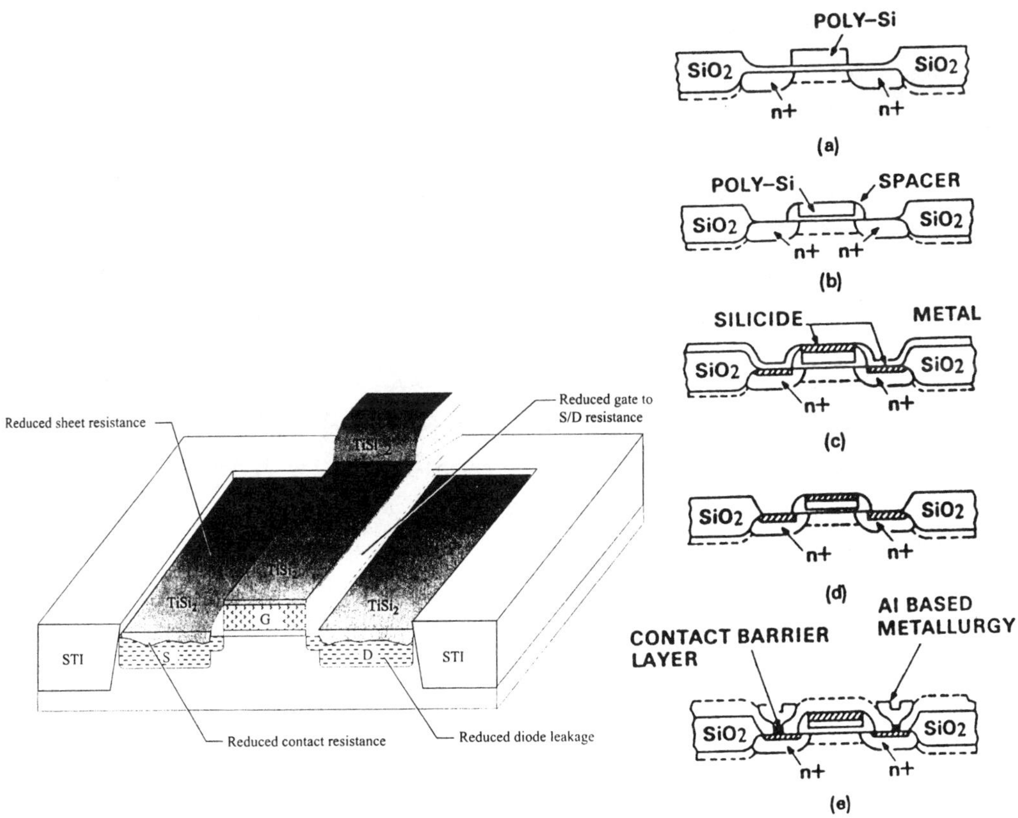

Fig. 13-4 Salicide processing steps and final structure.

2. Low junction leakage should exist on all source/drain geometries.

3. No bridging should occur across isolation regions or across gate sidewall spacers.

4. A low specific-contact-resistivity should exist for the silicide/Si contact.

The above contact-formation-method offers several advantages over the conventional contact structure:

1. The value of the parasitic-series-resistance component due to the bulk sheet-resistance of the source- and drain-regions R_{sh} becomes negligibly small because the ρ_{sh} value of the silicide is typically 1-2-Ω/sq, whereas that of the diffused junction region alone is 40-120 Ω/sq.

2. The contact area of the silicide and the Si is much larger than that of the metal-Si area in a conventional contact in a device of the same total area. Thus, the value of the contact resistance, R_{co} arising from the contact between the silicide and the Si substrate is smaller than in a conventional contact. That is, R_{co} is equal to the specific contact resistivity ρ_c divided by the contact area.

3. The contact hole is now used to make contact between the metal and the silicide, but the value of ρ_c exhibited by silicide-metal interfaces is typically $\leq 1x10^{-9}$-Ω-cm^2. This is two orders of magnitude lower than nominal metal-to-silicon specific contact resistivities, ~$1x10^{-7}$-Ω-cm^2, obtained at or near solid-solubility silicon-doping-limits. Hence, the contribution to R_{co} due to the metal-silicide contact resistance at the contact hole is also usually negligibly small.

As it became apparent that a self-aligned contact-structure would be beneficial for reducing the values of the parasitic-series-resistances of the conventional contact structure, some other processing trends also allowed a new group of materials to be considered as contact metals to Si. These materials were $TiSi_2$ and the group-VIII silicides $CoSi_2$ and NiSi. When it also became necessary to fabricate shallower junctions, processing technology had to be performed at lower temperatures to avoid the formation of deep junctions. When the maximum temperature after the silicide formation was thus restricted to less than 900°C, the door was opened to the use of these silicide films (since all of them are unstable at temperatures higher than 900°C).

There are several important reasons why $TiSi_2$ and the group-VIII metal silicides are especially attractive as candidates for self-aligned ohmic contacts and local interconnects to silicon. These are as follows:

1. Such silicides exhibit lower resistivities than the refractory-metal silicides of W, Ta, and Mo. That is, the respective resistivities are 13-20-$\mu\Omega$-cm for $TiSi_2$ (with C-54-phase crystal structure), 16-18-$\mu\Omega$-cm for $CoSi_2$, and 15-18-$\mu\Omega$-cm for NiSi.

2. All of the group-VIII metals react with Si at 600°C or less, and with the exception of Ti, there is no reaction of the metal with the masking oxide during the silicide formation. Even with Ti, the reaction is minimal below 600°C, and the reaction products are soluble in the Ti-removal etch.[8]

3. When the silicides are formed at such low temperatures, the metal atoms diffuse into the Si, and react with it. As the metal atoms diffuse into the SiO_2, no Si is available for reaction, and hence, the metal over device areas covered with SiO_2 remains unreacted. This also prevents lateral formation of the silicide over the narrow oxide-spacer-region during the silicide formation-step.*

4. For the cases of Co and Ni (but not for Ti), dopant atoms in the silicon are "snowplowed" into the silicon substrate as the $CoSi_2$ and NiSi are formed. Such snowplowing maintains a high doping concentration at the silicide-Si interface, which helps produce low-resistance contacts.

5. In all cases, after the silicide has been formed, the unreacted metal can be etched off from the oxide surface using chemical etchants that do not attack the oxide, the silicide, or the substrate silicon. This last characteristic allows self-aligned contacts to be formed through the use of these silicides.

The focus of most development for the salicide process has revolved around $TiSi_2$ and $CoSi_2$. Both of these silicides exhibit lower resistivities than silicides formed with the refractory metals, W, Ta, and Mo and they can withstand process temperatures in excess of 800°C (such temperatures are frequently used to reflow the doped glasses deposited over the salicided source/drain-regions).

The salicide process exhibits a limitation related to the fact that the gate and the source/drain-silicides are formed at the same time. On the gate, it is desirable for the silicide to have the lowest possible sheet resistance (so that the gate electrode will also possess a low interconnect-resistance). To achieve this, a thick

* There is an exception to this statement. During $TiSi_2$ formation at temperatures above 600°C, Si diffuses into the Ti. The diffused Si then reacts to form $TiSi_2$ over the oxide-spacer regions, resulting in shorting of the gate and source/drain regions. Procedures for avoiding this effect are discussed later.

Silicide-layer on the gate is needed. Over the source/drain-regions, however, the silicide can be only of limited thickness, in order to prevent excess consumption of the substrate-silicon by silicide formation. Thus, a thicker silicide, though favorable at the gate level, is detrimental to contact formation (and vice-versa). Going forward, this tradeoff between sheet-resistance and silicon-consumption represents a key scaling issue, in that thicker salicide films are needed to achieve low sheet-resistance-values, while shallower junction-depths require thinner silicide films.

Another problem is that it is difficult to metallize shallow junctions (≤ 200-nm-thick), because of the junction leakage that is observed when the silicidation process is used. The implant damage (caused during the source/drain junction-formation) may not be sufficiently annealed-out by the lower-temperature silicide-formation step. This damaged region may remain within the depletion-region of the shallow junction, giving rise to higher leakage. In addition, the RIE-process used to open the contact-holes in the doped oxide (deposited after silicide formation), may damage the silicide.[9] Techniques to reduce or eliminate this last set of problems are discussed in later sections. A final problem of salicide technology is that it degrades the effectiveness of some electrostatic-discharge-protection devices used with CMOS circuits.

13.4 THE EFFECTS OF SALICIDES ON MOSFET PERFORMANCE

A self-aligned-silicide on the source/drain-region lowers the contact-resistance because the metal-Si contact area is increased from the area of the contact window(s) to the entire source/drain area. For example, consider the contact resistance R_{co} (Ω) of an *n*MOSFET with a source/drain area of 1-μm x 1-μm and a contact window the size of 0.4-μm, with and without a silicide on the source/drain-region. Assume that the specific-contact-resistivity is identical for a silicide/n^+-Si contact and a metal/n^+-Si contact (e.g., in this case $1\text{x}10^{-7}$-Ω-cm^2), and $0.2\text{x}10^{-8}$-Ω-cm^2 for the metal/silicide contact. (Note that the value of the specific-contact-resistivity of metal-to-silicide is much smaller than the specific contact resistivity of silicide or metal to n^+-Si.)

Thus, if no silicide is formed on the source/drain region, then only the contact resistance R_{co} of the metal/n^+-Si contact exists, and its value is calculated from:

$$R_{co}([\text{metal}/n^+\text{-Si}]\ \text{contact}) = (\text{specific contact resistivity})/\text{contact area}$$

$$= (1\text{x}10^{-7}\text{-}\Omega\text{-cm}^2)/(0.16\text{x}10^{-8}\ \text{cm}^2) = 65\ \Omega \qquad \textbf{(13.3)}$$

On the other hand, if a silicide has been formed that covers the entire source/drain-regions, the contact-resistance will be the sum of the two contacts in series. That is, it will be sum of the contact resistance R_{co}([metal/silicide] contact), and R_{co}([silicide/n^+-Si] contact). The value of R_{co}([metal/silicide] contact) is found from:

$$R_{co}(\text{[metal/silicide] contact}) = (1\text{x}10^{-7}\text{-}\Omega\text{-cm}^2)/(1\text{x}10^{-8}\text{ cm}^2) = 10\ \Omega \qquad \textbf{(13.4)}$$

The value of R_{co}([silicide/n^+-Si] contact), is found from:

$$R_{co}(\text{[silicide/}n^+\text{-Si] contact}) = (0.2\text{x}10^{-8}\text{-}\Omega\text{-cm}^2)/(0.16\text{x}10^{-8}\text{ cm}^2) = 1.2\ \Omega \qquad \textbf{(13.5)}$$

The sum of these two contact resistances is:

$$R_{co}(\text{total}) = R_{co}(\text{[metal/silicide] contact}) + R_{co}(\text{[silicide/}n^+\text{Si] contact})$$
$$= 11.2\ \Omega \qquad \textbf{(13.6)}$$

Therefore, even for a nominal-sized MOSFET (i.e., in the case above, having a width of 1.0-μm), the self-aligned-silicide significantly decreases the contact resistance.

However, even a contact-resistance of 65-Ω is only a small fraction of the channel-resistance. Reducing this quantity even further does not significantly increase device performance of the nominally-sized MOSFETs of a circuit. On the other hand, for wider MOSFETs (used, for example, to supply more current to subsequent stages of the circuit, or to provide drive-current for off-circuit loads), the benefit can much more significant. Assume that a transistor is 10-μm wide (instead of being only 1.0-μm wide, as in the above example), and that this MOSFET is capable of delivering 4-mA of current when V_{GS} is 3.3 V. Let us see how the use of a self-aligned-silicide on the source- and drain-regions impacts the performance of such a device (compared to one that does not). We will do this by considering the effect of the silicide on each side of the device separately, namely first on the drain-side, and then on the source-side.

The effect of the contact-resistance R_{co} on the *drain-side* of such a wider MOSFET device is to reduce the drain voltage V_{DS} by the amount of $R_{co}I_D$. If the MOSFET is operating in the saturation region (as is the case in most digital circuits, when the MOSFET is ON), then the reduction in V_{DS} has little effect. If the transistor is operating in the linear region (as in analog circuits), then the output current I_D is reduced from:

$$I_D = V_{DS}/R_{channel} \qquad \textbf{(13.7)}$$

to:

$$I_D = (V_{DS} - R_{co}I_D)/R_{channel} \qquad \textbf{(13.8)}$$

or:

$$I_D = V_{DS}/(R_{co} + R_{channel}) \quad \textbf{(13.9)}$$

Therefore, the effective resistance in the linear channel has been increased by the amount R_{co}. For this MOSFET, the channel resistance in the linear region, $R_{channel}$, can be calculated as V_{Dsat}/I_D, or approximately 250-Ω. If no silicide exists, and only a single metal/n^+-Si, 0.4-μm contact is made to this drain region, the contribution of R_{co}, is about 25%.

The effect on the performance of the device due to the R_{co} on the *source-side* of the device is more significant. That is, a resistance on the source-side of the MOSFET not only increases the drain-voltage by the same amount as on the drain-side, but it also raises the potential at the source by $R_{co}I_D$. The effective gate voltage V_{GS} is then reduced by the same amount:

$$V'_{GS} = V_{GS} - R_{co}I_D \quad \textbf{(13.10)}$$

A reduction in effective gate voltage is equivalent to a reduction in transconductance:

$$g_m = \Delta I_D/\Delta V_{GS} \quad \textbf{(13.11)}$$

$$g'_m = g_m/(1 + R_{co}g_m) \quad \textbf{(13.12)}$$

A MOSFET with a g_m = 300-μS/μm and an R_{co} = 65-Ω, has a g'_m = 0.98-g_m, or only a 2% loss in drive current for a 1-μm-wide MOSFET. But for a 10-μm-wide MOSFET, a 20% loss for the drive current results. Therefore, for wide MOSFETs, either multiple contacts, or a self-aligned silicide is needed to reduce the parasitic resistance.

The contact-resistance will degrade the device performance more severely on MOSFETs with smaller minimum dimensions. At the 0.25-μm node, the contact window will be about 0.3-μm wide, making R_{co} ~140-Ω, compared to the 65-Ω in the example in which the contact window was 0.4-μm wide. At the 0.18-μm technology node, R_{co} will be about 250-Ω without a self-aligned silicide. Even for a 1-μm-wide MOSFET, the g'_m will be reduced to only about 0.9g_m. If a self-aligned silicide is used, R_{co} drops to about 18-Ω in the 0.25-μm node, and to about 25-Ω in the 0.18-μm node.

13.5 SALICIDES FORMED WITH TITANIUM SILICIDE ($TiSi_2$)

Salicides formed with titanium silicide $TiSi_2$ are currently being used extensively in the production of 0.25-μm CMOS devices. $TiSi_2$ is attractive for this salicide application because it exhibits low resistivity, and because it can reduce native-oxide layers (making it the only known refractory metal that can reliably form a silicide on both polycrystalline and single-crystal silicon through a thermal reaction). Furthermore, devices fabricated with titanium silicide on the gate electrode are more resistant to high-field-induced hot-electron degradation than are conventional poly-Si gate devices.[10] It is conjectured that the $TiSi_2$ is an effective getter for the hydrogen atoms introduced during the hydrogen anneal. Less hydrogen is thus incorporated into the gate oxide, and this improves the hot-electron reliability.

Some aspects of the $TiSi_2$ process are not as favorable, including:

1. The reactivity of Ti with SiO_2 can cause unwanted reactions with the oxide spacers during the silicide-formation process (a problem referred to as *bridging*).
2. $TiSi_2$ is less thermally stable than WSi_2 or $MoSi_2$ (i.e., it agglomerates at $T > 900°C$).
3. Because Ti films have a high propensity to oxidize, the silicide must be formed in ambients that are free of oxygen.
4. For line-widths smaller than about 0.25-μm it becomes difficult to form low-sheet-resistivity $TiSi_2$ films (and this phenomenon is termed the *$TiSi_2$ narrow-line effect*).
5. Silicon is consumed by the silicide reaction (i.e., about 30-nm of Si is consumed by 10-nm of Ti to form 35-nm of $TiSi_2$). Silicon consumption during silicide-formation becomes a major issue in the process of forming very shallow source/drain-junctions.
6. During silicide formation, dopants in the junctions redistribute themselves. This represents an especially severe problem with $TiSi_2$.

Each of these issues will be discussed in the upcoming sections.

13.5.1 Reaction Kinetics of $TiSi_2$ Formation (and the Problem of Bridging)

In practice, a two-step formation-process is typically used to form $TiSi_2$.[11] The first step (1st RTP step) is carried out at a temperature is between 620-680°C in nitrogen for 15-60 sec. This short RTP step transforms the Ti in contact with Si (or polysilicon) into the metastable, high-resistivity C-49 phase-$TiSi_2$:

$$Ti + 2Si \dashrightarrow TiSi_2 \quad (C\text{-}49) \qquad (13.13)$$

Following this step, selective etching is used to remove the unreacted Ti (e.g., in a room-temperature mixture of DI H_2O, 30% H_2O_2, and NH_4OH [5:1:1]). Then a second RTP-step at a temperature of 800°C in nitrogen for 30-60 sec is used to lower the $TiSi_2$ sheet-resistance by transforming the metastable C-49-phase into the low-energy, low-resistivity orthogonal C-54-$TiSi_2$ phase:[12]

$$TiSi_2\ (C\text{-}49) \dashrightarrow TiSi_2\ (C\text{-}54) \tag{13.14}$$

The resistivity of C-49-phase-$TiSi_2$ is about 3-4 times higher than that for C-54-phase-$TiSi_2$ (60-70-$\mu\Omega$-cm vs 13-20-$\mu\Omega$-cm). Phase-transformation occurs at 700-750°C, but it will be seen why the process is not carried out in a single step.

The reason why a two-step process is used to form C-54-phase-$TiSi_2$ involves the problem termed *bridging*. That is, oxide-spacers at the edges of the poly-silicon are formed to separate the silicided gate and source/drain-regions. But these spacers are typically only 200-300-nm wide. As a result, any lateral-formation of silicide can easily bridge the separation and cause the gate to be come shorted to the source/drain. Figure 13-5 illustrates the formation of $TiSi_2$ along the sidewall of the gate, resulting in a short between the gate and the source/drain. Such lateral-formation of $TiSi_2$ along the gate-sidewall oxide-spacer occurs because silicon is the moving species of this reaction. Thus, it diffuses into the Ti-regions that cover the spacer-oxide, where it subsequently reacts with the Ti. However, when annealing is performed at a lower temperature (and for short times), the Si does not diffuse as far, and this inhibits the lateral $TiSi_2$ reaction along the spacer. However, at least 600°C is needed to cause Ti to react with Si, even to form the C-49 phase. At this temperature, the Si-diffusion will already produce bridging - even if RTP is used.

It has also been reported that the surface of the SiO_2 sidewall spacers is covered with CF_x polymers after the dry-etch step used to from the spacers (carried out in a CHF_3/C_2F_6 plasma). The presence of these polymers enhances the lateral Ti-silicide growth over oxide-spacers and can result in bridging. Therefore a cleaning procedure (not based on a fluorine-based chemistry) should be used remove these polymers. A suggested post-spacer-etch cleaning-process in this report was an O_2 plasma + 2% diluted-HF in H_2O + APM (ammonia-hydroxide/hydrogen-peroxide mixture). This process was able to remove the fluorocarbon polymers while consuming less than 2.8-nm of the oxide-spacers.[13]

As a result, in addition to using the low-temperature RTP process, the 1st RTP step is also carried out in an N_2 ambient. This causes the Ti to absorb a significant amount of N_2 (e.g., more than 20atom%) before extensive Si diffusion occurs. The N_2 is absorbed preferentially at the Ti grain-boundaries, which "stuffs" the grain-boundary diffusion paths. This further reduces the diffusivity of Si in the Ti, and essentially suppresses the lateral silicide-reaction. Annealing in pure N_2 or pure forming gas (95% N_2 + 5% H_2), allows C-49-phase-$TiSi_2$ form-

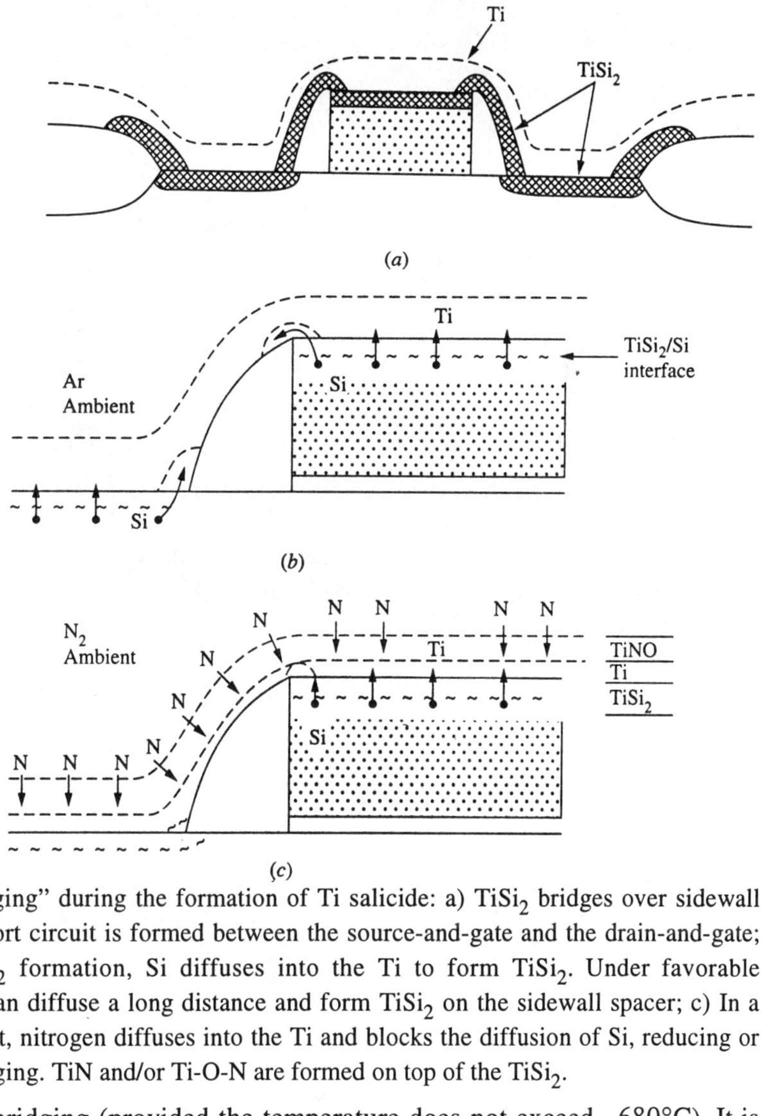

Fig. 13-5 "Bridging" during the formation of Ti salicide: a) $TiSi_2$ bridges over sidewall spacer and a short circuit is formed between the source-and-gate and the drain-and-gate; b) During $TiSi_2$ formation, Si diffuses into the Ti to form $TiSi_2$. Under favorable conditions, Ti can diffuse a long distance and form $TiSi_2$ on the sidewall spacer; c) In a nitrogen ambient, nitrogen diffuses into the Ti and blocks the diffusion of Si, reducing or eliminating bridging. TiN and/or Ti-O-N are formed on top of the $TiSi_2$.

ation without bridging (provided the temperature does not exceed ~680°C). It is also important to control the N_2 ambient so that it contains less than 5 ppma of oxygen or water to avoid unwanted oxidation of the Ti film.[15] The absorbed N_2 also reacts with the Ti to form TiN at the Ti surface.[14,15]

Another reason for performing the 1st RTP step at the temperatures below 700°C is that at higher-temperatures, Ti and the spacer SiO_2 will react rapidly enough to form significant amounts of titanium oxides. If any residues of this reaction remain on the oxide spacer, they can degrade device performance by compromising the oxide integrity or by producing bridging.

13.5.2 The $TiSi_2$ Narrow-Line Effect

As the polysilicon lines onto which the $TiSi_2$ is formed are made narrower, the sheet resistivity of the $TiSi_2$ is found to increase above the 13-20-$\mu\Omega$-cm exhibited by C-54 phase-$TiSi_2$ (Fig. 13-6). As a result, the resistance of the polycide-coated gate-poly lines no longer meet the resistance specifications for the corresponding technology nodes. The factor that causes this increase in $TiSi_2$ sheet resistivity on narrow polysilicon-lines is that the transformation of the C-49-phase into the C-54-phase is incomplete. It was observed that after the 1st RTP-step, small-grained (0.2-μm) C-49-$TiSi_2$ was formed on both wide (> 10-μm) and narrow (0.4-μm) polysilicon lines. After a 725°C anneal, the $TiSi_2$ on the wide polysilicon-lines transformed into large-grained (1 to 10-μm) C-54 phase. However, on narrow-lines only smaller-grained (< 1-μm) C-54-phase-$TiSi_2$ was observed, and only on a few lines. At higher temperatures (775°C anneal), C-54-phase-$TiSi_2$ was formed on both wide- and narrow-lines. At 800°C, $TiSi_2$ started to agglomerate, and the line resistance rapidly degraded. The incomplete transformation of C-49 to C-54 on narrow lines causes the line to have some regions along the line consisting of C-49 and others of C-54 (Fig. 13-7), resulting in an overall increase (and variability) in the $TiSi_2$ sheet-resistance.

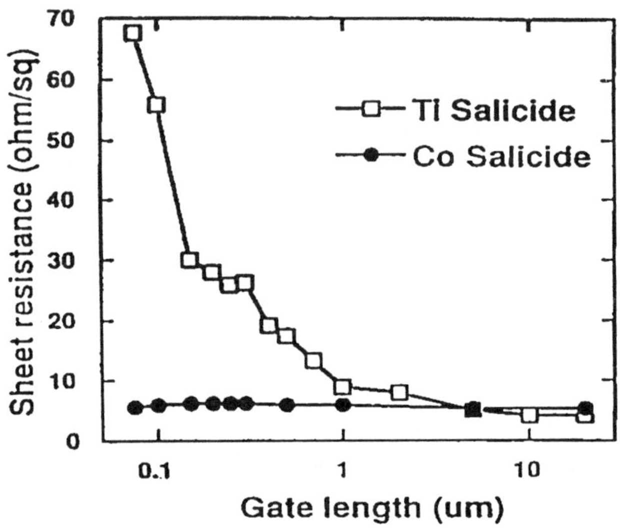

Fig. 13-6 The $TiSi_2$ narrow-width effect: As the polysilicon lines onto which the $TiSi_2$ is formed are made smaller (i.e., < 1 μm), the sheet resistivity of the $TiSi_2$ increases.

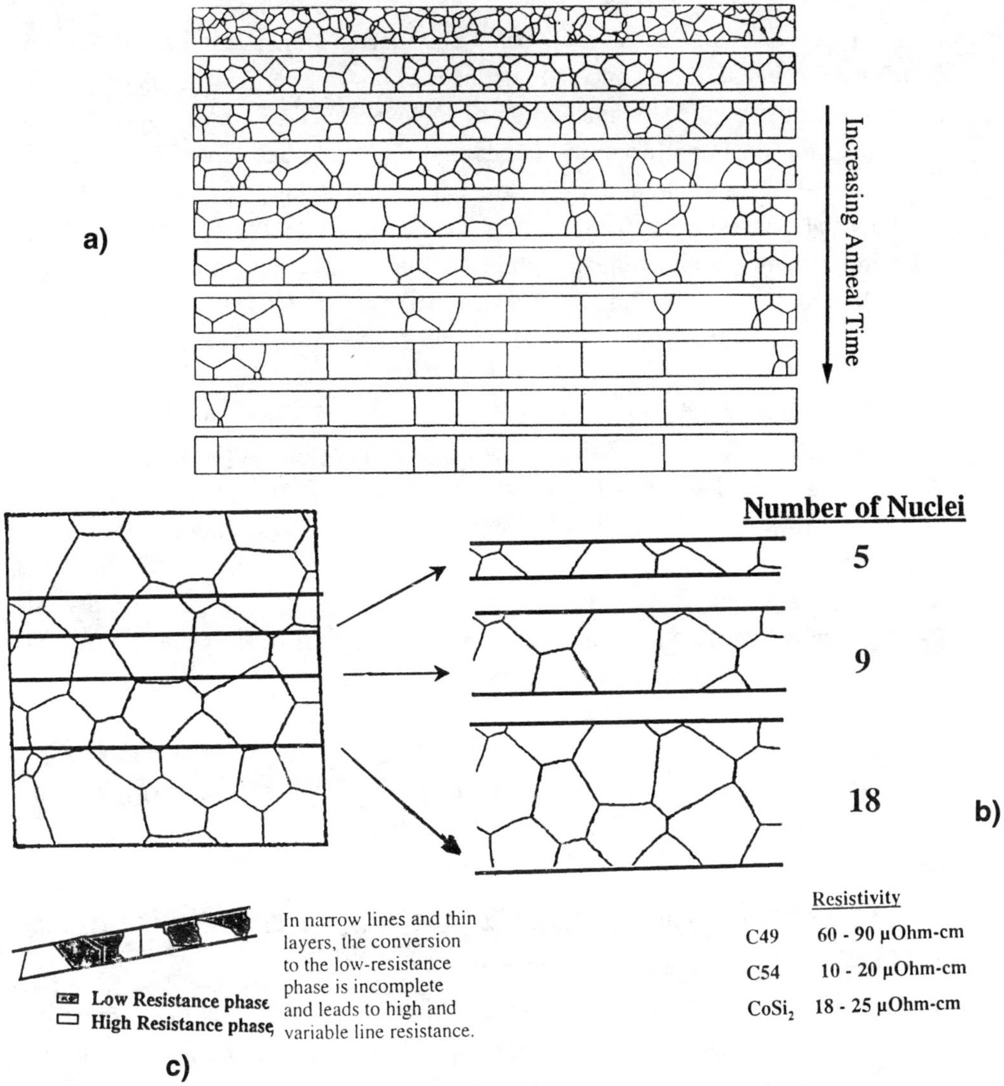

Fig. 13-7 a) Simulation of grain-growth within a line with increasing annealing time, starting with small grains; b) When the as-deposited grains are already large (relative to the line width) few nucleation sites for the C-54 phase exist. In fact, the narrower the line, the fewer the number of nuclei; c) Thus, the conversion to C-54-phase from C-49-phase is incomplete, and the final $TiSi_2$ sheet-resistance is a combination of low and high resistance phases, and is therefore higher (and variable).

The reason why there is difficulty in completely transforming C-49-phase into C-54-phase is based on the fact that on narrow polysilicon-lines the small grains of the C-49-phase are comparable to the width of the polysilicon lines.[16] As a result, there is a low-density of grain boundaries - in particular triple-grain boundaries - to function as nucleation sites for C-54-phase-$TiSi_2$ grain growth. (It has been shown by that the nucleation of C-54-phase-$TiSi_2$ preferentially occurs at triple-grain junctions [or grain-boundaries].)[17] In addition, on narrow-lines, the grain growth is limited by the linewidth, and only a fraction of the line is covered by the C-54-phase grown from the few nuclei available. On the other hand, on wide-lines, there are an adequate number of such nucleation sites per unit length of the line to allow C-54 grains to nucleate and grow to a size that transforms the entire $TiSi_2$ along the line to the C-54-phase (see Fig. 13-7). One way to solve the problem is to heat the $TiSi_2$ lines to > 750°C, to accelerate the nucleation of the C-54-phase. Unfortunately, thin films of $TiSi_2$ agglomerate at such temperatures, rendering this solution unusable. The linewidth and C-49 grain-size-dependence of the C-49-to-C-54 transformation-kinetics are in excellent agreement with a nucleation-density-model based on these assumptions.[18]

As shown in Fig. 13-8, as the polysilicon linewidth and $TiSi_2$ thickness decreases, it takes higher temperatures to transform the C-49-phase entirely into C-54-phase. Unfortunately, as the thickness of the film is decreased, the temperature at which it begins to agglomerate also decreases. As a result, the breadth of the process window (here expressed as the allowed variation in the time at which the 2nd RTP step is performed at the appropriate temperature) is also reduced. Without modifying the process conditions in some way, at some point it becomes no longer possible to produce $TiSi_2$ films with the sheet-resistance values required to meet the process specifications.

13.5.3 Proposed Process Modifications to Mitigate the $TiSi_2$ Narrow-Line Effect

Several techniques have been pursued as solutions to the narrow-line effect. One method is to replace $TiSi_2$ with $CoSi_2$, as $CoSi_2$ does not exhibit this effect. (Salicide formation with $CoSi_2$ is covered in Sect. 13.6.) However, to allow $TiSi_2$ to continue to be used on polysilicon lines below 0.25-μm, it has been proposed that the grain-size of the C-49-phase in the $TiSi_2$ be decreased. This will increase the density of C-54 nucleation sites in the film, allowing the C-49-phase to be transformed into C-54-phase even on narrower lines (Fig. 13-7a). The method that has been studied to decrease the grain sizes of the $TiSi_2$ C-49-phase is to perform a *pre-amorphization implant* (or *PAI*). That is, a shallow implant is made into the Si-substrate and poly to amorphize the top 30-nm of the silicon before the Ti is deposited (using either Ge, Si, or Xe as the implant species, Fig. 13-9).

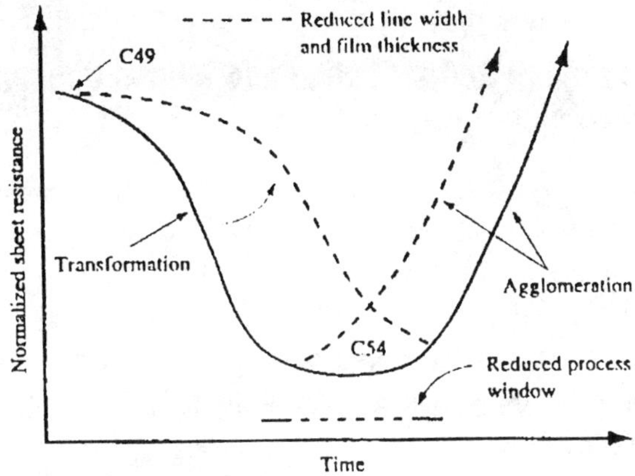

Fig. 13-8 For a thinner polysilicon line, the time required to transform the C-49-phase to the C-54-phase increases (i.e., at some chosen, constant temperature), but the time it takes for the $TiSi_2$ film on the line to undergo agglomeration decreases as the line width is decreased. Hence, the process-window for forming C-54-phase on thinner lines eventually closes.

This approach has been shown to induce a smaller C-49 grain size (~0.07-μm) compared to the 0.2-μm grain size found in the process without the PAI. This enables transformation of silicide films to the low-sheet-resistance-phase on structures down to 0.1-μm (Fig. 13-10).[19] It should be noted that such PAI steps can also cause increased junction leakage, due to damage by the implant of the Si substrate. Thus, the integration of PAI into a fabrication sequence involves careful design of the process conditions.[20]

An alternative approach to PAI which may allow the continued use of $TiSi_2$ on narrow polysilicon-lines is the implantation of refractory metal impurities into the Si surface before depositing the Ti, close to the Ti/Si interface.[21] Reports of using Mo implants for this purpose have been published.[22] This work indicates that the reaction between Ti and Mo-doped-polysilicon bypasses the C-49-phase, growing C-54-phase directly at low temperatures. On the contrary, however, the reaction between Ti and the Mo-doped single-crystal (100) silicon proceeds by the conventional sequence of C-49 growth first. To achieve a single-step growth of C-54 phase $TiSi_2$ on single-crystal-Si, it was found necessary to combine the Mo-implant into the Si-substrate with a PAI. Such a one-step Mo + PAI achieves low-sheet resistance on poly lines as small as 0.6-μm wide. Use of a single silicide-anneal step needed for the Mo + PAI process resulted in higher concentrations of B at the silicide *p*MOS interface, and nearly an order of magnitude improvement in the *p*MOS silicide-to-source/drain contact-resistance.

13.5.4 Junction Dopant Redistribution and Silicon Consumption During $TiSi_2$ Formation

Note also that $TiSi_2$ in contact with Si was originally thought to be capable of being subjected to temperatures up to ~900°C. ($TiSi_2$ agglomerates at higher temperatures.) It was observed, however, that if the temperature exceeds 800°C, the value of ρ_c to p^+ Si becomes too large.[23] However, the ρ_c of $TiSi_2$-to-n^+ Si can be less than 10^{-6}-Ω-cm^2 for anneal temperatures of 800°C. Even at anneal temperatures of 800°C, the value of ρ_c of $TiSi_2$-to- p^+-Si is 10^{-5}-Ω-cm^2, while at temperatures of 900°C, it increases to 10^{-3}-Ω-cm^2.

The cause of this problem is the rapid diffusion of the boron from the PMOS source/drain regions into the overlying silicide. Since silicon is the moving species in the course of the silicide reaction, during the formation of silicide, a large number of vacancies are generated in the Si substrate near the Si surface. This enhances the diffusion of dopants in the junction. This enhanced

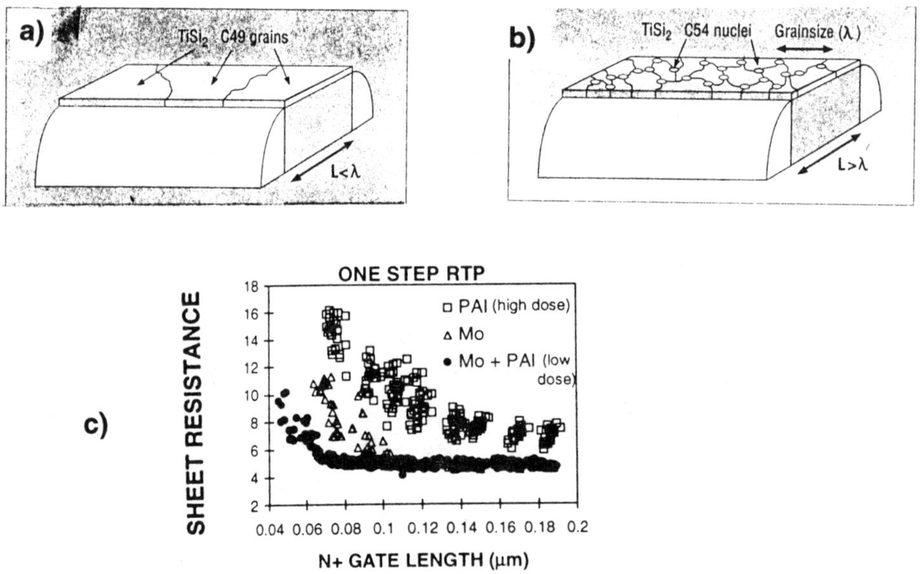

Fig. 13-9 The $TiSi_2$ C-49 (high resistivity) to C-54 (low resistivity) transformation-rate depends on the density of nucleation sites: a) Without a pre-amorphization implant (PAI) there are few nucleation sites on narrow polysilicon-lines; b) PAI reduces the C-49 grain size, increasing the number of triple-grain-boundaries and C-54 sites; c) By using PAI + MO the sheet–resistance of $TiSi_2$ lines remains low for narrower lines than when only PAI is used.

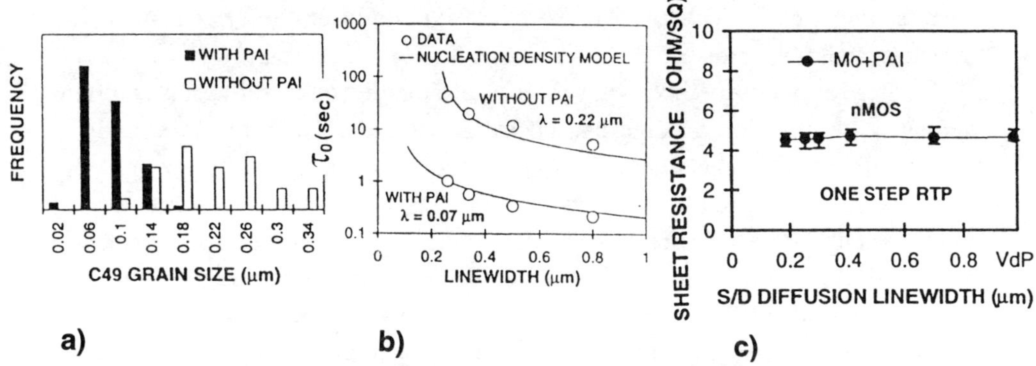

Fig. 13-10 a) C-49 grain size distribution with and without PAI. The C-49-grain size is significantly reduced by PAI, making it possible to achieve low sheet-resistance on 0.1-μm gates. b) Results and predictions of the nucleation-density-model for n^+ gates with- and without-PAI. The difference in C-49 grain size (λ) accounts for the large increase in transformation times. Transformation times τ_0 increase dramatically as the linewidth approaches λ, making the process without PAI incompatible with linewidths below 0.2-μm: c) Mo + PAI allows low-sheet resistance (4-Ω-sq) to be obtained for 0.2-μm-wide lines.

dopant diffusion has three deleterious effects. First, some of the dopants in the silicon diffuse deeper into the silicon, creating deeper junctions than would occur under normal diffusion. Second, the silicon containing the highest doping-concentration is consumed by the silicide. Third, the enhanced-diffusion causes more dopants from the silicon to diffuse into the silicide, further depleting the junction of dopants at the silicide/junction interface (and consequently, causing higher contact-resistance between the junction and silicide layer). Temperatures in excess of 800°C must thus be avoided following $TiSi_2$-formation in CMOS technologies. This can represent a problem if doped-glass-reflow is used in a CMOS process.[24] Reports have been published, however, that describe the use of nitrogen implants into the Si surface prior to the Ti-deposition to suppress the diffusion of B and As during the silicide formation.

Note, that one might think that a solution to the dopant-depletion problem would be to implant dopants into the *silicide* after it was formed, and then drive-out the dopants from the silicide to form the junction. Unfortunately, this solution is not effective for $TiSi_2$ because the dopants implanted into the $TiSi_2$ to produce the source/drain regions (e.g., B and As) form stable compounds such as TiB and TiAs.[25] Thus, the B and As species are tightly bound to the Ti and do not easily diffuse out from the silicide. This problem also contributes to the dopant depletion in the Si in contact with $TiSi_2$. That is, the B and As that diffuse into the silicide, do not back-diffuse into the Si, because once inside the silicide, the dopants bind tightly with the Ti. An extensive study of the effects of $TiSi_2$-formation on dopant redistribution is presented in Refs. 26 and 27.

Another serious issue involving the use of silicides on shallow source/drain-regions is the junction-leakage caused by the silicide. To form the silicide, a substantial portion of the silicon from the junction is consumed. For example, to form 50-nm of $TiSi_2$ requires a 20-nm-thick layer of Ti and the consumption of approximately 50-nm of silicon from the junction region. For a shallow-junction of about 150-nm deep (0.15-μm), this represents one-third of the junction-depth (assuming the silicide is uniformly produced). However, the silicide film tends to be not totally uniform in depth, and even a minor non-uniformity (say 20-nm), means that in some places 70-nm out of the 150-nm-deep junction are consumed - leaving a very shallow effective junction. As the diffused junction-depth beneath the silicide will continue to decrease with technology scaling, the silicide must also decrease in thickness so that the amount of silicon consumed is not greater than the junction-depth. In fact, it is typically desirable to leave about 50-nm between the bottom of the silicide and the junction, or else leakage current through the junction can become too large.

13.6 SALICIDES FORMED WITH COBALT SILICIDE ($CoSi_2$)

Although $TiSi_2$ has been the most widely implemented material for the salicide process, $CoSi_2$ has also received significant attention as an alternative candidate material. In fact, many companies are switching to Co at the 0.25/0.18-μm nodes. The switch to $CoSi_2$ relieves some of the constraints that exist for many established $TiSi_2$ processes used in production, widening a process window with good margins. On the other hand, it shifts the concerns from issues of lateral scaling (i.e., the narrow-line effect that afflicts $TiSi_2$), to issues related to the vertical scaling of junctions (such as diode leakage in junctions with $CoSi_2$).

13.6.1 Advantages and Drawbacks of Cobalt Salicides

$CoSi_2$ exhibits the following attractive characteristics for application in the salicide process:

1. It offers the benefits of low resistivity (16-18-$\mu\Omega$-cm) and high temperature stability. For example, 300-nm-thick $CoSi_2$ films exhibit a ρ_{sh} of 0.5-Ω/sq and remain stable when in contact with Si up to 900°C.[28,29] This demonstrates the compatibility of $CoSi_2$ with a 900°C glass-reflow step.

2. Unlike a conventional $TiSi_2$ process, nucleation does not limit the formation or transformation of Co-silicide phases, even for polysilicon linewidths smaller than 0.1-μm (see Fig. 13-9). This is due to the fact that

$CoSi_2$ has grain sizes ten times smaller than those of $TiSi_2$ (i.e., ~0.05-μm versus ~0.5-μm).

3. $CoSi_2$ is normally formed with a two-step RTP process:

$$\text{Co} + \text{Si} \dashrightarrow \text{CoSi} \quad (450°\text{C}) \tag{13.15}$$

$$\text{CoSi} + \text{Si} \dashrightarrow \text{CoSi}_2 \quad (700°\text{C}) \tag{13.16}$$

$CoSi_2$ can also be formed by RTP directly in a single annealing step. A wet-etch is then applied directly to the $CoSi_2$. Note that while Co is the species that moves during several stages of the Co-Si reaction, Si is still the main moving species during the formation of CoSi. Thus, bridging across the sidewall-spacer or isolation can still occur. The use of a Ti or TiN capping layer effectively reduces lateral growth and eliminates bridging for $CoSi_2$. In contrast with Ti, Co does not react with oxide or nitride films, alleviating some of the issues found for Ti salicide processing. An RTP system optimized for the formation of $CoSi_2$ films is described in Ref. 30.

4. Under optimized process conditions, the interface of the silicide and silicon can be smooth, and low-resistivity contacts to the silicon can be formed (as low as 15-$\mu\Omega$-cm^2 to both n^+ and p^+Si). The conditions that give rise to such smooth $CoSi_2$/Si interfaces will be described later.

5. $CoSi_2$ processes (unlike those for conventional $TiSi_2$) generally do not exhibit the strong dopant interaction and redistribution issues that can lead to interfacial dopant-depletion and higher silicide-to-source/drain contact-resistance. Instead, low-resistance-contacts to shallow As and B junctions can be successfully fabricated, and the dopant profiles in such junctions appear to be essentially unchanged after the silicide has been formed.

6. A selective chemical etch exists that makes it possible to remove the Co without affecting the $CoSi_2$ (e.g., with either a single-etch step using a hot H_2SO_4:H_2O_2 solution,[31] or with a two-step wet-etch process, consisting of an NH_4OH:H_2O_2:H_2O-etch followed by an HCl:H_2O_2:H_2O-etch.[32])

7. $CoSi_2$ is far less susceptible to removal by plasma-etching than $TiSi_2$. Hence, there is negligible silicide loss during the overetch time that may be needed when dry-etching the contact-openings through the glass-overlayer.

8. No competing reaction to the $CoSi_2$ formation reaction occurs, whereas when $TiSi_2$ is formed in nitrogen, TiN is simultaneously formed. In addition, lower shear forces (important to device integrity) are present in $CoSi_2$ than in $TiSi_2$ of the same thickness.[33]

9. It may be possible to form shallow junctions by implanting dopants into $CoSi_2$ after it has been formed, and then drive-out the dopants to form the shallow-junction. Unlike for the case of $TiSi_2$, good junctions have been reported by using this technique. This process is described in more detail in Sect. 13.9.

The use of $CoSi_2$, however, also has some drawbacks:

1. Cobalt silicide is sensitive to the surface condition of the Si regions prior to the Co deposition. That is, the presence of a native oxide or other surface contaminants can impact the silicide-formation-process and the resultant $CoSi_2$ film-characteristics. For example, both the $CoSi_2$ sheet-resistance and the leakage of the *pn*-junctions existing beneath the $CoSi_2$ salicide-film are influenced by previous processing steps that alter Si surface conditions, such as the spacer-etch, the screen-oxide-etch, and the subsequent surface-clean-and-preparation steps.

Unlike Ti, Co cannot reduce native oxides on Si, and therefore the formation of $CoSi_2$ is more sensitive than Ti to residual contaminants or to a native oxide on the Si-surface. However, several methods have been devised to reduce the sensitivity of $CoSi_2$ to surface conditions. In one of them, advantage is taken of the reactivity of Ti (and its ability to reduce native oxides), by incorporating Ti at the deposition stage. That is, by applying a capping layer of Ti onto the Co film (i.e., deposited *in situ* prior to the silicide formation), the sensitivity of the $CoSi_2$ formation process to interfacial contaminants such as oxygen is significantly reduced.[34] This allows the formation of $CoSi_2$ through native oxide layers. A similar effect was reported for a Co-Ti (10-20 atom.%) alloy deposition.[35]

Further discussion of the effect of surface-contaminants present on the Si surface prior to Co-deposition and silicide-formation on the leakage-current of $CoSi_2$ salicided *pn*-junctions is given in Sect. 13.7.3.

2. The silicon consumption during the $CoSi_2$ formation is about 25% higher than for the formation of $TiSi_2$.

3. Gate-oxide integrity is a concern when Co salicides are used because Co contamination will degrade GOI, and both the diffusivity and solubility of cobalt are high (significantly higher than those of Ti) at high temperatures. Degradation of film and interface morphology at high temperatures, as well as process conditions that induce rough, non-uniform silicide-Si interfaces, can effect GOI. These issues could limit the compatibility of Co salicide processes with aggressive scaling of polysilicon gate-thickness. GOI degradation can be avoided by a careful control of the thermal history.

4. $CoSi_2$ is sensitive to oxidation and contamination from the ambient. Measures must be taken to prevent this from occurring, and these are discussed in Sect. 13.7.4.

13.6.2 Details of the Cobalt Salicide Fabrication Process

The details of the process sequence for forming the $CoSi_2$ polycide and the self-aligned contacts to Si are as follows:

1. Since the formation of $CoSi_2$ is sensitive to the Si surface conditions, procedures must used to ensure that a clean Si surface exists prior to the Co deposition. One technique used to clean the silicon and polysilicon surfaces before the Co sputtering step is to dip the wafers in a dilute (100:1) solution of H_2O:HF just prior to loading them into the sputtering chamber. This is often followed by a short *in situ* rf-backsputtering procedure to remove ~10-nm of the silicon surface.[28,29] A prolonged rf-backsputtering process has the risk of allowing contaminant material sputtered from other parts of the wafer to redeposit onto the Si surfaces that are being cleaned. Nevertheless, the cleaning procedure is important, because cobalt is not able to reduce SiO_2 and because the presence of a native oxide can inhibit the silicidation reaction.

2. The cobalt is sputter-deposited to the desired thickness. The thickness is selected based on how much of the silicon in the source/drain regions can be consumed, as well as what value of silicide sheet-resistance is needed. For example, 35.9-nm of Si is consumed by 10-nm of Co to form 35.9-nm of $CoSi_2$. To form a $CoSi_2$ layer with ρ_{sh} = 1.5-Ω/sq, a 30-nm-thick layer of Co must be deposited. As noted earlier, the Si consumption during $CoSi_2$ formation is about 25% greater than that during $TiSi_2$ formation, constituting a disadvantage of $CoSi_2$ compared to $TiSi_2$.

3. The $CoSi_2$ is formed by reacting the Co with the Si and polysilicon. Although the reaction can be accomplished through either a furnace anneal or RTP, the latter has emerged as the more attractive approach.[36] Either a two-step anneal, or a single-step anneal can be employed, but the former is more commonly employed. An 80-nm-thick Co film will be completely converted to $CoSi_2$ at 700°C for 30 seconds. The gas ambient must be kept free of oxygen to ensure that oxide-free silicide films are formed.

4. The unreacted Co is etched away using the etchant mentioned earlier.

5. Finally, the wafer is covered with a CVD glass layer. Note that when Al is contacted to the $CoSi_2$, the maximum anneal temperature must thereafter be limited to 400°C (since Al will react with $CoSi_2$ at temperatures in excess of 400°C). To allow a higher anneal temperature to be used, a diffusion barrier layer (such as Ti:W), must be used between the $CoSi_2$ and the Al layers.[37]

13.6.3 Diode Leakage Issues Involving Cobalt Salicides

It was noted earlier that one of the major difficulties of using $CoSi_2$ is the issue of increased leakage of shallow source/drain-junctions when $CoSi_2$ is the salicide. A primary cause of such leakage is the formation of a rough interface between the $CoSi_2$ and the Si-substrate. Such interfacial roughness can arise if a native oxide or other contaminants are on the Si surface prior to the Co deposition. In that case, the silicidation reaction can be less uniform, proceeding faster in clean spots where the Co can diffuse faster, resulting in a rougher interface. As a result, an effective cleaning of the Si surface before sputtering the Co is mandatory for formation of smooth interfaces (and thus also low-leakage). As mentioned earlier, a dip in dilute-HF just before loading the wafer into the sputtering tool is thus normally performed. In addition, a brief *in situ* sputter-cleaning pre-Co-deposition-step is also used to remove contaminants from the Si surface.

The silicide interface is also rough if the silicidation is carried out at low temperatures. This, however, is the usual way of forming $CoSi_2$ films (i.e., a 1st RTP step is carried out at 450°C). At this temperature the formation of long 'needle-like' protrusions, or 'spikes' of some form of cobalt-silicide in the Si-substrate has been observed (with a length that is possibly great enough to penetrate shallow-junctions).[38] In contrast, high-temperature-processing promotes the formation of a smooth interface. At such higher temperatures the spikes formed at the lower temperatures are found to dissolve, and the $CoSi_2$/Si-interface becomes smoother. Such smoothing occurs because of the enhancement of diffusion kinetics and the increased solubility of Co in Si. Therefore, a very effective way to reduce diode leakage is the use of an optimized higher-temperature RTP, in the range of 800-850°C (Fig. 13-11). Because the growth of the low-resistivity Co-salicide-phase is easily achieved in deep-submicron structures, the second RTP step of $CoSi_2$ formation can be optimized in terms of its effects on diode-leakage, agglomeration, and source/drain-resistance, rather

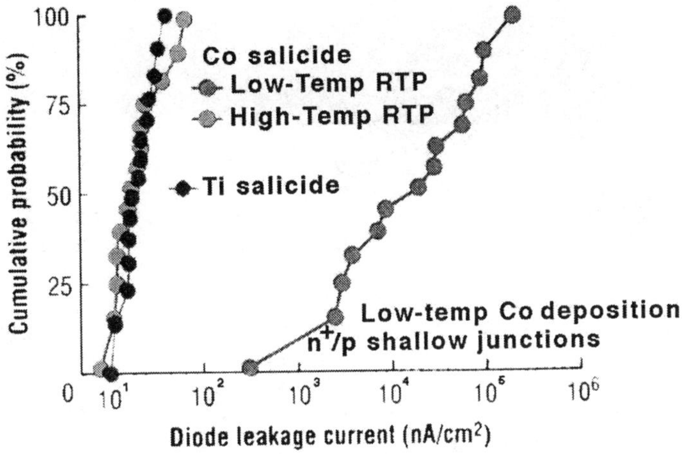

Fig. 13-11 High-temperature RTP is effective in improving diode-leakage in MOSFETs fabricated with $CoSi_2$ by eliminating high-leakage outliers.

than simply considering the growth of the low-resistivity phase (as must be done when using Ti- instead of Co-salicides). An upper-limit on the temperature that can be used during the second RTP-step is set by silicide agglomeration and the consequent degradation of the source/drain series-resistance. (An alternative technique for the reduction of diode-leakage is high-temperature cobalt deposition [i.e., the Co is sputter-deposited at 450°C].) In this process, the initial stages of the Co/Si reaction take place during the high-temperature cobalt-deposition-process.[39,40] $CoSi_2$-films deposited with such a high-temperature sputter-deposition-process were also reported to exhibit increased thermal stability, as compared to those sputter-deposited at room temperature.[41]

13.6.4 Sensitivity of Cobalt Salicide Films to Degradation by Oxygen Ambients: TiN-Capping of Cobalt-Salicide Films

$CoSi_2$ is also sensitive to oxidation and contamination due to the presence of oxygen or other contaminants in the ambient during the silicidation thermal steps. Such oxidation can cause the phase-transformation temperature from Co to $CoSi_2$ to change, and to produce films that exhibit higher sheet-resistance. A very effective approach to control ambient-contamination is the use of a capping layer, consisting of either Ti,[42] or TiN,[43] and deposited *in-situ* immediately following cobalt-deposition (Fig. 13-12). $CoSi_2$ processes with *in-situ* capping-layers have been shown to produce linewidth-independent, low-resistance films well below 0.1-μm dimensions. As also noted earlier, the use of a Ti or TiN capping layer

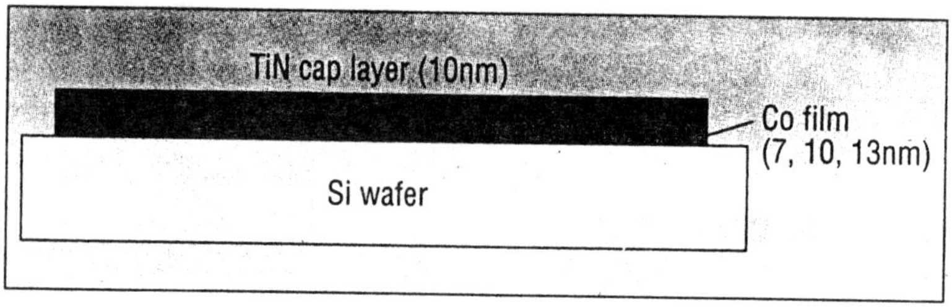

Fig. 13-12 Schematic cross-section of a $CoSi_2$ film with a TiN capping layer on a Si wafer.

also effectively reduces lateral growth of the $CoSi_2$ and thus eliminates bridging when carrying out the $CoSi_2$ salicide process.

For shallow junctions, the thickness of $CoSi_2$ films must also be scaled, to prevent excessive silicon-consumption of the substrate in which the shallow-junction is formed. The thermal-stability of $CoSi_2$ films decreases with thickness, and this becomes a major concern.[44] Deposition of a Ti capping-layer on the Co was found to increase the thermal stability of $CoSi_2$ when a Co-Ti-phase was formed on the surface.[45] According to another report, the Ti capping-layer also produced superior films than those having a TiN capping-layer. That is, the Ti-capped films formed on 0.15-μm-wide poly lines showed uniformly-low sheet-resistances and a thermal stability up to 900°C for 30 min. The TiN-capped films on the same lines showed a wide distribution of sheet-resistance values after annealing for 30 min at 800°C. The higher thermal-stability of the Ti-capped $CoSi_2$-layers was attributed to the fact that a large amount of Ti-atoms diffused into the $CoSi_2$ film along the grain boundaries, and that the presence of these Ti atoms slowed the agglomeration process.[46]

13.7 EPITAXIAL $CoSi_2$

The reaction mechanism that forms $CoSi_2$ layers on silicon-substrates that has been discussed to this point produces polycrystalline films of $CoSi_2$. However, it is also possible to nucleate and grow $CoSi_2$ epitaxially on (100)-Si. A potential advantage of epitaxial cobalt-silicide films is the achievement of smooth silicide-silicon interfaces that could result in lower diode-leakage-currents on ultra-shallow junctions. Thin epitaxial-silicide layers provide a shallow-junction contacting-technology that may open a path for further scaling of junction-depths.

Techniques have been developed for achieving epitaxial-$CoSi_2$-growth on (100)-Si that are compatible with salicide processing for IC manufacturing (Fig. 13-13).[47] One such approach consists of depositing a thin Ti-layer (~5-nm thick) before deposition of the Co film.[48] A different technique involves the growth of an ultrathin oxide layer (~0.5-nm-thick, which can be grown in a peroxide bath), and produces a similar effect.[49] In both cases, the mechanism appears to be solid-state epitaxy through a transport medium, with the main role of the interlayer being that of a diffusion barrier retarding the Co/Si reaction.

Several issues involved with the Ti-interlayer process (such as void formation at the silicide edges in patterned structures), significantly limits its applicability to salicide processing. The ability to control the characteristics of an ultrathin oxide layer may also limit the application of this process in a production environment. Epitaxial growth was also reported for a process that used a Co-Ti (10-20 atom%) alloy,[50] as well one that used a high-temperature sputter-deposition technique to deposit the Co-film.[51] An article comparing these techniques, and the influence of a BF_2 implant on the roughness of the $CoSi_2$/Si interfaces produced by each of these techniques, is listed in Ref. 52.

13.8 $CoSi_2$ AS A SORCE/DRAIN DIFFUSION-SOURCE

One technique for fabricating shallow-junctions that involves salicides is to implant the junction dopant after the salicide has been formed. To do this, the dopant would have to be implanted either: **1**) entirely into the silicide; or **2**) through the silicide. It is difficult to form very shallow junctions (< 0.15-μm) by

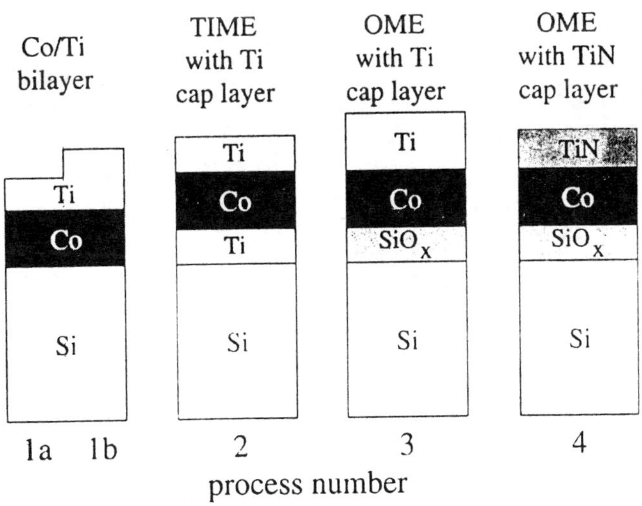

Fig. 13-13 Different silicide techniques; thickness are not scaled.

implanting through the silicide, since the peak of the implant has to be in the silicide and only the tail portion can be used to form the *pn*-junction in the Si. The junction-depth then becomes a sharp function of the silicide thickness, which is difficult to control precisely.

The second approach (implanting the junction-dopant into the silicide and driving it out) has been shown to successfully form good shallow-junctions with $CoSi_2$ but not with $TiSi_2$. This method has been dubbed the *Silicide as a Diffusion Source* (SADS) process.[53] As noted earlier, dopants such as B and As have high solubilities in Ti and also form stable compounds with Ti (i.e., TiB_2 and TiAs). As a result, if these dopants are implanted into Ti, it is very hard to get them to out-diffuse to form a junction.[54]

Forming junctions as shallow as 0.12-μm by implanting dopants into $CoSi_2$ were first reported in 1987.[55] In the SADS process that involves $CoSi_2$, the conventional S/D-implant-and-activation steps are skipped while a Co-salicide is formed, followed by S/D-implantation into the silicide-layer. A relatively low-temperature anneal is then performed to drive-out the dopants from the silicide film into the underlying silicon to form shallow junctions. Since B and As dopants diffuse rapidly in a $CoSi_2$ film, an ideal junction is expected to follow the silicide/Si interface-contour (i.e., conforming to any existing interface roughness), resulting in low-diode-leakage on the silicided ultra-shallow junctions. Two other benefits of the SADS process are that low source/drain series- resistance is achieved, due to the high dopant-concentrations at the silicide-Si interface, and that SADS is compatible with thin sidewall-spacers.

An example of an n^+/p-junction formed with the SADS process is shown in Fig. 13-14.[56] The *p*-doping concentration in the Si was $1x10^{18}$ cm^{-3}. It can be seen that a junction-depth of 40-nm beneath the $CoSi_2$/Si interface is obtained. Arsenic was implanted into the $CoSi_2$, followed by a low-temperature RTP-step. (Note that in Fig. 13-14, a baseline-temperature T_B is invoked (i.e., used in the conventional $CoSi_2$ process). Thus, the low-temperature-range referred to here is [T_B - 250°C] to [T_B - 50°C].) Junction-depths were observed to increase as the RTP-temperature was increased in this low-temperature-range. However, as the RTP-temperatures approached the baseline temperature ([T_B - 50°C] in Fig. 13-14), the SADS junction-depth became shallower (due to As evaporation from the $CoSi_2$ film). The minimum possible junction-depths in SADS process will probably thus be limited by the maximum temperature that can be endured by the thin $CoSi_2$ layers before they become thermally unstable (e.g., before they agglomerate). A capping-layer would aid in preventing evaporation at higher temperatures, allowing the formation of deeper junctions.

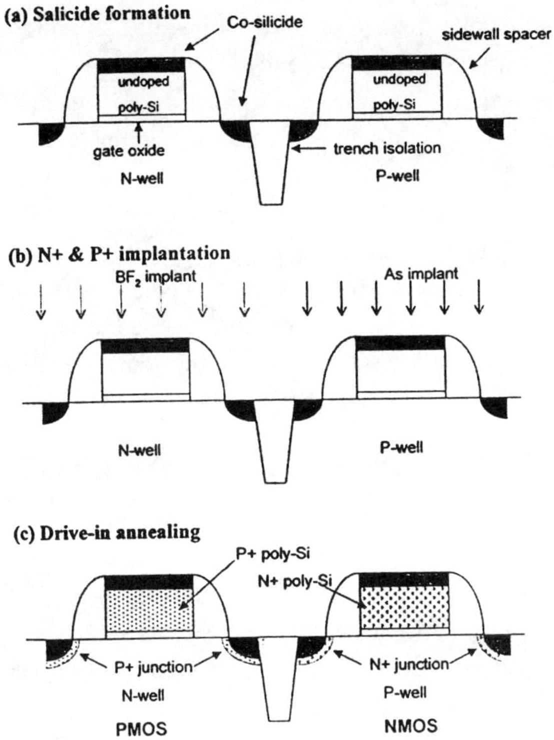

Fig. 13-14 Schematic of a process flow of 0.18-μm salicided dual-doped-gate CMOS using Co-silicide as the doping source.

Improvements in short-channel characteristics for thin (35-55-nm) spacers were verified for the SADS process in comparison with the conventional process, as better DIBL-values were observed (see Vol. 3, Chap. 5, *Silicon Processing for the VLSI Era*, for more information on DIBL).[57]

One of the main concerns associated with implementing silicided shallow junctions is diode-leakage. Good leakage characteristics, however, have been demonstrated with $CoSi_2$ as the diffusion source for both n^+/p and p^+/n junctions. However, most of the demonstrations were carried out in devices with low n-well and p-well concentrations that are not suitable for deep-submicron CMOS applications. For higher background-concentrations the depletion-region of the n^+/p-junction extends closer to the $CoSi_2/n^+$-Si interface. If the junction is too shallow, the depletion-region can extend through the whole n^+-Si-layer, and the shallow n^+/p-junction behaves as a Shannon-contact with high diode-leakage. Applicability of SADS is thus limited by the ability to obtain the appropriate junction-depths needed to achieve low diode-leakage, given the background-concentrations used in specific technologies.

A report that details the use of SADS for a full sub-micron, dual-doped-gate CMOS process was recently published.[58] As shown in Fig. 13-15, BF_2 and As were implanted separately into a Ti-capped, 35-nm-thick $CoSi_2$-layer that covered both the polysilicon and source/drain-regions of the MOSFETs. An As implant at 30-40-keV with a dose of $1x10^{16}$ cm^{-2}, was used to introduce the *n*-dopants, and a BF_2 implant at 30-keV with a dose of $5x10^{15}$ cm^{-2} was used to introduce the *p*-dopants. A single-step anneal at temperatures ranging from 800-1000°C for 30 seconds was used to drive-out the dopants to form the shallow junctions. These shallow-junctions exhibited good leakage-characteristics and low contact-resistance. The process also did not contribute to degradation of the gate-oxide, nor to B-penetration of the gate-oxide from the *p*-doped poly-gates.

13.9 NICKEL SILICIDE FILMS FOR SALICIDE APPLICATIONS

Ni silicides have also been investigated for use as salicide candidates, due to their low resistivity (~15-18-$\mu\Omega$-cm^2). A single RTP-step at temperatures in the 400-500°C range forms the NiSi film.[59] Like $CoSi_2$ the sheet resistance of NiSi films are not affected by the strong narrow-line-effects which plague $TiSi_2$ salicides. The slight increase in sheet resistance observed with decreasing linewidth has been attributed to thickening of NiSi films at gate edges. A potential advantage of NiSi over $CoSi_2$ is that silicon consumption is only 80% of that for $CoSi_2$.

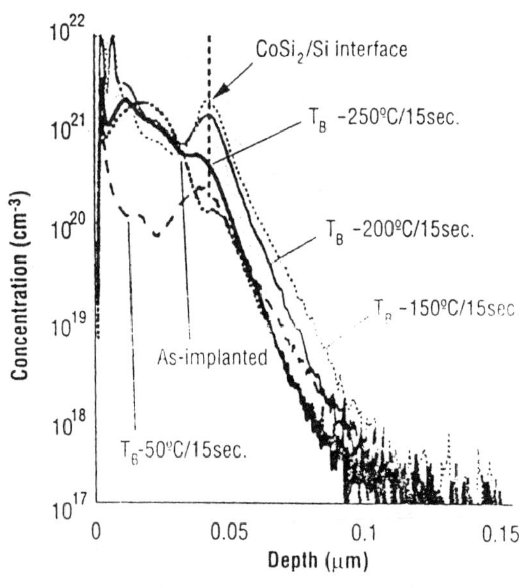

Fig. 13-15 Shallow junctions obtained using $CoSi_2$ as a diffusion source (SADS) process.

Several challenges, however, exist for implementing NiSi. First, NiSi (like $CoSi_2$) is sensitive to contamination on the Si surface, and this can lead to rough interfaces and higher-diode-leakage on shallow junctions. TiN capping-layers and nitrogen-doping can prevent oxidation from the ambient during silicide formation, and this reduces diode-leakage. High diode-leakage in the source/drain-edges of STI edge-intensive-structures has been reported, correlated to film-thickening and non-uniformity at these edges.[60] Furthermore, transformation into the high-resistivity $NiSi_2$-phase needs to be prevented during later processing steps. That is, NiSi films have poor thermal-stability, with an increase in sheet resistance observed at temperatures as low as 600°C.

REFERENCES

1. S.P. Murarka, *Sillicides for VLSI Applications,* Academic Pr., Orlando, FL 1983.

2. M.-A Nicolet and S. Lau, "Silicides," in N.G. Einspruch & G.B. Larabee,eds., *VLSI Microstructure Science*, Vol. 6, Academic Press, N.Y. 1983, p. 329.

3. K.C. Saraswat *et al*., "Properties of Low Pressure CVD Tungsten Silicide for MOS VLSI Interconnection," *IEEE Trans. on Electron Dev.,* **ED-30**, No. 11, November 1983, p. 1497.

4. M. Venkatesan and I. Beinglass, "Single-Wafer Deposition of Polycrystalline Silicon," *Solid State Technology*, March 1993, p. 49.

5. T. Hara et al., "Composition of Tungsten Silicide Films Deposited by Dichlorosilane Reduction of WF_6," *J. Electrochem. Soc.* , **137**, p. 2955 (1990).

6. S.G. Telford *et al.,* "CVD WSi_x Films Using Dichlorosilane in a Single Wafer Reactor," *J. Electrochem. Soc.* , **140**, 3689 (1993).

7. C. Y. Ting, "Silicide for Contacts and Interconnects," *Tech. Dig. IEDM*, 1984, p. 110.

8. R. A. Haken, "Applications of the Self-Aligned $TiSi_2$ Process to VLSI MOS CMOS Technologies," Workshop on Refractory Metals and Silicides for VLSI-IIII, San Juan Batista, CA, May, 1985.

9. R. Liu, D. S. Williams, and W. T. Lynch, "Mechanisms for Process-Induced Leakage in Shallow Silicided Junctions," *Tech. Dig. IEDM*, 1986, p. 58.

10. S.-T. Chang and K. Y. Chiu, *IEEE Electron Dev. Letts.,* May, 1986, p. 244.

11. *ibid. Ref. 7.*

12. C. Y. Ting *et al.,* "Interaction Between Ti and SiO_2," *J. Electrochem. Soc.* **131**, p. 2934, 1984.

13. Y.-B Kim *et al.,* "Characterization of the Post Dry-Etch Cleaning of Silicon Ti-Self-Aligned Silicide Technology," *J. Electrochem. Soc.* April 1999, p. 1549.

14. *ibid. Ref. 8,*

15. R. M. Vadjikar and R. P. Roberge, *Ext. Abs. Electrochem. Soc. Meeting*, Fall 1988, Abs. No. 467, p. 681, 1988.

16. J.B. Lasky *et al.,* "Comparison of Transformation to Low-Resistance Phase and Agglomeration of $TiSi_2$ & $CoSi_2$," *IEEE Trans. Electron Dev.,* 1991, p. 262.

17. Z. Ma *J. Appl. Phys.* 77, 4384 (1995).

18. J.A. Kittl *et al.,* "Novel One-Step RTP Salicide Process with Low Sheet Resistance 0.06 μm Gates and High Drive Current," *Tech. Dig. IEDM,* 1997, p. 111.

19. J.A. Kittl *et al.,* "A Ti-Salicide Process for 0.10 μm Gate Length CMOS Technology," *Symp. VLSI Tech* 1996, p. 14..

20. J. Kittl *et al.,* "Salicides for 0.10 μm Gate Lengths: A Comparative Study of One-Step RTP Ti with Mo Doping,, Ti with Pre-Amorphization, and Co Processes," *Symp. VLSI Tech.* 1997, p. 103.

21. R.W. Mann *et al., Appl. Phys. Letts.* 67, p. 975 3725, 1996.

22. J. Kittl, *Appl. Phys. Lett.,* 73, p. 900, 1998.

23. D. B. Scott *et al.*, "Titanium Disilicide Contact Resistivity and Its Impact on 1-μm CMOS Circuit Performance," *IEEE Trans. Electron Dev.*, March 1987, p. 562.

24.R. K. Shukla and J. S. Multani, "Thermal Stabillity of Titanium Silicide Thin Films," *Proceedings of the 4th Intl VMIC Conf.,* Santa Clara, CA, 1987, p. 471.

25. V. Probst, *J. Appl. Phys.* 70, 693, p. 1991.

26.C. M. Osburn *et al.,*"The Effects of Titanium Silicide Formation on Dopant Redistribution," *J. Electrochem. Soc.*, June 1988, p. 1490.

27. M.E. Lunnon, J. Chen, J.E. Baker, "Furnace and Rapid Thermal Annealing of p^+/n Junctions in BF_2^+-Implanted Silicon," *J. Electrochem. Soc.,* **132**, 2473 (1985).

28. L. C. Van de Hove *et al.,* " A Self-Aligned $CoSi_2$ Interconnection and Contact Technology for VLSI Applications," *IEEE Trans. Electron Dev.*, March, 1987, p. 554.

29. S. P. Murarka *et al.,* "Self-Aligned Cobalt Disilicide for Gate and Interconnection and Contacts to Shallow Junctions," *IEEE Trans. Electron Dev.*, Oct. 1987, p. 2108.

30. W.S. Yoo *et al.,* "Cobalt Silicide Processing in a Susceptor-Based LR-RTP System," *Solid State Technology,* July 1999, p. 126.

31. J. Kittl *et al.,* "Salicides and Alternative Technologies for Future ICs: Part I," *Solid State Technology* June 1999, p. 86.

32. D.K. Sohn *et al.,* "Effects of Ti-Capping on Formation and Stability of Co Silicide," *J. Electrochem. Soc.*, January 2000, p. 373

33. L. Van den Hove et al., *Electrochem. Soc. Abstracts*, Vol. 88-1, Atlanta, GA, May 15- 20, 1988, p. 312.

34. Q.F. Wang, *Dig. Symp.VLSI Tech.,* 1995, p. 17.

35. Iinuma *et al.,* "Highly Uniform Heteroepitaxy of Cobalt Silicide by Using Co-Ti Alloy for Sub-Quarter Micron Devices," *Symp. VLSI Tech.*, 1998, p. 188.

36. A. R. Sitaram and S. P. Murarka, *Ext. Abs. Electrochem. Soc. Meeting,* Fall 1988, Abs. No. 263, p. 376, 1988.

37. E.K. Broadbent *et al.,* "Application of Self-Aligned $CoSi_2$ Interconnection in sub-Micron CMOS Transistors," *Proceedings of the 5th IEEE VMIC Conf.*, Santa Clara, CA, 1988, p. 175.

38. K.-I. Goto *et al.,* "A New Leakage Mechanism of Co Salicide and Optimized Process Conditions," *IEEE Trans Electron Devices*, January 1999, p. 117.

39. Q.Z. Hong, "$CoSi_2$ With Low Diode Leakage and Low Sheet Resistance at 0.065 μm Gate Length," *Tech. Dig. IEDM* 1997, p. 107.

40. K. Inoue *et al.,* "A New Cobalt Salicide Technology for 0.15-μm CMOS Devices," *IEEE Trans. Electron Dev.,* November 1998, p. 2312.

41. H.Y. Huang, "Thermal Stability of $CoSi_2$ Thin Films on (001)Si by High Temperature Sputtering Deposition," *Mtg. ECS Spring 1999*, Abs 516.

42. *ibid. Ref.* 34.

43. T. Yamazaki *t al.,* "21 psec Switching 0.1 μm-CMOS at Room Temperature Using High Performance Co Salicide Process," *Tech Dig. IEDM,* 1933, p. 906.

44. K.Maex, *Mat. Sci. Eng. Rep.,* R11, p. 53 1993.

45. R.T. Tung and F. Schrey, *Appl. Phys. Lett.*, 67, 2164 (1995).

46. ***ibid. Ref. 32.***

47. J. Hohaus *et al.*, "Influence of BF_2-Implantation on Roughness of Cobalt Silicide/Silicon Interfaces," *Electrochem. & Solid State Letts.* March 2000, p. 147.

48. L.A Dass, *Appl. Phys. Lett.,* 58, p. 1308, 1991.

49. R.T. Tung *Appl. Phys. Lett.*, 61, 3461, (1996

50. *ibid.* Ref. 35.

51. *ibid.* Ref. 40.

52. ***ibid.*** Ref. 47.

53. J.C. Hu, *SPIE* Vol. 3506, p. 103, 1998.

54. C.Y. Lu *et al.,* "Process Limitation and Device Design Tradeoffs of Self-Aligned $TiSi_2$ Junction Formation in Submicrometer CMOS Devices," *IEEE Trans. Electron Dev.* **ED-38**, p. 246 (1991).

55. R. Liu *et al.,* "Formation of n^+/p Junctions with $CoSi_2$," in *Proceedings of the 1st International Symposium on ULSI Science and Technology,* S. Broydo and C.M Osburn, eds. Vol. 87-11, The Electrochemical Society, Pennington, NJ 1987, pp. 446.

56. J. A. Kittl *et al.,* "Salicidies and Alternative Technologies for Future ICs,": Part 2," *Solid State Technology*, August 1999, p. 55.

57. J.C. Hu, *SPIE* Vol. 3506, p. 103, 1998.

58. J.-S. Park *et al.,* "Simultaneous Formation of Shallow Junctions and Gate Doping for Dual Gate Structures Using Cobalt Silicide as a Dopant Source," *J. Electrochem. Soc.,* October 1999, p. 3856.

59. T. Ohguro *et al.*, "Analysis of Resistance Behavior in Ti- and Ni-Salicided Polysilicon Films," *IEEE Trans. Electron Dev.*, p. 2305, 1994.

60. K. Goto, *Proc. 36th Intl. Reliability Phys. Symp., IEEE* p. 363, 1998.

Chapter 14

LOW-*k* DIELECTRICS

For the purposes of IC microelectronic applications, dielectrics are categorized into two classes: **1**) *low-k dielectrics* and; **2**) *high-k dielectrics.* Those having a dielectric-constant k lower than that of thermal silicon dioxide (i.e., $k < 3.9$) are classified as *low-k dielectrics*, while those having a value higher than that of silicon nitride ($k > 7$), are classified as *high-k dielectrics.* Note that the minimum value of k is 1.0 for air. In this chapter low-k dielectrics are discussed, and in Chap. 4 high-k dielectrics are covered.

14.1 INTRODUCTION TO LOW-*k* DIELECTRICS

Low-k dielectric films are important in advanced IC applications because their use can significantly improve circuit performance characteristics, primarily with respect to interconnect delay. As detailed in Chap. 12, interconnect delay can be reduced not only by decreasing R of the conductor structures in ICs, but also by decreasing C of the dielectric layers (Fig. 14-1). The value of C, in turn, can be reduced by using dielectric materials with smaller permittivity values (i.e., *low-k dielectric materials*).[1] As semiconductor device geometries shrink below 0.25-μm, the parasitic capacitance in the intermetal-dielectric between metal lines becomes more and more important, in terms of the *RC*-delay in switching.

The primary dielectric film material for interconnects has been SiO_2. The choice of SiO_2 was based on its good dielectric and mechanical strength, as well as its ease of processing. However, SiO_2 has a dielectric-constant ranging from $k = 3.9$ to 4.5, depending on the method of its formation. This value is too high to be applicable in devices below about 0.18-μm. That is, as indicated in Fig. 14-1 only by using Cu as a conductor together with a low-k dielectric as the insulator, can the sufficiently-low values of *RC*-delay required for deep-submicron ICs be obtained. Thus, there has been a push to integrate new dielectric materials with k-values smaller than that of SiO_2 into IC fabrication. It should be pointed out that it has become apparent that the tacitly-hoped-for dielectric with $k \sim 1$ is not realizable, and that some of the qualities of SiO_2 will also no longer be attainable in actual low-k dielectrics. Evaluations of various dielectric materials for low-k thin-film applications will provide the empirical data needed to determine the trade-off limits of k-value versus mechanical performance.

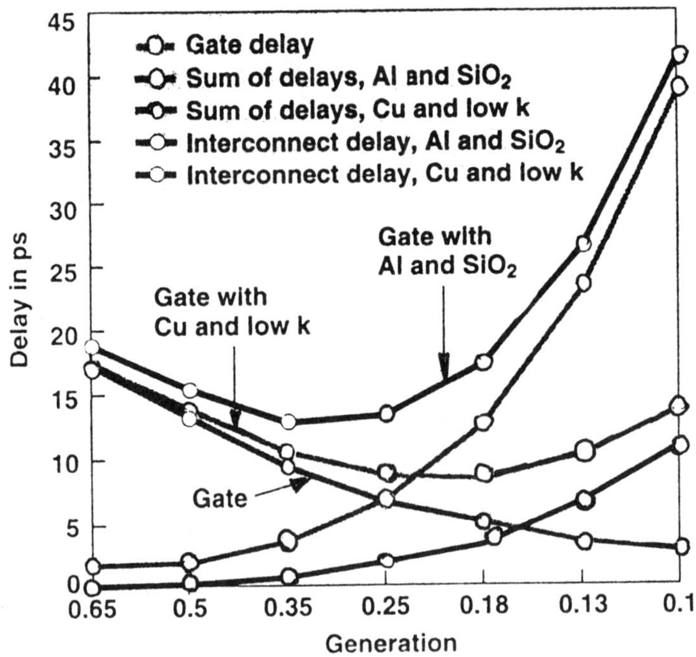

Fig. 14-1 *RC*-delays caused by interconnects are charted as a function of feature size for both Al/SiO_2 and Cu/low-*k* approaches. (© 1997 IEEE)

The conclusions from Fig. 14-1 can be justified by noting that the SiO_2 molecule is a pyramid-shaped tetrahedron (see Chap. 3), and its lattice structure thus fits together rather tightly. Although thermal SiO_2 does not possess long-range-order, it is nevertheless a relatively dense material (~2.3 g/cm^3). Since the goal is to find dielectric films with *k*-values smaller than that of oxide (~4.0), the factors that impact the dielectric constant should first be considered.

In general, no dielectric films as dense as thermal oxide can achieve *k*-values significantly smaller than 3.5. In fact, even CVD-oxides (which are also silica-based films) have about the same *k*-values, even though they are somewhat less dense. While CVD-oxides are more disordered than thermal-SiO_2 (which lowers their density), such disorder also makes them contain more silanol (OH) groups (i.e., at sites where Si-O-Si bonds are unable to form). Thus, they are prone to absorb more moisture than thermal oxides, and the absorbed water raises their *k*-value. Furthermore, materials that are less dense than thermal-oxide are also generally mechanically weaker. Nevertheless, to achieve a $k < 2$ in any dielectrics (including silica films), free space must be added, even though the strength of the material diminishes according to some power of the solids content.

The approaches used to produce dielectrics with *k*-values smaller than that of thermal-oxide and CVD-oxides therefore fall into three categories: **1**) reduction of the film density; **2**) reduction of ionization in the material, and **3**) reduction of the film polarization. Details of how these approaches are exploited in each low-*k* material-type are provided later, but a few general comments on them are given here.

First, the most dramatic technique for reducing the *k*-value is to *decrease the density* of the film material. Figure 14-2 shows the density of various low-*k* materials, as well as their *k*-values. It shows that the lower the material density, the lower the value of *k*. In all ultra-low-*k* dielectric-materials, the density is significantly lower than that of oxide films. This density reduction can occur either at the *microscopic level* (i.e., through the formation of a lattice structure that is more open than that of silicon oxide), or at the *macroscopic level* (i.e., through the formation of macropores in the film material - with sizes in the nm-range).

Second, *reduced ionization* and *reduced polarization* are only able to reduce the *k*-value to a much smaller extent. Reduced ionization is a common feature of the partially-organic low-*k*-materials. Such organic films contain carbon, often in the form of methyl groups (CH_3). These methyl groups replace oxygen atoms to create Si-CH_3 bonds in place of Si-O bonds. A smaller percentage of Si-CH_3 bonds ionize than do Si-O bonds (and they are less polarizable than are Si-O bonds). Reduced polarization is exploited in such dielectrics as fluorinated silicate glass (FSG). That is, the fluorine atoms introduced into the oxide lattice

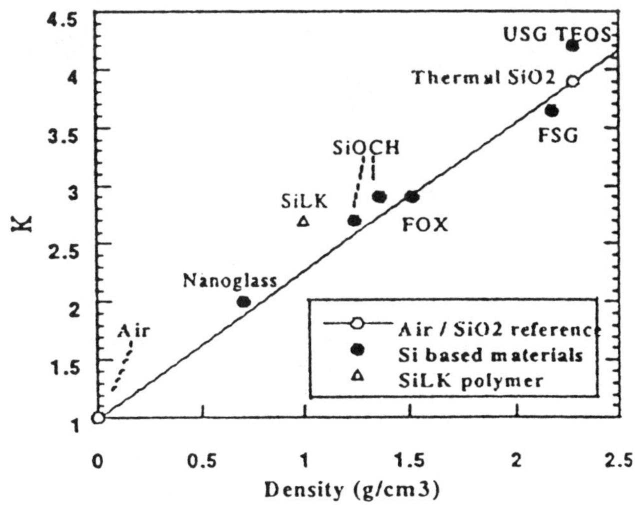

Fig. 14-2 Value of *k* is shown vs. density for various dielectrics in ULSI interconnects.

TABLE 14-1 ITRS Roadmap of the Effective Dielectric Constant Needed at Each Device Generation - Out to 2012

Table 1. SIA Roadmap								
1997								
Year of first product	**1997**	**1999**	**2001**	**2003**	**2006**	**2009**	**2012**	
Shipment technology node	250 nm	180 nm	150 nm	130 nm	100 nm	70 nm	50 nm	
Interlevel metal insulator effective k	3.0-4.1	2.5-3.0	2.0-2.5	1.5-2.0	1.5-2.0	<1.5	<1.5	
1999								
Year of first product		**1999**	**2001**	**2002**	**2005**	**2008**	**2011**	
Shipment technology node		180 nm	150 nm	130 nm	100 nm	70 nm	50 nm	
Interlevel metal insulator effective k		3.5-4.0	2.7-3.5	2.7-3.5	1.6-2.2	<1.5	<1.5	
2000?								
Year of first product			**2001**	**2003**	**2005**	**2007**	**2009**	**2011**
Shipment technology node			180 nm	150 nm	130 nm	100 nm	70 nm	50 nm
Interlevel metal insulator effective k			3.5-4.0	2.7-3.5	2.7-3.5	2.2-2.7	2.2-2.7	1.6-2.2

replace O atoms and form Si-F bonds instead of Si-O bonds. The shorter bond length of Si-F bonds makes the molecular structures containing a fraction of Si-F bonds less polarizable than those consisting of only Si-O bonds.

A few materials that were identified as good low-*k* on-chip insulators for ICs were first integrated into two-level-metal structures in the 1995-96 timeframe. These first-generation low-*k* films had *k*-values somewhat smaller than 4.0 ($2.8 < k < 3.5$). In 1999 the *International Technology Roadmap for Semiconductors* (ITRS-99) was published.[2] It urged IC manufacturers to identify, characterize, and integrate interlevel dielectrics with values of k = 2.5–3.0 for 0.18-μm devices, and k = 2.0–2.5 for 0.15-μm devices (Table 14-1). However, the industry proved to be more conservative. As Table 14-1 shows, there was a two-year delay in the implementation of the various low-*k* solutions in the years from 1997 to 1999. The delay in the adoption of low-*k* films was due to numerous technical challenges that had to be overcome before low-*k* materials could be successfully integrated into existing processes. As a result, the 0.25-μm and 0.18-μm technology generations were generally implemented with dielectrics having ranges of k = 3.0–4.0, and k = 2.5–3.0, respectively. As a result, ICs with low-*k* dielectrics did not appear in significant volume until about 1999. However, the consumption of materials used to produce low-*k* films is expected to gradually increase, and by 2003 the global low-*k* material market should reach $150 million and expand rapidly. In addition, it is quite possible that the expected change from a k range of 2.5-3.0 to a k range of < 2.0 will be stretched out beyond the 70-nm generation in 2005, due to a significant increase in integration challenges for materials with dielectric constants below 2.0.

As mentioned, there are many process-integration issues that must be satisfied before a new material is successfully incorporated into an IC fabrication sequence.[3] Candidate low-k materials must also satisfy these requirements to be useful in IC fabrication. Here a number of low-k materials being considered for use in deep-submicron IC processes are described, as well as some of the process integration challenges they offer.

14.1.1 Spin-On versus CVD Methods for Forming Low-k Dielectrics

The two most common ways of depositing low-k dielectric films are: **1**) spin coating followed by furnace annealing (*spin-on deposition*); and **2**) chemical vapor deposition (*CVD deposition*). As will be seen, there are many varieties of low-k films within the spin-on and CVD-deposition groups being studied for use in advanced IC interconnects. Note that spin-on films are all applied in liquid solution form, and are then cured in nitrogen at temperatures as high as 450°C. Curing is necessary for: **a**) driving-off solvent; **b**) completing chemical reactions; **c**) driving-off unreacted monomers and oligamers; and **d**) densification.

Proponents of the CVD deposition approach point to the reuse of existing toolsets and simpler integration due to the silicon-dioxide-like structure of CVD siloxanes as advantages of their method. However, the cost of depositing films by CVD is generally higher than the cost of a spin-on deposition. Hence, this is a factor that must also be considered when a choosing a low-k material. Manufacturers of CVD-tools for depositing low-k films include Applied Materials, Novellus, ASMI, and Trikon Technologies.

Manufacturers of spin-on materials and spin-on dielectric (SOD) tracks contend that the spin-on dielectrics allow better extendibility to future generations (especially in the sub-0.25-μm regime, when porous, low-k materials are likely to be used). Manufacturers of *spin-on dielectric* (*SOD*) coaters include: Dainippon Screen (DNS, Tokyo), Tokyo Electron (TEL America, Austin, TX), FSI International (FSI, Chaska, MN), Semix (Fremont, CA), and SVG (San Jose, CA). Spin-on dielectric material vendors include: Honeywell, Dow Chemical, Dow Corning, Hitachi Chemical, and Tokyo Ohka Kogyo. Each SOD equipment vendor has optimized the coating and curing modules specifically for SODs, making them significantly different from the photoresist tracks on which they were originally based. Note that one cost factor in SOD use is the quantitative dispense of the SOD. Due to the high cost (> $2000/liter) and potential sensitivity of some spin-on formulations to temperature, moisture, and oxygen, materials management could play a critical role in the successful integration of spin-on dielectrics. It should also be mentioned that use of fast heating and cooling rates in RTP reduces the spin-on-deposition processing time. Thus, implementation of rapid thermal processing tools into the spin-on low-k deposition process is being explored.

At present it is too early to tell which technology (spin-on or CVD), and which type(s) of low-k material will prevail. It seems that the whole industry will be characterized by uncertainty regarding this issue over the next few years. As one process integration engineer was said to remark in 1999, "My biggest problem right now is the confusion caused by the variety of material and process options that I face."[4]

14.1.2 Silicon-Based versus Carbon-Based Low-*k* Dielectrics

Low-k dielectrics are also categorized into two other groups: *silicon-based low-k dielectrics* (Si-based materials) and *carbon-based low-k dielectrics* (C-based materials). C-based materials typically have lower k-values, but some of their other characteristics may make Si-based materials the better overall choice, depending on the application. For example, Si-based materials usually have higher thermal stability and hardness compared to C-based materials, but Si-based materials tend to be more prone to moisture absorption.

Insofar as manufacturability issues are concerned, the Si-based materials are more compatible with existing device-fabrication flows. Adhesion to silicon, SiO_2, and silicon nitride (as well as to Al and Cu metallization) is better with Si-based materials. Also, the Si-based materials are easily etched in fluorine-based plasmas, whereas O_2-chemistries are needed for C-based materials. The selectivity of oxygen-based plasmas is not as high over organic photoresist. In addition, the Si-based materials are more compatible with CMP-processing because of their greater hardness.

14.1.3 Status of Low-*k* Dielectrics in 2001

Even by 2001 the news from the industry reflected the state of uncertainty as to which process will prevail. Early in 2000, IBM announced it would use SiLK (a spin-on film from Dow Corning with $k = 2.65$) in its 0.15- and 0.13-μm Cu/low-k technologies, while Texas Instruments stated it would use Black Diamond (a CVD process from Applied Materials which produces a film with $k = 2.7$) together with Cu, for its high-end DSP-products. Soon after, LSI Logic described how it would push the performance of its aluminum-based interconnects with

TABLE 14-2 List of Candidate Low-*k* Materials

Low-k Dielectric Candidates		
Material	Dielectric constant Low-k materials	Deposition method
Polyimides	3.0-3.6	Spin-on
Spin-on glasses	2.7-3.1	Spin-on
Fluorinated polyimides	2.6-2.9	Spin-on
DLC (diamond-like carbon)	2.8-3.0	CVD
Poly(arylene ethers)	2.6-2.9	Spin-on
Poly(arylenes)	2.6-2.8	Spin-on
Cyclotenes	2.6-2.8	Spin-on
Parylene N	2.6-2.8	CVD
Poly(norbornenes)	2.5-2.7	Spin-on
Polyimide-SSQ hybrids	2.7-3.0	Spin-on
Alkyl-silanes/N_2O	2.4-2.7	CVD
	Ultralow k	
Teflon-AF	1.9-2.1	Spin-on
Teflon microemulsion	1.9-2.1	Spin-on
	Porous dielectrics	
Polyimide nanofoams	2.2	Spin-on
Silica aerogels	1.1-2.2	Spin-on
Silica xerogels	1.5-2.2	Spin-on
Mesoporous silica	1.9-2.2	Spin-on

low-*k* FlowFill® CVD films from Trikon products on its G12 0.18-μm products. SONY said it had integrated FLARE® (a spin-on material from Honeywell) in its advanced devices. Yet, all of these announcements also appear to signal that the time of low-*k* dielectrics has come. In the discussion here, three ranges of low-*k* dielectric materials will be described: **1**) first generation (k = 2.8–3.5); **2**) second generation (k = 2.5–2.8); and **3**) ultra low-*k* (k = < 2.0). Table 14-2 is a list of various low-*k* dielectric material candidates.

As of mid-2000, the situation was described by Ku (of IBM), as[5] "Though leading candidates have emerged from among the dozen or so being considered a year ago, any choice appears to have downsides. No matter which material is selected, they are all fundamentally flawed in some major characteristic. Therefore, the criterion of low-*k* material selection for IC manufacturers is not purely technical; the question usually becomes which material shortcomings are we going to choose to integrate around." Table 14-3 shows the main commercial low-*k* products being used or evaluated on 2001.

14.2 DESIRED CHARACTERISTICS OF LOW-*k* DIELECTRIC FILMS

If a dielectric material is to be successfully used in an IC, it must not only exhibit the desired *k*-value, but it must also possess the following characteristics: **1**) It must exhibit adequate material properties (i.e., thermal, electrical, and mechanical); **2**) It must be able to work with the other materials of the interconnect structure; **3**) It must be compatible with the IC processes of cleaning, etching, CMP, and thermal treatments; **4**) It must be available in high purity form, and at low cost; and **5**) It must be able to operate reliably over the life of the product under the specified device operating conditions. Here the general characteristics of groups **1**) and **2**) are considered. Comments are made on how well the various dielectric materials meet these desired goals with respect to each type. The remaining issues are also discussed for each film in particular.

The most important material properties needed by low-*k* dielectric films are listed in order of importance in Table 14-4.[4] First, of course, is the *k*-value (in general, the smaller the better). However, next are the *thermal characteristics.* These include *thermal stability* (i.e., it should be stable up to temperatures of 425°C for at least short periods of time, to make it compatible with the thermal excursions of metal-deposition, assembly, and packaging). Next, it is desirable that these dielectrics have a *thermal conductivity* close to that of SiO_2 (12.0 mW/cm°C) for purposes of heat dissipation during device operation. It should be

TABLE 14-3 Some of the Main Low-*k* Products Being Used in 2001

Material	Manufacturer	Deposition	Category	k
FOx	Dow Corning	spin-on	inorganic (hydrogen silsesquioxane)	3.0
XLK	Dow Corning	spin-on	inorganic (porous hydrogen silsesquioxane)	2.0
SiLK	Dow Chemical	spin-on	organic (aromatic hydrocarbon polymer)	2.7
3MS-lowk	Dow Corning	CVD	C-doped oxide (Si-O-C)	2.7
Aurora	ASM-Japan	CVD	C-doped oxide (Si-O-C)	2.7
Black Diamond (BDII)	Applied Materials	CVD	C-doped oxide (Si-O-C)	2.0-2.7

noted, however, that the thermal-conductivity of typical low-k polymers is only about 2.4-3.7 mW/cm°C. This low thermal-conductivity is quite undesirable, since high-speed circuits generate significant amounts of heat that must be dissipated. The final thermal property a low-k dielectric should exhibit is a low thermal-expansion coefficient (to prevent cracking during thermal excursions (e.g., < 10 ppm/°C). It should be noted that k-value is generally traded off against thermal conductivity. For example, xerogel dielectrics (k = 2.4) possess thermal conductivity that is 10 times smaller than SiO_2 (which may therefore give rise to reliability problems from severe increases in circuit temperature).

The low-k films should also exhibit *little tendency to absorb moisture* and should also not exhibit an excessive permeability to moisture. Moisture can adversely impact the other materials of the interconnect structure, and can be outgassed during other processes (which can degrade the process ambient). The films should also have good *adhesion* properties. That is, these films must adhere well to the materials that are deposited on them, and they must also adhere tightly to the materials on which they themselves are deposited.

As far as *electrical characteristics*, such films should have leakage current and breakdown field behavior (at least 2–3 MeV/cm) similar to SiO_2 and low charge-trapping characteristics. They should also exhibit high reliability under normal operating voltages for the design life of the IC in which they will be used.

As far as *mechanical properties* are concerned, the films should exhibit low stress (< 10^9 Pa) and have the capability of being deposited to a thickness greater than 2-μm without cracking. From a *chemical perspective* the films must also possess chemical compatibility with various reactants (e.g., they must demonstrate good resistance to such reactants as acids, bases, solvents, and resist strippers). They should also exhibit good etch selectivity with respect to the photomasking material (e.g., resist), and with respect to the substrate (e.g., underlying metal). Note that a variety of plasma treatments, thermal treatments, or other steps may be added to stabilize the film during processing. These become a large part of a chip manufacturers expertise in bringing low-k materials into the production environment.

For non-damascene structures (i.e., in which the metal [Al] patterns are etched and then covered with a dielectric layer), the gap-fill characteristics of the low-k material will also be important (e.g., no voids should occur when filling a space that is 0.35-μm wide and has an aspect-ratio = 2). However, although the gap-fill characteristics of the low-k films are not a consideration for damascene structures (since the dielectric films are deposited on surfaces that have been planarized by CMP), they will need to be deposited in a single pass at a thickness become a

TABLE 14-4 List of Desirable Properties of Low-*k* Materials

1	Dielectric constant (out-of-plane, 1MHz @ room temperature)	<3.2
2	Thermal stability (C)	>425
3	Moisture absorption	<2.0
4	Adhesion (ILD/,metal, metal/ILD, ILD/ILD)	No peel after 450C cycles
5	Thermal expansion coefficient (ppm/C)	-10 to +50
6	Solvent resistance (acid, bases, solvents, PR strippers)	No weight change after dip
7	Etch rate (wet or dry)	
8	Etch selectivity (ILD/metal, metal/ILD, PR/ILD)	
9	Stress (Pa, tensile or compressive)	<1E9
10	Gap fill	No void @0.35μm and AR=2
11	Tensile strength (MPa)	<200
	Tensile modulus (GPa)	<1
	Elegation at break (%)	>5%
12	Thickness uniformity over topography (before and after curing)	
13	Cost	IBO
14	Corrosion to Al-Cu	
15	Environmental, health and safety	Pass

of about twice that used in dielectric films in non-damascene structures. Thus, they will need to exhibit crack-resistant characteristics for such film thicknesses. In addition, they will need to be compatible with CMP processing. Finally, use of spin-on dielectric films that evolve toxic vapors will be avoided.

From an economic perspective, the costs involved with both the materials needed to deposit the films and the CoO of the equipment required to fabricate the films should be low. In addition, the materials used to form the low-*k* films should be commercially available and have a reasonable shelf life.

14.3 GENERAL PROCESS INTEGRATION ISSUES INVOLVING LOW-*k* FILMS

There are several processes that need to be performed on the dielectric films after their deposition, and thus these films must be properly integrated into such processing. First, the films will need to undergo either etching or CMP.

As far as etching is concerned, vias (and perhaps grooves in a damascene technology) will have to be etched in the films. Issues of etch chemistry, etch selectivity to oxides, nitrides, and oxynitrides will need to be addressed. After etching, the resist-mask must be removed. The resist-removal processing (and any resist-residue-removal steps) should not damage the dielectric layers, nor undercut the low-*k* dielectric layer beneath the hard-mask (if one is being used).

When etching *organic* low-*k* films, O_2-based plasmas are generally used. However, since an O_2-based plasma will also ash the resist masking-layer while the low-*k*-layer is being etched, an oxide or nitride hard-mask must be used to transfer the pattern to the low-*k* film. The etch-process-sequence thus consists of opening the hard-mask (using the PR to define the pattern), and then proceeding with the low-*k*-dielectric etch. Depending on the type of low-*k* film, a wet-chemistry or proprietary dry-chemistry may be used to subsequently remove the resist. Since use of a pure-O_2 plasma will tend to undercut the hard-mask, such undercutting must be suppressed through the use of additives to the O_2-plasma.

When etching *inorganic* low-*k* dielectrics such as *organosilicate glasses, OSG* (which are oxide-based materials that also contain hydrogen and carbon), a common requirement is that the etch-process be selective to PR and to a silicon-nitride etch-stop-layer. (Silicon-nitride layers are used as embedded hard-masks in dual-damascene processes - see Chap. 15.) In general, oxide-etch-processes that are selective to nitride do not etch the OSG films due to the presence of the organic component in the film. Therefore, alternate process chemistries that include appropriate gases to etch the organic component in the OSG films and that maintain selectivity to photoresist and silicon nitride must be developed.

With respect to films that must be planarized by CMP, such materials will have to be polishing-process-compatible. As an example of the issues involved with CMP, polishing xerogel films is difficult due to the poor mechanical strength of the film. This necessitates the deposition of conformal oxide liners (both beneath the xerogel layer, and also as a capping layer on top of the xerogel). The issues of CMP involving low-*k* films are also covered in Chap. 8.

Another issue involving the use of CVD-deposited organosilicate glasses (OSG) is the possibility that they will poison the DUV-resists. That is, if the OSG contains any nitrogen, it could enter the DUV-resist and cause a "foot" to be produced in the DUV-resist layer that contacts the nitrogen-containing OSG interface. There are a number of "fixes" for this problem, such as: **1**) the use of a bottom anti-reflective coating (BARC); **2**) the deposition of a spin-on or CVD capping-layer on the OSG layer; or **3**) treating the OSG layer in an nitrogen-containing plasma (to form a thin nitride barrier on the OSG surface). More information on the DUV-resist-poisoning problem is given in Chaps. 6 and 15.

Another issue of integrating a low-*k* film into an interconnect process is the

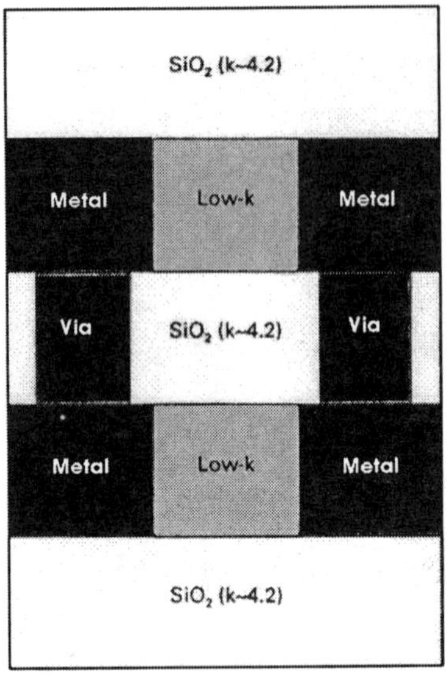

Fig. 14-3 Using low-*k* material between metal lines (but not at the via level) improves the mechanical strength and reliability of the stack at the expense of manufacturing complexity.

fact that such films lack the mechanical strength of oxides. One way to circumvent this difficulty is to place the low-*k* dielectric only *between* the metal lines, but not at the via levels (Fig. 14-3). Beyond strengthening the structure, this "embedded-approach" can reduce moisture-absorption, prevent via poisoning, and allow the fab to keep its conventional dry-etch and cleaning processes. The embedded-structure also makes for a more compatible CMP process, while allowing a company to learn from gradual integration of the low-*k* material. In addition, roughly 80% of the benefit is gained by using low-*k* material just between the metal lines. However, this approach requires that two ILD processes must be used. Figure 14-4 is a summary of the process integration issues involved with the use of low-*k* dielectrics as ILDs.

14.4 FIRST GENERATION LOW-*k* DIELECTRICS ($2.8 < k < 3.5$)

Dielectric materials with *k*-values moderately lower than SiO_2 were the first dielectric films to be adopted for this function. Three types of such moderately low-*k* films had found their way into IC production by 1999. The first two were the *spin-on dielectric* (SOD) films composed of *HSQ* and *MSQ,* with the third being CVD-deposited *fluorinated oxide films,** Their main use was reported to be in the 0.5-μm to 0.25-μm CMOS generations.

* These are also referred to as *fluorosilicated glass* (*FSG*), or *silicon oxyfluoride* (*SiOF*).

Film properties

Dielectric constant:
- — Bulk: k=2.5-3.0
- — Effective: k<3.0

Thermal stability:
- — High thermal conductivity
- — Tg>400°C, stable above 425° for short periods
- — Low expansion

Electrical properties:
- — High reliability
- — Leakage current: similar to SiO_2
- — Breakdown field: similar to SiO_2
- — Dissipation factor: <0.01
- — Low charge trapping

Film composition:
- — Low film stress
- — >2 μm thick cracking threshold

Manufacturing Integration
- —Good adhesion to metals (Ta, TaN, TiN, Cu), oxides/ nitrides
- —CMP compatible
- —Minimize need for liner/capping films
- — Etch selectivity to nitrides, oxides, oxynitrides
- — O_2 ash/solvent compatible
- — Avoid C_2H_6, C_3H_8 (CVD)
- — Avoid toxic solvents (spin-on dielectrics)

Fig. 14-4 Some of the many challenges of integrating low-*k* dielectrics in a copper dual-damascene process flow. Courtesy of Semiconductor International.

14.4.1 Hydrogen Silsesquioxanes (HSQ)

Films of the spin-on inorganic polymer, *hydrogen silsesquioxane* (*HSQ*) were initially investigated as replacements for siloxane spin-on-glass (SOG) films in non-etchback SOG planarization processes. When it was found that they also exhibit a lower *k*-value than do SiO_2 films (their *k*-value is < 2.9) as well as good gap-filling capability, they became candidates for low-*k* dielectric applications.[7]

HSQ is a subset of polyhedral oligomeric silsesquioxane macromers, a family of ordered three-dimensional polymers whose structure resembles a cage (Fig. 14-5a). HSQ can be considered a derivative of SiO_2, in which one of the four oxygen atoms bonded to every Si atom is replaced by hydrogen. That is, the caged structure of HSQ has an empirical chemical formula of $HSiO_{1.5}$. (The *sesqui* refers to the *one-and-a-half* in the stoichiometric bonding arrangement of $SiO_{1.5}$ to H.) In the SiO_2 lattice, all O-atoms act as bridges between the SiO_4 tetrahedra of the SiO_2 lattice (see Chap. 3 for details on the microscopic structure of SiO_2). However, when H (or a methyl group, CH_3), replaces an O-atom in the lattice, it no longer "bridges," but instead forms a terminal group. Thus, the tetra-

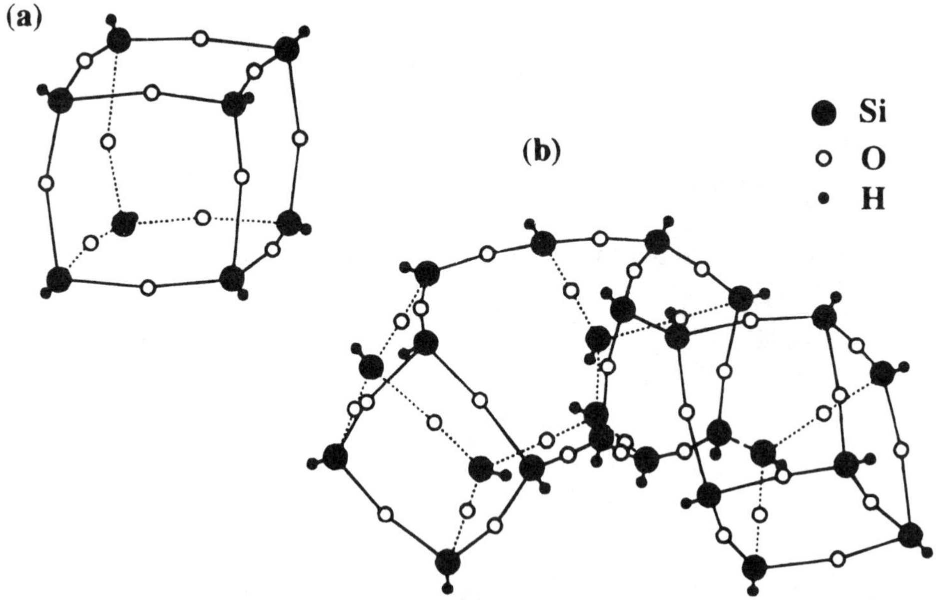

Fig. 14-5 Schematic representation of the molecular structure of HSQ: a) eight-corner oligomer; and b) resinous material.

hedron that contains the Si-H bond is then bonded only at three of its four corners. This results in a more open lattice structure (i.e.., it exhibits microporosity). Note that in HSQ films some of the Si atoms that would have bonded to O-atoms in SiO_2 are instead bonded to H-atoms. The incorporation of such Si-H bonds in spin-on dielectrics (SODs) is inherent to the spin-on material. In low-*k* Si-H bond-containing films formed by CVD, the Si-H bonds are formed by incorporating gases in a CVD-chamber that contain H.

In its uncured state, HSQ resin is composed of a mixture of Si-O networks and cage-like networks (Fig. 14-5b). This random mixture of networks is polymerized during a two-step thermal procedure consisting of : **1**) a bake at 150-250°C, and **2**) a cure at ~400°C. During these thermal treatments, a rearrangement occurs. That is, the cage structures are opened up, and there is an intermolecular redistribution of Si-O and S-H bonds. In other words, HSQ is cured from a near cage-like structure to a three-dimensional network which provides mechanical integrity to the film. The built-in microporosity of the cured HSQ network results in a reduced dielectric constant.

However, in addition to the rearrangement reaction during curing, other reactions which convert Si-H bonds to silica can also occur. If such conversion to silica is excessive, it will result in densification of porous HSQ (i.e., shrinkage of also increase the dielectric constant.[8] That is, low-k properties in cured HSQ films can only be achieved if the density of Si-H bonding is maintained at a high level and the formation of Si-OH bonds and the absorption or creation of water in the film is minimized. Thus, when curing HSQ films, the process must balance between promoting the favorable rearrangement-reaction and inhibiting the Si-H conversion-reactions.[9] To achieve this balance, the curing-temperature must not exceed 450°C, and an optimum curing-temperature was reported to be in the vicinity of 400°C.[10]

The HSQ material is available in solution from Dow Corning under the trade name of FOx® (Flowable Oxide).[11] HSQ was reported to have been integrated into products by Texas Instruments in the 0.5-μm generation, and in 1997 an article describing its use in a five-layer Al-interconnect structure was published.[3] An oxide cap was used on the HSQ film to improve the stability of its low-k value during W-CVD and Al-fill. That is, such HSQ-films must be capped with oxide to prevent moisture uptake in the film, and the integration sequence must be optimized to prevent cracking.

The built-in microporosity of the HSQ films, nevertheless, also reduces their mechanical strength. To overcome this drawback, a plasma treatment in NH_3 at 350°C for 20 min was performed. This was found to incorporate N atoms into the film. With the incorporated nitrogen, the HSQ films were found to be mechanically stronger. Specifically, scratching that was observed during CMP of non-plasma-treated HSQ films is mitigated.[12] In addition, such films could now be cured at temperatures up to 500°C without triggering the Si-H conversion-reactions.[13] Another benefit of such a plasma-treatment is that the thin nitride film formed on the HSQ-surface (35-nm thick) reduces leakage current through the film and also provides a good barrier against Cu migration. Leakage current was reportedly reduced by three orders of magnitude below that of an as-cured HSQ film, and evidence that Cu-migration was also suppressed was presented.[14]

14.4.2 Spin-On Methyl-Silsesquioxanes (MSQ)

Spin-on *methyl-silsesquioxane* (*MSQ*) films are similar to the HSQ materials described in the previous section, except methyl groups (CH_3) are used instead of hydrogen to replace some of the Si-O bonds (i.e., Si-CH_3 bonds are formed instead of Si-H bonds). Therefore, MSQ is often referred to as an organo-silicate material. With an empirical chemical formula of $CH_3SiO_{1.5}$, there are one-and-one-half (*sesqui*) stoichiometry of oxygen bound to silicon in a caged structure of MSQ. MSQ films have the benefit exhibiting excellent crack-resistance, and can thus be used to form thicker dielectric films. In addition, they show better moisture absorption properties than HSQ.

One example of a spin-on MSQ material is T-9,™ sold by Tokyo Ohka Kogyo. It's *k*-value is 2.7. The curing procedure is about the same as that for HSQ. Another spin-on film containing CH_3 is sold by Allied-Signal ($k = 2.9$). It's commercial name is Accuspin T-23.™ When cured at 380–400°C, it can tolerate temperatures as high as 450°C before decomposing. Note that development is being conducted to develop ultra-low-*k* dielectrics based on spin-on MSQs. One commercially available MSQ with sub-2.2 *k* value is Dendriglass.™

14.4.3 Fluorinated Silicate Glass (FSG) Films

FSG films are silicon oxyfluorides (F_xSiO_y) deposited by CVD. That is, FSG is produced by doping an oxide-film with F in an otherwise conventional CVD process. Fluorine is one of the most electronegative elements, and its presence in the oxide makes the oxide less polarizable (which causes the *k*-value to decrease). Specifically, fluorine atoms replace oxygen atoms in the oxide film to create some Si-F bonds instead of Si-O bonds. Because the attraction of fluorine to its valence electrons is greater, these Si-F bonds are less polarizable than the Si-O bonds they displaced. (Figure 14-6 shows the molecular building blocks of FSG.) In addition, fluorine-to-fluorine repulsion is stronger than oxygen-to-oxygen, and this makes oxide films containing fluorine less dense than those without incorporated fluorine. The FSG films used in ICs exhibit a minimum *k* value of ~3.5. This is because FSG films are formed in PECVD or HDP-CVD tools by adding a fluorine-source gas (e.g., SiF_4, CF_4, C_2F_6 , or NF_3, or other organo-silicate F sources), to the process-gas ambient used to deposit SiO_2 films by CVD (either silane and oxygen, or TEOS). Usually, SiF_4 or organo-silicate F sources are preferred because the incorporation of F into the silicate glass networks does not require breaking the strong Si-F bond (whose bond strength

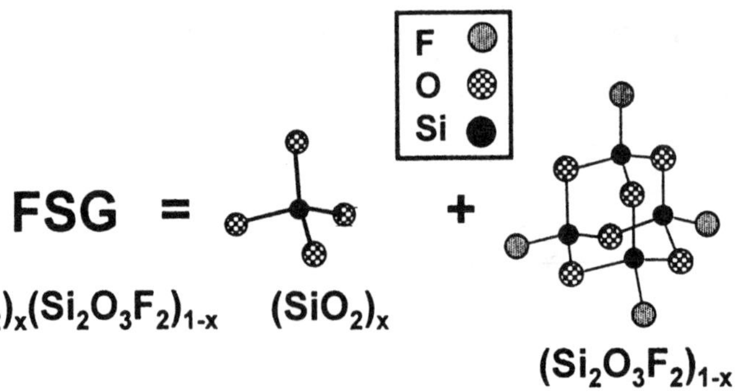

Fig. 14-6 Molecular structure of FSG materials.

is roughly 135 kcal/mole). By increasing the fluorine-source-gas flow-rate, more F is incorporated into the film. Higher concentrations of F cause the value of *k* to decrease, for the reasons explained above. However, a maximum of about 6% F is permitted to be incorporated into the FSG films because higher concentrations would make the FSG film unstable. That is, in such unstable films, F is evolved during RIE etching. This evolved F is highly reactive toward water. Aluminum interconnect lines embedded in this material would be corroded by the HF that is formed when the SiOF reacts with water. Because of this limitation, only a relatively small decrease in *k* values is possible with FSG. Thus, the benefits of replacing non-fluorinated CVD-oxides with FSG are modest.

However, FSG is relatively simple to integrate because of its similarity to the doped-CVD-oxides currently being used in the IC industry. In fact, FSG-films were reported to be successfully integrated into 0.35,- 0.25,- and 0.18-μm circuits by IBM.[15] The evolution of even small amounts of F from the FSG oxide has raised concerns about using this material in Cu-interconnect systems. The F readily attacks the Ta-based liners of Cu systems, leading to volatile TaF_2 formation, and the consequent loss of adhesion between the low-*k* film and the Ta. Despite such barrier stabilization problems, use of FSG has permitted extension of PECVD oxide processes to both subtractive-aluminum and copper-damascene interconnects. Note, however, that FSG solutions are different for the subtractive-aluminum process and the Cu-damascene process. That is, to obtain the gap-fill needed for the subtractive aluminum process, the sputter component encountered during the deposition of the FSG film bombards the depositing film. This breaks the bonds in the film and makes it less stable. Hence, even less F can be incorporated. In the damascene process, the dielectric film is deposited on a planar surface, and hence no gap-filling capability is needed. Thus, higher F-concentrations can be used. The *k*-value of FSG-films used with damascene interconnects is reduced by 0.2 to 0.3, compared to the *k*-values of FSG used with Al. As noted above, among the difficulties encountered when using FSG in a damascene process with Cu is the issue of keeping the fluorine tightly bonded in the CVD film (i.e., volatile TaF compounds can form at high temperatures).

Another possible application of FSG is replacing the conventional PECVD layer used to cap SOD low-*k* films. It has been demonstrated that use of FSG as a capping layer for an SOD resulted in 10% reduction in parasitic capacitance.[16]

14.5 SECOND-GENERATION LOW-*k* DIELECTRICS ($2.5 < k < 2.8$)

The largest variety low-*k* dielectrics is found in the second generation of such substances, which exhibit *k* values between 2.5 and 2.8. It is believed that these materials will begin to be incorporated into ICs at the 0.18-μm generation. Both *spin-on-* and *CVD-low-k* materials have been developed within this range of permittivities, and each of these types are described here.

As noted earlier, a heated battle at the $k = 2.8$ node between CVD- and spin-on-dielectric (SOD) films is taking place. It is staged on the one hand by the desire of wafer manufacturers to continue to use CVD-tool-sets for one more technology node, versus the potential lure of being able to extend the use of new SOD materials and process integration schemes for multiple generations.

That is, CVD-low-*k* processes can reuse existing tool sets and allow for simpler integration (due to the silicon-oxide-like structure of the CVD dielectrics). From a business perspective, CVD-tool-suppliers can also provide an integrated solution to the customer. For circumstances where there is a tight labor market for engineering talent (e.g., the Taiwan silicon foundries in 2000), the foundries need to rely on low-*k* solutions from their vendors that have been well characterized and can be easily introduced into production. Thus, adoption of CVD-low-*k* dielectrics is likely to be favored in such cases.

However, an alternative case for SOD-low-*k* dielectrics at the $k = 2.8$ node is made by the suppliers of SOD-materials and SOD-track-manufacturers. They claim that SOD processes are extendable to future technology nodes, especially those below 0.25-μm (when porous low-*k* materials will need to be used). Porous low-*k* SOD materials are available now for evaluation and testing. The SOD porous films may also produce films with smaller pore sizes than those deposited by CVD. This would make them mechanically stronger, allowing better film stability and better integration. The key for SOD processes is to develop an SOD solution that can be extended for multiple generations and can process 200-mm and 300-mm wafers on the same equipment. Some of the issues remaining to be resolved with organic SODs include poor adhesion and via-poisoning due to improper etch and resist removal. Spin-on-dielectrics also remain susceptible to cracking at thicknesses approaching one micron.

There is also a question at the 0.18-μm technology node as to which process should be implemented first - copper or low-*k*? It appears that the most cost-effective gain occurs with aluminum/low-*k*, particularly if $k < 3.0$ can be implemented. Hybrid structures with aluminum/low-*k* for the lower levels of metal, and copper at the very top levels of interconnect will allow manufacturers the ability to ship 0.15/0.13-μm products, forestalling the significant challenge of integrating copper and low-*k*. Some companies have elected to take this path, but others have decided to implement copper interconnects first, before incorporating a lower-*k* dielectric material.

14.5.1 2nd-Generation Spin-On-Dielectrics: (2.5 < *k* < 2.8)

Most spin-on *organic* low-*k*-polymers[17] are significantly different from their conventional SOG counterparts, primarily because they do not evolve moisture during curing. Since they exhibit better crack-resistance they can also be applied in a single pass to a greater thickness. Three organic spin-on low-*k*-dielectrics being considered are: **1**) *poly(alylene) ethers* (*PAE*); **2**) *fluoro-polyimides*; and **3**) *bis-benzocyclobutenes* (*BCB*). One other group of spin-on polymers (the *organo-siloxane polymers*), are more like their conventional SOG counterparts.

Spin-on-dielectric films are deposited without the use of an expensive CVD-tool. Instead, they are deposited with a less-costly track cluster-tool, consisting of spinners and hot plates. The cost of the spin-on liquid, however, is typically much higher per wafer than that of the CVD-reactant-gases.[18] In addition, a 100-nm-thick capping SiO_2 layer (deposited by CVD) will probably still be needed with spin-on low-*k*-layers. But an added benefit of using spin-on films is that they reduce the topology of any previous steps by more than 50%.

The *PAE materials* are aromatic structures with *k*-values of about 2.5 which can also withstand higher temperatures. A fluorinated PAE-product from Honeywell is named FLARE,™ and it has a *k* value of ~2.8. It was designed to have low-outgassing characteristics and high thermal and mechanical stability.[19] Dow Chemical has a similar material named SiLK (k = 2.62).™ Despite its name, SiLK is also a hydrocarbon that contains no silicon (and also no fluorine). A report by Fujitsu detailed their experience with integrating FLARE and SiLK into a two-level Cu-damascene-structure. These films showed little change in *k*-value after a five-hour anneal at 425°C.[20] Schumacher has also pioneered the use of non-fluorinated PAE-materials, with their latest such product being called VELOX.™ This material can be deposited without an adhesion promoter and does not require a capping layer. A *fluoro-polyimide* named FPI-136M with k = 2.7 is sold by DuPont.

In April 2000, IBM announced plans to integrate SiLK into its next-generation 0.15-μm and 0.13-μm Cu-11 process, intended to go into high-volume production in the first half of 2001. They claimed the interlevel dielectric layers made with SiLK demonstrated excellent resistance to Cu-migration, and the mechanical strength of SiLK films is high enough to make them compatible with CMP (and back-end processes such as wire-bonding). IBM said it chose to use a spin-on-dielectric film for two reasons. First, SiLK meets the critical thermal, mechanical and chemical requirementss. Second, it offers extendability to the next generation. "To get the effective dielectric constant down into the 2.2-2.3

range, there are no materials, and there is no way to get there with PVD or CVD. Our opinion is that the best way to achieve that is by the introduction of porosity. That would say, whether its an organic or an organosilicate, you're going to have to do a spin-on process."[21] One reported difficulty involving the use of SiLK is that if the film is immersed in HF solutions (even as dilute as 0.1%-HF), the HF will diffuse through the film and attack the oxide at the oxide/SiLK interface. This will lead to SiLK liftoff.[22]

BCB is a co-polymer of divinyl-siloxane and bis-*benzocyclobutene* (which is often referred to as *DVS-BCB*, or simply *BCB*). It is a thermoset silicon-containing resin that exhibits a *k*-value of 2.7, but which has limited thermal stability, even at temperatures of 350°C. Nevertheless, there are no by-products to the polymerization reaction and no corrosive or volatile by-products are evolved during the cure. BCB demonstrates excellent gap-fill, adhesion, and planarization capabilities. Dow Chemical offers a BCB product called Cyclotene™ for IC ILD-applications. A study in 1998 indicated that Cu drifts much more slowly through BCB films than through PAE.[23] Thus, use of BCB films could relax the requirements of metal diffusion barriers. But its limited thermal stability constrains its use to applications with a reduced thermal ceiling.

As noted above, there is also a class of SOG-like materials available in the 2nd-generation low-*k*-range. These are *hybrid-organo-siloxane polymers* (*HOSPs*) which have a *k*-value of 2.5 after curing. HOSPs have a 12% carbon-content, which allows them to be formed into layers up to 1.2-μm thick without cracking. Their *k*-value remains stable at temperatures up to 380°C. HOSPs demonstrate good gap-fill and planarization capabilities. However, the quality of HOSP films (and those of other carbon-containing spin-on-dielectrics) is degraded by damage from an oxygen plasma used when stripping resist. The conventional oxygen-based ashing-process can attack the low-*k*-dielectrics that contain any carbon, which will break down the film integrity. To overcome this limitation, a plasma treatment in H_2/N_2 or NH_3 (as discussed in previous sections), has proved to be effective. Honeywell markets such a low-*k* material under the name HOSP.™

Note that a plasma-treatment of organic SOD layers with an NH_3 gas reportedly improves some of the properties of these films (as was also discussed in the previous section the covered HSQ films). That is, such a treatment produces a thin nitride layer on the SOD surface, and this layer acts both as a barrier against Cu migration and a layer that improves the ashing resistance of such films.[24]

14.5.2 2nd Generation-CVD-Dielectrics: (2.5 < *k* < 2.8)

Three dielectric materials deposited by CVD and with a value of *k* in the range of 2.5–3.0 have been developed: **1)** *carbon-doped siloxanes,* also termed *organosilicates glasses* (*OSG*), which have a *k* value of 2.5; **2**), *fluorinated parylene* (*parylene-F*); and **3)** an *inorganic low-k material using organosilane/ozone.* One benefit of CVD deposition of low-*k* films is that the deposition process can be carried out in a multi-chamber tool that allows liner- and cap-layers to be CVD-deposited sequentially in a single tool. Another advantage is that the CVD processes use readily available and inexpensive gaseous precursors (in contrast to the expensive and proprietary spin-on liquid-precursors), which reduces the comparative chemical cost of the CVD approach. Furthermore, CVD does not require the handling of organic-solvent wastes.

14.5.2.1 2nd Generation Low-*k*-CVD Dielectrics: Organo-Silicate Glasses: The OSG dielectric films deposited by CVD basically have the same molecular structure as the HSQ- and MSQ-films deposited by spin-on methods. The CVD OSG films can be can be formed by PECVD or HDP-CVD, using a variety of precursors, including methylsilane, dimethylsilane, trimethylsilane (3MS), or tetramethylsilane, together with N_2O (or in some cases ozone). As noted earlier, in the discussion on HSQ- and MSQ-films, the methyl-groups form terminal Si-CH_3 bonds on the tetrahedral structures that make up the oxide lattice. These Si-CH_3 bonds contribute to reducing the *k*-value by increasing the microporosity of the film and by reducing its polarization tendency. In OSG-low-*k*-films formed by CVD, these methyl-groups come from the precursor gases used in the CVD process. In general, the CVD process must balance the film requirements for low-*k* with other desired film properties. That is, low methyl-content can lead to films that behave merely like standard oxide films (i.e., with *k* ~4.0). On the other hand, high methyl-content can lead to Si-C-Si bonding, which will degrade thermal stability and will hamper the etching of the film (i.e., through the production of carbon polymers during the etch process).

Precursors with more methyl-groups are capable of incorporating more carbon into the film, though the compounds are more difficult to dissociate. As a result, their deposition rates are smaller. For example, tetramethylsilane deposition rates are about 25% slower than trimethylsilane because it takes more energy to dissociate the tetramethyl molecule. Another significant difference lies in the delivery system, since trimethylsilane is a gas and tetramethylsilane is a liquid. (A trimethylsilane precursor gas is offered by Dow-Corning, which they have named Z3MS.™) An advantage of OSG films is that they have thermal conductivity values similar to those of silicon oxides. As high-speed devices will generate more heat, such relatively-high thermal-conductivity represents a significant benefit. Manufacturers of OSG-CVD-systems include Applied Materials and Novellus.

A CVD-low-*k* film using organosilane and ozone was also recently reported.[25] This film can be easily integrated into an interconnect scheme that employs W- plugs and Al-lines. In the CVD process, the Si-H-bond of the organosilane precursor is selectively broken by an active O_3-molecule to form a highly mobile $Si(CH_3)_x$-intermediate to achieve good step-coverage and gap-filling between metal lines. Void-free filling was achieved at spacings of 0.12-μm and an aspect-ratio of 5:1. A post-deposition-anneal at 400°C polymerizes and forms a low-*k* film. A cap PECVD-oxide film is deposited over the low-*k*-layer to avoid direct CMP of the low-*k*-material. With the capping oxide, the normal ashing and cleaning process can be applied to remove resist-residue and etch by-products. The W-via-filling process is straightforward after standard de-gas and liner-deposition. No via-poisoning was observed after the W-CMP step.

14.5.2.2 2nd Generation Low-*k*-CVD Films: *FlowFill*® CVD Films: The *FlowFill*® films from Trikon Technologies are OSG-films (silsesquioxanes) deposited with a thermal-CVD process using a mixture of methylsilane and hydrogen peroxide (H_2O_2) gases.[26] These react in the gas-phase to form methylsilicic acid, which condenses on a cool wafer (i.e., the wafer is kept cooled to 0°C during the deposition process). The condensed material takes the form of a methyl-substituted oxide-gel that contains a considerable amount of water. The high flowability and surface-tension of this gel enables it to seamlessly fill gaps and provide a significant degree of local planarization. After deposition, this gel is converted into an amorphous methyl-doped silicon-oxide by a higher temperature cure in a separate chamber. That is, annealing at 400°C drives-off the water and densifies the film, leaving a high-quality methyl-incorporated silicon-oxide-film (which has a porous structure and a low dielectric constant). Figure 14-7 shows the process flow of this CVD approach.

The FlowFill-films exhibit a *k*-value of 2.8 and can be used for gap-filling applications in a subtractive-aluminum interconnect-process, or a Cu-damascene-process. For gap-fill, the deposition sequence involves first putting down a plasma-deposited base-layer, then depositing the low-temperature flow-layer and performing a cure step (and finally capping this layer with a standard PECVD film). The overall *k*-value of this composite-film is ~3.1. FlowFill layers can also be deposited thicker (to > 1.5-μm) with a modified cure step. This allows a

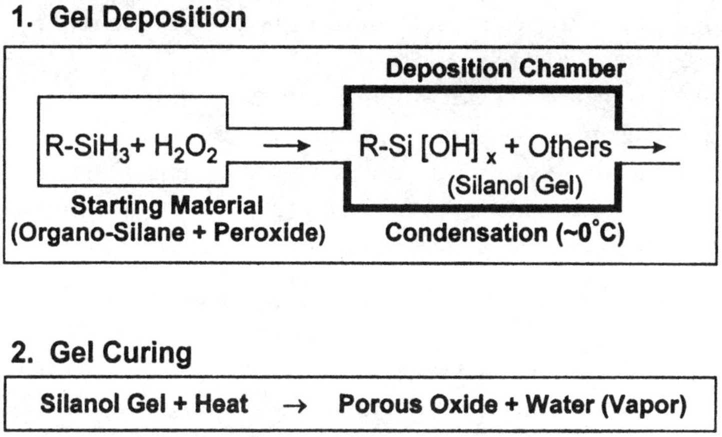

Fig. 14-7 Trikon *FlowFill®* CVD approach for depositing silsesquioxane films.

simple extension to timed-etch damascene structures using the same hardware and tool platform (i.e., the Trikon Planar® system). The film is attractive because its properties are very similar to spin-on glasses, whose process integration issues are well understood. However, its impact on circuit performance is limited by the need for encapsulation by the PECVD-SiO_2 capping layer.

14.5.2.3 2nd Generation Low-*k*-CVD Films: Black Diamond™ and Coral™ CVD Films: Applied Materials has developed a modification to the CVD OSG-films that they have termed Black Diamond.™ In this approach, a near-room-temperature PECVD-process is used to deposit a silsesquioxane film using the same organo-silane precursors, but with gaseous oxidizers like O_2 or N_2O instead of H_2O_2. By going from a cool deposition process (as is used in the *FlowFill* approach of Sect. 14.5.2.2) to a warmer one, Applied said it could eliminate the post-deposition furnace-cure step while positively influencing film-hardness and surface-roughness. This also produces an as-deposited film that already has a silsesquioxane type of porous structure, and which exhibits a *k*-value of about 2.7. Depending on the deposition temperature, curing may still be necessary to outgas the trapped reaction byproduct from the film. In addition, the high sticking-coefficients of the gaseous reactants result in poor step-coverage and gap-fill capability. Thus, Black Diamond films can be used as an inter-metal dielectric only in damascene interconnect schemes.

Applied Materials also claims that the same tool set used to deposit the first-generation FSG-films can also be used to deposit their Black Diamond films. This will allow the same equipment base to be used through several generations.

Applied Materials also announced that this material was validated for sub-0.18-μm devices by International SEMATECH.

Novellus calls their CVD OSG-films Coral.™ Their *k*-value is 2.7. Novellus claims that through process optimization, the hardness of the Coral film has been improved sufficiently so that it can be used in standard plastic packages. To protect the OSG-layer from attack during resist removal, a capping layer of SiO_2 is CVD typically deposited onto it (although SiC-layers are also being introduced).

14.5.2.4 2nd Generation Low-*k* -CVD Films: Parylene CVD-Films: Parylene (poly-*p*-xylylene)-based polymers are one of the most promising low-*k* CVD materials. Parylene polymers are high molecular weight (20,000-50,000 amu) linear chains of phenyl groups isolated by ethyl groups at the para positions. Figure 14-8 shows the different types of parylene polymers. The parylene precursors are low-melting-point solids that are composed of molecular dimers. The deposition consists of heating the solid in the presence of a carrier gas so that the dimers are carried to a cracking-furnace. The thermal energy in the furnace splits the dimers into the individual parylene monomers. The carrier gas then brings the monomers into a deposition chamber where they condense onto a wafer at temperatures below 0°C. Polymerization and cross-linking occur even at such low temperatures because the monomers are very reactive radicals. Parylene has a dielectric-constant of about 2.6. It does not absorb moisture, but is stable only up to about 300°C. Fluorinated parylene (parylene-F) exhibits a higher thermal stability (400°C). Parylene aliphatic tetrafluorinated poly-*p*-xylylene (parylene-AF4) demonstrates a *k*-value of 2.25 and even higher thermal-stability (greater than 450°C). Also it does not absorb moisture. A CVD process for depositing parylene AF4 is also commercially available from Novellus.[27]

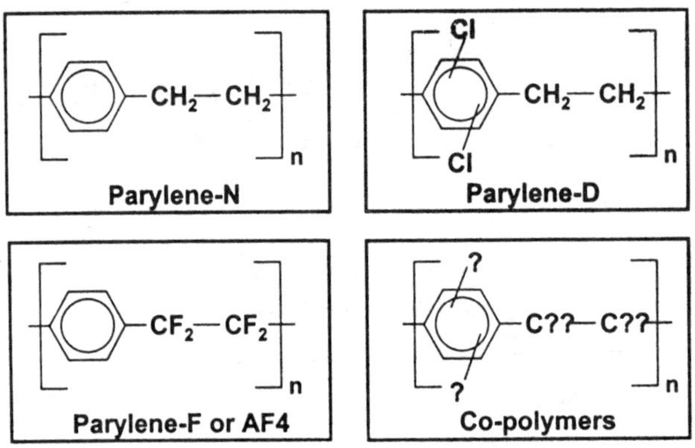

Fig. 14-8 Different types of parylene polymers.

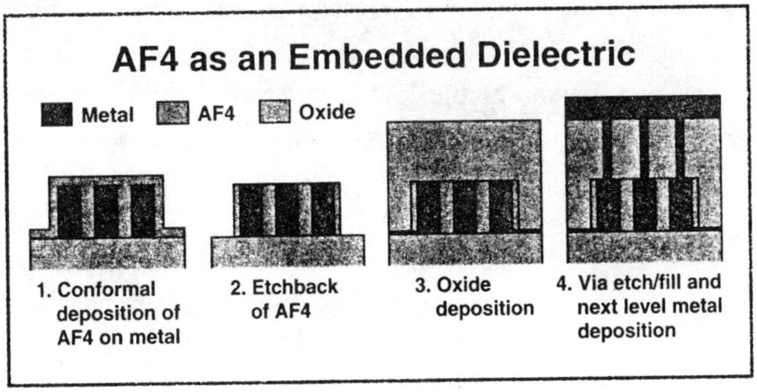

Fig. 14-9 In this integration scheme, the low-*k* material is encapsulated between the closely spaced metal lines where the low-*k* dielectric has its greatest influence on *RC*-delay. The interlevel dielectric is oxide, which has superior mechanical and thermal transport properties. Reprinted with permission of Semiconductor International.

In a subtractive Al-based interconnect structure, the parylene-AF4 layer is deposited first to conformally cover the Al lines. It is then etched back until the tops of the metal lines are exposed, at which point an oxide capping-layer is CVD deposited (Fig. 14-9). The parylene-AF4 remains in the spaces between the metal lines, where it does the most good in reducing side-by-side capacitance. The oxide above the metal lines is available as a layer through which via-etching takes place (a process that is mature). The chemical-vapor-deposition of parylene materials is a well-established technology, but the mechanical strength and applicability to via and damascene processes is still being researched. Adhesion is reportedly adequate for ULSI interconnect applications.[27]

14.6 ULTRA-LOW-*k* DIELECTRICS: ($k < 2.3$)

Ultra-low-*k* dielectric materials are also being investigated for their suitability as IC interconnect layers. Several different types were being considered, and are mentioned here. These types of ultra-low-*k* materials have not yet been proven for production-worthy devices. Specifically, according to the 2000 edition of the ITRS, it appears that such materials will not be included in commercial products until at least 2005.

One reason that ultra-low-*k* materials will be needed is that future high-performance circuits will use a copper damascene approach. Thus, the low-*k* dielectric likely be used in conjunction with an etch-stop (or polish-stop) dielectric layer (which will have a higher dielectric constant). Consequently the *k*-value of the ultra-low-*k*-layers will need to be decreased so that the overall dielectric constant of such ILD-stacks can meet the desired *k*-value specific-ations. Furthermore, adhesion of the low-*k* material to the other materials in the structure may also involve use of *dielectric-liners* with higher *k*-values. Many of these liners will probably be new to semiconductor processing.

Note that numerous ways of producing films with $k < 2.2$ are available. However, all of these methods invariably rely on reducing the density of the dielectric material. The most commonly used approach is based on a porous-silica structure (although silsesquioxane and aromatic-organic-materials have also been formed into porous films). These types of films, however, exhibit deterioration of the mechanical properties with decreasing *k*-values, as shown in Table 14-4.[28] In order to improve the mechanical properties of the porous films, a number of methods are being pursued. For films produced by the standard sol-gel technique (see later paragraphs in this section for more on sol-gels), particular attention must be given to the gelation and aging procedure, and to the termination of bonds on the interior pore walls (which typically requires an additional process step of some kind (e.g., thermal or plasma). However, silsesquioxane-based precursors reportedly can be treated with a sol-gel technique without requiring additional pore-wall passivation. In order to produce sol-gel films with more order and uniform pore size, self-organization stimulated by surfactant templating has been utilized. (Note that the information contained in this paragraph was presented at the Ultra-Low-*k* Workshop sponsored by SEMATECH in March 1999.)

Besides the sol-gel techniques, porosity has been introduced into dielectric films by burning out sacrificial material.[29] Another porosity producing method involves the interpenetration of a low-*k* constituent with a silica structure that is formed by standard SOG coat and bake procedures. One such approach involves depositing a multiphase film by CVD (that contains one film material that has a thermally-unstable phase). This is followed by an anneal (400°C for 8 hours), which removes this labile phase from the material. The dual-phase materials used in this study were deposited by PECVD from mixtures of SiCOH precursors with gases containing mainly carbon- and hydrogen-atoms. Following the anneal, the *k*-value was found to be reduced to 2.4 from its pre-annealed value of 2.8. Efforts are also being undertaken to produce porous dielectrics by PECVD using organosilicate glasses from a number of silane-based precursors.

The challenges of integrating such porous ultra-low-*k*-materials in copper damascene are formidable. Not only must they meet the low-*k* requirements, but the films should possess properties as close to the incumbent SiO_2-based materials as possible. These properties include low leakage-current, low thermal-coefficient-of-expansion, high dielectric-breakdown-voltage, low film-stress, low moisture-absorption, good adhesion to other materials, high thermal-conductivity, and mechanical-strength. Realistically, the materials properties of porous ultra-low-*k*-films will not match those of silicon dioxide, because they are compromised when pores are incorporated into a dielectric. Balancing low-*k* performance with sufficient mechanical properties (to survive mechanically intensive processes such as CMP), is one of the most significant hurdles facing ultra-low-*k* materials.

The first of the ultra-low-*k* materials considered here are the nanoporous-silica (SiO_2) xerogel materials whose *k*-value can be tuned from $k = 1.3$ to $k = 2.5$ (by varying the film porosity). Since the *k* of air is 1.0, by making a dielectric layer porous, more of it consists of air, and thus its overall *k*-value is lowered. In xerogels air is trapped as bubbles in a solidified gel. Xerogels are formed by evaporation drying, with some shrinkage. One commercial version is offered by Honeywell (formerly Allied Signal) under the name of Nanoglass.™ It was originally developed in 1996. In temperature testing of nanoporous-silica, feasibility studies indicated that such films are stable up to 500°C. Porous organic polymers are also being examined as low-*k*-layers. Note that all such porous materials (i.e., both oxide- and organic-based), must withstand such process steps as CMP, etching, and heat treatments without degradation of the pore structure. Other spin-on porous materials include porous-SiLK (from Dow Chemical), and porous versions of FLARE and HOSP (from Honeywell).

Dow-Corning has developed a spin-on film consisting of an interconnected pore structure based on HSQ-resin named XLK. In 2000 it could produce films with a *k*-value of 2.5, and the vendor claims it will be extendable to values of $k = 1.5$ in the future.[30,33] With a pore size of 2-3-nm, a 50% porosity, a tight pore-size-distribution, and a fully-interconnected network-structure, XLK appears to offer many of the characteristics needed for device integration. Pores are introduced through a three-step process on an SOD track system: **1**) gellation in the presence of a boiling-point solvent at ambient conditions; **2**) removal of the solvent without collapsing the resin structure; and **3**) formation of a strong network structure during cure near 400°C. Gellation occurs when the film is exposed to an NH_3 and H_2O environment for 30-120 seconds, which causes Si-H bonds to react with the H_2O to form SiOH. Then the SiOH bonds join together to form a SiOH gel. A plasma treatment in $H_2/N_2 + CF_4$ can be used to increase the mechanical properties of the XLK films.[31] In a single-level copper damascene

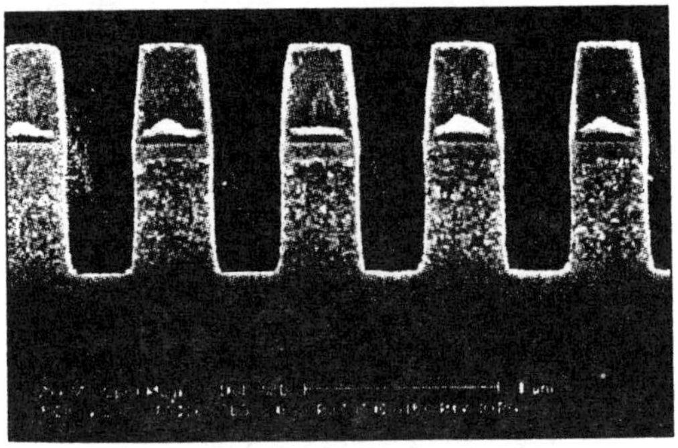

Fig. 14-10 Trench with 0.35-0.4-μm lines and spaces etched in 700-nm XLK film (ultra-low-k-dielectric) with 100-nm-thick oxide-cap-layer shows good sidewall etch-profiles.

integration study, 700-nm of XLK and 100-nm of an oxide-cap were deposited on a wafer with a 550-nm thermal oxide and a 100-nm nitride. A two-step recipe was used to etch the cap-oxide, then the low-k film, resulting in good etch-profiles (Fig 14-10). Following ashing in an HDP reactor, a 25-nm TaN barrier, with 100-nm Cu-seed and 1-μm of electroplated Cu were deposited. CMP of the structure showed no delamination. This work represented the first evidence of the feasibility of integrating an ultra-low-k film at the first-level of Cu-interconnect.

A second type of ultra-low-k-dielectric is formed with polyfluorotetraethylene (PTFE), a chemical related to Teflon (a DuPont tradename). PTFE is a chemical compound consisting of carbon and fluorine, made up of uncross-linked-CF_2-polymers (and so contains 67-at% F). A prototype spin-on version of a PTFE dielectric with a k = 1.9 has been introduced. The developer, W.L Gore and Associates gave it the name of SPEEDFILM. The material consists of an aqueous emulsion containing sub-20-nm PTFE particles and a surfactant. Films made with this material have good thermal-stability up to 400°C. Although Teflon does not adhere well to metals, SPEEDFILM has undergone adhesion testing (as well as CMP and etching studies). A thermal-CVD PTFE film first developed by Watkins-Johnson (now SVG Thermco Group), with a precursor from DuPont Electronic Gases Group, is also being offered.

Research continues on other ultra-low-k film materials. One possible class of candidates are porous organic-polymer materials. This is a fruitful pursuit, because if one starts with a k-value of 2.65 and makes it 20% porous, one can get

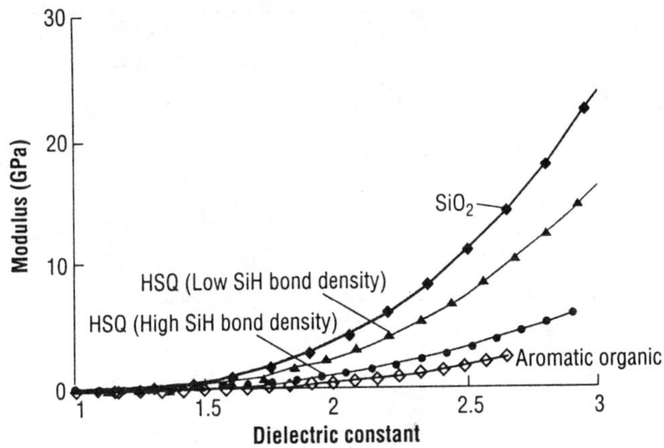

Fig. 14-11 Theoretical approximation of modulus of elasticity of porous ultra-low-*k* materials as a function of dielectric-constant indicates porous-SiO_2 offers the greatest mechanical stiffness.

a film with a *k*-value of 1.9. With an inorganic material like SiO_2 ($k = 3.9$), the film has to be 70% air to get *k* down to 2.0. Since the ultra-low-*k* films wont be commercially used until at least 2005, CVD companies still have time to develop a CVD film with a porous matrix.

In summary, the chief factor that determines the important characteristics of ultra-low-*k* materials is the pore structure. Each matrix material (whether it is silica-, silsesquioxane, or aromatic-organic based) has a unique pore structure.

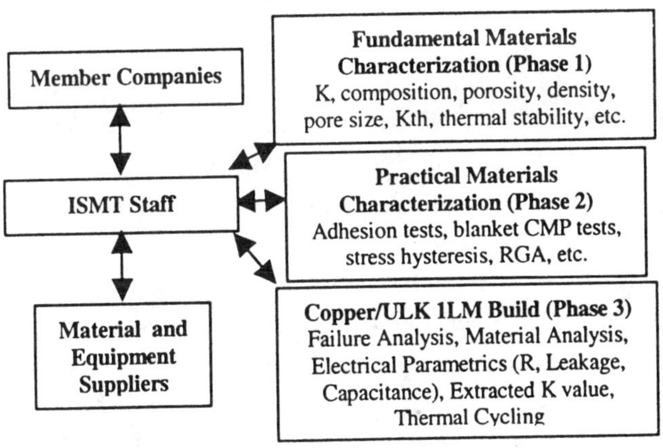

Fig. 14-12 Initial evaluation of candidate low-*k* materials at ISMT. The flow of information is meant to provide verification to material/equipment suppliers and probability of materials success in a two-level-metal integration.[32] (© IEEE 2001)

Material Property	Method	Specification	Guideline
K (relative permittivity)	MIS Capacitor 1MHz, RT, 40% RH	<2.5	<2.2
Thermal Stability (%loss/hr)	I-TGA In N2 and Ar, 425C	<1.0%	<0.5%
Film WIWNU, 1-sigma	Spectroscopic Ellipsometry	<3%	<2%
Young's Modulus (GPa)	Nanoindentation 1000nm thick film	>3	>4
Adhesion	ASTM Tape Pull	No Failures	No Failures
Maximum Pore Size	PALS/SANS+SXR	<25nm	<10nm
Film Quality	Visual Inspection Profilometry Light Scattering	No Striations, visual defects	Particles <100 @0.2um

Fig. 14-13 List of materials Specification and Guidelines. Specifications are the minimum expected material property, and Guidelines are the desired set of properties to ensure successful integration.[32] (© IEEE 2001)

For example, since air fills the pores, the *k*-value ultimately depends on that of the matrix material. In addition, the degradation of the modulus of elasticity E (which makes a material softer), is approximated by a power of the film porosity. Each type of porous dielectric-material exhibits a degradation of modulus of elasticity relative to fully dense silica (whose $E = 72$), as is shown in Fig. 14-11. Pore-size may also impact the continuity of Cu barrier-layer films, since these films are deposited with thicknesses comparable to the dimension of the pore size. Finally, compatibility with wet-chemical processing is influenced by pore structure. Open or connected pores are prone to absorb wet chemicals (such as water). Since the *k*-value of water is 72, any retained moisture in the pores degrades the dielectric constant. More details on this topic are given in Ref. 34.

One further change when ultra-low-*k*-dielectrics are used may involve going away from via-first dual-damascene interconnects (the approach most commonly used in 2000). With porous material, via-etching cannot be done first because after via opening the trenches are etched, and problems arise with resist absorption into the porous sidewalls of the via (see Chap. 15, Sect. 15.9.1.2).

Late in 2001 a report from Wetzel at International Sematech (ISMT) outlined a study that was being conducted there on candidate materials for ultra-low-*k*

ILDs.[32] Figure 14-12 shows the general scheme for the initial evaluation of such materials. The list of materials specifications sought for ultra-low-*k*-dielectrics is shown in Fig. 14-13. They suggested that integrating such low-*k* films into a dual-damascene process will face three main challenges: **1**) the ability to successfully survive CMP processing (especially with respect to the mechanical strength and adhesion of the films); **2**) the compatibility of ultra-low-*k* materials with DUV resist must be improved, especially when following a via-first dual-damascene approach; and **3**) porous-silica films must be able to maintain their initial *low-k* values through the processes that occur after the films are formed.

REFERENCES

1. L. Peters, "Pursuing the Perfect Low-*k* Dielectric," *Semiconductor International,* Sept. 1998, p. 64.

2. *International Technology Roadmap for Semiconductors* (ITRS-1999).

3. H. Meynen *et al.,* "The Integration of Low-*k* HSQ in Sub-0.35-μm Processes," *Proc. of the Electrochem. Soc. ,Symp. on Low- and High-k Dielectric Constant Materials,* May 1998, Abs. 157.

4. S.K. Chiang and C.L. Lassen, "The Market for Low-*k* Dielectrics," *Solid State Technology,* October 1999, p. 42.

5. V. Ku, "Low-*k* Integration Challenges," *What's Up from SEMI*, June 27, 2000.

6. K.A. Monnig, "Low-*k* Dielectrics for Integrated Circuit On-Chip Insulators," *Vacuum & Thin Film,* June 1999, p. 27.

7. M.J. Loboda *et al.,*"Properties of α-SiO_x:H Thin Films Deposited from Hydrogen Silsesquioxane Resins," *J. Electrochem. Soc.,* Vol. **145**, p. 2861 (1998).

8. J. Bremmer *et al., Spring Mtg. Materials Res. Soc.,* April 1997.

9. Y.K. Siew *et al.,* "Thermal Curing of Hydrogen Silsesquioxane (HSQ)," *Abs. Mtg. Electrochem. Soc.,* Spring 1999, Abs. 203.

10. Y.K. Siew, *et al.,* "Thermal Curing of Hydrogen Silsesquioxane," *J. Electrochem. Soc.,* Vol. 147, January 2000, p. 335.

11. H. Kojima *et al., Proceedings 5th Intl. IEEE VMIC Conf.* CA, p. 411, 1988.

12. T.-C. Chang *et al.,* "Ammonia Plasma-Passivation Effects on Properties of Post-CMP Low-*k* Hydrogen Silsesquioxane (HSQ)," *Abs. Mtg. Electrochem. Soc.,* Spring 2000, Abs. 240.

13. D.J. Kim, "Effect of NH_3 Plasma Treatment on Improvement of Reliability of W-B-N/HSQ Thin Films," *Abs. Mtg. Electrochem. Soc.,* Spring 2000, Abs. 230.

14. K.-M. Chang *et al.,* "Using NH_3 Treatment to Improve the Characteristics of Hydrogen Silsesquioxane for Copper Interconnection Application," *J. Electrochem. Soc.,* Vol. **147**, May 2000, p. 1957.

15. L. Peters, "IBM Reveals Key Copper/Low-k Integration Issues for Production," *Semiconductor International,* August 2000, p. 25.

16. C.-F. Yeh, "Comprehensive Investigation on FSG Prepared by Temperature-Difference-Based Liquid Phase Dep.," *J. Electrochem. Soc.,* January 2000, p. 330.
17. S. Murarka, "Low Dielectric Constant Materials for Interlayer Dielectric Applications," *Solid State Technology,* March 1996, p. 83.
18. E. Korczynski, "Low-*k* Dielectric Integration Cost Modeling," *Solid State Technology,,* October 1997, p. 123.
19. K.S. Lau *et al.,* "Characterization and Thin-Film Properties of FLARE," *Proc. 1996 VMIC Conf.,* p. 92.
20. M. Ikeda *et al.,* "Integration of Organic Low-*k* Material with Cu-Damascene," *IEEE International. Interconnect Conf., (IITC),* June 1998, p. 131.
21. P. Singer, "IBM to Use Dow's SiLK in Low-*k* Process," *Semiconductor International,* May 2000, p. 44.
22. G. Passemard and O. Demolliens, "Low-*k*: Damascene Integration Issues," *Abs. Mtg. Electrochem. Soc.,* Spring 2000, Abs. 228.
23. J. Gaynor and A. Ralston, "AF4 as a VLSI Interconnect Material," *Semiconductor International,* December 1997, p. 73.
24. L.-Q. Xia *et al.,* "Simplified Intermetal Dielectric Integration with Chemical Vapor Deposited Low-*k* Material Using Organosilane/Ozone Chemistry," *Abs. Mtg. Electrochem. Soc.,* Spring 2000, Abs. 229.
25. A.L.S. Loke *et al.,* "Copper Drift in Low-*k* Polymer Dielectrics for ULSI Metallization," *Dig. of Tech. Papers 1998 Symp. on VLSI Technology,* p. 26.
26. K.-M. Chang *et al.,* "A Novel Pretreatment Technology for Organic Low-Dielectric Material to Suppress Copper Diffusion and Improve Ashing Resistance," *J. Electrochem. Soc.,* Vol. **147**, June 2000, p. 2332.
27. M. Matsuura *et al.,* "A Highly Reliable Self-Planarizing Low-*k* ILD for Sub-Quarter Micron Interconnects," *Tech. Dig. IEDM,* 1997, p. 785.
28. B.T. Ahlburn, "Ultra Low-*k* Dielectrics, - Shortening the Time Line from R&D to Manufacturing ," *Abs. Mtg. Electrochem. Soc.,* Spring 2000, Abs. 218.
29. *Electrochemical and Solid-State Letts,* 2, p. 77, February 1999.
30. L. Peters, "IMEC, SEMATECH Show Feasibility of a Sub-2.0 Low-*k* Dielectric," *Semiconductor International,* July 2000, p. 33.
31. W. Chen *et al.,* "Plasma Treatment of Hydrogen Silsesquioxane (HSQ) Based Ultra-Low-*k* Films," *Abs. Mtg. Electrochem. Soc.,* Fall 2000, Abs. 535.
32. J. Wetzel *et al.,* "Evaluation of Material Properties & Performance of Ultra-Low-*k* Insulators for Inlaid Cu Metallization," *Tech. Dig. IEDM,* 2001, p. 73.
33. H.-J. Lee *et al.,* "Correlations Between Structural Characteristics and Process Conditions of HSQ-based Porous Low-*k* Thin Films," *Spring Materials Res. Conf., April 2001.*
34. J.N. Bremmer, "A New Class of Insulating Materials: The Emergence of Ultra-Low-*k*," *Solid State Technology,* September 2001, S3.

Chapter 15

DUAL-DAMASCENE INTERCONNECTS

As the submicron era evolved in the 1990's, several novel manufacturing methods were adopted to keep pace with Moore's law. One such approach, namely the formation of interconnect structures with *damascene processes* was chosen to fabricate IC interconnects. In this chapter we describe this technique.

15.1 DAMASCENE VERSUS SUBTRACTIVE INTERCONNECT STRUCTURES

The conventional way of fabricating each conducting layer of an interconnect structure of an IC is a so-called *subtractive process.* In such a subtractive process, a blanket metal film is deposited onto an insulator-covered wafer surface (with contacts - or vias - having been previously etched in the insulator layer - Fig. 15-1). Next this metal film (e.g., Metal-1) is patterned by etching - to create metal interconnect lines. Finally a dielectric film is deposited over these etched metal lines. Vias are then etched in the overlying dielectric film to allow connection passages for the next layer of metal that is to be deposited onto the dielectric film covering the Metal-1 lines. This subtractive approach requires several process capabilities: **1**) It must be possible to effectively etch the metal and; **2**) It must be possible to deposit both the dielectric and metal layers with a deposition process that has good gap-filling capability. In some interconnect schemes (especially as the ICs are scaled below 0.5-μm), it becomes difficult if not practically impossible, to meet these goals. For example, for technology nodes smaller than 0.5-μm it is desirable to replace Al with Cu as the intercon-

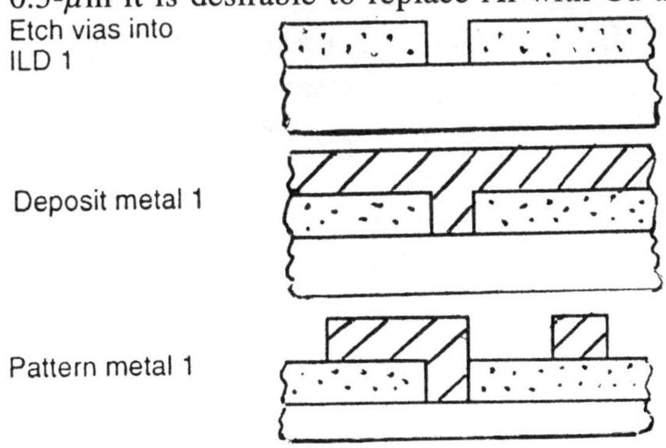

Fig. 15-1 Subtractive approach to forming interconnect structures for ICs.

nect conductor (i.e., because Cu has a higher conductivity than Al, and also exhibits superior electromigration characteristics). However, since it is difficult to etch Cu, an interconnect scheme that does not require Cu etching would be preferable. In addition, as the spaces between metal lines become smaller, it becomes progressively harder to fill such narrow gaps with an insulator film. Again, a significant advantage would be gained if it were possible to use a process that does not require gap-filling when insulator films are deposited. To achieve the above goals, an alternative interconnect fabrication technology to the subtractive method has thus been developed, called the *damascene method.*

15.2 DAMASCENE STRUCTURES AS ALTERNATIVE DEEP-SUBMICON INTERCONNECTS

In the damascene approach to forming interconnects on ICs, a trench is cut into the dielectric layer and then filled with metal. (The term *damascene* is derived from the decorative practice of creating metal inlays, as developed by the artisans of Damascus in the Middle Ages.) This method thus provides two major advantages over the subtractive method of forming interconnect structures:

1. In the damascene approach, no etching of the metal layers is required. The metal-etch process of the subtractive approach is replaced with a metal chemical-mechanical-polishing process (CMP). The metal-CMP step removes any metal deposited above the top surface of the trench (including the regions where the metal is deposited on the dielectric surface). This leaves a planarized dielectric surface with metal inlaid in trenches of the dielectric. Since the CMP process leaves a globally planarized surface, *the need for filling narrow gaps during the dielectric deposition process is eliminated.* The only gap-filling capability required by the damascene approach is during the deposition of metal into the recesses of the dielectric. Since metal etch (especially the etching of Cu) and dielectric gap-fill are viewed as two of the greatest challenges in the drive to smaller dimensions, elimination of these steps represents a major benefit. On the other hand, filling of even narrow, high-aspect-ratio trenches with metal (i.e., with W by CVD, or Cu by electroplating) is not as difficult.

2. There is only a bottom interface for Cu (versus top and bottom interfaces for Al-alloys with W plugs). This helps reduce overall via resistance.

We will also see that by using a *dual-damascene* process, the number of processing steps is smaller than for the subtractive approach. Hence, use of dual-damascene interconnects may also result in lower manufacturing costs. The damascene approach was pioneered by IBM, and began to appear in commercial IC products at about the 0.25-μm node.[1]

15.3 SINGLE-DAMASCENE INTERCONNECT STRUCTURES

Damascene interconnects can be formed with either a *single-damascene* or *dual-damascene* process. The dual-damascene process offers the advantage of reducing the number of process steps, and thus most of our discussion focuses on it. However, the single-damascene process is always carried out on at least the first level of a multi-level interconnect structure of an IC. For this reason, it is useful to begin with a description of the single-damascene approach.

In the *single-damascene process* (see Fig. 15-2 [and Fig. 12-10 in Sect. 12.1.4]), the interlevel dielectric layer (ILD) is planarized after it is deposited. Next, a contact (or via) is patterned with a lithography step, and then the contact opening is etched into the ILD layer. After completion of the contact-hole etch step, a metal plug is formed in this opening - typically by blanket deposition of the metal, followed by an etchback or CMP step to form the plug (e.g., a W-plug). Thus, the surface of the dielectric containing the metal plug after such an etchback or polish step is globally planar (see Fig. 8-92). The metal lines of the interconnect level are formed next. This can be done using a subtractive process, as described in Sect. 15.1, or with another single-damascene process.

Assuming a single damascene process is applied twice (first to create plugs, and second to create metal lines), the second time it is employed involves the following process steps (Fig. 15-2). Another dielectric layer is deposited over the

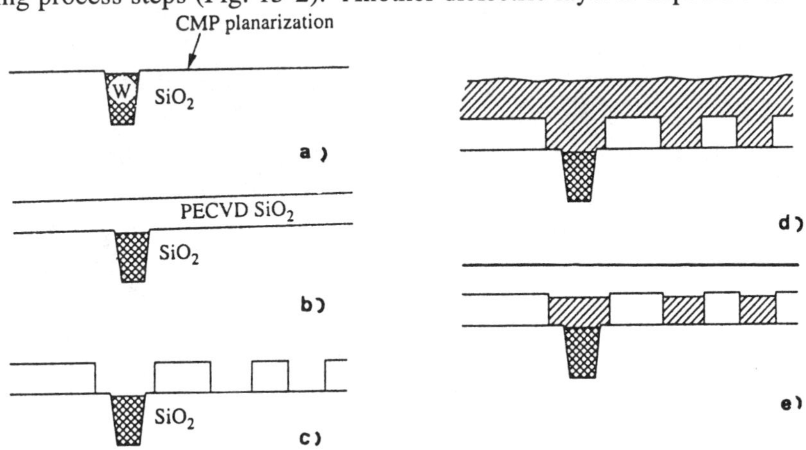

Fig. 15-2 A *single-damascene* process (applied twice) to form one level of an interconnect structure: (a) W-plugs formed after the SiO_2 is planarized; (b) IMD is deposited by CVD; (c) A trench is etched in the IMD; (d) Metal is deposited to fill trenches; (e) CMP metal.

underlying, plug-containing layer. This overlying dielectric film therefore also remains planarized. A second lithography step is then used to form the pattern of a trench, which is aligned to the previously fabricated via plug. With this aligned trench pattern acting as a template, a trench or groove is etched into the planarized overlying ILD. After the resist pattern is stripped, the trench is filled with a metal (such as aluminum or copper).[2] This metal is polished back by CMP to create a metal line embedded in a dielectric trench, and in contact with the underlying plugs. Finally, another dielectric layer is deposited onto the dielectric layer containing the metal-filled trenches. This dielectric layer will be the one through which the next level of vias (or holes to the contact pads) will be created.

It is worth emphasizing several important aspects of the single-damascene process: **1**) If CMP is used to form both the plug and the metal in the trench, no metal etch process is needed; **2**) No gaps have to be filled by the dielectric-film-deposition processes; **3**) The process still contains about the same number of steps as the subtractive interconnect process (i.e., the two polish steps replace the via and metal etch steps of the subtractive process, and the two metal-deposition steps replace the metal-plug-deposition step and main-metal-line deposition step of the plug/metal-line subtractive interconnect process); and **4**) Since the first-level interlevel dielectric layer is planarized before the contact holes are etched, *contact holes* will generally have different depths (e.g., those going to the substrate will be deeper than those going to the gates). Thus, to allow all contact holes to be opened, the shallower ones will expose the underlying material to the plasma etching reactants during the time the deeper contacts are being opened. To prevent the underlying material in the shallow contact holes from being attacked during this time, an etch-stop layer is often deposited after the salicide process is completed (typically using Si_3N_4 as such an etch-stop film). After the contact etch is completed, the nitride etch-stop layer is removed, Since the thickness of this layer is uniform on all regions, it can be etched away with a much shorter overetch time. This method allows all of the contact holes in a single-damascene process to be opened without damaging the underlying material on any part of the die.

15.4 INTRODUCTION TO DUAL-DAMASCENE INTERCONNECT PROCESSES

In a *dual-damascene* (*DD*) *structure*, only a single metal deposition step is used to simultaneously form the main metal lines and the metal in the vias. That is, both trenches and vias are formed in a single dielectric layer. The vias and trenches are defined by using two lithography steps (Fig. 15-3). Trenches are typically etched to a depth of 4000-5000-Å, and the vias are typically 5000-

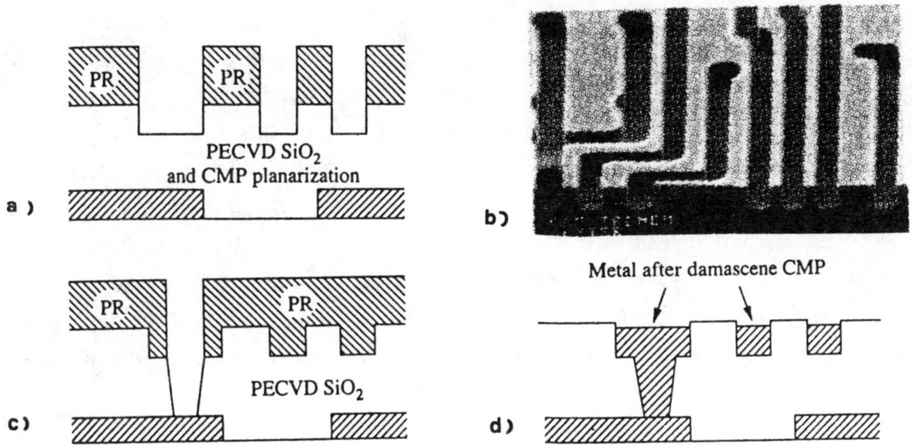

Fig. 15-3 *Dual-Damascene* process: (a) IMD is deposited by CVD and planarized by CMP. Trench is defined by PR #1 and then etched; (b) SEM of the etched trench after PR #1 has been stripped; (c) Vias are defined by PR #2 and then etched, using PR #2 to protect other regions of the IMD from etching; (d) Metal is deposited to simultaneously fill trenches and vias. CMP is used to remove excess metal. The sequence is repeated for the next level of metal.

7000Å-deep. After the via and trench recesses are etched, the via is filled in the same metal-deposition step that fills the trench. After filling, the excess metal that is deposited outside the trench is removed by a CMP process, and a planar structure with metal inlays is achieved. As in the single-damascene process, once a planarized surface is achieved, it is no longer necessary to perform CMP on the dielectric layers. Thus, one of the CMP steps needed for each metal level in the subtractive interconnect process is also eliminated.

It should also be mentioned that in early dual-damascene processes, the trench dimension was somewhat wider than that of the vias. This made the alignment of the via to the trench somewhat easier. However, with time, the via and trench widths have been made equal. (At the 0.18-μm node, the via and trench widths are 0.22-0.24-μm wide, while at the 0.13-μm node they will probably be ~0.18-μm wide.) Thus, modern dual-damascene processes are more sensitive to misalignment between the trench and via. However, this overlay sensitivity problem is mitigated to some degree by the nature of the damascene process, making it possible to maintain high interconnect packing density. That is, the alignment of the trenches to the vias in the dual-damascene sequences is an easier task than in the subtractive interconnect process. In the latter, aligning metal lines to the underlying contact holes (or vias) must be done after the opaque, shiny, blanket-metal film has been deposited. In damascene approaches, aligning the trenches to the vias is done through the transparent dielectric film, which allows tighter design-rule tolerances to be used when performing this alignment.

When comparing the single-damascene process to the simple dual-damascene process described above, one observes one metal-deposition step and one CMP step are eliminated in the latter (as well as one dielectric-deposition step). The reduction in the number of processing steps is another of the benefits that have driven the development of dual-damascene processes. Note that in single-damascene structures, plugs are typically filled with W, but in the dual-damascene-interconnects the metal that fills the holes is Al or Cu. The technology of filling of vias and trenches with Cu is described in Chap. 16. A dual-damascene process using Al was described by workers at IBM in 1998.[3]

15.5 THE THREE DUAL-DAMASCENE PROCESS SEQUENCES

Three different fabrication sequences have been developed to produce dual-damascene structures:

1. Trench-First Dual-Damascene
2. Via-First Dual-Damascene
3. Self-Aligned (or Buried-Via) Dual-Damascene

It should be noted that the *via-first* and the *trench-first* sequences are the processes currently being used in mainstream production. The trench-first sequence was probably the first one to be adopted, but the via-first sequence is currently the approach that has gained the greatest acceptance (and will likely remain so going forward). As a result, we will focus on these two sequences, and will point out the advantages and drawbacks of each one. While the third approach (the self-aligned process), has not been implemented in production (for reasons we will also explain), we will provide a brief description of it as well. However, whichever sequence is used, the final damascene structure is the same (see Fig. 15-4).

Before outlining the three dual-damascene processes, it should be mentioned that the dielectric layer in which the trenches and vias are etched actually consists of at least three layers of dielectric (see Fig. 15-8b). That is, SiO_2 (or a low-k dielectric is deposited to a depth of 5000-7000Å (and this will be the layer in which the *vias* will be etched). Next, a thin silicon nitride layer - or other material (which will serve as an *embedded etch-stop layer*), is deposited (about 30-nm thick). Finally, another SiO_2 layer (or a low-k dielectric layer) is deposited

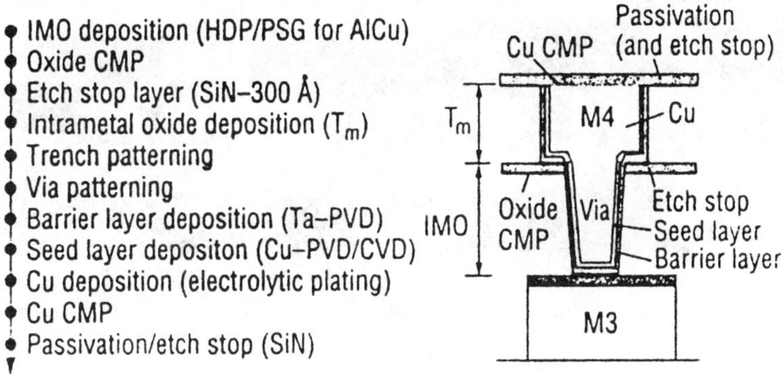

Fig. 15-4 A suggested dual-damascene-based Cu metallization flow and resulting cross-sectional structure.[8] Reprinted with permission of Solid State Technology. Published by PennWell.

(~4000-5000-Å thick). This will serve as the layer in which the *trenches* will be etched.

In the *trench-first* sequence (Fig. 15-5), the trench patterns are defined in the ILD first. That is, after spinning on the resist, the trench-pattern-mask is used to expose the resist. The trench is then produced by etching the dielectric down to the embedded etch-stop layer. After the trench-etching process, the first resist layer is stripped. A second resist layer is then spun on, and the via-pattern-mask is used to create openings in this resist layer - aligned to the trench that was etched previously. The resist protects the other parts of the wafer surface (including the etched regions of the trenches) so that the vias can be etched without further etching the dielectric in the trenches.

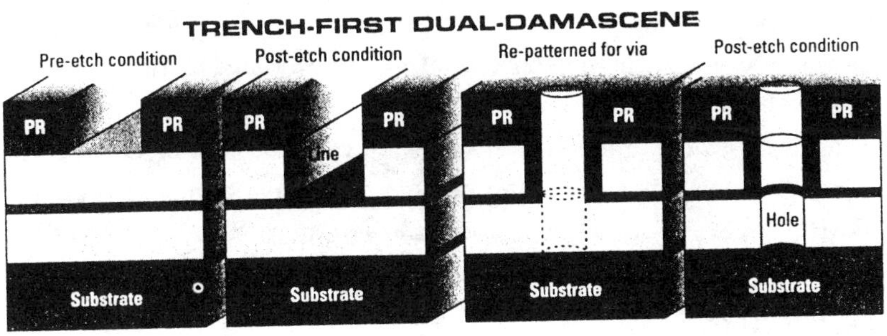

Fig. 15-5 In the trench-first approach, the vias are patterned & etched after trenches are etched.

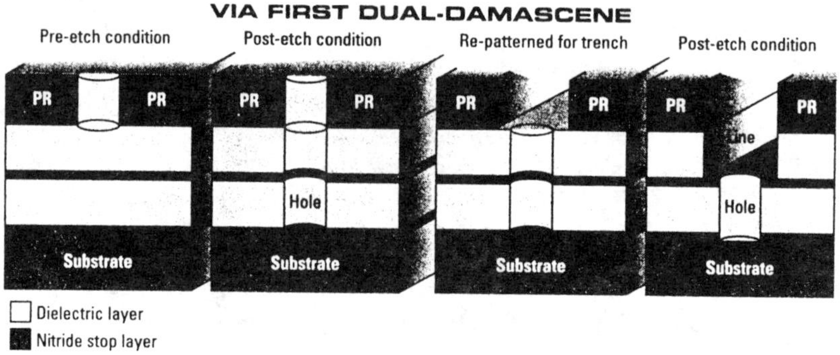

Fig. 15-6 In the via-first approach, the trench is patterned and etched after the via is formed.

The chief disadvantage of the trench-first sequence involves the fact that the vias must be patterned after the trench etch. That is, since resist is applied as a liquid onto the wafer surface, it fills recessed regions as water fills a lake. Hence, the top surface of the resist is planar. This means that the regions of resist over the damascene trenches is quite thick. Resolving fine features in thick resist is harder than in thinner resists. For features smaller than 0.25-μm the process latitude becomes too small for this to be a practical manufacturing process. Thus, while the trench-first sequence could be used for early dual-damascene processes (0.35-μm and some 0.25-μm processes), it is difficult to extend it to smaller technology nodes (i.e., in order to able to use it at such small dimensions, it may be necessary to implement a thin imaging-resist-layer process, as discussed later).

In the *via-first approach* the vias are defined first in the ILD, followed by patterning the trenches. As noted above, this is currently the most common method of fabricating dual-damascene interconnects (and is schematically shown in Fig. 15-6). The sequence of forming the damascene recesses in this approach begins by exposing the *via patterns* with the first mask. After etching the vias completely through the entire dielectric stack (except not through the barrier layer at the bottom of the dielectric stack) and stripping the resist, a second mask is used to pattern the trenches. The trenches are then created by etching the dielectric down to the embedded etch-stop layer. The barrier layer at the bottom of the vias is protected from further etching during the trench-etch either by resist or a BARC layer (applied for the trench-etch step) that floods the vias, or by using an etch process that is highly selective to the bottom etch-stop material (i.e., so that the etch-stop layer at the bottom of the via is not removed during the trench-etch step). After the resist is stripped and the etch-stop layer at the bottom of the via is removed by dry-etching, the metal that fills both the vias and the

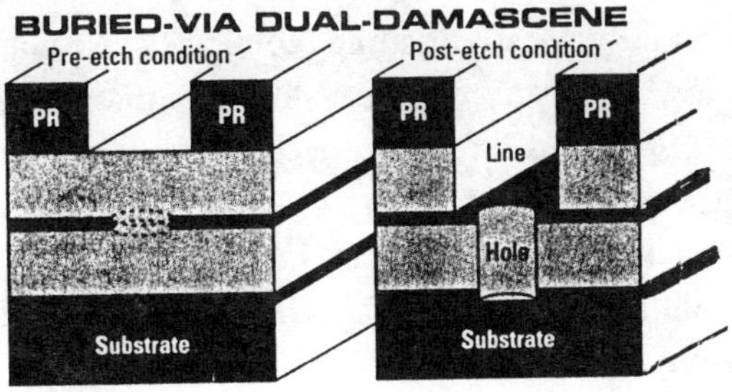

Fig. 15-7 In the self-aligned, or buried-via approach, a hard mask is patterned and etched. The trench and via can be etched in one step.

trenches can be deposited. After deposition, it is polished back to create the dual-damascene structure.

In the *self-aligned* dual-damascene process (Fig. 15-7), the via pattern is created in the embedded etch-stop layer with a lithography-and-etch sequence that is performed before the top dielectric layer of the stack is deposited. A special etch process is required for this step. After the via pattern is thus etched, the top dielectric layer is deposited. Finally, the trench mask is aligned to the via openings in the embedded nitride layer, and both the trench and via are opened with a single etch step. This sequence requires nearly perfect trench-to-via alignment, otherwise the via may no longer be round (but instead only half-moon shaped), resulting in high via resistance. In addition, an etch-process with high nitride-to-oxide selectivity is required. For these reasons this scheme has not found use in production.

15.6 THE EVOLUTION OF DUAL-DAMASCENE FROM SiO_2 TO LOW-*k* DIELECTRIC STACKS

In the first generation of dual-damascene processes (i.e., at the 0.25-μm node), Cu was used together with a conventional undoped PECVD-SiO_2 dielectric (with a PECVD-SiN layer beneath the SiO_2, which acts as both a Cu diffusion barrier and an etch-stop when the vias are etched, Fig. 15-8a).* The nitride and oxide layers may be deposited sequentially in the same tool (e.g., a PECVD reactor).

* This SiN layer is deposited first, and as noted, serves as a barrier against the diffusion of Cu between metal levels and also as an etch-stop in the dielectric etch that defines the vias. Because SiN has a high value of *k*, this layer should also be as thin as possible to avoid adding significantly to the interlevel capacitance. At the same time, the SiN layer must be low-stress, dense, and pinhole–free, so that its diffusion-barrier properties are not compromised. High-density is also critical to the nitride layer's role as an etch-stop.

This Cu/SiO_2 structure gave improved interconnect performance over the Al/W-plug subtractive process because Cu interconnects (i.e., with Cu in both the lines and vias) have a lower resistance than Al lines combined with W-plugs. Thus, the RC delay term was reduced through a decrease in R. A significant problem that occurs with using this dielectric structure is that the trench formation has no etch stop. Since the trenches are etched to an intermediate point part way through the dielectric layer, within-wafer etch-non-uniformity translates directly into within-wafer non-uniformity of the trench depths.

In second generation dual-damascene processes (down to about the 0.18-μm node), the CVD-SiO_2 material was replaced by a first generation low-k material, namely a PECVD-SiO_2 layer doped with fluorine (*fluorinated silicate glass, FSG*) whose $k = 3.5$. (FSG has many properties similar to SiO_2 and thus requires the smallest number of process integration changes.) A SiN layer, embedded halfway through the thick dielectric layer, was also added to the dielectric stack to serve as an etch-stop layer when the trenches are etched. However, since the k value of SiN is $k = 7.0$, the embedded-etch-stop SiN layer increases the net-k value of the resulting dielectric stack. In fact, the overall capacitance of the FSG dielectric stack with an embedded-SiN etch-stop layer was about the same as that of first generation SiO_2 layers - but it had the benefit of being able to ensure the trench is only etched halfway through the dielectric layer. The corresponding drawback is that fabricating a multi-layer dielectric stack adds cost and complexity to the manufacturing sequence.

The next step in the evolution of the dual-damascene dielectric stack was to incorporate a second generation low-k material ($k = 2.7$) into both the upper and lower segments of the dielectric stack, as this now allows the wiring capacitance to be reduced. (Note that the wiring capacitance is reduced by about 11% when a SiLK/SiN dielectric stack is implemented, compared to the wiring capacitance obtained when using a FSG/SiN stack.*) The first dual-damascene structure that incorporated a 2nd generation low-k dielectric in the stack (together with Cu), was announced by IBM in April 2000. It was intended for use in the IBM's Cu-15

* In a paper published in late 2001, an alternative dielectric stack to the SiLK/SiN stack was proposed. That is, it used FSG and a new embedded layer material (SiCN), whose k value of 5 is lower than that of SiN. The capacitance of this stack was about 9% lower than that of the FSG/SiN stack, making its wiring capacitance comparable to the SiLK/SiN stack. However, the authors claim that the SiLK/SiN stack is mechanically weak, and tends to fail during bonding and packaging processes, whereas the FSG/SiCN stack is more robust.[5]

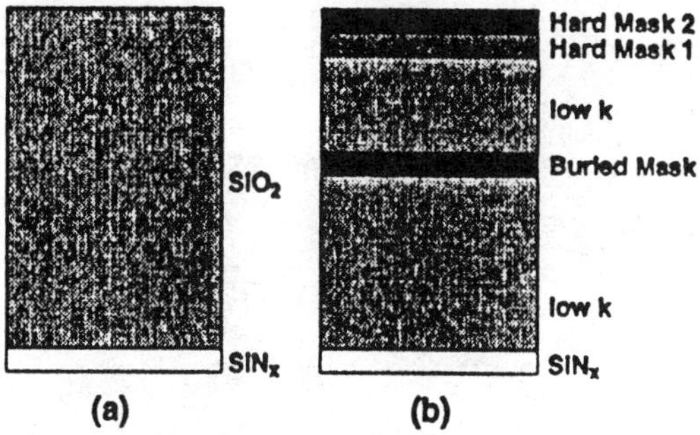

Fig. 15-8 a) First generation dual-damascene ILD stack, consisting of SiO_2 and a bottom SiN layer. b) Later generation dual-damascene ILD stack with low-*k* dielectric, embedded etch-stop layer, and hard-mask/capping layer (SiO_2).

0.13-μm CMOS technology.[4] The low-*k* dielectric material selected for this process was SiLK^{TM} from Dow Chemical (see Chap. 14). The dual-damascene process is used to form the #2 through #5 interconnect levels of the 5-level-metal interconnect structure used for this technology node.

It should also be noted that when a 2nd generation low-*k* dielectric replaces the FSG (or SiO_2) material in the stack, a hard mask must also be added to the top of the stack. The material chosen by IBM for this hard mask was SiO_2, giving the dielectric stack a structure as shown in Fig. 15-8b. SiO_2 was selected for this role because it has a significantly lower *k*-value than SiN. In addition, this SiO_2 hard mask remains in place during metal deposition, so it can later serve as a polish-stop during metal CMP. Since use of an SiO_2 hard mask maintains the same materials system encountered during the CMP of metal and a PECVD-SiO_2 dielectric stack (Cu/TaN/SiO_2), the developed consumables, equipment, and processes for Cu-CMP on plain SiO_2 should be extendable to Cu-on-SiO_2-capped low-*k*. However, since this oxide cap becomes a primary supporting structure against material loss by dielectric erosion during CMP, the allowable dielectric erosion (see Sect. 8.11.1) in a capped dielectric stack is dramatically decreased.

It should also be noted that in some cases different materials may be used for the dielectric at the via level than for the trench level. Since most of the capacitive coupling issues that a low-*k* material helps solve are encountered between adjacent metal lines, the *trench dielectric* will be where the lowest-*k*

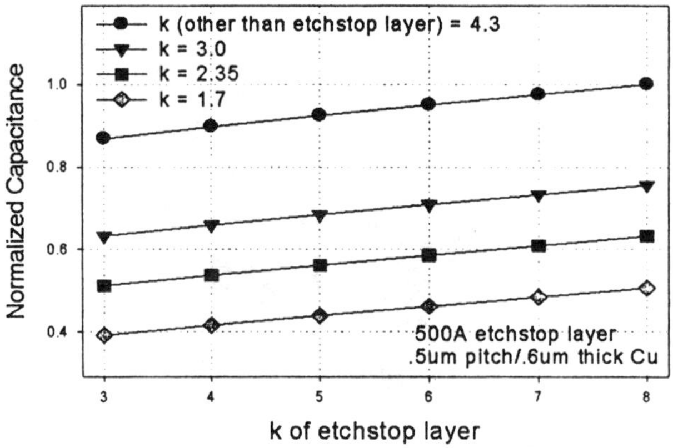

Fig. 15-9 Capacitance impact of the *k*-value of the etch-stop value in a dual-damascene structure for various dielectric constants.[32]

materials are first used. Since the via dielectric is less critical from a capacitance perspective, implementing a low-*k* material there may be delayed.

The use of SiN as an embedded etch-stop layer increases the capacitance. Figure 15-9 shows the effect of a 500-Å SiN on dielectric stacks with different low-*k* dielectrics.[32] To allow the capacitance of the dielectric stack to be reduced even further, a new material has been introduced to replace SiN for the embedded-etch-stop and hard-mask roles. This material is SiC (with $k = 4.5$). Figure 15-10 shows how much the use of Si-C instead of SiN reduces the effective *k*-value of the damascene dielectric stack.[33] Applied Materials introduced this material in 1999 under the name BLOk (barrier low-*k*).[6,7] In

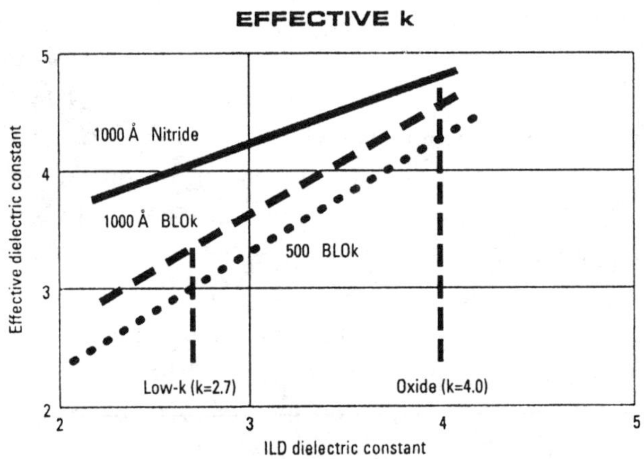

Fig. 15-10 The SiC-based dielectric barrier impacts the effective *k* value much less dramatically than the traditional nitride film. Courtesy Applied Materials.

addition, this material offers twice the etch selectivity of nitride to oxide-like films, has better Cu diffusion barriers than nitride, and shows good adhesion ot SiO_2, FSG, and organic-based low-*k* dielectrics. Furthermore, it exhibits the optical properties needed to make it behave like an anti-reflective coating in DUV imaging. The film is deposited in an HDP-CVD chamber using trimethylsilane at 350-400°C. The resulting Si-C film can also be etched using a "soft etch," which means it can also be used as the capping layer for Cu (i.e., at the bottom of the via of the via-first dual-damascene structure). Some companies have reportedly incorporated the Si-C films into damascene dielectric stacks which employ undoped SiO_2 or FSG, while others are using it with stacks containing lower-*k* films. Another vendor that supplies a tool and a process to deposit such an Si-C film is Novellus.

It is thought that the Si-C material will replace SiN as the etch-stop layer in low-*k* dielectric stacks for the 0.13-μm node. However, as devices are scaled even further, it is predicted that even the *k* value of 4.5 of Si-C will prove to be too large. Thus, it has been projected that at the 0.1-μm node, the embedded-etch-stop layer will be removed entirely. The ramifications of removing the embedded etch-stop layer will be discussed in a later section.

15.7 TRENCH-FIRST DUAL-DAMASCENE PROCESS FLOW

Here we present an example of a trench-first dual-damascene process flow, using undoped SiO_2 as the main dielectric material and Cu as the main interconnect metal. We assume that an STI isolation structure is used, so that the substrate is flat, up to the time of gate-stack formation. We will also see that both single-damascene and dual-damascene processing are used. Many variations of this example damascene-interconnect structure have been published.[8,9,10,11,12] Some advanced interconnect structures may elect to use both conventional W-plugs and etched Al-wire interconnects for some metal levels, and Cu-damascene for others. One such example is given in Ref. 13.

The fabrication sequence of this interconnect structure is begun by carrying out two single-damascene process sequences to form both the Metal-1 lines (Cu) and CVD-W plugs in the contact holes. The first process sequence is started by depositing a dielectric layer over the gate stack. This film serves as the *pre-metal dielectric* (PMD) layer in this structure, and it is polished flat with a CMP step. Vertical contact holes are etched into the PMD layer. A CVD-W process is then used to fill the contact holes. The W is then planarized, so that W-plugs are formed such that their top surface is flush with the top surface of the PMD.

The second single-damascene process is then performed. That is, an additional dielectric film is deposited onto the PMD layer, and trenches (aligned to the W-plugs) are etched into it, deep enough so that the tops of the W-plugs are cleared of dielectric. A liner film (e.g., TaN) and a Cu seed-layer are sequentially blanket-deposited deposited by PVD. The main Cu film is then blanket deposited by electroplating. The Cu film (and TaN liner layer) are polished off of the top surface of the dielectric film with CMP, leaving the Metal-1 Cu lines inlaid in the grooves. The main challenge here is to obtain the same polishing rate on the tantalum nitride (which is a hard refractory-metal-containing nitride), and the copper (which is comparatively soft). This completes the formation of the Metal-1 and contact plugs by two single-damascene processes. The surface of the wafer is flat for the next dielectric film deposition.

The subsequent layers of metal are formed by a dual-damascene process sequence. The sequence begins by depositing a thin, low-stress silicon nitride layer. This layer forms a top encapsulation layer on the level-1 inlaid Cu lines, and also acts as an etch-stop layer when the vias are etched in the dual-damascene sequence. Next, a dielectric layer (here, undoped SiO_2) is deposited by a CVD process to a thickness required for the via-hole depth. Since the surface is flat, gap-filling by this layer is not required. Then, the embedded etch-stop layer is deposited by CVD (typically PECVD-silicon-nitride with a thickness of 30 nm). Finally, the inter-metal-dielectric stack (IMD) is completed by depositing a second dielectric film, whose thickness is determined by the next-metal-level thickness. (Note that the multi-layer dielectric film, containing two nitride etch-stop layers adds cost and complexity, as well as increasing the interlevel capacitance of the interconnect stack.) Photoresist is applied to the wafer surface. The trench pattern in the resist is formed by exposing the resist through a mask containing this pattern (Fig. 15-5). An etch step is performed next, with the etch creating the trenches in the upper part of the IMD stack. The trench etch is stopped by the silicon-nitride etch-stop layer embedded in the dielectric stack. The resist is then stripped, and a second layer of resist is applied. Another mask (this one containing the via pattern), is used to expose the second resist layer, with the via aligned to the previously etched trench-pattern. The via pattern is then etched in the lower part of the IMD stack. This may be a two-step etch process, as etching the nitride etch-stop layer at the bottom of the via to expose the Cu of Metal-1 may require a separate plasma chemistry than the one used to etch the underlying oxide material. Note that in practice, the trench width may be only slightly larger than the via width.

After the second resist layer is stripped, the formation of the metal line is begun. Again, a TaN (or Ta or TiN) liner and a Cu seed layer are sequentially

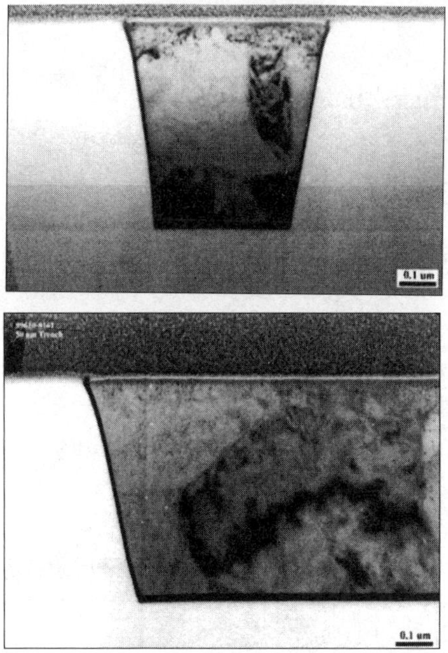

Fig. 15-11 SEM of an in-laid copper line in a dual-damascene interconnect-structure.[34] Reprinted with permission of Semiconductor International.

deposited. This may require an ionized-metal-plasma sputtering process, as the via hole is quite deep and has a high aspect ratio. The main part of the Cu interconnect is then deposited by electroplating. Note that both the via and the trench are filled simultaneously. A metal-CMP process is then used to remove the excess metal. A passivation/etch stop layer (silicon nitride) is deposited onto the top of the layer containing the grooves (trenches + vias) in the dielectric. This completely encapsulates the Cu with diffusion barriers. The entire interconnect structure is finally covered with an SiO_2/SiN passivation layer. Figure 15-11 shows an in-laid Cu-line in a dual-damascene interconnect structure.[34]

15.8 VIA-FIRST DUAL-DAMASCENE PROCESS WITH SiO_2 DIELECTRIC STACK

Many of the issues of forming the vias and trenches in the via-first process are common with the trench-first process and were described above. These include use of a single-damascene process to form the first layer of metal, and technologies for filling the damascene recesses with metal and polishing it back with CMP. We also assume that the via-first dual-damascene structure is being formed on the first level of metal, which has been covered with a SiN layer (that acts both as an etch-stop to the via etch and a barrier layer covering the top surface of the Metal-1 copper layer, to prevent its diffusion into the Si substrate).

However, there are also issues particular to the via-first sequence, and we will discuss them here.

For the purposes of opening this discussion, we assume that the dielectric stack consists of CVD-SiO_2/SiN/CVD-SiO_2. (The implications of using low-*k* material instead of SiO_2 in the stack were described in Sect 15-6.) The vias in this sequence are, of course, patterned first, using the via mask to align the vias to the Metal-1 lines produced previously (Fig. 15-6). The etch process used to open the via needs to be a multiple-step process because the first etch mechanisms must be *non-selective* (so that they can etch both SiO_2 and the embedded SiN etch-stop layer at about the same rates). However, the nitride capping layer at the bottom of the via should not be removed, and thus the etch mechanisms operating when the vias are close to being opened must be *selective* (to nitride).

After the via s are opened (but with the barrier-SiN layer at the bottom of the vias remains intact), the first layer of resist is stripped. Note that since SiO_2 is used in the dielectric stack, it can withstand the resist-stripping process (i.e., no extra hard mask is needed, as will be the case when most low-*k* materials are used in the ILD stack). The second layer of resist is then applied, and the trench mask is aligned to the opened vias. Note that the nitride at the bottom of the via must also remain intact throughout second (i.e., trench-etch) process.

Three methods are used to prevent removal of the nitride during trench etch:

1. The resist applied for the purpose of patterning the trenches will flood the open vias. If this resist is not exposed during the trench-patterning exposure step (i.e., because it is so deep beneath the top surface of the resist layer, that some may not develop after the exposure step. Thus, it remains to protect the via bottoms). However, due to pattern-density issues discussed later, such resist remaining in the vias may vary in thickness across the circuit. Thus, this approach will not always work.

2. A trench-etch process is used that is highly selective to the SiN that remains at the bottom of the open vias.

3. A bottom-anti-reflective coating (or BARC) is spun-on to partially (or completely) fill the open vias. This BARC remains after the resist is developed to protect the nitride at the via bottoms against the trench etch.[15] While this approach adds process control difficulties and process complexity (as the BARC must be applied, and separately removed later), this has been the method most widely adopted for preventing nitride-barrier attack during trench etching. We will continue the discussion of this approach in Sect 15.9.

In any case, the trench etch is carried out to create the trenches, stopping on the embedded SiN etch-stop layer. At that point the second resist layer is stripped.

After the second resist layer is stripped, the nitride at the bottom of the via must finally be opened. However, this procedure itself can cause some difficulties. First, the SiN-removal-step should be a "soft" etch. That is, the ions that bombard the surface being etched must not have enough energy to sputter Cu during the inevitable overetch time. If such sputtered Cu ends up on the via sidewalls it could be trapped on the wrong side of the subsequently deposited barrier-layer (exactly where you don't want it!). Another potential problem involves the possibility of slightly misaligning the via and the edge of the Cu film below. In this case a deep slit can be etched in the dielectric under the nitride. This slit would be difficult to fill with the barrier material, again leading to copper-diffusion problems. The way to avoid such slit-formation in the oxide, is to use a nitride-etch-process that is highly selective to oxide.

Another desired goal of the trench etching process is to reduce the handling of wafers between the various steps of the trench etching process (and the variability in waiting time between these steps). That is, the process actually contains four etch steps: **1**) organic ARC open (typically required for DUV resists); **2**) trench etch; **3**) resist strip; and **4**) nitride removal from the bottom of the via. Etching tools are now being made available that allow all four of these steps to be carried out in a single chamber, and use of such tools may be critical to achieving consistent device yield and lower overall cost-of-ownership.

15.9 PROCESS INTEGRATION ISSUES OF THE VIA-FIRST DUAL-DAMASCENE PROCESS WITH LOW-*k*-DIELECTRIC STACK

The process flow of forming each level of metal in a via-first dual-damascene interconnect system consists of the following steps:

- DEPOSIT Dielectric Stack (including the SiN barrier-layer over inlaid Cu);
- LITHO Step I (Via Patterning);
- Via ETCH;
- RESIST-STRIP I;
- Post-Via -Etch-Residue CLEAN;
- LITHO Step II (Trench Patterning);
- Trench ETCH;
- RESIST STRIP II;
- Post-Trench-Etch-Residue CLEAN;
- ETCH Barrier Film at Bottom of Via;

- Pre-Metal-Deposition CLEAN;
- Deposit Barrier Metal and "Seed" Cu Layer by PVD;
- Deposit Cu film by ELECTROPLATING;
- Post-Cu-Deposition ANNEAL;
- Remove Excess Cu/Barrier Metal (and do an SiO_2 Buff) by CMP;
- Post-CMP Residue CLEAN;
- CLEAN Top Surface of Cu Before Deposition of SiN Barrier/Etch-Stop Layer.

It can be seen that this process sequence contains: **1**) One Dielectric-Film-Stack Deposition sequence; **2**) Two Lithography processes; **3**) Two Resist-Strip processes; **4**) Three Etch processes; **5**) Five Cleaning processes; **6**) One CMP process; **7**) One PVD Deposition sequence; and **8**) One Electroplating (and Anneal) process.

Although many of these processes were mentioned earlier, there are some additional important process integration issues that should be discussed when considering the overall "via-first Cu/low-*k* dual-damascene" process flow. Thus, we will discuss these relevant process integration details here (Fig. 15-12). However, we will cover only those involved with Processes **2** through **5** above (since Process **1** is discussed in both Chap. 14 and in Sect. 15.6, and Process **6** is covered in Chaps. 8 and 16, and Processes **7** and **8** are described in Chap. 16).

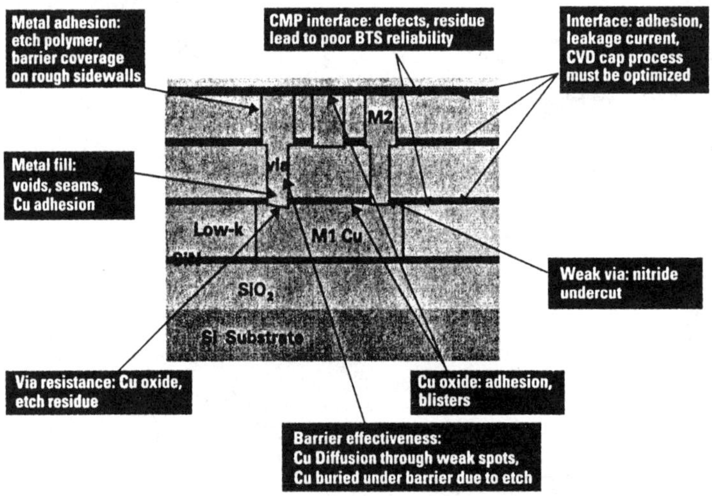

Fig. 15-12 Possible failure types associated with damascene processing pertain to film adhesion, barrier and metal fill capability, copper oxidation and CMP defects.

Besides the technical issues, there are those involving economics. In the face of increasingly difficult integration, the rise in the number of process steps per layer, and the 300-mm conversion, there will be the daunting economic challenges of containing interconnect processing cost, which in 2001 accounts for about 50% of the total wafer processing cost.[16]

15.9.1 Via-First-Damascene Integration Issues: Lithography

With the application of Cu/low-*k* dual-damascene integration, lithography and etch become more difficult because of the complex dielectric stack and intermediate topography. Here we consider the issues involved in carrying out lithography steps for the leading dual-damascene scheme, namely the *via-first approach*. There are two lithography processes needed in the via first approach, and each as its own challenges. They are: **1**) the *via-patterning lithography step*; and **2**) the *trench-patterning lithography step*.

Note that the issue of misalignment of the trench pattern to the via pattern is less severe for the via-first approach than for the trench-first approach. That is, as shown in Fig. 15-13b, even if the trench pattern and the via pattern are somewhat misaligned, since the via is already completely formed during the via etch process, it makes complete contact to the metal layer beneath it (provided it has been properly aligned to the underlying metal line). In the trench-first process, however, if the via is misaligned to the trench pattern, during the second etching step (via-etch) the entire via pattern will not be transferred to the bottom of the dielectric stack. This will cause the via resistance between the 2nd and first levels of metal to increase. Nevertheless, alignment of the via pattern to the underlying metal level below remain critical for both the via-first and trench-first processes, for reasons described in Sect. 15.9.2.3 (see also Fig. 15-13a).

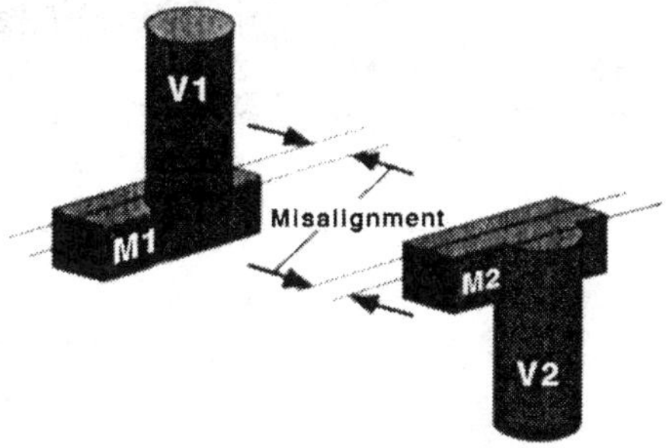

Fig. 15-13 Via-first damascene: a) Misalignment of the via to the underlying metal line is a problem; b) Partial misalignment of the trench to the previously formed via does not cause a problem because the via is completely etched through the ILD stack film.

15.9.1.1 Via-Patterning Lithography Issues: The major concern for the via-patterning lithography step is the ability of the resist to withstand the via etch process. This problem becomes worse as the vias are scaled because thinner resists have to be used to image smaller features, and the etch resistance of 193-nm resists is known to be inferior to that of 248-nm resists. The use of a bilayer resist process (as discussed in Sect. 15.9.1.3 below) may be required, since the imaging resist-layer contains silicon, making it more resistant to dry-etch attack than conventional DUV resists.

15.9.1.2 Trench-Patterning Lithography Issues - DUV Resist Poisoning: As is described in Chap. 6, DUV resists are sensitive to contamination by trace quantities of amines, a problem referred to as *DUV-resist poisoning*. While the most common source of such contamination is amines in the processing ambient, another source is amines in materials that are in direct contact with the resist. Some low-*k* dielectrics contain such amines (nitrogen compounds). For example, the SiOCH films (formed by PECVD using trimethylsilance and N_2O) entrap NH_3 products in the film as they are deposited. Such amines, if near the surface of the SiCOH films, can diffuse into and poison the DUV resist which is in direct contact with the dielectric surface. If an oxide hard mask is present on the top of the dielectric stack, the resist poisoning problem will not occur during the via-first etch. However, during the trench etch, the resist will come into contact with the exposed sidewalls of the SiOCH film under the hard mask, and resist poisoning can then occur during the trench-etch process. Performing an oxygen-plasma pre-treatment of the via sidewalls (perhaps as a part of the via-first resist stripping process) has been proposed as a way around this problem. Such a pre-treatment can leach out the amine compounds from the SiOCH surface on the sidewalls of the via. However, there is some concern that such a plasma pre-treatment may also take carbon out of SiOCH films and reduce their *k*-value.

As noted in Sect. 15.9.3, a plasma-treatment of the sidewalls of nanoporous low-*k* films with an NH_3 plasma has been proposed as a way to strengthen their surfaces against subsequent resist–stripping process of the via-patterning step. However, such a plasma-step can also trap amines in the film. These amines can later diffuse out and poison the resist during the trench-patterning step. (Figure 15-14 shows such a poisoned resist in the via after trench patterning.) Amines can readily diffuse in most low-density and nanoporous films, with diffusion lengths of hundreds of microns. The deposition processes, as well as etch and strip, must therefore be designed to avoid exposing the low-*k* film to any source of nitrogen. As mentioned above, if an SiO_2-capped low-*k* film is used, resist

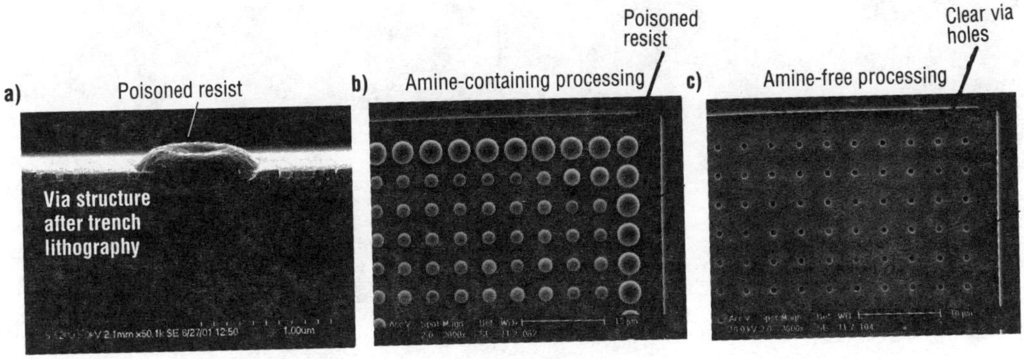

Fig. 15-14 DUV photoresist poisoning is indicated after trench patterning in a) cross-section, and in b) plan view. Poisoning was prevented in c) when the dielectric film was not exposed to nitrogen-containing plasma.[29]

poisoning may not be observed at the via lithography step, but poisoning can occur at the trench lithography step as a result of the trench-patterning resist coming in contact with exposed via sidewalls.[17]

15.9.1.3 Trench-Patterning Lithography Issues - Trench CD Control Issues: To be able to meet the stringent CD-control requirements (i.e., for trench width) for 0.18-μm technologies (and smaller), the resist film spun on a wafer should have the same thickness everywhere (to within ±100Å – see Vol. 1, 2nd Ed. Chap. 12). However, in the via-first damascene approach, since the resist floods the vias that were previously etched, the local thickness of the resist over the top surface of the oxide can be impacted by the presence of these vias. Namely, as shown in Figs. 15-15, and 15-16, the resist is considerably thinner in regions where a dense array of vias exists, compared to regions where only isolated vias exist. In addition, near the edge of such an array of vias, the resist thickness varies significantly, as also shown in these figures. Since higher resolution is generally

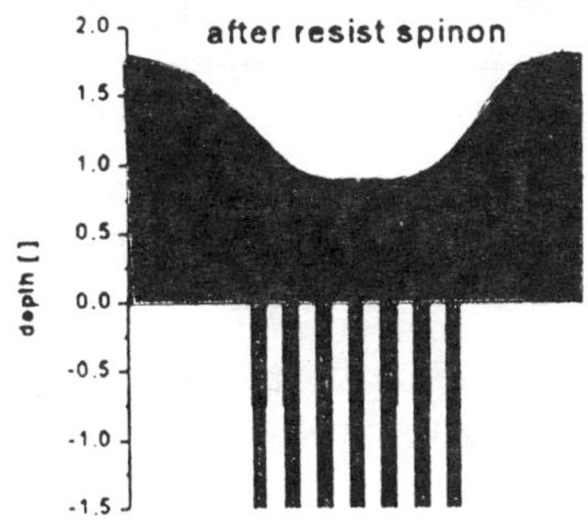

Fig. 15-15 Non-uniformity of resist layer within a chip due to pattern density variations.

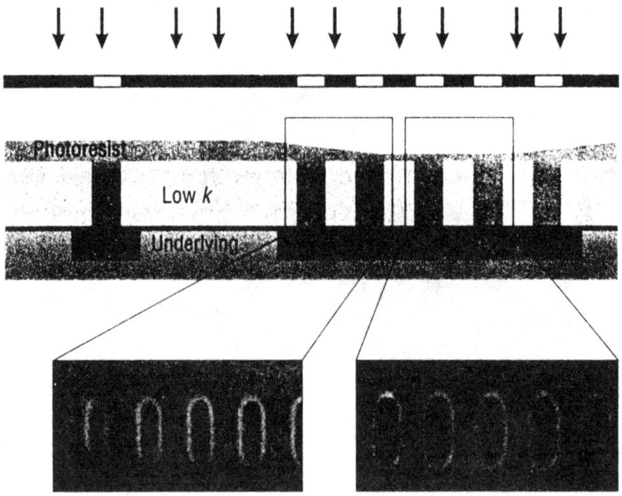

Fig. 15-16 Via-pattern density variations can lead to significant resist thickness non-uniformity during trench patterning in the via-first approach, and subsequent differences in trench widths. a) Trench widths vary at the edge of the via array. b) Trench widths do not vary at the center of the array of visas, where the local resist thickness is uniform.[20]

possible for thinner resist thicknesses, as the resist gets thicker the resolution degrades (e.g., the trench width in this case gets narrower). Figure 15-16 shows that the trench width in trenches patterned at the edge of an array of vias varies, compared to the trench widths patterned over vias at the center of the array. We will see in Sect. 15.9.1.5 that using bi-layer resist process (in which a planarizing underlayer produces more uniform resist thickness over the entire wafer) is a technique being explored to overcome this problem.

15.9.1.4 Trench-Patterning Lithography Issues - BARC Veils: Using a standard DUV lithography process for the trench patterning lithography step, a bottom antireflection coating (BARC) is applied. As noted earlier, the BARC serves two purposes. First, it improves pattern resolution by reducing substrate reflections. Second, the resulting partial fill of the vias with a BARC (Fig. 15-17b)[18] helps protect the SiN barrier layer from being attacked during the trench-etch step. This allows minimization of the barrier layer thickness. The BARC that partially fills the vias remains in place during trench etching and is removed later.

However, it is important to be able to control the depth to which the BARC fills the vias. (The BARC material that fills the via in this procedure is referred to a *via plug*.) That is if the via plug is not recessed below the buried SiN layer in

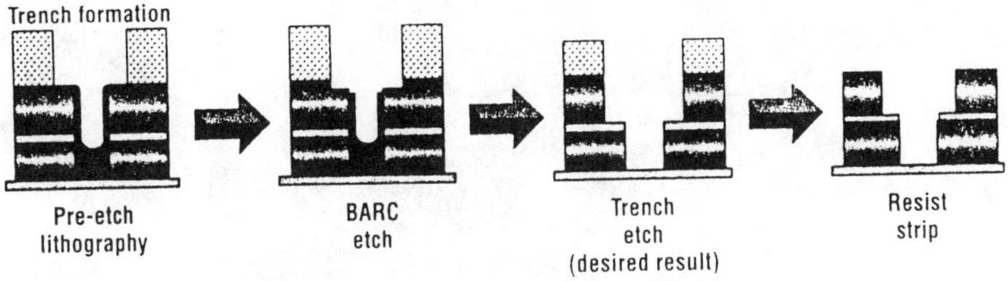

Fig. 15-17 Using a BARC to partially fill the vias prior to trench-PR deposition.[18]

the dielectric stack, "veils" (or "fences') may be created around the via opening at the bottom of the trench.

How such veils may arise is shown in Fig. 15-18. Here the BARC material (via plug) fills the via to a height that is above the top of the buried embedded-etch-stop SiN layer. As the trench is etched, the sidewalls of the via plug above the SiN layer are exposed to the etch ambient that removes the dielectric material to form the trench (Fig. 15-18a). The upwardly-protruding sidewalls of BARC can be chemically converted by an exposure to the plasma etching processes. This altered BARC-sidewall material may not be removable by the same resist stripping step that also removes unexposed BARC material. Thus, it may remain behind in the form of a "veil" surrounding the via opening. Figure 15-18b shows

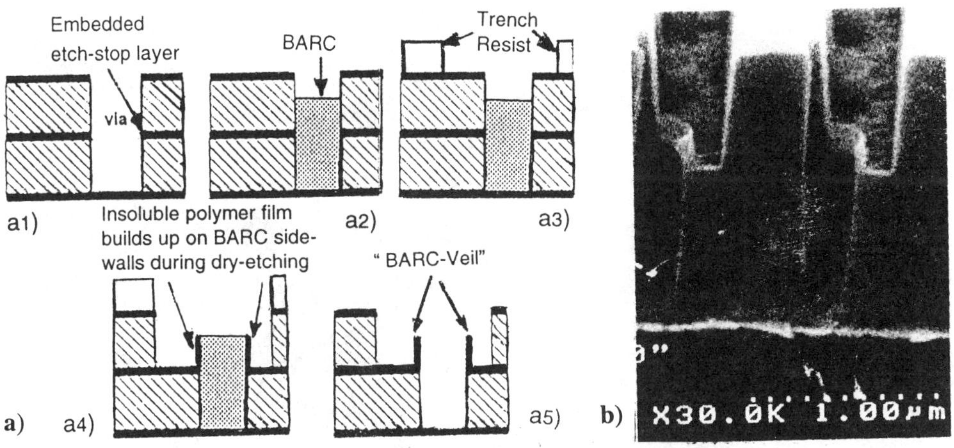

Fig. 15-18 a) BARC-veil formation mechanism in the Via-First Dual-Damascene process: **a**1) Etch Via, strip resist; **a**2) Via is filled with BARC to a level greater than that of the embedded etch-stop layer; **a**3) Deposit and pattern Trench photoresist; **a**4) Etch Trench. *(Note that during the dielectric etch step, neither the resist nor the BARC is attacked, BUT, an insoluble polymer builds up on the sidewalls of the BARC pillar that sticks up above the etch-stop layer.)*; **a**5) After Trench resist and BARC are stripped, the insoluble film that was created during the dry-etch on the BARC-pillar sidewalls, remains behind as "BARC-Veils." **b**) SEM of BARC via-veils after trench-patterning and resist-strip.[18]

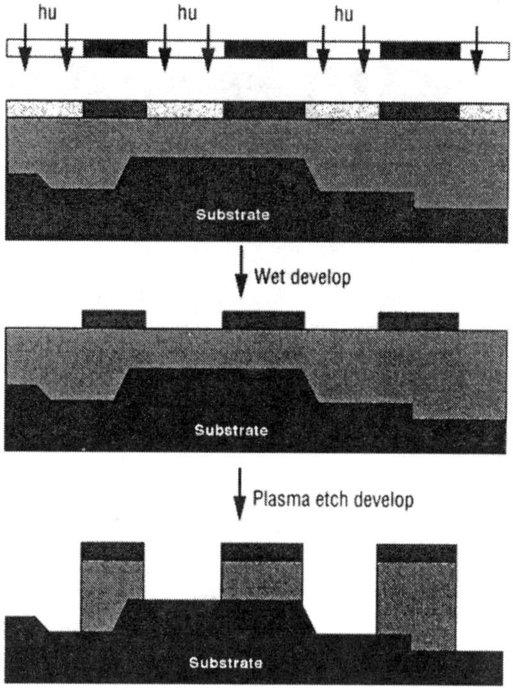

Fig. 15-19 TIS-2000 process flow with a thick planarizing underlayer and a silicon-containing imaging layer.

a SEM of such veils that remain in the trench around the via opening. Filling the trench/via openings with metal in the presence of such veils is problematic, and thus it is important that they be prevented from being allowed to form. To ensure that veils are not created by the trench etch step, it is necessary to control the degree to which the BARC fills the vias.

However, optimizing the BARC-fill of damascene vias is difficult, and often requires considerable process development and materials testing. It is possible to fill vias of different pitch equally by careful selection of the right BARC and coating conditions, but this only works if all the vias have the same size. A new BARC optimization also has to be done for each via diameter.

15.9.1.5 Bi-Layer Resist Process: To overcome the issues of resolution, critical-dimension control over arrays of vias, and the problem of via-plug depth-control that occur when the trenches are patterned in the via-first dual-damascene method, an alternative lithography approach has been proposed, namely a *bilayer, thin surface-imaging scheme* (Fig. 15-19) This method is based on a bilayer resist system developed by Arch Chemicals, called TIS-2000. The technique uses a

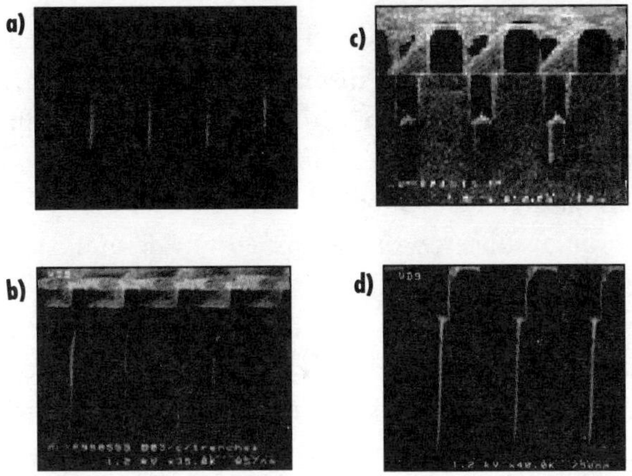

Fig. 15-20 SEMs of cleaved dual-damascene structures patterned with the TIS-2000 process: a) before exposure with vias filled; b) after wet development; c) after partial etch of underlayer and trench; d) after full underlayer etch (top resist still in place).[19]

thin silicon-containing top-imaging layer on top of a highly planarizing fully absorbing underlayer (chemically similar to a novolac resist). The processing sequence is similar to a process with an organic BARC. While this technique was developed for 248-nm lithography, it appears that it can be extended to 193-nm lithography (by using an alternate imaging layer and underlayer).

The thin imaging-layer on a planarized layer allows for very high resolution. The silicon present in the imaging-resist-layer also gives it a very-high etch resistance when the underlayer is dry-etched (developed). The underlayer effectively planarizes the topography of the via-containing dielectric stack (Fig. 15-20a).[19] In this figure (in which the underlayer fills 0.25-μm vias), the vias are completely filled and the top of the underlayer is flat without any topography caused by the vias underneath. The thickness of the underlayer over the substrate in the via-array region is about 350-nm, which is enough to insulate the imaging layer from any optical effects by substrate reflections (i.e., the underlayer also acts as an effective BARC layer). Figure 15-20b shows an array of vias after the *imaging layer* has been exposed and wet developed.

The next step is a dry-develop process done with an SO_2/O_2 plasma. This transfers the pattern created in the imaging-resist-layer to the underlayer. Dry-developing the underlayer to the dielectric surface below defines the trench pattern. At this point the vias are still completely filled with the underlayer material. If we stopped the develop step at this point, and proceeded with the etch step to remove the dielectric material which creates the trench, veils would again arise (in the same way they occur if the via plugs made from BARC material filled the vias too full). Thus, the dry-develop step is extended so via-plugs (now consisting of underlayer material instead of BARC) are recessed. (Since trench open-areas are large compared to via areas, an optical endpoint signal can be used to determine when the recess etching begins, providing good control.)

Figure 15-21 shows the patterned trench-resist and the recessed via-plug (whose level is below the top of the buried-nitride etch-stop layer). When the trench dielectric is now etched, no veils remain as residue within the trenches. Figure 15-22 shows the trench and via after trench-etch and resist-stripping. The trench and via profiles are vertical and veils are not present. Since the via-plug covers the bottom SiN barrier-layer during the trench etch, the trench-etch process does not have to be made selective to the nitride. This also allows a trench etch process with higher etch rates to be used.

Comparison of trench lithography at the edge of large via arrays shows another advantage of the top-surface bilayer-imaging technique. As is shown in Fig. 15-23, while trenches over vias at the edge of dense array vary considerably in width for standard DUV lithography (Fig. 15-23a), those patterned with TIS-2000 show no variation in width.[20] This is because the underlayer of the TIS completely fills the vias. Thus, the resist thickness is constant and there is little (or no) trench-width variation.

15.9.2 Via-First-Damascene Integration Issues: Dielectric Etching

The plasma etching and resist/polymer strip steps are among the most critical for the successful integration of a *low*-k dielectric. The etch process must be able to

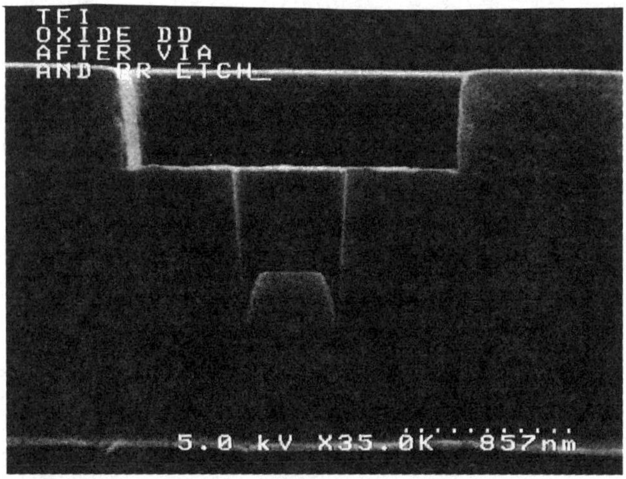

Fig. 15-21 Trench patterning using TIS-2000 bilayer resist. The dry-developed underlayer with recess etch brings the level of the via plug along the sidewalls below the nitride etch stop, which is important for the subsequent trench etch.[18]

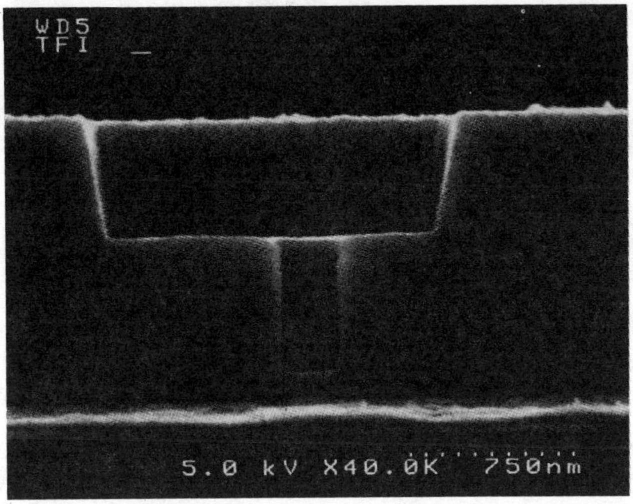

Fig. 15-22 Fully etched and stripped trenches over a via using the TIS-2000 bilayer resist process. The dual-damascene structure is nearly ideal in that there are no veils or faceting of the via due to trench etch.[18]

effectively perform the following tasks: **1**) Remove multi-layers of dielectric; **2**) Maintain a notch-free profile for optimum subsequent metal deposition; **3**) Achieve a high selectivity to resists (which is especially important as resists are getting thinner and 193-nm resists are less resistant to plasma etching than those used for 248-nm lithography); **4**) Retain the low-*k* value during the etch and strip processes so that ultimate device speed is not compromised; **5**) Achieve high selectivity to etch-stop layers, (especially since we want to make them as thin as possible to minimize the stack's effective *k*-value); **6**) Minimize microtrenching at the etch front; and **7**) Minimize aspect-ratio-dependent etch (ARDE) effects.

15.9.2.1 Via Etching for Low-*k* Processes: In the "via-first" dual-damascene process sequence, the vias are patterned and etched before the trenches. The via-

a)

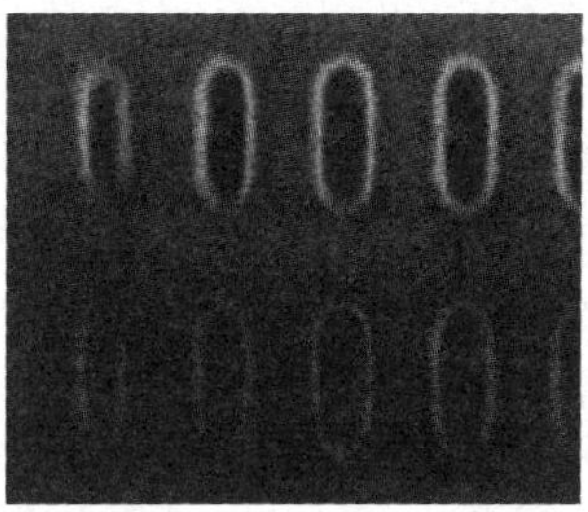

b)

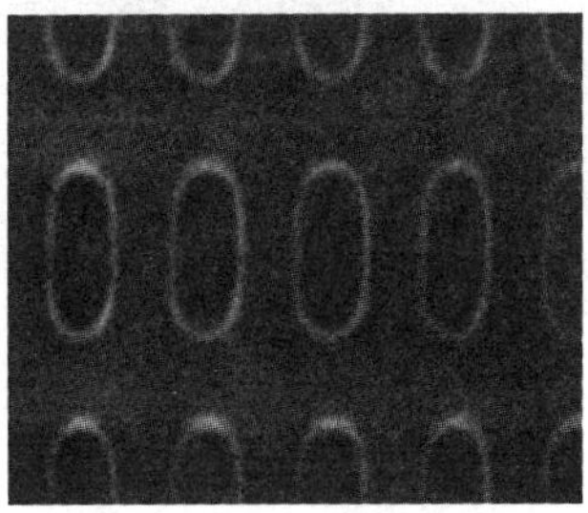

Fig. 15-23 Trenches over vias near the edge of a dense array: a) trench widths vary considerably for standard DUV lithography; b) the same area with TIS-2000 resist.[18]

etch process has some challenges, one of which was described in Sect. 15.9.1 (the resist must be able to withstand the deep via-etch process). To open the via, the following materials must be etched by the via-etch process in a low-*k* dielectric stack: **1**) BARC; **2**) SiO_2-hard-mask; **3**) upper low-*k* dielectric layer; **4**) thin SiN (or SiC) embedded etch-stop layer; and **5**) lower low-*k* dielectric layer. The etch step itself will therefore have to employ a multiple sub-steps (Fig. 15-17a).[18] The first such sub-step should be a non-selective etch, so that the BARC, SiO_2 hard-mask layer, low-*k* dielectric layers, and the embedded-etch-stop layers are all etched at about the same rate. This would ideally produce a via profile that is free of notching and bowing. This first sub-step would be terminated prior to removing all of the second low-*k* dielectric layer. The etching sequence would then switch to the second sub-step. The etch characteristic of this second-sub-step would be that it is highly selective to etching oxide and not nitride. This would allow it to remove the remaining lower low-*k* dielectric material from all of the vias without attacking the SiN etch-barrier layer at the bottom of the vias (i.e., a sufficiently long overetch time would be permitted to ensure that the lower low-*k* material was entirely removed from all of the vias).

While the above process is the desired goal, an even more complex etch process might be needed to produce a via-profile free of notching and bowing. That is, the etching characteristics of various low-*k* dielectrics are different from SiO_2, regardless of whether a spin-on organic-based material (such as SiLK® or FLARE®), or an SiCOH-CVD dielectric (such as Back Diamond,® Coral,® or Flowfill®) is used.

For example, a three-sub-step etch process might be needed for a SiO_2-hard-mask/SiLK stack. That is, fluorocarbon chemistries with high bias power might be used for the first-sub-step to etch the top SiO_2-hard-mask material and the embedded SiN layer (because this process can disrupt the strong Si-O and Si-N bonds), but when the SiLK layers are etched, a second-sub-step using oxygen as the active etchant is used. However, in an HDP tool, the high degree of gas-phase dissociation leads to high concentrations of atomic oxygen, which may produce some bowing of the via profile. Also, hard-mask undercut after via (and trench) etching can occur (Fig. 15-24), which would make it more difficult to achieve void-free metallization.[21]

When the low-*k* dielectric is a SiCOH-CVD film (having a higher porosity than normal CVD-SiO_2 films), the gas mixture for etching these low-*k* materials might be mixtures of CF_4 (the principal source of F atoms) and CH_2F_2 (a polymerizing gas for improving selectivity and providing sidewall passivation). Using such an etch procedure has reportedly allowed 0.3-μm vias to be etched in

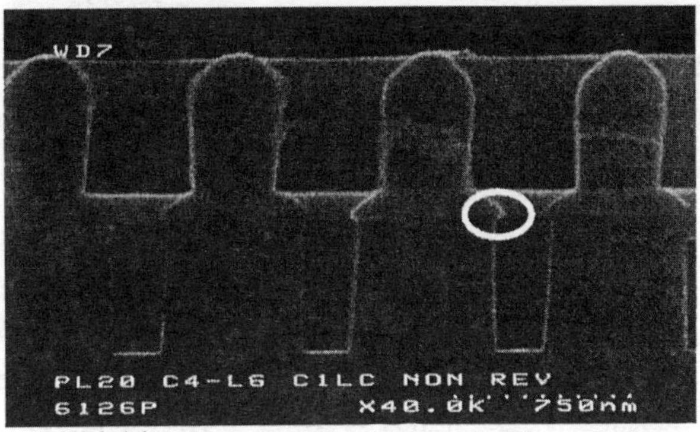

Fig. 15-24 Etched via profiles in SiLK® low-*k* dielectric showing hard-mask undercut.[21]

such stacks with the desired via profiles (Fig. 15-25).[22] The choice of gas mixture depends on the precise application being considered. That is, since these films contain some C and H, if their composition is changed (i.e., the more C, the lower the *k*-value), it will be necessary to re-tune the etch process in order to meet the selectivity and etch requirements.

15.9.2.2 Trench Etching: Following the via-etch-step in via-first, dual-damascene

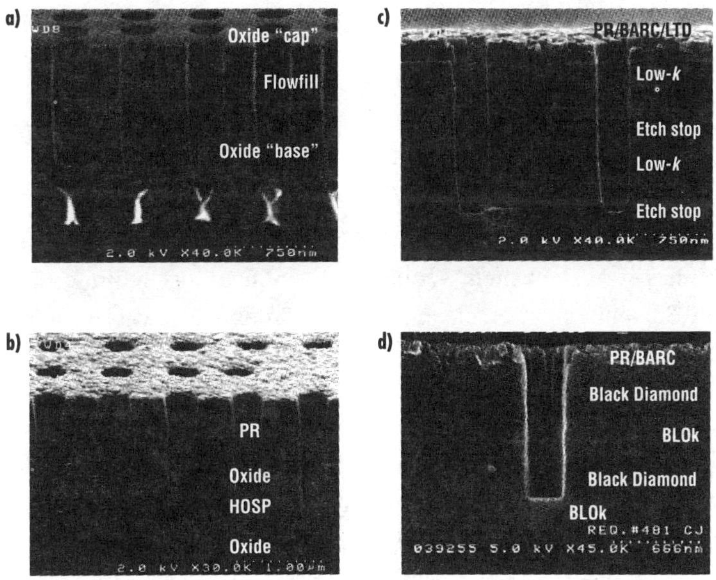

Fig. 15-25 Etching of: a) 0.3-μm vias through Trikon's low-*k* Flowfill ($k = 2.9$); b) 0.4-μm vias through Honeywells HOSP (k ~2.5); c) 0.28-μm vias through a Trikon dual-damascene stack (k ~2.8); and d) 0.29-μm vias through a dual-damascene stack containing Applied Materials Black Diamond film (k ~2.6) and BLOk (k ~4.5) etch stop.[22]

applications, trenches are etched down to the buried etch-stop layer. The issues involving protecting the barrier nitride at the bottom of the via (exposed during the previous via-etch step), from being etched during the trench-etch-step were previously considered in Sect. 15.9.1.

When a 2nd generation low-*k* dielectric is used, the top surface of the stack generally employs an SiO_2 hard-mask and a BARC to improve lithography resolution. Thus, the trench-etch process is also a multi-step procedure, involving: BARC-open; SiO_2-hard-mask etch; and low-*k* etch (down to the embedded etch-stop layer - Fig. 15-17b). Following the trench etch, the resist must be stripped, the BARC removed from the vias, and the barrier layer at the bottom of the vias etched away.

In early via-first, dual-damascene process flows, these steps were performed in several discrete tools. However, in the interests of improving cycle time (less handling and waiting), tool vendors are seeking to provide equipment that will allow an integrated-trench-etch process. That is, BARC opening, SiO_2-hard-mask etch, low-*k* dielectric etch, resist/BARC strip, and SiN barrier-layer open would all be carried out *in situ* in a single tool and a single process run.

15.9.2.3 Etching the SiN Barrier Layer at the Bottom of Vias: After the trench is completed and the second resist layer stripped, the nitride at the bottom of the via must be opened. As noted earlier, this etch process can cause some problems. First, the SiN etch step should be a "soft" etch. That is, the ions that bombard the SiN being etched should not have enough energy to sputter Cu during the inevitable overetch time. If such sputtering causes Cu to end up on the via sidewalls, it could be trapped on the wrong side of the subsequently deposited Cu-barrier-layer. Thus, any such sputtered Cu would have to be removed before the barrier metal was deposited. While a conventional dip in dilute HF is efficient for sidewall Cu contamination cleaning, this inevitably also cause some CD loss.

Another problem arises if the via is misaligned with the edge of the Cu line beneath it. In that case, during any overetch time of the nitride, a narrow slit would be etched in the dielectric alongside the Cu line (see Fig. 15-26).[31] This slit is hard to fill with barrier material, again leading to unwanted Cu diffusion. This issue can be addressed by employing a nitride etch process that is highly selective between nitride and oxide, and between nitride and Cu.

15.9.3 Via-First-Damascene Integration Issues: Resist Stripping

With the introduction of low-*k* and ultra-low-*k* porous dielectrics, existing resist-strip technology cannot be used without at least some modifications. Dry-etching in an oxygen plasma has become the most widely used stripping technology

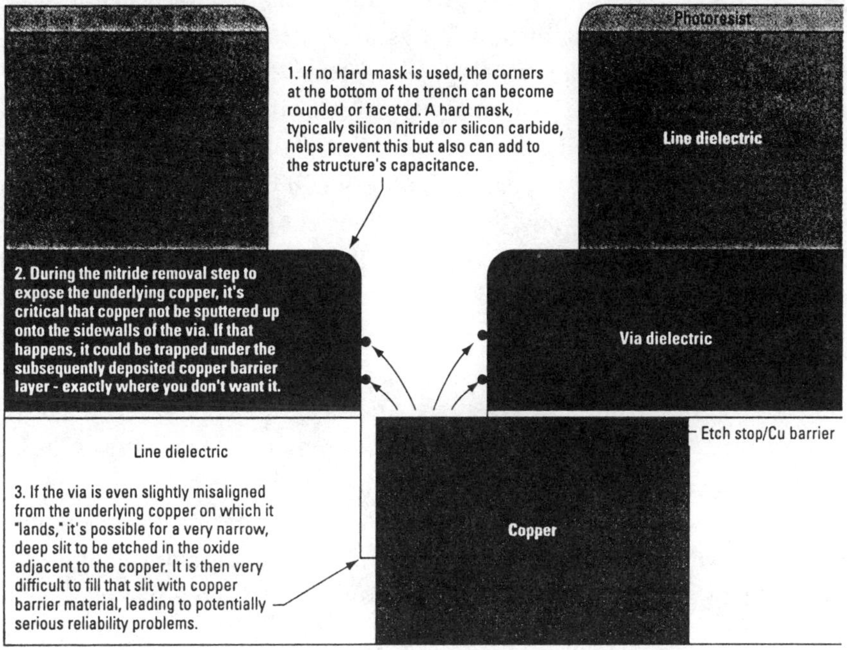

Fig. 15-26 Common problems in the via-first approach include rounding of the corners at the bottom of the trench; sputtering of the Cu onto the via sidewalls during nitride removal; and hard-to-fill slits etched next to the underlying Cu in the case of misalignment.[31]

(especially for back-end-of-line [BEOL] processes). However, many of the low-*k* dielectric materials are also etched by oxygen plasmas, and in some cases their *k* value may be degraded by reactive oxygen atoms as well. Here we discuss the issues facing resist stripping in such applications.

To begin, as noted earlier, SiO_2-hard-masks will typically be used on the top of the low-*k* dielectric stacks. These layers will actually serve multiple functions, including: **a**) as a hard-mask during via- and trench-etching; **b**) as a polish-stop layer for the Cu-CMP process; and **c**) as a dielectric anti-reflective coating (DARC). The use of a hard mask is mandatory if organic low-*k* films are used (since they are attacked by the same chemistries used when stripping the resists after an etch step). Even with a hard mask, however, difficulties arise when etching organic low-*k* materials. That is, if an etch-process is used that is based on an O_2 plasma, very reactive oxygen radicals are present in the plasma. These isotropically attack organic low-*k* dielectrics, causing bowing of the via and/or trench profiles (as shown in Fig. 15-27).[23] To stop such attack of the dielectric layer, sidewall-passivating gases must added to the O_2 etch gas. In one paper, ethylene was added to the O_2 etch gas when an organic low-*k* material was being etched. Undercutting of the hardmask was prevented, and good CD control was maintained, as the hardmask itself was not eroded.[24]

Even if an HSQ or MSQ-based low-*k* material is used instead of an organic film (see Chap. 14), the resist stripping process is still a challenge. That is, while the resist is being stripped, the oxygen radicals will react with the exposed low-*k* material of the trench sidewalls. These reactions convert the Si-CH_3 bonds in the MSQ, and the Si-H bonds in HSQ films, back to Si-O bonds. This causes the *k*-value of the dielectric to rise back toward that of SiO_2 (~4.0). As a result, an alternative resist stripping process is needed. That is, an alternative anisotropic ashing process must be available.

One method of making low-*k* materials more robust against attack by oxygen radicals is *plasma treatment of the exposed low-k material surfaces.* An example of such a plasma treatment was reported to have carried out on an organic low-*k* material. After exposing this dielectric to an NH_3 plasma for 10 minutes, a nitrided layer is formed on its surface. The nitrided low-*k* layer exhibited lower leakage current after an ashing plasma exposure than did the untreated material. The film also acted as a better barrier against Cu diffusion.[25] In another paper, three such NH_3-plasma treatments in were performed on a dual-damascene via-first etched dielectric stack consisting of HSQ/PECVD-SiO_2/HSQ. The first treatment was carried out prior to depositing the embedded etch-stop PECVD-SiO_2 layer on the first HSQ segment of the stack. This improved adhesion of the embedded etch-stop layer to the HSQ. The second treatment was carried out after the second HSQ layer was deposited and the trench etch was completed. This purpose of this treatment was to strengthen the exposed HSQ sidewall surfaces against erosion during the resist stripping step.[26]

Carbon-doped low-*k* materials such as Coral® and Black Diamond® are ozone

Fig. 15-27 Cross-section SEM picture of single damascene structure made in 3M's low-*k* dielectric after strip and etch in a low-pressure, low-temperature strip.[23]

resistant. Resist stripping in processes that use these dielectric materials might potentially be carried out with a DI ozone strip process.

15.9.4 Via-First-Damascene Integration Issues: Cleaning

Cleaning issues are critical to producing dual-damascene interconnects with high yield and reliability. The cleaning challenges are especially severe for dual-damascene structures compared to front-end-of-line (FEOL) processes, because more aggressive chemicals can be used on silicon and thermal oxides. That is, the strong acids and bases in the FEOL flow effectively remove contamination and resist residues after critical FEOL steps. However, metals in back-end-of-line (BEOL) cannot be subjected to such chemicals since they are susceptible to chemical corrosion. Fortunately, the CMP process used in dual-damascene interconnects tends to remove particles on the dielectric surface during polish, another inherent advantage over the subtractive interconnect approach. However, other residual films must still be removed, as will be discussed here.

15.9.4.1 Post-Dielectric-Etch-and-Resist-Strip Residue Removal: After photoresist is stripped with a dry-stripping step (ashing), residue from the etch and the ashing step often remains on the via and trench sidewalls, and must be removed. Typically this is done with wet-chemical residue cleaners, in a separate cleaning step after the dry-strip step. One goal (as a way to improve cycle time), is to integrate these two steps into a single tool.

As long as SiO_2 was the material used in the dielectric stack, the post-etch residue removal step was easier, and proprietary organic solvents sold by such vendors as EKC Technologies, ASCI, J.T. Baker, and Olin could effectively remove the residues (Fig 15-28). However, with the introduction of low-*k* dielec-

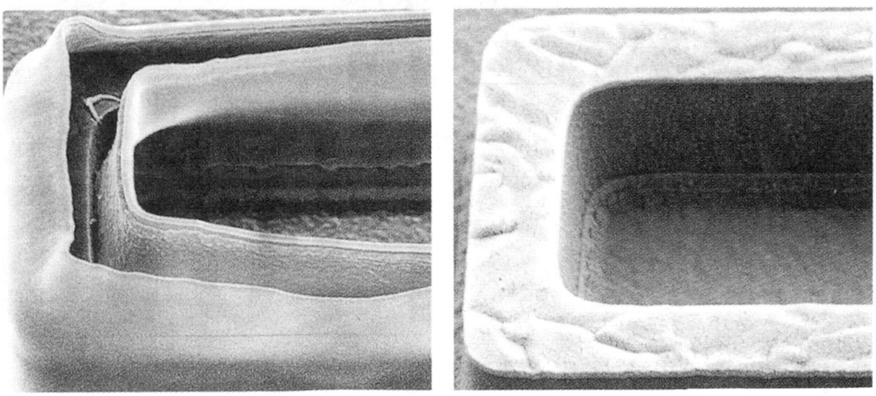

Fig. 15-28 Aqueous residue removers specifically target organometallic and inorganic residues left after ashing. Courtesy of J.T. Baker.

trics, the situation has changed. New cleaning chemicals must be used, and suppliers are working on chemistries for residue cleaning in Cu/low-k applications. However, a major issue in pursuing this goal is that it has not yet been decided exactly which low-k materials will be used for this application, and what are all of the material properties of the candidate materials. In addition, it appears that there will be a continuous transition from one material to another over the next few years, as ever lower-k materials are integrated into damascene structures. However, some general trends can be foretold.[27]

First, it must be ensured that residue-removal chemistries are compatible with specific the low-k dielectric and the metallurgy on the wafer. This is complicated by the lowering of the porosity of the new low-k materials and the thinning of the dielectric layers. Furthermore, since each low-k choice requires different plasma etching processes and chemistries to etch vias and trenches into the low-k materials, the nature of the residues can be different. Thus, no one product will likely to be able to meet all Cu/low-k applications. Nevertheless, it appears that some form of wet cleaning after etch and ashing will still be needed until below 0.13 μm. Fundamental barriers to wet cleaning approaches do not appear to be reached until well below 0.1 μm. At some point, however, there are concerns about liquids being able to go into and out of small, high-aspect-ratio features. At that time a dry-approach (or the use of supercritical fluids – see Sect. 15.9.4.5) may need to be used.

15.9.4.2 Pre-Metal-Deposition Clean: The pre-metal-deposition clean is carried out just before the barrier layer metal and Cu seed layers are deposited. The Cu film at the bottom of the vias must be free of organic residues and any native Cu oxide in order to ensure that low-resistance contact between the deposited barrier metal and the underlying Cu is fabricated with high yield. This clean step normally involves sputter etching of the wafer in the same tool in which the barrier layer metal and the seed Cu layer are sputter-deposited. Argon ion gas is used, and a dual-frequency rf sputtering module with a high-density plasma (HDP) source is required (so that a high degree of directionality can be obtained with the sputter-clean process).

During this pre-metal-deposition cleaning process, copper ejected from the lower layer can be re-deposited onto the via sidewalls, increasing the risk of line-to-line leakage. By 1999, however, the magnitude of this problem had not yet been quantified. It is hoped that the quantity of copper re-deposited will not be sufficient to degrade device performance. Nevertheless, if the problem becomes worse than expected, alternate via-clean methods will need to be developed that selectively remove the etch residue and copper oxides without removing any copper.

Another special problem occurs when trying to integrate Cu with a fluorine-containing dielectric material, such as FSG. That is, the F bonds in many FSG films are reactive (particularly with water), and they can be broken during other process steps to produce free fluorine. At the interface between the FSG film and the barrier metal, Ta can react with such free fluorine to form volatile by-products at elevated temperatures. The resulting weakened interface can fail when subjected to the stress of an overlayer (such as that produced by the SiN diffusion barrier layer). A number of solutions to this problem have been proposed, including: use of a TaN barrier material (which is less reactive than Ta); using a more stable FSG film; or using a hydrogen-plasma treatment before depositing the Ta to remove mobile fluorine from the surface.

15.9.4.3 Post-CMP Clean: The post-CMP clean step is a critical one for Cu CMP. The bevel and backside of the wafer will most likely be exposed copper-containing solutions during the CMP step. Any copper remaining on the bevel or backside of the wafer risks contaminating other tools within the fab. As in the electroplating process, simple cleaning methods are available to remove residual copper, but the risks associated with copper cross-contamination demand careful monitoring of the post-CMP clean (see also Chap. 8).[28]

15.9.4.4 Clean of Top-Surface of Cu Prior to Diffusion-Barrier-SiN-Deposition Process: After the CMP step of the metal, and post-CMP clean, a SiN layer is deposited on top of the damascene structure. As noted earlier, this SiN film serves as barrier to keep the Cu from diffusing into the dielectric layer above it, and also as an etch-stop layer for the via etching step. However, the SiN may fail to adhere to the Cu surface if that surface is not properly cleaned. That is, the SiN may not adhere to the Cu interface if even a small amount of residue is left on the Cu surface following CMP. It turns out that there are two classes of such contaminants that must be removed from the Cu-surface: 1) residual organics that are present in the Cu-CMP slurry; and 2) native oxide that has formed on the Cu surface.

The residual organics can be removed by a pre-treatment with an *in situ* exposure to an aggressive hydrogen plasma just before depositing the SiN layer. Another technique is to remove such organics (corrosion inhibitors, usually BTA) from the CMP process. In addition, Cu forms a native oxide that can interfere with SiN adhesion. That is, if this oxide is not completely removed from the Cu surface, the Cu/SiN interface will be weak. Thus, a cleaning step (consisting of a copper oxide reduction step using the same hydrogen plasma used to remove the residual organics) is recommended.[29]

15.9.4.5 Supercritical Fluids as Wafer Cleaners:* The temperature at which the vapor pressure above a pure liquid reaches one atmosphere is known as the normal boiling point – for water it is 100°C. In an open container, liquid water's temperature cannot be raised above 100°C because it would cause the vapor pressure to rise above one atmosphere, exceeding ambient pressure conditions. However, in a sealed container, liquid water can be heated to higher temperatures. As the water is heated, its density decreases (through thermal expansion), and simultaneously, the vapor density in the container increases.

This heating can be continued until the liquid's density has been so reduced – and that of the vapor so increased – that both densities are equal. This point is known as the *critical temperature*, and the pressure of the vapor at that point is called its *critical pressure.* Since the temperature and density of the water inside the sealed container are equal everywhere, thermodynamics dictates that the critical pressure inside the container also be equal everywhere. A fluid brought to conditions above its critical temperature and pressure (the *critical point*) is called a *super-critical fluid.*

Carbon dioxide (CO_2) is a material that is particularly attractive as a practically useful super-critical fluid. This is because it is non-toxic, non-flammable, and inexpensive. In addition, its critical conditions are easily achievable with existing process equipment. This unique set of characteristics has led workers to examine the possibility of using supercritical CO_2 instead of other solvents in such semiconductor processes as cleaning and photoresist stripping.

The attractiveness of using supercritical CO_2 as a replacement for traditional solvents in cleaning and stripping applications stems from the combination of gas- and liquid-like properties that it possesses. Table 15-1 compares the physico-chemical properties of a typical organic fluid in it's liquid, supercritical, and gaseous states. It can be seen that the supercritical fluid state possesses *gas-*

Table 15-1 PHYSICO-CHEMICAL PROPERTIES of TYPICAL ORGANIC FLUIDS

	Diffusivity (cm^2/sec)	Viscosity (mPa-sec)	Density (kg/m^3)
LIQUID	10^{-5}	1	1000
SUPERCRITICAL FLUID	10^{-3}	10^{-2}	700
GAS	10^{-1}	10^{-2}	1

* Written by James Rubin, Los Alamos National Laboratories. The material originally appeared in Semiconductor International. Reprinted with permission.

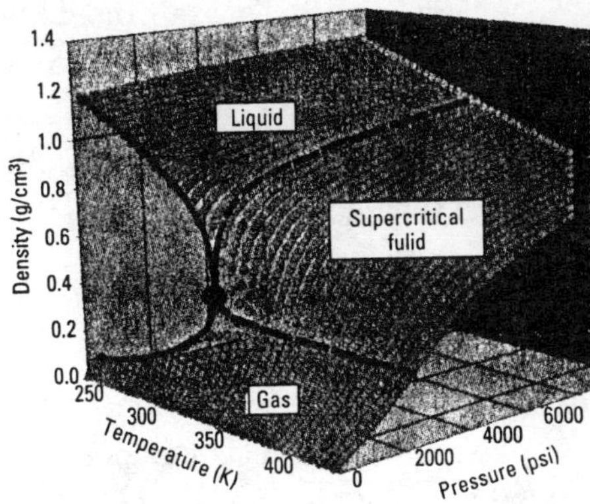

Fig. 15-29 Pressure-temperature-density surface for pure CO_2. The critical point is the small circle at the intersection of the contour lines.

like values of viscosity and diffusivity, and *liquid-like* density.

Figure 15-29 shows the density-pressure-temperature surface for pure CO_2. The critical point for pure CO_2 is 31°C (304K) and 1072 psi. To a first approximation, a fluid's solvent power is proportional to its density. Thus, high liquid-like densities achievable in supercritical fluids allow for liquid-equivalent solubilities. Figure 15-29 also indicates that small changes in pressure or temperature near the critical point result in large density changes. This implies that the density – and hence the solvent power – of a supercritical fluid is "tunable."

Figure 15-30 is a schematic pressure-temperature phase diagram for CO_2, incorporating the process flow for a closed-loop supercritical fluid treatment system. No uncontrolled waste streams can exit such a closed-loop system. All materials solubilized in the supercritical fluid are retained in the separation vessel

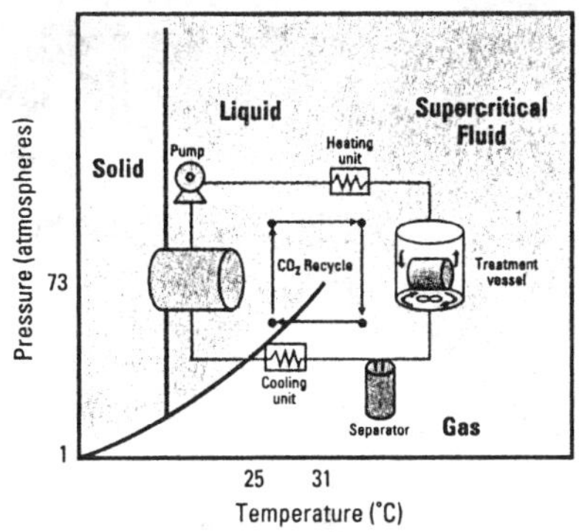

Fig. 15-30 Pressure-temperature phase diagram of pure CO_2 with a superimposed flow diagram for a closed-loop supercritical fluid treatment process.

for analysis, recycle, treatment, and/or disposal. In addition, these materials are concentrated in the separator, greatly reducing waste volume.

Several companies are working on commercializing tools that employ supercritical fluids in cleaning processes, including Supercritical Systems (a TEL-America company), and GT Equipment Technologies.

15.10 VIA-FIRST DUAL-DAMASCENE DIELECTRIC STACK WITH NO EMBEDDED ETCH-STOP LAYER

Eventually, the increase in the *k*-value of the damascene dielectric stack due to the embedded etch-stop layer (whether it is composed of SiN or SiC material) becomes too great. At that point, it may be necessary to use a damascene dielectric stack without an embedded etch-stop layer. Some predict that this approach will be needed for the 0.1-μm technology node.

The elimination of the etch-stop layer, however, produces some significant process integration issues. First, since there is no etch-stop layer, etch systems will have to have extremely good control of uniformity to ensure all trenches are of the same depth across the wafer, and from wafer-to-wafer. This could be difficult as there is no endpoint indicator, and the depth of the trench can only determined by the time of the etch step. Making the problem even more difficult is that low-*k* materials may not lend themselves to that kind of control. Second, if a hard etch-stop layer is not present, the top corner of the via can become rounded during the trench etch process (due to faceting, see Vol. 1, Chap. 11 - Physical Vapor Deposition, and also Fig. 15-26). A second issue involves the creation of microtrenches at the corners of the trenches themselves. This phenomenon is likely to be caused by enhanced ion-bombardment-flux at the corners of the trench (due to ions reflecting at shallow angles off of the sidewalls). Such microtrenches would provide locations that are poorly covered by the Cu diffusion barrier films. This could give rise to Cu diffusion through the barrier layer at such points where the barrier layer is thinned (or is even discontinuous).

15.11 PROCESS INTEGRATION ISSUES OF DUAL-DAMASCENE DIELECTRIC STACK WITH ULTRALOW-*k* DIELECTRICS

If damascene dielectric stacks are fabricated with ultra-low-*k* materials, additional process integration problems are encountered. That is, special techniques must be developed to protect these porous, less-robust materials during ashing and CMP. Two novel damascene processing steps have been proposed to handle such fragile films. The first is termed *half-via etch.* In this approach, the via is again etched first, but only half of the via is opened with this process, down to the depth of the trench. Then the rest of the via is opened during the trench-etch step. This provides a simpler etch and alignment process and makes the lithography problem of resolving small feature sizes easier.

Another change may involve returning to the trench-first etching sequence. With a porous material, the via cannot be etched first because, after the via is etched and you begin etching the trench, problems will arise because the resist that flooded the vias will begin to be absorbed by the porous low-*k* dielectric.

REFERENCES

1. P. Singer, "Making the Move to Dual Damascene Processing," *Semiconductor International,* August 1997, p. 79.

2. B. Luther *et al.,* "Planar Copper-Polyimide Back End of the Line Interconnections for ULSI Devices," *Proceedings of the 1993 VMIC Conf.,* p. 15.

3. R. Iggulden *et al.,* "Dual-Damascene Aluminum for 1-Gbit DRAMs," *Solid State Tech.*, November 1998, p. 37.

4. P. Singer, "IBM to Use Dow's SiLK in Low-*k* Process," *Semiconductor International,* May 2000, p. 44.

5. H. Aoki *et al.,* "Robust 130-nm-Node Cu Dual Damascene Technology with Low-*k* Barrier SiCN," *Tech. Dig. IEDM* 2001, p. 76.

6. "New Barrier/Etch-Stop for Copper Damascene," *Semiconductor International,* July 1999, p. 28.

7. S.-M. Cho *et al.,* "Low Dielectric Constant Material Deposited by HDPCVD for Barrier and Etch Stop Applications in Cu Damascene Structure," *Abs. Mtg. Electrochem. Soc.,* Spring 2000, Abs. 231.

8. X.W. Lin and D. Pramanik, "Future Interconnect Technologies and Copper Metallization," *Solid State Technology,* October 1998, p. 63.

9. B. Zhao *et al.,* "A Cu/Low-*k* Dual Damascene Interconnect for High Performance and Low Cost Integrated Circuits," *Dig. of Tech. Papers 1998 Symp. on VLSI Technology,* p. 23.

10. Y. Morand *et al.*, "Copper Integration in Self Aligned Dual Damascene Architecture," *Dig. of Tech. Papers 1997 Symp. on VLSI Technology,* p. 31.

11. S. Venkatesan *et al.*, "A High Performance 1.8V, 0.2 μm CMOS Technology with Copper Metallization," *Tech. Dig. IEDM,* 1998, p. 769.

12. D. Edelstein *et al.,* "Full Copper Wiring in a Sub-0.25 μm CMOS ULSI Technology," *Tech. Dig. IEDM,* 1998, p. 773.

13. M. Igarashi *et al.,* "The Best Combination of Aluminum and Copper Interconnects for a High-Performance 0.18-μm CMOS Logic Device," *Tech. Dig. IEDM,* 1998, p. 829.

14. P. Singer, "Tantalum, Copper, and Damascene: The Future of Interconnects," *Semiconductor International*, June 1998, p. 90.

Ramon CA, p. 183, 1999.

16. L. Peters, "Leaders Describe "Surmountable" Copper and Low-*k* Hurdles," *Semiconductor International,* September 2000, p. 25.

17. T. Mountsier *et al.,* to be published.

18. B. Gadsen *et al.,* "Top Surface Imaging Improves Copper Process Resolution," *Solid-State Technology,* March 2001, p. 77.

19. M. Neisser *et al.,* "Applying a Thin Imaging Resist System to Substrates with Topography," *Solid State Technology,* August 2000, p. 127.

20. M. Maenhoudt *et al.,* "248 nm and 193 nm Lithography for Damascene Patterning," *Solid State Technology,* April 2001, p. S15.

21. M. Moussavi "Advanced Interconnect Schemes Toward 0.1 μm," *Tech. Dig. IEDM* 1999, p. 611.

22. D.J. Thomas *et al.,* "Meeting the ITRS Roadmap Challenges with Low-*k* Dielectric Etching," *Solid State Technology,* March 2001 p. 108.

23. S. Vanhaelemeersch *et al.,* "Dry Etch & Clean Aspects for Advanced Integration of Low-*k* Dielectrics," *Mtg. Electrochem. Soc.,* Spring 2000, Abs. 224.

24. G. Rajagopalan *et al.,* "Development of Damascene Etch Processes for Organic Dielectric Materials," *Abs. Mtg. Electrochem. Soc.,* Fall 1999, Abs. 702.

25. I.-C. Deng *et al.,* "Using NH_3 Plasma Treated Spin-On Low-*k* Film as Barrier Metal-Free Dielectric to Suppress Copper Diffusion and Improve Its Ashing Resistance," *Abs. Mtg. Electrochem. Soc.,* Spring, 2000, Abs. 225.

26. T. Fukuda, *et al.,* "0.5-μm Pitch Copper Dual-Damascene Metallization Using Organic SOG (k = 2.9) for 0.18 μm CMOS Applications," *Tech. Dig. IEDM* 1999, p. 619.

27. A. Braun, "Photostrip Faces 300 mm, Copper and Low-*k* Convergence," *Semiconductor International,* September 2000, p. 88.

28. D. Hymes *et al.,* "The Challenges of the Copper CMP Clean," *Semiconductor International,* June 1998, p. 117.

29. V. Shannon, "Techniques to Improve Cu/Low-*k* Integration," *Solid State Technology,* September 2001, p. S22.

30. J. Rubin, "Supercritical Fluids as Wafer Cleaners," *Semiconductor International,* September 2000, p. 84.

31. P. Singer, "Dual-Damascene Challenges Dielectric Etch," *Semiconductor International,* August 1999, p. 69.

32. S. Russell *et al.,* "Integration Challenges of Ultra-Low-*k* Dielectrics," *Adv. Metallization and Integrated Systems for ULSI Applications,* 1997, p. 289.

33. L. Peters, "Low-*k* dielectrics: Will Spin-On or CVD Prevail?" *Semiconductor International,* June 2000, p. 108.

34. A. Braun, "CMP Battles Low-*k* Fundamental Barriers," *Semiconductor International,* October 2000, p. 66.

Chapter 16

COPPER INTERCONNECT PROCESS TECHNOLOGY

As described previously in Chap. 12, interconnects in deep-submicron ICs comprise a multilevel structure of fine wiring located on the top of the transistor circuitry of logic and memory chips. To avoid significant degradation of circuit speed, on-chip interconnections should permit rapid signal transmission among the various parts of the circuitry. Ever since the development of the integrated-circuit in the early 1960's, the materials most widely used for the wiring structures have been aluminum as the conductor (or eventually an aluminum-copper alloy for better reliability), and silicon dioxide as the insulator. In 1997 the transition to copper as the conductor was initiated with the announcement by IBM that they had developed a technology for forming copper interconnect lines and vias for IC applications. This was one of the most important changes in materials that the semiconductor industry has experienced since its inception. Since that breakthrough, AMD reportedly applied Cu to its microprocessor circuits at the 0.18-μm technology node, enabling it to introduce the first GHz-rated microprocessor in March 2000, ahead of competitors. In addition, circuit designers have also started to explore innovative layouts to take full advantage of the multiple benefits of Cu (as opposed to simply replacing Al with Cu in an existing design). Thus, it appears that the Age of Copper is dawning!

16.1 WHY COPPER FOR DEEP-SUBMICRON IC INTERCONNECTS?

As noted above, copper is now poised to take over as the main on-chip conductor for all types of ICs.[1,2] Early adopters of Cu technology put copper in production at the 0.25-μm node. Recent market-research studies indicate that for the 0.18-μm generation about 5% of the interconnect structures will be implemented with Cu. It also appears that most Cu in production (or pilot-line production) in 2001 was at the 0.18-μm and 0.13-μm technology nodes. For some manufacturers, 0.13-μm is a second-generation of Cu, for others 0.13-μm is the entry point for placing Cu into production (Fig. 16-1). It is predicted that Cu will represent about 60% of the interconnects for the 0.1-μm generation. The reasons why copper is displacing Al as the dominant conductive material for deep-submicron IC interconnects are explained next.

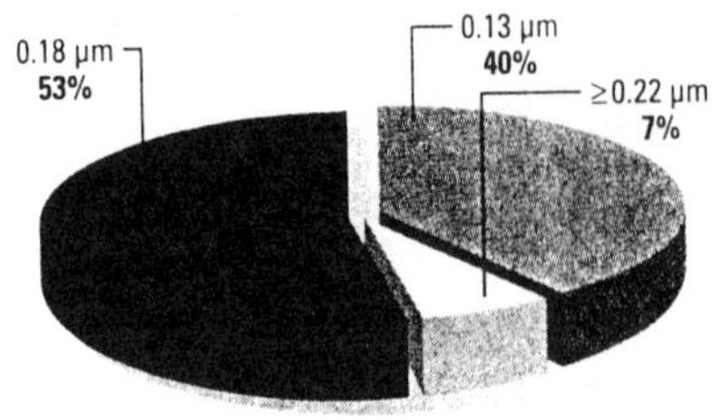

Fig. 16-1 Market analysis of device manufacturers using pilot production in 2001.[39]

16.1.1 The Advantages of Copper Interconnects

There are several critical advantages that copper interconnects provide over the traditional aluminum interconnects that have been used ever since ICs were invented in 1961:

1. Copper has a resistivity that is approximately 40% lower than Al (ρ_{Cu} = 1.7-$\mu\Omega$-cm versus ρ_{Al} = 2.7-$\mu\Omega$-cm), and 50% lower than the Al:Cu-alloys used in IC interconnects ($\rho_{Al:Cu(0.5\text{-at}\%)}$ = 2.9-3.3-$\mu\Omega$-cm). By referring to Eq. 12-7 for the interconnect *RC*-time-delay, it can be seen that *RC* scales directly with ρ. Specifically, if copper can be used instead of aluminum-alloys, the resistance of the wiring structures can also be proportionately reduced. That is, a reduction in the wiring resistance (and thus also the *RC*-delay) of as much as 45% can be obtained when copper replaces Al as the core-interconnect-metallurgy material.

   It is important to note that a more complete analysis of the wire resistance must take into account not only the core-material of the interconnect (Al:Cu or Cu), but also any other (and usually higher-resistivity) materials used to surround the core-conductor. A comparison between the effective resistivity ρ_{eff} of Al:Cu wires and Cu wires is illustrated in Fig. 16-2. That is, the Al:Cu lines in Fig. 16-2 are fabricated by etching an Al:Cu film that has both bottom and top Ti layers. During sintering the Ti-layers react with the Al to form $TiAl_3$ (whose resistivity is ~35-$\mu\Omega$-cm, and the thickness of the resulting $TiAl_3$-layers is about 3.5 times the original Ti-layer thickness). The Cu wires are formed by a damascene process (see Chap. 15). This requires that the bottom and sides of the Cu wires be clad with a thin layer of a diffusion-barrier material. While this barrier-layer film also has high resistivity, it does not react with the Cu (unlike Ti, which reacts with Al), and this material also serves as an adhesion-layer. In any case, it is clear from Fig. 16-2 that copper offers a very significant reduction in the effective resistivity for both present and future technologies compared to aluminum.

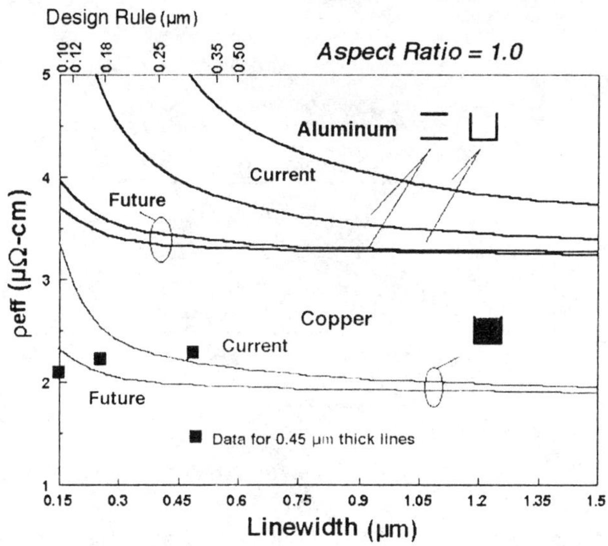

Fig. 16-2 Effective resistivity for aluminum and copper interconnects as functions of line-width and design-rule generation. Computational results (*solid lines*) have been obtained for various combinations of conductor material and barrier location shown in the right side of the figure. Experimental points correspond to published results from real structures.[46]

This lower effective-resistivity of Cu-wiring can be exploited in one of two ways to reduce the *RC*-time-constant of Cu interconnect-systems:

i) The total resistance R_{int} of an interconnect line can be reduced (by retaining the same metal thickness [and cross-sectional area]), or;

ii) The side-by-side capacitance C_{LL} of interconnect lines can be reduced by decreasing the metal-film thickness. This allows R_{int} to remain the same as it was for a thicker Al line, but it reduces C_{LL}. Meanwhile, the same current is carried in a smaller-cross-sectional feature, without fearing EM-failure (see following paragraphs).

The primary clock-speed benefit derived from Cu is gained by using it in the upper-interconnect-levels of a multilevel-metal structure (where the conductor lengths can be of the same order as the chip size). As was shown in Chap. 12, Fig. 12-1, by utilizing Cu together with a low-*k* dielectric to replace Al and SiO_2, the overall delay (and power consumption) in technologies below 0.25-μm is dramatically reduced.[3]

As also discussed in Chap. 12, the interconnect RC-delay is predicted to become larger than the *RC*-delay of MOSFET transistors for technology nodes below about 0.5-μm. As an example, Fig. 16-3 shows that the *RC*-

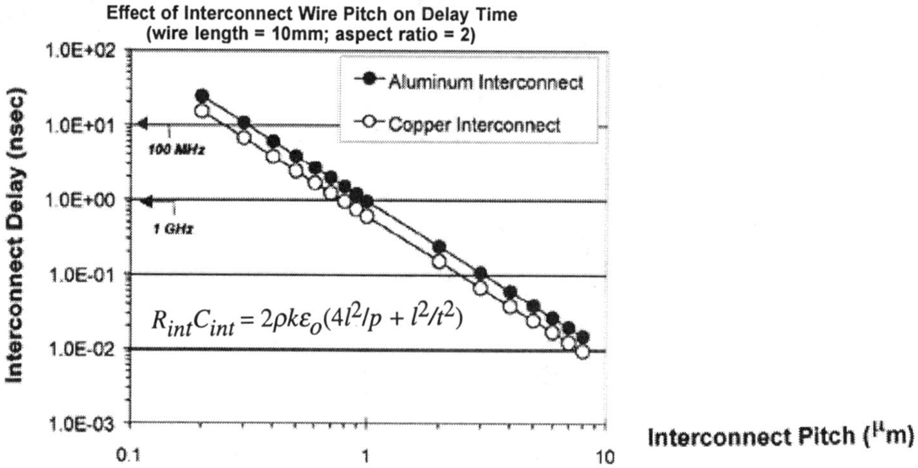

Fig. 16-3 The interconnect delay of both Cu and Al wiring has been calculated using Eq. 16-4 for *RC*. The delay-time in nanoseconds (1 nsec = 10^{-9} sec) is plotted versus the pitch of the wiring in microns (1 μm = 10^{-4} cm) for a line with an aspect-ratio = 2 and length l = 10-mm. The dielectric constant of the ILD has been set to k = 4.0, which is approximately equal to that of CVD deposited silicon-oxide. The graph shows that as the wire-pitch is decreased below 1-μm, the delay-time can become greater than the intrinsic switching-time of the transistors. The speed limit is increased by using Cu over Al.

delay of a dense array of 10-mm-long Al-alloy interconnect-lines with an aspect-ratio of 2 is longer than 1-nsec for pitches smaller than 0.8-μm. Since 1-nsec is equivalent to a frequency of 1-GHz, this implies that signal delays in IC wiring will limit circuit speed to about 1-GHz, (even though individual transistors may have faster intrinsic clock-rates). The interconnect thus represents a real limiter to high-performance ULSI chips – analogous to owning a sports car but only being allowed to drive it on a highway at rush hour. Switching to Cu is one way of improving this situation.

As was done in Chap. 12, Sect. 12-6, equations can be derived to calculate the interconnect delay-time $R_{int}C_{int}$. That derivation is repeated here, but for a simplified special case that exemplifies the impact of switching to copper. The following symbols apply to those aspects of the interconnect system depicted by the test-structure shown in Fig. 12-19.

That is, the metal-line *width* is w, the metal-line *thickness* is t, the *spacing* between the lines is s, the metal-line *pitch* is p ($p = w + s$), and the *thickness of the ILD between the top of the metal lines and the grounded planes* is d. Also the *length* of the metal lines is l, and the *resistivity* of the metal is ρ. Some equations can now be derived that allow the contributions made to R_{int} and C_{int} from: **1**) the materials properties; and **2**) the interconnect architecture, to be expressed. Using this approach, the resistance of a metal-line is calculated from:

$$R_{int} = \rho l/wt \tag{16.1}$$

Next, the capacitance between the metal lines and the grounded planes C_{LG} (separated by an ILD whose relative constant is k), is calculated from:

$$C_{LG} = k\varepsilon_o wl/d \tag{16.2}$$

and the intralevel capacitance C_{LL} (assuming the same ILD) is found from:

$$C_{LL} = k\varepsilon_o tl/s \tag{16.3}$$

where ε_o is the permittivity of free space. For the $R_{int}C_{int}$ time-delay-constant, it follows that:

$$R_{int}C_{int} = 2\,R_{int}(C_{LG} + C_{LL}) = 2\rho(l/wt)k\varepsilon_o(wl/d + tl/s) \tag{16.4}$$

To simplify the discussion, we assume that the pitch is equal to twice the line width ($p = 2w$), which also implies that $s = w$, and the metal-thickness is equal to the insulator thickness ($t = d$). In this case, Eq. 16.4 becomes:

$$R_{int}C_{int} = 2\rho k\varepsilon_o(4l^2/p^2 + l^2/t^2) \tag{16.5}$$

Equation 16-5 is the relationship that is used here to calculate the value of $R_{int}C_{int}$. As also pointed out in Chap. 12, the dependence on ρ, k and l in Eq. 16-5 shows that $R_{int}C_{int}$ can be reduced by choosing metals with low-resistivity, insulators with low-k-values, and/or making line-lengths as short as possible.

Using Eq. 16-5, it is also possible to calculate the dependence of $R_{int}C_{int}$ on metal-pitch and metal-thickness. The results of such a calculation is illustrated in Fig. 16-4, which shows that for a given metal-

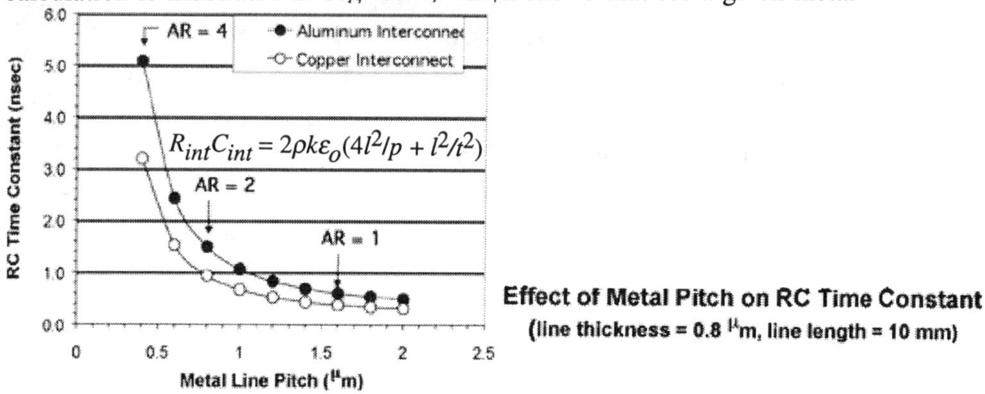

Fig. 16-4 The effect of Cu and Al metal-line-pitch on RC time-constant for a line thickness of t = 0.8-mm, a line-length of l = 10 mm for the case of both Cu and Al. Eq. 16-5 is shown on the graph and was used to calculate $R_{int}C_{int}$.

pitch and line-width that $R_{int}C_{int}$ can be reduced by increasing the aspect–ratio of the line (which of course reduces R_{int} due to the cross-sectional-area increase). However, for aspect-ratios greater than 2, this reduction in the $R_{int}C_{int}$-product is offset by a corresponding increase in the intra-line capacitance C_{LL}. This fact (and the practical difficulty of fabricating metal-lines with very-high-aspect-ratios), is consistent with industry roadmaps predicting metal-line aspect-ratios of less than about 2.7 through the 70-nm node.

After IBM debuted Cu interconnects in 1997, several other companies in 1998 announced the development of Cu-multilevel interconnect systems that they would implement in commercially-available ICs.[4,5] The resistivity of composite copper interconnects in these reports (with the diffusion-barrier in place) did not exceed 2.2-$\mu\Omega$-cm, a 40% improvement over that of comparable Ti/Al:Cu lines.

2. Copper films exhibit much better electromigration (EM) characteristics and stress-void resistance than do Al films. Electromigration is an atomic-scale phenomenon in which the electrons that constitute electric-current collide with the metal-atoms of an interconnect line and push them in the direction of the current-flow. If this "*electron wind*" is strong enough, a significant migration of the metal atoms will occur. Over time this transport of material can lead to the electrical failure of the interconnect in one of two ways: **1**) by a break in the line (causing an open-circuit) due to local depletion of material; or **2**) by the formation of a short-circuit to an adjacent line due to local accumulation of material. The strength of the electron wind increases as the current-density is increased, and the upper-limit of current-density to prevent electromigration in Al:Cu interconnects is 5×10^5 A/cm^2 (for a 20-year lifetime). This limit puts restrictions on circuit-design and chip-performance. (For example, a current of 5-mA in an interconnect-line of cross-section 0.5-μm x 1-μm (i.e., 0.5-μm^2 = 5×10^{-9} cm^2), represents a current-density 1×10^6 A/cm^2, which is twice as great as the current-density-limit for Al:Cu-lines. Scaling of interconnect wires to increase functional-density and chip-performance inevitably leads to increased current-densities (see also Chap. 12), making Al:Cu interconnects less attractive for future technology structures.

Since Cu-atoms are more massive and more tightly bound together than Al atoms, interconnects made of Cu are generally more resistant to electromigration than those made of Al or Al:Cu (i.e., they exhibit 10-100x

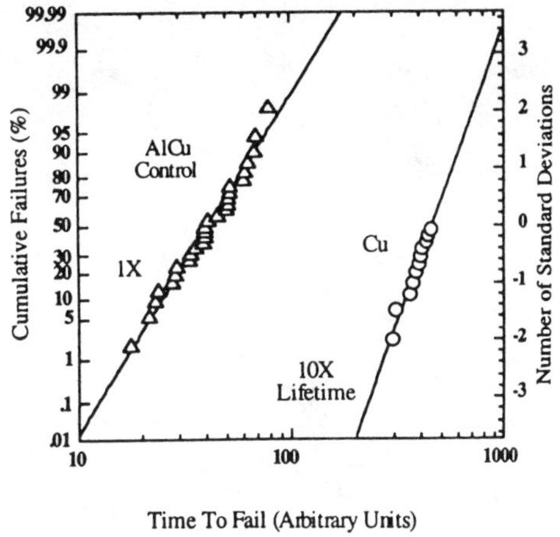

Fig. 16-5 Electromigration performance improvement using copper metallization.[5]

more EM-resistance). About an order of magnitude longer EM lifetimes are obtained for Cu lines of the same cross-section as Al lines, allowing Cu to deliver more reliability for a given interconnect design.[4] In addition to electromigration, stress-voiding (i.e., due to mass transport resulting from thermally-induced stress gradients), is another reliability concern.

Electromigration tests of 0.3-μm lines conducted at an elevated temperature (295°C) and a stress-current of $2.5\text{x}10^6$ A/cm^2 showed a mean-time-to-failure (MTF) that was more than 10 times longer for Cu than for Ti/Al(Cu)/Ti lines (see Fig. 16-5). Excellent results were also obtained for stress migration testing.[6] Both sets of results argue strongly in favor of using copper interconnections for higher usable current densities and better overall performance.

Another benefit of Cu interconnects with respect to electromigration failures involves the current-flow in vias. As shown in Fig. 16-6, a small diameter via serves as the electrical bridge for current flow from metal-level-n to metal-level-n +1. A problem known as "current-crowding" arises because the current from the rectangular interconnect is compelled to flow through the much smaller circular-cross-section of the via. Due to the change of geometry and reduction in cross-sectional area, the current-flow in the via can be highly non-uniform, with local current-density much higher than in the interconnect line. If the via current-density exceeds the design specifications, excessive heating and/or electromigration can result. In turn, this can lead to circuit-reliability problems. Replacing W-plugs in

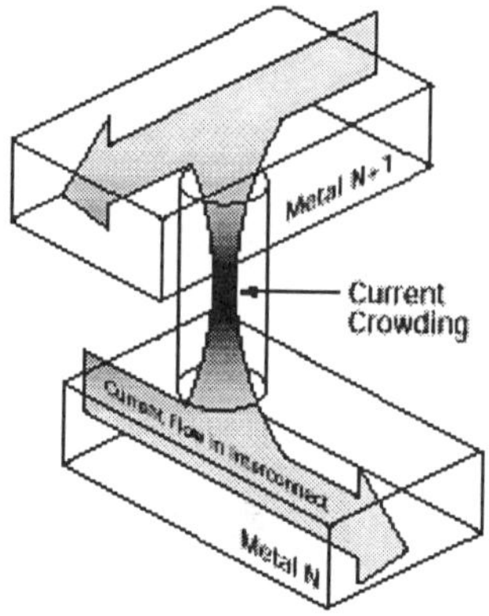

Fig. 16-6 Copper-filled vias can provide additional protection against structural failures associated with current crowding, thereby maintaining via reliability within designed specifications. In particular, the uncertainty in calculating the exact 3-D current-flow in the via is mitigated by the increased tolerance of Cu to electromigration at high current density.

the vias with Cu-plugs significantly reduces the electromigration problems observed in interconnects consisting of Al:Cu wires and W-plugs.

3. The third major benefit of employing copper (and some say it's the most important one) is that lower manufacturing costs can be obtained compared to using aluminum. The cost is lowered for two reasons.

First, because copper is difficult to etch, a new strategy called "damascene" or "in-laid" patterning was developed to form the Cu interconnect lines (see Chap. 15). IBM (which pioneered the process), claimed that the damascene approach requires 20-30% fewer steps than traditional subtractive patterning. Not only are there fewer steps, but dual-damascene eliminates or reduces the need for some of the most costly ones - including aluminum etch, and many of the tungsten-CMP and dielectric-CMP steps.

Second, using copper reduces costs because smaller lines can be employed to carry the same amount of current. Thus tighter packing-density can be achieved per level. This also implies that fewer levels of metal may be needed (see Fig. 16-7),[7] which significantly reduces manufacturing costs. This data indicates that for certain circuit applications only about half the number of levels are required if Cu is used instead of Al-lines and W-plugs.

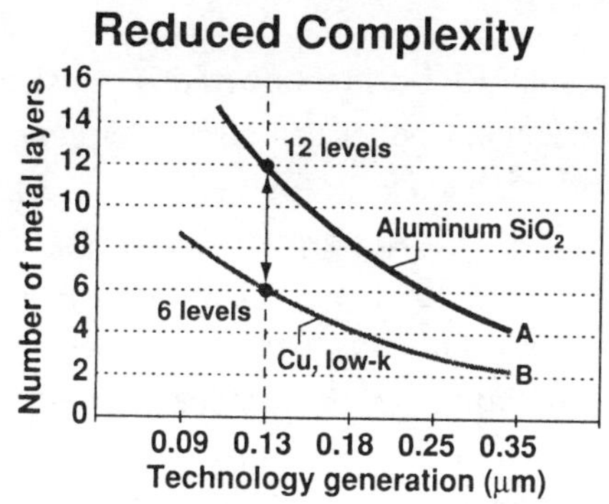

Fig. 16-7 One of the main benefits of copper, beyond the ability to increase chip speed and power consumption, is that the number of metal levels can potentially be reduced by as much as half.[1] Reprinted with permission of Semiconductor International.

4. The currently favored method for Cu deposition is electrolytic-plating, and this allows for much lower deposition temperatures that does Al sputtering or W-CVD. This enables the use of low-*k* dielectric materials whose thermal stability is lower than that of SiO_2.

16.1.2 Technological Challenges of Using Copper Interconnects

Although many powerful benefits can be gained if Cu is substituted for Al in the IC interconnect structure, a new set of technical challenges arise, including the following:

1. Cu does not form volatile copper fluorides and chlorides, making plasma etching of Cu impractical at temperatures below 200°C. Thus, another technique for forming Cu wires has been developed. Instead of patterning and etching Cu films, the Cu-films are first blanket-deposited over the entire wafer surface and into recesses in the dielectric layers. Then the Cu film is polished away by CMP in the flat areas, but is allowed to remain behind, in-laid in the recesses. This is the basis of *damascene* interconnect structure formation (as discussed in Chap. 15).

2. Unlike Al, Cu does not form a dense, self-passivating oxide on its surface. Thus, chemical reactions can continue to attack exposed Cu lines, making them more vulnerable to corrosion than Al lines.

3. Copper readily diffuses through SiO_2. If it is present in sufficient concentrations in a dielectric it can lead to shorts or leaky paths between conductors.

4. If the Cu manages to diffuse through the SiO_2 and then enters the silicon lattice, it will create deep impurity levels in the silicon bandgap. These levels act as *recombination/generation (R/G)* centers for free charge-carriers. The presence of such R/G centers degrades transistor performance and causes *p-n* junction leakage. Copper can also precipitate at (or near) the silicon wafer-surface, and such metallic precipitates are known to cause defects in the gate-oxide as it is being grown. Cu is also a fast-diffuser in single-crystal silicon (i.e., it exhibits the highest diffusivity in silicon of all metal elements), and thus will quickly distribute itself everywhere within the bulk of the wafer (including in the device-regions near the wafer surface, where it can do the most harm). For these reasons, Cu is viewed as "poison" if it is present in the silicon lattice. As a result, special Cu processing areas are also set up in wafer fabs to keep trace amounts of Cu from contaminating any other areas of the cleanroom.*

5. Unlike Al (which reacts with SiO_2 to form a strong interface with excellent adhesion), Cu adheres very poorly to SiO_2.

6. The deposited copper films are polycrystalline in nature, and the preferred orientation of the crystal structure is a high (111)-orientation. Control of this orientation is important because resistivity, stress, adhesion strength, and electromigration resistance of copper layers are largely affected by the films crystal orientation.[17,53]

To prevent Cu from diffusing to the device areas of an IC, the interconnect structure must have the built-in reliability of several protective measures against this possibility. The primary line of defense involves totally encapsulating the Cu wires and plugs at all locations and at all levels to prevent copper from escaping. In dual-damascene interconnect processes, this is done in two-steps. In the first one, a *conductive* diffusion barrier-layer (e.g., Ta, TaN, or TiN) is deposited over the wafer after the trenches and vias of each level of metal have been etched in the dielectric stack, (but before the copper layer is put down). Thus, the insides of the trenches and vias of the damascene dielectric-stack are lined with a diffusion-barrier layer prior to deposition of the Cu into them. Such conductive barrier-layers must be effective against Cu out-diffusion and must simultaneously be thin enough to maximize the cross-section of the Cu-wires and plugs in the recesses. The second diffusion-barrier-layer is formed after the Cu-layer is deposited and the Cu structures have been polished with CMP. That is, a *non-conductive* diffusion-barrier-layer is deposited onto the wafer to seal the top-side of in-laid Cu structures (typically a CVD-silicon-nitride layer is selected to serve in this

* The use of Cu as an alloy in Al:Cu interconnects generates a completely different set of requirements for containing Cu atoms. That is, Cu-atoms alloyed with Al are immobilized by the aluminum matrix, which acts as a very efficient getter.

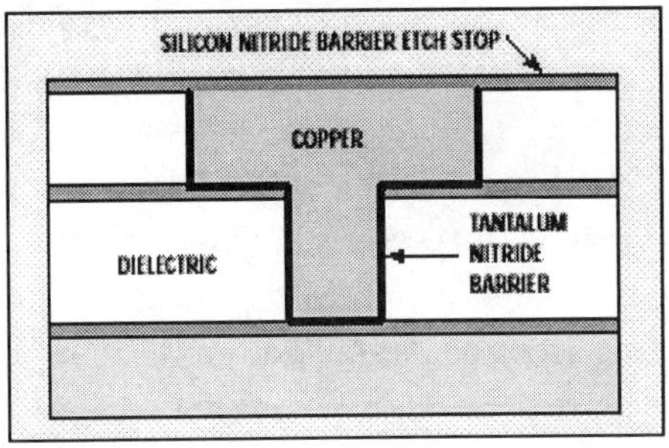

Fig. 16-8 Dielectric (silicon nitride) and conducting (tantalum nitride) barriers encapsulate copper in a damascene structure and protect against copper migration from interconnect structures to neighboring dielectrics and silicon.

role). Thus, Cu structures are ultimately enclosed on all sides by either a Ta (or TaN) or a silicon-nitride diffusion-barrier (Fig. 16-8).

The secondary on-chip line of defense is provided by W-plugs and phosphosilicate glass (PSG). The tungsten-plugs are used in the contact holes as a secondary barrier that keeps the first level of Cu-lines physically separated from the Si substrate (and this also means that a *single-damascene* process is used to form these W-plugs as well as the first level of Cu wires). PSG is known to getter metal atoms. The last level of protection on the chip itself is provided by a bottom silicon-nitride-layer, which further blocks the underlying silicon substrate from migrating Cu atoms.

While manufacturing protocols must also be modified to tightly control cross-contamination, the Cu-contamination problem does not seem to be as difficult as it first appeared to be. For example, fabrication tools cannot be shared between front-end and back-end processes, but copper manufacturing lines can be built in shared cleanroom facilities with aluminum interconnect manufacturing lines. Cleaning procedures must also be implemented to ensure that no detectable traces of copper are present on the bevels and backsides of wafers after processing.

Additional procedures have also been developed to heighten awareness of production workers and to contain copper contamination. For example, copper personnel in some fabs (especially maintenance technicians) wear cleanroom gowns that are used only in the copper bays, and they may not enter non-copper stations when wearing them. Specially colored wafer-carriers and conspicuous labels (indicating the presence of copper-products) are used to clearly identify wafers that have received copper processing and to segregate them from wafers that have not been subjected to copper processes.

Protocols also require that once a wafer has entered a copper bay it cannot return to a noncopper bay. In practice, however, wafers that leave the *front-end-of-line* (FEOL) after the contact holes to the silicon substrate have been opened, move to the *back-end-of-line* (BEOL), where the copper fabrication occurs. There is really no reason why the wafers should have to re-enter the FEOL part of the fab after that part of the process sequence is completed. Additional information on preventing cross-contamination from Cu is given in Sect. 16.7.

To counteract the problem of poor adhesion of Cu to oxide, the conductive diffusion-diffusion films under the Cu layers must also serve as adhesion promoters. The adhesion/diffusion layers used under the inlaid Cu features in damascene processes must therefore meet several criteria:

1. They must act as good adhesion-promoters for Cu onto SiO_2. That is, the adhesion/diffusion-layer materials should adhere well to oxide, and then the Cu films should adhere well to them.

2. They must be effective barriers against Cu diffusing through them under all post-copper-deposition temperatures.

3. They should exhibit low resistivity. That is, diffusion-barrier materials will be expected to exhibit resistivities of less than 1000-$\mu\Omega$-cm to meet the needs of advanced interconnect processes. This will keep the total resistivity of the composite Cu/barrier interconnect-structures below the resistivity of comparably-thick Al-alloy/barrier interconnect lines.

4. They should retain their diffusion-barrier properties when they are very thin, because they do not have the low-resistivity of Cu. That is, thinner diffusion barriers for Cu will be needed than for Al because in-laid-Cu interconnects require the barrier-layer to be formed on at least three sides of the Cu conductor (Fig. 16-9). To keep the lower-conductivity barrier-layers from impacting the resistance of the interconnect line, they must thus be relatively thinner for Cu lines.

5. They should protect the Cu from oxidation and corrosion.

6. They should be able to make a low-resistance contact to the copper layer of the metal-level under it, and this barrier-layer/Cu interface should exhibit good electromigration reliability characteristics.

7. They should provide a surface which will facilitate the nucleation of the Cu-seed layer deposited on it. Furthermore, the surface-condition and micro-

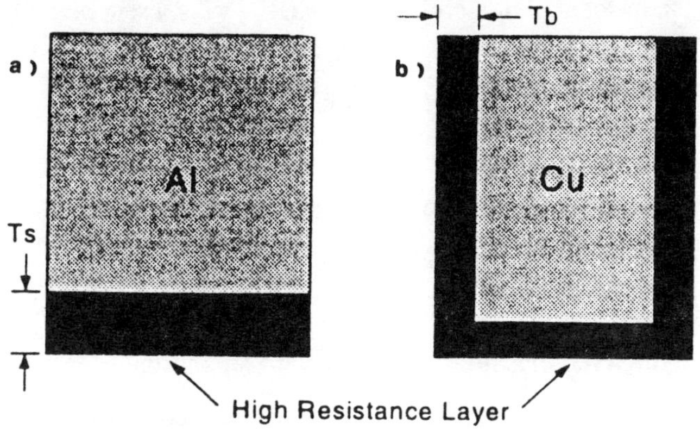

Fig. 16-9 Interconnect cross-sections showing the impact of higher resistance shunt or barrier layers: (a) Al interconnect; (b) Cu interconnect.[7] (© 1995 IEEE)

structure of the barrier-film can strongly affect the texture and grain-size of the deposited Cu (and these in turn, are critical in determining electromigration reliability).[50] Thus, the barrier-layer is also vital to seeding the desired Cu microstructure. Since the growth of Cu layers with a (111) preferred orientation provides Cu films with the best overall characteristics, the barrier-layer should enhance such Cu-structure.

In addition to the above materials-science issues associated with the use of copper, there are a number of difficult manufacturing challenges that will also have to be met when integrating Cu into an IC fabrication sequence. That is, getting a copper process (or any process) online and productive quickly is the key to profitability. The integration issues involved in bringing a Cu process to full-production are summarized in Fig. 16-10,[15] and they will be discussed in more detail in later sections of this chapter.

16.2 OVERVIEW OF COPPER PROCESS TECHNOLOGY

Before the details of depositing, polishing and cleaning copper IC interconnect structures are described, it is useful to give brief overview of the entire sequence of processes that are required to form a single-level of a copper dual-damascene structure. The sequence of process steps is as follows (and is also depicted in Fig. 16-11):

1. Non-Conductive Diffusion-Barrier-Layer Deposition
2. Dielectric Stack Deposition

3. Dual-Damascene Dielectric-Stack-Etch (Via, Trench, and Diffusion-Barrier-Layer Etch)
4. Conductive Barrier-Layer Deposition
5. Copper Seed-Layer Deposition
6. Copper Electrofill (and Annealing after Deposition))
7. Copper and Barrier-Layer CMP
8. Copper Clean
9. Non-Conductive Diffusion-Barrier-Layer Deposition.

The sequence starts by depositing a non-conductive diffusion-barrier-layer (e.g., a thin film of silicon nitride by PECVD) over the entire wafer (which has been planarized by a previous CMP step and may already contain one or more levels of metal). Next the remainder of the ILD stack is deposited (as described in detail in Chap. 15, Sect. 15.6). Using a lithography process, the vias and trenches of the dual-damascene structures are aligned to the previous layers of the circuit and are then etched with a dry-etch process. (This requires a sequence of two lithography and etch procedures, including stripping the resist after each etch step.) The SiN diffusion-barrier-layer at the bottom of the vias is also opened during the second of these etch steps, exposing the metal-layer of the previously formed level of the interconnect structure. Next, a conductive diffu-

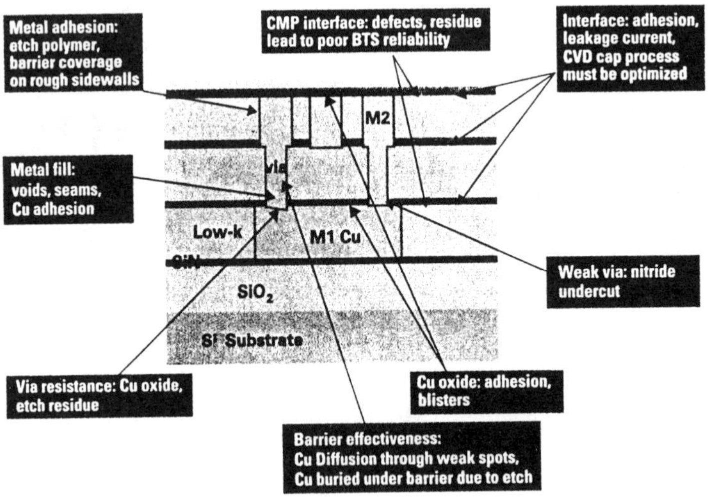

Fig. 16-10 Possible failure types associated with damascene processing pertain to film adhesion, barrier and metal fill capability, copper oxidation, and CMP defects.[9]

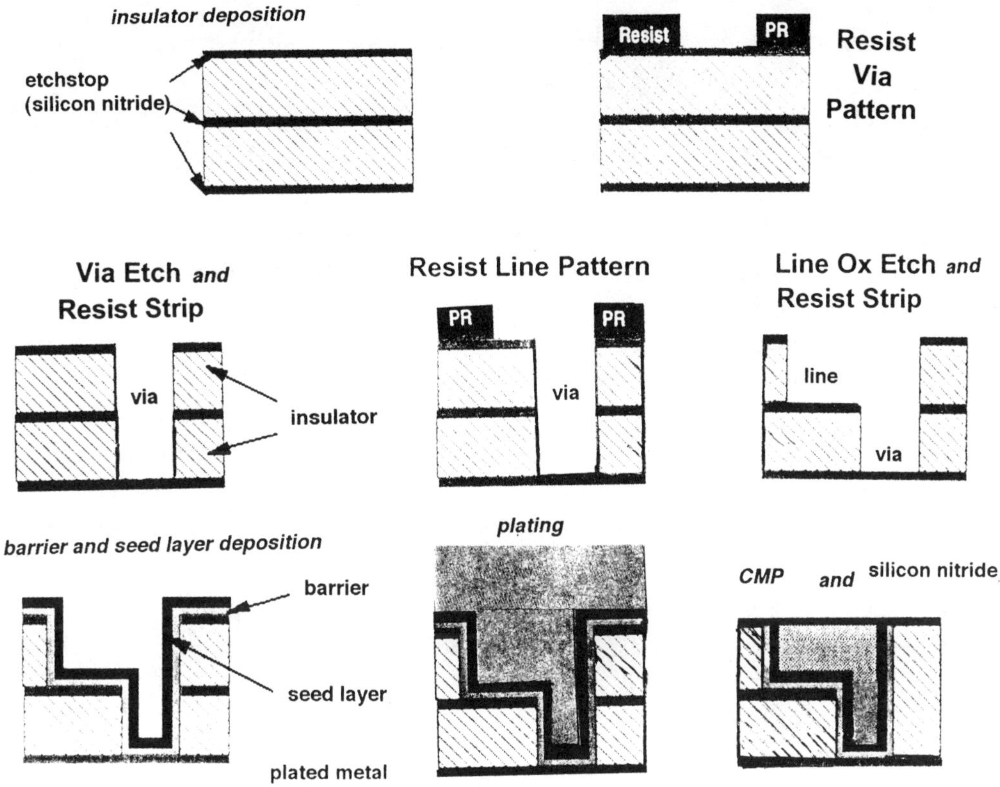

Fig. 16-11 Process sequence for forming a single-level of Cu metal in a dual-damascene interconnect structure (using a *via-first* approach).

sion-barrier/adhesion-layer is blanket-deposited onto the wafer, typically by PVD (with either a collimated-, a long-throw-, or [most typically] an ionized-metal-PVD process). Various materials have been selected to perform the role of this diffusion-barrier/adhesion-layer including Ta, TaN, TaSiN, TiN, TiSiN, WN, and WSiN (see Sect 16.3).

A copper seed-layer is deposited onto the diffusion-barrier/adhesion-layer to provide a low-resistance path for the copper electroplating process that comes next. The Cu-seed layer is also deposited *in situ* (i.e., without an air break) by one of same type of PVD-processes used to form the diffusion-barrier/adhesion layer film. Completing the process of filling the trenches and vias with Cu material is done with an electrolytic-plating technique. The key attributes of such Cu electrolytic-plating-processes include: **a**) the ability to fill the recessed structures without defects (i.e., no voids or seams should exist in the plated films after filling); **b**) the ability to achieve high-deposition rates; and **c**) the ability to obtain uniform deposition-thickness across-the-wafer, and from wafer-to-wafer (e.g., ±2% across a 200-mm wafer with a 5-mm edge exclusion).

A CMP process is carried out next to remove any excess copper above the vias and trenches (called the *overburden*), but allowing metal to remain in the damascene recesses. During the CMP process, the diffusion-barrier-layer must also be removed. A key technological challenge involves achieving the same polishing rate on the diffusion-barrier layer, because such materials as Ta or TaN are hard, refractory metals (or metallic compounds), while copper is a comparatively much softer metal. After the CMP process is finished, it must be ensured that the wafer is cleaned, such that no copper residue remains on the oxide on the frontside of the wafer (nor on the wafer backside or beveled edges). Finally, another non-conductive diffusion-barrier film (e.g., CVD silicon-nitride) is deposited onto the wafer surface, which serves to cap-off the copper (i.e., this completes the encapsulation of the recessed-copper structures, ensuring they are surrounded by a diffusion-barrier-material on all sides).

The details of the first three steps of this sequence are described in Chap. 15, Sects. 15.6 and 15.9 (and thus will not be repeated here). The discussion that follows therefore begins with the Step **4** (Diffusion-Barrier-Layer Deposition, and the properties of such materials) and then proceeds to cover the details of the remaining steps of the process sequence.

16.2.1 The Transition from Aluminum to Copper Interconnects

The process of incorporating copper into the interconnect process of an IC technology may be done in one of two ways. Some manufacturers have decided to use Al:Cu (and W-plugs) for several of the lower levels of a multi-level metal interconnect structure, and limit copper to the upper levels. In one such example, Al:Cu was used in the four lower levels of the 6-level metal interconnect structure and Cu for the upper two levels (Fig. 16-12).[10] Another process described by TSMC also uses four aluminum levels of metal and two of copper.[11] The interface between the top aluminum interconnect level and the lower Cu interconnect level can be implemented either with a W-plug or with a Cu-damascene via structure.[12]

On the other hand, other manufacturers decided to use copper for the entire interconnect structure (with the exception of W-plugs being used in the contact holes). An example of such a 6-level Cu-interconnect-structure (with W-plugs used to make contact to the Si-substrate) was described by IBM (see Fig. 16-13).[13] Figure 16-14 shows a SEM of a perspective view of a six-level metal Cu interconnect structure (see also Fig. 12-16 in Chap. 12).

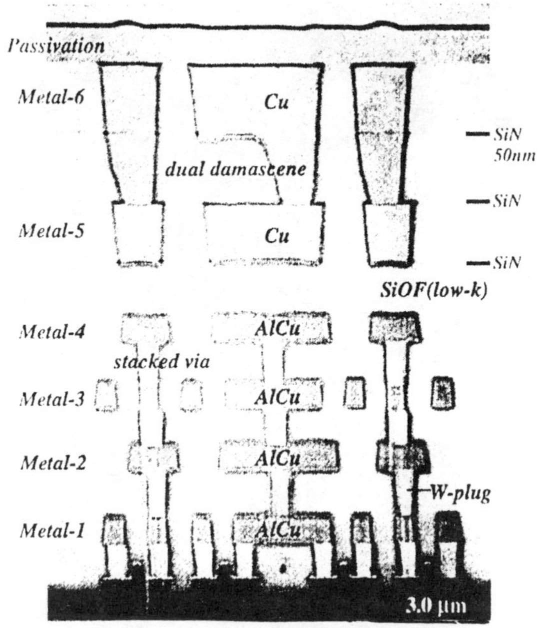

Fig. 16-12 Cross-section of six-level metal structure with the first four-levels being Al:Cu/W-plug interconnects and the top two levels of Cu metal.[10]

16.2.2 The Key Goal of Depositing Copper into Damascene Recesses

One of the most important aspects of depositing copper successfully into the recesses of a dual-damascene ILD stack is ensure it done without any voids or defects. The key to being able to do this is to fill the recesses with copper from

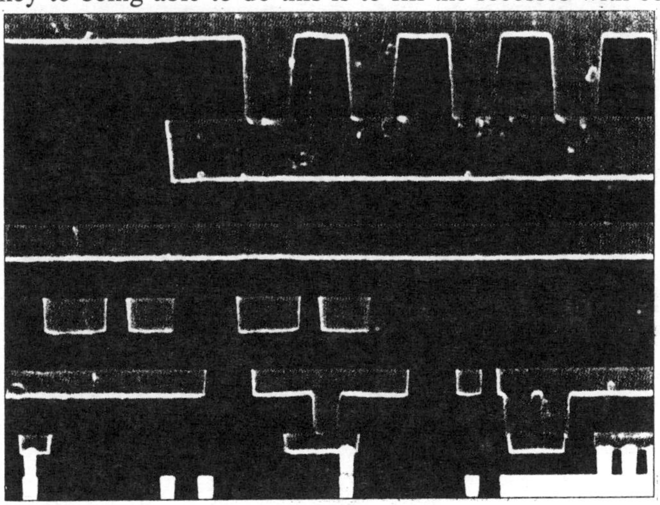

Fig. 16-13 Cross-section of six-level metal structure with all levels of Cu metal.[13]

Fig. 16-14 An SEM of a perspective view of a six-level metal Cu interconnect structure.[32]

the bottom up, a technique called "bottom-up" filling, or "superfilling." However, successful superfilling is affected not only by the electrofill process but also by the dielectric etch process, the barrier-layer deposition-process and the seed-layer deposition-process. Thus, this topic has to be addressed in a piecemeal fashion, as will be done in the three upcoming sections.

16.3 BARRIER LAYERS FOR COPPER INTERCONNECTS

As discussed previously, to stop Cu from diffusing into the dielectric materials in which it is embedded (or into the Si substrate), it must be surrounded by a barrier-layer on all sides. The underlying layer is deposited prior to depositing the copper (as noted in Sect 16.2). In this section, the technology for forming these conductive barrier-layers is described. The subjects that will be covered are:

1. The impact of the barrier-layer on the Cu-electroplating process;
2. The technology of depositing barrier-layers into damascene recesses by PVD;
3. The material properties of tantalum (and TaN & TaSiN) barrier-layers;
4. The other candidate materials for Cu-technology barrier-layers (including CVD processes for depositing such materials).

16.3.1 The Impact of the Barrier-Layer Deposition Process on the Filling of Damascene Recesses with Copper

To recap, a barrier-layer for Cu interconnects must prevent Cu diffusion into the ILD and must assure good adhesion between the Cu and the ILD. Furthermore, the barrier must be thin, conformal, and have a low electrical resistivity. Deposition should be at a low temperature with few particles being produced.

The barrier-layer material should also be compatible with CMP of Cu and with the Cu electroplating process. Finally, there a number of other aspects involved with the deposition of the barrier-layer that impact the ability to completely fill damascene recesses with copper without voids. These include depositing a barrier-layer that is continuous and adequately thick everywhere along the recess surfaces (typically 25-30-nm-thick), and with little (or better yet, no) overhang. The PVD-process used to deposit the barrier-layer impacts these last two factors, and the PVD techniques for depositing seed and barrier layer is discussed next.

16.3.2 Introduction to the Technology of Depositing Barrier-Layers for Cu Interconnects

The conductive barrier-layer deposited into recessed damascene features before the Cu is deposited can be done by PVD or CVD. While PVD has been the most widely used method down to at least the 0.13-μm technology node, there may be a change to CVD at some point. One reason for such a technology replacement involves the formation of a film within a trench or contact hole by a *physical vapor deposition process.* Assume a contact-hole exists with the top face being open with the dimensions of a cube of side **a**. This represents a contact hole with an aspect ratio of 1:1. In such a case there are five inner surfaces of the cube that must be covered by material which would otherwise deposit on area $\mathbf{a}^2$ on the top surface of the cube (i.e., if the top of the cube was not open, see Fig. 16-15). However, if the top of the contact hole is open, the total area inside the hole that must be covered now is $5\mathbf{a}^2$. Thus, if the material that would have covered $\mathbf{a}^2$

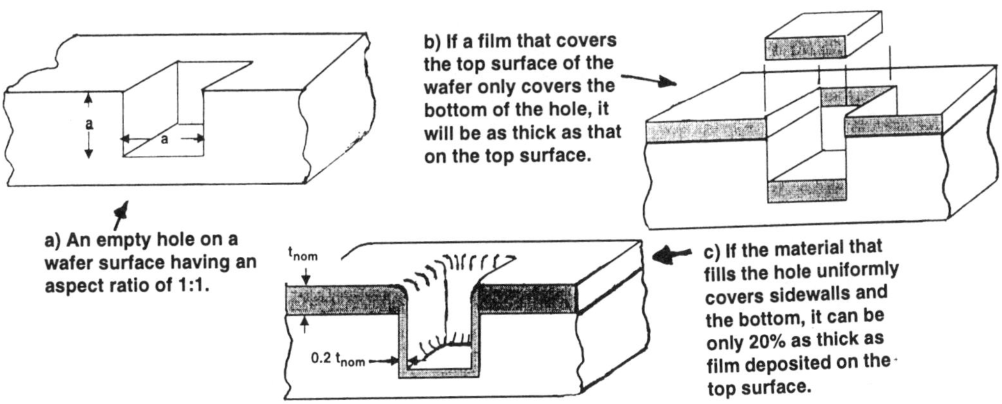

Fig. 16-15 One dilemma faced when forming films in high-aspect-ratio holes with PVD.

is evenly distributed onto all five insides of the contact hole, the film thickness will only be 20% as thick as that on the flat areas outside of the contact hole. Obviously, if the area of the hole-opening remains constant while its depth is increased, the aspect-ratio will also increase (e.g., if the square contact-hole opening-size remained at $\mathbf{a}^2$, but the depth became 2**a**, the aspect ratio will increase to 2:1, and the area that will have be covered by the material of $\mathbf{a}^2$ will now be $9\mathbf{a}^2$). Thus, for the case of a physical-vapor deposition process, as the aspect-ratio of a recessed feature is increased, the material entering the feature must be spread over a larger area, making the thickness per unit area decrease. Furthermore, if the film on the upper flat regions must be etched or polished away after the hole has been filled, a much thicker film now has to be removed (which makes the etch or CMP process task significantly more time consuming and difficult). Deposition by CVD may be able to avoid this problem because CVD processes can be reaction-rate limited (allowing deposition to be almost 100% conformal at a very high-aspect-ratio, see Vol.1, Chap. 6). Thus, CVD may be employed for depositing barrier-layer films for Cu in the lower levels of deep-submicron interconnects (where the aspect-ratios continue to increase).

In the previous discussion it was noted that it becomes progressively harder to fill contact holes by PVD as the aspect-ratio increases. A related problem further increases the difficulty of filling high aspect-ratio-recesses. Recall that the sputtered-atoms undergo collisions as they are transported in the gas-phase in the space between the target and the substrate. In addition, sputtered atoms leave a planar-target surface from each point at which sputtering occurs in directions that roughly exhibit a cosine distribution. As a result of these two mechanisms, in a conventional sputtering apparatus the sputtered-atoms will arrive at the upper corners of the trench or contact hole on the wafer from a larger arrival-angle than they will at surfaces inside the recess. Thus, the film deposition-rate at the top corners of the recess will be higher than along its sidewalls or bottom. As a result, the top opening will begin to close off (i.e., an *overhang* will start to form), as shown in Fig. 16-16a. Thus, for this case, as time goes on (Fig. 16-16b) the deposition-rate onto the sidewalls and bottom will continually decrease. If the overhang gets thick enough, the hole will close entirely, leaving a void or keyhole.

Even if deposition does not continue long enough to completely close-off the opening, the overhang will shadow the bottom corners of the contact hole, where the arrival angle for sputtered atoms is already the smallest (see Fig. 16-16b). Thus, the deposited film will be thinnest at the bottom corners of the hole. Since a barrier-film must have a minimum adequate thickness along the entire bottom

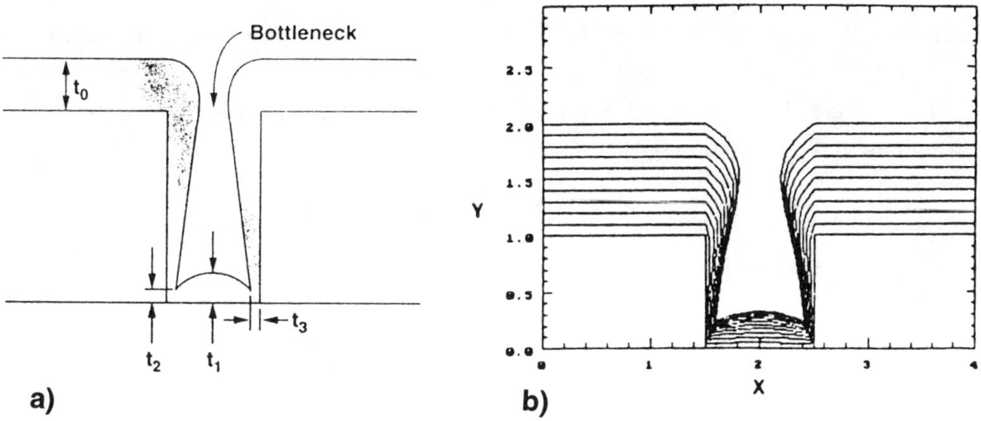

Fig. 16-16 a) Deposition of a sputtered film into a contact hole or via having a high aspect ratio. b) Development of a bottleneck due to the buildup of an overhang on the top corner.

of the hole in order to function effectively, such thin-spots at the corners can jeopardize the effectiveness of the barrier-layer.

Various methods have been explored to overcome this thinning of the film at the bottom corners of contact holes (including the use of CVD). However, modifications to the sputtering process have also been pursued. Two such early methods were *collimated sputtering* (in which a honeycomb baffle - i.e., a *collimator* - is inserted into the space between the target and the wafer to trap obliquely-directed vapor) and *long-throw-sputtering* (in which the target and wafer are separated by a distance 2–4 times longer than in a conventional magnetron source, and the plasma is also operated at a lower pressure). However, these two methods have proven to be inadequate for deep-submicron, dual-damascene applications. It is unlikely that they will even be commercially developed for 300-mm wafer processing. Thus, they will not be discussed further here (but more information on them is available in Vol.1, 2nd Ed, Chap. 11).

16.3.3 Ionized Magnetron Sputter Deposition

While the two sputtering techniques described above produce adequate barrier-layer films for contact-holes and vias for 0.35-μm (and some 0.25-μm) CMOS technologies, they appear to fall short for other 0.25-μm (and smaller) contact holes and vias when the aspect-ratio exceeds about 3:1. Thus, an alternative to collimated (or long-throw) sputtering, namely *ionized magnetron sputtering*, or *I-PVD* (also termed *ion metal plasma* [*IMP*] by Applied Materials), has been developed to deposit the next generation of ULSI barrier-layers with PVD. In this method, a large fraction of the neutral sputtered-metal-atoms are ionized while in flight between the target and the wafer. A bias on the wafer then causes an electrical acceleration of these ionic species to the wafer-surface with a controlled energy and direction (i.e., perpendicular to the wafer surface).[14,15,16] There have been several sputtering-source configurations developed to implement the ionized magnetron sputtering technique.

Figure 16-17a shows a schematic drawing of an ionized magnetron sputtering chamber used in the first commercially available I-PVD tools. A conventional magnetron source (powered with a dc power-supply) remains in place to sputter neutral metal-atoms from the target. (The magnetron produces a plasma and dark space that is confined near the target surface.) However, an rf-powered (and water-cooled) coil is now also located within the chamber (Fig. 16-17b). This coil inductively couples energy from a second 13.56 MHz rf-power supply into a secondary plasma downstream of the magnetron plasma. By increasing the pressure in the chamber to 20-40-mtorr (up from the 5-mtorr of a conventional magnetron sputtering process), and by increasing the depth of the secondary plasma to 8–10 cm (by increasing the distance between the cathode and the wafer chuck), the number of collisions experienced by the sputtered atoms before they

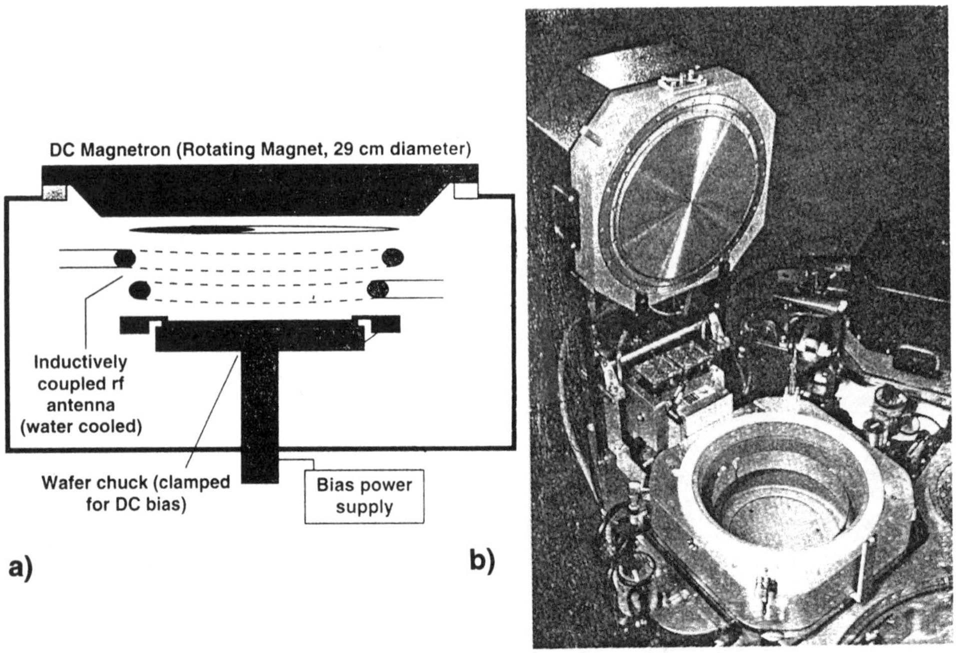

Fig. 16-17 a) Schematic of a magnetron sputter chamber fitted with an rf-coil which allows an ion magnetron sputtering process to be implemented. b) Photograph of an Applied Materials IMP system showing the internal coil. (Courtesy of Applied Materials)

arrive at the wafer (as well as their transport time through the gas phase), is increased. This longer transport time, and the fact that the ionization potential of metal atoms is much lower than that of Ar atoms, allows a much higher fraction of the metal atoms to be ionized by this secondary plasma (e.g., for a Ti-sputtering process, typically 50-70% of the sputtered Ti-atoms are ionized). A medium-density secondary-plasma is thus created in the region between the cathode and the wafer ($5x10^{11}$-10^{12} ions/cm^3).

The chief advantage of the I-PVD approach can now be exploited. That is, by applying a negative-bias to the wafers, the ionized metal-atoms can be accelerated toward the wafer surface (at voltages ranging from a few to several-hundred volts, depending on the specific application). Thus, the metal ions will strike the surface exactly at normal incidence to the wafer surface, regardless of their original trajectory. In addition, the third (substrate-bias) power-supply allows the kinetic energy of the metal ions striking the wafer to be controlled. Using an rf-power-supply to produce this bias overcomes the problems with insulating wafer surfaces (or backsides). Note that the ion-flux onto the wafer results in power dissipations ranging from about 25 watts to several hundred watts, which calls for some degree of wafer cooling.[16]

The bottom-coverage of high-aspect-ratio holes with I-PVD is better than that observed with collimated sputtering. That is, 40–80% bottom coverage has been achieved on 0.18-μm holes with aspect-ratios of up to 8:1.[17] Note that while the bottom-coverage is high, the sidewall-coverage is much thinner (although the film on the sidewall is continuous, see Fig. 16-18).

Several other advantages of I-PVD processes should be noted. First, much better target-utilization is obtained than with collimated sputtering (as almost all of the sputtered atoms arrive at the wafer surface, instead of some being intercepted by the collimator). Sputtering tool throughputs are therefore also increased. Second, the high flux of metal ions (as well as some Ar ions) directed perpendicularly onto the wafer surface appears to re-sputter the atoms that would otherwise form an overhang at the top-corners of a recess. This keeps the top opening of contact holes and vias from closing up. Figure 16-18 shows the coverage of the sidewall and bottom of a high-aspect ratio hole with I-PVD. Third, the negative-bias applied to the wafer not only causes the ionized metal atoms to preferentially strike the bottom of the via, but it allows the kinetic energy of these metal (and Ar) ions bombarding the wafer to be adjusted. For appropriate bias levels, re-sputtering from the feature base can be produced, which redistributes sputter-deposited material onto the lower sidewalls (Fig. 16-19). This is exactly where the lowest PVD coverage occurs in collimated and

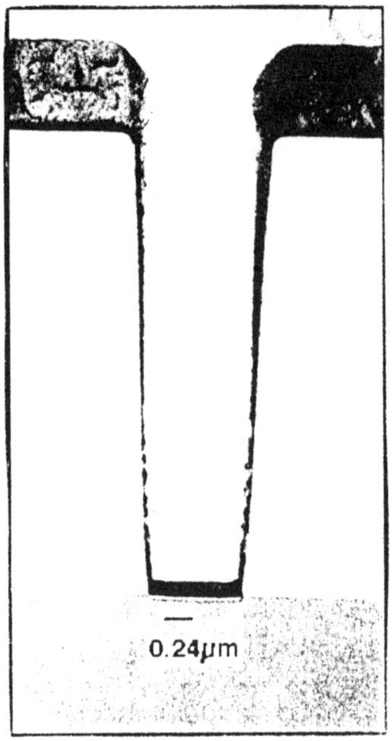

Fig. 16-18 SEM showing the coverage of the sidewall and bottom of a high-aspect ratio hole deposited by I-PVD. (Courtesy of Applied Materials)

long-throw PVD-processes. The process of local re-sputtering can also be used to tailor the conformality of the deposit to a greater degree, and this has extended the use of PVD for depositing barrier-layer and seed-layer films.

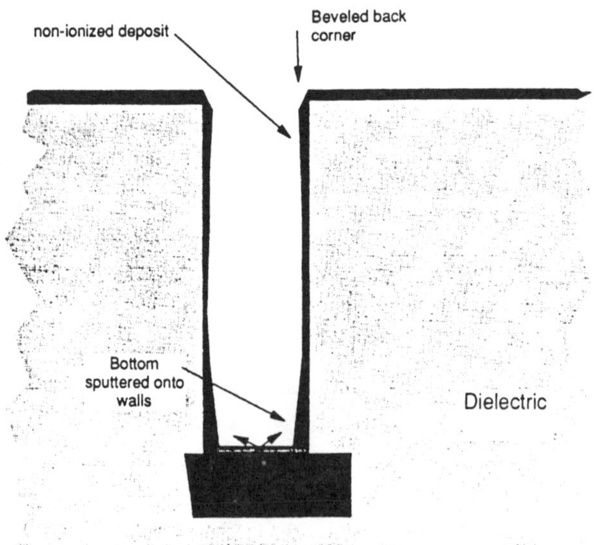

Fig. 16-19 Schematic drawing of I-PVD deposition of a high-aspect-ratio recess showing the effects of re-sputtering of the deposited film.

While the concept of I-PVD has the many advantages described above, the internal coil-configuration has some practical difficulties. First, the coil can get coated with the sputtered metal. If this gets too thick, the film may flake off and produce particles. Second, material from the coil itself can be sputtered, with some material ending up being re-deposited on the wafer. While a copper coil can be used to circumvent this problem when depositing Cu-seed-layers, this does not provide a fix when depositing Ta or Ti barrier-layers. Instead, a coil made of Ta (or Ti) would be needed, but forming such coils (which must also accommodate the flow of cooling water) is not a trivial task. As a result, two other I-PVD source configurations have recently been commercially introduced.

16.3.4 Hollow-Cathode Magnetron Sputtering Source

The *hollow-cathode magnetron* (HCM™) used in the Novellus INOVA™ sputtering system (Fig. 16-20) employs a similar technique to ionize sputtered atoms as they are transported to the substrate.[42,55] A uniform plasma throughout the volume of the cup-shaped hollow cathode target provides sputtering from a large fraction of the area of the target surface. Due to the geometry of the cup-shaped target a significant fraction of the sputtered neutrals are re-deposited on the target surface, where they can be sputtered again (instead of depositing on the chamber walls). This provides the benefit of high target utilization. In addition, no internal rf-coil is needed to generate the high-density plasma that ionizes the sputtered metal atoms. Instead, the orientation of the permanent magnets allows

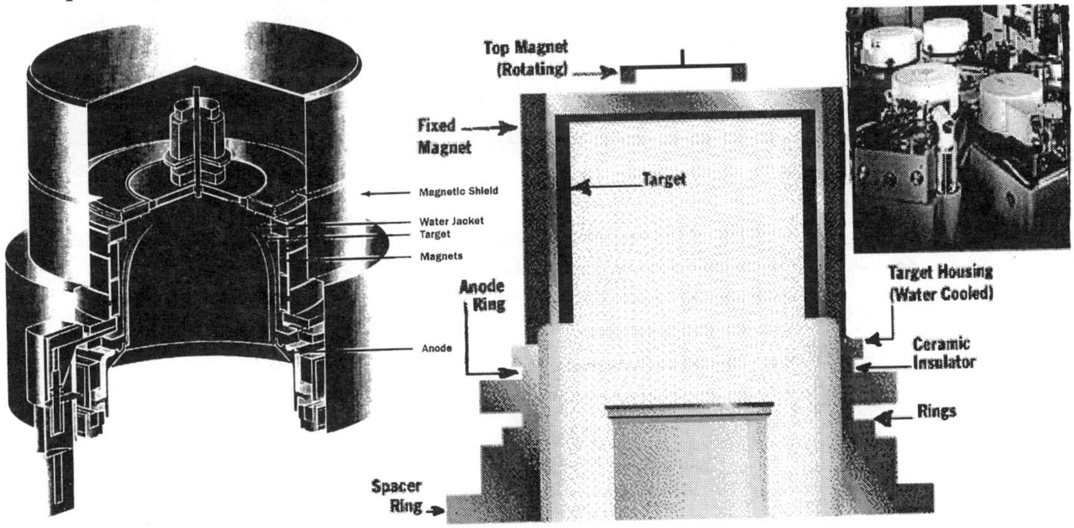

Fig. 16-20 Cutaway drawings of the hollow-cathode magnetron source used in the Novellus Systems's INOVA® sputtering system. Courtesy of Novellus Systems, Inc.

the electrons emitted by the target to efficiently ionize the sputtered neutrals (with the degree of ionization exceeding 90%). A bias on the wafer-chuck accelerates the ionized sputtered atoms toward the wafer in a direction perpendicular to its surface, allowing good coverage of high-aspect-ratio holes. The HCM can be used to sputter Ta, TaN, TiN, and Cu.

16.3.5 Self-Ionized-Plasma Sputtering Source

Another new sputtering source for implementing the I-PVD technique was recently introduced by Applied Materials. They refer to this source as *self-ionized-plasma* (SIP), and the internal rf-coil used in their IMP™ source to ionize the sputtered metal atoms is not needed. While few details about this source have been published, some of its aspects have been described.[18,19] The source operates at relatively low pressures (< 2 mTorr) with a target-to-substrate distance that is increased above the distance used in conventional magnetron sputtering sources. These measures increase the mean free path of the metal-ions and provide additional self-collimation to the metal-atom trajectories. The term *self-ionized-plasma* probably refers to the phenomenon in which some of the sputtered metal atoms that are ionized by the source, return to the target with enough energy to sputter other metal-atoms (and also produce additional electrons). These additional electrons, in turn, ionize more of the sputtered metal atoms, which keeps the plasma self-sustaining and produces more metal-ions. The phenomenon is apparently an efficient way of ionizing metal-atoms that have a small ionization energy (such as Cu). However, questions have been raised about whether it is effective for metals that have higher ionization energies (such as Ta).

16.3.6 Material Properties of Tantalum and Refining Tantalum Metal For Semiconductor Applications

The metal *tantalum* (or such tantalum-based compounds as TaN or TaSiN), have been the most commonly used barrier-layer materials for Cu-damascene interconnects. As such, it is useful to give some background about this element.

16.3.6.1 Material Properties of Tantalum: Tantalum is an uncommon (in nature) refractory metal, with an atomic-number of 73, an atomic-mass of ~181, a body-centered-cubic (BCC) crystal structure, and a steel gray color. It is an attractive Cu-barrier-material because of its high melting point (2996°C), and because it is immiscible with Cu. It is also a highly reactive metal that forms strong metal-metal and metal-oxygen bonds (much like titanium). Thus, it provides a low-resistance ohmic contact and good adhesion to the copper above (and to the dielectric material below) it in the Cu-damascene metal scheme. Doping the Ta with a few percent nitrogen blocks grain-boundary diffusion-paths. More heavily nitrided Ta (produced by reactive sputtering of Ta in the presence of nitrogen) can also be used as a barrier-layer. Such TaN barrier-layers will be discussed further later.

16.3.6.2 Refining and Forming Tantalum Metal For Sputtering Targets: Tantalum exists in nature as an oxide Ta_2O_5. The principle ore of tantalum is *tantalite* (Fe, Mn)Ta_2O_5. Mines in western Australia are the largest producers of high-grade tantalite, and such concentrates are shipped to the tantalum metal manufacturers for refining.[20] The main market segments for refined tantalum are capacitor-grade powder (about 50%), mill-products and wires (16%), chemical compounds (such as tantalum carbides, 14%), and use as a metal-alloy additive (6%). High-performance tantalum products such as *sputtering targets*, represent less than 10% of the tantalum mill-product and wire business.

Tantalite ore is not the only source of tantalum. Nearly 25% of the total Ta production is recycled. In fact, if the market for high-quality sputtering targets outpaces the rate of increase in tantalite production, it might become necessary for wafer manufacturers to return spent targets to the Ta manufacturers in order to be assured of future availability.

The extraction and refining processes for tantalum conclude with a chemical precipitation reaction that yields Ta metal in powder form. The high-density, metallurgical-grade powders (and recycled Ta scrap) are used as feed for *vacuum melting*. (Because tantalum exhibits a great affinity for oxygen, nitrogen, and carbon, thermal treatments must be conducted in a vacuum or inert gas atmosphere.) For demanding applications (such as sputter deposition), the refined tantalum must possess a number of material properties that can be obtained by subjecting the tantalum powders to such vacuum melting processes.

Typically, tantalum ingots are melted several times to purge most of the impurities from the metal. Because the melting point of Ta is much higher than most materials, this is effective in volatilizing the interstitial gases (such as O, N, C) and many metal impurities. However, the other refractory metal elements (i.e., niobium, molybdenum, and tungsten) are not volatilized during vacuum melting, but remain as soluble impurities in tantalum. However, these refractory metal contaminants can be minimized through specialized front-end chemical extraction processes, resulting in extremely pure-Ta ingots. For example, the Performance Materials Division of Cabot Corp. (Boyartown, PA) offers Ta-ingots that are 99.999% pure.

The pure ingots consist of poly-crystalline Ta with small-sized grains. These ingots can be cold-worked (i.e., at ambient temperature) to cause a re-crystallization of the grain structure in the finished annealed product. The desired microstructure for Ta sputtering targets is a fine, homogeneous polycrystalline structure. Tantalum crystal grains predominantly have a (111) or (001) type structure. It is preferable to have a Ta ingot with a preponderance of (111) grains. This is because Ta sputtering-targets which contain large (001) clusters exhibit lower sputtering yields. The reason is that in such body-centered cubic (BCC) metals as Ta, the erosion-rate of (111) incident planes is predicted to be far greater than for (001) crystals.[21] The proportion of (111) to (001) grains is governed primarily by deformation history.

In summary, attaining microstructural and textural homogeneity in tantalum sputtering targets through the use of proper thermochemical processing is critical to assuring the performance of such targets. It should also be noted, that all of the refining and thermochemical processes are expensive. In turn, this also makes Ta sputtering targets quite costly (reaching $40,000 or more for some target configurations).[22]

16.3.6.3 The Properties of Tantalum (and TaN & TaSiN) as Cu-Barrier Layers: To prevent the drift or diffusion of Cu atoms, a barrier-layer should not react with Cu. Such a barrier-layer should remain thermally stable up to the maximum temperature that the wafer will be subjected to once the Cu has been deposited. Since this temperature is not likely to exceed 500°C, the barrier-layers should be stable to this temperature. Ta and TaN films are observed to remain thermally stable up to about 550°C, and thus they are capable of meeting this requirement for back-end processes.[23]

However, as noted in Sect. 16.1, the barrier-layer materials must also serve as adhesion layers for Cu. There is typically a tradeoff between barrier integrity and adhesion performance with Cu. If a thin-film of the barrier-layer material does not react with Cu at all, it may exhibit poor adhesion. If the material reacts too easily, it may not function as a good diffusion-barrier. For good barrier properties and good adhesion with Cu, an ideal barrier would react with Cu to some self-limiting extent.

Tantalum films, sputter-deposited using a Ta target in an Ar ambient, have been investigated for such Cu barrier-layer applications. For the case of Ta, an as-deposited Cu-layer on a Ta film displays an abrupt interface with Ta. However, a 3-nm thick amorphous layer is observed at the Ta/Cu interface after annealing for 1 hour at 400°C.[24] The mechanism for the amorphous layer formation is not clearly understood because Ta has no equilibrium solubility with Cu at 400°C, and does not form compounds with Cu. Nevertheless, it is believed that the thin amorphous-layer improves the adhesion between Ta and Cu. In addition, the Ta film has the advantage of serving as a nucleation layer for Cu, enabling Cu films with (111)-orientation to be deposited.[25]

On the other hand, Ta barrier-layer films deposited by PVD are polycrystalline films. Thus, even though the reaction between Ta and Cu at 400°C occurs over only a short distance, the Cu may diffuse along the grain boundaries of the Ta film and escape into the neighboring dielectric material. Thus, by itself, Ta may not prove to be an adequate barrier-layer.

TaN is a related material to Ta, having a high melting point of 3087°C and a more dense microstructure. It is also typically deposited by PVD using a Ta target, but with nitrogen added to the argon gas in the sputtering gas ambient. Thus, as the reactive Ta is sputtered onto the wafer, it reacts with the nitrogen and forms a deposited film of TaN. The TaN material is more inert than Ta and does not react with Cu. In addition, TaN can be deposited as a polycrystalline film or an amorphous film by adjusting the nitrogen flow rate.

TaN films have been found to be better barrier diffusion-barrier films than Ta films. However, since TaN does not react with Cu, it exhibits worse adhesion behavior than Cu. For some Cu-CMP processes its adhesion properties are inadequate. To overcome the limitations that each film (i.e., Ta and TaN) exhibits when used alone, the implementation of bilayer films (TaN/Ta)[17] or trilayer films (Ta/TaN/Ta) have been investigated. These multilayer barrier-films reportedly offer better overall characteristics for Cu damascene interconnects.

TaSiN is another candidate for a Cu-barrier-layer material. It is formed by PVD of a TaSi target (with a low Si concentration) in an Ar/N_2 ambient. TaSiN is an amorphous refractory ternary material, and as such does not have any grain boundaries (which can serve as fast diffusion paths in polycrystalline films). It has been reported by Hara *et al.* that TaSiN exhibits better adhesion and barrier-layer characteristics than even TaN/Ta bilayers.[26,85] Furthermore, the same group reported that Cu-films deposited on TaSiN barrier-layers have a stronger (111) preferred orientation than do Cu films deposited on Ta or TaN layers.[27] A survey of over 75 papers from 1999-2000, indicated that TaSiN can maintain its diffusion-barrier properties for Cu to higher temperatures than Ta or TaN. However, since the Si containing ternary compounds (TaSiN and TiSiN) have higher resistivity, the Si industry tends to use TaN or Ta for Cu barrier-layers.[28]

16.3.7 Other Candidate Materials for Cu-Barrier Layers

Various other materials have been studied as Cu barrier-layers, and the most important of these (i.e., TiN, TiSiN, and WN) are discussed in this section. TiN is a good barrier-layer material for Al interconnects, and has been widely used for this application. While it also performs well as a barrier-layer for Cu, its behavior as an adhesion-layer and seed-layer for Cu is not as good as the Ta-based materials. CVD-TiN films have also been considered as Cu barrier-layer materials but they are reportedly less effective as diffusion barriers than PVD-deposited Ta and TaN films.[18,41] In addition, Cu films deposited on TiN have a weaker (111) structure than those deposited on Ta.[23] Hence, TiN deposited by either PVD or CVD has not been widely adopted for use in Cu interconnect metallurgy.[24] However, TiN deposited by ALD has been mentioned as a possible barrier-layer material for Cu. Work to deposit TiN by ALD has recently been reported.[29]

On the other hand, TiSiN-films deposited by CVD have been reported to exhibit significantly improved diffusion-barrier properties for Cu compared to TiN. The Si introduced into the TiN matrix bonds with the nitrogen (and this plays a role in producing improved barrier-layer behavior).[18] The TiSiN also has the additional benefit of good wetting to Cu, which has been correlated to improved Cu electromigration performance.[30] The CVD-TiSiN is deposited using TDMAT (tetrakis [dimethyl-amido] titanium) and silane as the precursors.[31,56] (Additional information CVD-TiN with TDMAT can also be found in Vol. 1, Chap. 6, Sect. 6.7.3.) The deposition of barrier-layers by CVD may become necessary for sub-0.1-μm technologies.

Tungsten nitride (WN) films deposited by CVD have also been considered as materials for Cu barrier-layers. Limited development of CVD-WN was described by Genus in 1999,[32] but they apparently decided not to commercialize this approach. Their decision may have been based on results indicating that such films are less effective than PVD-films of Ta or TaN as Cu diffusion-barriers.[18]

16.4 SEED-LAYER TECHNOLOGY FOR COPPER INTERCONNECTS

A *thin Cu seed-layer*, deposited by a method other than electroplating (e.g., PVD) must be deposited on top of the diffusion-barrier-layer.[9] This layer needs to be present to ensure that the subsequent Cu-electroplating operation can be performed successfully. Specifically, there are a number of requirements that the seed-layer must satisfy in order to enable the electroplating fill-process to be effectively carried out.[33] That is, if the seed-layer exhibits any one of a number of flaws, including poor morphology, inadequate sidewall- and bottom-coverage, excessive overhang, or surface contamination, the electroplating results will be unsatisfactory (even if the best electroplating process is available). The details of the characteristics that must be possessed by the Cu-seed-layer are described as follows:

1. *Low-Resistivity* - At the start of the electroplating process, the seed-layer is the primary-conducting layer that carries current from the edge of the wafer to the center (allowing the plating current-source to contact the wafer only near the edge). Thus, the seed-layer should provide a low-resistance path from the edge to the center of the wafer, to make the current-density during the Cu-electroplating process uniform across the entire wafer (especially when the plating process is being initiated). Uniform-current density across the wafer-surface helps maintain a uniform plating-rate across the wafer.

2. *Film Continuity Along the Sidewalls and Bottom of the Trenches and Vias* - The electroplating process must also occur within the recessed features on the wafer surface. Since, the seed layer carries current from the top surface into the bottom of vias and trenches, the electrical path to the bottom of such features must be continuous and adequately thick. If there is insufficient seed-layer thickness at the bottom or sidewalls of the vias, this will lead to a slowdown in the plating-rate, which, in turn can cause a void during the Cu-fill process (Fig. 16-21a). To get adequate Cu-seed-

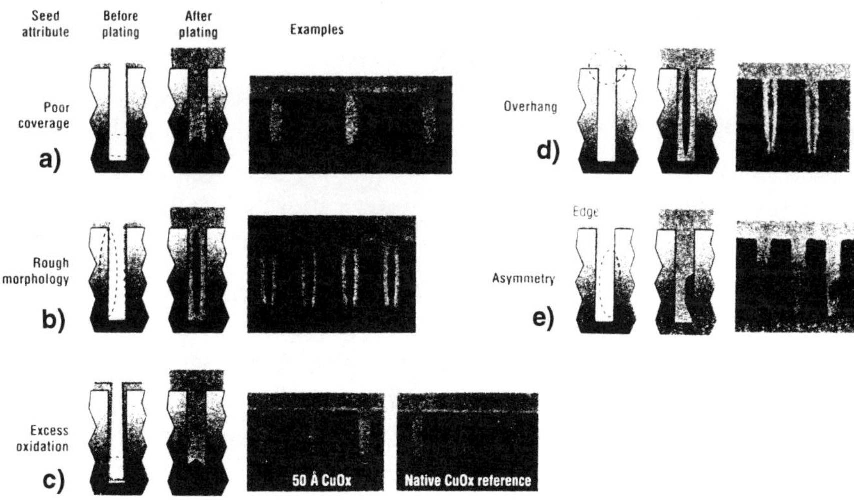

Fig. 16-21 Example set of illustrations and cross-sectional images showing key seed attributes and their impact on electroplating fill results.[33] Courtesy of Novellus Systems.

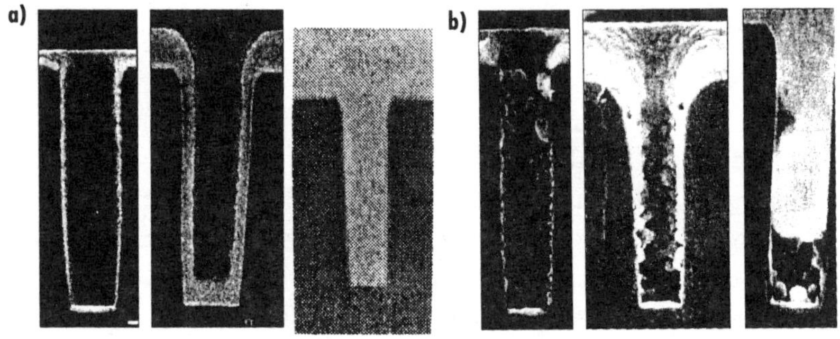

Fig. 16-22 Cross-sectional via images showing (left-to-right: a) Smooth seed coverage only, metal profile following partial fill, and complete fill by electroplating; b) Agglomerated seed coverage, metal profile following some electroplating, and resulting void at completion of electroplating.[33] Courtesy of Novellus Systems, Inc.

thickness along the sidewalls and bottom of deep-submicron, high aspect-ratio vias, either an ionized-PVD process or a CVD process is needed.

3. *Smooth Surface* - The seed layer must possess a smooth and continuous morphology. If the seed layer on the recess sidewalls is rough (or agglomerated) voids are also likely to form during plating. Rough sidewalls can lead to premature feature closure, and the discontinuity of an agglomerated film causes nucleation to fail at some locations. Either effect can give rise to creating a void (Fig. 16-21b). However, with a continuous, smooth seed-layer, plating readily occurs on all surfaces, thereby facilitating the subsequent as-desired, bottom-up filling (Fig. 16-22a).

It should be noted that the temperature of the wafer during the PVD process must be kept low to ensure that a smooth seed-layer surface is obtained (i.e., at room-temperature, or lower). Elevated wafer temperature during the seed-layer sputter-deposition process leads to a rough Cu surface (or even agglomeration, Fig. 16-23).

4. *Minimal Oxidation and Corrosion* – Oxidation and corrosion of the Cu seed-layer must be minimized both prior to plating and when the wafer is first immersed in the plating bath. Either effect can lead to seed-discontinuity and fill voiding. That is, the seed-surface must be largely free of oxide so that efficient charge-transfer can start to take place when the plating process is initiated. If the oxide on the PVD-Cu-surface is acid-soluble and sufficiently thin (i.e., in the thickness range of 10-20-Å), such initiation of the plating process is routine, and filling is complete.

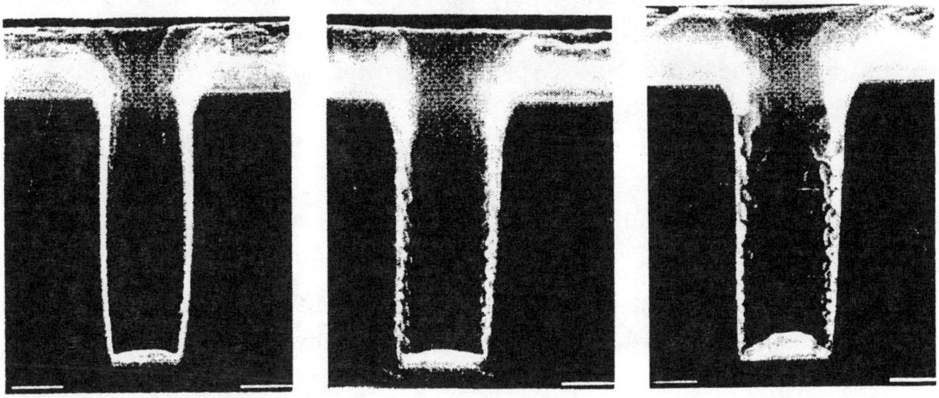

Fig. 16-23 Effect of wafer temperature on surface morphology of Cu deposited into vias by I-PVD. Courtesy of Applied Materials

However, when the cuprous oxides are thicker (i.e., 50-100-Å-thick), a greater degree of bottom voiding is observed (Fig. 16-21c).

5. *No Overhang at the Top Corners of Trenches and Vias* - Excessive overhang can lead to premature closure of the top of the recess during the plating process (Fig. 16-21d).

6. *No Asymmetry of Coverage Between Radially-Inward- and Radially-Outward-Facing Sidewalls* - Too thin a seed-layer on one side can impair nucleation and result in void-formation (Fig. 16-21e). That is, a much thinner seed-layer will form on that side than would be expected, and a void can thus result. This means that control of directional deposition during the PVD process must be maintained across the wafer plane from center-to-edge.

7. *Preferred Crystallographic Structure* – The crystallographic texture of the Cu seed-layer is influenced strongly by that of the underlying barrier-layer. This texture, in turn, determines the microstructure of the electroplated Cu deposited onto the seed layer. As noted previously, a (111)-orientation of the Cu is generally preferred because such films have the best overall film characteristics. A smooth Cu seed-layer-surface is also required to obtain this crystal orientation in the Cu layer that is subsequently deposited by electroplating.

In order to obtain a strong (111) Cu texture, it is also important that an *in situ* Cu-seed-layer deposition be performed. That is, the seed-layer deposition process should be carried-out in the same cluster PVD-tool in which the barrier-layer PVD process is performed. This ensures that the surface of the barrier-layer (typically consisting of a reactive metal, such as Ta), does not form an oxide. If the barrier layer is exposed to air prior to Cu deposition, the resulting (111) Cu texture is much weaker than without an air break. This is probably attributable to oxidation of the barrier surface, and the resulting elimination of the pseudohexagonal surface that would otherwise allow for heteroepitaxial deposition.

Controlling the via shape itself is also important to ensure repeatable feature-filling. That is, the via-shape influences the filling by impacting the continuity of the seed-layer. Problems with the shapes of the vias often occur when trying to integrate an organic low-*k* ILD and Cu. That is, the via profiles of such materials frequently exhibit undercutting and notching as a result of the via-etch process (see Chap. 14). Such notches in the via sidewall profile can cause a discontinuity in the seed layer, and adversely impact the ability to completely fill the feature.

To get adequate seed-layer thickness into along the sidewalls and bottoms of high aspect-ratio recesses, either an ionized-PVD-Cu or a CVD-Cu process may be needed. (Ionized-PVD technology for such applications is discussed in Sect. 16.3, and additional information about seed-layer deposition with CVD is given in later paragraphs of this section.) As also noted in Sect. 16.3.2, the ionized-PVD process offers an additional advantage provided by the fact that a negative-bias can be applied to the wafer. This not only causes the ionized metal-atoms to preferentially strike the bottom of the via, but it allows the kinetic energy of these ions bombarding the wafer to be adjusted. For appropriate bias levels, this can cause re-sputtering from the feature base, which redistributes sputter-deposited material onto the lower sidewalls. The improvement in the conformality of the seed-layer thickness helps enhance filling of the recess without voids during plating. The impingement of the energetic ions on the top corners also helps keep the top opening from forming an overhang. Since the aspect-ratios of mid- and upper-level lines and vias is not likely to exceed 5:1, it is probable that ionized PVD techniques will continue to find use for these applications in future technology generations.

Unfortunately, as the feature sizes get smaller without a corresponding reduction in the ILD thickness, the aspect-ratio of the recesses gets larger. This is a situation that is likely to be encountered for the lower metal-levels of sub-0.1-μm interconnect structures. As noted in Sect. 16.3.1, for a PVD process the amount of material entering the recess must shared among all of the recess

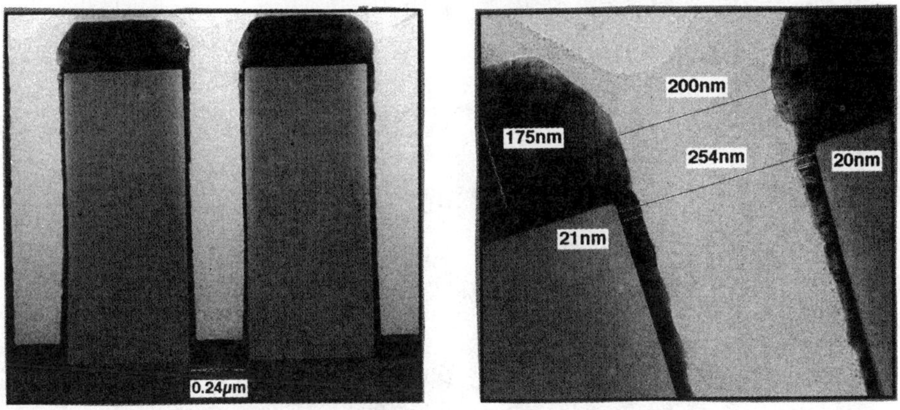

Fig. 16-24 SEM showing how thin the sidewall coverage is in high-aspect-ratio-features with I-PVD compared to the deposited layer thickness on the top surfaces of a wafer.

sidewall and bottom surface areas. For a via with an aspect-ratio of 5:1 the surface area of the sidewalls and bottom is 21x the surface area of the top opening. Thus, the material-thickness on all surfaces can only be 1/21 times as thick as the film deposited on the flat areas adjacent to the recess. Even at the 0.18-μm technology node, where the aspect ratio may be 5:1, a film-thickness of 2000-Å must be deposited in the flat regions to get a 200-Å-thick film on the via sidewalls (Fig. 16-24). Thus, at some time soon, it is predicted that PVD will no longer be capable of depositing an adequate seed-layer. Some have called the PVD-seed-layer the "weak-link" in the Cu-electroplating process.

At the point when the aspect-ratios for the lower levels of the interconnect system get to be too high to be adequately covered with any of the available PVD-processes (or if such a thick PVD-layer is needed that it causes too great an overhang), it may be necessary to switch to a CVD-Cu seed-layer. Some predict that this will occur for either the 100-nm or 70-nm CMOS nodes. Figure 16-25a shows a CVD Cu seed-layer on a CVD barrier layer deposited in a high-aspect-ratio recess that is only 0.14-μm wide. The CVD-Cu seed-layer is totally conformal and 500-Å-thick. The remaining 300-Å-wide slot was filled by electroplating without voids.[32] Figure 16-25b shows another example of a CVD-Cu seed-layer that is conformally deposited over high-aspect-ratio features.[34] One concern of using CVD-Cu seed-layers is the strength of their adhesion to the barrier-layer materials. Many freshly deposited barrier-layers (especially Ta-based), have highly reactive surfaces. These surfaces can degrade organic compounds resulting in interfacial contamination that prevents good adhesion.

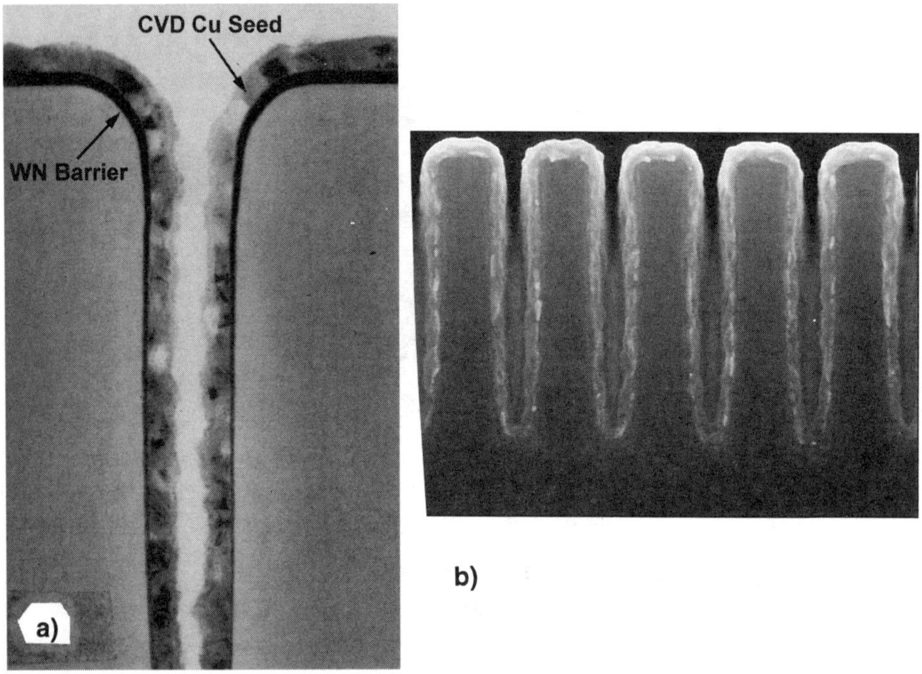

Fig. 16-25 a) SEM showing a CVD-Cu seed-layer integrated with clean and PECVD tungsten nitride (WN) barrier. (Courtesy Genus) b) SEM showing CVD-Cu seed-layer. (Courtesy Applied Materials)

This problem will need to be resolved by optimizing the barrier-layer choice, and improving the CVD deposition-process and precursor chemistry.

Two other alternative replacements for the PVD-Cu seed-layer have also been reported. The first is an electrochemical deposition process developed by Semitool. In their method, an ultra-thin Cu-layer (e.g., 200-Å-thick) is first deposited by PVD for adhesion purposes. This ultra-thin Cu-layer is then conformally thickened to provide a full-thickness seed-layer with an electrochemical process. Finally, the electrochemical process is switched over to a conventional electroplating chemistry to complete the feature fill.[35]

The other alternative replacement technology for PVD-Cu seed-layers is atomic-layer deposition (ALD - see Chap. 4, Sect. 4.6.1 for more background information on ALD).[36] ALD is a monolayer, stepwise growth-process that exploits the binding-energy-difference between chemisorption and physisorption. The growth proceeds by exposing the substrate-surface alternately to the precursors, and following each reaction, the excess species and by-products are removed with a nitrogen-pulse. The reactant-source pulses are thus separated by a pulse of an inert gas. As a result, ALD is capable of depositing conformal-films with thicknesses that can be determined accurately by the number of deposition-cycles. The Cu precursor chosen for the work in Ref. 36 was $Cu(hfac)_2{\cdot}XH_2O$ (see Sect. 16.5.3), and methanol, ethanol, and formalin were examined as the reducing agents. Smooth, conformal, near-bulk-resistivity Cu seed-layer films of 20-nm thickness with good adhesion were deposited by this ALD method.

As noted in Sect.16.3, another reported problem involving the PVD-seed-layer is its selective adhesion to barrier-layer materials. That is, the PVD-Cu seed-layers adhere well to Ta, but less well to TaN. Cu-films used with TaN barrier-layers can peel away during the later Cu-CMP process. Unfortunately Ta is not as good a diffusion barrier as TaN. To overcome this problem, use of a Ta/TaN bilayer has been reported.[37] Even better adhesion by the PVD Cu seed-layer was obtained by using a TaSiN barrier-layer.[26]

16.5 COPPER INTERCONNECT FILM DEPOSITION

Copper films for IC interconnect applications can be deposited in a number of ways, including: electrolytic-plating; electroless-plating; chemical-vapor-deposition (CVD); and physical-vapor deposition (PVD). Electrolytic-plating of Cu is the dominant technique used to fill recessed damascene features in deep-submicron ICs. Therefore, it is discussed in the greatest detail here (i.e., in Sect. 16.5.1). Electroless-plating of Cu and CVD-Cu are described more briefly in Sects. 16.8.2 and 16.5.3, respectively. PVD technology of Cu-films (used primarily to deposit Cu-seed-layers prior to the electrolytic-plating filling process), was covered in Sect. 16.4.

16.5.1 Electrolytic-Plating of Copper

As described in Chap. 15, dual-damascene processes have been developed for the multi-level Cu-interconnect structures of deep-submicron ICs. The damascene approach avoids the problems associated with having to etch Cu-films, but a challenge of a different sort arises from the need to fill the high-aspect-ratio damascene recesses without voids in the metal.

Line-of-sight PVD techniques are inadequate for filling such features because accumulation of deposits at the upper-corners of trenches closes-off the opening of the recess and this leads to void-formation. Chemical-vapor-deposition (CVD) has the potential for conformal coverage, but numerous chemical and hardware issues have hampered development of a manufacturable fill process for Cu. Thus, to the surprise of many in both the semiconductor and plating industries, electrolytic-plating has emerged as the deposition method of choice for damascene-Cu interconnects (although advanced sputtering techniques are still needed for depositing thin barrier- and seed-layers, as described in Sects. 16.3 and 16.4) before the filling of the recesses with a Cu-plating process.

Electrolytic-plating (also often called *electroplating*) and *electroless-plating* of metals are older technologies still used in a variety of industries including hardware, auto, and electronics. In the electronics industry these processes have been (and still are) widely used in the printed-circuit-board (PCB) industry to plate copper layers on both sides of plastic boards for PCB interconnects. By the late 1990's, work was begun to develop electrolytic-plating processes for IC fabrication technologies (especially for depositing copper films for IC interconnect structures).

The key reasons why development of electroplating processes for on-chip metallization was pursued stemmed from the future requirements of IC interconnects. That is, it became clear that in deep-submicron ULSI technologies, low-temperature, low-cost, metal-deposition-processes compatible with thermally fragile low-*k* dielectrics would be needed. Electroplating of metals (especially Cu) represented a very attractive process for such applications. This is because deposition of Cu by metal-plating offers a high deposition-rate process (0.5-1.0-μm/min), which in turn permits high-throughput (50 wafers/hr), and thereby also a potentially much lower cost-of-ownership than PVD or CVD. In addition, such processes can be carried out at temperatures $\leq 40°C$, and a well-tuned electroplating-process can fill high-aspect-ratio structures without defects (i.e., trenches with aspect ratios of 5:1 can be completely filled). The details of electroplating copper-films for deep-submicron ICs are described here. As noted earlier, since electrolytic-electroplating requires a closed-conducting-path for continuous electric current, it is not possible to plate directly onto an insulator such as SiO_2. In this case, an electrically-conducting contact-layer must first be deposited (i.e., in Cu electroplating, a low-resistivity Cu seed-layer is deposited on the insulator).

Electroplating processes utilize conductive solutions containing ions of the metal to be deposited.[13,38] In the case of copper, the solution typically contains copper sulfate ($CuSO_4$ in the concentration range of 17.5-60g/L), sulfuric acid (H_2SO_4 in the concentration range of 45-325 g/L), and water. The $CuSO_4$ in this acidified solution breaks up into Cu^{2+} (cupric-ions) and $SO_4{}^{2-}$ (sulfate-ions). The sulfuric acid added to the solution plays several roles: **a**) it dissolves any native copper-oxide on the copper seed-layer, so that the copper underneath is exposed, allowing the plating process to be reliably initiated across the entire wafer;* **b**) it

* Both CuO and Cu_2O (cuprous oxide) are unstable in a strongly acidic medium (pH < 3).

increases the conductivity of the electrolytic solution; and **c**) it improves the wetting capability of this solution. In commercial processes, proprietary organic additives are also added to the plating solution in order to improve the filling of recessed features by the deposited Cu film (as is described in more detail in Sect. 16.5.2.3). As with other types of deposition, the success of the process is measured by factors such as: deposition-rate, film-morphology, trench- and via-filling-capability, deposited-film-thickness-uniformity across-the-wafer and from wafer-to-wafer.** Plating chemistries were evolved from the plating bath formulations used in the PCB industry, and these were made to work for semiconductor applications.

In the Cu-electroplating-process for on-chip interconnects, wafers with a conductive top surface (which in this case is a PVD-Cu seed layer, as mentioned earlier) are immersed face-down in the electrolytic solution. The wafer-surface is also electrically connected to the negative-side of an external dc power-supply (by a conducting sealing-ring at the outer edge of the wafer, see Fig. 16-26a[50]), making it the *cathode* in this circuit. Electric-current is passed so that a chemical reaction occurs at the wafer surface which causes Cu plating to occur (with the cupric-ions that carry electric charge constituting the current in the electrolytic solution, see Fig. 16-26b). That is, as the positive cupric-ions arrive at the negative-biased-cathode they adsorb on its surface. They also acquire two electrons and are reduced to Cu metal (which plates out), with the reaction for Cu being:

$$Cu^{2+} + 2e^{-} \rightarrow Cu^{0} \quad \textbf{(16.6)}$$

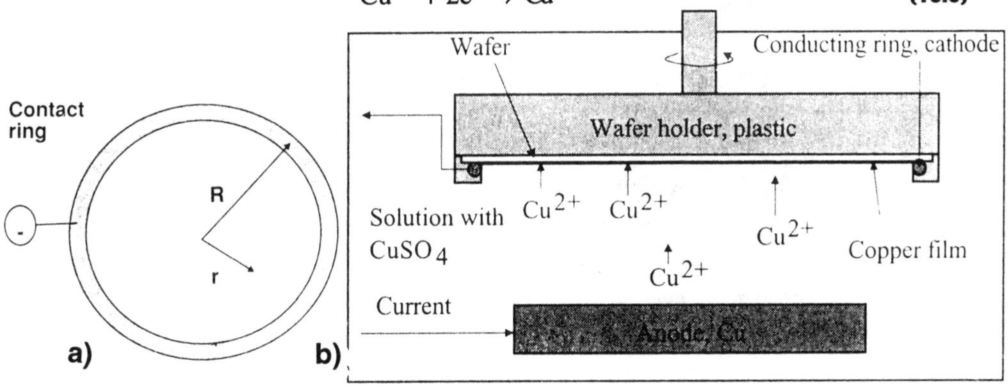

Fig. 16-26 a) Contact ring of a copper plating cell. b) Copper electroplating process.

** While the semiconductor industry has predominantly deposited Cu films electrochemically using acid baths, alkaline baths have also been considered. This is because alkaline baths can plate directly onto TiN diffusion barriers without a copper seed-layer and without the need for organic additives. On such bath studied by Semitool used an Cu-EDTA-complexed bath and the deposited Cu exhibited superior adhesion to TiN.[49]

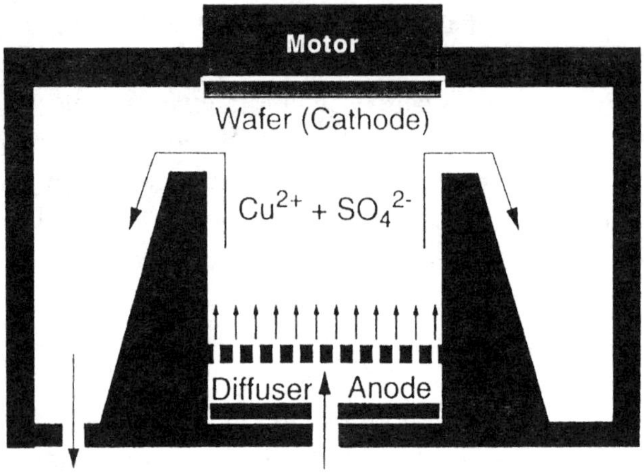

Fig. 16-27 a) Schematic drawing of a cross-section of a Cu electroplating cell.

Another electrically active surface (a solid-copper electrode which serves as the *anode*), is present in the conductive solution to complete the electric circuit (i.e., it is connected to the positive side of the power-supply). At the anode, the oxidation reaction occurs that balances the current flow at the cathode, thus maintaining the electrical neutrality in the solution. That is, all of the cupric-ions removed from solution by the plating of Cu onto the cathode are replaced by dissolution from a solid-copper anode immersed in the solution. (The copper atoms of the anode are oxidized to form the Cu^{2+} ions. These ions are released from the anode and dissolve into the solution.) The metal-ion formation-reaction is thus sustained by passing an electrical current through the bath.

If the applied potential is reversed, the cathode and anode are also reversed, and so are the locations where Cu deposition and dissolution occur. Alternating between deposition and dissolution (a process known as *pulse-plating*) may be useful in order to improve the filling capability of a plated film (see Sect. 16.5.2.5 for more details on pulse-plating). To enhance the uniformity of deposition across the wafer, the electrolyte is usually pumped against the wafer during the plating process, and the wafer may also be rotated at the same time (see Fig. 16-27).

The electroplating rate is a direct function of the current-density. To obtain uniform deposition across a wafer, the plating cell must be designed to give uniform potentials across the wafer surface. (Because of concerns about Cu contamination, it is also critical that the plating cell be designed so that no copper is deposited on the backside of the wafer.) The current delivered to a conductive surface is directly proportional to the quantity of metal deposited (according to

Faraday's Law of Electrolysis). By using this relationship, the mass deposited can be readily controlled through variations of plating current and time.

If no potential is applied across the interface between a metal and the electrolytic solution, an equilibrium potential exists between the two (see Fig. 16-28). Once the potential is shifted by an external power supply away from the equilibrium potential, a current will be driven across the interface. Under conditions typical of most plating processes, this current-flow is approximated by an exponential relationship known as the *Tafel equation*.

Figure 16-28 shows a current-vs.-potential curve of a copper electroplating process (also referred to as a *Tafel plot*). As the potential is increased from equilibrium to more negative values, the current increases in an exponential manner (Tafel region). In this region of operation the overall deposition-rate is determined largely by reaction-rate kinetics at the cathode. Ordinarily, conversion of Cu^{2+} ions to plated Cu-atoms occurs with nearly 100% efficiency, so that the electric current-density is a direct measure of the plating-rate. This also implies that ohmic effects (primarily through the uniformity of current distribution) are one of the important parameters that impact plating-thickness uniformity across a wafer. The strong dependence of current upon potential implies that the plating cells must be designed to enhance the uniformity of the potential across the wafer surface. The resistance of the barrier/seed-layer also impacts such ohmic effects, as will be described in later paragraphs.

As the potential continues to increase, mass-transfer effects become predominant and a limiting current-plateau is reached. This occurs because the Cu^{2+} species reacting at the cathode no longer reach the interface at a rate

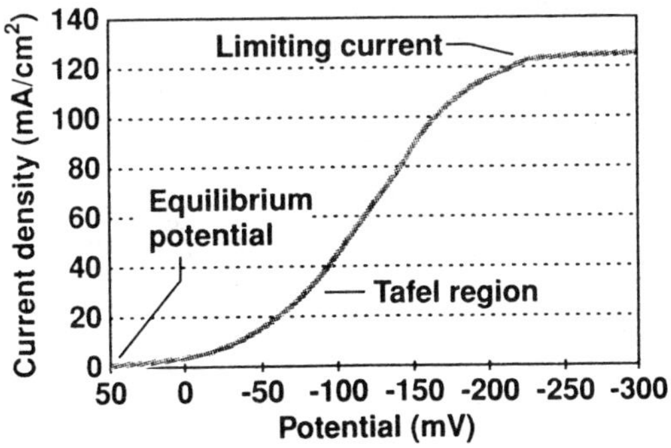

Fig. 16-28 A current-potential curve (Tafel plot) typical of a copper electroplating process.

sufficient to sustain the high reaction-rate. As a general guideline, Cu plating processes are operated at currents no greater than 30-50% of the limiting current to avoid undesirable film-deposition-characteristics.

To ensure that the rate of mass-transfer of Cu species to the interface is large compared to the reaction-rate (and that the mass-transfer rate is uniform across the wafer), the rates of migration, diffusion, and convection must be well understood and controlled. Convection is the most important mode of mass transfer in a liquid, and can vary from stagnant to laminar (or turbulent) flow. (See Vol. 1, p. 72 [Chap. 3] and p. 156 [Chap. 6] for a discussion of turbulent and laminar flow, and p. 157-8 [Chap. 6] for a discussion of mass-transfer in regions where movement by convection is zero [i.e., in stagnant regions].) In some electroplating systems, the wafer is also rotated while it is immersed in the electrolyte to get a uniform boundary-layer and to ensure that the fluid mechanics and material balance for the copper-diffusion match well everywhere on the wafer.

Generally, electroplating-processes have high deposition-rates (e.g., 400-1000 nm/min) and form desirable metallurgy. However, as noted in Sect. 16.4, electroplating requires a thin seed-layer of Cu (about 50-nm-thick) which must first be deposited by some other method (e.g., CVD or PVD) onto the barrier/adhesion-layer. As noted in Sect. 16.4, the characteristics of the seed-layer are another of the critical factors that impact the ability to completely-fill recesses by electroplating without voids or other defects. It would be ideal if a single layer could serve as both a barrier- and seed-layer, but currently two materials are used. An adhesion/barrier layer like Ta, TaN, Ta/TaN or TaSiN is first deposited, with a Cu seed-layer applied in a second step.

Note that to obtain uniform deposition across the entire wafer, the current-density should be equal at all locations. Since the conducting layer has some resistance, there will be a voltage-drop from the edge-to-center of the wafer. This means that the edges of the wafer will plate faster than the wafer center (and the deposition-rate non-uniformity is greatest at the initiation of the electroplating process). The problem also gets more severe as the wafer diameter is increased, and thus is worse for 300-mm- than for 200-mm-wafers. To minimize such deposition-rate non-uniformity-effects, the voltage-drop from edge-to-center must be kept tolerably small. Since the barrier-layer is very thin and typically has a relatively high resistivity, a Cu seed-layer must be formed on top of the barrier-layer to increase the conductivity to a sufficiently high degree. (It has also been reported that reverse pulse-plating [as discussed later], will help reduce the non-uniformity of deposition, as the copper also dissolves more rapidly during the part of reverse-plating cycle in places where it is most heavily plated.[25])

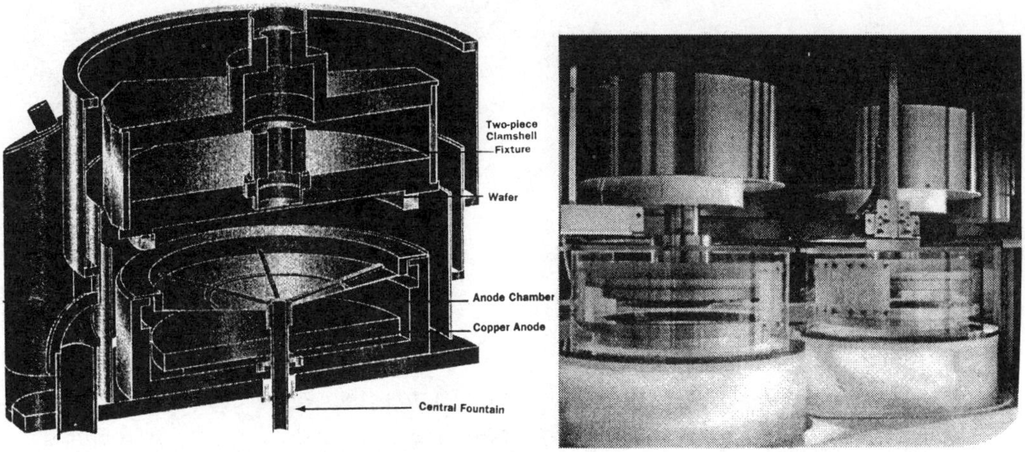

Fig. 16-29 Schematic drawings of plating cell of the Novellus SABRE^{TM} electroplating tool. Courtesy of Novellus Systems.

Copper electroplating tools are commercially available from a number of companies. In 2000 the three companies with the largest market share were Novellus Systems, Semitool, and Applied Materials. The market for copper tools is expected to continue to grow significantly in the decade starting at 2001. Figure 16-29 shows illustrations of the Cu electroplating cell in the Novellus Systems SABRE^{TM} elecroplating tool. Figure 16-30 is a drawing of the overall SABRE tool layout. This electrofill system enables void-free copper in extremely narrow high-aspect-ratio structures (up to 10:1) without backside contamination. Throughputs of up to 75 wafers are possible, with manufacturing availability above 90%.

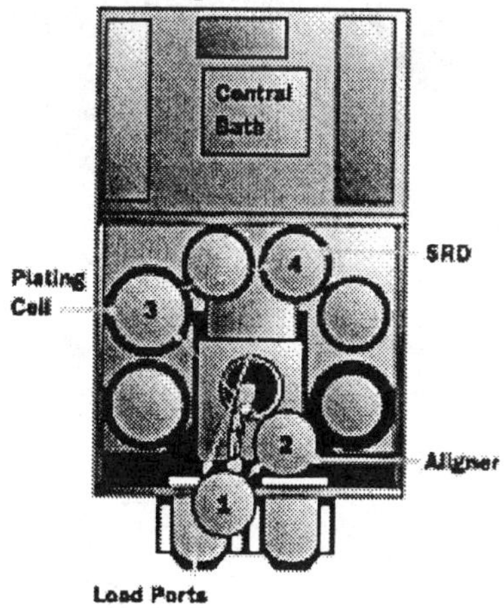

Fig. 16-30 Schematic drawing of the layout of the Novellus SABRE^{TM} electroplating tool. Courtesy of Novellus Systems.

Figure 16-31 depicts the Applied Materials Electra™ Cu electroplating system. It has dual process cells and dual wafer handling blades for high throughput and reliability. Wafers are loaded into the twin plating cells. After plating they are processed in an edge-cleaning station followed by a spin-rinse-dry. The wafers are then moved to the integrated anneal chambers before being delivered back to the cassettes.

16.5.2 Filling High-Aspect-Ratio Recesses Without Voids Using a Cu Electrolytic Electroplating Process

The deposition process of Cu films for dual-damascene IC interconnects must not only have a high and uniform deposition-rate across the wafer, but it must also be capable of producing *void-free fill* of high-aspect-ratio damascene features. There a number of factors that impact the ability of the Cu electroplating process to fill such high-aspect-ratio recessed structures without the formation of voids or seams, including: **1**) the incoming feature profile; **2**) barrier-layer and seed-layer attributes; and **3**) key electroplating process parameters that are optimized to encourage acceleration of deposition near the base of damascene features ("bottom-up" fill). That is, for deep-submicron IC applications an accelerated plating rate is desired within submicron recessed features to achieve void-free fill. The attributes of the barrier- and seed-layers needed for permitting void-free fill were discussed in Sect. 16.3 and 16.4, respectively. The void-free filling factors involving the electroplating process are considered in the following subsections. They include: **1**) the electroplating solution must provide *rapid uniform wetting of the wafer surface*, especially within the area of high-aspect-ratio features; **2**) the deposition process must be *properly initiated*; **3**) appropriate

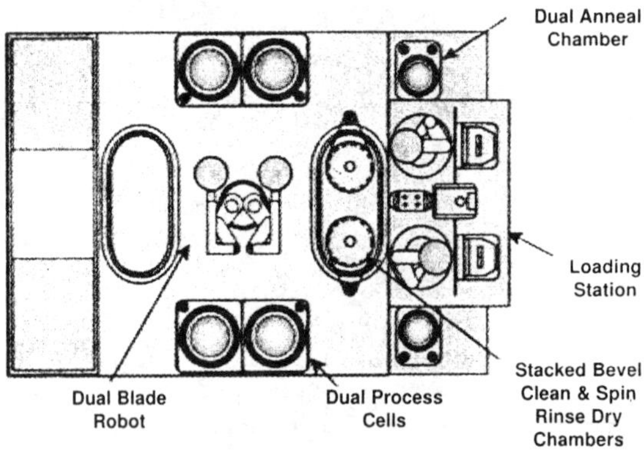

Fig. 16-31 Schematic drawing of the layout of the Applied Materials Electra™ Integrated ECP system. Courtesy of Applied Materials.

organic additives are added to the electroplating baths to facilitate void-free filling; and **4**) the *electric-current* waveform used for the deposition process can impact the filling characteristics. Topics **1**) and **2**) involving the electroplating process itself are described in Sect 16.5.2.1, while the effect of additives in the plating bath is covered in Sects. 16.5.2.2 and 16.5.2.3. The topic of plating-bath chemistry-management is discussed in Sect. 16.5.2.4 and the subject of pulsed plating in Sect. 16.5.2.5. The issue of topography in plated Cu-layers is addressed in Sect. 16.5.2.6.

16.5.2.1 The Electroplating Process Sequence: The electroplating process sequence typically consists of the following events. The wafers are immersed face down in the plating cell containing the electrolytic bath. Adequate surface wetting prior to initial deposition of the copper must occur. That is, rapid uniform wetting of the wafer surface and the inside of the high-aspect-ratio features is another critical requirement for good filling performance. For a solution with good wetting properties, capillary action should force solution into submicron features. Wetting is highly dependent on plating chemistry, seed-layer structure, composition, surface oxidation, and contaminants, as well as how the wafer is immersed into the plating bath and how the plating is initiated.[39]

The time the wafer is being wetted is referred to as the "zero-current-induction- time," or "induction-time" (see Fig. 16-32). During this time no external potential is applied, and thus no cathodic current flows. It represents the time period over which the wafer is exposed to the plating electrolyte at an uncontrolled (floating) potential. If the wafer is not polarized during this time, copper corrosion and cuprous oxide dissolution can occur.[40] To avoid the possibility of corrosion during the initial immersion period, the wafer may instead be introduced into the bath with an immediate small current flowing (or with a very-small cathodic potential being applied). In this way these corrosion reactions can be reduced or reversed. Immersing the wafers while applying a small cathodic potential is applied is thus referred to as *cathodic protection.*[33]

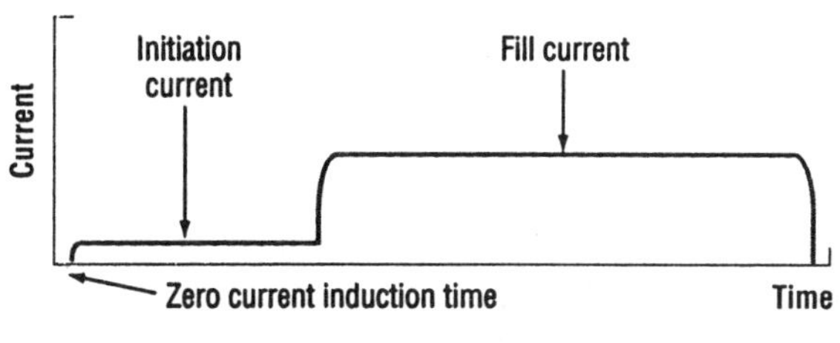

Fig. 16-32 Example of a two-step dc plating-waveform with three key features identified.[33]

After the induction time has passed, an "initiation current" is applied to the wafer (see again Fig. 16-32). During this time the current level is small, the deposition rate is slow, and the deposition occurs conformally. Thus, the seed-layer thickens uniformly during this period (allowing it to become continuous if a discontinuity in its coverage had occurred). In some cases, a thicker Cu seed-layer is needed to support the nucleation and high current density required in the subsequent fill step.

At the conclusion of the initiation-current stage, the "fill-step" stage is performed (Fig. 16-32). Since the damascene recesses are typically about 0.5-μm deep, and the Cu deposition-rate during the electrofilling stage of the plating process is ~1.0-μm/min, the fill-step lasts 20-40-sec, depending on bath conditions and feature size. Smaller features tend to fill earlier because they develop the conditions necessary for bottom-up fill (as will be described in the next sections). Since the larger features (i.e., with aspect-ratios much smaller than unity) fill more slowly, the copper layer must typically be deposited to a thickness somewhat greater than that of the damascene dielectric film stack. This issue will be considered in more detail later.

16.5.2.2 Additives in Plating Baths for Damascene Cu Interconnects: As noted earlier, the most important challenge for depositing Cu-films by plating for damascene applications is to completely fill vias, holes, trenches, and their combinations without voids or seams. For such defect-free filling, a higher deposition-rate along the sides and the bottom of the feature is needed than on the flat, top-wafer surface. As early as 1990 at IBM, it was discovered that plating from certain electrolytic solutions that contain additives leads to this kind of "bottom-up" filling. Since that time, work has continued to understand the role of such additives in plating solutions that will give the best filling results.

Additives to electroplating baths were originally developed to enhance the brightness and hardness of metallic deposits. Both of these attributes can be related to decreased grain-size in the deposits and reduction (or leveling) of the topography over rough surfaces. When the Cu-plating for submicron IC interconnects was being developed it was also observed that without additives the grain-size of the Cu was large (roughly 1-μm) and the film surface was rough. Additives led to a much finer grain-size (<1000-Å) and a smooth, highly reflective surface that had significantly less topography than the underlying substrate. From that work it was surmised that perhaps additional additives could be found that would also permit the damascene recesses to be filled without voids. It was discovered that a number of different additives (i.e., four or more) need to be added to the plating solution to give the best filling results. (It also appears that the effects of these multiple additives are interactive.) Here such additives and their characteristics are described. In the next section a model is given for how these additives allow recesses to be filled from the "bottom-up."

The plating baths, as formulated by a number of specialty suppliers, include low concentrations (in the millimolar or micromolar range) of a variety of organic agents (generically referred to as *additives*) which serve to modify the properties of the deposited films as compared to results otherwise obtained from additive-free copper-sulfate/acid solutions. The manufacturing processes generally employ Cu plating-solutions purchased commercially, and the identity of the additives is proprietary. The examples of the various additives listed in this section come from the patent literature, as well as from various research articles that describe the additives being investigated. At the outset of the discussion the various classes of plating-bath additives are described. These classes include: suppressors, accelerators, and levelers (which, as will be noted, are also referred to by other names).

Accelerators are additives that speed or enhance deposition of the Cu during electroplating. Accelerators adsorb strongly on seed surfaces, where they also compete with suppressors for surface adsorption sites. Accelerators are usually unsaturated small organic molecules containing a polar sulfur, oxygen, or nitrogen functional group. In addition to increasing the Cu deposition-rate, they promote dense nucleation (and consequently the growth of films with fine grains). As a result, accelerators are also referred to as *brighteners* (i.e., because they cause the Cu-film to have a smooth, well-textured [i.e., bright] finish). Accelerators are added to plating baths in low concentration-levels (1-25 ppm). Examples of accelerators reported in the literature include: bis(3-sulfopropyl)disulfide (SPS);[41] sulfonated acetylthiourea;[42] 3-mercapto-1-propanesulfonate (MPSA);[33] and dibenzyl-dithio-carbammate.[43]

Suppressors (also called *inhibitors, carriers, surfactants or wetting-agents*) are plating-bath additives that reduce the Cu plating rate. Suppressors are usually present in the plating bath at higher concentrations (200-2000-ppm), so that their concentration at the interface is not strongly dependent on their rate of mass-transfer or diffusion to the wafer surface.

The suppressors are generally polymeric surfactants (with 1500 molecular-weight polyethylene glycol [PEG] reportedly being the most widely-used material for this purpose).[33] The suppressor molecules slow down the deposition-rate of Cu by adsorbing on the wafer surface, where they form a monolayer which offers a diffusion barrier to the cupric ions. In addition, they enhance the cathodic polarization needed for fine-grain structure. In order for the polyethylene glycol to act as an effective suppressor, however, chlorine ions must also be added to the bath (in the form of HCl), and thus the chloride ion Cl^- can be considered a co-suppressor. In fact, HCl is the only non-organic additive in plating solutions.[33]

The role of chloride ions should be discussed in more detail. Chloride ions are believed to adsorb on the wafer surface, where they attract PEG molecules, and cause them to be more strongly adsorbed there as well. In fact, it has been observed that low chloride concentration leads to weak or negligible PEG adsorption.[44] This effect results in poor suppression of current at the wafer surface, and therefore the deposition on the wafer surface is not suppressed (which impairs the desirable condition of a differential deposition-rate between the top surface and inside the recesses that assists the complete filling of recesses). On the other hand, as the Cl^- concentration is allowed to increase to the appropriate level, good suppression of Cu deposition on the wafer surface is obtained. The levels of Cl^- within the recesses, however, might not be sufficiently high to maintain a high level of PEG adsorption, allowing accelerator molecules to be adsorbed on those surface instead. Thus, the conditions for good "bottom-up" filling are not disrupted. But too high a concentration of Cl^- concentration can interfere with "bottom-up" filling, because Cl^- can compete successfully with the accelerator molecules for surface sites in the recesses, and subsequently attract suppressing PEG molecules within the recess.

Levelers are plating-bath additives used to reduce surface roughness and to smooth the topography of the plated-film. They perform this role by locally reducing the Cu deposition-rate (i.e., by adsorbing on surface sites, blocking them from serving as locations where cupric ions can plate). Hence, levelers can also be considered as kin to suppressor additives. However, unlike suppressors, levelers are added to plating baths only in low concentrations (i.e., 1-25 ppm, comparable to the accelerator concentration). Hence, unlike suppressors, the concentration of levelers at the interface is mass-transport dependent and the large size of the leveler molecules also impedes their migration into trenches. In these ways, isolated locations (such as the insides of recesses, where mass-transport is limited) have fewer leveler species adsorbed on their surfaces. Thus, these regions experience less deposition-reduction effects caused by leveler adsorption than do the top, flat-surface regions of the wafer.

On the other hand, protruding-surfaces on wafer flat areas, or trench corners (to which mass-transport by diffusion or surface-migration is more efficient), experience a greater suppression of the deposition-rate. That is, the number of leveler molecules adsorbed on wafer surfaces is continually being depleted during the plating process – by incorporation into the film or by electrolytic decomposition. The suppression of plating at protrusions on the wafer flat-areas and the upper corners of the trenches is maintained by constant replenishment of the leveler species that block surface sites (provided the leveler is present at the correct concentration). The suppression of the growth of protrusions on the wafer surface results in a more *level* film surface on the top-surface of the wafer (hence the name *leveler*). The suppression of deposition on the corners of the trenches allows the tops of the recesses to remain open while the "bottom-up" filling proceeds to completion. Examples of leveler additives listed in the literature are Janus Green B (JGB) a kind of dye,[45] and quatenary ammonium salt (QAS).[43]

For Cu-plating in dual-damascene applications, the composition and concentration of accelerators and suppressors are selected such that the accelerator surface-concentration dominates that of suppressors on the interior surfaces of trenches and vias. Local deposition-rates are thus suppressed on the flat surfaces on top of the wafer relative to the insides of recessed features, leading to the desired "bottom-up" deposition and void-free metal filling.

As a result, when these various additives are correctly balanced in the plating bath solution, it is possible to get a correct fill without voids and seeds. However, high precision is required to balance the roles of these materials. This topic is addressed in Sect. 16.5.2.4, where the issues of managing plating-bath chemistries are covered.

It should also be noted that in addition to void-free filling of recesses by Cu layers, it is also desirable to end up with a smooth (not rough) and topography-free (i.e., level) film after plating. This will allow a CMP process without an extensive overpolish time (i.e., that would be needed to get rid of the high spots that accompany rough and non-planar films) to be used. If an extensive overpolish time is required, this will lead to more severe dishing & erosion problems. Adding levelers to the plating bath and employing reverse-pulse plating waveforms are measures used to try and planarize Cu-layer topography.

16.5.2.3 Models of How Additives in Plating Baths Impact "Bottom-Up" Filling: A model that describes the effects of the plating-additives on the "bottom-up" filling of damascene recesses was published by Reid in 2000.[33] Reid's model is based on a large volume of experimental work, and it proposes that the mechanism for "bottom-up" fill is primarily due to the accumulation of *accelerator species* within damascene features. It supercedes a model by

Andricacos in 1998 that held that the "bottom-up" filling was driven primarily by the establishment of a gradient of *suppressor polymers* in the recesses.[46]

Before presenting the mechanisms that Reid proposes are responsible for "bottom-up" recess-filling, a few preliminary remarks are in order. First it has been observed by others that if the co-suppressors PEG and Cl^- are the only additives present in a plating bath, feature filling is poor. If PEG and Cl^- and an accelerator (e.g., SPS) are the additives present, better filling is seen to occur. If all three additive types (i.e., suppressor, accelerator and leveler) are present, the best filling results are obtained.[47] Furthermore, the concentrations in the bath must be correctly balanced to obtain the best filling effects. This implies that the effects of the three types of additive are interactive, as is also proposed in Reid's model.

Figure 16-33 shows the possible ways for a profile of plated copper to evolve during plating. In conformal plating, a deposit of equal thickness at all points of a

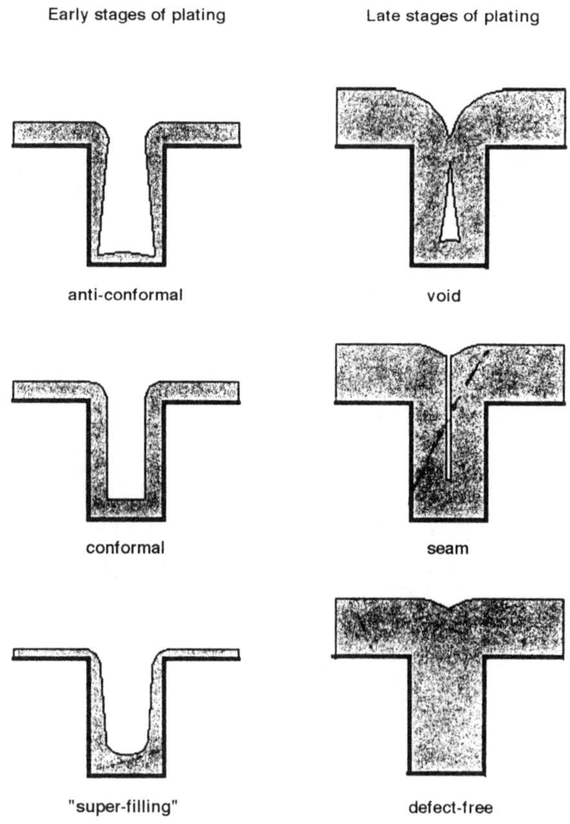

Fig. 16-33 Types of profile evolution in damascene plating.

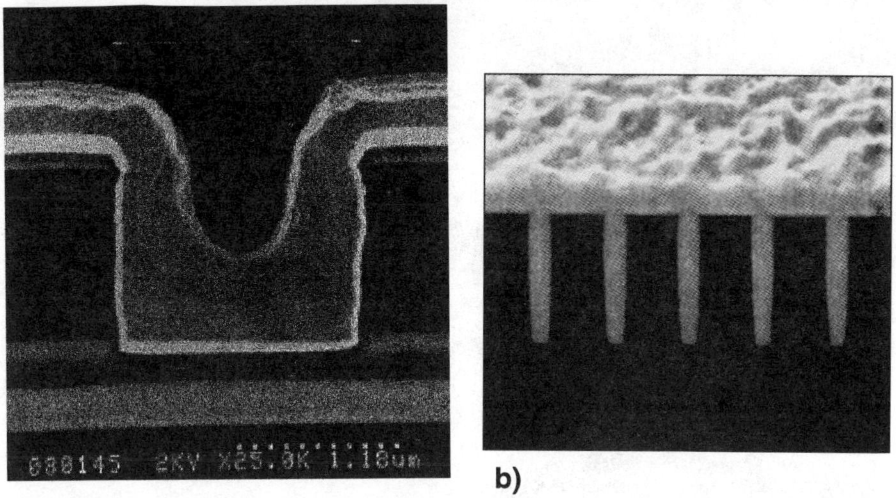

a) b)

Fig. 16-34 a) Cross-section of a partially-filled line showing the profile of electroplated copper that exhibits the "bottom-up" filling phenomena.[48] b) Vias completely filled with Cu.

feature leads to the creation of a seam. If the shape of the feature is re-entrant, a void is formed. Subconformal plating also leads to void formation, even in straight-walled features. Subconformal plating occurs when substantial depletion of the cupric ions in the plating solution leads to significant concentration overpotentials, which in turn, cause the current to flow preferentially to more accessible locations outside of the feature. Thus, for defect-free filling, an increasing deposition-rate along the sides and bottom of the feature is desired ("bottom-up" filling). Fig. 16-34 shows a SEM of partially-filled line showing the profile of electroplated copper that exhibits such a "bottom-up" filling phenomenon.[48] Figure 16-35 shows SEMs of the time-evolution of filling a trench with a Cu-layer in a "bottom-up" manner.[49]

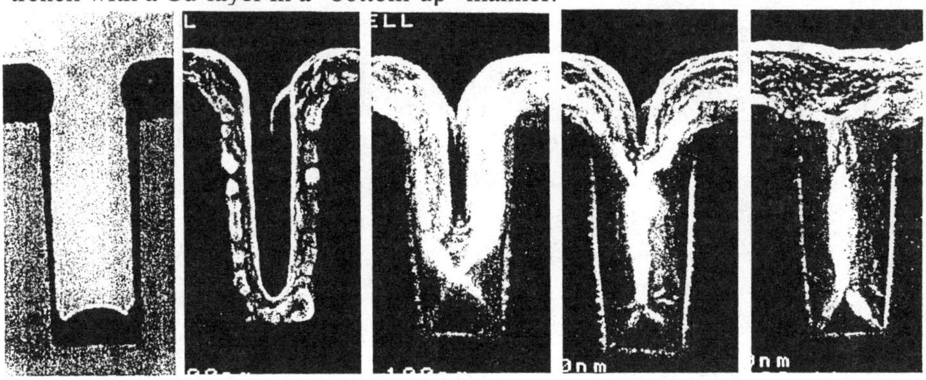

Seed 10sec 20sec 30sec 40sec

Fig. 16-35 SEMs of how "bottom-up" filling of Cu damascene recesses occurs versus time.[49]

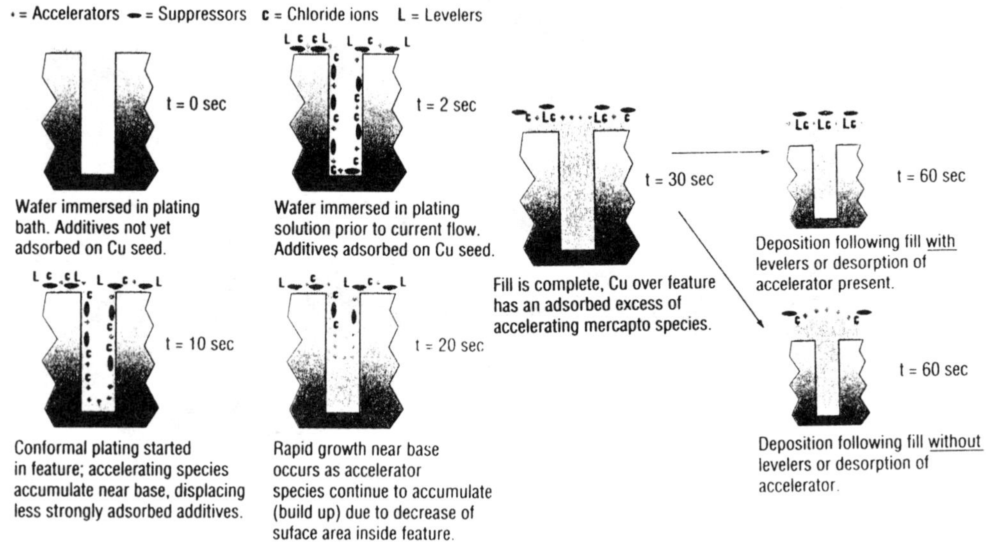

Fig. 16-36 Illustration of additive adsorption behavior during plating. A mechanism for the establishment and propagation of "bottom-up" fill is suggested. By permission of Solid State Technology. Published by PennWell.[33]

Reid's model of the sequence of events that accompanies "bottom-up" filling during Cu-plating is shown using Fig. 16-36. It begins by assuming that when a wafer is initially immersed in the plating bath (t = 0 in Fig. 16-36), that no additives are adsorbed on the Cu seed. However, within a few seconds (i.e., t = 2 sec in Fig. 16-36) both the suppressor and accelerator additives have been adsorbed on all areas of the wafer (including the insides of recesses) at their equilibrium levels. Thus, when the plating process is initiated by applying an external voltage to the cathode, the initial currents on all surfaces should be approximately equivalent on all surfaces. This leads to a deposition-rate that is constant everywhere. This perception is supported by the observed conformal deposition seen in the first 5-10 sec of a filling process (see also Fig. 16-35b - 10 sec). As was shown in Fig. 16-32, the plating current is kept low during this time (initiation current), to keep this conformal deposition thin.

After some time (which depends on current density, chemical concentrations, and feature size - e.g., t = 10 sec in Fig. 16-36), two phenomena are thought to begin to contribute to turn the fill process from a *conformal* filling-mode to a "*bottom-up*" filling-mode. First, the accelerator molecules start to accumulate at higher concentrations at locations near the bottom of narrow recesses. The increase in accelerator concentration near a recess bottom occurs because the surface area within the recess decreases as it fills (even if the fill is conformal), and the accelerator molecules are neither incorporated in the depositing Cu nor are they desorbed into solution. (Recall from Sect. 16.5.2.2 that accelerator species adsorb strongly on seed-layer surfaces.) Meanwhile the PEG and Cl^- species are held less tightly on the seed-layer surface, and so they are displaced by the accelerator molecules and desorb into the solution. Thus, with a higher concentration of accelerator molecules in the areas of geometric concentration, the current-density (and the Cu-deposition-rate) is increased (t = 20 sec in Fig. 16-36).

The concentration of accelerator species in the solution must remain within an appropriate range. Too much accelerator in a bath disrupts fill because the accumulation of accelerating species also begins to take place on the top wafer surface between the recesses, and the differential deposition-rate with respect to the bottom of recesses is lost.

16.5.2.4 Electroplating Bath Control: Copper plating-baths are complex mixtures of chemicals, whose compositions contain various components that must be maintained at correct concentration levels. This is a challenging requirement because the composition of the bath changes as the plating process proceeds. The most dynamic ingredients of electroplating solutions are the organic additives. During normal electroplating the organic chemicals may be consumed in stoichiometric reactions (or they may break down in a variety of oxidative side reactions, some of which are non-stoichiometric). In any event, for consistent performance of the plating-bath, the original concentrations of these organics must be maintained, and their breakdown products must be removed. Even a small imbalance between components of the additives can cause defects in the filling of the trenches and/or degradation of the properties of the electroplated copper. As higher aspect-ratio-recesses are encountered, the ability to fill them properly will thus increase the importance of electrolyte management systems. (It should be noted that the bath solutions themselves are supplied by specialty-chemical vendors, such as Shipley Ronal.)

To enable the plating tools to be used in a high-volume production environment, the complex plating-bath chemistries have to be continuously monitored and controlled. Rigorous analysis of the plating bath chemistry has to be performed to determine how the bath composition changes with time as a consequence of plating, and then appropriate measures to counteract these changes must be taken. Such real-time analysis and feedback control of electrolyte are key to operating plating processes in a manufacturing mode. A self-adjusting, self-monitoring electrolyte is necessary to ensure precise control of the plating chemistry to guarantee repeatable lot-to-lot fill. This has to be done on a large number of plating tools in a wafer fab that may be producing 20,000 wafers per month (with six or more levels of copper interconnect).

In some cases a separate, standalone, closed-loop analysis-and-dosing-system for multiple plating tools is used. In other cases the tool is designed with an on-board closed-loop electrolyte management system (e.g., as is done in the Applied Materials Electra ECP Deposition System™).[50] In the Electra tool, the bath management system measures (and then doses) the bath on-board at the system level, enabling precise control of suppressor, accelerator, leveler, copper, and acid concentrations. The system has a 200-liter tank with filtration on-board and six dosing channels. If the monitoring unit detects that, for example, suppressor x is under-concentrated, it immediately calculates how much is needed to dose the tank, and then does it. Feedback is on the order of minutes, rather than hours.

One method for monitoring the electroplating bath is a technique called *cyclic voltametric stripping* (CVS). The CVS principle has been widely used in the PCB industry, and was subsequently implemented for monitoring the electrolyte baths for IC Cu-plating. In the CVS method, a Cu film is deposited on an inert electrode and then this film is stripped by anodic dissolution. During stripping, the current is integrated over time to quantify the electric charge required for complete film dissolution. This charge is directly related to the molar quantity of the metal stripped (and thus to the amount initially deposited). From this information, calibration curves can be generated that provide reasonably accurate quantification of the total additive concentration in a plating bath. The stripping charge is monitored instead of the deposition charge because it is less sensitive to changing electrode surface states and impurity currents. Monitoring a process that proceeds to a well-defined endpoint also has inherent advantages.

The small plating cells used for plating Cu onto silicon wafers create very dynamic changes in chemical component activity, and this has generated a demand for a continuous, on-line, closed-loop analytical system. ECI Technology (Rutherford, N.J.) has developed such a CVS system for monitoring Cu-plating processes for silicon fabrication,[51] the Quali-Line QLC-5000.™ The CVS bath controller samples multiple plating tanks as well as a standard solution and then outputs the results.

The main limitation of CVS is its sensitivity to "matrix effects." Many bath components and their breakdown products display convoluted electrochemical interactions, so that the stripping-charge response is ambiguous if several components simultaneously change concentration. Thus, it is difficult to precisely correlate observed changes of electrochemical activity to variations in individual bath components over extended times. These attributes make CVS less than ideal if the user's goal is individual monitoring of each bath component. However, under conditions where the principal requirement is to detect and correct changes in one or two electrochemically active components that have a rapid rate of consumption compared to others, the CVS method may be adequate for a closed-loop control system. In addition, CVS can provide rapid measurements with low-hardware investment, and little added waste-disposal burden.[52]

Another more elaborate plating-bath monitoring technology involves metal-free and ion liquid chromatography (also termed *high performance liquid chromatography* [HPLC] and *ion chromatography* [IC]). This method works on the principle of separating each of the individual components in the bath prior to detecting and quantifying them. It can provide information on the concentration and proportional balances of components in the plating solutions, as well as information on contaminants and degradation products.[53] A bath monitoring HPLC system for IC Cu-plating applications is being developed by Dionex Corp.[63] A short-coming of the chromatography methods include waste generation and the fact that the time required to perform the analyses can be long. It appears that a workable approach may be one in which both CVS and chromatography are integrated into a hierarchical method that exploits their relative strengths and minimizes their shortcomings.

When it has been determined that a plating bath needs to be adjusted (or replaced), several strategies can be employed. The simplest (and crudest) approach is to dump the bath when the plating results no longer meet the required specifications (and this is associated with a condition where the organic contaminant level exceeds a specified value). This technique is referred to as *batch dump,* or *bleed-and-dump-to-drain.* The level of contaminants in the bath is determined by correlating the Cu deposition tool throughput (and its related current-density) to the concentration of organic contaminants produced by breakdown of the organic additives. The increase in contaminant concentration is correlated to the amount of charge passed through the plating bath, and this relationship is expressed in units of *mg/amp-hr* (see Fig. 16-37). Thus, if this quantity is known (and the maximum acceptable contaminant-level is specified), the length time that a bath can be used before dumping is known. However, this method generates large quantities of waste that must be treated due to the copper and acid content. There is also no control (in either the general, or statistical sense), over the bath chemistry between dumps.

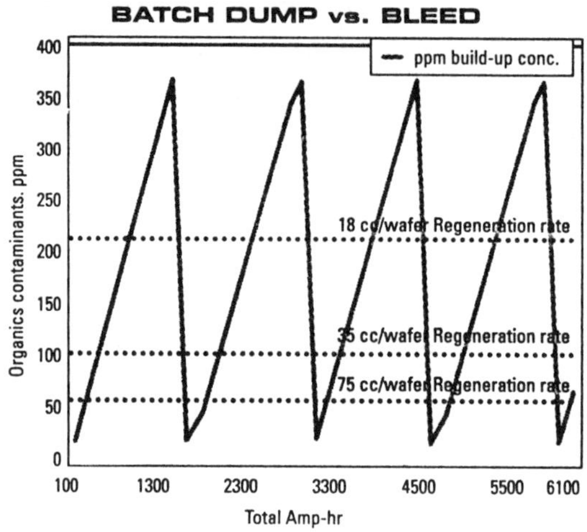

Fig. 16-37 "Batch dump" control versus "Bleed" control (0.3 amp-hr/wafer used to plate 1-μm thickness).[54] By permission of Solid State Technology. Published by PennWell.[33]

In a more refined technique, some fraction (typically about 10%) of the bath is removed daily, and it is replaced with new electrolyte solution. This is called the *bleed-and-feed* method. By analyzing a sample of the removed bath solution (e.g., with CVS), the concentration of additives in the new solution added to the bath can be adjusted to keep the total additive concentration within specified limits. However, as shown in Fig. 16-37, only a higher bleed-and-feed flow-rate can actively control the bath contaminant build-up that adversely impacts the bottom-up filling mechanisms. In addition, the bleed-and-feed method generates about the same amount of waste over time as the batch-dump method. Finally, while it does remove some of the contaminants that arise from breakdown of the organic additives, it does not completely remove them. It only dilutes them somewhat to a steady-state concentration. These drawbacks have led to a new approach to managing bath chemistry, referred to as a *remove-and-reconstitute* technique.

In this closed-loop method, the bath chemistry is continuously regenerated by stripping all organics (active and contaminant) from the recirculating bath. Then new organic additives (with the original concentration), are added back in. The bath solution is removed with flow-rates that are high enough to stabilize the concentration of contaminants at low levels. This flow rate is adjusted automatically, based on such inputs from the process tool as: amp-hours; wafer throughput; tool status, etc. The organic components are removed from the electrolyte by filtration, followed by their destruction using oxidizing agents combined with heat and UV-light exposure. The organics are photochemically decomposed to carbon dioxide by exposure to the UV light - a clean process that involves no direct contact with the process fluid.

Process stabilization of both by-product and active organics are the benefits of this remove-and-reconstitute approach. The cost of consumables is also greatly reduced as only the organic additives are being removed (so only the organic additives need to be replaced). By using a closed-loop system, costly treatment of waste copper and acid is eliminated. A system in which this process is carried out is built by BOC Edwards. They call it the CuBIS™ (Copper Bath Integrated System), and it is shown in Fig. 16-38.[54] Its centralized monitoring system can simultaneously track five plating tools (from different equipment manufacturers).

16.5.2.5 Pulsed Plating: Utilizing pulsed-current waveforms (see Fig. 16-39b) during plating (instead of a dc-current) is a technique that complements the use of additives to obtain complete Cu filling of recessed features without voids. By periodically ceasing the current-flow, the plating reaction at the surface is halted in that time interval. The Cu filling process is improved for the following reason:

During electroplating a diffusion boundary layer is established between the bulk of the electrolyte solution and the wafer surface. In this boundary layer the Cu concentration trails off towards the wafer surface (since cupric ions are plated

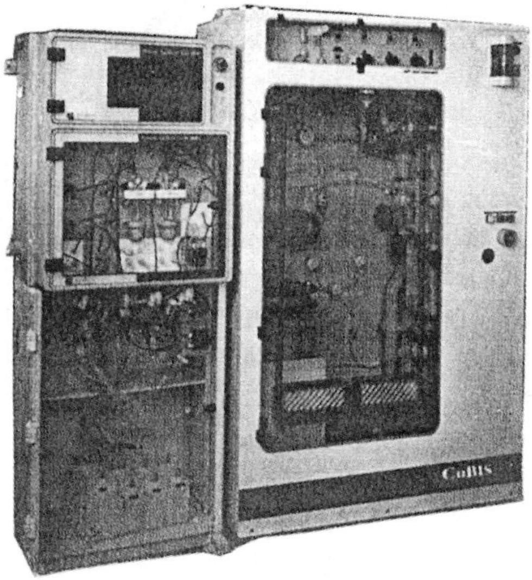

Fig. 16-38 Chemistry management is moving away from costly bleed-and-dump-to-drain approaches to bleed-and-feed. BOC's CuBIS Autorator uses a "remove-and-reconstitute" technique, which recycles the bleed stream, replenishing the bath at a rate automatically determined by plating conditions and by-product formantion. Courtesy of BOC Edwards.[54]

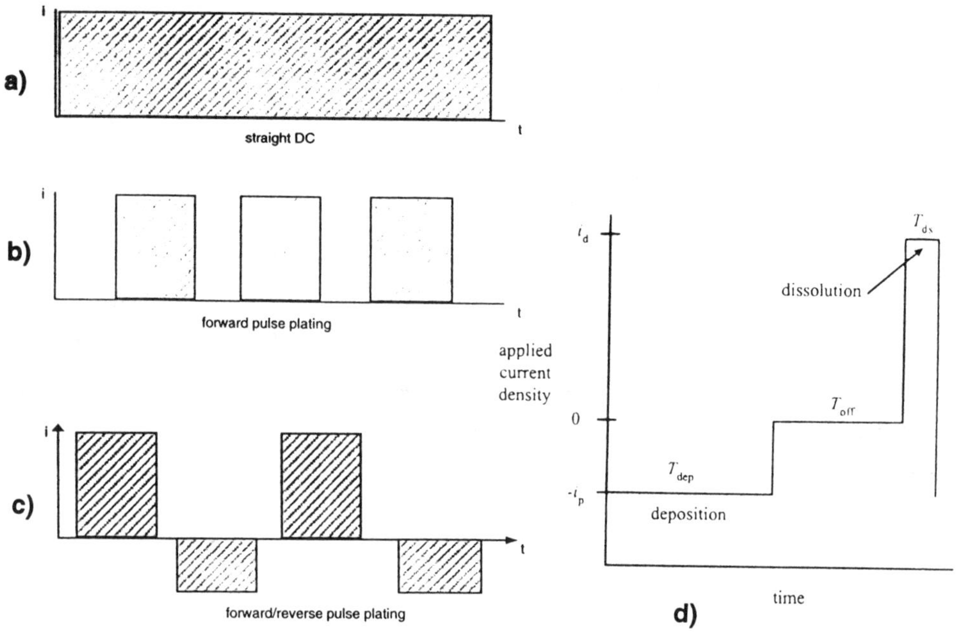

Fig. 16-39 a) dc current plating waveform; b) pulsed plating waveform; c) reversed-pulse-plating waveform; d) Example of reverse-pulse-plating waveform with optimized current amplitudes and pulse durations.[57]

out of the solution onto wafer surface). This boundary-layer is typically several microns thick. For a boundary-layer thickness that is larger than the pattern dimension, when the plating process reaches a steady-state condition, the copper-ion density is higher at the top-edges of trenches than at the bottom, Thus, for a dc-current-plating-process, a higher deposition rate is expected at the top of the trenches than at the bottom - and plating in this manner can lead to void formation.

However, by using pulsed-mode current, the concentration of cupric-ions depleted deep in the damascene recesses by the plating reaction can be restored to a higher level by Cu-ion diffusion to the bottom of the recess during such "rest-periods." This allows the deposition-rate at the bottom of the recess to equal the deposition-rate at the top when the current is restored during the next pulse of the waveform. This type of plating process is thus called *pulse-plating*. With an optimally shaped voltage pulse (both in amplitude and temporal shape), Cu filling of the recesses can be significantly improved.

Not only can the dc plating-current waveform be replaced with a pulse-waveform, but the polarity of the pulses can also be periodically reversed (Fig.

16-39c). The effect of reversing the polarity is to cause the Cu-film to undergo dissolution during the time polarity is reversed. Under the proper waveform conditions, the dissolution occurs more selectively at the top-corners of the trenches, thus allowing the trenches to remain open during the deposition-part of the plating waveform. Figure 16-39d shows an example of pulse-reverse waveform.[57] In one report, the application of periodic polarity reversal, together with adequately formulated and dosed surface-active additives allowed high aspect-ration via/trench filling capability. [58] One report observed that using small pulse periods gave better filling characteristics inside the trench. Using a small duty cycle and current density along with small pulse period appeared to give the best results.[59]

16.5.2.6 Copper Thickness Variation Within a Wafer After Electroplating: The copper topography across a wafer after the completion of a conventional electroplating process is not planar. Such within-wafer non-planarity is due to several effects. First, as discussed in Sect. 16.5.1, the higher current-density at the edge of the wafer next to the ring electrodes causes the Cu deposition-rate to be higher there than at the wafer-center. Thus, there is a buildup of copper thicker than at the center. One of the significant issues of achieving tight dishing and erosion specs after Cu-CMP is that these thicker Cu-regions require extra polishing time to get rid of them. This tends to increase dishing and erosion on other areas of the wafer.

Second, topography variations are caused by the feature-size-dependence of the filling process that is a characteristic of conventional electroplating. For example, the additives present in the electroplating baths used to enhance "bottom-up" filling of recesses with Cu can cause copper mounds to grow above regions containing dense arrays of narrow features (Fig. 16-40a). This occurs because the concentration of additives which accelerate the Cu-deposition rates in the recesses remain adsorbed on the top of the copper layer after the recesses

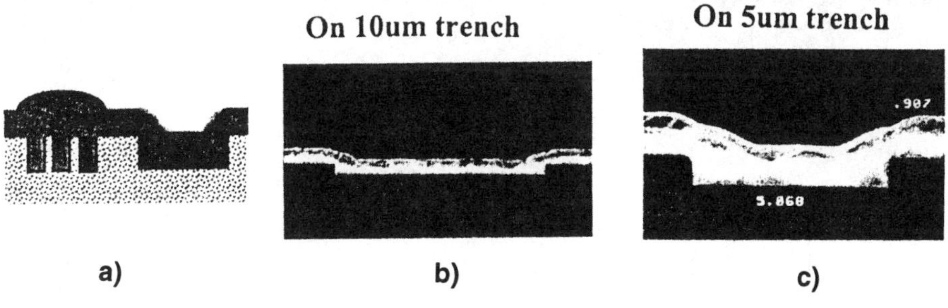

Fig. 16-40 a) Topography of a Cu-layer after plating into damascene recesses of varying widths; b) SEM of Cu plating into 10-μm-wide trench; c) into a 5-μm-wide trench.[61]

filling has been completed in those areas. This accelerates copper growth with respect to other areas of the wafer. If plating is done using electroltye solutions in which no additives are present to inhibit this effect, the copper layer over dense features can be twice as thick as the copper layer on wafer regions with no recesses. This reflects a continuation of copper growth beyond the time of filling completion, and the phenomenon is sometimes referred to as *overplating* (or *overburdening*, see also Fig. 16-36, t = 30 sec, and t = 60 sec).[60] The presence of mounds of excess copper in such regions again requires additional CMP over-polish. As a result, levelers are added to the plating bath (and reverse-pulsed waveforms are also used) to suppress the degree of such overburdening.

On the other hand, wide features do not fill as rapidly with Cu as narrow ones. Thus, there are steps in the copper layer at the edge of wide features. Figure 16-40a also shows such a step at the edge of a wide recess. Figures 16-40b and 16-40c show SEMs of Cu layers plated onto wide recesses, again showing the topography that results from a slower filling of wide recesses by plating relative that at which narrower recesses are filled.[61]

In summary, the topography that results from conventional electroplating processes exacerbates the problems of dishing and erosion that occur in a Cu-CMP process. Modifications to the plating process (such as those described in Sect. 16.5.4) have been proposed to alleviate this difficulty.

16.5.3 Post-Electroplating Cu Annealing and Recrystallization

When work began on copper-electroplating for IC interconnects, it was discovered that Cu-films are unstable after being deposited by electroplating, in that they self-anneal at room temperature (with the transient film-changes occurring over a period of hours to days). Furthermore, the transformation time is a strong function of film thickness. Thinner films take substantially longer to transform than do thicker films. In addition, the Cu seed-layer thickness also appears to play a role in the time and the degree to which annealing reduces the resistivity of the plated Cu-films. Greater resistivity reductions were observed as the plated films self-annealed on thinner Cu seed-layers than on thicker ones.[62]

In any case, if these variables are not controlled, each plated wafer headed for CMP may be different, depending on how much time has elapsed before it reaches the CMP step (a period that can vary depending on load size or tool selection). This can lead to wide distribution in electromigration behavior due to grain-size differences. To prevent such lack of control in plated-Cu-film characteristics, a post-ECP anneal at an elevated temperature was instituted. Such

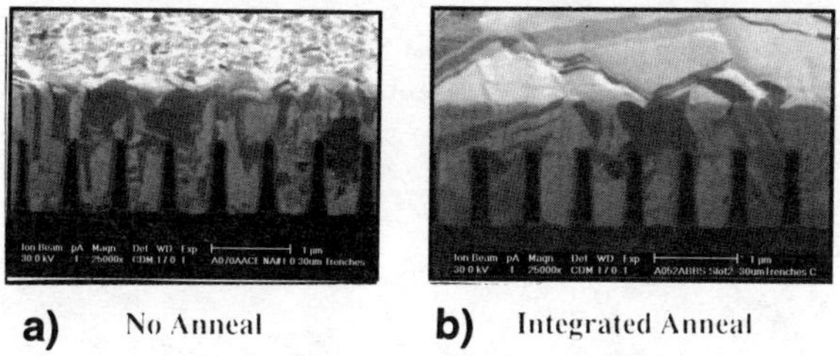

Fig. 16-41 Comparison of grain size in damascene trenches: a) without; and b) with anneal.

post-plating anneals have since become a standard part of the Cu-deposition sequence.

The annealing process causes several changes in the properties of the as-plated Cu-films. First, grain-growth and strong (111)-texture is observed after 400°C annealing (Fig. 16-41). The as-plated grain-size of 1000 Å increases to several microns.[77] The increase in grain-size is significant because the film resistivity is also a strong function of grain-size. (Thin films with larger grains generally have a smaller resistivity than do those with smaller grains. This is because in films with large grains there are fewer grain boundaries – which cause carrier scattering, and thus degrade conductivity.) Thus, annealing of plated-Cu films also reduces their resistivity by up to 20%.[50]

Furthermore, annealing before CMP stabilizes both the Cu-grains and the Cu film-density (i.e., they no longer undergo changes with time at room-temperature). The result of such stabilization is not only a more controllable (and repeatable) Cu-CMP polishing-rate, but this rate also tends to be somewhat higher (by about 20-30%).[64] However, the temperature and time-duration of the annealing process can have an impact on the reliability of the Cu lines, and this must be considered when specifying the annealing conditions (as is discussed in more detail in Sect. 16.8.2).

To avoid having to raise the temperature of the Cu annealing process beyond the point where it makes no further improvements, studies of changes in the sheet resistance versus anneal-temperature and time were conducted. In one study it was found that a temperature of 200°C for 30 seconds would adequately reduce the resistivity and improve the Cu CMP removal-rate (Fig. 16-42). Thus, by designing a resistively-heated-pedestal-based annealing chamber within an electroplating tool enclosure, an annealing process can be integrated into the plating system. A chill-plate is also used in this arrangement to quickly cool the wafer to 50°C after annealing before exposing it to atmosphere (to keep the hot copper from oxidizing, as it would do if it were allowed to cool in air). The wafer

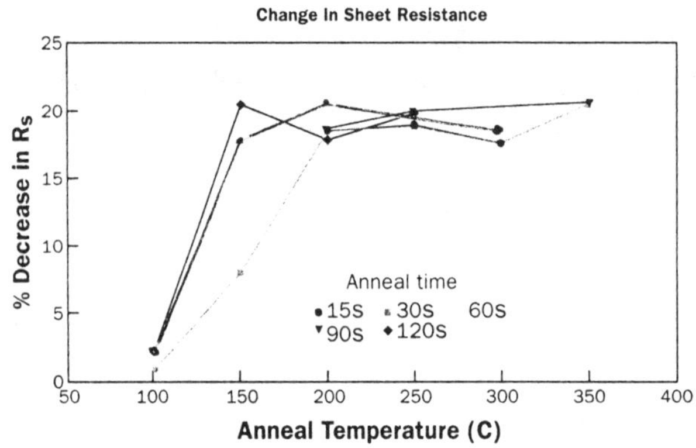

Fig. 16-42 Sheet resistance of Cu film as a function of anneal time and temperature after electroplating.[50]

is actually moved close to the chill plate (i.e., proximity cooling is used). To prevent the copper film from oxidizing during annealing, the ambient in the annealing chamber is maintained with < 5-ppm oxygen.

The Cu self-annealing effect can be explained as a recrystallization and purification process.[65] That is, an as-deposited plated-Cu-film is highly defective and the resistivity of the as-deposited film is high. The defects in the film are likely to be impurity inclusions caused by additives in the baths. (The large-grained deposits from additive-free baths do not undergo this recrystallization.) The inclusion of impurities at grain boundaries from the additives or their electrolyte by-products would be expected to cause pinning that would suppress grain growth. The as-deposited electroplated Cu film is not stable because of this high defect density. The film tends to reduce the defect level by recrystallization. During recrystallization, new grains are formed with fewer defects. These grains grow larger by consuming defective grains, leaving defects mainly in the grain boundaries and film interfaces. The stress in electroplated copper also changes from compressive-stress to tensile-stress after self-annealing. This is the macro evidence of film contraction caused by defect reduction.

16.5.4 ElectroChemical Mechanical Deposition (ECMD™)

As noted in Sect. 16.5.2.2, the Cu electroplating deposition-rate within recessed regions of a wafer exhibits a feature-size-dependence. That is, the Cu-deposition-rate is higher on areas of the wafer where there is a high-density of narrow, recessed damascene-features (and the plating process continues beyond the point where the recess is filled). This results in mounding of the copper to a local thickness above those regions that is greater than in the flat regions devoid of recessed features (overburdening). In wide recessed-features, the deposition-rate is close to that on the flat-areas devoid of features, and hence a downward step can occur in the copper film when going from a flat area adjacent to a wide-recessed area. This has several detrimental consequences, especially with respect to the CMP process. That is, the thickness of the plated film must be increased to ensure that the wide recesses are completely filled. Thus, the process time of the CMP to polish the thicker Cu-film and the overburdened areas must be longer (which reduces the CMP throughput). There is also apt to be more Cu dishing in the wider-recessed areas. This problem would be overcome if the Cu could be deposited so that its surface after plating was globally planar. While techniques such as the use of leveler additives and pulsed plating can reportedly reduce the non-planarity to some degree, further improvement in the global planarity is still being pursued.

A novel process and deposition tool has been introduced by NuTool Inc. which seeks to address this issue directly. That is, in a process they call *electrochemical mechanical deposition* (ECMD™), as a copper layer is deposited in their tool (NuTool 2000™), a polishing pad is in physical contact with the wafer. This mechanical polishing coupled with the NuTool process chemistry simultaneously deposits copper and removes it. A differential growth rate between Cu deposition on the wafer flat areas and Cu deposition in the features is obtained. The speed, pressure, and mechanical path of the pad can all be controlled with software, eliminating the common problems of polishing copper, such as dishing and erosion.[66]

The ECMD process solves the problem of *excess-overburdening* (i.e., the buildup of excess copper over small densely arrayed features) and the downward steps that occur at the edges of wide recessed-features, by providing *aspect-ratio-independent filling* of features. Because the copper surface is not only globally planar but it is thinner in the field area than in a conventional Cu electroplating deposition process, the ECMD technique permits the CMP polish time to be reduced.

The tool and process were introduced in 2000, and thus information about its use in production is not available at the time of this book's publication. Some observers have commented that the ECMD process combines electroplating with some form of CMP on the wafer. Whether this combination of phenomena can be made to work simultaneously in a manufacturing environment will have to be demonstrated.

16.5.5 Electroless-Plating of Copper

Copper films can also be deposited by an electroless plating process, and this technology is already used in electronic packaging. Some work has also been done to develop Cu electroless plating processes for IC interconnect films. However, while this method has not gained the acceptance of electrolytic electroplating for forming Cu films, it is also briefly described here.

An *electroless* process involves the formation of a thin-film from an electrolytic solution without externally applied voltage. The deposition is due to an electrochemical reaction among the metal ions, reducing-agent, complexing-agents, and pH-adjusters on catalytic surfaces. The chemical process involves a CuO_4 and formaldehyde reduction:[67]

$$Cu^{2+} + 2HCHO + 4OH^- \rightarrow Cu^o(\text{solid}) + 2H_2O(\textit{liquid}) + H_2(\textit{gas}) \qquad \textbf{(16.7)}$$

An electroless Cu plating process for ICs was described in 1993 to give films with resistivities of 1.9 $\mu\Omega$-cm, close to that of bulk-Cu (1.7 $\mu\Omega$-cm).[68] Later work demonstrated filling of vias with aspect ratios of greater than 4:1 onto diffusion-barrier films of TiN.[69]

Although potentially a low-cost process because of its batch nature, electroless plating lost ground to electrolytic plating because of concerns due to hydrogen evolution, complexity of process control, low deposition rates, and potentially adverse environmental impact. However, it is possible that electroless plating will offer significant advantages for future wiring structures with improvements in sophistication and understanding. For example, some think that electroless plating may be used to deposit the Cu seed-layer at the 0.7-µm technology node.

16.5.6 CVD of Copper

While electroplating is the dominant method used to deposit Cu films for IC interconnects, work has been done on CVD-Cu deposition methods as well. One reason why interest remains in Cu-CVD processes is that it becomes increasingly difficult to deposit a uniform, conformal Cu seed-layer in the vias and trenches as their aspect-ratio continues to increase. At some point, it may not be possible to deposit a Cu-PVD-layer with sufficient thickness and/or continuity into such narrow features. A more conformal CVD-process could potentially produce a better Cu seed-layer in such high-aspect-ratio features. Thus, CVD-Cu technology approaches are briefly described here. (It should be noted however, that a copper PVD-process which deposits a very-thin nucleation layer will still be still needed before the Cu-CVD process.)

Organo-metallic sources are the most widely used precursors of CVD copper. These include Copper(II) {bis-hexa-fluoro-acetyl-acetonate copper} [$Cu^{II}(hfac)_2$] and Copper(I) {hexa-fluoro-acetyl-acetonatoL copper} or [Cu^{I}(hfac)L], where L is a neutral ligand weakly bonded to the Cu. One such ligand that has been used is vinyl-tri-methyl-silane (or *vtms*). $Cu^{II}(hfac)_2$ is a solid at room temperatures and must be dissolved in isopropyl alcohol in order to form a liquid precursor that can be delivered as a vapor (with hydrogen used as the carrier-gas). High quality Cu-films with low resistivity can be formed by the hydrogen reduction of $Cu^{II}(hfac)_2$. However, to get the lowest-resistivity Cu-films, the deposition temperature must be between 350–450°C. Since Cu is likely to be used with thermally-fragile low-*k* dielectrics, such high deposition-temperatures are undesirable. A low-pressure plasma process has reportedly been developed which allows the deposition temperature to be reduced to about 170°C, and which produces films with resistivities of 2.2 $\mu\Omega$-cm.

The Cu^{I}(hfac)L precursor (where the L = vtms) is commercially available as a product called Cupra Select,^{TM} developed by Schumacher.[70] It is a liquid at room temperature and is fed into the process chamber by direct-liquid-injection (i.e., the liquid is pumped to a vaporizer and then injected). This delivery method is needed for such precursors as Cu^{I}(hfac)-vtms, which decompose after long periods of heating in a bubbler. Low-resistivity Cu films can be formed at deposition temperatures of 150°C. The Cu is then deposited through a disproportionation mechanism:

$$2Cu^{I}(hfac)\text{-}vtms\ (gas) \rightarrow Cu^{0}(solid) + Cu^{II}(hfac)_2\ (gas) + 2vtms\ (gas) \qquad \textbf{(16.8)}$$

Both the $Cu^{II}(hfac)_2$ and the vtms ligand are stable compounds and readily desorb from the surface at the deposition temperature, leaving behind a Cu film relatively free of residues. Since the reaction is reversible, vtms can be used to dry clean the deposition chamber. By removing the Cu deposited on the parts inside the chamber, particle contamination by the film's cracking and flaking can be prevented. However, an issue that must also be addressed with CVD-Cu is wafer-backside Cu-deposition.

16.6 WET-ETCHING AND ELECTROPOLISHING OF CU FILMS

Work has also been carried out to develop processes that would replace (or shorten the process time) of Cu-CMP. Two basic approaches have been pursued: **1**) chemical *spin-etch planarization* (SEP) - in which the Cu film is etched with wet-chemicals using a special spin-processing tool; and **2**) electrochemically removing the Cu film with a process called *electropolishing*.

16.6.1 Spin-Etch Planarization of Cu Films

A SEP process was jointly developed by SEZ and Honeywell.[71] The SEP process removes layers of Cu material at a controlled rate by exposure to a reactive chemical solution while the substrate is spinning. Its goal is to remove the copper overfill material after a plating process with the same planarization results as CMP (but with no risk of damage to the low-*k* materials in the structure), and at much lower cost than CMP. Damage to the low-*k* is avoided and low-erosion of the dielectric is also achieved because there is no mechanical downforce as occurs in CMP.

During SEP the wafer is suspended horizontally on a nitrogen cushion above a rotating chuck (utilizing the Bernoulli principle). The substrate is held in place laterally with locating pins about the wafer perimeter. As the chuck and wafers are spun, wet-etchants are dispensed onto the wafer from a radially oscillating nozzle. An appropriate etching solution and spinning of the wafer result in a planar final surface (with the simultaneous radial and tangential etchant flow greatly enhancing planarization over simpler static etch processes).[72] DI water and nitrogen are then applied to the wafer to achieve rapid cleaning and drying. Etch rates of copper layers can range from 500-2000-nm/min, depending on the chemistry selected, and almost infinite selectivity exists with respect to etching of the underlying Ta and TaN barrier-layers.

By 2000, SEP was still in an early stage of development. However, it was demonstrated that the SEP process could provide a high degree of planarization for small features (0.35-0.5-μm-wide), and work was proceeding to improve the planarization of larger features in the 5-100-μm range. One way that SEP may become more practical is if a short CMP process (with a small downforce) could be used to knock off the large mounds of excess overburden present on parts of the wafer where high densities of narrow features are present. This would make the surface substantially more planar, making the SEP process more effective.

16.6.2 Electropolishing of Cu Films

Electropolishing of the overfill material after Cu electroplating is another process that has been pursued to eliminate or shorten the Cu CMP process. That is, after the electroplating step is completed, the current flow can be reversed, and the Cu film will then de-plate. The idea is to use such de-plating to remove the Cu overburden while leaving the Cu in the damascene recesses. Since reverse electroplating will remove Cu uniformly across the wafer, at first glance the concept appears feasible. However, it proves not to be practical as long as the problem of aspect-ratio-dependent filling-effects occur during the initial electroplating process (see Sect. 16.5.2.6). That is, the non-uniform topography of conventional electroplating processes will result in overpolishing in the wide trench-regions and under-polishing in the circuit regions where there are dense-arrays of small-features. However, with the advent of a globally-planar Cu-deposition-process such as described in Sect. 16.5.4, electropolishing is once again being investigated.

In 2001 a Cu-electropolishing process was described in which the Cu-CMP step was either eliminated or significantly shortened.[61] A globally-planarized Cu-film was then deposited (such as was described in Sect. 16.5.4). With a globally planarized Cu film, the electropolishing can remove copper across the wafer without thinning the metal in wide trenches. In the Cu-CMP-free process that was described in Ref. 61, the Cu-overburden was electroplolished down to the barrier-layer. The barrier-layer was then removed by dry-etching with a fluorine-based plasma (that does not etch the exposed Cu remaining in the trenches). The wafer was then cleaned with brush scrubbing (Fig. 16-43). In the CMP-less process, the Cu-overburden was partially-removed by electropolishing and the remainder by conventional CMP. Less dishing could be achieved with the CMP-less process. A smooth Cu surface and highly (111) texture was also obtained by selecting the proper current density during the electropolish step.

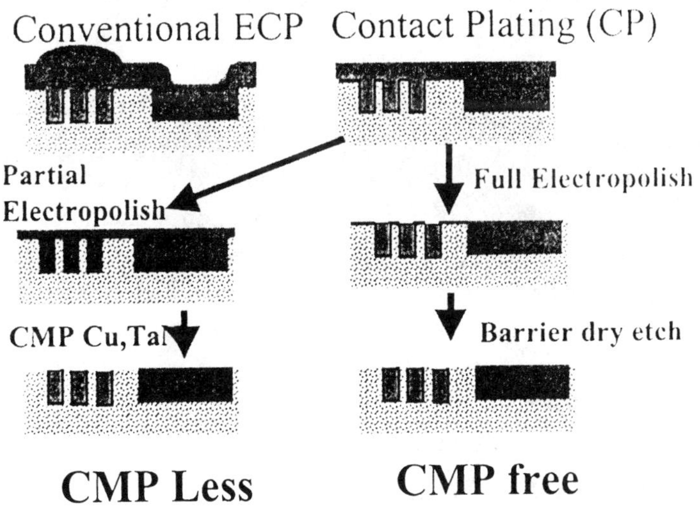

Fig. 16-43 Schematic diagram of CMP-less and CMP-free process flows by using a process that produced planar-Cu films after the plating process.[61]

A tool designed for precise electropolishing was introduced by ACM Research Inc. They refer to the process as *stress-free polishing* (SFP), and tool is called the ACM Ultra SFP.^{TM} It is claimed that the SFP Ultra can polish Cu supported by ultra-low-*k* dielectrics (i.e., with a *k*-value less than 1.8).

16.7 CONTAMINATION CONTROL IN COPPER TECHNOLOGY

In Section 16.1 the issue of copper-contamination was introduced. Also briefly listed were the on-chip lines-of-defense and Cu-manufacturing protocols developed by wafer fabs to guard against this occurrence. Here additional information is provided about methods used to control copper-contamination in wafer manufacturing environments.

16.7.1 Sources of Copper Cross-Contamination in a Wafer Fab

In a completed device with a dual-damascene interconnect (and during many stages of the wafer manufacturing sequence), the embedded copper lines and vias are fully encapsulated with barrier-layers to prevent Cu out-diffusion into the surrounding dielectric and silicon areas. In those processes the likelihood of copper contaminating either other parts of the chip, or the wafer-fab environment, are small. However, in several process steps, Cu is exposed or is present outside of these sealed boundary areas (e.g., during those steps that occur from Cu-deposition to Cu-CMP). The most obvious risk of Cu-contamination occurs when large open areas of Cu are present on the wafer, or when Cu-containing chemicals or waste products - such as electroplating solutions or Cu-CMP slurry - are present. Another critical mode of Cu-contamination results from the high-energy bombardment of exposed Cu surfaces during plasma processing (e.g., via-open etch, and resist-strip). The following wafer processing steps (in roughly descending order of importance), can thus be considered to carry the highest risk of spreading Cu contamination - either into the deeper layers of an individual wafer or from wafer-to-wafer: **a**) Cu-CMP; **b**) Cu-deposition (including PVD seed-layer-deposition and electroplating); **c**) post-plating anneal: **d**) cap diffusion-barrier layer deposition; and **e**) via opening.[73]

During Cu-seed-layer deposition, electroplating, and post-plating anneal processes, the front side of the wafer surface is almost completely covered with a blanket Cu-film (and some Cu may find its way to the bevel and wafer backside during one or more of these steps). Therefore, cleaning of the wafer backside and bevel at certain critical steps during this stage may be necessary to prevent the spread of Cu particles from wafer-to-wafer by way of wafer-handling equipment (see Sect. 16.8.3). A barrier layer should also ideally extend over the whole wafer surface, including around the bevel and on the backside to prevent Cu from diffusing into the wafer. This could be accomplished by using a wafer frontside barrier-layer (typically the conductive barrier-layer deposited before Cu), which overlaps a diffusion-barrier layer present on the wafer backside (e.g., CVD-SiN). Thus, the wafer bevel would also be completely covered by one or both of these barrier-layers. In many cases, the backside SiN diffusion-barrier-layer is not present. The backside and bevel cleaning is even more important in such cases. In some cases, if there is no backside diffusion-barrier layer, a silicon etch is performed after the backside and bevel cleaning step.[74] Its purpose is to remove a thin layer of silicon from the wafer backside that contains any copper that was deposited by the previous processes.

As noted above, contamination from copper can arise from tools and equipment involved in the deposition and handling of wafers. For example, a deposition tool that coats wafers with copper may cause it to be deposited on the bevel of the wafer. This wafer with copper on its bevel may be sent to a metrology tool equipped with a wafer handler that manipulates the edges of the wafers. (Such metrology tools are commonly shared among various process steps because manufacturers generally find that the acquisition of dedicated metrology equipment for copper processing is cost-prohibitive. These tools may include in-line SEMs for critical-dimension measurement, film-thickness monitors, and optical defect inspection devices.) The metrology-tool wafer-handler is now contaminated with copper, and can then cross-contaminate other wafers (for example, those that are on their way to an etch tool). The etcher thus eventually becomes contaminated with copper, and this contamination can spread throughout the fab (accumulating in plasma process chambers, wet benches, and lithography steppers). Contamination can also be spread by way of the wafer backside, as many wafer-handlers and robots hold and lift wafers by their backsides.

Wafer-handling by transport in cassettes is another likely source of copper cross-contamination. If copper is left on the wafer bevel as it exits a process tool, the carrier will invariably be contaminated with copper. Great care must thus be taken to either limit the use of wafer carriers to a specified area or to switch to a clean carrier once the copper is removed from the bevel. Wafer carriers may also need to be cleaned more frequently than otherwise required just for particle control. Furthermore, to prevent wafer-handlers and wafer-carriers from being contaminated by Cu particles on the wafer edge, it is preferable to maintain a sufficient edge exclusion of the Cu seed-layer or electroplated film. However, a full-faced seed-layer is typically required to allow electroplating with an edge-exclusion region of less than 4-mm. If dedicated cassettes are used to move the wafers from the PVD-tool to the plating tool, and a cleaning step that removes the Cu from the bevel (and wafer backside) after electroplating is used, this problem can be circumvented.

Another source of possible copper cross-contamination is wafer breakage. Although the incidence of wafer breakage in processing tools has dramatically declined in recent years, it still occasionally happens. Thus, protocols for cleaning tools in which a wafer that contains copper has broken, are crucial to preventing factory-wide cross-contamination. A decontamination plan is required for each area and tool, which mandates appropriate cleaning methods and wipedowns in the immediate area in which the breakage occurred. However, there is no easy test for demonstrating the cleanliness of a tool after such breakage and cleanup, short of analyzing a test-wafer for copper-contamination.

Copper can reach the wafer bevel or backside during a deposition process. However, these surfaces can be protected by clamping and sealing the wafer hermetically during the plating operation so that only the face of the wafer is exposed during the process as required. The SABRE Electrofill™ tool from Novellus Systems uses a clamping method known as a *clamshell* to prevent the wafer backsides from being exposed to copper-containing solutions during the electroplating process. Surface analysis over a period of time during which 100,000 wafers were processed through a SABRE tool showed that a consistently low copper concentration of $< 10^{11}$ atoms/cm^2 were present (which was similar to the copper concentrations on control-wafers that did not undergo the plating process).[75]

The plating tool from Applied Materials (Electra™) uses a different approach. It contains a separate backside- and bevel-cleaning station integrated into the tool to treat each wafer after the electroplating step such that any copper that is deposited is removed (see Sect. 16.8.3 for more details).

16.7.2 Effects of Copper-Cross Contamination at the Back-End-of-Line (BEOL)

The concern for copper cross-contamination at the front-end-of-line (FEOL) stages of wafer processing is that the copper can diffuse into the active device regions and cause degradation of the gate-oxide-integrity or *pn*-junction leakage. The method of ensuring that copper-contamination is not occurring in the FEOL bays of a wafer fab is to monitor the MOS-C characteristics of wafers passing through the process line (much in the same way that is also done to safeguard against other forms of metal contamination - e.g., sodium).

However, the concerns at the BEOL phase of the wafer processing sequence are different from those feared at FEOL. That is, the solid solubility of Cu at

BEOL processing temperatures is relatively low, so not much Cu is expected to be incorporated into the silicon substrate from Cu on the wafer surface as long as the process temperature does not exceed 400°C. A test at SEMATECH took wafers with Cu plated on the backside, and put them into a furnace at 400°C.[76] Subsequent tests for leakage and threshold-voltage-shift showed no deviation from control wafers. Apparently the gettering mechanisms that protect devices from other contaminants can also handle levels of copper-contamination from such an event. As a result, worries about Cu entering the Si wafer at BEOL are probably somewhat exaggerated. On the other hand, the concern that copper will diffuse into the interlevel dielectric and cause leakage during BEOL processing is more real. The failure due to such a degradation of interconnect-leakage from copper contamination could also be one that develops slowly during the life of a part.

16.7.3 Cu Removal from the Wafer Bevel/Edge and Backside

Isolating Cu production from the rest of a wafer fab's manufacturing area is one approach to control copper contamination. But, many fabs share such critical and sophisticated tools as metrology and lithography equipment. Thus, such tools must accept Cu-processed wafers but it must be ensured that no contamination transfers to non-Cu wafers. While backside hermetic sealing and clamp rings are used to provide electrical contact and to mask the wafers exclusion zone from Cu deposition, they may not be 100% effective in preventing Cu deposition in the exclusion zone and on the wafer backside. This condition is further exacerbated by trends that minimize the exclusion zone to 3-mm (and even 2-mm), effectively putting the deposited Cu blanket onto the beveled edge of the wafer (Fig. 16-44a). Furthermore, since CMP cannot remove copper from the bevel and

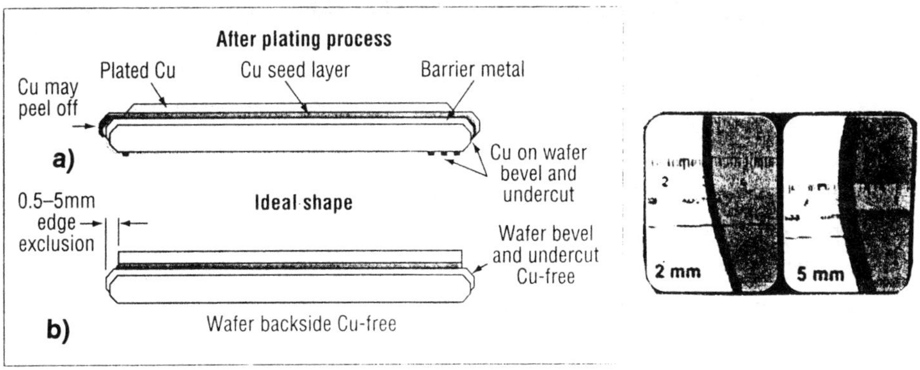

Fig. 16-44 a)Full coverage electrochemical Cu plating; b) The ideal Cu condition on the wafer after SPCE. By permission of Solid State Technology. Published by PennWell.[74]

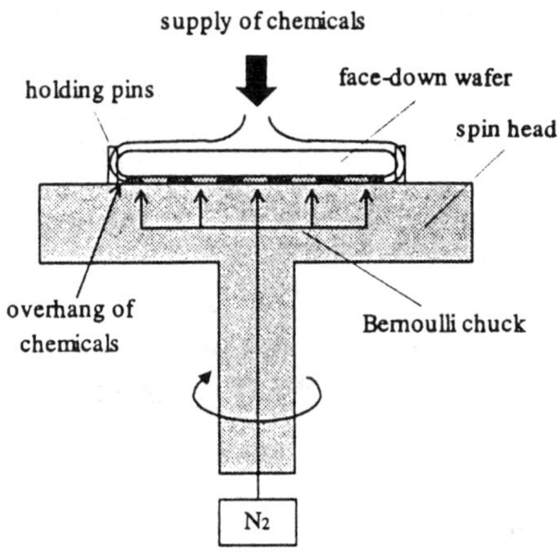

Fig. 16-45 Schematic diagram of the single-wafer cleaning system with a Bernoulli chuck.

backside, after a few layers, Cu-peeling (and resulting Cu cross-contamination) due to film buildup may occur.

The approach used to overcome this problem is to remove any copper that may have accumulated on the wafer-backside, bevel/edge, and exclusion-zone with a spin-etching process. Such a process is carried out immediately after the electroplating step is completed, and may be conducted either on a standalone spin-processor or on spin-processor that is integrated into the platform of the elctroplating tool.

A *standalone, single-wafer spin-cleaning-system* for backside Cu removal has been developed by SEZ Inc. They call their technology *spin-process contamination elimination* (SPCE^{TM}). During SPCE the wafer is suspended horizontally and frontside-down on a nitrogen cushion above a rotating chuck, utilizing the Bernoulli principle (Fig. 16-45). The substrate is held in place lateral with locating pins about the wafer perimeter. As the chuck and wafers are spun, the wafer is lifted about 1-mm. Wet-etchants are dispensed onto the wafer backside from a radially oscillating nozzle (e.g., modified HNO_3 chemicals are used for Cu removal), and the N_2 flow protects the entire front of the wafer (except the edge) from the chemicals. The degree of wraparound of chemicals onto the front of the wafer is a function of the N_2 flow-rate, the rotation speed of the chuck, and the viscosity and composition of the chemicals. This enables contamination and film removal from the back, edge/bevel, and front of the wafer without damaging the devices and without the use of masking materials on the frontside of the wafer. The width of the exclusion ring is tunable from 0.5-5-

Fig. 16-46 Edge-clean chambers in the Applied Materials Electra electroplating tool. (Courtesy of Applied Materials)

mm.[74] DI-water and nitrogen are applied to the wafer for rapid rinsing and drying. Fig. 16-44b shows ideal Cu coverage on a wafer after an SPCE process.

A backside and edge *Cu-removal spin processor* has also been *integrated* into the Electra^{TM} electroplating tool of Applied Materials. After the electroplating process, the wafer is moved to the cleaning chamber, where it is loaded face down by a robot onto the self-centering wafer hoop. A wafer chuck holds each wafer, which is then lowered into a cavity with three chemical channels below and three above the wafer (Fig. 16-46). While the wafer is rotated, a DI-water rinse first rinses the electrolyte from the wafer surface. As the wafer is spun dry at high rotation speeds, the chemical dispensing arm moves into position to release its etching/cleaning chemical precisely at the edge of the wafer. Three sets of nozzles (120 degrees apart) enable simultaneous chemical dispensing for very short edge clean times. While the chemicals remove any edge and backside copper, DI-water is sprayed on the front side. By rotating the wafer and balancing the flows, it is possible to control how much material from the backside overwraps, either on the side or the bevel, or just above the bevel, or whether its kept from coming to the front side at all. Figure 16-47 shows the programmable edge exclusion with this edge clean process.[50]

An integrated bevel and backside cleaning process is also offered in the Novellus SABRE^{TM} electroplating system. In this approach, the Cu is removed

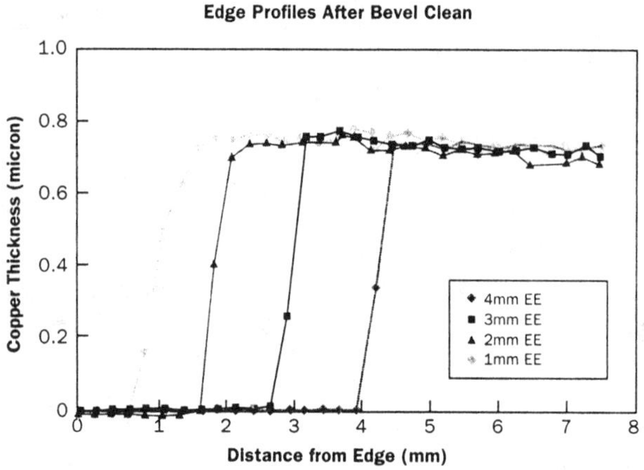

Fig. 16-47 Programmable edge exclusion with the applied Materials Electra^{TM} edge-clean process. (Courtesy of Applied Materials)

from the bevel after the Cu seed-layer is deposited (i.e., prior to the electroplating process). The removal is carried out in the cleaning station of the tool.[78]

16.8 RELIABILITY OF COPPER INTERCONNECTS

As described in Sect. 16.1, copper lines offer the benefit of improved reliability over Al lines, primarily due to their superior electromigration lifetimes. However, the issue of copper interconnect reliability has other aspects as well, and some of these are described in this section.

16.8.1 Reliability of Copper Interconnects: "Missing-Metal" Defects

One Cu reliability issue involves post-CMP defects in narrow lines, which are referred to as "missing-metal" defects (Fig. 16-48). Such defects occur as a result

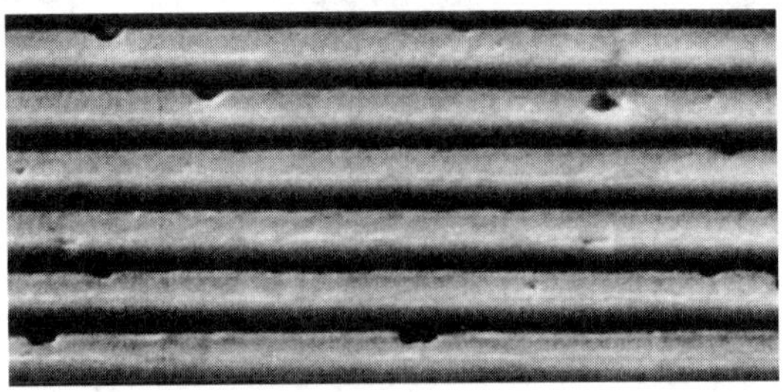

Fig. 16-48 Example of "missing-metal"-type defects. These defects arise from grain pull-out or corrosion and are exposed during the CMP process.[39] Reprinted with permission of Semiconductor International.

of corrosion or *grain-pullout* of lines exposed to a CMP-process. While the mechanism for these defects is poorly understood, it appears that their occurrence is influenced by a number of factors, including: plating-chemistry; plating-waveform; composition of the barrier-layer; and thermal-treatment. The latter effect appears to be most significant, and is thus discussed further here.[39]

Copper-lines are annealed after electrodeposition with the Cu overburden still in place. High thermal-excursions can cause large stress in the Cu films. That is, as the temperature is increased, the thermal expansion of the copper is greater than that of the surrounding dielectric material. This Cu expansion thus induces compressive-stress in the copper film that causes the film to re-crystallize and relax through dislocation motion. However, upon cooling the copper contracts and goes into tensile stress that is much higher under ambient temperature conditions than in the as-deposited film, and such stressed crystalline-films are prone to grain-pullout during CMP.

To reduce the incidence of defects related to thermal treatment it is necessary to minimize the temperature excursion of the post-deposition-anneal process (i.e., while the Cu overburden remains in place). Temperatures should be kept below the Cu plastic-flow limits (~200°C) so the copper can expand and contract elastically. Figure 16-49a shows the sensitivity of via-chain yield to the post-electroplating annealing temperature. Some have said that controlling the defects related to the copper process is perhaps the most important reliability issue facing the drive to achieve higher yields for copper in production.[39]

16.8.2 Reliability of Copper Interconnects: Plated vs. CVD Cu Films

Another reliability aspect observed in copper interconnects is that electroplated copper is more reliable than CVD-Cu with respect to electromigration failures.

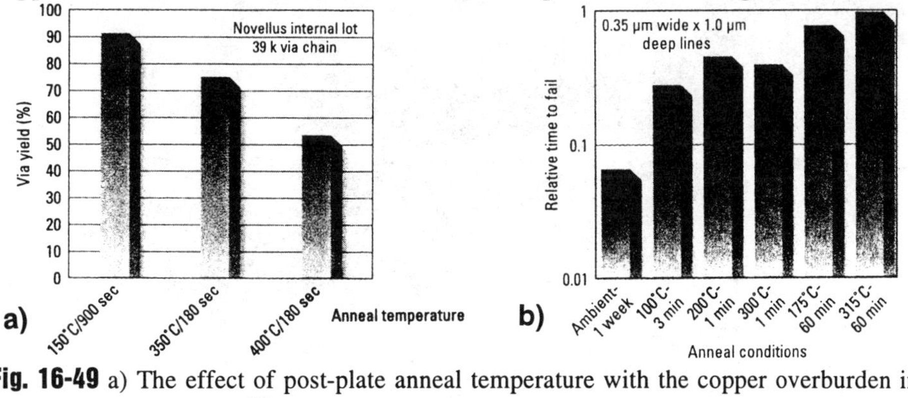

Fig. 16-49 a) The effect of post-plate anneal temperature with the copper overburden in place on via-chain yield.[39] b) Electromigration lifetimes for narrow lines (0.35-μm) with various pre-CMP anneals.[39] Reprinted with permission of Semiconductor International.

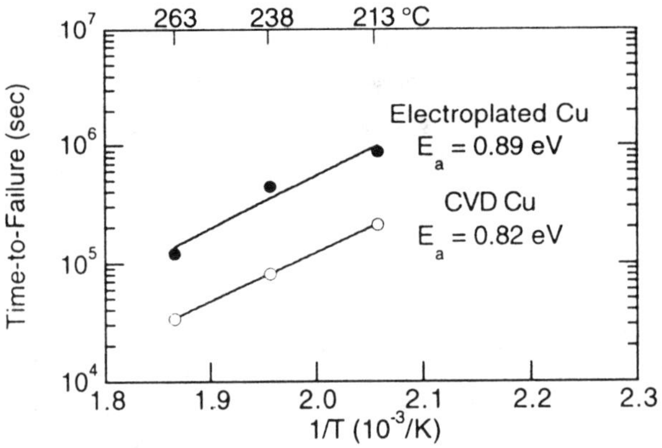

Fig. 16-50 Plot of time-to-failure of CVD-Cu and electroplated-Cu lines.[80] (© IEEE 1998)

As shown in Fig. 16-50, this is especially true for deep-submicron Cu linewidths. (The 0.4-μm-thick Cu-lines in Fig. 16-50 were tested for electromigration behavior after deposition, annealing [at 400°C for one hour], and CMP.[80]) Copper-lines deposited by electroplating exhibit longer lifetimes under electromigration-stressing than do CVD-Cu-lines because CVD-Cu has a smaller as-deposited grain size. In the narrow trenches of deep-submicron dual-damascene interconnects, electroplated-Cu has a bamboo grain-structure, which is conducive to superior electromigration behavior (whereas CVD-Cu has smaller grains after annealing than Cu deposited by electroplating, as can be seen in Fig. 16-51).[80,81]

As far as the electroplated-Cu-lines themselves, their electromigration reliability is impacted by the line-width and post-plating anneal-cycle. Wide lines anneal fully at the lowest anneal-temperatures, and when fully annealed they

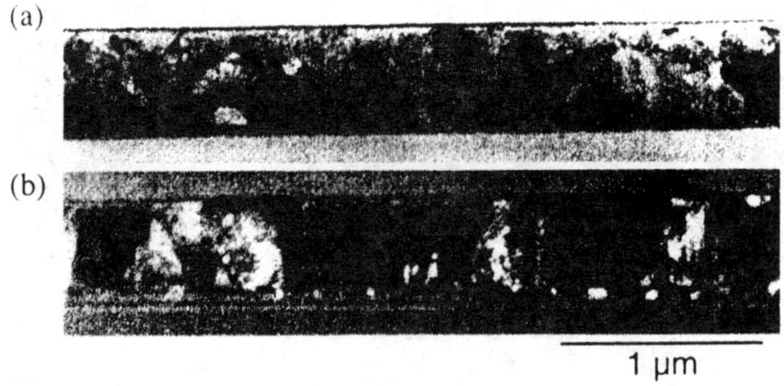

Fig. 16-51 Plan-view TEM micrographs of Cu lines in trenches after annealing: a) CVD-Cu; and b) electroplated Cu. (© IEEE 1998)

exhibit the longest electromigration lifetimes. Narrower linewidths require a higher-temperature anneal-cycle to achieve the same reliability behavior (see Fig. 16-49b, which shows the electromigration lifetimes of narrow copper-lines versus anneal-time and temperatures).[39] However, higher-temperature annealing cycles lead to greater "missing-metal" defect levels in narrow Cu-lines. Thus, a tradeoff must be made between deteriorating defect-performance and improved electromigration-performance as the anneal-temperature is increased.[39]

It was also reported that by incorporating a small amount of magnesium into the copper, various characteristics of the Cu interconnect were improved. That is, by forming a CuMg/Cu/CuMg stack, it was observed that the Mg diffuses through the copper and aggregates on the Cu surfaces. At the interface between the Cu seed-layer and the TiN barrier-layer this Mg improves the adhesion of copper to the TiN. At the top surface of the Cu-layers, the Mg forms a self-limiting oxide that hinders copper oxidation. Electromigration characteristics of such Cu/Mg stacked layers showed no difference relative to pure copper layers.[79]

16.9 ENVIRONMENT, HEALTH, AND SAFETY ISSUES OF COPPER TECHNOLOGY

The implementation of copper technology in IC fabrication introduces new chemistries, tools, and waste streams into the manufacturing arena. Before copper processes are put into production, it is important to address the *environment, health, and safety issues* (EHS) involved with them. Copper brings five new technologies to wafer fabrication, and their EHS impacts are discussed here: barrier-layer deposition; copper seed-layer deposition; Cu-fill by electroplating; copper CMP; and copper wafer reclaim.[82]

The first two of these processes, *barrier-layer deposition* and *Cu seed-layer deposition* are currently PVD processes which use solid sources, and hence do not involve any special EHS considerations. *Electrolytic plating for Cu fill,* however, is a new technology for the semiconductor industry, which also involves new chemistries, tools, and wastes. Nonetheless, commercially available electroplating chemistries used in the printed-circuit wiring board (PWB) industry have been evaluated and found to have no particular hazards or worker exposure issues. That is, all materials are TSCA listed (Toxic Substance Control Act). The primary concern is the two waste streams generated by the process: rinse water and a concentrated bath.

Copper CMP involves the use of a slurry that may or may not contain any chemicals that are environmentally hazardous. However, the process effluent is a potential environmental hazard, insofar as the it contains all the copper removed from the wafer by the CMP process. Development and manufacturing fabs generate a large number of *test wafers that are reclaimed.* When these have a copper film, they cannot be reclaimed using existing methods. Nevertheless, relatively environmentally friendly techniques have been developed for this purpose (e.g., ammonium persulfate/sulfuric acid),[83] which is also compatible with electroplating waste. Nevertheless, the wafer reclaim process, like the electroplating process, generates the same two waste streams.

Thus, the two key steps in a copper dual-damascene structure – electroplating and CMP – are both wet-chemical processes. They use large amounts of water and create large amounts of acid and heavy-metal waste. They require large quantities of slurries, plating solution, and other liquid feedstocks. Bulk chemical distribution systems and waste-water treatment facilities for these materials can be built into new fabs. Accommodating them in an existing fab is more difficult. Typical fab delivery systems are designed around gases and DI-water, not slurries and plating chemicals.

16.9.1 Copper Waste Handling

The liquid chemical wastes from the various copper process steps described above must be evaluated for their toxic substance concentrations, and treated appropriately. It has been determined that the rinse water from the copper plating and wafer reclaim process ordinarily contains relatively low-levels of copper and contribute only minor increases in the total copper concentration of the industrial waste-water discharge. However, copper CMP contributes more than 80% of the discharged copper on a mass basis, and is likely to require treatment before being discharged to the industrial waste-water.

16.9.1.1 Handling Concentrated Waste from Plating and Wafer Reclaim: The liquid chemical wastes and slurries containing concentrated copper must be treated to remove the copper. A process similar to the electroplating process (called *electrowinning)* is one method for treating the concentrated plating baths and wafer reclaim wastes. That is, copper is reduced and plated out of solution at the cathode, and water is oxidized (split) to form oxygen at the anode. This process is capable of reducing the copper concentration in these waste liquids to levels below the current concentration based discharge limit for *publicly owned treatment works* (POTW). While, the time to reach such limits with the electrowinning process is 20-40 hours, it is still a cost-effective method and it does not generate other waste streams (e.g., sludge).

16.9.1.2 Handling Copper CMP Effluents: The treatment of the Cu-CMP effluent is done in another way. The primary goal of such treatment is to remove the copper from the effluent. Recycling of the water is a secondary goal. In many CMP effluents the copper is entirely soluble and is not incorporated into the abrasive particles of the slurry residue. Thus, the treatment first involves microfiltering the effluent to remove the nonhazardous abrasive solids (which are then discharged into the fabs industrial wastewater and ultimately to the POTW). Next the water containing the slurry chemicals and copper is treated. The slurry oxidizer (e.g., H_2O_2) is removed by appropriate chemical addition. Then the copper can be removed by several different methods:

1. An ion exchange process can be employed, using cation-exchange canisters. These require periodic on-site or off-site regeneration (which again involves an electrowinning procedure). Such a treatment scheme can achieve very high levels of copper removal from the relatively dilute CMP effluent.

2. Coagulation and microfiltration is another method that can effectively reduce the copper content to less than 0.1 ppm.[83]

3. To insure copper removal to sub-0.1 ppm levels due to stringent POTW discharge requirements, it may be necessary to use water-soluble heavy-metal removal agents. These materials typically contain sulfide functionality, which bonds strongly to transition metals and forms insoluble precipitants.[84]

REFERENCES

1. P. Singer. "Tantalum, Copper, and Damascene: The Future of Interconnects," *Semiconductor International*, June 1998, p. 90.

2. R.L. Jackson *et al.*, "Processing and Integration of Copper Interconnects," *Solid State Technology*, March 1998, p. 49.

3. S.C. Sun, "Process Technologies for Advanced Metallization and Interconnect Systems," *Tech. Dig. .IEDM,* 1997, p. 765.

4. T. Takewaki *et al., Symp. on VLSI Technology, Dig. of Tech. Papers*, (1995), p. 31.

5. S. Venkatesan et al., "A High Performance 1.8 V, 0.20 μm CMOS Technology with Copper Metallization," *Tech. Dig. IEDM*, 1997, p. 769.

6. D. Edelstein *et al.,* "Full Cu Wiring in Sub-0.25-μm CMOS Technology," *Tech. Dig. IEDM*, 1997 p. 773.

7. M. Bohr, "Interconnect Scaling: The Real Limiter to High Performance ULSI," *Tech. Dig. IEDM,* 1995, p. 241.

8. C. Ryu *et al., International Reliability Physics Symp. Proc.,* 1997, p. 201.

9. L. Peters, "Leaders Describe 'Surmountable' Copper and Low-*k* Hurdles," *Semiconductor International*, September 2000, p. 25.
10. M. Igarashi *et al.*, "The Best Combination of Aluminum and Copper Interconnects for a High Performance 0.18-μm CMOS Logic Device," ***Tech.*** *Dig. IEDM,* 1998, p. 829.
11. A.E. Braun, "Aluminum Persists in a Copper Age," *Semiconductor International*, August 1999, p. 58.
12. P.K.K. Ho *et al.*, "Direct Integration of Damascene Copper onto Aluminum Interconnect," *Abs. Mtg. Electrochem. Soc.,* Fall 2000, Abs. 525.
13. P.C. Andricacos, "Copper On-Chip Interconnections,," *Electrochemical Society Interface,* Vol. 8, No. 1, Spring 1999, p. 32.
14. S. Rossnagel and J. Hopwood, *Applied Physics Letters,* **63**, 3285 (1993).
15. S. Rossnagel and J. Hopwood, "Metal Ion Deposition from Ionized Magnetron Sputtering Discharges," *J. Vac. Sci Technol.,* **B12**, 449 (1994).
16. S.M. Rossnagel, "Ionized Magnetron Sputtering for Lining and Filling Trenches and Vias," *Semiconductor International,* February 1996, p. 99.
17. G.A. Dixit *et al.,* "Ion Metal Plasma Deposited Titanium Liners for 0.25 μm/0.18-μm Multi-level Interconnections," *Tech Dig. IEDM*, 1996, p. 357.
18. B.L. Chin *et al.*, "Barrier and Seed Technologies for Sub-0.1 μm Copper Chips," *Semiconductor International*, May 2001, p. 107.
19. P. Gopalraja and J. Forster, *Appl. Phys. Letts*., Vol 77, 2000, p. 3526.
20. C. Michaluk *et al.,* "Tantalum 101: Economics and Technology of Ta Materials," *Semiconductor International*, July 2000, p. 271.
21. H.E. Roosendaal, "Sputtering by Particle Bombardment I," in *Physical Sputtering of Single-Element Solids,* Behrisch (Ed.), Springer-Verlag, Berlin 1981, pp. 219-56.
22. D. Tracy, "The Emerging Copper Revolution Will Impact ICs with Critical Dimensions Smaller than 100-nm," *Chip Scale Review,* March 2001, p. 83.
23. C. Ryu *et al., Proc. Mtg. Material Res. Soc.,* Spring 1998, p. 75.
24. C. Ryu *et al.,* "Barriers for Copper Interconnections," *Solid State Technology,* April 1999, p. 53.
25. S.S. Wong *et al,. Proc. International Interconnect Conf.* (IEEE) 1998, p. 107.
26. T. Hara and K. Sakata, "Stress in Copper Seed Layer Employed in Copper Interconnects," *Electrochem. Soc. Letts.,* March 2001, G77.
27. T. Hara *et al.,* "Control of the (111) Orientation in Copper Interconnection Layer," *Electrochem. Soc. Letts.,* March 2002, C41.
28. A.A. Istratov and E.R. Weber, "The Importance of Understanding the Physics of Copper in Silicon for Development and Characteristics of Copper Diffusion Barriers," *Abs. Mtg. Electrochem. Soc.,* Fall 2000, Abs. 522.

29. M. Juppo *et al.*, "Atomic Layer Deposition of Titanium Nitride Thin Films Using *tert*-Butylamine and Allylamine as Reductive Nitrogen Sources," *Electrochem. Soc. Letts.*, January 2002, C4.
30. S. Hirao *et al., Dig. Tech. Papers Symp. VLSI Technology,* 1997, p. 57.
31. J.-S. Min *et al.*, "The Mechanism of Si Incorporation with the Digital Control of Si Content during the Metallorganic Atomic Layer Deposition of Ti-Si-N Thin films," *J. Electrochem. Soc.* **147**, October 2000. P. 3868.
32. A.E. Braun, "Copper Electroplating Enters Mainstream Processing," *Semiconductor International*, April 1999, p. 58.
33. J.D. Reid *et al.*, "Factors Influencing Damascene Feature Filling Using Copper PVD and Electroplating," *Solid State Technology,* July 2000, p. 89.
34. A.E. Braun, "ECP Technology Faces Chemical, Dielectric Hurdles," *Semiconductor International*, May 2000, p. 60.
35. I. Chen and T. Ritzdorf, "Seed-Layer Enhancement in Ultra-Thin Seed for Inlaid Copper Metallization," *Abs. Mtg. Electrochem. Soc.* Spring 1999, Abs. 275.
36. R. Solankey and B. Pathangey, "Atomic Layer Deposition of Copper Seed Layers," *Electrochem. Soc. Letts.*, October 2000, p. 479.
37. T. Nogami *et al., Proceedings of the IITC*, p. 298 (1998).
38. J.D. Reid, "Fundamentals of Copper Electroplating," *Semiconductor International*, June 1998, p. 94.
39. V. Shannon and D.C. Smith, "Copper Interconnect for High-Volume Manufacturing," *Semiconductor International,* May 2001, p. 93.
40. S. Mayer *et al.*, "Integration of Copper PVD and Electroplating Processes for Damascene Feature Filling," *Abs. Mtg. Electrochem. Soc.*, Fall 1999, Abs. 732.
41. J. Horkans and J.O. Dukovic, "Monitoring of SPS-Based Additives in Cu Plating," *Abs. Mtg. Electrochem. Soc.*, Spring 2000, Abs. 360.
42. J.C. Hu *et al.* "Investigation of the Leveling Effect on Electrodeposited Cu Films for ULSI Applications," *Abs. Mtg. Electrochem. Soc.*, Spring 2000, Abs. 388.
43. P.L. Cavalotti *et al.* "Copper Electrodeposition for On-Chip Interconnections: Role of Different Additives," *Abs. Mtg. Electrochem. Soc.* Spring 2000 Abs. 359.
44. M. Yokio *et al.*, "Adsorption Behavior of Polyethylene Glycol on a Copper Surface in Acid Copper Sulfate Bath," *Denki Kagaku*, Vol. 52, p. 218, 1994.
45. J.L. Kelly *et al.*, "Leveling and Microstructural Effects of Additives for Copper Electrodeposition," *J. Electrochem. Soc.*, **146**, July 1999, p. 2540.
46. P.C. Andricacos *et al.*, "Damascene Copper Electroplating for Chip Interconnections," *IBM J. Res. & Develop.* **42**, 567 (1998).
47. J.J. Kelly and A.C. West, "Experimental and Numerical Study of Leveling of Submicron Features by Organic Additives," *Abs. Mtg. Electrochem. Soc.*, Spring 1999, Abs. 262.
48. P. Andricacos *et al.*, "Damascene Copper Electroplating for Chip Interconnects," *Abs. Mtg. Electrochem. Soc.*, Spring 1998, Abs. 254.

49. W.Y. Hsieh, "Interconnect Manufacturing Issues," *Advanced Interconnects: Design, Process, Integration, IEEE Short Course,* IEDM 2000.
50. S. Subrahmanyan ***et al.,*** "Integrated Electroplating – Key to Chip Production," ***Vacuum Technology & Coating***, March 2000, p. 50.
51. G. Chalyt and P. Bratin, "On-Line Analysis of Organic Additives in Damascene Copper Processes, *Semiconductor International,* April 1999, p. 66.
52. T. Taylor *et al.,* "Electroplating Bath Control for Copper Interconnects," *Solid State Technology* November 1998, p. 56.
53. "Analysis of Copper Plating Baths," Dionex Corp. Website: www.dionex.com.
54. O. Blachier and J. Clark, "Taking a New Approach to Copper ECD Bath Management," *Semiconductor International*, July 2000, p. 229.
55. M. Biberger *et al.,* "Low Pressure Sputtering of Copper and Related Barriers for Seed Layers and Complete Planarization," Novellus Systems Website.
56. C. Marcadal *et al.,* "Metallorganic Chemical Vapor Deposited TiN Barrier Enhancement with SiH_4 Treatment," *J. Electrochem. Soc.* January 2002, p. C52.
57. A.C. West *et al.,* "Pulse-Reverse Copper Electrodeposition in High-Aspect-Ratio Trenches and Vias," *J. Electrochem. Soc.*, **145**, September 1998, p. 3070.
58. S. Lopatin, "Extendability of Electrochemical Deposition for High-Aspect-Ratio Cu Interconnects," *Abs. Mtg. Electrochem. Soc.,* Spring 1999, Abs. 260.
59. D. Varadarajan *et al.,* "A Mass Transfer Model for the Pulse Plating of Copper Into High-Aspect-Ratio sub-0.25 Micron Trenches," *Abs. Mtg. Electrochem. Soc.,* Spring 1999, Abs. 268.
60. P. Singer, "Copper CMP: A Question of Tradeoffs," *Semiconductor International*, May 2000, p. 73.
61. M. Tsai *et al.,* "CMP-Free and CMP-Less Approach for Multilevel Cu/Low-*k* BEOL Integration," *Tech. Dig. IEDM,* 2001, p. 80.
62. B. Baker *et al.,* "The Interactions Between PVD Copper and ECD Copper as Observed in the Self-Annealing Process," *Abs. Mtg. Electrochem. Soc.* Spring 2000, Abs. 364.
63. E. Kaiser *et al.*, "Improved Determination of Additives in Acid Copper Plating Baths," *Abs. Mtg. Electrochem. Soc.,* Fall 2001, Abs. 1303.
64. T. Hara *et al.,* "Chemical Mechanical Planarization of Copper and Barrier Layers by Manganese (IV) Oxide Slurry," *Electrochem. Soc. Letts.,* December 2001, G109.
65. E. Shan *et al.,* "Effect of Deposition Temperature on ECD Copper Self-Annealing," *Abs. Mtg. Electrochem. Soc.* Spring 1999, Abs. 274.
66. J. Chappell, "NuTool Unveils ECMD Tool," *Electronic News*, 12/18/2000.
67. V.M. Dubin *et al.*, "Selective and Blanket Electroless Cu Platingby Contact Displacement for Deep Submicron Via and Contact Filling," *Proc. VLSI Multilevel Interconnect Conf.,* 1995, p. 315.

68. J. Van der Meerakker, "On the Mechanism of Electroless Plating - I - Oxidation of Formaldahyde at Different Electrode Surfaces," *J. Appl. Electrochem.*, **11**, p. 387 (1981).
69. A. Contolini *et al.*, "Copper Electroplating Process for Sub-Half Micron ULSI Structures," *Proc. VLSI Multilevel Interconnect Conf.*, 1995, p. 322.
70. J.A. Norman *et al.*, "New OMCVD Precursors for Selective Copper Metallization," *Proc. VLSI Multilevel Interconnect Conf.*, 1991, p. 123.
71. D. DeBear *et al., Solid State Technology*, March 2000, p. 53.
72. J.A. Levert *et al.*, "A Novel Spin-Etch Planarization Process for Copper Dual-Damascene," *Abs. Mtg. Electrochem. Soc.* Fall 1999, Abs. 731.
73. D. Dornisch *et al.*, "Cu Contamination Control for Advanced Interconnect Manufacturing," *Solid State Technology*, May 2000, p. 137.
74. P.S. Lysaght and M. West, "Addressing Cu Contamination Via Spin-Etch Processing," *Solid State Technology,* November 1999, p. 63.
75. T. Cacouris, "Preventing Cross-Contamination Caused by Copper Diffusion and Other Sources," *Micro*, July 1999.
76. L. Peters," SEMATECH Explores Copper Contamination," *Semiconductor International*, May 2000, p. 89.
77. M.E. Gross *et al,* "Implications of Damascene Topography for Electroplated Copper Interconnects," *Solid State Technology*, August 1999, p. 47.
78. For more information on this process visit www.novellus.com.
79. L. Peters, "Following a Steep Learning Curve for Copper," *Semiconductor International*, October 2000, p. 25.
80. C. Ryu *et al.*, "Electromigration of Damascene Copper Interconnects," *Dig. Tech. Papers VLISI Symp.* 1998, p. 156.
81. H.S. Rathore *et al., 16th International VLSI Multilevel Interconnect Conf.* (1999), p. 89.
82. L. Mendocino and P.T. Brown, "The Environment, Health, and Safety Side of Copper Metallization," *Semiconductor International*, June 1998, p. 105.
83. R.D. Letterman (Ed.) *Water Quality and Treatment*, McGraw-Hill, New York, 1999, 5th Edition.
84. J.H. Golden *et al.*, "Evaluating and Treating CMP Wastewater," *Semiconductor International*, October 2000, p. 85.
85. T. Hara *et al.*, "Improved Barrier and Adhesion Properties in Sputtered TaSiN Layer for Copper Interconnects," *Electrochem. Soc. Letts.*, March 2002, G36.

INDEX

Abruptness of doping profile in SDE, 508
Accelerators (Cu plating additive), defined, 757
Accumulation-layer resistance (R_{ac}), 198-200, 509
additives to Cu plating baths, 756-63, 764-9
Aluminum, CMP of, 418, 711-3, 715-7, 726
  Electromigration behavior of, 716-7
  Hybrid Al + Cu interconnects, 726
  Resistivity of. 712-3
Aluminum oxide (Al_2O_3), 146, 172-3
  ALD of, 172-3
Amine contamination of DUV resists, 230-3
  Filtering to prevent, 232-3
Annealing of Plated Cu films, 770-2
  Sheet resistance vs. annealing temp., 771
Anode-hole model of oxide breakdown, 94-7, 97-8
Anti-reflective coatings (ARCs), 237, 244-51
  Bottom ARCs, 245, 246-50
    Inorganic, 249-50
    Organic, 248-9
    PVD ARCs, 250
  Top ARCs (TARs), 245
  UARLs, 251
ARC-open step, 248
ArF excimer lasers, 265
Aspect-ratio, defined, 316
Atomic-layer deposition (ALD), 170-2, 172-3
Atomic-layer epitaxy (ALE), 170, 508

Backbone polymers in resists, 238, 242
  In 193-nm resists, 238
  In 157-nm resists, 242
Back-gate in FD-SOI MOSFETs, 524
BARC in dual-damascene process, 678, 686, 692-4, 698, 700
BARC veils, in dual damascene lithography, 692-4
Barrier layers for Cu interconnects, 720, 725, 728-40
$BaSrTiO_3$, 145, 146, 152, 162-9
  Deposition of BST films, 163-5
  Integration issues, 167-9
  Microstructure of, 166-7
Bath (Cu plating) control, 763-9
Beveled edge, wafer,
  Cu cleaning from, 403
  Polishing of, 59
BiCMOS, Si:Ge, 479, 491, 496, 498
Bilayer resist process, 237, 694-6. 697
  in dual-damascene process, 694-6. 697
Binary imaging mask (BIM), 285
Bird's beak, in LOCOS, 433
Bleed & dump-to-drain (Cu plating bath Management, 765
Bleed & feed (Cu plating bath management), 766
Body contact in PD-SOI MOSFETs, 520-22
Bond-and-Etch-Back (BESOI), 541-2
Boron penetration of gate oxide, 112, 138, 139
"Bottom-up" filling of recesses with Cu plating, 728, 756, 759-63, 769
  Models for, 759-63
BOX (see Buried oxide)
Bridging, in $TiSi_2$ formation, 615
Bright-field inspection systems, 410, 412
Brighteners (in Cu plating baths) def., 757
BST (see BaSr TiO_3)
Buff polish step, in CMP, 404, 416
Buried oxide (BOX)
  in STI, 435-6, 7
  in SOI, 502, 530, 532-3, 536-7

Calcium fluoride (CaF_2), 267, 267, 269, 270-2, 289-90
  Extrinsic birefringence, 271
  Intrinsic birefringence, 271-2
CARL resist process, 237
Catadioptric optical system, 269
Cathodic protection, 755

C-49 phase of $TiSi_2$,
615, 618, 619, 620, 621, 622
C-49 phase of $TiSi_2$,
615, 618, 619, 620, 621, 622
Chained implants, 186-7
"Champagne-glass" via, 579
Charge-to-breakdown (Q_{BD}), 98-9
Charge storage capacity in DRAM cells, 150
Chemically-amplified resists (CA), 228-9
Chemical mechanical polishing, 313-428
of Aluminum, 418
Cleaning issues, 392-406, 407, 408
Brush scrubbing, 399-404
Buff polish step, 404-5
Corrosion during, 418
Equipment, 406-7, 408
Megasonic cleaning, 404, 405
Metallic contaminants, 405
Steric Stabilization of particles, 398-9
VanderWaals force, 393-4
Vibrational scrubbing, 404
Zeta potential, 394-6
of Copper, 417-8, 726
Dishing, 421-26
Direct polishing, to prevent, 424-5
Dummy-active regions, to prevent, 423-4
in STI, 462-4
Nitride cap, to reduce, 423, 467
RIE/CMP, to prevent, 424, 464-5
Economics of CMP, 427
Edge effect in CMP, 364-5
Efficiency of planarization (EOP), 326
Endpoint detection in CMP, 385-9
Frontside laser interferometry, 388
In-line, defined, 385-7
In-situ, 387
Motor-current monitoring, 391
Optical monitoring, 390
Spectroscopy based, 389
History of, 333-5
of Interlevel dielectrics, 414
of Low-*k* Dielectrics, 420-1
Metrology of CMP,
Degree of planarity, 408
Thickness measurement, 407
Modeling of the Mechanisms of, 335-52
Metal removal, 349-52
Oxide removal, 341-8
Pads, 375-8, 382-4, 414
Slurry-free pads, 382-4
Pad conditioners, 378-82
Pattern dependence of planarization rate, 346-9
of Photoresist, 419
Polishing tools, 352-8
Fixed Table, 353, 356-7
Linear, 353, 355-6
Orbital, 353, 355
Reliability of, 411-2
Rotary, 353, 354
of Polysilicon, 418-9
Process integration issues of, 412-3
Removal rate, defined, 325
Slurries, 366-71
For metal CMP, 370-1
For oxide CMP, 367-9
Ceria, 369
Colloidal silica, 368-9
Fumed silica, 367, 368-9
Slurry distribution systems, 371-4
Slurry-free pads, 382-4
in STI, 438-9
of Tantalum (Ta), 417
of Tungsten (W), 415-17, 500-2
Wafer-to-wafer non-uniformity (WTWNU), 330-1
Within-die non-uniformity (WIDNU), 327-9
Within-wafer non-uniformity (WIWNU), 329-30, 339-40
Child's Law, 237
Cleaning Issues of CMP, 392-406
CMP (see chemical mechanical polishing)
Cobalt Silicide ($CoSi_2$),
203-4, 604, 610-11, 620, 624-34
Diode leakage in, 628-30, 633
Epitaxial $CoSi_2$, 630-1
Process sequence for forming, 625, 627-8

Silicon consumption during formation, 626-7
as Source/Drain diffusion source, 631-2
TiN capping layer, 629-30
Collimated sputtering, 731, 733
Contact region of drain, 10, 203, 516
Resistance of (R_{co}), 203-4, 208, 610, 612-14
Contact resistivity (ρ_c), 610
Salicide to silicon, 610
Copper (Cu), 711-89
Advantages of, 712-9
Barrier layers for, 720, 725, 728-40
CMP of, 417-8, 726,
CMP cleaning of, 403
Contamination control of Cu processes, 778-81
Effects of Cu on IC reliability, 780-82
Removal of Cu from wafer bevel and backside, 781-4
Sources of Cu contamination, 778-80
Corrosion during CMP, 418
CVD of Cu, 774-5
Difficulties of using Cu, 719-23
Diffusion barriers for, 728-40
Conductive, 720, 722-3, 728-40
Silicon nitride, 721
Tungsten-plugs, 721
Dishing during CMP of, 425
Effect of Cu on the reliability of ICs, 719-20, 723-4
Electroless plating of Cu, 774
Electrochemical mechanical deposition (ECMD), 772-3
Electromigration behavior of, 716-7
Electroplating of Cu, 747-69
Accelerators, def. 757
"Bottom-up" filling, 728, 756, 759-63, 769
Models for, 759-63 756, 759-63, 769
Brighteners, def. 757
Basics of, 747-51
Levelers, def. 758
Plating-bath control, 763-9
Cyclic voltametric stripping (CVS), 764-5
High-performance-liquid-chromatography (HPLC), 765
Plating tools, Cu, 753-4
Process sequence, Cu plating, 755-6
Pulse-plating, 752, 767-9
Reverse-pulse-plating, 768-9
Supprssors, def. 757-8
Electropolishing of Cu films, 775-8
Spin-etch planarization of Cu films, 776
Environmental, Health, and Safety issues of Cu processing, 787-9
Manufacturing protocols for, 721-2
Propagation delay in, 713-16
Resistivity of, 712, 713
Seed-layers of Cu, 725-6, 740-7
ALD of, 746-7
CVD of, 745-6
Desired characteristics of, 740-1
Preferred crystal orientation, 742-3
PVD of, 743-4
Self-annealing of plated Cu films, 770-2
Supprssors, def. 757-8
Topography of Cu films after plating, 769-70
Wet etching of Cu films, 776
COPs (see crystal-originated particles)
Corner effect in STI, 434, 451, 469-72
Corner parasitic MOSFETs in STI, 436
Corner Rounding in STI, 451, 458
Critical dose in high-dose SIMOX, 532
Crystal-originated particles (COPs) 21-32, 27-9, 47, 48, 64-5
Annihilation of, 27-9, 30
Defined, 23
Formation of, 23-5
In epitaxial wafers, 48
In fast-pulled wafers, 47
Current crowding in Cu vias, 717-8
Current-gain cutoff frequency (f_T), 481, 490, 494, 498
Cyclic order skipping, in film thickness measurements, 407

Cyclic voltametric stripping (CVS), 764-5

Damascene (see Dual Damascene)
DARC, 251
Dark-field inspection systems, 410, 412
Decel in low-energy implanters, 207
Deep-trench isolation, 490
Deep-UV resists, 228-44
Amine contamination of, 230-2
248-nm, 228-37
193-nm, 237-41
157-nm, 241-4
Poisoning, 690-1
T-top formation in, 230-2
Defects in CMP, 408
Types, 408
Degree of planarization, 313-5, 326
Measurement of, 408
Denuded zone, 39, 40-1, 45, 50-2
In "New" wafers, 48-50
Magic Denuded Zone, 50-2
Dielectric isolation, 517
Direct polish process to prevent dishing in STI, 424, 467-8
Direct tunneling leakage current in gate oxides, 101-4
Dishing, 421-26
in Cu CMP, 417
in CMP, 421-6
in STI, 462-4
Direct polishing, to prevent, 424-5
Dummy-active regions, to prevent, 423-4
Nitride cap, to reduce, 423, 467
RIE/CMP, to prevent, 424, 464-5
DNQ/novolac resists, 207-8
Dose-loss during shallow-junction formation, 214-7
Out-diffusion, 215-6
Screen oxide effect, 216-7
Self-sputtering, 214
Double-level metal interconnects, 576-7, 580
Dual damascene interconnects, 582, 583, 584, 671-709
BARC veils, 692-4
Buried via, 676, 679
Defined, 672, 674-5
Dielectric stack in, 679-80
Embedded etch stop layer, 676, 681-2
Silicon carbide, 682-3
Silicon nitride, 681-2
Hard mask in dielectric stack, 681
Process flow, basic, 584, 675-6, 687-8
Process integration issues, 687-706
Cleaning, 703-8
Post-CMP clean, 705
Pre-metal deposition, 704-5
Resist-strip residue, 703-4
Supercritical fluids, using, 706-8
Etching, 697-700
SiN layer, 700
Trench, 699-700
Via, 698-99
Lithography, 689-96
BARC veils, 692-4
Bi-layer resist process, 694-6, 697
DUV-resist poisoning, 690
Trench CD control, 691-2
Resist stripping, 700-3, 703-4
Ultra-low-*k* dielectrics, 708-9
Self-aligned process, 676, 679
Trench-first process, 676, 677-8, 683-5
Via-first process, 676, 678-9, 685-7, 687-8, 708-9
Dual-gate MOSFETs, 1, 524-5, 656-6
Dual-gate oxide thickness IC, 121
Dual-doped poly, in CMOS, 12, 13
Dummy active regions to prevent dishing in STI, 423-4, 464, 465-6

Edge effect in CMP, 364
Edge exclusion, in CMP, 356-6
Edge polishing of wafers, 59. 60
Edge-rounding of wafers, 59, 60
Electrochemical mechanical polishing (ECMD), Of Cu, 772-3
Electromigration behavior of Cu interconnects, 716-7
Electron beam projection lithography (EPL), 397-9

Electroplating of Cu (see Cu Interconnects)
Electroless plating of Cu, 747, 748, 774
Electropolishing of Cu films, 775-8
Elevated source/drains, 203, 205-6
ELTRAN, 547-50
Endpoint-detection in CMP, 385-9, 407
- Frontside laser interferometry, 388
- In-line, defined, 385-7
- In-situ, 387
- Motor-current monitoring, 391
- Optical monitoring, 390
- Spectroscopy based, 389

Epi-optimized wafers, 12, 35, 45
Epitaxial layers, Si:Ge, 485-6, 487, 492
- Compatibility with ion-implantation, 493
- Compatibility with thermal oxidation, 493-4
- On Patterned wafers, 492
- RTP-CVD of Si:Ge, 489-90
- Selective Si:Ge, 490
- Thermal stability of, 492
- UHV-CVD of Si:Ge, 488-9

Epitaxial wafers, 11
EPL (see Electron-beam projection lithography)
Equivalent oxide thickness, 130, 146-7, 175
Erosion, in CMP, 425
Etalon, 262, 264
EUV (see Extreme UV lithography)
Excess overburden, in Cu plating, 770, 773
Excimer, defined, 260-1
Excimer lasers for DUV lithography, 259-65
- 248-nm (KrF), 262-5
- 193-nm (ArF), 265
- 157-nm (F_2), 265
- Spectral bandwidth of, defined, 261

Exposure tools for DUV lithography, 266-73
- For 248-nm lithograhy, 266-7
- For 197-nm lithograhy, 267-8
- For 157-nm lithograhy, 268-9
- Mix-and-match lithography, 274, 299
- 300-mm Exposure tools, 272-3

Extinction coefficient in BARCs, 246
Extreme UV lithography, 301-7
- Masks for EUV, 305-6
- Multilayer reflector for, 305

Fast-pulled silicon ingots, 30-2, 45, 47
Fe-B (iron-boron) pairs, 12, 38-9
Ferroelectric dielectrics, 151, 163
Ferroelectric RAMS (FeRAMs), 154, 155, 163
Fixed oxide charge (Q_f), 82
Floating-body effects in Partially-depleted SOI MOSFETs, 519-20, 559-61, 562-3
- Transient, 562

Fluorinated oxides, 420, 650, 654-5, 680
- CMP of, 420

Fluorine (F_2) excimer lasers, 265
FOSB, 62-3
Fowler-Nordheim tunneling, 100-1
Front opening shipping box, 52-3
Fully-depleted SOI MOSFETs, 518, 522-6

Gallium-arsenide, 476-7, 478
Gas immersion laser diffusion, (GILD), 508
Gate current, 99-104
Gate oxides
- Breakdown of, 90-9
  - Anode-hole Model, 94-5
  - Current-driven model, 94-5
  - E-field driven model, 93-4
  - Fluorinated oxides, 120-1
  - Growth models of thin oxides, 104-9
  - Interfaces charges, 82-90
  - Microstructure of, 80-2
  - Nitrided oxides, 112-20
  - Rapid thermal oxidation (RTO), 109-12
  - Scaling of, 76-8
  - Soft-Breakdown of SiO_2, 126-8

Gate oxide integrity, 12, 22, 31, 47, 48, 64-5
Gettering, 32-45
- Basic methods, 34-6
- Extrinsic, 35, 36-7
- Impurity pairs, 38-9
- Intrinsic, 35, 37-9, 41-4, 51, 66
- Mechanisms of, 34-6
- Oxygen precipitates, 37-8, 39
- Phosphorus gettering, 35

Global planarization, defined, 313-4, 323
GOI (see gate oxide integrity)
Grain noise, in post-CMP processing, 409

Hafnium oxide (HfO_2), 146, 175-6

HALO implant, 7
HBT (see heterojunction bipolar transistors)
HDP-CVD of trench dielectrics, 459, 461-2
Hemispherical silicon grain (HSG), 151, 159
Heterojunction bipolar transistors (HBTs), 477, 478, 480-86, 498
  Large-bandgap emitters, 482-4
  Small bandgap emitters, 484-6
  Si:Ge, 478-80
HF-de-glaze steps in STI, 452
Hi-Wafers, 28
High-density plasmas for depositing trench dielectrics for STI, 460-2
High-*k* Dielectrics, 145-178
High-dose SIMOX, 531-5
High-performance-liquid-chromatography (HPLC), 765
Hollow-cathode magnetron sputtering, 735-6
Hot-carrier injection, 8
Hybrid-organo-siloxane-polymer (HOSP), 658
Hydrogen annealing of Si/SiO_2 interface, 85-6
Hydrogen annealing of Si wafers, 27-9
Hydrogen annealing to achieve corner rounding In STI, 458
Hydrogen silsesquioxanes (HSQ), 651-3

Implant shadowing effects, 188-90
Indium channel implants, 193-4
Induction time in Cu plating, 755
Indium implants for HALO applications, 221-2
Interface trapped charge (Q_{it}, D_{it}), 82, 83-9
  Effect of interface traps on ICs, 86
Interconnects for ICs
  (see Multilevel interconnects)
Interstitials, self, in Si, 22, 50-2
Initiation current in Cu plating, 755-6
Intrinsic point defects, in Si, 22
Ionized magnetron sputtering (I-PVD), 731-4, 744
Ionized metal plasma (IMP), 731
ITOX (internal thermal oxidation) 536-7, 538
ITRS Roadmap, 505, 506, 514, 642

"Kink effect" in STI, 437, 438, 469-72
"Kink effect" in SOI, 520, 559-60
Kooi effect in STI, 454
KrF excimer lasers, 262-5
KrF (see Krypton fluorine lasers)

Laser marking of wafers, 56
LATIPS, 8, 219-20
LDD (see lightly doped drain
Leakage paths in STI structures, 441-2
Levelers (Cu-plating-bath additive), 757
Lightly-doped drain, 9, 195-6
Line jogs, 282
LLDs (localized light scatterers), 24
LOCOS isolation, 13, 14, 433, 434, 436
  Mini-LOCOS in STI, 458
  Recessed LOCOS for fabricating thin-body FD SOI MOSFETs, 525-6
Long-throw sputtering, 731, 734
Low-angle tilt-implants to dope STI trench sidewalls, 444-5
Low-dose SIMOX, 530, 535-7
Low-energy implants, 206
  Decel in, 207
  Tools, 207
Low-*k* Dielectrics, 639-59
  BCB materials, 657, 658
  Carbon-based materials, 644, 649
  CMP of, 420-1
  CVD materials, 643, 654-5, 656, 659-62, 662-3
  Defined, 639
  Density impact on *k*-value, 640-1, 664-5
  Desired characteristics of, 646-8
  Embedded low-*k* films 650,
  Fluorinated oxide films (FSG), 650, 654-5, 680
  Hybrid-organo-siloxane-polymers (HOSP) materials, 658
  Hydrogen silsesquioxanes (MSQ), 651-3
  Inorganic materials, 649
  Ionization impact on *k*-value, 641
  Methyl silsesquioxanes (HSQ), 651-3
  Organic materials, 649
  Organosilicate glasses (OSG), 659-62, 698

PAE materials, 657, 658
Parylene materials, 662-3
Polarization impact on *k*-value, 641
Porous ultra-low-*k* films, 668-9
Process integration isues of, 648-9
Silicon based materials, 649
Sol-gel materials, 664
Spin-on materials, 643, 650, 651-3, 653-4, 656, 657-8
Ultra-low-*k* materials, 663-9
Xerogel materials, 665
Low-temperature epitaxy, 487-8, 508
of Si:Ge layers, 487-8
LSTDs, 23

Magic Denuded Zone, 50-2
Magnesium oxide (MnO_2), CMP slurry, 421
Mask-error factor (MEF), 292-5
Maximum frequency of oscillation (f_{max}), 481, 490, 494, 498
MEF (see Mask error factor)
MEEF (see Mask error factor)
Megasonic cleaning, 404, 405
Metal-insulator-metal capacitors (MIM) in DRAMS, 153, 160
Metal-insulator-silicon capacitors (MIS) in DRAMS, 152, 155
Methyl silsesquioxanes (MSQ), 653-4
Mini-LOCOS process for STI, 458
Mix-and-match lithography, 274, 299
Mobile ionic charge (Q_m), 82
Molecular airborne contamination, 137
Oxide growth problem, 137
Molybdenum silicide ($MoSi_2$), 604, 605
Multilevel interconnects for ULSI, 573-601
Contact resistance in, 598
Crosstalk in, 596-7
Double-level metal interconnects, 576-7
ESD protection in, 599
Geometrical aspects, 585-8
Orthogonal tier architecture, 585-7
Global interconnects, def., 585
Local interconnects, def., 585
Power dissipation in, 595-6
Propagation delay (RC) defined, 588-9
Repeaters in, 593-5
Single-level metal interconnects, 575-6
Trends in, 599-600

Nanocleave, 544-7
Narrow-line effect in $TiSi_2$, 618-22
Next-generation lithography (NGL), 301-9
EPL, 307-9
EUV, 301-7
NGL (see Next-generation lithography)
Nickel silicides (NiSi), 610, 611, 634-5
Nitride cap process to reduce dishing in STI, 423, 467
Nitrided oxides, 112-20
Nitrogen doping of Si ingots, 30-1
Nitrogen implantation of Si to inhibit growth of oxide, 121
Non-optical lithography, 301-9
Notches, in Si wafers, 53-4

Octohedral voids in Si wafers, 23
Off-axis illumination, 275-9
Annualr apperture, 278
Quarupole apperture, 278
Offset-spacer to form SDE overlap, 217
157-nm DUV resists, 241-4
193-nm resists, 237-42
Line edge roughness, 240
Outgassing, 241
Resist collapse, 240-1
Shrinkage from e-beam, 241
Onium salts, 228-9
Optical proximity correction (OPC), 279
Organo-silicate glasses (OSG), 659-62, 698
OSF ring (see Oxidation Stacking Fault ring)
Out-diffusion, dopant loss during shallow-junction formation, 216-7
Overburden, 726, 770, 773
Defined, 726
Excess, 770, 773
Overlap of SDE and gate edge, 216-7
Overplating, 770
Oxidation-induced stacking faults (OSF), 26
Oxide/nitride/oxide (ONO), 151
Oxide trapped charge (Q_{ot}), 89-92

Effects on ICs, 90
Oxygen precipitates, 37, 38-9, 41-4
Oxygen stacking faults (OSF) ring, 26-7, 30
Oxynitridation of Si in N_2O, 113-6
Oxynitridation of SiO2 in N_2O or NO, 116-7

Pads, CMP, 375-8, 382-3 (see also CMP)
Contribution to dishing, 422, 463-4
PAE low-*k* materials, 657, 658
PAG (see photoacid generator)
Paraelectric dielectrics, 163
Partially-depleted (PD) SOI MOSFETs, 518, 519-20, 559-61, 562-3
Body contacts in, 520-2
Floating-body effects in 519-20, 559-61, 562-3
Particle adhesion mechanisms, 393-5
Parylene low-*k* materials, 662-3
"Perfect" silicon, 29-30, 45, 47
Perovskites, 163, 165, 166, 168
Phase-shift masking (PSM), 283-92
Alternating PSM (ALT-PSM), 286-7
Attenuating-PSM, 288
Half-tone PSM, 288
Rim-shift PSM, 288
Sidewall-chrome alternating aperture, 287
Strong, 288, 291, 292
Weak, 288, 292
Photoacid generator, 228-9, 239, 241
Photo-islands, 252-6
Photoresist processing systems, 251-4
Track systems, 252-4
Physical-vapor deposition (PVD), 729-36
Collimated sputtering, 731, 733
Filling recesses with, 729-33
Hollow-cathode magnetron sputtering, 735-6
Ionized magnetron sputter deposition (I-PVD), 731-4
Ionized metal plasma (IMP), 731
Long-throw sputtering, 731, 734
Self-ionized plasma sputtering (SPS), 736
Planarization, partial (in double-level metal), 578-9
Planarization factor, 315
Plasma immersion ion implantation, 531
Plasma-induced damage of oxides, 134-5
Platinum, 153, 167
Platinum group of metals (PGM), 153, 161
"Plus-One" approximation of TED, 209
Pocket implant, 8, 219-21
Polycides, 4, 603, 605-8
Polyfluorotetraethyl, low-*k* material, 666
Polysilicon depletion, 128-33
Polysilicon, CMP of, 41
Pores in low-*k* films, 668
Post-exposure bake (PEB), 230-2, 235
Pourbaix diagram, 351
Praseodymium oxide (Pr_2O_3), 176-7
Pre-amorphization implant (PAI) in $TiSi_2$, 620-2
Preston's constant, 337
Preston's Law, 336-8, 340-2
PREVAIL, 309
Propagation delay (RC) in IC interconnects
defined, 588-95, 639-40
in Cu interconnects, 713-6
PSM (see Phase-shift masking)
Pseudomorphic-epitaxial layers in Si:Ge, 405-6
Puller, Si ingot, 20
Pull rate, Si ingot, 20, Pull-speed, Si ingot, 20
Pulse plating of Cu interconnects, 750, 767-9
Reverse-pulse plating of Cu, 768-9
PZT, 145, 154

Quad Implants, 188-90, 221
Quadruple off-axis illumination, 278

RELACS, 235-6, 470-1
Reliability issues of SOI, 567-8
Replacement gate MOSFETs, 1
Remove-and-reconstitute (Cu-plating-bath management), 767
Resist stripping in dual-damascene process, 700-3
Resolution-enhancement technologies (RETs), 274-91
Off-axis illumination (OAI), 275-9
Optical proximity correction (OPC), 279-84
Optical proximity correction features, 280-84
Line jogs, 282
Scattering bars, 283
Serifs, 282

Phase-shift masking (PSM), 283-92
Alternating PSM (ALT-PSM), 286-7
Attenuating-PSM, 288
Half-tone PSM, 288
Rim-shift PSM, 288
Sidewall-chrome alternating aperture, 287
Strong, 288, 291, 292
Weak, 288, 292
Rapid-thermal oxidation (RTO) process, 109-14
Wet RTO processes, 111-2
Rapid-thermal annealing of shallow junctions, 210-11
Retrograde wells in CMOS, 12, 182-91
Reverse-narrow-channel effect, 471
RIE/CMP process to reduce dishing in STI CMP, 423, 464-5
Ruthenium, 153, 158, 167, 168, 169
Ruthenium oxide (RuO_2), 158, 160, 167, 168

Sacrificial gate oxide , 454,
Sacrificial layer etchback, 320-1
SADS (see silicide as diffusion source)
Salicide, 4, 203, 603, 604, 605, 608-35
$CoSi_2$ (see Cobalt silicide)
Defined, 605
Nickel silicide, 610, 611, 634-5
$TiSi_2$ (Titanium silicide)
Sawing wafers, 53-7
Scanning capacitance microscopy (SCM), 219
Scanning tunneling microscopy (SCM), 219
SCALPEL, 307
Scattering bars, 283
SDE (see source/drain extension)
Seed-layers of Cu (see Cu Interconnects)
Selective epitaxial growth (SEG) of Si:Ge films, 490, 508
Self-annealing of plated Cu films, 770-2
Self-heating effects in SOI MOSFETs, 566
Self-interstitials in Si, 22, 25
Self-ionized plasma sputtering (SPS), 736
Self-sputtering during low-energy implants, 313-4
Separation by Implanted Oxygen (SIMOX), 518, 529-37
High-dose, 530, 531-5
Low-dose, 530, 535-7
ITOX, 536-7, 538
Serifs, 281-2
Shallow-junction formation, 206-10, 213-7, 506-7
Dopant loss during, 313-4
Out-diffusion, 215-6
Self-sputtering, 213-4
Screen oxide effect, 216-7
Doping profile abruptness, 508
RTA of, 210
Shallow-trench isolation, 433-72
Baseline process flow, 446-55
Corner rounding in, 451, 458
in CMOS, 422
CMP, polish stop layer, 447
Dishing in, 422
"Kink effect" in STI, 437, 438, 450
Leakage paths in, 441-4
Si trench depth issues, 457
in SOI, 556, 557
Trench filling process, 458-62
Sheet resistance, 509, 611, 618, 624
$CoSi_2$, 611, 624
$TiSi_2$, 611, 618
Short-channel effects in SOI MOSFETs, 564-6
Silicide, 603
Silicon
COPs in, 21-32, 27-9. 47. 48, 64-5
Epitaxy optimized wafers, 12, 32, 45, 46, 47-8
Grown-in defects in, 21-32
"Perfect" Si, 29-30, 45, 47
300-mm Si wafer growth, 17-21
Silicon carbide (SiC), 682-3
Silicon as diffusion source (SADS), 631-4
SiLK, low-*k* dielectric, 657-8, 680-1
Silicon:Germanium (Si"Ge), 475-98
Base-after gate process, 495, 496
Base-during gate process, 495-6
Base layer, 486
HBTs, 478-80
Process integration issues, 491-4

Process technology for making Si:Ge HBTs, 486-90
Vertical scaling of, 494-5
Silicon-on-Insulator, 14, 501-68
Fabricating SOI wafers, 526-50
BESOI, 541-2
ELTRAN, 5467-50,
Nanaocleave, 544-7
SIMOX, 518, 529-37
Smart-Cut, 542-4
Silicon-on-Sapphire (SOS), 517, 527-8
Wafer Bonding, 518, 537-50
Fully-depleted SOI MOSFETs, 518, 522-6
History of SOI, 516-8
Isolation issues in, 502-5
in CMOS, 504-5
Lateral isolation in, 550-6
LOCOS isolation, 551-4, 555
Mesa isolation, 55-1, 552, 553
STI, 556
Partially-depleted (PD) SOI MOSFETs, 518, 519-20, 559-61, 562-3
Process technology of fabricating FD SOI MOSFETs, 525-6
Reliability Issues of SI, 501, 567-8
Self-heating effects in SOI, 566
Short-channel effects in SOI, 565-6
Thick SOI, 518
Thin SOI, 518
Silicon-on-Sapphire (SOS), 517, 527-9
Silicon dioxide (see Gate oxides)
SIMOX (see Separation by Implanted Oxygen)
Single-damascene process, 673-4, 676
Single-event upset (SEU), 517
Slurry, CMP, 366-70 (and see CMP)
Slurry distribution systems, 371-3
EHS issues, 373-4
Smart-Cut, 542-4
Soft-Breakdown of SiO_2, 126-8
Soft-error rate (SER), 150, 513
Sol-gel technique, 664
Solid-phase epitaxial (SPS) growth of SOS films, 529
Double SPS (DSPS) films, 529
SOI (see Silicon-on-Insulator)
Source/Drain Extension, 10, 208, 216-7, 219-20, 516
Abruptness of doping profile, 218
Offset spacer in, 217
Overlap with gate edge, 216-7
Resistance of (R_{DSE}), 196
Spacers, offset, to form SDE-overlap, 217
Spacers, sidewall, 9
Spreading resistance of drain (R_{sp}), 198-203, 508-9
SDE doping profile abruptness, 218-9
Spike anneals, 208, 211-2, 507
Spin-etch planarization (SEP) of Cu films, 776
Spin-process contamination elimination of Cu (SPCE), 782-3
SSR (see Super-steep retrograde channel doping)
Stacked gate dielectric films, 138-9
Steric stabilization of particles in a solution, 398
STI (see Shallow trench isolation), 13
Stress-induced leakage current, 126
Subtractive interconnect process, 671
Super-steep-retrograde channel doping (SSR), 6, 7, 191-95
Surface-channel MOSFETs, 13
Subtractive interconnect process, 575
Supercritical fluids, 664
"Superfilling" of recesses with Cu plating, 727-8
Suppressors (Cu plating bath additive), 770-2
Surface grinding, Si ingot, 53-4
TAETO, 155-6
Tafel equation, 751
Tafel plot, 751
Tantalum, 417, 720, 725-6, 728, 735-6, 738-9, 740, 747, 752, 776
CMP of, 417
Diffusion barrier properties of, 738-9, 740, 747
Material properties of, 736-7
Refinng of, 737-8
Tantalum nitride (TaN), 684, 720, 725-6, 737-40, 747, 776
Diffusion barrier for Cu, 739-40, 747, 752

Tantalum silicide ($TaSi_2$), 604, 605
Tantalum silicon nitride (TaSiN),
725, 728, 736, 739, 752
Tantalum pentoxide (Ta_2O_5),
145, 146, 148, 149, 151, 154-61
as DRAM capacitor dielectric, 154-60
Oxygen annealing of, 157
as MOSFET gate dielectric, 160-1
Thick SOI MOSFETs, 501, 518, 520
Thin SOI MOSFETs, 518
(311)-defects, 209
Time-dependent-dielectric breakdown (TDDB),
91, 97, 125
Time-to-breakdown (t_{BD}), 93, 94, 98-9
Titanium aluminide ($TiAl_3$), 712
Titanium arsenide (TiAs), 623, 632
Titanium boride (TiB), 623, 632
Titanium nitride, as Cu diffusion barrier,
720, 725, 736, 740
Titanium silicon nitride (TiSiN), 725, 739-40
Titanium silicide ($TiSi_2$), 203, 204, 604, 610,
611, 614-24, 625, 627, 632
Bridging during formation, 615-18
Nitrogen stuffing to prevent, 616-7
C-49 phase of $TiSi_2$,
615, 618, 619, 620, 621, 622
C-49 phase of $TiSi_2$,
615, 618, 619, 620, 621, 622
Dopant redistribution during formation,
622-4
Molybdenum implant into, 621
Narrow-line effect, 618-22
Pre-amorphization implant (PAI), 620-22
Process sequence for forming, 622-4
Silicon consumption during formation,
622-4
Track systems, photoresist, 252-6
Transient-enhanced diffusion, 209-10, 211-2
Trench filling processes for STI, 458-62
Trench-first dual-damascene process,
676, 677-8, 683-5
T-top formation in DUV resists, 228-9
248-nm DUV resists, 228-37
Post-exposure baking of, 235-6
RELACS, 235-6
Tungsten
CMP of, 415-17
W-Plugs, 580-2, 683
and Al lines, 583
Tungsten hexafluoride (WF_6), 606, 607
Tungsten nitride (WN) as Cu diffusion barrier,
725, 740
Tungsten silicide (WSi_2), 604, 605-6
Deposited with SiH_4 and WF_6, 606
Deposited with SiH_2Cl_2 and WF_6, 607
Tunneling currents in oxides, 99-104
Direct, 102-4
Fowler-Nordheim, 100-1

UARL, 251
UHV-CVD of Si:Ge layers, 488

Vacancies, in Si, 22, 25, 50-2
Vacancy-rich region in Si ingot, 26, 45
Van der Waals force, 393-4
Vertical gate MOSFET, 1
Via-first dual damascene process
(see Dual damascene interconnects)
Void formation during dielectric film deposition,
322-3
Void formation during trench dielectric
deposition in STI, 460-2
Void-free filling of recesses with Cu, 754-6

Wafer Bonding, 518, 537-50
BESOI, 541-2
ELTRAN, 5467-50
Generic, 538-41
Nanaocleave, 544-7
Smart-Cut (Hydrogen implantation), 542-4
Wafering, 52-66
Economics of, 66-8
Edge rounding of wafers, 59
Edge polishing of wafers, 59
Flat grinding, 53
Lapping and grinding of wafers, 57-9, 61
Laser marking, 56
Notch grinding, 53
Sawing of wafers, 55-7
Shipping wafers, 62-3

Specifications of 300-mm wafers, 63
Well dopants in CMOS, 6
"Well engineering," 184
Well formation in CMOS, 181
Retrograde wells, 182-191
Wire sawing of wafers, 56-7
Within-wafer non-uniformity from CMP, 408, 426-7

Xerogel low-*k* materials, 665

Zeta potential, 394-6
Zirconium oxide (ZrO_2), 146, 173-5
Yttrium stabilized (YZT), 174

ORDER FORM

LATTICE PRESS
POST OFFICE BOX 340
SUNSET BEACH, CA 90742, U.S.A.

www.latticepress.com
Phone: (714) 840-5010
FAX: (562) 592-1976

Please send the number of copies indicated below of **SILICON PROCESSING FOR THE VLSI ERA - Vols. 1, 2, 3,** and/or **4 (please specify!)** A check, money order, or Credit Card payment for the full amount for the number of copies ordered is enclosed. Make checks payable to: **LATTICE PRESS.** I understand that I may return any book within 30 days for a full refund if not completely satisfied!

____
NAME (and COMPANY - if part of your address)

____
ADDRESS

____
CITY

____
STATE ZIP CODE COUNTRY

______________ ______________
Daytime Phone Number e-Mail Address

____________________ __________
Credit Card Number Expiration Date

____ copies **Vol. 1** @ $199.95 each, plus $10.00 for each copy - shipping-handling fee.
(*California Residents*: Please add $15.00 per copy for Sales Tax.)

____ copies **Vol. 2** @ $139.95 each, plus $10.00 for each copy, shipping-handling fee.
(*California Residents*: Please add $11.00 per copy for Sales Tax.)

____ copies **Vol. 3** @ $149.95 each. plus $10.00 for each copy, shipping-handling fee.
(*California Residents*: Please add $12.00 per copy for Sales Tax.)

____ copies **Vol. 4** @ $199.95 each, plus $10.00 for each copy shipping-handling fee.
(*California Residents.* Please add $15.00 per copy for Sales Tax.)

International Ordering Information:

Enclose full payment in Travelers Checks (in $US), International Money Order, or by a Check payable to **Lattice Press** in **US dollars drawn on a U.S. bank.** Or send Credit Card payment.

Include a $12.00 US shipping/handling fee per copy for Overseas Orders shipped via SURFACE MAIL. (Allow **8-10 weeks** for Surface Mail delivery.)
Include a $30.00 US shipping/handling fee per copy for Overseas Orders shipped by AIR MAIL, with tracking capability. (Allow **2 weeks** for Air Mail delivery.)

VISIT OUR WEBSITE AT: ***www.latticepress.com***

ORDER FORM

LATTICE PRESS
POST OFFICE BOX 340
SUNSET BEACH, CA 90742, U.S.A.

www.latticepress.com
Phone: (714) 840-5010
FAX: (562) 592-1976

Please send the number of copies indicated below of **SILICON PROCESSING FOR THE VLSI ERA - Vols. 1, 2, 3,** and/or **4 (please specify!)** A check, money order, or Credit Card payment for the full amount for the number of copies ordered is enclosed. Make checks payable to: **LATTICE PRESS.** I understand that I may return any book within 30 days for a full refund if not completely satisfied!

____
NAME (and COMPANY - if part of your address)

____
ADDRESS

____
CITY

____
STATE   ZIP CODE   COUNTRY

________________   ________________
Daytime Phone Number   e-Mail Address

________________   ________________
Credit Card Number   Expiration Date

____ copies **Vol. 1** @ $199.95 each, plus $10.00 for each copy - shipping-handling fee.
(*California Residents*: Please add $15.00 per copy for Sales Tax.)

____ copies **Vol. 2** @ $139.95 each, plus $10.00 for each copy, shipping-handling fee.
(*California Residents*: Please add $11.00 per copy for Sales Tax.)

____ copies **Vol. 3** @ $149.95 each. plus $10.00 for each copy, shipping-handling fee.
(*California Residents*: Please add $12.00 per copy for Sales Tax.)

____ copies **Vol. 4** @ $199.95 each, plus $10.00 for each copy shipping-handling fee.
(*California Residents.* Please add $15.00 per copy for Sales Tax.)

International Ordering Information:

Enclose full payment in Travelers Checks (in $US), International Money Order, or by a Check payable to **Lattice Press** in **US dollars drawn on a U.S. bank.** Or send Credit Card payment.

Include a $12.00 US shipping/handling fee per copy for Overseas Orders shipped via SURFACE MAIL. (Allow **8-10 weeks** for Surface Mail delivery.)
Include a $30.00 US shipping/handling fee per copy for Overseas Orders shipped by AIR MAIL, with tracking capability. (Allow **2 weeks** for Air Mail delivery.)

VISIT OUR WEBSITE AT: ***www.latticepress.com***